The Routledge Handbook of Translation and Cognition

The Routledge Handbook of Translation and Cognition provides a comprehensive, state-of-the-art overview of how translation and cognition relate to each other, discussing the most important issues in the fledgling sub-discipline of Cognitive Translation Studies (CTS), from foundational to applied aspects.

With a strong focus on interdisciplinarity, the handbook surveys concepts and methods in neighbouring disciplines that are concerned with cognition and how they relate to translational activity from a cognitive perspective. Looking at different types of cognitive processes, this volume also ventures into emergent areas such as neuroscience, artificial intelligence, cognitive ergonomics and human–computer interaction.

With an editors' introduction and 30 chapters authored by leading scholars in the field of Cognitive Translation Studies, this handbook is the essential reference and resource for students and researchers of translation and cognition and will also be of interest to those working in bilingualism, second-language acquisition and related areas.

Fabio Alves is Professor of Translation Studies at Universidade Federal de Minas Gerais (UFMG) and a research fellow of the National Research Council (CNPq), Brazil.

Arnt Lykke Jakobsen is Professor Emeritus of Translation and Translation Technology at Copenhagen Business School.

Routledge Handbooks in Translation and Interpreting Studies

Routledge Handbooks in Translation and Interpreting Studies provide comprehensive overviews of the key topics in translation and interpreting studies. All entries for the handbooks are specially commissioned and written by leading scholars in the field. Clear, accessible and carefully edited, *Routledge Handbooks in Translation and Interpreting Studies* are the ideal resource for both advanced undergraduates and postgraduate students.

The Routledge Handbook of Audiovisual Translation
Edited by Luis Pérez-González

The Routledge Handbook of Translation and Philosophy
Edited by Piers Rawling and Philip Wilson

The Routledge Handbook of Literary Translation
Edited by Kelly Washbourne and Ben Van Wyke

The Routledge Handbook of Translation and Politics
Edited by Fruela Fernández and Jonathan Evans

The Routledge Handbook of Translation and Culture
Edited by Sue-Ann Harding and Ovidi Carbonell Cortés

The Routledge Handbook of Translation Studies and Linguistics
Edited by Kirsten Malmkjaer

The Routledge Handbook of Translation and Pragmatics
Edited by Rebecca Tipton and Louisa Desilla

The Routledge Handbook of Translation and Technology
Edited by Minako O'Hagan

The Routledge Handbook of Translation and Education
Edited by Sara Laviosa and Maria González-Davies

The Routledge Handbook of Translation and Activism
Edited by Rebecca Ruth Gould and Kayvan Tahmasebian

The Routledge Handbook of Translation, Feminism and Gender
Edited by Luise von Flotow and Hala Kamal

The Routledge Handbook of Translation and Cognition
Edited by Fabio Alves and Arnt Lykke Jakobsen

For a full list of titles in this series, please visit www.routledge.com/Routledge-Handbooks-in-Translation-and-Interpreting-Studies/book-series/RHTI.

The Routledge Handbook of Translation and Cognition

Edited by Fabio Alves and Arnt Lykke Jakobsen

LONDON AND NEW YORK

First published 2021
by Routledge
2 Park Square, Milton Park, Abingdon, Oxon OX14 4RN

and by Routledge
52 Vanderbilt Avenue, New York, NY 10017

Routledge is an imprint of the Taylor & Francis Group, an informa business

British Library Cataloguing-in-Publication Data
A catalogue record for this book is available from the British Library

Library of Congress Cataloging-in-Publication Data
A catalog record for this book has been requested

ISBN: 978-1-138-03700-7 (hbk)
ISBN: 978-0-367-50339-0 (pbk)
ISBN: 978-1-315-17812-7 (ebk)

Typeset in Bembo
by Newgen Publishing UK

Contents

Contributors

Fabio Alves is Professor of Translation Studies at Universidade Federal de Minas Gerais (UFMG) and a Research Fellow of the National Research Council (CNPq), Brazil. He has published widely about translation process research and expertise in translation in journals such as *Target, Meta, Across Languages and Cultures*, and in book series by John Benjamins, Routledge and Springer. He serves on the editorial board of *Target* and *Translation, Cognition & Behavior.*

Gerrit Bayer-Hohenwarter holds a PhD from the University of Graz with her thesis on the development of translational creativity (2011). She started her career as a language professional in 1999, gained several years of experience as an in-house translator and interpreter as well as a translation teacher and has been working as a freelance translator, technical writer and editor since 2013. Recently, she has been increasingly enjoying working as a copywriter and social media professional.

Michael Carl is Professor at Kent State University/USA and Director of the Center for Research and Innovation in Translation and Translation Technology (CRITT). He has published widely on machine translation, natural language processing and Cognitive Translation Studies. His current research interest is related to the investigation of human translation processes and interactive machine translation.

Andrew Chesterman retired in 2010 from his post as Professor of Multilingual Communication at the University of Helsinki. His research interests have been in contrastive analysis; translation theory, norms, universals and ethics; and research methodology. He was CETRA Professor in 1999 (Catholic University of Leuven), and has an honorary doctorate from the Copenhagen Business School. His most recent book is *Reflections on Translation Theory. Selected Papers 1993–2014* (John Benjamins, 2017).

Paul Chilton is Professor Emeritus of Linguistics at the University of Lancaster and a Visiting Fellow in the Centre for Applied Linguistics at the University of Warwick. His research interests are in cognitive linguistics and the analysis of discourse in various practical domains. He is also a published translator.

Agnieszka Chmiel is Assistant Professor in the Department of Translation Studies, Faculty of English, Adam Mickiewicz University, Poznań, Poland. Her research interests include interpreting studies, lexical processing and memory in interpreting, reading in sight translation, audiovisual translation and audio description. She also works as a freelance conference interpreter.

Igor A. L. da Silva is Assistant Professor at Universidade Federal de Uberlandia (UFU), Brazil. He holds a Master's and a PhD degree in Linguistics from Universidade Federal de Minas Gerais (UFMG), Brazil. His main fields of research comprise translation expertise and human-computer interaction. He is an associate member of LETRA, the Laboratory for Experimentation in Translation at UFMG.

Maureen Ehrensberger-Dow is Professor of Translation Studies in the Institute of Translation and Interpreting at Zurich University of Applied Sciences (ZHAW). She has written about the ergonomics of translation in both academic and professional publications. She serves on the board of the *European Society for Translation Studies* and also as the book reviews editor of *Target*.

Adolfo M. García serves as Scientific Director of the Laboratory of Experimental Psychology and Neuroscience (INCYT), Assistant Researcher at CONICET and Professor of Neurolinguistics at UNCuyo, Argentina. He also works at the Departamento de Lingüística y Literatura, Universidad de Santiago de Chile, Santiago, Chile. He leads research teams in over a dozen countries and has produced more than 130 publications, including books, chapters and articles in leading journals on neuroscience, language and translation.

Daniel Gile is Professor Emeritus at Université Sorbonne Nouvelle, Paris. He had his initial training in mathematics and sociology. He holds a PhD in Japanese and a PhD in linguistics. Conference interpreter and former technical translator, conference interpreter and translator trainer. He is a Co-founder and former President of the European Society for Translation Studies, the Founder of CIRIN (www.cirinandgile.com) and an author of numerous publications.

Sandra L. Halverson is Professor at the University of Agder, Norway. She has published extensively on various issues related to translational cognition, the epistemology of Translation Studies and research methodology. Professor Halverson is a member of the Translation Research, Empiricism and Cognition network (TREC). She was co-editor of *Target* for eight years and currently serves on the editorial boards of several international journals. She was appointed CETRA Chair Professor in 2018.

Silvia Hansen-Schirra is Full Professor of English linguistics and Translation Studies at Johannes Gutenberg University Mainz in Germersheim, Germany. She is the director of the Translation & Cognition (TRACO) Center in Germersheim and co-editor of the online book series *Translation and Multilingual Natural Language Processing*. Her main research interests include specialized communication, text comprehensibility, post-editing, translation process and competence research.

Amparo Hurtado Albir is Full Professor at the Universitat Autònoma de Barcelona. She is the team leader of a number of research projects on translation pedagogy and the acquisition of translation competence and head of the PACTE group. She is the author of more than 100 publications on the theory and pedagogy of translation.

Kristian Hvelplund is Associate Professor of English and Translation Studies in the Department of English, Germanic and Romance Studies at the University of Copenhagen. His research is mainly experimental using eye tracking and keylogging to study translator expertise and translators' interaction with texts and resources.

Alina Karakanta is a PhD candidate at the Machine Translation (HLT-MT) group at Fondazione Bruno Kessler. She was a research assistant at the Department of Language Science and Technology, Saarland University for the SFB-funded project Modelling Human Translation with a Noisy Channel. She is a co-organizer of LoResMT.

Haidee Kotze is Professor and Chair of Translation Studies in the Department of Languages, Literature and Communication at Utrecht University, and holds an appointment as Extraordinary Professor at the North-West University in South Africa. Her research focuses on language contact, language variation and language change, with a particular emphasis on translation. She is the co-editor of *Target: International Journal of Translation Studies* and co-editor of the series *Translation, Interpreting and Transfer* (Leuven University Press).

Jan-Louis Kruger is Head of the Department of Linguistics at Macquarie University in Sydney, Australia, where he also teaches in audiovisual translation (AVT). His main research interests include studies on the reception and processing of language in multimodal contexts. He is on the editorial board of the *Journal of Audiovisual Translation* (JAT), as well as on the advisory boards of a number of audiovisual translation conferences.

Paul Kußmaul is a pioneer of translational creativity research and research into the cognitive aspects of translation. From 1971 to 2005 he trained translators at the University of Mainz. He is a member of the advisory board for the series *Studien zur Translation* and the journal *Translation, Cognition & Behavior.* Among his numerous publications is his most recent book, *Verstehen und Übersetzen* (3rd edition, 2015).

Caroline Lehr is Research Associate at the Institute of Translation and Interpreting at Zurich University of Applied Sciences (ZHAW). She received her PhD from the University of Geneva in 2014. Her research focuses on the affective aspects of translation.

Victoria Lei is Associate Professor at the University of Macau, Macao, holding a PhD in English Literature. In addition to training translators and interpreters, she is also a conference interpreter. Originally trained as a journalist and literary historian, she formerly worked as a journalist, media translator and news anchor. In recent years her research has been dedicated to cross-disciplinary collaboration looking at cognitive processes in translation and interpreting.

Arnt Lykke Jakobsen is Professor Emeritus of Translation and Translation Technology at Copenhagen Business School. In 1995, he invented Translog, and in 2005 he established the CBS Centre for Research and Innovation in Translation and Translation Technology (CRITT), which he directed until his retirement in 2013. From 2006 to 2009 he was a principal investigator in the EU Eye-to-IT project, and he has been the President of the European Society for Translation Studies since 2016.

Kathleen Macdonald has a background in environmental community development in remote Queensland and Western Australia, and currently works in Social Programs for the Martu-run organisation, Kanyirninpa Jukurrpa. She obtained her PhD from the Hong Kong Polytechnic

University under a University Grants Committee Scholarship. This enabled her to explore the Linguistic Relativity Principle using contrastive description of English and Korean language use and comparative analysis using Korean–English translation of texts about reinventions of music culture. Her research interests include linguistics, translation, anthropology, aesthetics, and music analysis.

Kirsten Malmkjær has taught at the universities of Birmingham, Cambridge, Middlesex and Leicester. Recent publications include the *Routledge Handbook of Translation Studies and Linguistics* (2018), the collection of articles, *Key Cultural Texts in Translation* (John Benjamins), co-edited with Adriana Serban and Fransiska Louwagie (2018), and *Translation and Creativity*, Routledge (2019).

Celia Martín de León is Associate Professor of Translation at the University of Las Palmas de Gran Canaria, Spain. Since 2002, she has been a member of PETRA Research Group, devoted to empirical research into the cognitive aspects of translation and interpreting. She is also a member of the TREC network (Translation, Research, Empiricism, Cognition).

José Martínez Martínez is a PhD candidate at the Department of Language Science and Technology at the Universität des Saarlandes, Saarbrücken, Germany. His career in academia has revolved around human translation, contrastive studies and corpus linguistics. He currently works as an NLP (natural language processing) data scientist at Datamaran in Valencia, Spain.

Christian M. I. M. Matthiessen is Chair Professor of Language Sciences at the Hong Kong Polytechnic University and an honorary professor at the Australian National University. He has published extensively in systemic functional linguistics since 1981, covering general theory, description of particular languages and language typology, and a number of particular areas, including Translation Studies, discourse analysis, rhetorical structure theory, register studies, healthcare communication studies, computational linguistics and educational linguistics.

Edinson Muñoz is a Professor of Linguistics and a Japanese–Spanish translator. He is currently the director of the Department of Linguistics and Literature at Universidad de Santiago de Chile, where he teaches Japanese–Spanish Contrastive Grammar, Japanese–Spanish Translation and Japanese Graphemics. He has also supervised undergraduate theses on those topics. His research work deals with cognitive linguistics, neurolinguistics and translation.

Ricardo Muñoz Martín is Professor of Translation Studies at the University of Bologna at Forli, Italy. He is the coordinator of PETRA Research Group (Expertise and Environment in Translation, Spanish acronym) and a member of the international research network TREC. Muñoz Martín is the editor of the journal *Translation, Cognition & Behavior.*

Stella Neumann is Professor of English Linguistics at RWTH Aachen University. She obtained her PhD from Saarland University, Germany. Her research interests include quantitative register analysis across languages and varieties and the empirical modelling of translation. She currently holds a grant for empirical translation research funded by the German Research Council. She was previously the review editor of the journal *Languages in Contrast* and is now on the editorial board of the LangSci Press book series *Translation and Multilingual Natural Language Processing.*

Jean Nitzke received her PhD in Germersheim at the Johannes Gutenberg University Mainz, where she also had a teaching and research position until March 2018 and is still a post-doc. Currently, she is substituting a full-professor position in Hildesheim.

Sharon O'Brien is Professor of Translation Studies at the School of Applied Language and Intercultural Studies in Dublin City University (DCU). She teaches and writes about translation technology. She is a funded investigator in Science Foundation Ireland's Research Centre ADAPT and a member of the Centre for Translation and Textual Studies at DCU.

Adriana S. Pagano is Professor of Translation Studies at Federal University of Minas Gerais (UFMG), Brazil, where she advises doctoral theses in the Graduate Programme in Linguistics and Applied Linguistics and conducts research at the Laboratory for Experimentation in Translation (LETRA). Her research interests include meaning modelling in translation and multilingual tasks.

Anthony Pym teaches at the University of Melbourne, Australia, and is Distinguished Professor of Translation and Intercultural Communication at Universitat Rovira i Virgili in Spain and Extra-Ordinary Professor at Stellenbosch University in South Africa. His current research focuses on risk management as a frame for understanding cross-cultural communication.

Hanna Risku is Professor for Translation Studies and head of the research group *Socio-Cognitive Translation Studies: Processes and Networks* at the University of Vienna, Austria. Prior to her work in Vienna, she was professor at the University of Graz and at the Danube University Krems, Austria, guest professor at Aarhus University, Denmark, and lecturer at the University of Skövde, Sweden. Her main areas of interest are the cognitive foundations of translation, translation networks, ethnographic workplace research on translation and translation as computer-supported cooperative work.

Regina Rogl holds an MA in Interpreting and is currently working as a graduate research and teaching assistant at the Vienna Centre for Translation Studies. She is a member of the research group *Socio-Cognitive Translation Studies: Processes and Networks*. Further research interests include translators'/interpreters' networks, socio-technical issues of translation/interpreting, non-professional translation/interpreting and workplace research.

Christina Schäffner is Professor Emerita at Aston University, Birmingham. Her main research interests are political discourse in translation, news translation, metaphor in translation and translation didactics. Her recent co-edited works include *Political Discourse, Media and Translation* (2010) and *Interpreting in a Changing Landscape* (2013).

Tatiana Serbina is a postdoctoral research assistant in English linguistics at RWTH Aachen University. Her PhD thesis combined insights from empirical Translation Studies and construction grammar. She was on the scientific committee of several Translation in Transition conferences. Her recent publications include articles in the journals *Pragmatics and Society* (9:1), *Translation, Cognition & Behavior* (1:2) and *Target* (31:1).

Gregory M. Shreve is Professor Emeritus of Translation Studies at Kent State University and Adjunct Professor of Translation at New York University. At Kent State, he founded the first comprehensive Translation Studies programme in the United States. Shreve's research interests include Cognitive Translation Studies, translation expertise and text linguistics in translation.

Erich Steiner holds the Chair of English Translation Studies, Department of Language Science and Technology, University of Saarland, Saarbrücken. His major research interests include translation theory and comparative linguistics, functional and empirical linguistics. Major publications include *Exploring Translation and Multilingual Text Production: Beyond Content* (edited 2001 with Colin Yallop), *Translated Texts: Properties, Variants, Evaluation* (2004) and *Cross-linguistic Corpora for the Study of Translations. Insights from the Language Pair English–German* (with Silvia Hansen-Schirra and Stella Neumann, 2012).

Elke Teich is Full Professor of English Linguistics and Translation Studies at Universität des Saarlandes, Saarbrücken, Germany. She is head of the Saarbrücken Collaborative Research Center (SFB 1102) *Information Density and Linguistic Encoding* funded by the German Research Foundation (DFG) and principal investigator in the CLARIN-D (*Common Language Resources and Technology Infrastructure*) project and has published two monographs and over 70 peer-reviewed papers.

Introduction

Arnt Lykke Jakobsen and Fabio Alves

Aim

The Routledge Handbook of Translation and Cognition aims at bringing together in a single volume a full overview of the current state of research-based thinking about Translation and Cognition in 30 invited chapters, all written by leading scholars in their respective specializations. Cognitive Translation Studies (CTS) is a term introduced by Halverson (2010) to refer to a variety of "process-oriented" translation research activities. CTS has developed rapidly in the last three and a half decades in interaction with disciplines like anthropology, philosophy, psychology, computer science, artificial intelligence, neuroscience and linguistics, in full accordance with the vision of the pioneers of the "cognitive turn" in the 1950s (see Miller, 2003, and Chapter 3 of this volume). The details of the development of CTS, especially since the middle of the 1980s, are fully covered in the Handbook, theoretically, methodologically and by topic, with a particularly strong focus on written translation, as a Routledge Handbook devoted specifically to interpreting and cognition is being prepared.

Despite the rapid development of CTS, or possibly because of it, CTS is not at present a unified discipline with a shared grand theory, methodology, epistemology or even ontology. The very nature of what is the subject or object studied is contested, and therefore a variety of research methodologies are used, some of which explore translational language as a way of getting to know about the mind, while others attempt to explore the mind from behavioural observation of either body or brain. In both approaches, computational models are also invoked. Our aim in this regard has been to openly acknowledge this lack of unity by admitting space to reflect the variety of conceptualizations, approaches and topics that collectively represent the current state of CTS as the editors see it.

Translation and cognition

The word "and" in the title of the Handbook gives wide scope for discussing the relationship between the two concepts it connects, but it should also signal that the Handbook is not about all aspects of translation. Nor is it about cognition in general. Translation can be, and has been, studied from historical, sociological and other approaches as well as from functional, technological or linguistic perspectives, without any consideration of cognition. Likewise, there are psychological, physiological and medical interests in cognition that we need not be particularly concerned with in CTS. The focus in the Handbook is on what translation is cognitively and what it means to study translation from a cognitive perspective. What qualifies all the chapters for inclusion in the volume is that they consider translation and Translation Studies from a cognitive perspective.

The chapters have been organized into the following four parts:

Part I, Foundational aspects of translation and cognition, tackles questions of the kind just mentioned, presents two current, influential theoretical views of CTS, and offers an overarching theory under which Cognitive Translation Studies can perhaps be subsumed.

Part II, Translation and cognition at interdisciplinary interfaces, surveys concepts and methods in neighbouring disciplines that are all concerned with cognition, from which inspiration has been drawn in an effort to develop CTS.

Part III, Translation and types of cognitive processing, is concerned with different types of cognitive processes, what triggers them, how they are managed and measured, and how knowledge about translation as a whole is acquired.

Finally, Part IV, Taking Cognitive Translation Studies into the future, offers five bids for the future direction in which CTS is moving, complementing the chapters in Part I.

Cognitive Translation Studies—an introductory overview

Representing meaning in language

Wonderland 2016[1] is a video recorded by Erkan Özgen, which shows a 13-year-old deaf and dumb Kurdish boy, Muhammed, as he apparently attempts to share his experience of war in Syria. Muhammed has vision and body movement and can utter inarticulate sounds, but, as he is both deaf and dumb and has not been taught sign language, he has no verbal language. As part of his motivation for making the recording, Özgen has said that "perhaps only Muhammed could truly communicate this unseen and unheard brutality [...] The power of his body language made any other language form insufficient and insignificant" (quoted from the Tate Modern presentation material). Özgen's message with the video has been interpreted differently, but what it very movingly illustrates is the loneliness of a communicatively handicapped individual and our inability to know the boy's mind. The boy's physical situation very severely constrains his ability to engage socially and interpersonally with others, except emotionally, and illustrates the dependency of communicative verbal skills on physical and biological human factors, with respect to both representing our meaning to others and making sense of other people's meaning representations.

There are similar physical, social and interpersonal dimensions to translation. It is possible to translate mentally without attempting to represent translation in a medium, but if we want to share a translation with others, our mental translation will have to be represented in a physical code to which others can assign relevant and probable meaning.

The big issue in the history of translation ever since antiquity has been whether to translate from the word (*verbum e verbo*) or from the meaning (*sensum de sensu*), often formulated as a distinction between translating literally and freely. At different times this debate has been prescriptive, pedagogical, author oriented, reader oriented and/or text-type oriented. In the cognitive paradigm, the issue goes by other names and is modelled differently, e.g. by reference to Kahneman's distinction between slow and fast thinking (2011), but it remains unresolved and a bone of contention. The reason for this is the inextricable interdependence and interactivity between experienced meaning and the language we use to articulate and share meaning. The association is so close and routinized that we are not sure there is such a thing as "deverbalized" meaning, for without language there is no shareable translatable meaning beyond that demonstrated by Muhammed. On the other hand, can we say that he was not attempting to convey meaning? Several of the ensuing chapters tackle the issue, and some will present a different view of the relationship between language and meaning. Is language a way to understanding the mind? Do we construe experience through linguistic meaning? Or do we construe experience as meaning pre-linguistically?

Construing reality as meaning in language and translation

In CTS it is widely assumed that we cognitively construe or translate experience into meaning. It is this meaning that we learn to articulate in language, and this is the meaning that translators translate. Our lived meaning-making process is initially an individual affair, and philosophers and psychologists tell us that we cannot know exactly what a person's experienced meaning is, but learning a language and speaking a language involve processes of cognitive alignment in which an individual's experience of reality is constantly modulated by language use and adapted to the way members of the language community perceive and construe reality. Knowing a language thereby gives us reason to believe that our construal of the world is both similar to that of other people and shareable with them if we speak the same language, for knowing a language is having learned how to conventionally represent our meaning, our perception of reality, in speech or writing (or some other physical medium).

There is no doubt that the language a person speaks has considerable impact on how the person construes meaning, how an episode will be reported and how the report will be understood by others. The language a speech community develops collectively is bound to have signs for concepts that are particularly important, salient and functional in the community's physical, social and cultural environment. Therefore, every language will have a unique lexical-conceptual profile articulating a bias reflecting the blend of what a community cognitively foregrounds and what it is less attentive to or ignores. This is well known from anthropology and commonly identified as a special challenge in Translation Studies, for regardless of whether our point of departure is cognition and meaning or language, the challenge is there. All members of a culture and all languages have "unique items" (Tirkkonen-Condit, 2004).

From a cognitive perspective, meaning can never be exhaustively represented in a language or other representational system. Symbolic representation is vague, ambiguous, polysemous and indeterminate (Quine, 1960, pp. 73ff.). Therefore, every language depends on language users' ability to activate relevant meaning, but since it is based on "interpretation", i.e. on the meaning a language user assigns to an utterance, a huge variety of potentially relevant meaning can be activated, and reports can be endlessly retold, represented and re-represented from new perspectives and new focuses. For the same reason, the meaning a translator construes from the representation of meaning in a source text is necessarily creative. As there is no meaning in the material source-text representation itself, meaning always has to be supplied by a listener/reader familiar with the code. In CTS, this process is traditionally referred to as a process of comprehension, and it is understood as no less creative than the process of articulating meaning. Comprehension is a very active and creative semiotic, meaning-making process (Pickering & Garrod, 2013).

On the production side, similarly, translation (and interpreting) require(s) a creative meaning-representation effort, not just when it comes to translation of text with language rich in imagery and other rhetorical effects, although it is most clearly seen here (see Chapters 17 and 18). Historically, it has often been argued that translation of texts of this kind requires a certain relaxation of such norms as equivalence, adequacy, accuracy or loyalty (Chapter 19), but theoretically the constraint that most often applies is the one that applies to all translation, i.e. that of interpretive resemblance (as proposed in relevance theory; see Gutt, 1991, p. 100; see also Chapter 7).

When understood meaning is re-represented, it may become the subject of public negotiation in the community and part of the process of constant adaptation by which a language community's norms for regulating the connection between language, experience and meaning are maintained. The meaning construed by the translator is the translator's theory of what was on the mind of the source-text author, and this theory of an author's meaning can be represented, revised and re-represented in a thousand different articulations, constrained to some extent by the translator's (or interpreter's) assessment of the relevance and possibility of achieving similarity of meaning in specific cases. Whether strict similarity in the form of equivalence or some form of transcreation is aimed for will depend on multiple social and professional considerations, but in the end listeners/readers will be assigning *their* meaning to the chosen representation, based on their meaning-making ability.

Cognitive processes in translation

The distinct characteristics of CTS are that it locates the centre of interest in the mental operations performed by translators when they translate and that it sees translation as inseparably linked to human meaning making.

Since the beginning of the cognitive turn in the 1950s, the concept of a cognitive translation process has changed quite radically from being understood until the 1980s as a primarily intellectual and rational process by which problems were solved and decisions made in a logical way that could be simulated computationally to being construed from the 1990s on as a far more complex and full-bodied process, at once rational and irrational, calculated, emotional and intuitive, dependent on processes both in the brain and in the body from its interaction with the environment and other agents in the environment.

Perhaps the main change brought about by the cognitive orientation of Translation Studies was the shift of focus from translation product studies to process-oriented research. Research in the 1980s and 1990s often used the think-aloud protocol (TAP) method (see e.g. Chapter 12) with a view to discovering by what labyrinthine processes, described as language operations, translation students arrived at their final translation solutions. Much of this research was strongly classroom oriented, inspired by the tradition for contrastive linguistics and by ideas then current in foreign language learning research (e.g. the concept of "interlanguage"; Chapter 6). As a by-product of the method, which successfully identified numerous language operations, a lot of what was verbalized was found to indicate that translation was far from being a neat, controlled process. It was not very linear, often not a process driven by conscious decision making; problem solving was often random, and verbalizations were often only indirectly related to the actual performance of the task; they were questions to the experimenter, comments on the situation and expressions of emotions.

The TAP method remains a strong method, as there may not be a better way of getting information about a person's mind than by having the person tell us about it in words. The problems with relying on verbal reports, in daily life and in research, are well known: people may not know what motivated a certain decision, may not give a true report and may not give a full report. From a research perspective, the TAP method is also cumbersome. After a session, recorded data have to be transcribed. This involves interpretation and a risk of data misrepresentation. Then, data have to be categorized, which involves further interpretation, and *only then* analysed as data, not as ordinary statements.

The difficulty of interpreting verbal reports as data was part of the motivation behind the invention of the *Translog* program in 1995, which has been an important research tool in translation process research. The program aimed at producing a complete and very accurately timed data record of a translator's typing process while performing a translation task. The timed records produced in the program not only showed all edits, deletions, typos and changes and exactly when they were made, permitting measurement of production speed and editing effort, but also showed pauses before identifiable trouble spots and the rhythm with which solutions were typed, which could arguably be related to cognitive processing characteristics. It may be true that much of the research done with keylogging was technology driven rather than theory driven. The technology was invented to enable accurate exploration of translational behaviour as a means of getting underneath language and making inferences about cognition from another type of data. The aim was to access behavioural data that could be taken as evidence of cognitive phenomena related to meaning-making and meaning-representation processes.

The same was true of the addition of eye tracking, which made it possible to combine gaze and key data and obtain data both about reading/comprehension processes and from writing/typing processes and to observe how they were coordinated. Analysis of combined key and gaze data, especially the way attention was shifted between reading and typing and (visual) monitoring of the whole process, clearly confirmed what keylogging had already demonstrated: that written translation proceeds in "bursts" or "chunks", i.e. chunks of words, such as phrases. Phrase-level translation has been found to produce more idiomatic solutions than word-level translation in machine translation, but the fact that translation is observed to proceed behaviourally and presumably also cognitively in this manner is evidence that humans translate the meaning signalled by phrases. They do not translate from a mental phrase book but know how to represent the meaning they have construed from phrases (see e.g. Chapters 12 and 15).

If cognitive translation research was technology driven in the early phases, the most recent development has been more clearly driven by theories of mind, cognition and meaning developed by cognitive scientists who have insisted that cognition is situated and distributed as well as embodied, embedded, enacted, extended and affective (Chapters 2, 3, 27 and 29). Just as cognitive linguistics has been seen as essentially a recontextualization movement (Geeraerts, 2002), especially as a reaction to formal generative grammar, so "translatology" was launched as an attempt to recontextualize or "reembed" translation process research (Muñoz Martín, 2016). The aim was to broaden the scope of CTS to include consideration of extended and affective dimensions, to pull CTS away from computational models and mechanical information processing thinking, and to philosophically bury Cartesian ideas of certainty and dualism by uniting mind and body and grounding mind in matter. Several chapters contribute to the discussion of what methodological consequences follow from this recontextualization and how it affects the scope of CTS.

Epistemological considerations

Thinking about translation takes us to the core of some of the toughest philosophical questions about how we experience and know the world, how we build the assumptions by which we interact with other people and our environment, how we develop cognitive skills like communicating and speaking, and how we manage to understand each other across language barriers and cultural and personal differences by means of translation. Translation makes us think about what truth is, and how we can have certain knowledge of anything (Chapter 1), seeing how much our perception of the world varies across cultures and languages (Chapter 5) and how much members of the same cultural environment, even family members, may differ and disagree. And yet, despite differences and a lurking suspicion that we may never hope to achieve certain knowledge, we carry on our lives on the—possibly false—assumption that we are part of one world, physically and biologically, although we do not fully understand it cognitively. We interact with this perceived reality, mostly quite confidently, and assume that there is a reachable common humanity that makes it possible for us to have similar, explicable and translatable thoughts and experiences despite the differences in the reality we construe, within and across languages and cultures. Without being certain that the sense we get from another person's communication to us is a reliable and fair reflection of what this person intended to communicate, we carry on and enjoy mediating our thoughts and understanding to others in our environment and find that there is wonderful life to be lived, reported and translated without epistemic certainty, just as there is valuable understanding to be gained from imperfect translations, as long as the meaning they activate in us is perceived to be relevant and interesting.

Ontological considerations

To Jerome Bruner (1960), a central figure in the cognitive turn, our construal of reality begins as an "act of meaning", an idea similar to ideas propounded earlier by phenomenologist philosophers like Husserl, Bergson and Merleau-Ponty. From the perspective of language and cognition, the basic ontological question is whether there is a world that we begin by perceiving and construing pre-linguistically and then perceive in greater culture-specific detail as we learn a language, or whether we construe what we call reality through the language(s) we learn to speak. Do we need a language to know reality, or do we know reality through acts of meaning, which we learn later to report in language?

From the perspective of CTS, the ontological question is whether translation is fundamentally a matter of meaning, separately from how it is represented, or whether meaning and meaning representation in language are so inextricably connected that they cannot be teased apart. E. M. Forster's famous "How can I tell what I think till I see what I say?" (1927, p. 99) suggests that meaning and its articulation in language are inseparable, that there is no knowable meaning until it has been articulated in language. This view is epitomized in the Saussurean concept of the linguistic sign as at once form and meaning (expression and content; see also Chapters 11 and 29). On the other hand, we all interact wordlessly with other people in our environment by interpreting meaning from their face, appearance and behaviour, showing that there is a lot of non-verbalized meaning in our lives. This is related to why two competing construals or conceptualizations of how the mind construes reality have dominated cognitive science since its start. One construal sees the brain as a large computer working on symbolic representations of the world (associated with linguistic signs) or on so-called sub-symbolic representations and outputting consciousness and meaning. This goes back to the information processing theory initiated by Shannon and Weaver, Allen, Miller and others (Chapters 3 and 28) and to the later branch of cognitive science known as connectionism. Both these construals are sometimes referred to as cognitivist, and the associated theories as representing the computational theory of mind.

The other, recent construal focuses on the complexity and embodiment of experience and mind and on seeing the jungle-like, biological and neuro-physical processes in the brain as connected to what we experience consciously in the mind rather than as belonging in a qualitatively different realm. In this construal, perception and cognition are preliminaries to making sense of the world—both our outside world and our inside world. We partake of the world and interact with our environment, with our senses, our minds and our bodies, and we construe reality from this interaction. The claim from this perspective is that what goes on in the brain is quite different from what goes on in a computer's processor. Symbols (such as words) can only very sketchily represent meaning, it is claimed, which is a reason why meaning cannot be computed by processing of symbols. Meaning is not *in* symbols. Meaning is in people. It has its origin in perception but can also be activated by symbols in humans who know how to generate meaning from them. This second construal, now often referred to as situated or 4EA cognition, rejects the mechanical and finite nature of the computer as a model of the mind and prefers metaphors that highlight translation as a living, biologically grounded activity, somewhat like a complex eco-system in which there are all kinds of organic systems in constant interaction and adaptation. As will be apparent from several chapters in the volume, this construal is still under development. In one version, that of cognitive linguistics (Chapter 2), language use is proposed as a means of cognitive discovery. In another, the process of meaning construal is seen as a phenomenon we need to understand before we can hope to understand how meaning is represented in language and translation (Chapter 3). The computer is still used as a metaphor of the brain, especially in the neural version, as in neural machine translation (Chapter 28). Giant networks are also used as metaphors referring both to the brain and to the way cognition is argued to extend beyond individual minds into the environment, particularly into the social networks which translators and the rest of us increasingly operate in. According to Edelman (1992, p. 162), whose metaphor for the processes in human minds was a jungle (see Chapter 1), "[m]inds do not exist disembodied", but several contributors to this Handbook make reference to the view that minds are not necessarily embodied in individual organisms but extend into giant networks or complex adaptive eco-system-like structures (Chapters 4 and 27). This is where the discussion stands at present, and we leave it to readers to contribute to the debate.

Translation and multimodal/multimedia communication

The concept of translation is still mostly associated with translation of a written text into another text written in a different language. Sometimes the notion of "text" is expanded to include a spoken or signed source (as e.g. in Chapter 14). The widespread use of multimodal communication using multimedia (Chapter 24) is expanding the scope of what is translated even further and has made it abundantly clear that translation is not a monomodal language operation. For instance, in audiovisual subtitling, it is customary to include titles for certain audible non-verbal features in the audio track to give people with reduced or no hearing access to what may be important contextual cues to making sense of a situation or narrative. Similarly, in an effort to allow persons with reduced or no eyesight (but with hearing) access to visual information, e.g. in the cinema, audio description has been developed (see Chapter 24). This is widely accepted as a form of translation, although the "text" that is translated is not text as traditionally understood, but visually represented action in a filmed environment. Roman Jakobson's (1959/2000) famous categorization of translation into intralingual, interlingual and intersemiotic translation has been a reminder that the concept of translation could be understood broadly, but intersemiotic translation was often categorized as adaptation, and it is only now beginning to be fully appreciated that there is no clear demarcation between interlingual and intersemiotic translation. "No text is, strictly speaking, monomodal", as Gambier rightly stated (2006). But if text is always multimodal, and translation can include rendition of visual cinematic narrative into a verbal narrative, then we no longer need to distinguish theoretically between interlingual and intersemiotic translation. All translation is intersemiotic. A sports journalist's radio narrative reporting on a Bundesliga football match is an act of translation. The journalist has an immediate and direct visual and auditive experience of the action. As we listen to the spoken report, we are able to invest the description with an idea of the action and share the journalist/translator's emotional excitement, probably aided by the so-called "paralinguistic" features of the reporter's narrative brought into play to help listeners visualize and sculpt the experience: quality of voice, loudness, stress, speed and effort imitation. In the widest interpretation, as already stated, the meaning we make of reality is created in acts of translation and can be mediated in different modes.

The preceding explains why we are always able to reformulate an utterance if asked what we mean. We can engage in intralingual translation because we know what we want to say, we have the meaning we wish to communicate, and there are a thousand different ways in which this meaning can be represented. Therefore, knowing how to use language is knowing how to translate experience, intention, memories, emotions and the rest. In that sense, all language users are natural (intralingual) translators. In Chapter 3, the question is asked why it is that some bilingual language users with no training in translation or interpreting nevertheless have acceptable translation or interpreting skills, but this is perhaps not very surprising, for having an idea, having something you want to say and being bilingual means that you can represent your idea in either language. What we learn in learning a language or several languages is how to materially represent meaning in it or them.

Translation, technology and cognition

The increasing use and availability of translation technology as an automatic service for Internet users in many countries and in several social media as well as in dedicated smartphone apps, some offering speech-to-speech machine translation, has already changed the face of global communication. It has vastly extended our ability to translate across several languages and to make sense of texts in many languages. It has extended our range of meaningful interaction and enabled us to better navigate the ocean of available information. Working methods and conditions in the translation industry have been very deeply affected by the increasing use of machine translation, computer-assisted translation (CAT) tools and other technological solutions. In Chapter 17, the development is described in the following way: "What production robots, autonomous vehicles, surgery robots, telemedicine and legal technology are for these professions [...], Google translate, post-editing of machine translation and translation vendor platforms are for the translation business." This technological revolution has created new areas for CTS to explore and given rise to numerous studies of cognitive effects on users, in ergonomics (Chapter 8), translator–computer interaction (Chapter 21), risk evaluation (Chapter 25) and post-editing of machine-translated text (several chapters).

The use of technology in experimental research and its potential effect on participants' performance is a topic taken up in several chapters, e.g. in discussions of the reactivity of concurrent think-aloud and the consequent potential skewing of data, the ecological validity (or invalidity) of data from keylogging and gaze data collected in a laboratory environment, and the degree of invasiveness of technologies like eye tracking, electroencephalography (EEG), functional near-infrared spectroscopy (fNIRS) or functional magnetic resonance imaging (fMRI). Advantages and disadvantages are weighed in several chapters, but the general observation one can perhaps suggest is that in each research situation, the observer's paradox applies. Each situation has its own ecology, which it is up to the researcher to explicate as far as possible.

Using computational methods to explore translational phenomena does not commit one to a computational theory of mind. Corpus analysis is a proven method for exploring translational language, also from a cognitive perspective. In the present Handbook it is used to explore a cognitive phenomenon like "shining through" (Chapter 23), notions like expectancy, surprisal and entropy (Chapter 20), and ideas that combine artificial intelligence with findings in user-activity corpus data to build an up-to-date computational view of translation that integrates the most recent ideas of extended and distributed cognition (Chapter 28).

Brief presentation of individual parts and chapters

Part I

The four chapters in Part I, Foundational aspects of translation and cognition, are all foundational in the sense that each of them offers reflections on the epistemological and ontological bases on which CTS is currently conceptualized as standing. The opening chapter asks the most fundamental questions of all about the epistemological basis of human knowledge: Can we really obtain certain knowledge about translation and cognition? If so, with what methods; and with what degree of certainty can we hope to obtain knowledge? The next two chapters ask questions about the ontological nature of translation. Is translation essentially a language operation, something that people (and machines) do with words, or is it primarily an operation based on meaning? Is language ontologically primary in our conceptualization of translation, or is meaning? One answer, given in Chapter 2, prioritizes language, as understood in cognitive linguistics. Here, a concern with language is posited as the ontological core of our discipline and cognitive linguistics as the theoretical and methodological centre. A different, more purely cognitivist answer is given in Chapter 3, where translating is said to be a matter not of language but, rather, of meaning or communication. Throughout the volume, readers will find this oscillation between the primary focus being placed on language as a means of discovery about translation and cognition and the focus being on cognition and meaning as the primary source of knowledge about language and translation. The last of the foundational chapters offers a theoretical framework for viewing translation as a complex adaptive system (CAS).

Chapters 2, 3 and 4 should be read together, as they have all been written with the positions in the other chapters in mind. They are also theoretically and thematically close to two of the chapters (27 and 28) in Part IV.

Chapter 1, by Andrew Chesterman, examines our attempts to know about cognitive aspects of CTS from an *epistemological* standpoint, asking how we can extract reliable information from introspection or from subjective verbalizations recorded in a TAP and how we can analyse such verbal data as reliable evidence or even as explanations of cognitive phenomena. Chesterman also asks, quite critically, how we can make reliable inferences about cognition from recorded keystrokes and the intervals between them during the typing of a translation, regardless of how accurately they have been recorded. The answer given can perhaps best be described as sceptically optimistic. Chesterman affirms belief in the value both of conceptual and empirical research and of hypothesis testing, whereby some real knowledge can be found—at least about some things, such as the things we need to know in order to be able to navigate our immediate environment. Although he adopts a "fallibilistic" standpoint, meaning that what we think we know today may turn out not to be true tomorrow, he firmly repudiates nihilistic relativism. There are things that we *can* know, or think we can.

In Chapter 2, Sandra Halverson makes a plea for grounding CTS in a *linguistic commitment*. She also takes up fundamental ontological and epistemological questions concerning cognitive translation and looks for "clearer lines of demarcation between major approaches to the study of cognition in general and to the study of translational cognition in particular". One major approach is the information processing approach, which sees the mind as a symbol manipulating engine, much like a computer, and is firmly based on language as the dominant representational system. Halverson sides with the other major approach, the translatological approach, but argues that a clearer commitment is required to a view of language as "central to our fundamental ontology in all cognitive approaches to translation". She argues that cognitive linguistics can be the unifying element and rewrites its cognitive commitment to what is generally known (in cognitive science) about the mind and brain into a commitment that cognitive translatology should accept to ensure that its account of translation should always accord with what is generally known about language from cognitive linguistics.

In Chapter 3, Ricardo Muñoz Martín and Celia Martín de León trace the beginnings of *cognitive science* in the 1950s and survey several strands in Translation Studies that contributed to developing CTS. In interpreting studies, the influential concept of deverbalization suggested that meaning could be held in an interpreter's mind independently of a fixed verbal form or other symbolic representation. One main model of communication was Shannon and Weaver's information processing model, which represented the human brain as a computer operating "intelligently", i.e. logically, on symbols and problem solving. This is largely the model that continues to underlie computational attempts to model cognition and cognitive translation. In the 1990s, new ideas began to develop, jointly leading to the construal of 4EA cognition (embodied, enacted, embedded, extended and affective), which involves a radical recontextualization and socio-cognitive expansion of the scope both of cognitive science and of CTS and challenges the importance of language as a means of understanding cognition. Cognitive science is appealed to as the science that will give us proper insight into the sub-symbolic processes in the mind by which meaning is generated, the meaning that we construe in listening and reading and the meaning we articulate and represent in translation (and in interpreting).

In Chapter 4, Gregory Shreve takes up the challenge of exploring the theoretical implications of the key proposals from information processing, cognitive linguistics and cognitive science put forward in the previous two chapters. Rather than attempting to explore and integrate all key concepts, Shreve's proposal is to include CTS (translatology) under the umbrella of CAS theory. Arguments are given in support of viewing translation as a CAS. If conceptualized in this manner, translation is multi-scale and can be studied at all levels, from the level of cell activity or below to the social and cultural level or beyond. It is therefore important that the scope of a piece of cognitive translation research is always specified. This is particularly important when notions of distributed and extended cognition are brought into play, for any "system we study is linked to systems both above and below it in a multi-scale hierarchy". As translation is understood as inevitably communicative and functional, goals should be specifiable at all hierarchical levels. The question of who or what can have goals leads to the question of agency. Shreve concludes that because words can signal their producer's goal orientation, they embody agency, thereby accepting a view of embodiment and agency that extends beyond living bodies.

Part II

In Part II, Translation and cognition at interdisciplinary interfaces, all nine chapters deal with the interfaces between CTS and such disciplines as anthropology, contact linguistics, pragmatics, ergonomics, language ontologies, corpus linguistics, linguistics, psycholinguistics and neuroscience. CTS has overlapping research interests with all of them, shares a number of epistemological and ontological assumptions with them, uses the same or similar methods, and attempts to explore and model translation in ways that constantly adapt to what is known in these interface disciplines. Translation and cognition penetrate every human activity, every knowledge field and every institutionalized discipline and are therefore eminently interdisciplinary. The disciplines we have selected for special attention illustrate some of this variety, but they have historically been disciplines with which CTS has interacted most vigorously. Sometimes the contact has been chiefly mediated through conceptual overlaps (exchanges, borrowings and lendings); at other times the contact has mainly resulted in methodological inspiration.

Where *anthropology* (Chapter 5) is focused mainly on studying people in their cultural environment and from their language, "the most articulate code of human experience", *ergonomics* (Chapter 8) is particularly aimed at studying the environments that people work in, with the central focus on how well working environments are in tune with human bodies and minds and how everything in the environment affects employees' cognition. Chapter 6 studies what happens when speakers of different languages have contact. The phenomenon of translational interference has been studied intensely at the level of individual translations or corpora of translations, but there is a broader way of approaching linguistic change, not as a result of translation but in the context of *language contact*. Chapters 7 and 9 through 12 all concern interfaces with various ways of studying language, *pragmatics* (Chapter 7), *domain ontology* (Chapter 9), *corpus-based CTS* (Chapter 10), *linguistics* (Chapter 11) and *psycholinguistics* (Chapter 12). Whatever our view on the relative importance of cognition, meaning and language in CTS, language and the various branches of linguistics remain crucial. The final chapter concerns the increasingly important interface with *neuroscience* (Chapter 13).

To Kathleen Macdonald (Chapter 5), "*Anthropology*, as a discourse of disparate cultural communities, has always been concerned with translating the minds and behaviours of individuals and collectives within one culture, across to those from another context." Anthropologists are intensely aware of the interplay between thought, culture and language. To Sapir, language was an instrument that made thought possible, while thought, in turn, helped refine the language instrument. Whorf's view of the inextricable tripartite of thought, language and culture led him to identify isolates of experience and isolates of meaning from both explicitly (overtly) and implicitly (covertly) signalled meaning to identify the distinctive particularism of a culture. His celebrated linguistic relativity principle is not representative of an essentialist view of culture and meaning (making translation impossible), nor is it a mere compromise between universalism and essentialism. In the words of Penny Lee, Whorf assumes "that there are universal configurations of experience upon which different linguistic schemes of classification operate in a variable way" (Lee, 1996, p. 96). "Thick" descriptions of linguistic (or other semiotic) articulations are a way of bringing out both the specificity and the complexity of culture-specific meaning making.

Where anthropologists have tended to study cultures without much interaction with other cultures or languages, Haidee Kotze (Chapter 6) looks at the kinds of language change that result from different languages being in close contact, both socially and in the minds of language users, not only translators. Her focus is on the interface of CTS with *contact linguistics*, particularly on (socio-)cognitive aspects of cross-linguistic influence (CLI) of the kind sometimes referred to (in translation) as interference or shining through. The cognitive effects of simultaneous activation of two languages in a translator's mind may produce instances of CLI, which may in turn propagate into non-translational uses of language and may—in the long perspective—contribute to language change. The process has both an individual and a social dimension and is viewed as embedded in two simultaneous socio-cognitive processes: text production and reception. Whereas many CLI effects and other contact effects are well documented in language production, especially by means of corpus studies, the wider effects through reception to more permanent effects on monolingual production in the recipients' language are difficult to establish. To explore them, corpus methods have to be supplemented with process-type or experimental methods in order to establish longer-term language change effects and what (socio-)cognitive and social factors mediate such changes. Relevant factors may be language users' perceived relative power, the dominance and prestige of languages, and the relative strength of purism or openness to change with which a language is socially supported.

In Chapter 7, on translation, *pragmatics* and cognition, Fabio Alves, quoting Morris (1938), defines pragmatics as "the science of the relationship of signs to their interpreters", i.e. the science of how humans use language to exchange thoughts. The perspective from which Alves approaches pragmatics is first through the Anglo-American tradition of language philosophy with its strong philosophical and increasingly cognitive orientation, as well as its discovery of the many ways language users "do things" with words in so-called speech acts, and the subtle ways in which intended and understood meaning depends on context and situation and implied meaning as much as on explicitly stated meaning. The main focus of the chapter is on the approach of relevance theory (RT) to pragmatics as applied to translation. It revisits the axioms of RT and expounds the way RT describes comprehension and the notion of cognitive environment. RT sees translation as an act of interlingual interpretive language use, based on Gutt's concept of interpretive resemblance. The chapter also looks at processing effort and cognitive effects in translation as well as instances of metarepresentation (how communicators interpret each other's minds) and higher-order representations. It further looks at the semantics–pragmatics interface and issues related to explicitation and explicitness in translation. Here the difference between explicatures and implicatures is clearly expounded. The chapter ends by reporting on experimental pragmatics as a current and future development.

In Chapter 8, Maureen Ehrensberger-Dow examines the interface between translation, cognition and *ergonomics*, placing translators physically in a workplace environment and studying the way in which external factors, including other humans, impact on them as well as on their translation performance and their own (cognitive) perception of their work and work situation. Field study methods are preferred in order to get as close as possible to where the real action is, with screen recording as a possible source of supplementary data. Although the primary concern of ergonomics has been aimed at seeking to optimize behaviour by designing the tools used in production, including desks, chairs, keyboards, computer screens and interfaces, and other elements in the work environment, ergonomics also takes an interest in softer factors like social organization, and there is a concomitant concern for health, well-being and cognitive ergonomics. From the point of view of program designers, much effort has gone into developing ergonomically attractive screen and keyboard layouts, especially now that nearly all translation is done in interaction with computers.

In Chapter 9, Adriana Pagano tackles the concept of *domain* in theories of translation and theories of competence and expertise. After criticizing code-oriented models of translation that separate language, thought and reality by distinguishing between language "form" and language "content", she adopts an approach through systemic functional linguistics (SFL), which offers a language-based approach to human cognition that sees "language as inextricably embedded in social context" and a domain as "construed through language", whether in translation or not. A language-based ontology of a domain can help translators know "how meanings are worded in a particular domain". Text mining in a domain makes it possible to build an ontology in which conceptual categories are all based on natural language expressions, e.g. by nurses. Building an ontology on natural language use makes it easier to extend ontologies multilingually. Artificial intelligence routines applied to natural language ontological data can successfully identify connections, make inferences, formulate new axioms and suggest translations, and can also assist with making accurate diagnoses and suggest treatment. Ontologies are stronger than bilingual dictionaries by providing a richer linguistic modelling of the action and the full social experience in a domain by drawing on the natural language used in the domain. There is therefore great potential in multilingually oriented ontologies to assist translators.

Chapter 10 by Stella Neumann and Tatiana Serbina concerns the kinds of contribution *corpus linguistics* techniques can make to Cognitive Translation Studies. One way is by testing hypotheses about cognitive phenomena. Text corpus data do not give access to process data but can nevertheless lead to assumptions about cognitive causes of phenomena manifested in the product data. For instance, machine learning programs are very good at discriminating between translated and non-translated texts, which makes it fair to assume that translated texts (often) share certain recognizable features that may have a cognitive origin. Explicitation, normalization, simplification and shining through are familiar examples of features that can be documented by corpus methods to be frequently found in translations. More subtle frequency effects that have no known cognitive explanation can also be detected by corpus analysis. Cognitive phenomena like chunking, schematization and entrenchment are more difficult to demonstrate with corpus linguistic methods, although attempts have been made to relate translation shifts to different levels of entrenchment in the translators' minds. Assumptions arising from frequency observations in corpus data illustrate the usefulness of multi-method approaches. If combined with experimental research, and if assumptions are supported by experimental evidence, speculative assumptions can grow into strong hypotheses. This multi-method (triangulating) approach is strongly advocated by the authors.

In Chapter 11, Kirsten Malmkjær, writing about the interface between translation, cognition and *linguistics*, points out the fundamental difference between rationalism and empiricism as two ways of understanding human reason. Rationalism "ascribes considerable innateness of cognitive phenomena to people" and tends to assume that being human involves sharing universal cognitive abilities and constraints. Empiricism, by contrast, "considers socialization and cognitive development to be mainly a question of acquiring and learning from experience"—and not necessarily arriving at the same understanding or speaking about experience in similar ways. Contemporary rationalist thinking in linguistics is associated not only with Chomsky, who has had very little influence on Translation Studies, but also with the cognitivist tradition. The main influence on Translation Studies before the cognitive turn was from empiricists like Boas, Sapir and Whorf, whose main observation was that we do not develop identical cognitive pictures of the world across different languages and cultures, and from descriptivists like Halliday and Toury. Contemporary cognitivist approaches to language, as represented e.g. by Langacker, Dirven and Fillmore, and also relevance theorists, like Sperber, Wilson and Gutt, are categorized as continuing the rationalist tradition, but at the end it is clear that rationalists, empiricists and descriptivists need one another. As the author states, "there can be no language use without cognitive engagement, just as there cannot be [...] thought that is not expressible in language".

In Chapter 12, Agnieszka Chmiel examines the interface of translation and cognition with *psycholinguistics*, itself a discipline at the interface of psychology and language. This is undoubtedly where there has been the closest and most intense interaction in CTS since the 1980s. The aim of psycholinguistics is to "explain how language is processed in the human mind", which resonates perfectly with CTS's aim to explain how translation works in the human mind. The chapter gives a full account of psycholinguistic methods used (from TAPs, keylogging and eye tracking to EEG and other neuroscience methods) in the analysis of lexical and syntactic processing, reading and writing patterns and pauses, memory and executive functions, and the question of directionality. Psycholinguistically motivated models of translation (horizontal/vertical, monitor and recursive models) are also treated, as well as Halverson's gravitational pull hypothesis. The chapter also deals with how to design relevant tasks using authentic or manipulated materials; how to find a proper balance between control of variables and an ecologically valid simulation of experienced reality; what data to elicit and what measures to make; and how to triangulate findings based on analysis of qualitative and quantitative data. Finally, a strong plea is made for the use of mixed-method approaches applied to well-powered experiments and the use of strong statistics with reports of effect sizes.

In Chapter 13, Adolfo García and Edinson Muñoz study activity in the brain that can be associated with translation and cognition from a *neuroscience* perspective. In their terminology, translation is "interlingual reformulation" in either the "forward" or the "backward" direction. Scientific interest in neuroscientific approaches has developed so rapidly that the authors declare it a full-blown sub-discipline of CTS. Neuroscience employs methods like EEG, positron emission tomography (PET), fNIRS and fMRI to study electrical and haemodynamic activity in the brain to identify which areas are particularly active during the execution of specific translation tasks. Research in this paradigm is experimental and laboratory oriented and includes a battery of behavioural and linguistic tasks. Most methods are non-invasive, but certain methods involve insertion of electrodes into brain regions to assess a specific region's role in the execution of a task. Important findings include knowledge about each brain hemisphere's contribution to translation skills, neurocognitive routes, temporal task dynamics, and the impact of interpreting expertise on neuroplasticity. Although activity is widely distributed in the whole brain during translation, the left hemisphere appears to dominate in the vast majority of the population in the execution of translation tasks. Broca's area and neighbouring areas appear to be particularly involved, with word translation and sentence translation engaging different brain regions, associated with different kinds of processing (faster or slower) along different neurocognitive routes. Evidence has been found that interpreters' grey matter increases in volume with growing expertise, demonstrating the functional adaptability and biological plasticity of the human brain.

Part III

In Part III, Translation and types of cognitive processing, all 12 chapters look at different types of cognitive processes related to translation task execution. They are concerned with what triggers them, how they are managed and measured, and how knowledge about translation as a whole is acquired. The chapters in Part III concern different aspects of cognitive processing involving greater or lesser *effort* (Chapter 14), *attention* (Chapter 15), *emotion* (Chapter 16), *creativity* (Chapter 17), challenges of translating *metaphor* (Chapter 18), achieving *equivalence* (Chapter 19), applying *information theory*, e.g. to the study of expectancy effects (Chapter 20), dealing with *human–computer interaction* (Chapter 21) and, generally, how to *acquire translation competence* (Chapter 22); and further, how a combination of *process and product data* analysis can throw light on translational cognition (Chapter 23), how *multimodal communication* is processed (Chapter 24) and how *translational risk* is managed (Chapter 25). Collectively, the chapters in this part of the volume provide an overview of internal and external factors potentially affecting cognitive translational processes, ways of measuring and managing those effects, and how to acquire and build the necessary translation competence to handle all these factors.

In Chapter 14, Daniel Gile and Victoria Lei discuss the emergence of interest in *effort* (also known as workload) as a cognitive construal related to the assumption that the central processing capacity in the working memory is limited and to the subjective experience of translation being sometimes more effortful and sometimes less. Cognitive overload has a negative impact on performance, but the relationship between the amount of effort invested in a translation task and the success or quality of the performance is not straightforward. Some research indicates that there is both slow and fast translation, associated with two distinct processing modes, horizontal and vertical, depending on the degree of conceptual mediation required. In simultaneous interpreting, effort is often challenged to the limit, especially in adverse conditions caused e.g. by a speaker's fast delivery of information-dense speech, perhaps in an unfamiliar accent and in an acoustically load environment. A frequently used metric of effort is processing time per unit, as reflected in typing speed (for written translation) and in patterns of eye movements on source information. Hesitations and omissions in interpreters' spoken output are other potential indicators of high-intensity effort. Pupillometric and brain imaging techniques are also used, sometimes in combination with subjective reports to underpin the construct. A limitation pointed out by the authors is the current inability of research to relate specific levels of effort to specific comprehension or formulation issues.

Chapter 15, by Kristian Hvelplund, discusses *attention* as a cognitive process by which specific environmental objects are attended to while others are ignored. The main methods used in CTS to know what a translator or interpreter attended to at any given point in time include verbal reports in the form of concurrent think-aloud verbalizations or retrospective reports and behavioural measurements like keylogging and eye tracking (including pupillometry). Physiological data from brain imaging and EEG are especially useful for identifying what brain regions were particularly active when a certain item was attended to and with what speed and intensity. What a translator attends to visually (mostly one or a few words) is referred to as an "attention unit", but gaze data can only identify what graphic representation of meaning was attended to, not what meaning the translator was cognitively attending to and working on. Therefore, what is identified as being visually attended to may be considerably displaced relative to what was being cognitively processed at the time. The fact that attention can be split between multiple concurrent tasks and very quickly switched from one task to another also contributes to the difficulty of knowing what exactly is attended to, when and for how long. Better technologies will hopefully advance the accuracy with which a translator's true focus of attention can be identified.

In Chapter 16, Caroline Lehr takes up the theme of the influence of *emotions* on thinking, which has only recently begun to be researched empirically and systematically integrated into translation theory. Where the emphasis in the classical cognitive paradigm was on viewing cognitive processes as fundamentally rational, increasingly, particularly since the 1990s, emotion and affect have come to be seen as crucially important factors in shaping human cognition and consciousness and in generating situational meaning and understanding. Emotion can be triggered by both external and internal stimuli. It is understood as an episode involving a change in such states as an organism's action tendency, appraisal of an important event or a subjective feeling, and also has a physiological and expression component. Emotions may be provoked by the source text, the translator's performance, external working conditions and the translator's personal well-being and will affect a translator's attention, judgement, problem solving and decision-making activity. Empirical research has demonstrated how emotions arising from positive or negative feedback have a regulatory effect on a translator's behaviour, e.g. towards greater creativity or stricter monitoring of accuracy. Lehr ends by stating that emotional competences are an integral part of translation competence and should be part of the training of future translators.

Chapter 17, by Gerrit Bayer-Hohenwarter and Paul Kußmaul, examines the role of *creativity* in Cognitive Translation Studies. Adopting a functionalist *skopos*-oriented approach, which they see as an overall framework for creativity in translation, they emphasize creativity as a human translational prerogative, which cannot be emulated by computers. Creativity manifests itself both in the translation process and in translation products. In translations, creativity is manifested in elements of novelty or even uniqueness, which emerge in a process characterized by periods of preparation, incubation, illumination and evaluation. Visualization of comprehended meaning has been found to be helpful in producing creative solutions based on more complex analogical reasoning. Other findings indicate that professional translators are better at switching between a routine mode of production and a cognitively more demanding creative mode. A social media case is used to illustrate the way in which creative translation may result from a translator's interaction with many individuals, exploiting their crowd knowledge and creativity to find both the most likely contextual meaning of an unfamiliar expression and an adequate creative translation of it.

In Chapter 18, Christina Schäffner and Paul Chilton look specifically at *metaphor* translation, which has traditionally been approached in terms of whether or not to translate a linguistic metaphor into another linguistic metaphor. Metaphorical expressions have frequently been considered a challenge to translation and have stimulated discussions of translatability. Schäffner and Chilton approach metaphor translation from a cognitive perspective, looking at metaphors from the perspective of Lakoff and Johnson's conceptual metaphor theory, which sees metaphor not primarily as a language operation but as a mapping across conceptual domains, which can be represented in a variety of language manifestations. The theory makes a distinction between primary and complex metaphors, with primary metaphors being potentially universal, as they relate to mappings related to bodily experience. Manifestations of primary conceptual metaphors generally translate more easily than complex metaphors, which tend to make use of culturally based conceptual frames. Empirical research tends to agree that metaphor translation is more cognitively demanding than translation of non-metaphorical expressions, but much remains to be empirically investigated, multimethodologically and in real-life, technology-supported contexts.

Chapter 19, by Erich Steiner, tackles the question of *equivalence*, perhaps the most hotly debated concept in the history of Translation Studies. According to Steiner, "it is not easy to define translation entirely without it". He uses the concept to distinguish between paraphrase and variation by truth-conditional rather than cognitive criteria, with translation as an "approximation to a multi-functional paraphrase of the ST [source text] by some TT [target text]". Translation is "text production under the constraints of a ST". Translation of multimodal genres is said to be in need of a motivated notion of equivalence. Models of translation and cognition are also still lacking in detail and concreteness. Equivalence can be applied at any level of textuality, but linguistic representation at any level is full of ambiguity before it is read, interpreted and disambiguated by a reader/translator. It is stated that this interpreted and understood meaning is what the translator represents ("instantiates") in the TT, but existing models of translation need to be more explicit about their commitment to either text or reading (interpreted meaning), Steiner argues. He is more comfortable working with proven naturalistic language corpus data (within an SFL theoretical framework) than with venturing into research based on what he sees as problematic process data recorded by immature methods. However, if product and process studies are integrated, linguistic evidence is likely to emerge showing that translation involves a search for equivalence.

Chapter 20, by Elke Teich, José Martínez Martínez and Alina Karakanta, deals with the interface between cognition, *information theory* and cognition. Using a computational calculation of predictability in (a linguistic) context to model human translation based on an information processing ("noisy channel") theory of communication, they seek to relate probabilities of occurrence of words or phrases to cognitive notions like "expectancy", "surprisal", "choice", "difficulty", "world knowledge" and "assumptions about the addressee". Mathematical entropy applied as a measure of the complexity of a translational choice has been found to correlate positively with cognitive uncertainty, surprisal and processing effort. The success of modelling translation in this manner is measured by equivalence (fidelity of TT to ST) and adequacy (conformity of TT with TL norms), and translation is operationalized as "a search task in a space of alternative linguistic options". "Context" is defined in terms of ambient text and by reference to register, and equivalence and adequacy are established from probabilities of linguistic occurrences in corpora. It will be interesting to see if future machine translation (MT) solutions based on a model of human translation as information processing will be able to further optimize

MT quality without needing to implement a language-independent, cognitive construal of situation and context.

In Chapter 21, Sharon O'Brien discusses cognitive aspects related to *human–computer interaction* in translation. In less than 30 years, technology has radically changed the way most professional translators work, regardless of whether they work with regular text or e.g. with audiovisual translation, "with the translator in some circumstances becoming almost symbiotic with the 'machine'". Interaction with the machine is not just with dedicated translation software like translation memory and machine translation systems, which are increasingly integrated, but with the Internet and the resources to which it gives access. This makes new demands, both physical and cognitive, on translators and interpreters, but it also has the potential to reduce those demands. The benefits of technological aids are evident: translation is speeded up, throughput volume is up and quality is up. There may also be wider social and ethical benefits. More problematically, cost is down, and there are other challenges, many of them cognitive. Technology does away with plenty of routine work, but many translators feel dehumanized by the machine and robbed of their creativity by having to post-edit errors they would never have made themselves. To reduce cognitive friction, translation software should be interactive and adaptive; it should learn from the user and function more as an extension of the translator's mind. This can be achieved if the translation software of tomorrow is created to allow tailor-made personalization.

In Chapter 22, Amparo Hurtado Albir offers a comprehensive survey of attempts to define *translation competence* and *translation competence acquisition*. Most studies have concluded that the construct can be broken down into a number of components such as bilingual competence (at some level), including reception and production competence, strategic competence and transfer competence, but cultural competence, world knowledge, theoretical knowledge and many more have also been claimed as components. "Competence" is widely understood as a scale, which is one source of difficulty in reaching agreement on a definition. Other problems arise from competence being in competition with terms like ability, skill, proficiency and expertise, all of which are also scalable, and all of which can be understood as having both a competence and a performance interpretation, with both a knowledge and a behaviour dimension. Furthermore, there have been relatively few empirical studies of competence seeking to validate a competence model, although a few have been large-scale projects with many participants.

The discussion of translation competence acquisition (TCA) departs from Harris's notion of "natural translation" followed by Toury's socialization idea and Shreve's description of TCA as proceeding from natural translation to professional ("constructed") translation following an expertise trajectory, perhaps along Chesterman's proposed five stages, leading to increasing automatization and higher-level, more holistic decision making, but along an itinerary that is not linear but spiral, both progressive and regressive, and unevenly distributed across subcompetences. Large-scale, empirical, longitudinal and replicable research targeting TCA in well-defined contexts is called for.

In Chapter 23, Silvia Hansen-Schirra and Jean Nitzke argue that we need to study the *process–product interface* to know about translational cognition. Corpus methods are good at discovering linguistic patterns that distinguish translated from non-translated language, but they lack a method for identifying what causes such "universal" patterns as normalization and shining through. Process-oriented research operates with concepts like priming and monitoring. If such patterns and concepts could be convincingly linked and, for example, establish a link between monitoring and normalization and between priming and shining through, our understanding of translation would be improved. Methods for linking data obtained with the two methods, either from different materials or from the same materials, are suggested. Another example concerns translation of cognates. Corpus analysis demonstrates that lexical variety in translation of non-cognates is higher than for cognates. This effect has been shown to be modulated by many factors, including translation mode, level of expertise and others, but also by priming and/or monitoring. A product-process approach would be able to show if such normalization was caused more by priming or by monitoring.

In Chapter 24, Jan-Louis Kruger focuses on the cognitive demands associated with translation of multimodal genres (especially subtitling and audio description) and on the impact of *multimodality* on recipients' cognitive processes. After centuries of conceptualizing translation as text translation, it is now apparent that there is more to translation than translating words. In multimodal communication "linguistic content is always framed in, supplemented by, and informed by context, co-texts and non-verbal modes that determine the translation as well as the reception of the translated product". In audio description the "source text" is not linguistic at all, and translation is not interlingual, but intermodal. What is translated (and represented in speech) is the action shown visually in a movie. Studies of the way multimodality affects the way viewers/listeners/readers construe meaning and how much cognitive load is imposed on them have found that subtitles may be found distractive, because they automatically attract visual attention, but may also help an audience become more fully immersed in a movie. Similarly, redundancy effects in multimodal communication may have both positive and negative effects. One important methodology used in this research is eye tracking, which shows that the way e.g. living pictures are "read" is very different from the way text is read, showing that we engage differently with the new, more complex representations of reality mediated by multimodal communication.

Chapter 25, by Anthony Pym, is about the concept of *risk management*, which may be applied socio-economically, and also to cognitive concerns a translator may have concerning a translation. In particular, risk can be related to cognitive concepts like confidence and uncertainty in the handling of ambiguity and indeterminacy. Professional translators appear to display greater tolerance of uncertainty than novice translators because of their superior ability to assess risk and to discriminate between high-risk and lower-risk situations. This allows professional translators to more effectively allocate cognitive effort, deal routinely with low-risk translation and devote more time, attention and reflection to high-risk challenges. If translation is done collaboratively or in interaction with a CAT tool and a machine translation engine, assessment of the risk scene depends crucially on the trust a translator is prepared to invest in contributions from collaborators (including the machine) and the credibility a translator aims at having with colleagues and clients. If a translation is a text that clients receive and accept as a translation, then credibility is a translator's most important asset. Managing risk involves reducing, transferring, mitigating and avoiding risk or taking it. Most studies of translational risk have suggested that risk aversion dominates. This tendency could be culturally induced. Recognition "that risk taking can be justified, pleasurable and socially rewarded" could help translators break away from subservience, fear and inadequacy.

Part IV

The five chapters in Part IV, Taking Cognitive Translation Studies into the future, conclude the Handbook and point to important new avenues for CTS. Part IV closes the circle by means of a dialogue with the four chapters in Part I and by offering bids for future directions and reflections on where CTS may be moving.

Chapter 26 discusses *expertise and expert performance* as alternative concepts to competence and attempts to construct a trail to finding ways of better assessing translation performance. If this could be achieved, it would have very important consequences for translation training and for the translation profession. Chapter 27 undertakes to describe some of the far-reaching consequences for CTS, in particular the scope of CTS, of an acceptance and implementation of the ideas of *situated, embodied, distributed, embedded and extended cognition*. Chapter 28 is an ambitious forward-looking attempt to model human translation computationally in a way that attempts to include several of these notions, in particular embedding and extension. Chapter 29 fits Translation Studies into the fully developed SFL system, offering SFL as an alternative theoretical framework to the CAS framework presented in Chapter 4. Chapter 30 summarizes the editors' reflections of the overall direction of CTS.

In Chapter 26, Igor da Silva gives an account of ways in which *expertise* and *expert performance* have been employed to describe a very high level of consistently superior translation performance, sometimes in an effort to replace the competence construct with a hopefully more robust and theoretically enlightening construct. Like competence, expertise is an acquired skill, but long experience in a domain does not necessarily lead to expertise. According to expertise theory, it takes years of deliberate practice to achieve it. When achieved, it is manifested in consistently excellent superior performance in a particular domain. Cognitively, experts have superior working memories, have better knowledge and produce faster solutions than others, but only within an often quite narrow domain. This makes it very difficult to devise a performance model with well-defined tasks by which expertise can be measured. Expertise is sometimes construed as an absolute concept. In empirical research, it has often been construed somewhat loosely as a relative concept, e.g. in research comparing the performance of a group of "professional translators" (hypothetical translation experts) with a group of "novices" (hypothetically not yet translation experts) and a group of "domain professionals" (professionals in a domain, but non-translators). This type of research has demonstrated considerable difficulty in pinning down translational expertise and devising relevant tasks and metrics.

In Chapter 27, Hanna Risku and Regina Rogl offer their view of translation from the evolving perspective of *situated, embodied, distributed, embedded and extended cognition*. They first present a historical overview of first-generation (computational information processing) and second-generation (connectionist) cognitive science theories of the human mind and track the itinerary from a view of the mind as a computer-like manipulator of symbols to a view of mind as a giant network of experience-based and action-oriented sub-symbolic activations in neurons organized in a parallel and distributed architecture. Second-generation ideas inspired e.g. prototype and scenes-and-frames semantics and brought contextuality, creativity and individual experience to the fore. The third generation is dominated by situated 4EA cognition, which sees cognition as extending beyond the human mind and brain, into the whole body, into social interaction, and even into the environment and human artefacts. The five varieties of third-generation theories listed in the title of the chapter are described separately, although they admittedly overlap considerably, e.g. in that they all share an enactive view of cognition. One main consequence of the situated perspective is that meaning always emerges in a specific context. Translation, similarly, always takes place in a specific physical and social environment with which translators interact and which impacts on their cognition. The consequences of this for the scope of CTS will have to be clarified, but the third-generation views of cognition should inspire research agendas "that are committed to a non-computational understanding of the cognitive processes in translation".

In Chapter 28, Michael Carl describes the different ways in which *artificial intelligence* solutions have been variously conceived of as replacing, simulating or extending human cognitive processes. In the early cognitive science assumption, the human mind was a syntactically driven symbol-processing machine-like organ which generated meaning and intelligent solutions to problems by manipulating symbols. In the strong version of the artificial intelligence (AI) hypothesis, the machine could be claimed to have intelligence. In a weaker interpretation, the machine would only be said to simulate intelligent human behaviour. Subsequent connectionist modelling of human cognition was based on the idea of cognition as happening in neural networks whose distributed and parallel processing activity was assumed to operate not on symbols but on sub-symbolic representations. Several recent neural MT and AI solutions have suggested architectures that seek to embed the translator in an interactive computational, technological environment in which the computer will engage translators when necessary and learn from them, making user and machine operate like a single agent. In a different architecture, the interactivity between translators and the machine operates as a cognitive extension of the translators' cognitive powers. Cloud- and crowd-based solutions across distributed tasks and platforms have also been implemented. All such recent solutions provide strong interactive cognitive support to translators, but none of them makes a claim to imitate human cognitive processes.

In Chapter 29, Christian M. I. M. Matthiessen first presents a grand outline of what makes systemic functional linguistics systemic. Then he defines SFL's conceptualization of translation and *multilingual text production* as a semiotic "4th order" phenomenon foregrounding meaning, not as a cognitive phenomenon foregrounding knowledge. The two perspectives are complementary, but there are many advantages, says Matthiessen, to viewing translation as a semiotic (primarily linguistic) phenomenon construed as recreation of meaning in context. Together, the two perspectives ideally provide a more holistic view of translation. A semiotic system is an immaterial social system with meaning embedded in it. A system like language both carries ("embeds") and creates meaning, and it is embodied in semiotic beings, e.g. in "multilingual meaners" like translators. SFL construes language as organized in a multidimensional network of relations in which phenomena get their meaning from the many global and local relations they enter into. As translators recreate meaning, they constantly make choices in different modes of meaning (ideational, interpersonal and textual). Particular interest is devoted to choices leading to "shifts", instances of "grammatical metaphor" leading to different levels of (un)packing of meaning. Being a semiotic process and a metalinguistic phenomenon operating on existing texts in context, like summarizing, being also "enacted socially (in groups), embodied biologically (in organisms) and ultimately manifested physically", translation should not be construed and enacted as a distinct discipline, leading potentially to fragmentation of knowledge. SFL is offered as an overarching framework for (Cognitive) Translation Studies.

Chapter 30 is the editors' envoi, in which they reflect on where CTS is headed. The authors look at the four foundational chapters in Part I in relation to the four concluding chapters in Part IV, suggesting the likelihood of a unified framework for CTS that is epistemologically, ontologically, theoretically and methodologically grounded. The chapter ends with some tentative considerations about how the intricacies concerning translation and cognition are situated in relation to interpreting and cognition. In view of the forthcoming publication of a Routledge Handbook of Interpreting and Cognition, they suggest that further reflections on the nature of translation and interpreting with respect to cognition may give rise to an overarching framework for cognitive translation and interpreting studies (CTIS).

Note

1 https://vimeo.com/354654462

References

Bruner, G. (1990). *Acts of meaning*. Cambridge, MA: Harvard University Press.

Edelman, G. (1992). *Bright air, brilliant fire. On the matter of the mind.* New York: Basic Books, Harper Collins.
Forster, E. M. (1927). *Aspects of the novel.* London: Edward Arnold.
Gambier, Y. (2006). Multimodality and audiovisual translation. *EU-High-Level Scientific Conference Series MuTra 2006—Audiovisual Translation Scenarios: Conference Proceedings.*
Geeraerts, D. (2002). Decontextualizing and recontextualizing tendencies in 20th-century linguistics and literary theory. In E. Mengel, H.-J. Schmid, & M. Steppat (Eds.), *Anglistentag 2002* (pp. 369–379). Bayreuth. Trier: Wissenschaftlicher Verlag. Retrieved August 20, 2019, from http://wwwling.arts.kuleuven.be/qlvl/PDFPublications/03Decontextualising.pdf.
Gutt, E. A. (1991). *Translation and relevance. Cognition and context.* Blackwell: Oxford.
Halverson, S. (2010). Cognitive translation studies: Developments in theory and method. In G. Shreve, & E. Angelone (Eds.), *Translation and cognition* (pp. 349–369). Amsterdam: John Benjamins.
Jakobson, R. (1959/2000). On linguistic aspects of translation. In L. Venuti (Ed.), *The translation studies reader* (pp. 113–118). London: Routledge.
Kahneman, D. (2011). *Thinking, fast and slow.* London: Allen Lane.
Lee, P. (1996). *The Whorf Theory Complex: A critical reconstruction.* Studies in the History of the Language Sciences, Series III. Amsterdam: John Benjamins.
Miller, G. A. (2003). The cognitive revolution: A historical perspective. *Trends in Cognitive Sciences,* 7(3), 141–144.
Morris, C. H. (1938). Foundations of the theory of signs. In R. Carnap et al. (Eds.), *International encyclopedia of unified science.* Chicago: University of Chicago Press.
Muñoz Martín, R. (Ed.). (2016). *Reembedding translation process research.* Amsterdam: John Benjamins.
Pickering, M. J., & Garrod, S. (2013). An integrated theory of language production and comprehension. *Behavioral and Brain Sciences, 36,* 329–392.
Quine, W.V. O. (1960). *Word and object.* New York and London: MIT & Wiley & Sons.
Tirkkonen-Condit, S. (2004). Unique items—over- or under-represented in translated language? In A. Mauranen, & P. Kujamäki (Eds.), *Translation universals: Do they exist?* (pp. 177–184). Amsterdam: John Benjamins.

Part I

Foundational aspects of translation and cognition

1
Translation, epistemology and cognition

Andrew Chesterman

1.1 Introduction

The human brain, writes Nobel Prize winner Gerald Edelman (1992, p. 23), is "the most complicated material object in the known universe". Note the careful "known". After all, it seems that currently we can only perceive about 5% of the universe, the rest being mysterious dark matter and equally mysterious dark energy (so we lay folk are told, at least). Edelman's qualification "known" is thus eminently justified, and a useful reminder of the virtue of modesty in making claims about the world. But in what sense, and to what extent, do we even really "know" that 5%?

How much do we know about the human brain? Less than 5%? Do we even know how much we don't know?

And further: what do we think we know about "knowing"? From Plato on, philosophers have usually started from the assumption that "knowledge" is justified true belief; but such a definition carries a number of problems and has given rise to a good many debates. (For a relatively reader-friendly survey, see Steup, 2018.) The debated issues include the questions of how we can know—for certain—anything at all, what counts as justification (and for whom), and of course, what is truth. These are problems of epistemology, the philosophical study of knowledge. In the present context, my rather sceptical focus will be on the limitations of knowledge and of our methods of acquiring it. My position is thus a fallibilistic one: I assume that when we think we know something, we may be mistaken.

So, when Krings (1986) asked "What goes on in translators' heads?"—in one of the most influential early publications in what was to become the "cognitive turn" in Translation Studies (TS)—we can at least answer: well, it is unlikely that we will ever know completely. But we might get to know something, and this something might eventually "evolve" into greater knowledge, better descriptions, better explanations and better understanding.

Suppose I introspect for a minute: what happened in my head when I just wrote *what goes on in translators' heads*? For this is not what Krings actually wrote, which was: *Was in den Köpfen von Übersetzern vorgeht*. Actually, I first translated this as *what happens* .... But then I paused, and changed *what happens* to *what goes on*. Why? I'm not sure: perhaps it sounds more informal? Perhaps I was unconsciously influenced by the morphology of the German *vor-geht*? Would this interference hypothesis be strengthened by other examples of me being influenced by German morphology when translating into English? What about my translations from (or indeed into) other languages? Tests could be done .... Or was the change just due to intuition? The same kind of intuition that prompted me to write *in translators' heads* rather than *in the heads of translators*, which would have followed the German form more literally? I just don't know, however hard I try to introspect and to retrospect on what I just did. I do know that in the first line of this paragraph I wrote *happened in my head* rather than *went on* precisely because in the previous paragraph I had just shifted *happens* to *goes on* and did not want to repeat the same lexical item so close to the previous mention. But if I had not made this shift, would I have written *goes on in my head*? Again, I don't know.

As I continue my retrospective introspection, the thought occurs that there is actually a semantic difference between *happens* and *goes on* (when neither is used in a progressive form), such that *happen* co-occurs more with events and *go on* with states or processes. (*What happens when I press this switch? What goes on when the light is out?*) I find this thought supported by a quick Internet search. But I was certainly not consciously aware of it when I made my change. Was this just my native language intuition, working below the level of consciousness? Perhaps. But why, then, did I write *happens* in the first place? Because I have lived for several decades outside an English-speaking culture and have lost touch with some of the niceties of my native English? And further: how do *you* know that I am introspecting and retrospecting, and reporting, truthfully? On what grounds do you trust me (or not)?

This trivial example illustrates just some of the problems faced by cognitive research into the translation process: problems of causality and explanation; the reliability of different data sources; the plausibility of conclusions about decision making and choices, and about the possible influence of earlier choices on later decisions; accessing the possible roles of the unconscious and of intuition; and even perhaps the influence of personal life history.

1.2 Historical perspective

A sense of the historical development of research on translation and cognition can be framed as a series of epistemological assumptions. Initial assumptions are always necessary, of course; but they may turn out to be misleading or mistaken, or only partially justified. We can start with the idea of the translator's mind as a black box, with an input arrow on one side and an output arrow on the other, and a question mark in the box itself. This box model itself reveals an initial assumption: that we can learn something about the mind by considering it in isolation from its environment and from its own history, both phylogenetic and ontogenetic. Such an assumption can of course be defended as a necessary heuristic preliminary at the early stages of an investigation, a useful simplification. True, if it is borne in mind that it is just such an assumption; otherwise, there is the risk that the simplistic map is actually taken to be the complex territory it represents.

Looking at this box, first, we just wonder and speculate—surely where all science starts. We can observe what is going in and coming out, and attempt to draw inferences from what we see. The "interpretive school" of interpreter training (based at L'École supérieure d'interprètes et de traducteurs (ESIT) in Paris) arose from the initial inference that good interpreting (and translating) requires the deverbalization of the incoming message, so that its "sense" can be separated from its form and then reformulated in the target language (Seleskovitch & Lederer, 1984). This view of the cognitive process involved was based mainly on experience and intuition, and led to successful training methods that produced excellent professional interpreters. But it was not supported by empirical research on cognition, and the status of the deverbalization phase was challenged, e.g. by research supporting the "literal translation hypothesis". This claims that interpreters and translators are highly influenced by the form of the source text and tend first to select target forms that match those of the source when possible. If this is the case, it would imply that deverbalization does not take place—or at least not always. Interesting evidence supporting the literal translation hypothesis comes from studies of interim drafts and the revision process (see e.g. Englund Dimitrova, 2005). The difference between these two positions may be partly explained by the ESIT focus on what should happen in ideal interpreting, and therefore on what would be pedagogically relevant. But inferences may also be mistaken.

Continuing to ponder our black box, we might assume that our translator's mind contains a number of smaller, interconnected boxes (modules, components) with different functions: one deals with comprehension, for instance, another with target-language formulation, yet another is perhaps engaged in some kind of quality checking, etc. So we assume that the mind is modular, or at least that it is useful to regard it as such at the initial stages of an investigation. The image is of the mind as some kind of machine, with various components all doing their own jobs, converting input into output.

Then we try manipulating the input and see how the output changes: i.e. we adopt an experimental method. We also begin scratching the surface of the box, with the idea of getting below the surface, at least to some small degree. We thus assume that by entering the outer layer of the brain we can begin to glimpse the workings of the mind: we can, for instance, use EEG (electroencephalography) or PET (positron emission topography) and examine which bits of the brain appear to be activated under different conditions (e.g. Kurz, 1994; Tommola, 1999). Some later research, however (e.g. García et al., 2016), has cast doubt on the idea that the brain has areas that are dedicated specifically to translation or interpreting.

But do these methods give us any knowledge about the mind? Is it reasonable to assume that the brain is a window on the mind? Not a transparent one, at least, it would seem. The huge mind–brain issue (or more generally, the mind–body issue) has not been resolved to general satisfaction, although a strict dualist position seems now to be out of fashion. The jury is still out and may well remain out for some time to come. Indeed, perhaps this issue is actually not solvable at all. (But see later.) There is, however, the risk that we too easily assume that new information about the brain tells us something about the mind. It may, but it may not: reservations are needed.

From the mid-1980s on, Krings and others experimented with a new way of collecting data on what were assumed to be the cognitive processes involved in translation. This was introspection in the form of spoken think-aloud protocols (TAPs). The method had previously been used in psychology (see Ericsson & Simon, 1993). Part of my brief introduction earlier was a kind of thinking aloud, not spoken in this case, but slightly delayed and to some extent monitored before being expressed in writing. In my case, I was retrospecting; but in the protocols used by the early TAP scholars the idea was to have the subjects talk aloud at the same time as they translated (usually working on a computer, with a separate recording machine).

It was widely recognized that the TAP method had limitations (it was obviously not appropriate for research on interpreting) and flaws. Assumptions were made that now look hard to justify. Critics pointed out that the very act of talking aloud would surely interfere with the normal cognitive process that was allegedly being studied. Indeed, the very act of being observed may have interfered (the observer effect). How reliable would the protocols be? Wouldn't they be rather selective, focusing only on what can be verbalized, and within that, on translation problems at the expense of routine flows? How much of what goes through the subject's mind *would* actually be verbalized? And wouldn't the TAPs actually concern the results of cognitive processes rather than the processes themselves? Such criticisms were often discussed, and often acknowledged, but TAP research nevertheless went ahead. Attempts were made to rectify some of the problems, for instance by setting up TAPs with pairs or small groups of subjects, on the assumption that this would allow a more natural dialogue rather than a possibly unnatural monologue. TAPs were also combined with retrospective interviews. Yet despite these developments, the numbers of subjects studied in each project remained very small—for the obvious reason that this kind of research is extremely time-consuming. (See Jääskeläinen, 2002, for a survey.)

Further refinements added new kinds of data gathered from new kinds of technical tools, the most noteworthy being Jakobsen's Translog keystroke logging program, and later the application of eye-tracking technology. (See e.g. Jakobsen, 2017.) Here too there are interesting assumptions. Now, for instance, it is the eye that is assumed to serve as a window on the mind. There is also the assumption that when the subject pauses, i.e. stops typing for a few seconds or minutes, this absence of observable activity is a significant indication of non-observed cognitive activity, such as grappling with an awkward translation problem (rather than, say, day-dreaming). One big advantage of these new methods was their use in triangulating results, e.g. cross-checking the results of a TAP with those of keystroke logging and/or eye-tracking data and/or a retrospective interview. These innovations allowed more specific hypothesis formulation and more robust hypothesis testing (for instance of the above-mentioned literal translation hypothesis). TAPs have suggested significant differences in the way professionals or experts translate in comparison with students or untrained translators. They have also generated promising new lines of research, such as on the relevance of attitudinal and emotional factors, which have further complicated our original image of the black box. Much of this research has been methodology driven (we do it because we now can) rather than theory driven (e.g. to test theoretically significant hypotheses) or problem driven (e.g. to solve translation quality problems). It has often been exploratory: let's do this, and see what we can find. This is not to say that such research does not bring new knowledge, but such knowledge may appear fragmentary if its theoretical framework is unclear. Observation is used to test theory, yes, such as when hypotheses are tested; but theory is also needed to make sense of what is observed.

Research into the cognitive processes involved in translation has been increasingly influenced by aspects of the related fields of psycholinguistics, cognitive linguistics and cognitive science. Translation is a complex activity, true, but on what grounds might we assume that the nature of translators' cognitive processes would be significantly different from those of electricians, or dentists? Apart from the factor of the two languages, of course. More recently, "Cognitive Translation Studies" has been launched as a cover term for a broader framework of study that would open up to other relevant fields such as bilingual research and language acquisition, and more broadly still, neuroscience. (See e.g. Muñoz Martín, 2014, 2016a.)

We started with the assumption that the translator's mind could be initially conceptualized as a black box, and it has seemed that the study of translation cognition also started with a similar image of itself, i.e. as a self-contained field. In both cases, such assumptions have now been largely abandoned or become superseded by others. This in itself is of epistemological significance. As noted earlier, it is well known that we need concepts in order to interpret what we observe. It is also well known that our concepts and models, including our heuristic metaphors, influence the way we make sense of what we observe. If you see the brain as a machine, you will be disposed to think of it in terms of components and functions, etc. If you see it as an information processing computer, you will visualize algorithms and programs. Indeed, several of the first models of the translator's mind were pictured as flow diagrams that were algorithmic in form, from Krings onward. More recent work influenced by computational linguistics also attempts to model human translation processes in computational terms. (See e.g. Carl & Schaeffer, 2017.)

But suppose, like Edelman (1992, p. 29), you see the mind as a jungle, a rainforest? What would then follow? At least this: that everything has evolved in connection with everything else, in a complex environment including emotions, history, evolution, material factors, social and cultural factors, and much else. Such an image would be consistent with the position known as embodied cognition. (See further later in this chapter.)

This is an empirical position: research on translation and cognition has indeed been dominated by a general empirical paradigm, as opposed to the postmodern, relativist paradigm, which has become prominent in some other areas of Translation Studies (see Delabastita, 2003). The former view takes TS to be an empirical human science rather than a branch of philosophy or literary theory. On this view, it is at least possible to strive for neutral and objective knowledge, although we know (or assume) that this ideal is ultimately unattainable.

1.3 Core issues

1.3.1 Different concepts of theory

Knowledge evolves. This implies that even our concepts of what knowledge is, or what a theory is, presumably also evolve. If you have a broad view of what a theory can be, you can even see myths as a kind of pre-scientific theory, offering at least some kind of explanation for phenomena such as the cycle of the seasons or the causes of volcanoes (cf. the myth of Vulcan's forge). Metaphors too can be seen as a kind of primitive theory: recall the many metaphors that have been proposed for the act of translation, from the dominant but problematic transfer metaphor to hundreds of others, as a "way of seeing" translation (cf. the Greek etymology of the word *theory*). In the general empirical paradigm, one can take hypotheses, too, as mini-theories, hopefully related to a larger framework of descriptive and explanatory propositions that together make sense and perhaps allow prediction.

But how far can we go without such a larger framework? The so-called "new experimentalist" position argues that progress can be made without claims being necessarily embedded into a large-scale theory. Experiments that work, i.e. that have the predicted effect and are corroborated when they are replicated, remain valid evidence even though no-one can explain why they work. (See Chalmers, 1976/1999, Ch. 13, for a philosophical discussion of this point.) This view does not claim that we do not need theory, but that good progress can sometimes be made before theory can catch up, as it were. This seems to be what is happening in some contemporary work on translation cognition (e.g. Carl et al., 2015). On this view, then, tentative progress is first made inductively, only later being integrated into a theoretical structure from which further hypotheses can be formulated deductively.

1.3.2 Interpretive hypotheses

Another question concerns the role of purely conceptual analysis in empirical research. Definitions (e.g. of "cognition"), categorizations, and interpretations of X *as* Y can all be taken as hypotheses of a kind, sometimes known as interpretive hypotheses (see further Chesterman, 2008a). I have just claimed that definitions etc. can be taken *as* hypotheses: this itself is an interpretive hypothesis, in my view at least. Not all scholars agree with this characterization, which indicates that interpretations can always be argued over. As interpretations, however, they cannot be falsified (cf. Popper's criterion for scientific knowledge, e.g. Popper, 1959). They are not empirical claims but conceptual ones. However, they can certainly be tested: not on the true-or-false criterion but on the pragmatic one of usefulness. A useful interpretive hypothesis has added value in that it brings clarity, generates significant empirical research, stimulates further questions and so on. An interpretive hypothesis that turns out not to be useful simply fades away.

This testing process is epistemologically just as important for conceptual claims as for empirical ones; however, this testing is often neglected. Conceptual proposals are often made, and supportive illustrations given, without adequate account being taken of counter-examples to which the proposal does not seem to apply, and without specification of the added value of the proposal in question—its advantage over competing proposals, for instance. One advantage of the "interpretive hypothesis" proposal is that the term "hypothesis" relates these conceptual claims to empirical hypotheses and hence to the necessity of testing them, albeit on different criteria. That is, the term underlines their status as precisely hypotheses, not facts or truths. There is always the risk of presenting a conceptual distinction (such as "there are three kinds of translation") in a way that makes it look like an empirical fact when it is really a conceptual suggestion that is hopefully a useful way of thinking about something.

A particularly important kind of interpretive hypothesis concerns the notion of a category and the process of categorization, without which we cannot formulate or use concepts, for a concept is itself the result of a categorization. Ellis (1993) shows how categorization is absolutely central to knowledge, because without categories we cannot generalize. Indeed, the whole point of a category is to create the possibility of generalization. A category groups together individual instances that can be counted as relevantly similar for a given purpose—to allow a generalization. Instances within a category are thus counted as equivalent, for a given purpose, although, as individual instances, they are in fact all different in some respects. Ellis (1993, p. 117) cites Peirce on the idea that to know something is precisely to categorize it, i.e. to place it in relation to other things. Categories may or may not be "natural categories" (translation, for instance, does not seem to be a natural category); categories may be classical (Aristotelian), fuzzy, prototypes, clusters, continua, etc., but the point to bear in mind is that they are all conceptual tools, like all interpretive hypotheses, to be used or revised or discarded, as seems appropriate.

As noted earlier, research into translators' cognition is empirical research. But the object of this research is not like the objects of research done in the natural sciences, because we are dealing with the mental behaviour of human beings, which is (we assume) much more unpredictable, messy and varied than, say, the behaviour of atoms and molecules. This assumption too may turn out to be false as more is revealed about the role of chance and unpredictability at the quantum level of physical matter: the unknown unknowns are multiple. However, given what seems still to be a significant difference between the natural sciences and the human sciences, we are justified in enquiring about the status of our explanations in cognitive research. The standard view has long been that the natural sciences look for explanation, whereas the humanities aim for understanding (see von Wright, 1971). Cognition research perhaps lies somewhere between these two alternatives, given its significant experimental input. And perhaps the relation between explanation and understanding is not one of opposition but, rather, complementarity. When we think we have a good explanation, we say that we understand. And it is only to the extent that (we think) we understand that we can (we think) validly explain.

1.3.3 Causes and other explanations

Yet there does remain an important epistemological difference vis-à-vis the natural sciences, and it has to do with causality. In the natural sciences explanations are typically causal. More specifically, in Aristotelian terms they are typically material causes, based on law-like (nomic) generalizations. (True, one may then wonder why nature should seem to obey laws at all—see e.g. Chalmers, 1976/1999, Ch. 14.) On the other hand, although explanations of human behaviour, including cognitive behaviour, may partly be based on material causes, they are also based on final causes (purpose, intention), formal causes (e.g. socio-cultural norms), and efficient causes (e.g. pertaining to features of the agent(s) in question). Consider history as a humanities discipline: the causes of historical events such as a war may be a mixture of all four cause types. As historical explanations, these causes are not law-like, and they do not allow precise predictions. Some scholars would hesitate to use the term "cause" at all in such contexts, and speak instead of influencing conditions, quasi-causes or the like.

Now consider the translation process. All four Aristotelian causes are relevant here too, but in a non-nomic way. In any given translation event, the process has a goal: say, a decent translation of a particular kind for a particular purpose (final cause). The process is influenced by expectations and norms (what will count as a decent translation, etc. in the given situation: formal cause). It is influenced by the linguistic form of the source text and by the material constraints of the target language (material causes). And it is influenced by the mood, experience, competence, attitudes, habitus, etc. of the translator (efficient cause); and perhaps by other things too, such as the situational constraints of time, resources and so on. Within TS, all these cause types have been explored both conceptually and empirically. Historically, the focus was first on the material causes, i.e. the contrastive relation between the two languages in question, but within the past few decades there has been a huge expansion into research on cultural, social and psychological factors as well. (See e.g. Chesterman, 1998.)

Establishing firm evidence for any of these causes is far from unproblematic. Final causes imply intentions, which are notoriously difficult to pin down. Formal causes such as norms imply both textual evidence (regularities) and extratextual evidence (such as an authoritative statement that X is indeed a norm), since textual regularities may also have other causes themselves (such as cognitive constraints). The range of potential efficient causes appears to be enormous. And the material causes may intertwine with any or all of the others .... The assumption of a *single* cause, in any given case, would rarely be justified.

There are also other possible types of explanation, not just causes or influencing factors. Here, too, scholarly opinions differ. On one view, generalizations also rank as a kind of explanation. This is because generalization allows us to relate a single puzzling event, for which an explanation is needed, to other similar puzzling events: we generalize from an event to a class of events. And then we can perhaps generalize across classes in terms of a higher class. The lower-level events can then be said to be "explained" by appeal to a higher generalization—even though the causal mechanism is not known. The distinction between description and explanation thus becomes blurred. (See Croft, 1990/2003, pp. 284–285; Halverson, 2003.) One problem in the human sciences, however, is that proposed generalizations may be based on rather few instances (such as in TAP studies) or be too vague to allow testable predictions. Generalizations are often formulated as tendencies, but information on exactly how a given tendency is defined in terms of probability is often lacking, which means that the claim in question cannot be adequately tested, let alone falsified. (Does the phenomenon occur in at least x% of potential cases? More often than some opposing tendency? Always?) And the scope of a generalization also needs to be stated if the claim is to be tested. (Under what conditions, exactly, is the phenomenon claimed to occur?)

Yet another kind of explanation is what Salmon has called "unification" (Salmon, 1998, pp. 69–70). This works by relating the *explanandum* to a wider context so as to bring together a variety of facts and synthesize them under a single structuring concept. The classic example is Darwin's concept of natural selection (ibid., p. 360), which made excellent sense of a great deal of data, even though he was not aware of the underlying genetic mechanisms.

Any explanation can be framed as an answer to an explanation-asking question. And any explanation may be wrong, even though it may satisfy the questioner, at least for a time. (What's that smoke, Mummy?—That's just Vulcan, heating up his forge.—Oh, I see.) An interesting question is then: what do people accept as an explanation, or as a good explanation, and why? Opinions will often differ. (For further discussion, see Cattrysse, 2014, Ch. 10; Chesterman, 2008b.)

1.3.4 Models

Another problematic issue concerns the kinds of models that are proposed when scholars wish to represent the translation process. In terms of a distinction made by Toury (1995, p. 249), some of these models seem to be of the translation *act* (cognitive, not directly observable) and others are of the translation *event* (observable, sociological). A translation act is thus embedded in a translation event, although in Toury's view the two cannot be completely separated.

Chesterman (2013a) related three kinds of process model to Toury's discussion of types of translation problem. A "virtual process" is defined as one that outlines "ideal" translation strategies, either in general or with reference to the translation of a given source-text item or item type. Such a process model usually has a pedagogical function and is often formulated as a flow diagram marked with decision points and available options. An early example was Hönig (1995). A "reverse-engineered process" model starts with a translation solution and aims to reconstruct the decision chain that led to it. These models are proposed or implied e.g. in research on errors, when possible causes and sources of errors are investigated. They have also been used in attempts to plot the source of unusually successful, creative translation solutions (e.g. Kußmaul, 2007). These models also usually have a pedagogical function. A third type of model seeks to represent "actual processes" in real time, as is done in TAP research. Such models have many functions, including the aim of developing better CAT (computer-assisted translation) tools for translators.

Muñoz Martín (2016b) responded to Chesterman's proposal with a detailed and clarifying critique. A key point in this critique is that the original distinction between cognitive acts and sociological events is a misleading oversimplification of the relation between the human brain and its environment—i.e. an assumption that needs reassessing. However, models too are conceptual tools. For some research questions, an albeit simplifying distinction between translation act and event may be useful, but not for others. The perceived usefulness of the distinction also depends on the initial theoretical position taken by the researcher. Proponents of the embodied cognition view will obviously not find it so relevant. But a view is just that: a view, a perspective; other perspectives are also possible, including ones no-one has thought of yet. In any case, any proposal about a model of the translation process needs to be as explicit as possible about precisely what is being modelled, and for what purpose, so that it can be appropriately tested. What does the model predict, if anything? What evidence or argument would suggest that the model needs revision? In what respects is it better than alternative models?

1.4 Some recent debates

The great linguist Sapir famously said that "all grammars leak" (1921, p. 39). It now appears that the mind leaks too: it is "a leaky organ" (Clark, 1997, p. 53, cited in Risku, 2014, p. 335). This view implies that we cannot understand the mind adequately if we consider it in isolation from the physical environment in which it is "embodied"—not only the brain but also the rest of the body, plus the wider context, both temporal and spatial. It is also assumed that these processes are highly contingent on local circumstances. Risku summarizes thus (2014, p. 335): "we will actually also have to study translators in their authentic, personal, historically embedded environments and translation situations if we want to be able to describe the cognitive process". Edelman's rainforest metaphor seems to take on renewed relevance.

If this is indeed the direction in which cognitive research is heading, the methodological challenges are daunting. If everything is connected to everything else, and the agent is seen as embedded in a complex environment, it will not be easy to isolate elements for close study; and if one has to study the whole network, where does the network end? To totally understand a single atom, does one have to study the universe? Multiple methods will need to be combined and triangulated, as Risku says. One recent suggestion is that more use should be made of "mixed methods" research (e.g. Meister, 2017). This pragmatic approach aims to integrate qualitative and quantitative methods, and it uses both induction and deduction, back and forth by turn, in an iterative cycle. However, it does not yet seem clear what kind of developed and coherent epistemology might underlie such a mixture, where subjective and objective are also merged.

And what about the relation between cognition and the unconscious? On the rainforest view, presumably these too are connected? Venuti (2002) attempted to illustrate the influence of the translator's unconscious on translation decisions by offering some anecdotal evidence. But we are a very long way from any systematic knowledge of what this relation might be, or even how to study it.

Mention was made earlier of the well-known observer effect, evident not only in sociology but also in quantum physics, we are told. A not dissimilar problem arises when the results of empirical research on the translation process become filtered into translator training. If trainee translators become aware of what professional translators typically do, and if some of these tendencies are, in the trainees' view or that of their teacher, not desirable ones, then one result might be that these trainees decide to act differently. For example, after conscious exposure to certain so-called "universals" such as the under-representation of target-language-specific items (Tirkkonen-Condit's "unique items hypothesis", 2004), students might decide to resist this tendency in their own work. In so doing, they will themselves be weakening the given tendency that scholars have described, and thus undermining the predictability of the original description, or at least reducing its generality. The very fact of describing, and then publishing and publicizing the description, could in principle even end up falsifying the original description. In this way, translation practice—and the cognitive processes underlying it—may evolve, and thus not remain constant. This in turn would imply that the whole object of cognitive translation process research could be a moving target. (See further Chesterman, 2013b.) The target is also moving because of the increasing use of CAT tools such as translation memory systems. Here, the technological progress that has been one applied goal of research on translators' cognition is itself affecting the processes that are the object of the research. (For an example, see Dragsted, 2004.)

As noted above, our metaphors and categories influence how we think, including what we think we know. So do our concepts, and so does the very language we use. Consider the term "process". This nominalization seems to have some inbuilt sense of a linear direction, of going from a starting point to a goal. But could this be a misleading idea? In the rainforest of the mind, would it be more appropriate to imagine cognitive activities that were simply "going on" without much sense of direction? At least, such "stuff going on" would be a significant background against which something more goal oriented could be set.

Mind, brain, context, environment—these are all grammatically nouns (in e.g. English): they are conceptually reified. To what extent might this fact mislead us? The naturalist Richard Dawkins (2004, p. 232) comments on the human tendency to think in terms of discrete entities—nouns—rather than continua: he calls this "the tyranny of the discontinuous mind". Perhaps this is one consequence of the traditional image of the mind itself as a discontinuous organ—precisely the view that embodied cognition rejects. What effect might this "tyranny" have on our thinking about translation and cognition, about the very concepts we use? Do we just accept them "without thinking", as it were? It was no less a thinker than Wittgenstein who wrote that "'concept' is a vague concept" (1956/1978, #70).

1.5 Concluding remarks

In the light of the reservations and limitations mentioned earlier, it seems that we would do well to be rather cautious about what we think we know about translators' cognition. So many different kinds of data sources are now being made use of, with so many different methodologies, that hesitation is warranted when colligating all the various interpretations into some kind of a coherent theory, whatever that might look like. There is always a gap between data and interpretation, of course, but perhaps also between the interpretation and the linguistic form in which we choose to express it. And there is a gap between laboratory experiments and real life. Generalizations are hard to make in the face of a wealth of variation; and their validity is constrained by the number of cases or subjects concerned. Potential explanations are many and diverse.

More generally, there will often remain a gap between "is" and "seems". It is not always clear what counts as a fact. It was once apparently believed to be known, as a fact, that the Earth is flat. This was believed to be quite adequately justified, e.g. by common sense and the fact [sic] that no-one had ever fallen off the edge. But as we know, knowledge evolves, and most of us now believe differently. Today's knowledge may have a different status tomorrow (Is butter good for you, or bad for you? The "facts" appear to change quite often!). But this does not mean we just have to give up and descend into a nihilistic relativism. One can justifiably be more certain of some facts than others. I am 100% certain that, as I write, Helsinki is the capital of Finland. And in general, I am pretty certain of a great many large and small facts within a kind of medium range of size and time. I am much less certain about what little I know of the very small (subatomic quarks and the like) or the very large (gravitational waves, the edge of the solar system …) or the distant past. And I am not too certain about what I think I know about cognition.

As Delabastita puts it (2003), empirical researchers in the humanities (including TS) are epistemologically utopian. They acknowledge that totally objective knowledge is an unattainable ideal, but an ideal that is nevertheless worth striving towards. None of our sources of knowledge are infallible: perception can be deceptive, introspection may be selective, memory can be false, the testimony of others may be unreliable, and even rational induction may be misleading. All this can be admitted. Yet, realistically, we can always seek for better understanding by testing our claims and hypotheses: not in order to confirm or prove them—the utopian ideal—but to check whether they are supported or not, or need revising, or are perhaps false after all; and, of course, by testing and replicating the tests themselves. We can aim for more truthlikeness, even though ultimate truths seem [sic] to be out of reach.

With respect to knowledge about cognition, Edelman (1992, p. 162) proposes a "biologically based epistemology", which recognizes that knowledge must remain "fragmentary and corrigible", precisely because we are human, we evolve, and so does knowledge. So there is always room for doubt. Pym (1993, p.17) starts his first epistemological seminar on translation from "the primacy of doubt". Indeed, to paraphrase Descartes: "Dubito, ergo cogito, ergo sum" (I doubt, therefore I think, therefore I am).

Further reading

Cattrysse, P. (2014). *Descriptive adaptation studies: Epistemological and methodological issues*. Antwerp and Apeldoorn: Garant.

Chapter 10 in particular is a thought-provoking discussion of epistemological issues pertaining to adaptation and translation, issues that are eminently relevant also to research on translation and cognition.

Delabastita, D. (2003). Translation Studies for the 21st century. Trends and perspectives. *Génesis. Revista científica do ISAI, 3*, 7–24. www.academia.edu/4222581/Translation_Studies_for_the_21st_century_Trends_and_perspectives

A clear presentation of the basic epistemological divide in TS, between empirical and postmodern positions, and the implications of this for the increasing fragmentation of the field.

Pym, A. (1993). *Epistemological problems in translation and its teaching*. Calaceit: Caminade. www.academia.edu/35864334/Epistemological_problems_in_translation_and_its_teaching

Written version of an entertaining and stimulating series of seminars "for thinking students".

Risku, H. (2014). Translation process research as interaction research: from mental to socio-cognitive processes. *MonTI Special Issue—Minding Translation*, pp. 331–353.

An accessible overview of the most recent development in cognition research, towards embodied cognition.

References

Carl, M., Bangalore, S., & Schaeffer, M. (Eds.). (2015). *New directions in empirical translation process research*. Cham: Springer.

Carl, M., & Schaeffer, M. J. (2017). Models of the translation process. In J. W. Schwieter, & A. Ferreira (Eds.), *The handbook of translation and cognition* (pp. 50–70). Hoboken, NJ: Wiley.

Cattrysse, P. (2014). *Descriptive adaptation studies: Epistemological and methodological issues*. Antwerp and Apeldoorn: Garant.

Chalmers, A. F. (1976/1999). *What is this thing called science?* (3rd ed.). St. Lucia, Queensland: University of Queensland Press.

Chesterman, A. (1998). Causes, translations, effects. *Target, 10*(2), 201–230.

Chesterman, A. (2008a). The status of interpretive hypotheses. In G. Hansen, A. Chesterman, & H. Gerzymisch Arbogast (Eds.), *Efforts and models in interpreting and translation research* (pp. 49–61). Amsterdam: John Benjamins.

Chesterman, A. (2008b). On explanation. In A. Pym, M. Shlesinger, & D. Simeoni (Eds.), *Beyond descriptive studies. Investigations in homage to Gideon Toury* (pp. 363–379). Amsterdam: John Benjamins.

Chesterman, A. (2013a). Models of what processes? *Translation and interpreting studies, 8*(2), 155–168.

Chesterman, A. (2013b). The descriptive paradox, or how theory can affect practice. *Tijdschrift voor Scandinavistiek, 33*(1), 29–40.

Clark, A. (1997). *Being there: Putting brain, body and world together again*. Cambridge, MA: MIT Press.

Croft, W. (1990/2003). *Typology and universals*. Cambridge: Cambridge University Press.

Dawkins, R. (2004). *The ancestor's tale*. London: Weidenfeld & Nicolson.

Delabastita, D. (2003). Translation studies for the 21st century. Trends and perspectives, *Génesis. Revista científica do ISAI, 3*, 7–24. www.academia.edu/4222581/Translation_Studies_for_the_21st_century_Trends_and_perspectives.

Dragsted, B. (2004). *Segmentation in translation and translation memory systems: An empirical investigation of cognitive segmentation and effects of integrating a TM system into the translation process*. Copenhagen: Copenhagen Business School.

Edelman, G. (1992). *Bright air, brilliant fire. On the matter of the mind*. New York: Basic Books.

Ellis, J. M. (1993). *Language, thought and logic*. Evanston, IL: Northwestern University Press.

Englund Dimitrova, B. (2005). *Expertise and explicitation in the translation process*. Amsterdam: John Benjamins.

Ericsson, K.-A., & Simon, H. (1984; 2nd ed. 1993). *Protocol analysis: Verbal reports as data*. Cambridge, MA: MIT Press.

García, A. M., Mikulan, E., & Ibáñez, A. (2016). A neuroscientific toolkit for translation studies. In R. Muñoz Martín (Ed.), *Reembedding translation processing research* (pp. 21–46). Amsterdam: John Benjamins.

Halverson, S. (2003). The cognitive basis of translation universals. *Target, 15*(2), 197–241.

Hönig, H. (1995). *Konstruktives Übersetzen*. Tübingen: Stauffenberg.

Jääskeläinen, R. (2002). Think-aloud protocol studies into translation. An annotated bibliography. *Target, 14*(1), 107–136.

Jakobsen, A. L. (2017). Translation process research. In J. W. Schwieter, & A. Ferreira (Eds.), *The handbook of translation and cognition* (pp. 21–49). Hoboken, NJ: Wiley.

Krings, H. P. (1986). *Was in den Köpfen von Übersetzern vorgeht: eine empirische Untersuchung zur Struktur des Übersetzungsprozesses an fortgeschrittenen Französischlernern.* Tübingen: Narr.

Kurz, I. (1994). A look into the black box—EEG probability mapping during mental simultaneous interpreting. In M. Snell-Hornby (Ed.), *Translation studies—An interdiscipline* (pp. 199–207). Amsterdam: John Benjamins.

Kußmaul, P. (2007). *Verstehen und Übersetzen.* Tübingen: Narr.

Meister, L. (2017). On methodology: How mixed methods research can contribute to translation studies. *Translation Studies, 11*(1), 66–83.

Muñoz Martín, R. (2014). A blurred snapshot of advances in translation process research. *MonTI, Special Issue 1—Minding Translation*, 49–84.

Muñoz Martín, R. (Ed.). (2016a). *Reembedding translation processing research.* Amsterdam: John Benjamins.

Muñoz Martín, R. (2016b). Processes of what models? On the cognitive indivisibility of translation acts and events. *Translation Spaces, 5*(1), 145–161.

Popper, K. R. (1959). *The logic of scientific discovery.* London: Hutchinson.

Pym, A. (1993). *Epistemological problems in translation and its teaching.* Calaceit: Caminade.

Risku, H. (2014). Translation process research as interaction research: From mental to socio-cognitive processes. *MonTI Special Issue (1)—Minding Translation*, 331–353.

Salmon, W. C. (1998). *Causality and explanation.* New York: Oxford University Press.

Sapir, E. (1921). *Language. An introduction to the study of speech.* New York: Harcourt, Brace and Company.

Seleskovitch, D., & Lederer, M. (1984). *Interpréter pour traduire.* Paris: Didier Érudition.

Steup, M. (2018). Epistemology. In E. N. Zalta (Ed.), *The Stanford encyclopedia of philosophy* (Winter 2018 ed.). https://plato.stanford.edu/archives/win2018/entries/epistemology/

Tirkkonen-Condit, S. (2004). Unique items—over- or under-represented in translated language? In A. Mauranen, & P. Kujamäki (Eds.), *Translation universals. Do they exist?* (pp. 177–184). Amsterdam: John Benjamins.

Tommola, J. (1999). New trends in interpreting research. Going psycho—or neuro? In A. Álvarez Lugrís, & A. Fernández Ocampo (Eds.), *Anovar / Anosar estudios de traducción e interpretación*, vol. 1 (pp. 321–330). Vigo: Universidade de Vigo.

Toury, G. (1995; revised ed. 2012). *Descriptive translation studies and beyond.* Amsterdam: John Benjamins.

Venuti, L. (2002). The difference that translation makes: The translator's unconscious. In A. Riccardi (Ed.), *Translation studies: Perspectives on an emerging discipline* (pp. 214–241). Cambridge: Cambridge University Press.

Von Wright, G. H. (1971). *Explanation and understanding.* Ithaca, NY: Cornell University Press.

Wittgenstein, L. (1956/1978). *Remarks on the foundations of mathematics* (revised ed.) (G. H. von Wright, R. Rhees, & G. E. M. Anscombe, Eds.; G. E. M. Anscombe, Trans.). Oxford: Blackwell.

2

Translation, linguistic commitment and cognition

Sandra L. Halverson

2.1 Introduction

Cognitive translation and interpreting studies (henceforth CTIS), though it goes by different names and is construed with quite considerable differences in scope, has broadly concerned itself with questions pertaining to "the translating mind" (Jääskeläinen & Lacruz, 2018, p. 2). In a recent publication, Jääskeläinen and Lacruz state that "[c]ognitive research in translation and interpreting has reached a critical threshold of maturity that is triggering rapid expansion along several innovative paths" (2019, p. 1). This maturity is witnessed by the rapidly growing number of publications of various kinds, including handbooks of the present type, as well as the introduction of dedicated journals, conferences and competitively funded projects. An interesting parallel to this growth is the slow emergence of clearer lines of demarcation between major approaches to the study of cognition in general and to the study of translational cognition in particular.

In a number of publications, Muñoz Martín (2010, 2016; see also Muñoz Martín & Martín de León, this volume) has articulated the key differences between three main approaches, which he refers to as the "cognitivist approach" (also sometimes referred to as a "computational" approach), the "connectionist approach" and "cognitive translatologies" (see also Alves, 2015; Marín García, 2017; Risku, 2013). The differences between the different programmes are quite fundamental, as they bear upon such basic issues as the following: What are the roles of mind and body in cognitive processes? What is the relationship between general and domain-specific cognitive processes? What does linguistic processing involve? and How should "(linguistic) meaning" be conceptualized? The two predominant positions, the first and third on the list, are diametrically opposed on most of these questions.[1]

As summarized by Muñoz Martín, the two main positions, which follow the distinction in cognitive science between traditional cognitivism and so-called "situated cognition", or sometimes "4EA (embodied, embedded, enacted, extended, affective) cognition", line up as follows. For cognitivist/ computational approaches:

> Thought is (mostly) conscious, rational and logical. The mind is modular, with independent modules in charge of, or (almost) exclusively devoted to different tasks or faculties. One such faculty is language, which allows us to engage in linguistic behaviour by combining finite

> sets of symbols. By and large, the mind works serially, as a neutral problem-solving device that manipulates symbols. [...] Linguistic symbols carry stable, self-contained meaning or content. The purpose of language is to build or to prompt internal representations of messages. Denotative or conceptual meaning is objective but arbitrary, because it does not resemble the object or concept it stands for (e.g. Monday, green). Pragmatic, affective, or connotative meanings consist of modifications and additions of meaning induced by the context.
>
> *Muñoz Martín, 2017, p. 561*

The view from cognitive translatology is quite different. As Muñoz Martín outlines, within this approach, the brain is viewed as a "huge, plastic (i.e. modifiable) learning machine that slowly adapts its wiring to steady demands" (2017, p. 564). Moreover, rather than consisting of separate modules, the assumption is that higher cognitive functions (planning, decision making, attention control, etc.) rely on and are integrated with more elementary cognitive functions such as perception and motor functions. In short, "thinking is thinking-for-action, and it is done not only *in* the brain but *by* the brain in interaction with the body and the environment" (ibid., author's emphasis). From this perspective, language is seen as a set of prompts to activate mental routines, which may be more or less well established. Language processing involves spreading activation, and meaning is "a process, not a thing" (ibid.): it involves structured networks covering in principle unlimited encyclopaedic information of a wide variety of types (see also Langacker, 1987/1991, 2008).

It is difficult to do these positions justice in a brief chapter such as this, and their characteristics and content have already been explicated in previous work. Readers are referred to Lakoff (1987) and Langacker (1987/1991) for the linguistics perspective, and the two alternatives are also discussed within the context of CTIS by Muñoz Martín and Rojo López (2018), in addition to the references already cited. This brief introduction of the two key approaches serves a specific purpose in the context of the present discussion. As has been pointed out by Muñoz Martín (2017), CTIS is at a critical juncture in its development, and there is an obvious imbalance in the degree of development and internal coherence within the newer, cognitive translatological approach as opposed to its more established computational counterpart. It has been claimed that the translatological approach is fragmented and incomplete and that much work is required before it can present a true challenge to the more well-developed computational approaches (Muñoz Martín, 2017, pp. 564–565). This chapter may be taken as one response to this situation, and the arguments put forward here are intended to serve as a basis for the further development of this alternative and at the same time to enable more targeted contrasting and testing of the two approaches, as called for by Marín García (2017).

Keeping this motivation in mind, this chapter has one overall objective, which is to argue that all cognitive translational research programmes must build on a clearly articulated commitment to a view of language and language processing in translation; or, in other words, that language must be central to our fundamental ontology in all cognitive approaches to translation. In considering the status quo, it will be suggested that some cognitive translatological approaches, in their eagerness to establish a new research programme and to distinguish it from the old, have in some ways neglected, or may even be unwilling, to give adequate space or priority to language. I will argue that this constitutes a risk to the programme as such, for reasons that will be given. The argument will be developed by looking first at how the two dominant approaches to translational cognition do or do not incorporate an ontological position regarding language. The next step is to outline the epistemological problems that follow from the current situation in cognitive translatology. In the closing remarks, the overarching argument is considered in light of discussions of ontology and epistemology in TIS more broadly.

2.2 CTIS and the position of language

This discussion adopts Laudan's approach to epistemology (Laudan, 1977) as a starting point. Laudan outlines a non-positivist, sociological philosophy of science in which scientific progress is conceived of as a continued process of solving conceptual and empirical problems. The identification and resolution of such problems is a central task, and more specifically, progress is achieved through the continued comparison of alternative theories, and the theory that provides the better solution to a problem or problems is considered to be the better theory. Within this view, the function of a theory is to "resolve ambiguity, to reduce irregularity to uniformity, to show that what happens is somehow intelligible and predictable" (ibid., p. 13). In scientific practice, however, Laudan emphasizes that it is important to distinguish between "theory", which in his view should denote

> a very specific set of related doctrines (commonly called "hypotheses", or "axioms" or "principles") which can be utilized for making specific experimental predictions and for giving detailed explanations of natural phenomena.
>
> *Laudan, 1977, p. 11*

and "research traditions", a term intended to capture "more general, much less easily testable, sets of doctrines or assumptions" (ibid.). In Laudan's terms, a "research tradition" provides both an ontology, a specification of, "the types of fundamental entities which exist in the domain or domains within which the research tradition is embedded" as well as a set of methodological commitments regarding how best the domain may be investigated (1977, pp. 78–81). Within a research tradition, there may be any number of more specific theories linked to specific domains of enquiry. To add to Laudan, we might also suggest that theories, in turn, may comprise any number of "constructs", which are understood as conceptual elements of the theory that have been operationalized such that they may be empirically investigated.

Laudan's approach has been used to compare theories within the two alternative approaches to translational cognition in the dissertation by Marín García (2017). Like Marín García, I consider cognitivist/computationalist and cognitive translatological approaches to constitute two distinct research traditions. It is not within the scope of this discussion to compare all the ontological or methodological commitments that characterize the two traditions. Some of the ontological ones are evident in the descriptions cited from Muñoz Martín. The methodological commitments are not often overtly articulated as such and must be derived from the evidence represented by common practices. It is also important to note that many of the fundamental assumptions are imported into CTIS covertly, through the adoption of concepts or methods from cognitive psychology or psycholinguistics. It is not always clear whether such covert, imported assumptions have been critically examined or fully appreciated, as is also pointed out by Marín García (2017, p. v). At the same time, it is only fair to recall that, as Laudan puts it, "a theory, taken abstractly, does not have its 'parent' research tradition stamped all over it" (1977, p. 86). The bottom line is that while specific theories or hypotheses within CTIS may not have flagged all their philosophical loyalties, at a certain point in the evolution of a research tradition, it becomes urgent that some scholars within that tradition do so. The present discussion aims to contribute to that project.

In the following two sections, both ontological and methodological assumptions are used diagnostically in studying the content and function of language constructs within the two research traditions.

2.2.1 Cognitivist approaches

In a recent publication, Shreve and Diamond (2016) provide a detailed presentation of the cognitivist/computational, information processing approach to Cognitive Translation Studies. This comprehensive account provides a detailed summary and overview of the main elements of this research tradition, and will thus be taken as representative of this approach. As the authors state, this approach "has been wholeheartedly adopted by cognitive translation scholars" (ibid., p. 141). In the following, the language construct inherent in this tradition will be defined first, as a key element of its ontology. Secondly, some of the methodological practices related to this view are presented.

From the perspective of Shreve and Diamond, this tradition, like others within the broader field of cognitive science, studies "the mental processes that organisms with brains can carry out. These processes underlie our ability to acquire, store, generate, manipulate and use knowledge as well as our ability to act on that knowledge in the world" (2016, p. 143). In their view, cognition, including translational cognition, is best understood as "the result of 'computational' procedures that operate on representational structures in the mind" (ibid.). As far as language is concerned, it must be assumed that it is the representational structure that is the subject of mental computations. The translation process is described as consisting of the stages of comprehension, transfer and production (ibid., p. 150), and "language use" is conceived of as one of many "primary mental processes" that make up those three stages, along with attention, memory, perception, metacognition and problem solving (ibid.). Here is clear evidence of the modularity that was mentioned in the introductory remarks (language is separate from other cognitive processes).

Another body of work that is typical of this research tradition is exemplified in Carl (2013) and Schaeffer and Carl (2017). In Carl (2013), two different computational alternatives are considered: the ACT-R model and a statistical model. The former models human translation through a set of five "production rules". These function in an algorithmic procedure operating on words. The statistical model is offered as a means of calculating probable processing sequences on the basis of data for reading and writing processes (which are constituents of the composite translation process). Here, too, translation is seen as a decomposable, algorithmic process.

In Schaeffer and Carl (2017), the authors use translation and post-editing data to derive an interactive model of these two related activities. Both processes are modelled as a "transition network" in which four alternative states are connected in a series of iterative state transitions. The authors then calculate the distribution of the four states—source-text reading, target-text reading, target-text writing and pausing—in a dataset including keystroke and gaze data, and use this distribution to model a process consisting of potential state changes. This too is a typical algorithmic approach, building on a fully computational language construct.

Shreve and Diamond (2016) provide a comprehensive overview of the methodological consequences of this view of translational cognition, mentioning the adoption of keystroke logging and eye-tracking methods to study the time course of translating activity, including pauses and regressions as indicators of processing effort or a focus of attention and the identification of stages in the process. Strikingly, the authors state:

> Indeed, one could argue that the advent of more precise methodologies for collecting and analyzing behavioral data accelerated a process of integration with the cognitive sciences and the dominant information-processing model that had already begun years before.
>
> *Shreve & Diamond, 2016, p. 149*

Language is central to the cognitivist approach to cognition, as it is often considered to be the quintessential representational system. Similarly, algorithmic operations on symbols are a recognizable and commonly adopted description of syntactic processes, among other things. Given the dominance of this research program in CTIS, this particular view of language is commonly adopted, either explicitly or implicitly.

2.2.2 Cognitive translatology

A commitment to a view of language is much less evident within cognitive translatological approaches. A good deal of research emphasis within this research tradition has been on implementing broader models of cognition, in which language is only one of many elements, and not necessarily a prominent one (e.g., Risku et al., 2013; Risku & Windhager, 2013). In Risku et al.'s network model of translatorial cognition and action, for instance, communicative action, including translation, is modelled as the dynamic interplay of cognitions, actions, interests, social structure, material and artefacts in integrated environments (ibid.; see also Risku & Windhager, 2013). Knowledge of languages is taken to be part of cognition, though it may also possibly be involved in mental artefacts such as "checklists" or "guidelines" (2013, p. 163). Language or linguistic processing as such is not highly visible in the network model.

Ontological/methodological commitments to 4EA cognition are also in evidence in a body of work that focuses on the interaction of the translator and her physical environment (Ehrensberger-Dow, 2017; Risku, 2014;, Risku & Windhager, 2013). In some studies of this type, ethnographic methods are added to the repertoire and complement experimental or quasi-experimental methods designed to collect data on cognitive processes. The ethnographic complement integrates observations of translators' interactions with other people, with various elements of their physical environment and with a range of artefacts, including translation technologies. Interview data can also contribute to the investigation of translator environments and interactions.

In another strand within this research tradition, the non-modular view of cognition (fundamental to this tradition) has led to increased research interest in other areas of translator psychology, i.e. areas that are often excluded if linguistic processes are taken to be cognitively distinct, as is claimed by modularity. Examples of this work include studies of such psychological phenomena as emotion, creativity, tolerance of ambiguity and self-confidence as related to translators and translational action, also including responses to translations (Bontempo et al., 2014; Del Mar Haro-Soler, 2018; Hubscher-Davidson, 2013, 2016, 2018; Lehr, 2014; Rojo López, 2017a, 2017b; Rojo López & Ramos Caro, 2018). In much of this work, linguistic production is tested in investigations of the possible translational effects of various types of personality traits or emotional states of varying duration and type. Language constructs are not often problematized, and in some work the linguistic nature of the task is not in focus at all, as the variables are primarily psychological ones, such as those mentioned earlier, or additional variables such as expertise or translator competence.

Another strand of research within cognitive translatology incorporates the fundamental assumptions of cognitive linguistics. The language construct adopted here is of the type described in the introductory remarks and is exemplified in a volume of papers edited by Rojo and Ibarrexte-Antuñano (2013). In addition to the editors' own account of the potential of cognitive linguistics for the study of translation (2013, pp. 1–30), the papers illustrate the use of cognitive theoretical tools for the investigation of lexical and grammatical patterns, metonymy and a range of other themes related to meaning construction in translation. The ontological commitments that underlie this type of work are clearly articulated in the introduction to the 2013 volume:

- "language is an integral part of cognition, and thus a product of general cognitive abilities" (2013, p.11)
- "human language is symbolic in nature" (ibid.)
- language is usage based (ibid.)

In their discussion of these commitments and the potential of cognitive linguistics (henceforth CL) in the study of translation, the authors claim that the CL approach is commensurable with ("can be bridged to") translation studies (TS); that CL has the tools required for TS; that CL tools and concepts can help investigate underexplored areas and issues within TS; and that insights from cognitive TS can also enrich CL. In these authors' view, CL can serve as the linguistic foundation for a broad cognitive theory of translation (2013, pp. 18–26).

Finally, it has also been argued within this research tradition that the role of language in translation has been overstated, and that cognitive theorizing on translation is better served by giving pride of place to constructs such as "meaning" or "communication" (Muñoz Martín & Martín de León, this volume). The authors' view is quite clearly stated in the following:

> if we agree that meaning is encyclopedic; that the mental organization of stored information is sub symbolic; that each language symbol may trigger the activation of a network of sub symbolic nodes that will never be identical due to the interaction with other activated nodes through spreading activation; if we agree that language underspecifies meaning in that what we experience mentally is far richer than what is meant with language symbols; that meaning is an active process that will use any kind of inputs and stored information: then we will necessarily conclude that translating is not a matter of language, but rather a matter of meaning or of communication.
>
> *ibid.*

The view of language, meaning and communication sketched out here is a fair representation of the view taken by most scholars within the cognitive translatological research tradition. However, the conclusion drawn by these scholars is not a necessary conclusion at all. In fact, the exact opposite position may also be logically taken. The reasoning behind this position is outlined in Section 2.3.

The difference in ontological assumptions within the cognitive translatological research tradition is, as expected, linked to a difference in the research methods used. Investigations of translational ergonomics (see Ehrensberger-Dow, 2017, for survey) have led to an increase in the use of ethnographic methods, including various observational techniques, interviews, etc. Interest in translator psychology has led to the introduction of various psychometric techniques and instruments. Cognitive linguistic studies have investigated corpus data and elicited data of different types. Some of the production data used in computational studies are also used here (e.g. keystroke logs and eye-tracking data), sometimes in combination with a range of other data types (see Halverson, 2017). The sketch provided here demonstrates some of the fragmentation that currently characterizes the cognitive translatological research program in CTIS as regards central theoretical concepts and theories. In some general models, a language construct is not visible or is identifiable only as an inferred component of another construct (e.g. Risku et al., 2013). In work on the impact of various translator characteristics or states, a language construct underlies certain variables, such as "accuracy" or "creativity" or "ambiguity", but is not explicitly developed or referred to in individual studies. In one central volume of work, cognitive linguistic theory is put forward as an appropriate grounding language theory, and the potential of that theory is demonstrated in a collection of studies. In a quite recent paper (Muñoz Martín & Martín de León, this volume), there is a call for a change in priorities and a move away from language as a central concern. Clearly, if this research programme is to become more fully developed and internally consistent, some work is required to either articulate the coherence that is there or to pinpoint areas of disagreement or inconsistency in order to facilitate the resolution of conceptual and/empirical problems.

At the end of this outline of the two main research traditions within CTIS, it is important to note that while there are obvious differences in the *content* of the two traditions, these differences are rooted in fundamental ontological commitments, and it is these commitments that are of primary interest here. In principle, it is not impossible to model translational cognition in a computational manner and also include both situational parameters and personality-related ones. Conversely, it is not impossible to investigate core linguistic processes from a cognitive translatological (situated cognition) starting point (for discussion and research findings, see Spivey & Richardson, 2009). However, the causal role played by environmental factors or parameters related to the translator psychology is less obvious in a modular view of language processing and is external to the central translational processes. In the cognitive translatological view, on the other hand, translational cognition fundamentally includes translator psychology and extends into, and is embedded in, the translator's environment. The two research traditions require different concepts and theories, and they investigate different data in different ways. In order for the cognitive translatological programme to consolidate into a fully-fledged alternative to computationalism, it is imperative that the precise relationship between translational linguistic processing and other cognitive psychological and environmental factors in translational task settings be theorized in more specific detail. It is also imperative that the foundational constructs that unite the research tradition be clearly articulated and debated. An additional issue is that of scope: there is a clear contrast between the specific investigations of translator psychology and/or ergonomics, for instance, and more general models such as Risku et al.'s (2013). The point of interest here is the need for some means of dialoguing across study types and of ensuring at least a minimum of coherence within the research tradition. A principled view of language and linguistic processing could serve as the necessary hinge or nexus.

2.3 Epistemological challenges for cognitive translatology

This section is concerned with the epistemological challenges that follow from the current situation of fragmentation in cognitive translatology. In Section 2.3.1, ontological assumptions from 4EA cognition are put forward to argue for the centrality of a language construct, and Section 2.3.2 provides an illustration of how a language construct can ground work towards resolving conceptual or empirical problems within this research tradition. The final part of this section, Section 2.3.3, presents a call for a linguistic commitment for cognitive translatology.

2.3.1 Why language: Arguments from ontology

In their chapter in this volume, Muñoz Martín and Martín de León refer to the distinction in cognitive science between "microcognition" and "macrocognition". The former term is used to refer to actions and activities on a small time scale and is usually investigated in experimental, laboratory settings. "Macrocognition" refers to "complex cognitive functions (...) in natural settings" (ibid.: p. 57) and often employs ecologically situated investigations of situated decision making. The two are sometimes seen as antagonistic and sometimes as complementary: there is no consensus. Nor is there complete agreement on the utility of the distinction itself. Muñoz Martín and Martín de León, following Flach (2008), argue that CTIS would be best served by rejecting the distinction and adopting an integrated, ecologically situated approach to cognition, i.e. the view that macrocognition is primary. This is the philosophy that underlies most work within cognitive translatology today.

The "ecological/situated cognition" view that Flach supports "argues for one reality and claims that the reality is grounded in the physics of ecologies or situations" (2008, p. 34). On this view, also laboratory experiments (which remain viable) must be grounded in the ontologically prior ecology of real-world situations. Flach continues to outline three implications of adopting this view. The first of these is of key interest in the present context. Flach states that research on cognition must "start with the phenomena as they are lived" (2008, p. 36). In other words, in Flach's view, epistemology must start with a human-based, naturalistic approach to lived reality. We might add translation and interpreting to his list of "the natural phenomena of cognition", examples of which include firefighting, meteorology, piloting, ship navigation, military command, etc. (ibid.)

If translation and interpreting are also "natural phenomena" of cognition, and if we are to investigate these phenomena "as they are lived", then it seems obvious that language must be an absolutely central element of the ecology. Indeed, the bulk of the activity that is referred to by the terms "translating" or "interpreting", or their counterparts in other human languages, involves phenomena that must be included in a language construct; for example, the raw material of language (graphological encodings, sound waves), the mental actions involving linguistic material, and the linguistically related communicative objectives and settings in which translation is embedded. This seems almost too obvious. Evidence of the ubiquity of language in translating and interpreting is found everywhere in empirical studies of translational/interpreting activity. The raw material of language is visible in anterior texts, whether they are written or spoken. The translator/interpreter attends to this material and also produces new linguistic material in a selected production mode while engaging in mental activity involving language.[2] Linguistic elements are selected, rejected, added or deleted often in response to environmental stimuli, either real or inferred. To the extent that the translator/interpreter verbalizes his/her process, s/he refers to linguistic elements, processes or relationships (see Halverson, 2018 on metalinguistic awareness in translation). To the extent that the translator reflects upon the nature of the task itself, reference is made to linguistic elements or relationships and the communicative situations in which they are or can be employed (Presas, 2017).

If the aim is to consolidate and develop a situated/4EA research tradition for CTIS, built on the ontology that is being developed within this tradition within cognitive science, then the cognitive translatological research tradition must have a central language construct if it is to investigate "the phenomena as they are lived". It might be helpful to recall that within a research tradition, numerous theories may aim to account for different phenomena and/or at different levels of abstraction or with differences in scope. Thus, a theory of translational ergonomics, or of the relationship between translation performance and affect, would probably have quite different contents (cf. Laudan, 1977, pp. 81–86). On the other hand, in a viable research tradition, different theories would share assumptions and tenets regarding the fundamental phenomena in their segment of reality as it is lived. These shared foundations are the bedrock on which a research tradition must be built, and it seems counterproductive to try to build a research program that does not recognize the central role that language plays in the reality of translation and interpreting.

As an addendum, however, it is important to point out that the approach to language represented by CL builds on assumptions that are shared with a broader approach to cognitive semiotics (see Zlatev, 2011). In other words, the fundamental meaning-making apparatus devised for language in the CL view is also valid for other semiotic systems that might be of interest. This means that a research tradition based on a 4EA view of translational cognition, with a centrally positioned, appropriately conceived set of assumptions regarding language, is perfectly capable of accommodating other semiotic systems within its remit. Translation and interpreting can be conceptualized with reference to meaning negotiation across any codified semiotic system. This issue will be discussed also in the concluding remarks in Section 2.4.

2.3.2 Grounding for resolution of conceptual and empirical problems

A consensus on the central role of language in our fundamental conceptualization of translation could play a central role in consolidating cognitive translatology as a fully-fledged research tradition (a programmatic position most thoroughly articulated in Muñoz Martín, 2010). In arguing this case, the current discussion adopts Laudan's view of scientific progress, as outlined in his 1977 volume. In that work, Laudan articulates the differences between broad research traditions, as mentioned in Section 2.2, and the theories that are associated with such traditions. In order to take a closer look at the epistemological need for a shared and centrally positioned view of language, we shall consider the relationship between the research tradition itself (and its fundamental ontology) and the theories that constitute it. According to Laudan, there are two specific "modes" in which broad research traditions and the theories that constitute them may be related: historically (through the specific allegiances and views of particular scholars working within established or emerging traditions) and conceptually. Conceptual relationships are best understood by looking at the ways in which research traditions and theories interact (Laudan, 1977, pp. 81–93). Laudan describes four ways in which research traditions influence the theories that constitute them: i) by "strongly influencing [...] the range and the weighting of the empirical problems with which its theories must grapple" (1977, p. 86) or the conceptual problems they must resolve, ii) by constraining the domain of the theories, iii) by serving as a heuristic for the development of new theories, and iv) by justifying or rationalizing theories.

The discussion in Section 2.2 demonstrated how the two research traditions in cognitive science, and subsequently within CTIS, are constituted by quite different sets of empirical problems and research domains. As was discussed, the domain of cognitive translatology, given its non-modular view of cognition, is much broader than that of the cognitivist tradition. The development of the construct of "default translation" by the present author (Halverson, 2019) is an example of the research tradition serving as a heuristic for the development of new theories or constructs. Of course, it was a particular view of language that served the heuristic purpose in this effort.

It is also important to recall, as mentioned in Section 2.2, that in Laudan's view, scientific progress is achieved through careful comparison of alternative theories to see which provides a better solution to a conceptual or empirical problem. Such comparisons may take place within or across research traditions. It seems reasonable to assume that the more coherent and fully developed a research tradition is, with clearly articulated foundational tenets, the more robust and meaningful the comparison will be.

One example of a comparison of this type is Marín García's (2017) comparison of the constructs of translational competence and expertise. In his analysis, the focus is on the two versions of the construct as applied within cognitive translatology, and the analysis made use of a conceptual performance model developed by the author. This analysis is a prime example of how research traditions might progress through careful comparative analysis and grounding.

In the current context, however, the point is to illustrate the role of foundational (ontological) assumptions in such comparisons. To reiterate: in Laudan's view, the comparison of contending accounts must take place at the level of the theory or construct, not the research tradition as such. However, uncertainty about the underlying world view, or ontology, may impact the feasibility of carrying out comparisons of this type, especially if the fundamental element that is either missing or underdeveloped is one that the specific theory itself does not articulate. To take a fictitious example, but a realistic one: it has been proposed that translators choose alternatives in their production that are simpler, more explicit, "equalizing" (non-extreme), as a means of avoiding risk (Pym, 2015, p. 76). Aggregate tendencies such as these (and other, similar ones) have long been recognized as being within the remit of Translation Studies, and the cognitive basis of this behaviour has long been sought. Cognitive theories to account for this behaviour must propose a risk construct that manifests both socio-cultural and cognitive aspects and must also posit a causal mechanism to account for the ways in which "risk" affects the cognitive processes resulting in specific linguistic choices. To the best of this author's knowledge, no cognitive theory has yet been proposed for this purpose. Moreover, it is quite clear that cognitivist and 4EA views of cognition would have to go about this task in quite different ways. For CTIS, it is not possible to build theories to capture socio-cognitive entities such as "risk", or "norm" or "ideology" for that matter, without a fundamental position on language to make these ideas relevant for translation. If the underlying view of language is dubious or unclear, theories and hypotheses cannot be compared, new theories cannot be developed and existing ones cannot be justified.

A final challenge for cognitive translatology has more to do with interdisciplinary relationships and less to do with the internal development of the research tradition itself. As has been made amply clear in the work by Muñoz Martín cited in the introduction, the two main approaches to translational cognition share many assumptions, theories and constructs from their counterparts in cognitive science. This was also pointed out for the cognitivist tradition in the article by Shreve and Diamond (2016), cited in Section 2.2.1. Of particular concern in the present discussion, related to the diagnosis of fragmentation and immaturity given in Muñoz Martín (2017), is the fact that the research program of 4EA cognition is itself engaged in debates about the interpretation and implementation of some of its own fundamental concepts. Both "embodiment" and "extended cognition" are the subject of disagreement, and the programme as such is thus not fully developed with regard to these ideas (see e.g. Kiverstein, 2012; Wilson & Golonka, 2013; Wheeler, 2014). At the very least, cognitive translatology must acknowledge these uncertainties; it must not proceed as though the 4EA programme were *pret-a-porter*. On a more positive note, this status quo makes it distinctly possible that CTIS can contribute to the investigation and possible resolution of these issues.

Investigating translation and interpreting, particularly in a highly technologized environment such as the one most translators work in today, using an appropriately framed language-based framework, would provide a highly relevant test ground for key issues related to both cognitive extension and embodiment. For instance, a study by Schaeffer et al. (2019) investigated the impact of removing the visual representation of the emerging translation on the computer screen. One of the interesting findings was that translators, in certain circumstances, fixate on the empty screen (or rather on the area of the screen where the emerging text would be, as indicated by the cursor). Though this paper did not discuss an interpretation in terms of spatial indexing (as a form of external memory support), this is a relevant notion for the further investigation of such phenomena. The basic idea here is that language processing can involve spatially indexing meaning content (for instance in a location on a computer screen) as a means of supporting memory (Spivey & Richardson, 2009). Studies such as Schaeffer et al.'s represent an area in which translation research could contribute to the debate on what it means for cognition to be "extended" into the environment. A language construct must be implicated in such studies.

Another issue which cognitive translatology could address is that of mental representations. Martín de León (2017) has discussed this issue at some length, pointing out the various positions within cognitive science. As she points out, there are also differences within the 4EA tradition regarding the need for, and possible form of, mental representations (see also Rowlands, 2009). Martín de León outlines the cognitive translatological position on mental representations, stating:

> However, translation takes place decoupled from the communicative processes that it makes possible, and it involves representation-hungry tasks, such as text comprehension or addressee profiling, which may require the support of dynamic, action-oriented mental representations. These can be regarded as internal scaffolding that supports meaning construction and translation processes in a dynamic, situated, even distributed way.
>
> *Martín de León, 2017, p. 120*

The "representation-hungry" constituents of translational cognition, for example text comprehension or addressee profiling, mentioned in the extract are linguistic tasks and must be theorized with reference to linguistic processing. The differences between cognitivist and cognitive translatological views of representation can be clearly illuminated through contrastive theory testing, which must be linguistically grounded.

2.3.3 A linguistic commitment for cognitive translatology

Cognitive linguistics (of the type advocated for CTIS by Rojo and Ibarrexte-Antuñano as well as the present author) emerged in the 1970s as a reaction to formalist linguistics, including the cognitivist programme of generative linguistics. In the early stages of its development, the founding scholars expended considerable effort in positioning the project relative to its competitors and in building a foundation for the programme in a viable ontology and epistemology (Johnson, 1987; Lakoff, 1987; Lakoff & Johnson, 1980, 1999). As part of that effort, Lakoff articulated two key commitments for the field: the cognitive commitment and the generalization commitment (1990). The cognitive commitment deals with the relationship between linguistics and other areas of cognitive science and enquiry. The generalization commitment articulates the aim to elaborate general principles that apply to all aspects of human language. The cognitive commitment is relevant in the current discussion: it requires cognitive linguists to "make one's account of language accord with what is generally known about the mind and brain from disciplines other than linguistics" (Lakoff, 1990, p. 50). This commitment is not only an expression of desired disciplinary relationships or practices: it is also fundamentally an expression of how linguistic cognition is conceived of within this research tradition. The commitment is entailed by the adopted view of linguistic cognition: if linguistic cognition is believed to build on and be integrated with other general cognitive processes, then this commitment is a necessary consequence.

An adapted version of Lakoff's cognitive commitment has been proposed as one of the foundational tenets for cognitive translatology (Muñoz Martín, 2010, p. 174). This proposal is firmly endorsed here. The current argument is an extrapolation of that position. As suggested in Section 2.2, a language construct has always been a central element of the cognitivist tradition, and this is also true of the cognitivist tradition within CTIS. In this author's view, however, the cognitive translatological tradition is in need of a firm commitment to a fundamental language construct, for the reasons outlined in Sections 2.3.1 and 2.3.2. This call may be seen as the converse of Lakoff's original cognitive commitment; in other words, a linguistic commitment for cognitive translatology might be articulated in a reversal of Lakoff, as follows:

> A linguistic commitment requires cognitive translation scholars to make their account of translational cognition accord with what is generally known about language and other signifying systems from cognitive linguistics and semiotics.

A commitment of this type would constitute a step towards consolidating a clear alternative to the cognitivist research tradition in CTIS. It would establish a necessary part of the shared foundation for this research tradition, in line with the characteristics outlined by Laudan (1977). Moreover, a commitment to a foundational language construct is not at all at odds with the view of meaning, communication and language referred to by Muñoz Martín and Martín de León (this volume). Though meaning and communication are not limited to language, they are intertwined with language in most of the forms of translation and interpreting activity that remain of predominant interest in the field. A centrally positioned, cognitive semiotic/linguistic view of language[3] is exactly what is needed to ground and further a research tradition that views translation as an embodied, embedded, enacted, extended and affective cognitive activity.

2.4 Concluding remarks

The case being made in this chapter is that a view of language such as that proposed by CL (commensurate also with broader cognitive theories of semiotics) can function as a unifying and consolidating element in the cognitive translatological research tradition. The claim is that the tradition remains weak and fragmented without it. The arguments for making language central are based on ontological and epistemological considerations.

Several scholars have called for stronger epistemological foundations for translation and interpreting studies. The argument has been made recently by Marais (2014, pp. 77), by Gambier and van Doorslaer (2016, p. 4) and by Marín García (2017). The case made in the present discussion is narrower in scope and is directed primarily at the emerging research tradition of cognitive translatology. The articulation of a specific view of language and a particular position for language in a broader ontology is one that will serve this tradition first, but it will also serve the discipline at large in that it will facilitate better dialogue and comparison across research traditions, cognitive and otherwise. Regardless of whether one adopts Laudan's view of scientific progress or not, dialogue on fundamental philosophical issues is almost always illuminating.

There is one final issue that must be returned to, as it is important to mitigate the risk of this argument being misunderstood. It may be reasonably expected that the call to prioritize language will be viewed as retrograde and counterproductive by some. It has become commonplace in translation and interpreting studies to dismiss linguistic approaches to translation as positivist, overly empiricist or, at best, extremely limited. Part of the response to this in the current discussion has been to emphasize the need for a phenomenological take on translational phenomena: a call to investigate translation and interpreting "as they are lived". The second response is to point out that the view of language advocated here builds on a seamless link between language and other semiotic systems in terms of the cognitive processes involved. In the view pursued here, linguistic meaning and non-linguistic meaning are of the same stuff, and the representational material at our disposal is oftentimes very rich. Giving priority to language in the cognitive investigation of translation does not mean that meaning stemming from rich settings or other representational codes is not factored in: of course, it is.

Considering this issue from the perspective of Translation Studies at large, an interesting question is how cognitive explorations based on the view of language, semiotics and translational ecology advocated here might be fruitfully situated within a broader Translation Studies as envisaged by Marais and Kull in their discussion of translation and biosemiotics (2016). It would seem that the approaches share an overarching interest in forms of and conditions for meaning making in living organisms and systems. There is no contradiction in calling for a linguistically/semiotically oriented cognitive translatology to study human translational processes and phenomena, while at the same time wishing to situate these processes within a broader, more encompassing world. The question is not whether to do so, but how.

Notes

1 The "connectionist" position is in some ways related to both the other two and thus constitutes something of a middle position. It does not view cognition as algorithmic symbol manipulation but as an "experiential, cultural act" (Risku, 2013), and thus departs from the cognitivist view on a very important issue. It is not concerned with the integrated role of environmental phenomena in cognition, as is the latter view, and thus cannot be entirely subsumed under this position either (see also Muñoz Martín, 2010).
2 Obviously, source and target texts also incorporate non-linguistic material in the form of visual elements of various kinds as well as contextual and encyclopaedic information that is in principle unlimited. This does not change the fact that most translational/interpreting activity is language dominant. This only means that we must have a language construct and theoretical frameworks that allow the integration of all of these elements into the meaning-creation process: it does not mean that language is less important.
3 While this chapter advocates a cognitive linguistic/semiotic approach, this is a specific realization of a broad group of linguistic approaches that are often characterized as "usage based". Most of these approaches share a fundamental set of ontological and epistemological commitments, which is the important factor in the current discussion. In other words, cognitive translatology need not put all its eggs in the cognitive linguistic basket for philosophical purposes.

Further reading

Marín García, Á. (2017). Theoretical hedging: The scope of knowledge in translation process research [PhD dissertation]. Kent State University.

A dissertation presenting a rare analysis of translation process research from a philosophy of science framework.

Muñoz Martín, R. (2017). Looking toward the future of Cognitive Translation Studies. In J. W. Schwieter, & A. Ferreira (Eds.), *The handbook of translation and cognition* (pp. 555–572). Hoboken, NJ: John Wiley & Sons.

A recent overview with insightful commentary on expected developments in CTS.

Shreve, G. M., & Diamond, B. J. (2016). Cognitive neurosciences and cognitive translations studies. About the information processing paradigm. In Y. Gambier , & L. van Doorslaer (Eds.), *Border crossings. Translation studies and other disciplines* (pp. 141–167). Amsterdam: John Benjamins.

A detailed presentation of the computationalist view of translational cognition.

See also the chapter on *Translation and cognitive science* by Ricardo Muñoz Martín and Celia Martín de León in the present Handbook for a state-of-the-art review which clearly identifies and situates cognitive scientific thinking within TS.

References

Alves, F. (2015). Translation process research at the interface. Paradigmatic, theoretical, and methodological issues in dialogue with cognitive science, expertise studies and psycholinguistics. In A. Ferreira, & J. W. Schwieter (Eds.), *Psycholinguistic and cognitive inquiries into translation and interpreting* (pp. 17–40). Amsterdam: John Benjamins.

Bontempo, K., Napier, J., Hayes, L., & Brashear, V. (2014). Personality matters: An international study of sign language interpreter disposition. *International Journal of Translation and Interpreting Research*, *6*(1), 23–46.

Carl, M. (2013). A computational framework for a cognitive model of human translation processes. In S. Bandyopdhyay, S. Kumar Naskar, & A. Ekbal (Eds.), *Emerging applications of natural languages processing: Concepts and new research* (pp. 110–129). Hershey, PA: IGI Global.
Del Mar Haro-Soler, M. (2018). Self-confidence and its role in translator training: The students' perspective. In I. Lacruz, & R. Jääskeläinen (Eds.), *Innovation and expansion in translation process research* (pp. 131–160). Amsterdam: John Benjamins.
Ehrensberger-Dow, M. (2017). An ergonomic perspective of translation. In J. W. Schwieter, & A. Ferreira (Eds.), *The handbook of translation and cognition* (pp. 332–349). Hoboken, NJ: John Wiley & Sons.
Flach, J. (2008). Mind the gap: A skeptical view of macrocognition. In J. M. Schraagan, L. G. Militello, T. Ormerod, & R. Lipshitz (Eds.), *Naturalistic decision making and macrocognition* (pp. 27–49). Aldershot: Ashgate.
Gambier, Y., & van Doorslaer, L. (2016). Disciplinary dialogues with translation studies. The background chapter. In Y. Gambier, & L. van Doorslaer (Eds.). *Border crossings. Translation Studies and other disciplines* (pp. 1–21). Amsterdam: John Benjamins.
Halverson, S. (2017). Multimethod approaches. In J. W. Schwieter, & A. Ferreira (Eds.), *The handbook of translation and cognition* (pp. 195–212). Hoboken, NJ: John Wiley & Sons.
Halverson, S. (2018). Metalinguistic knowledge/awareness/ability and translation: Some questions. *Hermes, 57*, 11–28.
Halverson, S. (2019). "Default" translation: A construct for cognitive translation and interpreting studies. *Translation, Cognition and Behavior, 2*(2), 169–193.
Hubscher-Davidson, S. (2013). Emotional intelligence and translation studies: A new bridge. *Meta: Translators' Journal, 58*(2), 324–346.
Hubscher-Davidson, S. (2016). Trait emotional intelligence and translation: A study of professional translators. *Target, 28*(1), 129–154.
Hubscher-Davidson, S. (2018). Do translation professionals need to tolerate ambiguity to be successful? A study of the links between tolerance of ambiguity, emotional intelligence and job satisfaction. In I. Lacruz, & R. Jääskeläinen (Eds.), *Innovation and expansion in translation process research* (pp. 77–104). Amsterdam: John Benjamins.
Jääskeläinen, R., & Lacruz, I. (2018). Translation—cognition—affect—and beyond: Reflections on an expanding field of research. In I. Lacruz, & R. Jääskeläinen (Eds.), *Innovation and expansion in translation process research* (pp. 1–16). Amsterdam: John Benjamins.
Johnson, M. (1987). *The body in the mind*. Chicago: University of Chicago Press.
Kiverstein, J. (2012). The meaning of embodiment. *Topics in Cognitive Science, 4*, 740–758.
Lakoff, G. (1987). *Women, fire, and dangerous things*. Chicago: University of Chicago Press.
Lakoff, G. (1990). The invariance hypothesis: Is abstract reason based on image-schemas? *Cognitive Linguistics, 1*, 39–74.
Lakoff, G., & Johnson, M. (1980). *Metaphors we live by*. Chicago: University of Chicago Press.
Lakoff, G., & Johnson, M. (1999). *Philosophy in the flesh*. New York: Basic Books.
Langacker, R. W. (1987/1991). *Foundations of cognitive grammar. Vol. I and II*. Stanford, CA: Stanford University Press.
Langacker, R. W. (2008). *Introduction to cognitive grammar*.
Laudan, L. (1977). *Progress and its problems. Towards a theory of scientific growth*. Berkeley: University of California Press.
Lehr, C. (2014). The influence of emotion on language performance: Study of a neglected determinant of decision-making in professional translators [unpublished thèse de doctorat]. Univ. Genève.
Marais, K. (2014). *Translation theory and development studies. A complexity theory approach*. London: Routledge.
Marais, K., & Kull, K. (2016). Biosemiotics and translation studies: Challenging 'translation. In Y. Gambier, & L. van Doorslaer (Eds.), *Border crossings. Translation Studies and other disciplines* (pp. 169–188). Amsterdam: John Benjamins.
Marín García, Á. (2017). Theoretical hedging: The scope of knowledge in translation process research [unpublished PhD dissertation]. Kent State University.
Martín de León, C. (2017). Mental representations. In J. W. Schwieter, & A. Ferreira (Eds.), *The handbook of translation and cognition* (pp. 106–126). Hoboken, NJ: John Wiley & Sons.
Muñoz Martín, R. (2010). On paradigms and cognitive translatology. In G. M. Shreve, & E. Angelone (Eds.), *Translation and cognition* (pp. 69–187). Amsterdam: John Benjamins.
Muñoz Martín, R. (2016). Of minds and men–computers and translators. *Poznań Studies in Contemporary Linguistics, 52*(2), 351–381.

Muñoz Martín, R. (2017). Looking toward the future of cognitive translation studies. In J. W. Schwieter, & A. Ferreira (Eds.), *The handbook of translation and cognition* (pp. 555–572). Hoboken, NJ: John Wiley & Sons.

Muñoz Martín, R., & Martín de León, C. (2020). Translation and cognitive science. In F. Alves, & A. L. Jakobsen (Eds.), *The Routledge handbook of translation and cognition* (pp. 52–68). Routledge.

Muñoz Martín, R. & Rojo López, A. M. (2018). Meaning. *Routledge handbook of translation and culture* (pp. 61–78). London: Routledge.

Presas, M. (2017). Implicit theories and conceptual change in translator training. In J. W. Schwieter, & A. Ferreira (Eds.), *The handbook of translation and cognition* (pp. 519–534). Hoboken, NJ: John Wiley & Sons.

Pym, A. (2015). Translating as risk management. *Journal of Pragmatics, 85*, 67–80.

Risku, H. (2013). Cognitive approaches to translation. In C. Chapelle (Ed.), *The encyclopedia of applied linguistics*. doi.org/10.1002/9781405198431.wbeal0145

Risku, H. (2014). Translation process research as interaction research: From mental to sociocognitive processes. In R. Muñoz Martín (Ed.), *MonTi Special Issue: Minding translation* (pp. 331–353).

Risku, H., & Windhager, F. (2013). Extended translation: A sociocognitive research agenda. *Target, 25*(1), 33–45.

Risku, H., Windhager, F., & Apfelthaler, M. (2013). A dynamic network model of translatorial cognition and action. *Translation Spaces, 2*, 151–182.

Rojo, A., & Ibarrexte-Antuñano, I. (2013). Cognitive linguistics and translation studies: Past, present and future. In A. Rojo, & I. Ibarrexte-Antuñano (Eds.), *Cognitive linguistics and translation. Advances in some theoretical models and applications* (pp. 3–30). Berlin: Walter de Gruyter.

Rojo, A., & Ibarrexte-Antuñano, I. (2013). *Cognitive linguistics and translation. Advances in some theoretical models and applications*. Berlin: Walter de Gruyter.

Rojo López, A. M. (2017a). The role of creativity. In J. W. Schwieter, & A. Ferreira (Eds.), *The handbook of translation and cognition* (pp. 350–368). Hoboken, NJ: John Wiley & Sons.

Rojo López, A. M. (2017b). The role of emotions. In J. W. Schwieter, & A. Ferreira (Eds.), *The handbook of translation and cognition* (pp. 369–387). Hoboken, NJ: John Wiley & Sons.

Rojo López, A. M., & Ramos Caro, M. (2018). The role of expertise in emotion regulation. Exploring the effect of expertise on translation performance under emotional stir. In I. Lacruz, & R. Jääskeläinen (Eds.), *Innovation and expansion in translation process research* (pp. 105–129). Amsterdam: John Benjamins.

Rowlands, M. (2009). Situated representations. In P. Robins, & M. Ayded (Eds.), *The Cambridge handbook of situated cognition* (pp. 117–133). Cambridge: Cambridge University Press.

Schaeffer, M., & Carl, M. (2017). A minimal cognitive model for translating and post-editing. *Proceedings of MT Summit XVI* (Vol. 1), 144–155. http://aamt.info/app-def/S-102/mtsummit/2017/conference-proceedings/

Schaeffer, M., Halverson, S. L., & Hansen-Schirra, S. (2019). "Monitoring" in translation—the role of visual feedback. *Translation, Cognition and Behavior, 2*(1), 1–33.

Shreve, G. M., & Diamond, B. J. (2016). Cognitive neurosciences and cognitive translations studies. About the information processing paradigm. In Y. Gambier, & L. van Doorslaer (Eds.), *Border crossings. Translation Studies and other disciplines* (pp. 141–167). Amsterdam: John Benjamins.

Spivey, M., & Richardson, D. (2009). Language processing embodied and embedded. In P. Robins, & M. Aydede (Eds.), *The Cambridge handbook of situated cognition* (pp. 382–400). Cambridge: Cambridge University Press.

Wheeler, M. (2014). Revolution, reform, or business as usual? The future prospects for embodied cognition. In L. Shapiro (Ed.), *The Routledge handbook of embodied cognition* (pp. 374–383). Abingdon: Routledge.

Wilson, A. D., & Golonka, S. (2013). Embodied cognition is not what you think it is. *Frontiers in Psychology, 4*, art. 58.

Zlatev, J. (2011). What is cognitive semiotics? *SemiotiX* XN-6 (Not paginated).

3

Translation and cognitive science

Ricardo Muñoz Martín and Celia Martín de León

3.1 Introduction

The inception of cognitive science might be traced back to 11 September 1956 (Miller, 2003, p.142), the second day of the first "Dartmouth College Summer Research Project on Artificial Intelligence". On that day, several researchers from different disciplines found that the ideas they presented converged into creating what was, for them, a distinct new world view. One by one, they displayed the belief that everything was falling into place. Under the influence of John von Neumann's thoughts, John McCarthy had coined the term "artificial intelligence" (AI) one year earlier. McCarthy was one of the convenors of that brainstorming summer school that welcomed Newell and Simon's (1956) presentation of their *Logic Theorist*, the first AI program ever. Afterwards, and inspired by the work of Claude Shannon, Noam Chomsky (1956) sketched many ideas he would publish in *Syntactic Structures* one year later. Finally, George Miller (1956) gave a presentation on the limitations of memory and bottlenecks in mental processing.

These presentations may seem disconnected, but they had a common thread: for Newell and Simon, mental processes consisted of applying search and planning rules onto internal representations. Chomsky suggested the existence of an inner, universal, mathematical grammar with a finite set of rules that would nevertheless allow people to be creative when constructing novel sentences. Also, drawing on Shannon's work, Miller showed that there were memory limits that could be tackled in terms of units. Such units might be complex and, in turn, consist of other, minor units—hence, they entailed some internal organization and structure. In brief, all presenters had different proposals that closed onto the existence of internal mental states. It was, in other words, an interdisciplinary return to mentalism. Only now it would be called *cognition*, following Jerome Bruner's preference for a term that avoided the stigma of a *mind*, all too often likened to soul in the 19th century and dubbed "unscientific" by psychologists at the time.

These and other moves crystallized into a so-called "cognitive revolution" that entailed sweeping changes in established scientific assumptions. Among other consequences, it brought about the end of behaviourism in psychology. Behaviourism had fallen short of explaining complex behaviour and, anyway, symbols were not behaviour but internal representations. In linguistics, the change called for an end to structuralism, because now the focus was on grammar rather than on the lexicon. Grammar rules were developed, according to Chomsky (1957), with minimal external input thanks to an innate language faculty. The mind emerged from the physical properties of the brain. This notion, originally from the Berlin Gestalt School of psychology, was used to separate brain and mind in order to make it possible to liken computers and humans as cognizing agents.[1]

The mind, like a computer program, would be composed of several interacting parts or modules. In this view, cognition would amount to flows of information within complex systems controlling diverse processes to (1) sense the environment; (2) symbolize it for internal handling; (3) transform, reduce, elaborate, store, recover and use such symbols; and (4) yield controlled behaviour as a consequence (cf. Neisser, 1967/2014, p. 4). Thought was reduced to problem solving, and personal, social and cultural variations were disregarded, because the human mental machine was one, and universal. Indeed, psychologists still think that humans' minds are basically alike, but nowadays they try to take on board all that was left out before, as we will see later.

The nascent cognitive sciences—note the plural, then more common—underpinned each other but were far from being cohesive.[2] This would be crucial in the development of Cognitive Translation Studies (CTS, henceforth), because cognitive psychology and the first version of cognitive linguistics (i.e. generative linguistics) took quite diverse epistemological paths. The father of cognitive psychology, Ulric Neisser, made it clear that behaviour was still the only acceptable evidence to sustain hypothesized mental processes (1967/2014), whereas linguists felt quite comfortable with introspection. With cognitive psychologists trying to discern the structure, modules and properties of the mind, and generative linguists busy trying to develop a universal grammar (see later), there was still room to focus on the interaction between mind and language.

The psychology of language may be traced back to Wilhelm Wundt and Karl Bühler, but the immediate trigger of a renovation and leap forward in psycholinguistics was the cooperation between psychologists and linguists, e.g. between Osgood and Sebeok (1954), and then between Chomsky and Miller, who inaugurated the trend of sentence processing. Many of the first empirical publications on human translation were carried out by psychologists who studied "simultaneous translation", e.g. Oléron and Nanpon (1965), Barik (1969), Chernov (1971), Gerver (1971) and Goldman-Eisler (1972). The term "psycholinguistics" had been coined by Jacob R. Kantor (1936), but such studies on interpreting epitomize a new take on psycholinguistics that would last 30 years: it would be a branch of psychology centred on the interface between mental skills and faculties and decontextualized language use in laboratory settings, within a clearly generative approach. Their efforts focused on manipulating an input variable, such as speech delivery rate, and then watching its effects on the interpreters' output (e.g. input segmentation, ear-voice span, pauses). In general, simultaneous interpreting was for them a peculiar task to trigger differential effects. Even language was deemed a secondary research topic when compared with message processing, attention and memory.

Meanwhile, written translation had languished over the centuries as a minor concern within the realm of languages, literatures and cultures, until machine translation (MT) jump-started it elsewhere in academia. Chomsky had been working in Yehoshua Bar-Hillel's MT team, and researchers in this budding field soon realized that their goal was not as easy to tackle as expected. MT research then forked out into applied, trial-and-error projects, looking for language rules, and a "pure" line that focused on the very nature of translation and language in the mind and paved the way to "theoretical linguistics". Noam Chomsky was very influential in the first years of the movement, mainly because of the publication of his 1959 review of *Verbal Behavior* (1959), an all-out attack on Skinner's (1957) radical behaviourist manifesto. Then, the Automatic Language Processing Advisory Committee (ALPAC; 1966) report nearly killed MT research in the USA, and it also called for research to focus on computational linguistics and human translation. Many founders of modern Translation Studies were scholars in humanities departments who had started working on problems and concepts inherited from MT views, now applied to human translation (Muñoz, 2016a, pp. 5–6).

3.1.1 First attempts to develop Cognitive Translation Studies

The events summarized previously were the academic milieu in which the Karl Marx University of Leipzig founded the Institute of Interpreters in September 1956 under the direction of Albrecht Neubert. Otto Kade, the deputy director, obtained his PhD in 1964 with work seeking to find out "whether solutions for translation problems could be derived from rules or were rather the result of individual accomplishments" (1964, p. 7, our translation).[3] Inspired as they were by Chomsky, the Leipzig translation scholars—mainly Gert Jäger, Otto Kade, Albrecht Neubert, Heide Schmidt and Gerd Wotjak—never saw themselves as a school of thought proper (Wotjak, 2003, p. 7, note 1). They shared, nevertheless, some important basic tenets, such as their deductive approach and their search for translation grammars that would apply to both translation and interpreting, for which they coined the German term *Translation* [transla'tsjo:n] as a hypernym of *Übersetzen* (translating) and *Dolmetschen* (interpreting).

In spite of a restrictive, objectivist approach that led them to proscribe literary translation, the contributions of the Leipzig scholars were far richer and varied than commonly acknowledged. They soon recognized a "double nature" in translation (linguistic and communicative) and developed the notion of "communicative equivalence" (Jäger, 1977, pp. 16–17), which put an end to the extreme formalism inherited from MT. Kade realized that the study of translation and interpreting needed to go beyond linguistics; Neubert was very pragmatic and included socio-cultural aspects first and text linguistics later; Wotjak slowly drifted towards cognitive linguistic approaches (see also Muñoz Martín, 2016a, p. 6). This was to little avail, because some basic assumptions inherited from MT and the information processing paradigm of cognition—the one that emerged from the cognitive revolution—remained unshattered in their work. By the mid-1980s, the Leipzig model of a science of translation based on generative linguistics had already broken down. For instance, Höhlein (1984) had to acknowledge several important drawbacks both in standard generative theory and in case theory when applied to translation. In its last years, however, the Leipzig School scholars turned to West Germany and were instrumental in fostering the development of Translation Studies there, where many of their basic insights were continued by e.g. Wolfram Wilss and, more recently, Gutt (e.g. 1989, 2000).

At the same time, the Paris School came into being after Danica Seleskovitch joined the Sorbonne in 1957 and created a PhD programme, which would become the mainstay of the *théorie du sense*, later known as the "interpretive theory of translation". The Paris School mistrusted previous experimental research by psycholinguists. Observation—but mainly introspection, much in the Chomskyan fashion she otherwise criticized—became centre stage. Hence, being an interpreter was very important to doing this kind of research, and a whole generation of researchers led by Seleskovitch would be dubbed "practisearchers". The Paris School also disliked the mechanistic approach of the Leipzig School but were not ready to criticize the then prestigious generative views (see Muñoz Martín, 2016a, pp. 6–7). Instead, they tried to enlarge, complement and reconcile heterogeneous sources—from semiotics to cognitive psychology—to propose an alternative account of interpreting that would later be generalized to (non-literary) translation.

The cornerstone of the interpretive theory of translation was the highly contested concept of "deverbalization", an intermediate step between comprehension and reformulation where words would lose "their form". Interpreters would then just find the words in another language to "repack" such deverbalized, pure meaning. Seleskovitch's was a very abstract and deductive model, so it was very difficult to falsify with anything but logical argument. Furthermore, its correlation with contemporary (and evolving) models of the mind was sometimes doubtful. Lederer, Déjean Le Féal, Delisle (who added insights from discourse analysis) and García Landa, among others, made an effort to streamline Seleskovitch's model and to flesh it out. Even so, many basic concepts, such as comprehension, remained highly idealized. In spite of this, the Paris School were first at many things that today are more or less widely accepted in CTS, such as the focus on translators and interpreters rather than on a disembodied mind; the importance given to an individual construction of meaning, where personal experience was crucial; and the notion that translating is not a matter of language but of communication.

The Paris School faded away after Seleskovitch passed away. By the end of the 1980s, Translation Studies was becoming a university discipline in its own right thanks to scholars such as Snell-Hornby, Toury and Holmes, who opened it to approaches within the tradition of the humanities. Some scholars—e.g. Fraser, Guerloff, Kußmaul, Lörscher and Tirkkonen-Condit—embraced psychological concepts and methods and focused on the empirical research of translation processes, but most of them shared notions of cognition, meaning, language and communication similar to those held by the Leipzig and Paris Schools. Towards the turn of the 21st century, some scholars—e.g. Shreve, Halverson, Risku and the authors of the present chapter—were promoting a change in the theoretical foundations of CTS, taking on board advances mainly in cognitive linguistics, cognitive psychology and the philosophy of mind. Such advances may shed new light on received approaches to translation process research (TPR) against the backdrop of a changing landscape in cognitive science.

3.1.2 The changing landscape in cognition

Jerome S. Bruner co-authored *A Study of Thinking* (Bruner et al., 1956), the first landmark of the cognitive revolution. Later on, however, Bruner did not seem very happy with the way the cognitive revolution had evolved: "at least in my view, that revolution has now been diverted into issues that are marginal to the impulse that brought it into being. Indeed, it has been technicalized in a manner that even undermines that original impulse" (Bruner, 1990, p. 1). As early as 1962, Bruner (1962/1979, pp. 129–130) wrote:

> Man does not respond to a world that exists for direct touching. Nor is he locked in a prison of his own subjectivity. Rather, he represents the world to himself and acts on behalf of or in reaction to his representations. The representations are products of his own spirit as it has been formed by living in a society with a language, myths, a history, and ways of doing things.

For Bruner, who became a proponent of "cultural psychology", meaning making was the cornerstone for cognition, and the aim of the cognitive revolution should have been

> to discover and to describe formally the meanings that human beings created out of their encounters with the world, and then to propose hypotheses about what meaning-making processes were implicated. It focused upon the symbolic activities that human beings employed in constructing and in making sense not only of the world, but of themselves
>
> *Bruner, 1990, p. 2*

Bruner, thus, may also be seen as having contributed to a second cognitive revolution that started in the 1990s. Neisser also became disappointed with information processing theories and simplistic laboratory-based methods because they could not apprehend the way people think in natural settings, and argued that research should be designed to explore how people think in real-world tasks and environments. Neisser also contributed (e.g., 1976) to this new intellectual revolution by becoming an advocate for "ecological" cognitive research. Behaviourism had not been as dominant in Europe, and Bruner sympathized with the work of European psychologists such as Bartlett (e.g., 1932), Vygotsky and Piaget. In Europe, the second cognitive revolution felt rather more like a merger of new American and existing European traditions.

Theoretical differences had been there from the start. In the meeting at the Dartmouth College, Nathaniel Rochester et al. (1956) presented a paper drawing on the work of Marvin Minsky (1954). They had tested Donald Hebb's (1949) theory of learning through cell assemblies—later known as artificial neural networks (ANN), which would pivot on the links between word features rather than on whole concepts. This trend was called "parallel distributed processing" (PDP; Rumelhart et al., 1986), or "connectionism", and would claim that the brain is a network of simple units working together. Thought and memories would be stored not as full-blown representations (symbols) but, rather, as patterns of activity throughout the network linking symbol features, i.e. below the level of symbols. Such patterns would be determined by the strengths of the connections between the units, which would vary as a function of exposure. In brief, mental representation was portrayed as "sub-symbolic", and this entailed an important correction of the by now classical cognitive paradigm of information processing. Neural networks became very successful, and they lie at the core of the current advances and hype with Big Data and also with "neural machine translation".[4] However, Minsky himself first curbed connectionists' enthusiasm by showing some limits of ANNs' capacities (Minsky & Papert, 1969) and also by challenging the difference between rational thought and emotions (Minsky, 2006).

The moves by many pioneers echoed parallel developments in other realms of cognitive science. For example, de Groot (1992a, 1992b) suggested decompositional conceptual representations in bilingual memory that amount to distributed memory representations very much in the PDP fashion. Damasio (1994) studied the neurological foundations of emotions and saw in them bodily reactions to stimuli that play a crucial role in decision-making processes. In other words, emotions and rational thought cannot be separated in human experience. The challenge to Cartesian mind/body dualism is now one of the new foci—see Muñoz Martín (2016a) and Risku, this volume—for researchers in various strands of cognitive science. Varela et al. (1991) showed that categorization is based on experience and is graded; that is, that knowledge is embodied and situated. Rowlands (2010, pp. 51–84) summarizes the new convergence in describing cognitive processes as (1) the product of the interaction with the environment, rather than being internal and self-sufficient; (2) conditioned and mediated by the body, rather than being autonomous; (3) oriented to action, rather than neutrally (aimlessly, objectively) processing information; and (4) supported by tools and environmental affordances that blur the internal/external distinction. In other words, cognition is embedded, embodied, enacted and extended. Since we have seen that emotions also belong to it, an A, for affective, is often added to the acronym: "4EA cognition".

In the last decades, there has been a call to ensure the "psychological reality" of constructs and tenets. As Johnson-Laird (1980, p. 110) put it for cognitive science in general, in building CTS we may be torn apart by two divergent tendencies: an empirical pedantry where only facts count, no matter how limited their purview, and a concept of truth leading to systematic delusion, where all that counts is internal consistency, no matter how remote it is from reality. Furthermore, we need to strive for coherence between our views and the tenets of different branches of cognitive science, now mainly in the singular (cf. Lakoff's cognitive commitment; 1990, pp. 54–55). In the next sections, we will try to put a few issues in perspective, and we will use the distinction between "microcognition" and "macrocognition" to weave a topical thread. Other scopes are equally appropriate, but this one has drawn much attention recently, thanks to an article by Chesterman (2013) in which he developed Toury's (2012, pp. 67–68) notions of "translation act" and "translation event" in ways that parallel extended views on microcognition and macrocognition.

3.2 Microcognitive and macrocognitive approaches to cognition and translation

In the cognitive sciences, microcognition and macrocognition are usually described as theoretically, epistemologically and methodologically opposite approaches. While microcognitive approaches focus on cognition "within" the individual mind and tend to use experimental, quantitative methodologies, macro-approaches focus on action and social interaction, and they mostly use ethnographic, qualitative methods. Furthermore, quantitative and qualitative approaches usually rely on different epistemological positions. While quantitative (micro-) approaches are deductive and claim universality for their results, qualitative (macro-)approaches are more inductive, and their results tend to be contextual or contingent (Halverson, 2017, p. 199). For these reasons, both kinds of approaches have often been considered incommensurable. Let us consider the differences between micro- and macro-approaches to cognition and translation and the possibilities of combining them.

3.2.1 Getting the big picture or zooming in on details?

The term "macrocognition" was coined by Cacciabue and Hollnagel (1995) to refer to the study of complex cognitive functions—such as decision making and problem detection—in natural settings, while performing realistic tasks, in contrast to traditional, information processing approaches, which operated in artificial laboratory settings and focused on single cognitive functions, such as perception and memory. Microcognitive research usually works at time scales of seconds or less, while macrocognitive approaches focus on complex processes extending over time that often can only be tackled with qualitative methods, because their time scale is too long and too noisy to use quantitative methods such as reaction time measures (West et al., 2013). Macrocognition does not study what goes on in the individual mind while performing a task, but the overall performance of the whole complex socio-technical system (Cacciabue & Hollnagel, 1995; Schraagen et al., 2008).

Some specific questions about human cognition seem to have a better fit in microcognitive, experimental approaches, while the findings of these approaches can only be applied to real-world situations using a macrocognitive perspective (Smieszek & Rußwinkel, 2013). Thus, micro- and macro-approaches to cognition are not always considered as antagonistic frameworks. Staszewski (2008), for instance, argued for the search of convergent findings between both approaches in order to strengthen their validity, and Klein et al. (2000) claimed that micro- and macro-approaches can be complementary. Smieszek and Rußwinkel (2013) described three attempts to bridge the gap between them that relate to three different understandings of the term "macrocognition".

The first follows the line initiated by Cacciabue and Hollnagel (1995) within cognitive "modelling" and focuses on the overall complex human–machine system. Here, the strategy to connect microcognition and macrocognition amounts to downscaling the model to reach a finer granularity. This is done by integrating microcognitive functions into the macrocognitive model; for example, Smieszek et al. (2013) implemented a micro-model of limited working memory within a macrocognitive model of air traffic control with the purpose of predicting more accurately the overall performance of the whole human–machine system. In order to gain domain knowledge, macro-modellers rely on interviews with experts and on observations of their interaction with the working environment, which are ethnography-inspired methods.

The second take on macrocognition is based on the "macro architecture hypothesis", first proposed by Newell (1990) with the aim of building a complete cognitive architecture that could model all macro-level tasks in all knowledge domains (West et al., 2013). According to this hypothesis, human intelligence can be described at different levels, each of which relies on the next lower level: macro-level architectures are built on micro-level architectures, which, in turn, are built on neural architectures. Like macrocognitive models, cognitive architectures allow researchers to model complex, real-world behaviours, although in this case the gap between microcognition and macrocognition is overcome by upscaling from microcognitive architectures to model macrocognitive functions, such as expert performance in chaotic, multiagent environments (West & Nagy, 2007). Another difference between macrocognitive models and architectures is that macrocognitive architecture is hypothesized to exist in the individual brain, and it is supposed to enable people to perform complex real-world tasks using information processing abilities that emerge from micro-architectures (West et al., 2013, p. 427).

The third understanding of macrocognition, also labelled "shared cognition", stems from research on teamwork. Here, macrocognition refers to the cognitive processes of a group of people collaborating in complex socio-technical contexts (e.g. Fiore et al., 2010; Grauel et al., 2013). Studies of team cognition coincide with distributed and extended cognition approaches (e.g. Clark, 2001; Hutchins, 1995) in the view that cognitive processes take place not only "within" individuals but also "between" individuals, in their cooperation and in their interactions with the environment. Macrocognition in teams describes how individual internalized knowledge is transformed into externalized team knowledge in order to build a shared understanding of the task at hand (Fiore et al., 2010). For these approaches, microcognitive aspects of individual cognition are only relevant inasmuch as they can contribute to team performance, and the connection between micro- and macro-levels takes place through knowledge internalization and externalization processes.

In a broad sense, the first and third understandings of macrocognition described previously may be related to situated and distributed approaches to cognition, while the second may be nearer to classical views of cognition, in particular to Newell's goal of reducing intelligent behaviour to simple micro-processes of symbol manipulation. So, it would seem that macrocognition could be addressed from both a symbol-manipulation and a situated/distributed perspective. And yet, the gap between the classical assumption that cognition—mainly understood as information processing and problem solving—happens in the individual mind, and the view of cognition as taking place "in the wild" (Hutchins, 1995), may be too deep to ignore.

3.2.2 Minding the gap: Cognition in the wild

Can macrocognition be understood from a symbol-manipulation perspective? Are experimental, laboratory research and ethnographic studies incommensurable? Before trying to answer these questions, it may be helpful to take a closer look at the differences between micro- and macro-approaches. Flach (2008) identified three interrelated gaps between them:

The first gap may be said to be implicit in the methodological approaches of traditional experimental research, on the one hand, and the more ethological, situated studies of human cognition, on the other. This gap has to do with the dynamics of the investigated processes. While laboratory research mostly implies some linear dynamics, i.e. a series of isolated, simple processes that take place step by step, ethological observations suggest some non-linear dynamics of coupled processes interacting in a flexible, adaptive way (Flach, 2008, p. 29).

The second gap pertains to the locus of cognition. While micro-approaches—and, as we have seen, macro-architectures—locate cognition in the individual mind, situated and distributed approaches describe cognition as taking place in the interactions between individuals, and between them and their socio-technical environment (Flach, 2008, p. 29). Hutchins (1995), for example, approached navigation from a distributed perspective, using an ethnographic methodology, and described the cognitive and computational properties of a complex socio-technical system—a ship entering a port—in terms of the cooperation of crew-members and their interaction with the technical equipment of the ship.

The third gap described by Flach (2008, p. 30) relates to the traditional disconnection between rationality and emotion. Micro-approaches to cognition have tended to consider affective, motivational aspects of experience as epiphenomena of cognition that disturb pure rationality. Enactive approaches, on the contrary, view cognition as essentially affective, since it depends on the cognizer's evaluation of the objects of cognition (Ward & Stapleton, 2012). Emotions are both mental and bodily (Damasio, 1999) and, as such, they may play an essential role in the attempts to bridge the traditional dichotomy between body and mind (Colombetti, 2007).

This dichotomy of physical and mental events characterizes the metaphysics guiding Western science and causes a deeper rift of which the three gaps sketched here are only the visible part. Flach (2008, p. 30) rejected the term "macrocognition", arguing that all cognition is macro, and proposed two alternatives to the body/mind dichotomy: (1) the perspectives of ecological psychology (Gibson, 1979) and distributed/situated cognition (Hutchins, 1995; Suchman, 1987); and (2) radical empiricism (James, 1912). Common to these approaches is the focus on situated action and the idea, also formulated by Varela et al. (1991), that cognition takes place in the dynamic coupling of organism and environment. Gibson's (1979) notion of "affordance" is the epitome of this relational approach. Affordances are the opportunities for action perceived by an organism in the environment. They are not completely objective, since they are configured by the agent's perception and exploratory behaviour; but they are not completely subjective either, because they emerge from the history of dynamic interactions between the organism and the environment. Affordances are both physical and mental (Gibson, 1979, p.129).

Now we may attempt to answer the first question posed earlier about approaching macrocognition from a symbol-manipulation perspective. Approaches inspired by ecological psychology, distributed/situated cognition and radical empiricism envision cognition from a relational, dynamic perspective. They focus on situated action and interaction, so they may be best suited to approach macrocognition. And, since all cognition takes place in the wild—since all cognition is macro (Muñoz Martín, 2016b)—these situated approaches may be the best way to address cognition in general.

In order to answer the second question about the commensurability of laboratory and ethnographic research, it is worth considering the methodological implications of ecological, situated approaches for experimental research. Flach (2008, pp. 36–37) identified three implications. The first pertains to the common distinction between basic and applied research. Psychologists Hoffman and Deffenbacher (1993, pp. 323–324) claimed that this distinction had been inadequately specified, since it was understood as a one-way road from basic to applied research, although the applied-to-basic direction was not an exception in the history of experimental psychology. New research paradigms can emerge from applied research, and existing theories can be transformed by their applications. Basic and applied research depend on each other and cross-fertilize one another. Flach (2008, pp. 36) argued that an ecological or radical empiricism approach to cognition implies "to start with the phenomena of human experience as it is lived", that is, to start with what traditionally have been considered "applied" questions.

The second implication addresses the focus of research, which should not be placed on the "internal" mental processes of the participants but on the whole socio-technical system. The "context" should not be considered an "external" factor, but intrinsic to the phenomena under study. The third implication refers to the relevance of laboratory research when generalized to the real world, i.e. to external or ecological validity. Flach (2008, p. 37) did not argue for taking research "out" of the laboratory but, rather, for taking "in" the ecological context in the research design and in the generalizations made from experimental studies.

This last implication is particularly relevant to the question of the compatibility of laboratory and ethnographic research. For Gibson (1979, p. 3), the laboratory "had to be" like life. In the context of the psychology of perception, Brunswik (1955, p. 199) originally defined ecological validity as the correlation between two variables, one of which can function as a probability cue for the other. For instance, some points of light moving in the dark can be perceived as a person. In this first definition, ecological validity was not a property of an experiment but a property of a cue. This early definition was then extended to mean the correlation between the features of experimental research—materials, tasks, conditions and setting—and the natural conditions of human cognition. Hoffman and Deffenbacher (1993, pp. 329–331) further expanded Brunswik's (1955) concept into four ecological dimensions for analysing experimental research:

1. Validity: the correlation between experimental materials, tasks, conditions and setting, and the natural conditions of real-world human cognition.
2. Relevance: the pertinence of experimental tasks to actual human experiences and activities.
3. Salience: the connection between experimental tasks and important human activities.
4. Representativeness: the connection between experimental tasks and conditions, and frequent human activities and environments.

In sum, if laboratory research starts with "applied" questions, if the focus is on the whole socio-technical system in which cognition takes place, and if it preserves natural forms and is relevant to actual, salient and frequent human activities, then it can be said to be compatible with ecological, situated approaches to cognition and ethnographic methodologies.

3.2.3 Macro-approaches to translation: Workplaces, classrooms and laboratories

The first methodological implication of adopting an ecological, situated perspective on cognition as suggested by Flach (2008)—i.e. that research starts with applied questions—is fulfilled in cognitive approaches to translation, since their object of study is a culturally informed, socially embedded human activity. As an applied science, cognitive translatology can both help to test and contribute to creating new hypotheses in the cognitive sciences.

As for the second implication—that research focuses on the whole socio-technical system—Halverson (2014) argued that adding cognitive approaches to the study of translation had prompted a transition from linguistic micro-approaches, centred on the text and the comparison between language systems, to broader views focused on translators and their situated work. However, and despite what may be a growing tendency, not all cognitive approaches to translation can be considered "macro", since not all of them envision the whole socio-technical context as an integral part of translation processes. The classical view of cognition as a series of rational, disembodied and isolated processes—and of contexts and situations as something external to cognition—was also initially adopted in CTS. Muñoz Martín (2016a) identified various simultaneous efforts taken in the 1990s to go beyond this inherited view of cognition, drawing on the theoretical frameworks of expertise studies (Shreve, 2002, 2006), cognitive linguistics (Halverson, 1996; Muñoz Martín, 1994), and situated cognition (Risku, 1994, 2002). Common to these approaches was the focus on different aspects of cognition neglected by micro-approaches, such as the social embeddedness of expertise acquisition, the embodied nature of language, and the situatedness of cognition and translation.

These efforts crystallized in some theoretical proposals that adopted a macrocognitive perspective. Risku (2000, 2002, 2010) proposed applying the findings of situated and embodied cognition to translation research, broadening its object of study to include, for example, the history of use of artefacts, the social organization of work, and communication between experts. The new approach should focus on real-life translation and use qualitative research methods "to study basic practices central to the organization of translation" (Risku, 2010, p. 104). The members of PETRA Research Group (see Muñoz Martín, 2006, 2010a, 2010b) started developing a theoretical framework for a "cognitive translatology" drawing from cognitive linguistics, situated and embodied cognition, and social constructivism. PETRA's research methods included non-invasive data collection and ecological validity checks.

Adopting a theoretical macro-perspective paved the way to introducing ethnographic methods in translation process research (reviews in Risku, 2017 and Risku, Rogl, & Miloševic, 2017). Ethnographic methods include participant observation—with researchers taking part in the observed activity in some way—and interviews with translators at the workplace. Although ethnographic research projects usually approached new aspects of the translation process, such as translation project management (Risku, 2004; Olohan & Davitti, 2015) or ergonomic needs (Ehrensberger-Dow, 2014), some projects have thrown new light on aspects frequently addressed by microcognitive approaches to translation, such as creativity (Risku, Miloševic, & Rogl, 2017) and literary translation (Kolb, 2017), adopting a broader perspective that includes not just the individual translator process, but also interactions at the workplace and the organization of work. In other cases, the results of macrocognitive approaches converge with the findings of microcognitive research. For example, Risku, Miloševic, and Pein-Weber (2016) compared the processes of translating and writing in an ethnographic case study and confirmed the findings of Flower and Hayes (1981) and Immonen (2006).

The projects developed by Ehrensberger-Dow and her colleagues (e.g. Ehrensberger-Dow, 2014; Ehrensberger-Dow & Hunziker Heeb, 2016; Ehrensberger-Dow & Massey, 2017) addressed translation from a situated perspective and combined methods in the ethnographic tradition—observation of the workplace and semi-structured interviews—with techniques commonly used in more controlled settings, such as computer logging, screen recording and retrospective verbalization. Moreover, their corpora of professional translators' processes include data collected at the workplace and in the controlled setting of a laboratory. The use of both sources of information allowed them to obtain data about the constraints placed on the situated activity of professional translators by comparing the interviews conducted in the workplace with those carried out in the lab (Ehrensberger-Dow & Massey, 2017, pp. 107–108). Combining workplace and laboratory research methods is a promising path to paint a broader picture of translation processes based on robust sets of data underpinned by the ecological validity of field research.

Let us now turn to the ecological validity of classroom and laboratory research. Cognitive approaches to translation have developed in universities and, from the start, they have been concerned with translators' education. Thus, empirical research on translation processes has frequently taken place in the classroom. Although classroom and laboratory research have often been placed together, as belonging to one pole, and in opposition to field research, there are important differences between them. Classrooms are authentic settings where social and human–computer interactions take place "in a natural way". In traditions such as action research, teachers-as-researchers are often involved in the situation and participate in the activities of the class as a community of practice. This sets classroom research closer to ethnographic approaches than to experimental settings. On the other hand, classroom conditions are more flexible and can be more controlled than the circumstances of the workplace; for example, the same text can be translated by many translators, allowing comparisons that are not possible in the workplace (Ehrensberger-Dow, 2014). For these reasons, classroom research can be placed somewhere between ethnographic approaches and laboratory settings, depending on the research interests and methodologies, and it can exhibit high ecological validity and relevance when the objects of study are novices or advanced students.

Experimental research "involves the specific manipulation of conditions or variables in a controlled environment" (Mellinger & Hanson, 2017, p. 6) and, accordingly, it tends to have less ecological validity than ethnographic and classroom research, although some efforts can make the laboratory (more) lifelike. Beyond the ecological validity of the experimental methods and setting, the possibilities of combining laboratory research with macro-approaches to translation also depend on their ecological relevance—the pertinence of experimental tasks to actual human activities. From a 4EA perspective on cognition and translation, translation processes encompass the interactions with, and the cognitive processes of, all the people involved, including the addressees. Experimental research on reception from an embodied perspective that includes emotional response (e.g. Ramos, 2016; Ramos & Rojo, 2014; Rojo et al., 2014) is an example of how laboratory research can contribute to expanding the study of translation from a micro- to a macro-perspective.

In CTS—in particular, in translation process research—the use of multi-method research designs has increased in the last decades (Halverson, 2017). Very often, multi-method research takes the form of triangulation of quantitative and qualitative data of the same event (Jakobsen, 2014). Expanding this kind of triangulation to combine micro- and macro-approaches may contribute to further strengthening the foundations of our research. Compound projects balancing laboratory and ecological data collection instances may contribute to blurring the theoretical dichotomy between micro- and macro-approaches in the future, in the same way as the dichotomy between process and product has become outdated in recent years (Halverson, 2017).

3.3 Concluding remarks: Tearing down walls for a cognitive translatology

Cognitive science is today no more cohesive than it was in the 1950s, even though many research strands tend to share basic assumptions, which can be the backbone for any '"cognitive translatology", i.e. for any framework within CTS that draws from 4EA cognition (Muñoz Martín, 2010a). We would like to close by summarizing several notions we have discussed that feel outdated and that should simply be left behind. We cannot keep entertaining naïve notions of a mental lexicon that looks like a dictionary, in view of de Groot's and Minsky's contributions (see earlier) and others such as Elman's (2004, 2011) notion that words are stimuli that operate directly on mental states and Langacker's (1987) basic contentions that (a) grammar forms a continuum with the lexicon in our minds and that (b) any symbolic unit—not only words—is in our minds a point of access to a network or cognitive routines with nodes that work much in the connectionist fashion (1987, pp. 163–164, 182).

The differences between the classical cognitivist paradigm and the (common assumptions in) current trends in cognitive science are too general and fundamental for a small set of experiments to directly test between them: they are not rival theories, but differing philosophical paradigms (van Dijk et al., 2008, p. 299). This also means that we could go on and on reviewing every single notion and aspect in the CTS landscape, but there is not enough room for this here. Besides, this whole volume offers a wide array of perspectives on cognition, and the reader is thus invited to consider their philosophical foundations as an exercise in critical reading.

First, and foremost, if we agree that (a) meaning is encyclopaedic; (b) the mental organization of stored information is sub-symbolic; (c) each language symbol may trigger the activation of a network of sub-symbolic nodes that will never be identical due to the interaction with other activated nodes through spreading activation; (d) language underspecifies meaning, in that what we experience mentally is far richer than what is meant with language symbols; and (e) meaning is an active process that will use any kind of inputs and stored information, then we will necessarily conclude that translating is not a matter of language but, rather, a matter of meaning or of communication.

This is indeed what the members of both the Leipzig and the Paris School sensed and hinted at, and what Kokkola and Ketola (2015) and Ketola (2016) have argued and proved. Language is a cognitive tool that will help us perspectivize knowledge, organize and develop ideas, and learn. But when it is separated from its communicative purposes, it drifts away from psychological reality. Language is also a social tool, and humans simply cannot use it in a pure, self-sufficient form. It always comes with an intonation, a font size, an addressee; it is always multimodal. Only abstractions of some of its regularities can give us the false impression that it is an autonomous system. This is why both psychology and linguistics, cognitive or otherwise, are not adequate referential frameworks to explain how we translate. Cognitive science is.

Using cognitive science as the referential framework for our attempts to explain how we translate and interpret would also put an end to the traditional divide between Cognitive Translation Studies and (cognitive) Interpreting Studies. If we are trying to account for a set of social behaviours that we can overarchingly describe as multilectal mediated communication, then we need to come to terms with the fact that our prototypical opposition between simultaneous interpreting and book translation are but two instances of a wide array of activities whose features used to be neatly clustered around oral and written language (use) but today can be easily separated, as in remote interpreting, asynchronous interpreting, sight translation, respeaking, post-editing, trans-creating and the like: new labels for old tasks, sometimes activities at the fuzzy borders, but well within our research scope.

This, of course, does not mean that there are no differences between translation and interpreting but, rather, that we need to seek to accommodate current and future forms of multilectal mediated communication as instances along continuums with a common core in the structure of the mind, its neurological roots, and the way language and communication are socially and culturally nurtured and gradually modified throughout a lifetime. There is also an undeniable, prominent role for (purely, merely) linguistic and psychological topics within cognitive translatology, but they will always be only partial explanations of phenomena that go well beyond their scopes.

We also need to expand our research in different directions. From a distance, CTS has a distinct Western-centric flavour and a clear bias towards professional forms of multilectal mediated communication. But "professional" is a social, not a cognitive concept. Away from some more developed countries, the most usual forms of translation and interpreting are not necessarily similar to those in complex, rich, post-industrial societies. What is it that we learn when we learn how to translate? How is it that some people will reach acceptable translation or interpreting skills on their own, and, crucially, why are we not studying them?

We have perhaps also been blindly focused on mediators. After all, we started out by focusing on the minds of interpreters and translators. But what about the other agents who participate in multilectal mediated communication? Do we behave in exactly the same way when we communicate "through" an interpreter with a third party? Do translators and interpreters take that into account? Are translation addressees constructed in translators' minds in fashions similar to those used by writers and speakers to construct their readers or audiences in monolingual settings? Are such addressee constructions intuitively rooted in communication? If so, we should include audiences, interlocutors, and addressees into the picture. We cannot afford to leave part of the picture out. These are some of the main cornerstones that cognitive science can provide for Cognitive Translation Studies.

Notes

1 According to Pinker (2002, pp. 31–32):

> *The mental world can be grounded in the physical world by the concepts of information, computation, and feedback.* A great divide between mind and matter has always seemed natural because behavior appears to have a different kind of trigger than other physical events. Ordinary events have *causes,* it seems, but human behavior has *reasons* [italics in the original].

2 A certain leeway or vagueness would even be welcome:

> It often does more harm than good to force definitions on things we don't understand. [...] The things we deal with in practical life are usually too complicated to be represented by neat, compact expressions. [...] In any case, one must not mistake defining things for knowing what they are. You can know what a tiger is without defining it. You may define a tiger, yet know scarcely anything about it.
>
> (Minsky, 1985, p. 39)

3 "ob die Lösung von Übersetzungsproblemen einer gewissen Gesetzmäßigkeit unterliegt oder ausschließlich als individuelle Leistung zu betrachten ist"

4 The use of mathematical neural networks to develop and learn statistical models to translate texts based on regularities abstracted from large corpora of examples. See some caveats for the notion of ANNs as mimicking human brains and the use of Big Data in Muñoz Martín (2017, pp. 562–563, 565).

Further reading

Clark, A. (2015). *Surfing uncertainty: Prediction, action, and the embodied mind.* Oxford: Oxford University Press.
Clark draws from empirical evidence to argue, sometimes densely and somewhat dryly, that the brain is proactive and that it self-organizes to minimize error in its constantly updated predictions. Such predictions shape perception and lead action, but an embodied, symmetrical bottom-up correction mechanism renders the mind a hierarchical set of mechanisms for Bayesian inference.

Gallagher, S. (2017). *Enactivist interventions: Rethinking the mind.* Oxford: Oxford University Press.
A clear introduction to enactivism, which distinguishes it from the other Es and tackles the role of intersubjective social engagement in language, while it shows that the construction of meaning and the acts of communication are multimodal.

Mellinger, C. D., & Hanson, T. A. (2017). *Quantitative research methods in translation and interpreting studies.* New York: Routledge.
This guide to quantitative research methods in translation and interpreting studies provides information and advice on the different stages of the research process, from survey design to the interpretation of results and reporting. Although this methodological book is not exclusively devoted to Cognitive Translation Studies, it offers useful insights for microcognitive approaches to translation and interpreting.

Risku, H. (2017). Ethnographies of translation and situated cognition. In J. W. Schwieter, & A. Ferreira (Eds.), *The handbook of translation and cognition* (pp. 290–310). Malden, MA: Wiley Blackwell.
Risku describes the developments that led to the study of cognitive processes in their authentic environments, and provides examples on the use of ethnographic methods in cognitive translation and interpreting studies. She also describes the methodological commitments of macrocognitive approaches and the challenges they pose to translation and cognition research.

References

ALPAC (1966). *Languages and machines: Computers in translation and linguistics.* Washington, DC: National Academy of Sciences, National Research Council.

Barik, H. C. (1969). A study of simultaneous interpretation [Unpublished PhD dissertation]. University of North Carolina, Chapel Hill.

Bartlett, F. C. (1932). *Remembering: A study in experimental and social psychology.* Cambridge: Cambridge University Press.

Bruner, J. S. (1962/1979). *On knowing. Essays for the left hand* (2nd ed.). Cambridge, MA: Belknap Press.

Bruner, J. S. (1990). *Acts of meaning.* Cambridge, MA: Harvard University Press.

Bruner, J. S., Goodnow, J. J., & Austin, G. A. (1956). *A study of thinking.* New York: John Wiley & Sons.

Brunswik, E. (1955). Representative design and probabilistic theory in a functional psychology. *Psychological Review, 62*(3), 193–217.

Cacciabue, P. C., & Hollnagel, E. (1995). Simulation of cognition: Applications. In J.-M. Hoc, P. C. Cacciabue, & E. Hollnagel (Eds.), *Expertise and technology. Cognition & human–computer cooperation* (pp. 55–73). Hillsdale, NJ: Lawrence Erlbaum.

Chernov, G.V. (1971). Eksperimental'naya proverka odnoy modeli [An experimental test of a model]. *Tetradi Perevodchik, 8*, 55–65.

Chesterman, A. (2013). Models of what processes? *Translation and Interpreting Studies, 8*(2), 155–168.

Chomsky, N. (1956). Three models for the description of language. *IRE Transactions on Information Theory,* 2(3), 113–124.

Chomsky, N. (1957/2002). *Syntactic structures* (2nd ed.). Berlin: Mouton de Gruyter.

Chomsky, N. (1959). Review of Skinner's *Verbal Behavior. Language, 35*, 26–58.

Clark, A. (2001) *Mindware.* Oxford: Oxford University Press.

Colombetti, G. (2007). Enactive appraisal. *Phenomenology and the Cognitive Sciences, 6*, 527–546.

Damasio, A. R. (1994). *Descartes' error: Emotion, reason, and the human brain.* New York: Avon.

Damasio, A. R. (1999). *The feeling of what happens.* New York: Harcourt Brace.

de Groot, A. M. B. (1992a). Bilingual lexical representation: A closer look at conceptual representations. In R. Frost, & L. Katz (Eds.), *Orthography, phonology, morphology, and meaning* (pp. 389–412). Amsterdam: Elsevier.

de Groot, A. M. B. (1992b). Determinants of word translation. *Journal of Experimental Psychology: Learning, Memory, and Cognition, 18*, 1001–1018.

Ehrensberger-Dow, M. (2014). Challenges of translation process research at the workplace. In R. Muñoz Martín (Ed.), *Minding translation. MonTI 1 (special issue)*, 355–383.

Ehrensberger-Dow, M., & Hunziker Heeb, A. (2016). Investigating the ergonomics of a technologized translation workplace. In R. Muñoz Martín (Ed.), *Reembedding translation process research* (pp. 69–88). Amsterdam: John Benjamins.

Ehrensberger-Dow, M., & Massey, G. (2017). Socio-technical issues in professional translation practice. In H. Risku, R. Rogl, & J. Miloševic (Guest Eds.), *Translation practice in the field: Current research on socio-cognitive processes.* Special issue of *Translation Spaces, 6*(1), 104–121.

Elman, J. L. (2004). An alternative view of the mental lexicon. *Trends in Cognitive Sciences, 8*(7), 301–306.

Elman, J. L. (2011). Lexical knowledge without a lexicon? *Mental Lexicon, 6*(1), 1–33.

Fiore, S. M., Rosen M. A., Smith-Jentsch, K. A., Salas, E., Letsky, M., & Warner, N. (2010). Toward an understanding of macrocognition in teams: Predicting processes in complex collaborative contexts. *Human Factors: The Journal of the Human Factors and Ergonomics Society, 52*(2), 203–224.

Flach, J. M. (2008). Mind the gap: A sceptical view of macrocognition. In J. M. Schraagen, L. G. Militello, T. Ormerod, & R. Lipshitz (Eds.), *Naturalistic decision making and macrocognition* (pp. 27–40). Aldershot: Ashgate.

Flower, L., & Hayes, J. (1981). A cognitive process theory of writing. *College Composition and Communication, 32*(4), 365–387.

Gerver, D. (1971). Aspects of simultaneous interpretation and human information processing [unpublished PhD dissertation]. Oxford University.

Gibson, J. J. (1979). *The ecological approach to visual perception.* Boston: Houghton Mifflin.

Goldman-Eisler, F. (1972). Segmentation of input in simultaneous translation. *Journal of Psycholinguistic Research, 1*, 127–140.

Grauel, B. M., Kluge, A., & Adolph, L. (2013). Analyse vorausgehender Bedingungen für die Unterstützung makrokognitiver Prozesse in Teams in der industriellen Instandhaltung. *Kognitive Systeme,* 1. https://duepublico.uni-duisburg-essen.de/servlets/DocumentServlet?id=31344

Gutt, E. A. (1989). Translation and relevance [unpublished PhD dissertation]. University of London. http://discovery.ucl.ac.uk/1317504/1/241978.pdf
Gutt, E. A. (2000). *Translation and relevance: Cognition and context* (revised ed.). Manchester: St Jerome.
Halverson, S. L. (1996). Conceptual categories in translation studies: Moving from classical to prototype. Paper presented at the 2nd International Conference on Current Trends in Studies of Translation and Interpreting. Budapest, 5–7 September 1996.
Halverson, S. L. (2014). Reorienting translation studies: Cognitive approaches and the centrality of the translator. In J. House (Ed.), *Translation: A multidisciplinary approach* (pp. 116–139). Basingstoke: Palgrave Macmillan.
Halverson, S. L. (2017). Multimethod approaches. In J. W. Schwieter, & A. Ferreira (Eds.), *The handbook of translation and cognition* (pp. 195–212). Malden, MA: Wiley Blackwell.
Hebb, D. O. (1949). *The organization of behavior: A neuropsychological theory.* New York: Wiley and Sons.
Hoffman, R. R., & Deffenbacher, K. A. (1993). An analysis of the relations between basic and applied psychology. *Ecological Psychology, 5*(4), 315–352.
Höhlein, H. (1984). *Die Relevanz von Tiefstrukturen bei der sprachlichen Translation* [The relevance of deep structures for translation]. Tübingen: Narr.
Hutchins, E. (1995). *Cognition in the wild.* Cambridge, MA: MIT Press.
Immonen, S. (2006). Translation as a writing process. Pauses in translation versus monolingual text production. *Target, 18*(2), 313–336.
Jäger, G. (1977). Zu Gegenstand und Zielen der Übersetzungswissenschaft [On the object and goals of translation science]. In O. Kade (Ed.), *Vermittelte Kommunikation, Sprachmittlung, Translation* [Mediated communication, language mediation, and translation] (pp. 14–26). Leipzig: VEB Verlag Enzykoplädie.
Jakobsen, A. L. (2014). The development and current state of translation process research. In E. Brems, R. Meylaerts, & L. van Doorslaer (Eds.), *The known unknowns of translation studies* (pp. 65–88). Amsterdam: John Benjamins.
James, W. (1912) *Essays in radical empiricism* (R. B. Perry, Ed.). New York: Longman Green and Co.
Johnson-Laird, P. N. (1980). Mental models in cognitive science. *Cognitive Science, 4*, 75–115.
Kade, O. (1964) Subjektive und objektive Faktoren im Übersetzungsprozess. Ein Beitrag zur Ermittlung objektiver Kriterien des Übersetzens als Voraussetzung für eine wissenschaftliche Lösung des Übersetzungsproblems [Subjective and objective factors in the translation process. A contribution to the establishment of objective criteria for translation as a precondition for a scientific solution to the problem of translation] [unpublished PhD dissertation]. Karl Marx Universität Leipzig.
Kantor, J. R. (1936). *An objective psychology of grammar.* Bloomington, IN: Indiana University.
Ketola, A. (2016). An illustrated technical text in translation: Choice network analysis as a tool for depicting word-image interaction. *Trans-Kom, 9*(1), 79–97.
Klein, D. E., Klein, H. A., & Klein, G. (2000). Macrocognition: Linking cognitive psychology and cognitive ergonomics. *Proceedings of 5th International Conference on Human Interactions with Complex Systems* (pp. 173–177). Urbana-Champaign: University of Illinois at Urbana-Champaign.
Kokkola, S., & Ketola, A. (2015). Thinking outside the "methods box": New avenues for research in multimodal translation. In D. Rellstab, & N. Siponkoski (Eds.), *Rajojen dynamiikkaa, Gränsernas dynamik, Borders under Negotiation, Grenzen und ihre Dynamik* (pp. 219–228). Vaasa: Vakki Publications.
Kolb, W. (2017). "It was on my mind all day". Literary translators working from home—Some implications of workplace dynamics. In H. Risku, R. Rogl, & J. Miloševic (Guest Eds.), *Translation practice in the field: Current research on socio-cognitive processes.* Special issue of *Translation Spaces,* 6(1), 27–43.
Lakoff, G. (1990). Cognitive versus generative linguistics: How commitments influence results. *Language and Communication, 1*(1), 53–62.
Langacker, R. (1987) *Foundations of cognitive grammar, vol. 1: Theoretical prerequisites.* Stanford: Stanford University Press.
Mellinger, C. D., & Hanson, T. A. (2017). *Quantitative research methods in translation and interpreting studies.* New York: Routledge.
Miller, G. A. (1956). Human memory and the storage of information. *IRE Transactions on Information Theory, 2*(3), 129–137.
Miller, G. A. (2003). The cognitive revolution: A historical perspective. *Trends in Cognitive Sciences,* 7(3), 141–144.
Minsky, M. (1954). Theory of neural-analog reinforcement system and its applications to the brain-model problem [unpublished PhD dissertation]. University of Princeton.
Minsky, M. (1985). *The society of mind.* New York: Simon & Schuster.

Minsky, M. (2006). *The emotion machine*. New York: Simon & Schuster.
Minsky, M., & Papert, S. (1969). *Perceptrons – An introduction to computational geometry*. Cambridge, MA: The MIT Press.
Muñoz Martín, R. (1994). La lingüística cognitiva, la teoría de la traducción y la navaja de Occam [Cognitive linguistics, translation theory, and Occam's razor]. Paper presented at I Taller Internacional de Traducción e Interpretación. Universidad de La Habana, 6–9 December 1994.
Muñoz Martín, R. (2006). Expertise and environment in translation. *Mutatis Mutandis, 2*(1), 24–37.
Muñoz Martín, R. (2010a). On paradigms and cognitive translatology. In G. M. Shreve, & E. Angelone (Eds.), *Translation and cognition* (pp. 169–187). Amsterdam: John Benjamins.
Muñoz Martín, R. (2010b). Leave no stone unturned. On the development of cognitive translatology. *Translation & Interpreting Studies, 5*(2), 145–162.
Muñoz Martín, R. (2016a). Reembedding translation process research. An introduction. In R. Muñoz Martín (Ed.), *Reembedding translation process research* (pp. 1–19). Amsterdam: John Benjamins.
Muñoz Martín, R. (2016b). Processes of what models? On the cognitive indivisibility of translation acts and events. In M. Ehrensberger-Dow, & B. Englund Dimitrova (Eds.), Cognitive space: Exploring the situational interface (special issue). *Translation Spaces, 5*(1), 145–161.
Muñoz Martín, R. (2017). Looking toward the future of cognitive translation studies. In J. W. Schwieter, & A. Ferreira (Eds.), *The handbook of translation and cognition* (pp. 555–572). Malden, MA: Wiley-Blackwell.
Neisser, U. (1967/2014). *Cognitive psychology*. New York: Psychology Press.
Neisser, U. (1976). *Cognition and reality*. San Francisco, CA: Freeman.
Newell, A. (1990). *Unified theories of cognition*. Cambridge, MA: Harvard University Press.
Newell, A., & Simon, H. A. (1956). The logic theory machine. A complex information processing system. *IRE Transactions on Information Theory, 2*(3), 61–79.
Oléron, P., & Nanpon, H. (1965). Recherches sur la traduction simultanée. *Journal de psychologie normale et pathologique, 1*, 73–94.
Olohan, M., & Davitti, E. (2015). Dynamics of trusting in translation project management. Leaps of faith and balancing acts. *Journal of Contemporary Ethnography, 46*(4), 391–416.
Osgood, C. E., & Sebeok, T. A. (Eds.) (1954). *Psycholinguistics: A survey of theory and research problems*. Bloomington: Indiana University Press.
Pinker, S. (2002). *The blank slate*. New York: Penguin.
Ramos Caro, M. (2016). Testing audio narration: The emotional impact of language in audio description. *Perspectives, 24*(4), 606–634.
Ramos Caro, M., & Rojo, A. (2014). "Feeling" audio description. Exploring the impact of AD on emotional response. *Translation Spaces, 3*, 133–150.
Risku, H. (1994). Aktive Expertenroutine oder reactive Verhaltensautomatik? Überlegungen zum Begriff der Übersetzungsfertigkeit bei Wilss [Active expert routine or reactive, automated behaviour? Reflections on Wilss' concept of translation skill]. *TextconText, 9*, 237–253.
Risku, H. (2000). Situated Translation und Situated Cognition: ungleiche Schwestern. In M. Kadric, K. Kaindl, & F. Pöchhäcker (Eds.), *Translationswissenschaft* (pp. 81–91). Tübingen: Stauffenburg.
Risku, H. (2002). Situatedness in translation studies. *Cognitive Systems Research, 3*(3), 523–533.
Risku, H. (2004). *Translationsmanagement: Interkulturelle Fachkommunikation im Informationszeitalter*. Tübingen: Narr.
Risku, H. (2010). A cognitive scientific view on technical communication and translation. Do embodiment and situatedness really make a difference? *Target, 22*(1), 94–111.
Risku, H. (2017). Ethnographies of translation and situated cognition. In J. W. Schwieter, & A. Ferreira (Eds.), *The handbook of translation and cognition* (pp. 290–310). Malden, MA: Wiley Blackwell.
Risku, H., Miloševic, J., & Pein-Weber, C. (2016). Writing vs. translating. Dimensions of text production in comparison. In R. Muñoz Martín (Ed.), *Reembedding translation process research* (pp. 47–68). Amsterdam: John Benjamins.
Risku, H., Miloševic, J., & Rogl, R. (2017). Creativity in the translation workplace. In L. Cercel, M. Agnetta, & M. T. Amido Lozano (Eds.), *Kreativität und Hermeneutik in der Translation* (pp. 455–469). Tübingen: Narr Francke Attempto.
Risku, H., Rogl, R., & Miloševic, J. (2017). Translation practice in the field. Current research on socio-cognitive processes. In H. Risku, R. Rogl, & J. Miloševic (Guest Eds.), *Translation practice in the field: Current research on socio-cognitive processes.* Special issue of *Translation Spaces, 6*(1), 3–26.
Rochester, N., Holland, J. H., Haibt, L. H., & Duda, W. L. (1956). Test on a cell assembly theory of the action of the brain, using a large digital computer. *IRE Transactions on Information Theory, 2*(3), 80–93.

Rojo, A., Ramos, M., & Valenzuela, J. (2014). The emotional impact of translation: A heart rate study. *Journal of Pragmatics, 71*, 31–44.

Rowlands, M. (2010). *The new science of the mind: From extended mind to embodied phenomenology*. Cambridge, MA: Bradford Books.

Rumelhart, E. D., McClelland, J. L., and the PDP Research Group (Eds.) (1986). *Parallel distributed processing: Explorations in the microstructure of cognition*. Cambridge, MA: MIT Press.

Schraagen, J. M., Klein, G., & Hoffman, R. R. (2008). The macrocognition framework of naturalistic decision making. In J. M. Schraagen, L. G. Militello, T. Ormerod, & R. Lipshitz (Eds.), *Naturalistic decision making and macrocognition* (pp. 3–25). Aldershot: Ashgate.

Shreve, G. M. (2002). Knowing translation: Cognitive and experiential aspects of translation expertise from the perspective of expertise studies. In A. Riccardi (Ed.), *Translation studies: Perspectives on an emerging discipline* (pp. 150–171). Cambridge: Cambridge University Press.

Shreve, G. M. (2006). The deliberate practice: Translation and expertise. *Journal of Translation Studies, 9*(1), 27–42.

Skinner, B. F. (1957). *Verbal behavior*. New York: Appleton-Century-Crofts.

Smieszek, H., Manske, P., Hasselberg, A., Russwinkel, N., & Moehlenbrink, C. (2013). Cognitive simulation of limited working memory capacity applied to an air traffic control task. In R. West, & T. Stewart (Eds.), *Proceedings of the 12th international conference on cognitive modeling*. Ottawa: Carleton University. https://pdfs.semanticscholar.org/7b5f/8da45a6858df2abc43ebbe8e667945ad6885.pdf

Smieszek, H., & Rußwinkel, N. (2013). Micro-cognition and macro-cognition—Trying to bridge the gap. Paper presented at the 10 Berliner Werkstatt Mensch-Maschine-Systeme: Grundlagen und Anwendungen der Mensch-Maschine-Interaktion, October 2013. www.researchgate.net/publication/260436847_Micro-cognition_and_macro-cognition_trying_to_bridge_the_gap/overview

Staszewski, J. J. (2008). Cognitive engineering based on expert skill: Notes on success and surprises. In J. M. Schraagen, L. G. Militello, T. Ormerod, & R. Lipshitz (Eds.), *Naturalistic decision making and macrocognition* (pp. 317–347). Aldershot: Ashgate.

Suchman, L. A. (1987). *Plans and situated actions: The problem of human–machine communication*. Cambridge: University Press.

Toury, G. (2012). *Descriptive translation studies—and beyond* (revised ed.). Amsterdam: John Benjamins.

van Dijk, J., Kerkhofs, R., van Rooij, I., & Haselager, P. (2008). Can there be such a thing as embodied embedded cognitive neuroscience? *Theory & Psychology, 18*(3), 297–316.

Varela, F. J., Thompson, E., & Rosch, E. (1991). *The embodied mind: Cognitive science and human experience*. Cambridge MA: MIT Press.

Ward, D., & Stapleton, M. (2012). Es are good: Cognition as enacted, embodied, embedded, affective and extended. In F. Paglieri (Ed.), *Consciousness in interaction: The role of the natural and social environment in shaping consciousness* (pp. 89–104). Amsterdam: John Benjamins.

West, R. L., Hancock, E., Somers, S., MacDougall, K., & Jeanson, F. (2013). The macro architecture hypothesis: Applications to modeling teamwork, conflict resolution, and literary analysis. In R. West & T. Stewart (Eds.), *Proceedings of the 12th international conference on cognitive modeling*. Ottawa: Carleton University.

West, R. L., & Nagy, G. (2007). Using GOMS for modeling routine tasks within complex sociotechnical systems: Connecting macrocognitive models to microcognition. *Journal of Cognitive Engineering and Decision Making, 1*(2), 186–211.

Wotjak, G. (2003). La escuela de traductología de Leipzig [The Leipzig school of translatology]. *Hyeronimus Complutensis,* 9/10, 7–26. https://cvc.cervantes.es/lengua/hieronymus/pdf/09_10/09_10_007.pdf

4

Translation as a complex adaptive system

A framework for theory building in cognitive translatology

Gregory M. Shreve

4.1 Introduction

Cognitive Translation Studies is a generic cover term referring to a research tradition within Translation Studies that focuses on explaining the cognitive foundations of translating and other language mediation tasks like interpreting. Such studies have also been referred to as *translation process research* (an older term describing a variety of studies during the last three decades) and more recently as *cognitive translatology*, a paradigm for the study of translation and cognition that has moved forcefully to articulate the nature and principles of contemporary Cognitive Translation Studies (see Muñoz Martín, 2010, 2015). Some of the issues we take up in this chapter are, in fact, an exploration of the implications of some of the key proposals of cognitive translatology for building cognitive theories of translation within the jurisdiction of a specific theoretical framework—namely, complex adaptive systems (CAS) theory. In this chapter I will use the term cognitive translatology preferentially because I believe that it labels the most useful current framework for understanding translational phenomena from a cognitive perspective. Cognitive translatology has the scope to usefully encompass all forms of theory building in cognitive translation research.

The study of translation and cognition is by nature interdisciplinary and intersects with cognitive science, neuroscience, psychology, philosophy, linguistics, anthropology, artificial intelligence and other disciplines that illuminate our understanding of the mental and social processes that underlie the highly complex observable behaviour of translation. Theoretical constructs have entered our interdiscipline quite opportunistically from a wide variety of sources. As a result, multiple definitions or interpretations of basic theoretical constructs often co-exist and have not yet been conceptually aligned and integrated (Muñoz Martín, 2010, p. 180). Similarly, the natures of the theories emerging from cognitive translatology and other prior cognitive study frameworks have varied widely, influenced by their multiple disciplinary wellsprings.

For instance, Sandra Halverson's theoretical work (Halverson, 2013) derives much from cognitive linguistics. Gregory M. Shreve and Barbara Moser-Mercer have explained the development of the translation skill through the lens of expertise studies in psychology (Moser-Mercer et al., 2000; Shreve, 2002). Tymoczko (2012) and others (Annoni et al., 2012) have examined the implications of neuroscience for understanding translation. Hanna Risku explored the implications of the situated cognition model from general cognitive science for Translation Studies and has examined, along with her colleagues, notions of extended and distributed cognition (Risku & Windhager, 2013). Our cognitive literature is rife with other examples.

Cognitive translatology can be situated within the boundaries of cognitive science generally. Cognitive science, of course, is also interdisciplinary, and has itself struggled (and is still struggling) with integrating constructs that arose from multiple disciplinary frameworks. Indeed, many of the most pressing issues of cognitive translatology today are reflections of issues facing cognitive science broadly; for instance, how to integrate more traditional computationalist (information processing) models of cognition with both connectionist and extended cognition frameworks such as distributed and situated cognition.

It is not my intent here to review all of these different, and sometimes competing, constructs, but to outline the possible nature of a more coherent framework for building cognitive translation theories and integrating conceptual frameworks that appear, at first glance, to be incommensurate. To that end, we propose looking at translation as a *complex adaptive system*, detailing how some of the properties of such complex systems could be extrapolated as general principles for reinterpreting cognitive translatology.

4.2 Core topics

4.2.1 Cognitive translatology

"Cognitive translatology" was first coined as a term and articulated as a disciplinary stance by Ricardo Muñoz Martín in a seminal 2010 paper. The paradigm is evolving rapidly, especially as it encourages interdisciplinary relationships and "borrowing" (under certain principled conditions) from the other cognitive disciplines. But, as O'Brien points out,

> cognitive translatology has matured over the last few years, but it is arguably still in its infancy. There are many ways in which further development could take place by borrowing even more from more established disciplines.
>
> *O'Brien, 2013, p. 13*

What O'Brien suggests is most certainly true, but one could extend that argument. Our disciplinary borrowings might be even more fruitfully integrated and their utility for theory building in cognitive translatology even further enhanced if they were placed in the context of an integrating theoretical framework that has garnered wide support in most of the disciplines we currently borrow from.

Of course, borrowing the complex adaptive systems framework to extend cognitive translatology only makes sense if there is some conceptual consonance between the two. If we lay out in a brief fashion the main arguments of cognitive translatology, we can perhaps then proceed to see where CAS might provide theoretical support. Muñoz Martín articulated a total of ten postulates to define cognitive translatology (2010, pp. 173–179):

1. "The aim of translatology is to offer a realistic, detailed account of a set of special, complex communicative events and their products."
2. "Translatology should include cognitive approaches to account for translation and interpreting."

3. "Cognitive translatologies need to be based on scientific, empirical research."
4. "Cognitive translatologies need to make their account of human translation and accord with what is generally and currently known about the mind and brain."
5. "Cognitive translatologies are functional by definition."
6. "It is interpretations, not texts or discourses, which are translated or interpreted."
7. "Translating is an interpersonal activity."
8. "Translation is a form of creative imitation."
9. "Translation expertise implies the continuous development of natural cognitive skills."
10. "Cognitive translatology should focus on the interaction between translators and their environment."

Taken together, these postulates lay out some important properties of translation from the perspective of cognitive translatology. I take the liberty of abstracting them in a deliberate fashion in Table 4.1.

Even if we consider Table 4.1 simply a foil for advancing my thesis, a close reading of the author's paper will find that these ideas (expressed in a variety of ways) pervade his argument. While not mentioning the CAS paradigm directly, it is our argument here that Muñoz Martín has nevertheless presciently laid out a framework perfectly consonant with that paradigm and has therefore led us to a closer disciplinary integration with the social and behavioural sciences (and an emerging paradigm in linguistics).

To give credit where it is due, I am not the first translation scholar to recognize the value of complex systems theory in Translation Studies. Kobus Marais (2014) had already detailed the implications of complexity thinking for Translation Studies and focused on translation as an emergent phenomenon. However, in a fortuitous case of parallel thinking, I had not yet read his work when drafting this paper—yet we have reached many of the same conclusions.

4.2.2 Complex adaptive systems

Complex adaptive systems have been a topic of discussion in the philosophy of science, cognitive science and physics, and other disciplines for many decades. Ladyman et al. (2013), Heylighen (2008), and Beckner et al. (2009) provide an excellent overview of the central concepts—the third citation focusing specifically on language as a complex adaptive system. Ladyman (2013, p. 27) proposes a useful starting definition, stating: "A complex system is an ensemble of many elements which are interacting in a disordered way, resulting in robust organisation and memory."

Complex systems are composed of *multiple ensembles* (collections) of *multiple similar elements*. Thus, a social organization is a grouping of many human individuals; a neural network is a

Table 4.1 Properties of translation from the stance of cognitive translatology

Property	*Postulate*
Translation is complex, communicative and interactive	1, 6, 7
Translation involves multiple agents	7, 10
Translation is cognitive, adaptive and continuously develops	2, 4, 9
Translation is goal oriented	5
Translation is both ordered and disordered, but patterns emerge	6, 8
Translation is multi-scale and hierarchical	4, 10
Translation involves response to the environment	9, 10
Translation can be understood empirically with proper methods	3, 4

collection of a great number of neurons. These multiple similar elements are organized *hierarchically*; that is to say, they exhibit *different levels of organization.*

> Ensembles of similar elements at one level form a higher-level structure, which then interacts with other similar higher-level structures. As an example, consider a society. Many cells make up a human body; many human bodies make up a group; many groups make up a societal structure.
>
> *Ladyman et al., 2013, p. 28*

The elements and ensembles must be able to interact directly via some mechanism. In neurons, this interaction occurs via some sort of biochemical information transfer; in human beings, gesture, facial expression, language and artefacts enable interaction. The interaction of elements and ensembles is continuous—and the interaction "affects all the other variables contained in the system and thus also affects itself" (Van Geert, 1994, p. 50). Thus, there is always a strong and active *interdependence* both between elements and between elements and other ensembles in the system. Heylighen (2008, p. 4) makes the observation that in complex systems the "components are both distinct and connected, both autonomous and to some degree mutually dependent".

The unitary and collective elements of the system are generally referred to as *agents*, and this term reflects their *active* role. Agents *act upon their environment* in response to the changes or events *they experience in that environment.* Due to the high connectivity of agents at all levels of organization, the actions they take (e.g. in their responses to local conditions) may affect the actions of other agents. The impact of any given response may propagate outwards from local agents to more distant ones:

> Such interactions are initially local: they start out affecting only the agents in the immediate neighbourhood of the initial actor. However, their consequences are often global, affecting the system of agents as a whole, like a ripple produced by a pebble that locally disturbs the surface of the water, but then widens to encompass the whole pond.
>
> *Heylighen, 2008, p. 4*

In a complex adaptive system, agents are also assumed to be *goal directed* and *adaptive*. Holland and Miller (1991, p. 365) explain:

> An agent in such a system is adaptive if it satisfies an additional pair of criteria: the actions of the agent in its environment can be assigned a value (performance, utility, payoff, fitness, or the like), and the agent behaves so as to increase this value over time.

Agents can act upon and change their environment, but they may also change themselves in keeping with the goal of maximizing their (let us call it for now) *utility*. As Odell says, such "agents must be able to react to their environment and possibly change their behaviour" (2002, p. 37). Changes in behaviour can be precipitated by inputs from the environment but can also be mediated by mechanisms such as positive and negative feedback to their own actions. A change in any agent can cause a very small change in the overall system, a very great change, or perhaps no change at all.

Thus, the dynamics of change in complex adaptive systems are *non-linear*—that is to say, "the effects may not be proportional to the causes" (Heylighen, 2008, p. 4). The non-linearity of such systems results from the amplification effects of positive feedback, where small changes (perturbations) are reinforced as they propagate through the systems. Similarly, negative feedback can lead to a dampening effect (2008, p. 5). Ultimately, this means that we cannot easily (or at all) predict system outcomes from initial conditions. Complex adaptive systems are quite sensitive to initial conditions: "small differences in the values of the initial conditions may make for radically different macrostates" (Ladyman, 2013, p. 5).

This entails the idea that while complex adaptive systems exhibit structure, changes to that structure are generally not explainable by recourse to simple rules and cannot be accounted for without looking at other levels of the system—that is to say, at lower and higher levels of agents. So far, we have argued that the "changes" effected by the actions of local agents, by dint of the dense interconnectivity of complex systems, affect other agents more globally; we have also emphasized that complex systems are in continuous interaction—this dynamic aspect balances such systems between order and disorder.

Order and disorder are both present in complex systems, but "processes of self-organization literally create order out of disorder. They are responsible for most of the patterns, structures and orderly arrangements that we find in the natural world" (Heylighen, 2008, p. 2). A completely disordered system would show no patterns—the relative states of the agents or elements would be random; a completely ordered system would show no change—the states of the agents would always be identical.

Complex systems are always balanced on the turbulent "edge of chaos" (Heylighen, 2008, p. 4), but structures, patterns and organization (e.g. some sort of stable spatial-temporal configuration of agent states and agent relationships) can arise via the processes of *emergence* and *self-organization*. Emergence is a property of the collective interactions of system elements at the micro-scale. Their interactions produce a macro-level (higher-level) pattern that cannot be explained by examining the constituent agents alone. Self-organization and emergence are related concepts, and there is no need to debate any further theoretical distinctions here. Heylighen provides a definition that equates the two usefully enough for our purposes:

> Self-organization can be defined as the spontaneous emergence of global structure out of local interactions. "Spontaneous" means that no internal or external agent is in control of the process: for a large enough system, any individual agent can be eliminated or replaced without damaging the resulting structure. The process is truly collective, i.e. parallel and distributed over all the agents. This makes the resulting organization intrinsically robust and resistant to damage and perturbations.
>
> *Heylighen, 2008, p. 6*

Emergence and self-organization can create patterns that appear to be purposefully adaptive—goals are met, intentions are realized. The constant action of agents in maximizing their utility seems to provide evidence of central control, of planning and direction, but it is instead sometimes just the result of the constant jostling of agents to find utility or fitness in their local environment.

Sometimes the patterns (structure, macro-level behaviours) that arise from complex systems persist for a time. Heylighen called these patterns "robust", and Ladyman (2013, p. 29) uses this expression also but adds the condition that "on an appropriate time scale the order is robust". The notion of robustness implies that even as individual elements continue to "interact in a disordered way", some of the patterns that have emerged from the collective action may persist *for a time*. Although order arises, any particular pattern or structure is not necessarily permanent—it persists for a time and will potentially alter as the elements of the system continue to interact and respond to one another.

Thus, it is important to understand the temporal nature of order in complex systems, an idea captured in the notion of *dynamism*. Ladyman also termed the persistence of pattern in complex systems *memory*. Memory or robustness also plays out in another way, in that previous patterns of the complex system, because they are retained for a time, can influence the way the system will behave in the future. The *past history* of a complex system becomes another critical factor in its future behaviour.

4.2.3 Translation as a complex adaptive system

There is certainly much more we could say about complex adaptive systems; this is a brief introduction to some of the main concepts we want to align with cognitive translatology. The first assertion to be made is that cognitive systems, of which translation would be an example, are complex adaptive systems. Many cognitive scientists share this point of view (see Jordan et al., 2015). A unifying characteristic of a complex systems view of cognition is that it challenges the idea that "cognition, perception, and action constitute system functions that can be isolated and measured independently of the rest of the organism-environment system" (Jordan et al., 2015, p. 316). A cognitive system would share the general properties of other complex systems—but it remains to specify exactly what merits the modifier *cognitive*. Rather than review the (copious) literature arguing this point, I advance a different practical argument, pertinent to cognitive translatology.

4.2.3.1 Cognitive translation processes

A fundamental notion in cognitive translatology, and one with great longevity, is that of *translation processes*. The term is generally taken to refer to a variety of mental operations that engage during a translation task and enable its completion. Cognitive science has generally seen cognitive processes as (relatively) discrete mental operations that have some sort of circumscribed scope, although there is not, as yet, a single commonly accepted definition of what such processes are. Newen (2015, p. 8) has argued that cognitive processes are *information transfer processes* that "connect multiple (or complex) informational inputs to form a minimally flexible cognitive system with a spectrum of minimally flexible behavioural outputs". What makes these processes *cognitive*, Newen argues, is primarily the fact that they involve the paradigmatic process domains of the cognitive sciences: perception, memory, learning, emotion, intentionality, self-representation, reasoning, problem solving and so on.

Newen's argument is quite pragmatic and has some utility for theory building in cognitive translatology, since it claims, in part, that what *cognitive* means is, at least at this stage of development of the cognitive sciences, circumscribed by what constructs those disciplines have adopted as objects of study. Thus, the notion of cognitive processes in our area of study is greatly dependent on "actual practice in the typical cognitive sciences, the development of specific research methods, etc." (Newen, 2015, p. 17). In our attempts at building translation process theories, we in cognitive translatology have basically enacted Newen's approach. As we imported certain constructs from our sister sciences, we implicitly demarcated them as cognitive because the disciplines we borrowed them from had already done so. We did not provide a grounding definition of our own and, more importantly, perhaps we didn't need to. Nevertheless, as we import ideas like "cognitive process", we need to be clear about their disciplinary origins and entailments, and where they might fit into our own schemes for theoretical integration.

If we want to retain the notion of cognitive processes for a complex adaptive systems view of translation, then how would we explicate such processes from that theoretical position? Process would be understood primarily as the response of a collection of agents to conditions in the environment (implying the receipt of information), resulting in a change of state in the system. That changed state, in turn, is an information unit. From a complex systems point of view, processes are not just actions or responses of individual unitary agents, but a term to describe changes emerging as a pattern, behaviour or structure at a macro-level due to the continuous interaction of agents at lower levels. Thus, processes have to be understood as multi-scale (they can apply at different levels of system organization) and dynamic.

By classifying cognitive processes as information transfer operations, as Newen does, some other insights emerge. First, cognitive operations are *transformative*. They take inputs, whether from the outside environment (including from other agent collectives) or from precedent processes, and always modify them in some fashion. Theoretically speaking, it makes no sense to posit a process that simply "passes through" information unchanged. The process perforce generates outputs, or modified information, passing them to other downstream processes, some of them emerging eventually from motor processes as visible behaviour. The type of input, the type of output and by implication the nature of the transformation provide a basis for understanding the scope of a process and demarcating it from other processes. Complex systems theory also requires us to specify the level of organization at which the process is operating—the scale at which a structure or behaviour emerges and what agents are involved.

This notion is particularly useful for looking at translation, where we have (certainly) complex informational inputs, in the form of the source text and its context of situation, and a recognizable behavioural output in the form of the serial production and revision of target-text segments. However, from a cognitive science perspective, the notion of a singular discrete translation process seems unlikely. Rather, it is more probable that what we call translation processes are actually more "paradigmatic" processes that engage and interact in a flexible way during a translation task—and at different levels in the system hierarchy. In building cognitive process theories of translation, we need to, as Muñoz Martín (2010, p. 174) also argues, "be consonant with the findings of the other cognitive sciences". This means, minimally, that we shouldn't introduce novel process constructs that can't be integrated with the existing process constructs of our sister sciences—but we should take it as also implying that we need to be clear about systemic questions: what are the agents, how do agents interact, what is the environment to which the agents are responding, and how do changes in the collective states of agents become interpreted as processes, structures or behaviours at higher and lower levels of scale?

Shreve and Lacruz (2014) conceptualized translation as a "higher-order process"—meaning that multiple lower-order (e.g. paradigmatic) constituent processes are engaged serially and in parallel during a given translation task within a finite time course. These processes, brought online during the task, interact with one another, comprising what Newen termed a "minimally flexible cognitive system". This means that the system output is not rigidly determined by the input but that the system of interdependent "processes" is flexible enough to produce variable outputs. This flexibility is due in great part to variability in the interaction of constituent processes—collections of agents don't exchange information in a predetermined way. Further, the transformations they produce are dependent on the current environmental conditions to which the system is responding. Complex systems are dynamic, and that dynamism derives not only from environmental variability but also from the intrinsic interactive nature of the cellular agents of the body (neuronal, motor, organ). It also derives from the unpredictable way in which system outcomes are associated with variations in initial conditions. I can't help thinking here of Muñoz Martín's postulate number 8 concerning imitation and creativity.

The source text comprises some of the initial conditions for the engagement of a temporally circumscribed local "translation system". We can't predict the outcome of that system to a high degree of precision—the contrastive "rules" of language don't alone suffice. There is some "imitation" (constraint on the outcome) but also, quite necessarily, "creativity" (divergence, novelty). This is in the nature of the propagation of the signal through the complex system. We don't know exactly where we will end up.

Viewing cognition during translation from a CAS perspective has, we feel, great value in theory building. Cognitive translatology needs to directly confront the systemic nature of cognition during the translational act. In this chapter we will argue that useful theories of translation and cognition are, to a great extent, complex adaptive systems theories. This implies, at least, that when building cognitive translation theories, we need to deal with systemic questions.

4.2.3.2 Cognitive system boundaries

One of the explicit entailments of cognitive translatology is a questioning of the boundaries of cognition; this critical stance is quite consonant with a CAS view. From a complex system point of view, we must consider that the boundaries of any "complex" cognitive translation system under study are quite possibly flexible from a theoretical view. If translation is a complex system of structures or processes at multiple scales, then the initial boundaries of the system *could be set* at the initiation of the first information transfer, that is, with the processes of visual perception that accompany reading. The terminal boundary would be a terminating behaviour, such as a final text production or revision act or a final reading of a source or target-text segment, after which no further task-relevant behaviour is deemed to occur. The cognitive system is demarcated at one temporal pole by an initial perception of stimuli from the environment and at the other by a terminating act.

However, this demarcation is at least partially a matter of disciplinary tradition and of theoretical convenience. In the first case, we study the systems as our colleagues have studied them (for commensurability) and in the second case, because our research questions often assume an object of study that is presumably contained within the system boundaries we have set.

We could set the boundaries of the system *differently*, and if we do so, then we extend the cognitive system beyond the traditional boundaries of what is thought to be cognitive: that is to say, the mental operations of an individual mind. Theories of *situated cognition* do not so abruptly set the system boundary at individual perception, because the relationship between the perceiving agents and the situation, environment or context is conceived of quite differently. Cognition, in this view, must explicitly include the immediate context in which an intelligent agent acts; there is a dynamic interchange between the so-called *affordances* present in the situation and the processes of perception and later cognition that subsequently occur (Gibson, 1977). This view of translation processes fits very well with several new research directions emerging in Translation Studies and with the principles of cognitive translatology.

Similarly, cognitive systems need not be constrained to an individual mind. *Distributed cognition* (Hollan et al., 2000; Hutchins, 1995) argues that cognitive systems can extend to include not only other minds but also artefacts in the environment. A distributed cognitive system is one "whose structures and processes are distributed between internal and external representations, across a group of individuals, and across space and time" (Zhang & Patel, 2006).

It is beyond the scope of this paper to describe these *extended cognition* frameworks in detail (see Clark & Chalmers, 1998, on the "extended mind"), but if we are building cognitive translatology theories, it is important to specify the scope of the systems we are trying to explain. Where we set the boundaries of the system determines how we conceptualize the (arbitrary) initiating and terminating states of the system and which processes (and their associated input conditions and output phenomena) are presumed to be actively interdependent during a given time course—and therefore of presumed relevance to the explanation we are seeking.

From a practical standpoint, complex systems theory reminds us that any "discrete" system we study is linked to systems both above and below it in a multi-scale hierarchy. However, we cannot hope to study the entirety of a system's entanglements with other systems—this is why we set specific system boundaries in theory building. If we do not, then we have a practical "theory management problem"—the object of study becomes too diffuse, the inputs and outputs too diverse and too many; the patterns and structures are at an unwieldy diversity of scales.

Also, the scope of any theory is necessarily coextensive with the observational limits of the phenomena under study and the observational methods available in the discipline. We can't study all the interactions multiple processes in any given system might have with other processes—and all the inputs and outputs. To some extent, what we do is only study processes that seem to be of explanatory relevance—that is, that pertain to the research question asked, and where the chosen methods of observation and analysis can yield some grounds for claiming that some output is the complex result of the active interdependence of specifiable collections of agents.

4.2.3.3 Hierarchy in the cognitive translation system

When building theories in cognitive translatology, we have to "situate" the explanatory objectives of any theory at the appropriate level of the cognitive system. We have argued, to this point, that a cognitive process is a label for some condition-action event associated with a collection of elements at a certain scale; changes of system state occur, and information is transformed. Because complex systems are multi-scale, our process explanation can also be at different levels of scale.

For the sake of argument, and setting the system boundaries at the individual mind, we could consider translation as involving collections of agents interacting to enable comprehension, transfer and production processes. These processes are at a very high level of scale; for some explanations, for certain research questions, this tripartite granularity may be quite sufficient. But comprehension itself may be viewed as involving both reading and memory retrieval; thus, a more granular explanation of comprehension in translation—a more specific research question—would involve specifying the systemic nature of those two processes and how they interact.

Reading may further be functionally decomposed into even more granular processes of signal perception and processing. Composition and decomposition, from a complex systems perspective, is a theoretical process of moving up and down the hierarchy, or as Ladyman argues (2013, p. 32), recognizing that "the different levels of the hierarchy are made up of regroupings of lower levels, of a redefining of the system's boundaries".

Cognitive translatology has historically concerned itself with rather broad higher-order "meta" processes that take place online during translation. The term "meta" here signifies that such processes are convenient constructs for discussing events that, in fact, involve large numbers of interacting elements at different levels of structural consideration. We "collapse" all these details into a construct that encapsulates the details and focuses on compositional abstractions—we abstract the relevant input conditions, emphasize certain changes of state, and focus on certain outcomes.

The further down we go to the "bottom" of the systemic hierarchy, the closer we get to the individual and more unitary agents that are presumed to enable the operation of the entire complex system. In an embodied (human) cognitive system, the lowest level could be claimed to be neurons—although, in fact, one could make the case that at the ultimate level of decomposition, we are left looking at collections of specialized body cells—of the eye, of the brain, of the muscles with which we type our translations. As we travel further "up" the system hierarchy, we extend the system boundaries outside the individual mind and include not only other individuals but also the artefacts their own system processes have produced.

Of course, while all these levels are within the scope of theory building, we always delimit ourselves (Shreve, 2018, p. 106):

> When we seek explanations of how any task can be carried out, we can go quite possibly very much further down in levels of explanation; the neuron networks of the brain are not the lowest level of consideration. We could proceed to the cellular, the molecular, the atomic, and, perhaps, the quantum. As we move up in levels of explanation from the situated task, we could also go quite a lot further up, from the social and cultural perhaps all the way to the cosmological. But, as a matter of necessity, we circumscribe our levels of explanation, restricting our interests to those task-adjacent levels that contribute materially to the robustness of the explanation we wish to furnish.

4.2.3.4 Translation as task

Theory building in cognitive translatology must be concerned with where we set the boundaries of the translation system. Those boundaries can be flexible, but, in reality, they generally reflect the research questions we ask, the character of our observations, and to a great extent the observational methods we can use.

We can argue, as externalism does, that cognition extends beyond the individual mind to include external actors and artefacts—and therefore, by implication, that the time course of a set of cognitive system processes is not necessarily initiated at individual perception and terminated in a final individual act. Yet, it seems unworkable—with respect to practical research questions that we might hope to answer—that we should extend cognition outwards to include all possible influencing processes, artefacts and actors. On what basis should we constrain the system we wish to explain?

In their discussion of extended cognition, Clark and Chalmers (1998, p. 8) offered a useful basis for limiting the scope of a cognitive system as an object of study: "all the components in the system play an active causal role, and they jointly govern behaviour in the same sort of way that cognition usually does". This formulation generally considers that we would constrain the system by being able to demonstrate which processes were *active* and therefore had some *demonstrable* influence on subsequent or concurrent processes. Whether they are internal (of the mind) or external (of the environment—or even outputs of the behaviour of others) is in this view irrelevant—an external process is a cognitive process if its presence or absence influences the "system's behavioural competence"—its ability to achieve its functional goal. External and internal resources are temporarily "coupled" during the time course of any given cognitive task.

Any given act of translation is a transient system, brought online for goal-oriented reasons emerging out of a social-communicative context. The "functional" behaviour of the system, potentially involving external actors and processes, is objectively directed to the completion of a particular task. In theory building in cognitive translatology it is imperative to recognize how the notion of a *task* is critical to theory building. Muñoz Martín (2010, p. 173), articulating the principles of cognitive translatology, argues that the "aim of translatology is to offer a realistic, detailed account of a set of special, complex communicative events and their products" (postulate 1); a few paragraphs later, he adds "cognitive translatologies are functionalist by definition" (postulate 5). If we take these two statements together, they underscore an important underpinning of theory in cognitive translatology; it is *communicative*—and therefore must at some point account for the influence of external entities and extend itself into the social—and it is *functional*.

Both of these properties of cognitive translatology align very well with complex systems theory. We will briefly explore these two notions before turning to the question of task.

The communicative property speaks directly to the nature of interaction between collections of elements (in this case stakeholders in the translation process); as such, translation can be seen at the social system level as a complex behaviour that emerges, at least initially, from the collective behaviour of individuals acting together in a social structure. We can understand intent, purpose, function and even the setting of task conditions (e.g. schedule, deadlines) as properties emerging from the social system. The fact that we *can* communicate—and that we share certain language knowledge, world knowledge and domain knowledge—can be seen as a property of the extended social system. Each of these systems can play a role in cognition—properly extended via the right research question—although it does not follow, if we circumscribe the scope of our explanation properly, that we *must* deal with those systems. Still, the point here for scholars in our discipline, like Hanna Risku, Maureen Ehrensberger-Dow and Ricardo Muñoz Martín, is that cognition can and does extend to those systems—and theories developed in cognitive translatology must account for that extension.

The functional property of translation also aligns quite neatly with complex system theory. Earlier we discussed the property of agency and its relationship to goal orientation, and here Heylighen, once again, provides a useful perspective:

> To explain the appearance of organization, we need to make one further assumption, namely that the outcome of interactions is not arbitrary, but exhibits a "preference" for certain situations over others. The principle is analogous to natural selection: certain configurations are intrinsically "fitter" than others, and therefore will be preferentially retained and/or multiplied during the system's evolution. When the agents are goal directed, the origin of this preference is obvious: an agent will prefer an outcome that brings it closer to its goals.
>
> *Heylighen, 2008, p. 7*

Functionalism has a long history in Translation Studies, going back to Hans Vermeer's 1978 articulation of *Skopostheorie*, with its emphasis on the functional purpose of a translation as the primary determinant of translational strategy. It is time for us to recognize (as Muñoz Martín exhorts us to do) that we can't ignore the fact that translation, at whatever level of system explanation, is goal directed. All of how translation behaviour plays out in real life and real time is inextricably related to the goals the translator sets for the activity—or perceives as having been set. Goals are notoriously slippery to grasp, however. We can talk about the goals of a translation at the social system level—e.g. with words like "purpose" or "function" and terms like "translation brief". But we must also consider that system elements are goal oriented at levels further down in the systemic hierarchy. So, what are the "goals" of lower-level agent ensembles—or, better yet, how do the goals at greater system levels *scale down* to explain lower levels? Similarly, there are goals further up the systemic scale—the translator's objectives for the act of translation don't appear *sui generis* but derive from the outputs of external social agents.

Let me give an example here of scaling down and take the case of a translator proceeding through a translation sentence by sentence. During translation the translator appears (from the empirical evidence) to work on discrete "chunks" of textual material. A specific and specifiable (via eye-tracking and keystroke logging technologies) segment of source text appears to be focused on by the translator and its meaning extracted. That meaning is then "transferred" into a target-language production, a target segment taken to be a re-expression of the message of the originating segment. This chunk, a so-called translation unit, is demarcated from preceding and succeeding units by pauses.

We have, apparently, during the time course of the entire translation, a recursive process whereby several successive cycles of reading, text comprehension and text production occur. Each of these chunks is initiated (we have empirical evidence of such) by an initial eye movement and terminated by final production behaviour before another cycle begins. The theoretical question here is this. At this level of granularity, how would we characterize the goal of this particular condition-action sequence? Certain broad conditions were set, certainly, before the entire sequence was kicked off—but are these conditions specific enough to explain why the transition between one chunk and another was made, or, rather, why the decision to move from one chunk to the next was made? (See here the notion of "tipping point" in CAS theory (Gladwell, 2000)—the point at which a system moves away from one state to another.)

Functional considerations had to play a role in the decision to continue, did they not? Yet not all the goal setting was external and broad—the translator must have set some more specific internal goals him- or herself; and therefore, some goals pertain to the chunk rather than to the textual process as a whole. What I am suggesting here is that "previous states" of the translator's cognitive system (the history of the system—its memory) may have contributed additional internal goal conditions. The translator may have had certain expectations—set additional micro-goals—that needed to be met. These were extrapolated both from the initial conditions and from the memory of past translations. Such a conception fits not only with what we know of metacognition but also with complex systems theory. We could continue our consideration of goal orientation all the way down to examining why a particular neural array activates and deactivates—should our research focus require this. Interestingly, recent work by Carl, Tonge and Lacruz (forthcoming) in human and machine translation provides some avenues for investigating the micro-scale emergence of structure and changes of state in translation processes using the framework of "dissipative systems" and "entropic gravity". Their work may allow us to better understand how and when translational tipping points are reached.

Now we turn our attention to the notion of the translation task—a construct, I believe, central to cognitive translatology. The translation task is functionally motivated, arising from the constraints placed on it by other agencies in a social network; it is communicative, because at some point, like other agents, during its time course, processes engage that accept information from external sources, transform it and then pass it to others before moving beyond the translator's individual scope. But, additionally, we have to consider whether the construct of the task really represents the way the external aspects of cognition impinge upon the internal aspects. In Translation Studies and in functional linguistics we have long understood the role of ideas like "context of situation". The pragmatic turn in Translation Studies was a widespread recognition that translation, as a system, is exquisitely conditioned by its external parameters. This orientation of Translation Studies is perfectly consonant with how complex system theory understands the relationship of system and environment and applies directly to theory building within cognitive translatology.

4.2.3.5 Translation, system experience and control

Complex systems are dynamic, but also adaptive. Cognitive systems not only adapt but also can preserve aspects of the adaptation in memory to influence future adaptation. A long-standing interest in cognitive translatology is how the ability to translate develops. Various constructs such as competence, and more recently, expertise, have been used to look at what *accumulation* of experience is most efficacious in translational success. It should be noted here that both competence and expertise necessarily involve not only acquiring experience but also transforming it in some way so as to benefit the future performance of the system in meeting its goals. The cognitive system transforms raw experience, preserving that which has proven efficacious—a sort of fitness or utility function—in achieving goals. Expertise is an adaptation, a construct that captures not only the goal orientation of the translation system but also the results of competition between alternative experiences. Which ones get preserved—stored in system memory? There has to be a mechanism that explains what is retained and what is discarded—or, rather, what leaves a more persistent cognitive trace rather than a weaker one. Here, of course, theoretical notions stemming from cognitive linguistics, such as entrenchment and salience, would come into play as possible explanatory mechanisms.

Thus, all sorts of constructs we use in cognitive translatology, from decision making to problem solving, proceduralization, schematization, automatization and translation strategy (among others), could be seen as various aspects of the pattern-creation properties of complex systems. Each of these ideas involves the longer-term persistence of certain patterns in the system—stored in some way for later use in achieving the goals of similar tasks.

Some constructs we use in cognitive translatology, such as planning and metacognition, relate to the issue of control in the system. Cognitive translation scholars have long understood that translation is a composite systemic activity, bringing together within the scope of a specific translation task an array of cognitive resources activated by constituent processes within the scope of the task; we've done this implicitly even if we haven't used the language of complex system theory to express the ideas.

We assume that the constituent processes are coordinated under the constraints of a given translation task for the fulfilment of a specific communicative goal. The notion of coordination is an important one; it implies that there is control, e.g. that the constituent activities are, somehow, both integrated and directed to the completion of the task. Complex systems theory can give us another perspective on this, suggesting that there is no "hierarchy of command and control", as Kaisler and Madey (2009) claim: "there is no planning or managing, but there is a constant reorganizing to find the best fit with the environment". From this perspective, much of what we think of as control is, in fact, an aspect of self-organization within the confines of a task—with the consideration that previous patterns (a redacted history of system states preserved, literally, in our memory) may serve to help certain patterns of action to emerge rather than others.

As an example, *task awareness* is a central concept in metacognition. But we could conceive of it as simply an emergent property of the current state of the system whereby the awareness of the current task is conditioned and precipitated by our precedent experience of the task. Translation strategies, similarly, are a stored pattern extracted from previous experience, represented ultimately as a network of interconnected neurons that activate under the right conditions. Even our decision making and problem solving aren't the simple acts of will we might imagine them to be, but a result of our history of experience interacting dynamically with the currently engaged system.

4.2.3.6 Transformation, language and artefact

In dealing with translation we cannot, of course, avoid dealing with *language*. Language, too, is a complex adaptive system. This perspective has gained significant disciplinary traction in recent years (see Beckner et al., 2009, and Baicchi, 2015, for an overview). We cannot treat with this subject extensively here, but simply make the claim that the patterns of language, from the phonological to the textual, are patterns emerging from the interpenetrating actions of systems of agents at multiple scales.

We have used the word "pattern" to refer to a persistent, stable configuration of system states over a finite time course. We can also give another interpretation to the term (Baicchi,

2015, p. 19): "In complexity theories the term information is synonymous with the presence of distinguishable patterns and relationships." Thus, one way we could conceive of the stock-in-trade of the translator—words, sentences, paragraphs and texts—would be to see them as persistent emergent patterns, structures arising from large numbers of interacting agents at the social level influencing the formation of patterns at multiple levels in individual minds—this would also alter our view of the notion of mental representation. Or, as Baicchi describes it (2015, p. 22),

> If we [...] observe the internal structure of language, we realize how the different levels, or sub-systems, of its organization constantly compete, merge and adapt so as to produce a coherent text. In other words, the textual requirements involve the interactions of various sub-systems (the phonological, morphological, syntactic, lexical sub-systems), which feed into one another in ways that are instrumental to produce a text that is efficient and appropriate to the situational context. The different language levels can be conceived of as multiple agents that co-adapt their behaviours in order to synergistically produce an optimal text for the specific situation of occurrence.

From this point of view, we can understand a text as an emergent structure arising from the interaction of multiple systems—some biological, some cognitive, and some social. At the point at which it becomes written down, the text becomes an *artefact*. Artefacts might be seen as an end result of interacting cognitive and social systems—patterns that take on a higher level of persistence by being deliberately manifested or realized as (or in) material objects because they have greater utility in that state. Artefacts have a central role in theories of distributed cognition because they are an element of extended cognition, "external representations" (Zhang & Patel, 2006) of knowledge and structure. And when they are externalized, they also become part of a broader system with which an individual mind interacts—and through which it influences other minds at greater spatial and temporal dimensions.

At this point an interesting question arises concerning agents and agency. We have implied, of course, that translators and their clients, among others involved in any given translation interaction, are agents. Further, we have argued that neurons and collections of neurons are potential agents. But are words, sentences or texts agents? Baicchi (2015) argues that the different levels of language can be seen as agents—but I think it is useful to deconstruct the implications of that statement.

First, it is clear from a CAS perspective that the structures of language are the products of agent interaction. We can conceive of a mental representation of a word as a persistent ensemble of agents. The transient representation of a sentence in memory is also an ensemble of agents. Yet, once uttered or written and sent into the environment, are these system outputs still agents? Can they be the product of agents and also agents themselves? One possible solution to this conundrum is to focus upon issues of goal orientation and functionality. As Muñoz Martín argues, translations (and indeed all produced language structures) are not just communicative; they are functional. They are *intended to do something*: to influence, to persuade, to inform, to entertain and so on. This is an insight that has long informed Translation Studies, discourse analysis, text linguistics and systemic functional linguistics.

Thus, to the extent that language structures carry, for lack of a more useful word, *intent*, I would argue that they also carry agency. Because they are released into the environment to affect other social agents in some deliberate way, they become part of external ensembles that interact with other ensembles to influence future social and communicative outcomes. In their very shape and nature, they are tools, and, as such, carry with them signals of their producer's goal orientation or intent and markers of their producer's purpose. They embody agency, and therefore become agents.

Source texts are artefacts, external material representations of the results of an author's cognition and purpose—of patterns that emerged from competition to be captured in writing (or digitally) and to some extent frozen. During the time course of a translation, this artefact—and specifically its constituent elements—enter into interaction with other systems. First, the active social system provides a social context—with the intents, purposes, desires and needs of a group of social actors exerting influence on the translation. Other relevant artefacts in the environment—from dictionaries to search engines to software tools—will also enter into the equation. Target texts will also become artefacts when the time course of an active translational system ends. They will persist to influence others and to be of value in achieving future goals.

4.2.3.7 Translation units

Explaining what happens during the time course of an active translation system has been the traditional objective of cognitive translatology. Yet, as we have argued here, we cannot, if even by failure to clarify, assume that all of what happens next is internal—nor is it as linear as we might suppose. The "translation task" begins externally and manifests as a specific temporal progression of system interactions—or, rather, multiple systems at multiple levels of scale interact. It ends when, for all practical purposes, the translator fails to return to modify the text in any way. Establishing the temporal boundaries is necessarily arbitrary—and conforms generally to the research questions posed.

We do seem to know with some degree of certainty that translation proceeds in a chunk-wise fashion; we have already introduced the notion of the translation unit. What this means from a system's point of view is that there is a constant oscillation between creating internal patterns and processing external patterns. For instance, if we take reading as an example, several internal systems (and their component elements) must engage to process the perceived graphemes and then engage lexical systems. Grammatical systems must engage, including conceptual systems, to work out contextualized meaning. There are undoubtedly competing patterns, but it is the one that emerges successfully that is then committed to writing. But even here, there is microstructure; there are micro-patterns within the unit, chunks within the chunks. The emergence (and I use the word here both metaphorically and in the complex system sense) of a target segment is by fits and starts—as can be seen from eye gaze patterns and keystroke logs. There are pauses internal to the chunk—micro-pauses—and revision behaviours, deletions, retyping, and so on. The translator may move her eyes to the source text again and again and then watch her own typing as the target text takes shape. All of this takes time; but in some sense, the time it takes, the number of pauses, the extent of keystroking, the looking here and there, are all constituents of the temporal space it takes for the engaged systems to resolve upon a sufficient determination of fitness so that the translator moves on to the next chunk. And, all this activity in the time course necessary to determine fitness is the behavioural manifestation, certainly, of what we normally call cognitive effort in the discipline. A multi-scale view of translation would, however, also extend the notion of effort out beyond that manifesting in the individual mind—what about interpersonal effort, social effort?

I'd also like to point out that as the translator proceeds chunk by chunk, he or she is externalizing internal patterns—creating another artefact. This is at least partly the result of the cognitive constraints of working memory (a system limitation); the information created in the mind has to be offloaded—distributed to an external artefact to preserve it. Once externalized, it becomes another environmental influence and provides new inputs into the system to affect subsequent processing. From a distributed cognition and complex systems view, it is a memory—just not a neurologically realized one.

4.2.3.8 Translation and transfer

A consistently bothersome and persistently fuzzy concept in cognitive translatology has been the nature of "transfer", a kind of black box construct that links the reading of the source-text segment to the production of a target-text segment. It seems to me that at least one useful way for us to address the nature of transfer is to define it as an "outcome" of the utility-driven competition of competing systems and patterns. There is no specific control mechanism (e.g. something like Green's language task action schema (Green, 1998)) that links certain kinds of language input (in the L2) to certain kinds of language output (in the L1), as I used to believe. Rather, a pattern emerges from the interaction of the relevant systems and element collectives and survives the fitness test.

There is no transfer in any simple sense. There is no mysterious undefined "something" existent in the source text that is "carried over" to the target. Discussions about the transfer of "message" or "sense" only work because we simplify and abstract those constructs so that we can talk about them to students and colleagues. Rather, "transfer" is a convenient label for a non-linear pattern that emerges dynamically during the time course of a translation. Transfer is initially a locally emergent phenomenon—localized in discrete translation units. A source-text production and its precedent context enter the translation system and are influenced by pertinent agent collectives stored in the translator's memory. A translation unit emerges into a target-text production. As new productions are added to an also emerging text, past productions become new influencing factors. Those productions may themselves later be revisited and modified. Why? Because re-reading of the source text and the progressively greater coherence of the target provide new information that stimulates new patterns to emerge. Choices made affect future choices—both elaborating and constraining the possibilities of outcome. By the time the entire emergent pattern is released back into the environment, we have a target text: one that we say, quite simplistically, has "transferred" the meaning of the source. But the transference itself was anything but simple.

The source text was an artefact—but it was also a representation of a state of affairs (what Kintsch (1988, p. 180) called a situation model). One of the things we have to account for is exactly how the situation model is "transferred" or, rather, reconstituted for the target reader. The translator hasn't just produced a system of words on paper but produced an output that will be the environmental input to someone else's act of cognition—and there propagate, in unpredictable ways, through the constellation of systems it touches.

4.3 Concluding remarks

To a great extent, much of what I have argued here requires that we accept a complex adaptive systems view of translation—from that basis, when I use terms like perception, reading, comprehension, production, even "transfer", it is with the understanding that each of these is a label for information transformation operations—emergent pattern and structure building—realized by multiple multi-scale systems of agents. It is not my objective to "prove" that reading or writing, or language in general, can be understood via CAS—others have already begun to make more convincing arguments than I. I have accepted this perspective as a fruitful one and offer it here as a useful way of understanding and elaborating some central constructs of cognitive translatology. Cognitive translatology made a call for a paradigmatic shift in previous approaches to Cognitive Translation Studies, and one way to accomplish this is to harness our local disciplinary shift to a greater one.

Given the constraints of space, there are necessarily many theoretical constructs important to Translation Studies that have not been given attention here. For instance, we interpreted translation strategies earlier as an example of a kind of persistent emergent pattern—particularly in the context of the individual translator's mind. But, of course, one could extrapolate to the notion of procedural norms at the social level—and contextualize translational norms of all kinds as robust emergent patterns that have self-organized in the community of translation practice. Still, my objective was not to address every critical issue in our discipline—although I have touched on many—but to argue that Muñoz Martín's innovative cognitive translatology can be supported by a broader disciplinary shift in the physical, natural, social and behavioural sciences. If we accept that argument, then the corollaries of complex adaptive system theory will greatly and gradually inform theory building in our discipline.

Earlier I mentioned the work of Kobus Marais. Where his work concentrates on the general implications of complexity thinking for Translation Studies broadly, mine focuses more specifically on its utility in understanding translation and cognition. Yet, because of the very nature of complex adaptive systems, CAS theory may serve to unite cognitive theories of translation with more general theories and point the way to a unified theoretical framework. The different levels of scale, in particular, help us deal with issues not just of translation process, but of translation product and function.

Translation—the act of translating—is a complex adaptive system. The complexity engages systems of agents from the level of the neuron upwards to include social systems. A translation—the product—is also a "memory"—a persistent ordering of linguistic elements at various levels. Some of the order that emerges during a particular time course (translation options that arise only in the mind and are not preserved in the target text) may be persistent for only a short time—during the time course of a translation unit. However, the final translation product is an artefact of greater longevity that preserves an order and intent that arose from the unpredictable yet organized interaction of many concurrent and precedent system states. It is a material memory. Similarly, the act of translating, though transient, affects the patterns of neurons. The individual stores some record of the experience—memory in a more direct sense—to influence future translation and translations.

Further reading

Baicchi, A. (2015). Complex adaptive systems: The case of language. In A. Baicchi, *Construction learning as a complex adaptive system* (pp. 9–31). New York: Springer. doi: 10.1007/978-3-319-18269-8_2

Baicchi makes the case for viewing language as a complex adaptive system. Individual agents (people) interact in speech communities. Over time, their interactions produce emergent linguistic regularities. The author makes the case that grammar and lexis are the emergent "sedimentation" of frequently used forms. We can view translations and translating from a similar perspective.

Beckner, C., Blythe, R., Bybee, J., Christiansen, M., Croft, W., Ellis, N., Holland, J., Ke, J., Larsen-Freeman, D., & Schoenemann, T. (2009). Language is a complex adaptive system: Position paper. *Language Learning, 59*, Suppl. 1, December 2009, 1–26.

The authors argue the position that processes of human interaction collaborate with domain-general cognitive processes to create the structures of language. These social and cognitive processes are interdependent facets of a larger complex adaptive system with multiple agents, evidence of adaptive behaviour and speech productions that are the consequence of competing factors ranging from perceptual constraints to social motivations. According to the authors, the structures of language emerge from interrelated patterns of experience, social interaction and cognition. The same case could be made for the products of translation.

Heylighen, F. (2008). Complexity and self-organization. In M. J. Bates, & M. Niles Maack (Eds.), *Encyclopedia of library and information sciences* (pp. 1–20). Oxford: Taylor and Francis.

This article introduces some of the main concepts and methods of the science of studying complex, self-organizing systems and networks. A complex system is a hierarchically organized collection of interacting agents at different levels of scale. Because of the nature of the interactions, the systems behave in a non-linear fashion and are, to some degree, unpredictable and uncontrollable. However, Heylighen argues, such systems tend to self-organize, and out of a multitude of local interactions, global coordination and persistent regular patterns emerge. In this chapter we use many of Heylighen's conceptions to understand translation as a self-organizing system with different levels of scale and thereby address some current issues in cognitive translatology, such as distributed cognition.

Ladyman, J., Lambert, J., & Wiesner, K. (2013) What is a complex system? *European Journal for Philosophy of Science, 3*, 33–67. doi.org/10.1007/s13194-012-0056-8
Many disciplines, including our own, Translation Studies, are beginning to use the notions of complexity and complex systems. Ladyman points out that there is no concise, commonly accepted definition of a complex system. This article reviews various attempts to characterize complex systems and considers a core set of features we could ascribe to such systems. Ladyman's work provides a basic architecture for considering whether translation can be successfully understood as a complex system.

Marais, K. (2014). *Translation theory and development studies: A complexity theory approach.* New York and Abingdon: Routledge.
Marais uses complexity theory to try to reframe translation theory, arguing that translation in all its aspects, especially the social, is a complex system. Marais argues that the complex systems view addresses some fundamental epistemological problems in Translation Studies and, as in the case of this work, offers some elegant solutions to persistent thorny issues in the discipline. While Marais does not specifically address cognitive translatology, nevertheless his monograph is among the first to look at translation from a complex systems point of view.

References

Annoni, J., Lee-Jahnke, H., & Sturm, A. (2012). Neurocognitive aspects of translation. *Meta, 57*(1), 96–107. doi:10.7202/1012743ar

Baicchi, A. (2015). Complex adaptive systems: The case of language. In A. Baicchi, *Construction learning as a complex adaptive system* (pp. 9–31). New York: Springer. doi: 10.1007/978-3-319-18269-8_2

Beckner, C., Blythe, R., Bybee, J., Christiansen, M., Croft, W., Ellis, N., Holland, J., Ke, J., Larsen-Freeman, D., & Schoenemann, T. (2009). Language is a complex adaptive system: Position paper. *Language Learning, 59*(Suppl. 1), December 2009, 1–26.

Carl, M., Tonge, A., & Lacruz, I. (forthcoming). An information philosophy perspective on human and machine translation. *Entropy,* 1–17.

Clark, A., & D. J. Chalmers. (1998). The extended mind. *Analysis, 58*, 7–19.

Gibson, J. J. (1977). The theory of affordances. In R. Shaw, & J. Bransford (Eds.), *Perceiving, acting, and knowing. Toward an ecological psychology* (pp. 67–82). Hillsdale, NJ: Lawrence Erlbaum Associates.

Gladwell, M. (2000). *The tipping point: How little things can make a big difference.* London: Little, Brown.

Green, D. (1998). Mental control of the bilingual lexico-semantic system. *Bilingualism: Language and Cognition, 1*, 67–81.

Halverson, S. (2013). Implications of cognitive linguistics for translation studies. In A. Rojo, & I. Ibarretxe-Antuñano (Eds.), *Cognitive linguistics and translation: Advances in some theoretical models and applications* (pp. 33–74). Berlin: Mouton de Gruyter.

Heylighen, F. (2008). Complexity and self-organization. In M. J. Bates, & M. Niles Maack, *Encyclopedia of library and information sciences* (pp. 1–20). Oxford: Taylor and Francis. Retrieved September 2, 2019 from http://pcp.vub.ac.be/Papers/Elis-complexity.pdf

Hollan, J., Hutchins, E., & Kirsh, D. (2000). Distributed cognition: Toward a new foundation for human–computer interaction research. *ACM Transactions on Computer-Human Interaction,* 7(2), 174–196.

Holland, J. H., & Miller, J. H. (1991). Artificial adaptive agents in economic theory. Paper #: 91–05–025, Santa Fe Institute. Retrieved September 2, 2019 from www.santafe.edu/research/results/working-papers/artificial-adaptive-agents-in-economic-theory

Hutchins, E. (1995). *Cognition in the wild.* Cambridge, MA: MIT Press.

Jordan, J. S., Srinivasan, N., & van Leeuwen, C. (2015). The role of complex systems theory in cognitive science. *Cognitive Processing, 16*, 315–317.

Kaisler, S. H., & Madey, G. (2009). Complex adaptive systems: Emergence and self-organization. Tutorial presented at HICSS-42. (n.p.). www3.nd.edu/~gmadey/Activities/CAS-Briefing.pdf

Kintsch, W. (1988). The role of knowledge in discourse comprehension: A construction-integration model. *Psychological Review, 95*, 163–182.
Ladyman, J., Lambert, J., & Wiesner, K. (2013). What is a complex system? *European Journal for Philosophy of Science, 3*, 33–67. doi: 10.1007/s13194-012-0056-8
Marais, K. (2014). *Translation theory and development studies: A complexity theory approach*. New York and Abingdon: Routledge.
Moser-Mercer, B., Frauenfelder, U., Casado, B., & Künzli, A. (2000). Searching to define expertise in interpreting. In K. Hyltenstam, & B. Englund Dimitrova (Eds.), *Language processing and simultaneous interpreting* (pp. 107–132). Amsterdam: John Benjamins.
Muñoz Martín, R. (2010). On paradigms and cognitive translatology. In G. Shreve, & E. Angelone (Eds.), *Translation and cognition* (pp. 169–187). Amsterdam: John Benjamins. doi: 10.1075/ata.xv.10mun
Muñoz Martín, R. (2015). From process studies to cognitive translatology. *EST Newsletter, 46*, 20–21.
Newen, A. (2015). What are cognitive processes? An example-based approach. *Synthese, 194*. doi: 10.1007/s11229-015-0812-3
O'Brien, S. (2013). The borrowers: Researching the cognitive aspects of translation. *Target. International Journal of Translation Studies, 25*(1), 5–17.
Odell, J. (2002). Agents and complex systems. *Journal of Object Technology, 1*(2), 35–45.
Risku, H., & Windhager, F. (2013). Extended translation: A sociocognitive research agenda. *Target. International Journal of Translation Studies, 25*(1), 33–45.
Shreve, G. M. (2002). Knowing translation: Cognitive and experiential aspects of translation expertise from the perspective of expertise studies. In A. Riccardi (Ed.), *Translation studies: Perspectives on an emerging discipline* (pp. 150–171). Cambridge: Cambridge University Press.
Shreve, G. M. (2018). Levels of translation expertise: Expertise as an emergent phenomenon. *Hermes, 57*, 97–108.
Shreve, G. M., & Lacruz, I. (2014). Translation as higher-order text processing. In S. Bermann, & C. Porter (Eds.), *A companion to translation studies* (pp. 107–118). Oxford, UK: John Wiley & Sons, Ltd. doi: 10.1002/9781118613504.ch8
Tymoczko, M. (2012). The neuroscience of translation. In E. Brems, R. Meylaerts, & L. van Doorslaer (Eds.), *The known unknowns of translation studies* (pp. 83–102). Amsterdam: John Benjamins.
Van Geert, P. (1994). *Dynamic systems of development: Change between complexity and chaos*. New York: Harvester.
Vermeer, H. (1978). Ein Rahmen für eine allgemeine Translationstheorie. *Lebende Sprachen, 23*(3), 99–102.
Zhang, J., & Patel, V. L. (2006). Distributed cognition, representation, and affordance. *Pragmatics & Cognition, 14*(2), 333–341.

Part II

Translation and cognition at interdisciplinary interfaces

5

Translation, anthropology and cognition

Kathleen Macdonald

5.1 Anthropology: Translation of cultures, collectives, minds and modalities

Anthropology, as a discourse of disparate cultural communities, has always been concerned with translating the minds and behaviours of individuals and collectives within one culture, across to those from another context. Therefore, its inherent attribute is the translation of perception and social interaction (Ingold, 1994; Rubel & Rosman, 2013). As a first step, such mediation occurs between individuals within one social code and others within that same circle (Bateson, 1972). The web of complexity in the description of even a single society was early articulated by sociologist David Emile Durkheim (1858–1917), who said that individual consciousness is steeped in and manifests collective social consciousness, which depends on and derives from the enabling language of a people (Darnell, 1998; Rapport & Overing, 2000, p. 71). From the anthropological tradition, particularly in its nascent years, descriptive mediation was done between worlds having unrelated and entirely unfamiliar linguistic codes for which there might not even be written forms. The process involved the interpretation of the world of users of one code in the modes and forms available to the world of users of another code, with language as the most articulate code of human experience. Standing at the interface of cultures that have evolved entirely different ways of interacting with each other and viewing the world, the challenge for anthropologists was to enable access codes to bi-directional portals between these divergent modes of experience.

Cultural anthropologist Franz Boas (1858–1942) expounded the potential of contrastive language description for reinterpreting the human condition as a whole:

> The investigation of the laws governing the growth of human culture is carried out by means of comparative methods .... Their great value for the study of the human mind lies in the fact that the forms of thought which are the subject of investigation have grown up entirely outside of the conditions which govern our own thoughts. They furnish, therefore, material for a truly comparative psychology. The results of the study of comparative linguistics form an important portion of this material, because the forms of thought find their clearest expressions in the forms of language.
>
> *Boas, 1899, p. 96*

As Boas suggests here, very different histories and ecologies led to very different ways of thinking, and a useful way in to experiencing such alternative ways of perception was comparative linguistics. In fact, Boas defined his project as "ethnolinguistics": with ethnology being the science of the human mind and social customs within a culturally specific setting, as opposed to the more general science of anthropology (Barnard, 2004, p. 2); and with language as the most important manifestation of such ethnographies (Boas, 1899, p. 94; Boas, 1911, p. 63). His methodology was one which aimed to give the culture its own voice: whether that was through developing his own personal linguacultural fluency, working closely with native members of communities he studied to develop their writing systems and describe their grammars, or compiling texts that could speak for themselves of the narratives, myths, ritual and songs of the people. One distinctive contribution of anthropology to translation and cognition was that it was transdisciplinary and collaborative, as is typified by the two-volume work *Ethnology of Kwakiutl* (Hunt & Boas, 1921). Through such means, ethnolinguists worked towards pattern-based generalizations on the whole of the human condition (Benedict, 2005/1934), which in turn contributed to more specialized fields like psychology (Sapir, 1921), in a bi-directional process.

In the social anthropological tradition, anthropologists and their benefactors have been interested in speculating on the processes of human evolution, civil development, the transfer of innovations across societies and, more recently, interactionalism and auto-ethnography. This tradition was concomitantly concerned with interpreting diverse histories across distant regions of the world, in synergy with following common threads in social behaviour, collective motivation and phenomena that typify and unite us. In its infancy, social anthropology tended to approach the understanding of humanity from the dominant analogic discourse, which has resulted in affiliations with colonial hegemonies and indiscriminate versions of evolution and Eurocentric models of social progress, as well as xenophobic biological theories such as eugenics. The miasma from these associations lingers over the reputation of anthropology and socio-cultural theories even today. However, these earlier theoretical assumptions were refined in the face of important field-based work, such as Malinowski's ethnography of Kiriwinan in *Argonauts of the Western Pacific* (1922). Significant in Malinowski's work was the role of translation in developing frameworks for authentic discourse on divergent cultures in ways that are also meaningful to outsiders. The generalizations which emerged through such ethnographies helped to reorient much of the work in both social analysis and linguistics, towards more holistic and dynamic approaches to translating and imagining humanity (Barnard & Spencer, 2002).

The specific contribution of anthropology comes from its attention to the social and cultural aspects of humanity. While it is important to have a working definition of both society and culture, such definitions are notoriously hard to pin down: from dichotomies between mental representations and performance (such as *cognitive: phenomenological*, Ingold, 1994; or *ideational: behavioural*, Harris, 1999); to backgrounding filters (culture) against which the immediate environment of interaction (society) can be interpreted (Halliday & Hasan, 1985, p. 46). Consistent among these is the role of linguistic mediation between individuals and communities in a complex interweaving of mind–language–society (Rapport & Overing, 2000). The consequences of positioning on even such basic shared experiences as society and culture are often more clearly worked out within a focused setting. To give a cross-institutional example, in differentiating the often competing definitions of "society" in legal and anthropological terms in the outworking of National Native Title (NNT) in the Australian context, Palmer suggests that the legal interpretation recognizes society as a repository of traditional laws and customs held by groups with common language use and land stewardship; while the anthropological one implies sets of relationships that are uncertain, malleable and defy reification (2018, p. 37). This intercultural space between Indigenous Australian and scientific practice is illustrative of social and cultural entanglements, so is continued in Section 5.3 to explore translations of spatial experience as temporal conceptualization. Before going there, it is helpful to review the formative dialogue within cultural and social anthropology that relates to translation and cognition.

5.2 Historical contributions of ethnolinguistics to translation and cognition

Dealing with alternative linguacultures at some remove from those of the anthropologists themselves and their audiences, meant that the tension between sameness and difference was intrinsic to their narrative. This tension has long been recognized in intercultural interaction, even where the cultures concerned were quite closely related, as on the European continent. The translation of minds and societies is inevitable and indelible in the translation of languages. Despite how innocuous, or even well-meaning, these contemplations might appear, it is a question that has been polarizing scholars since the period of Romanticism (Leavitt, 2011; Penn, 1972), and continues to incite invested debate (Chomsky, 2000; Jackendoff, 2002; Lakoff, 1987; Langacker, 1987; Pinker, 1994; Pullum, 1991). This is somewhat because agency is at stake in the question of the role that language might play in shaping our thoughts and realities.

The debate is often formalized as an orientation to languages that is either *essentialist* or *universalist* (for a comprehensive overview, see Leavitt, 2011). Those holding the former view prioritize that each language has a distinct essence, which may be irreconcilable with others, and which is only arbitrarily connected to reality.[1] By contrast, the latter position holds that languages are more similar than different, potentially deriving from a common history and mediating experiences in relation to a common "truth".[2] While these ideas extended much further back to the Enlightenment with Locke[3] and Leibnitz on the arbitrary and innate nature of language, respectively, neither specifically identified thought with language (Penn, 1972, p. 47), so the early Romantic thinking is a useful point of departure. It is also the period of the emergence of anthropology in parallel with expanding global travel and trade. In the past, the essentialist and universalist positions have been used to rank and discriminate, minimize and homogenize speakers and communities (Koerner, 1999, 2000; Leavitt, 2011; Risjord, 2007), despite both being untenable in real-world communicative practice, implying an immutability that does not align with actual experience.

5.2.1 *Essential matters: A particular interpretation*

Franz Boas is known as the "Father of American Anthropology". His extensive work on AmerIndian languages empowered him to caution against viewing language difference as reflective of disparity between mental capacities. Commenting on an apparent absence of linguistic categories for drawing generalizations in Kwakiutl (spoken on Vancouver Island), he observed that this had no impact on their ability to form such structures if they became relevant. Likewise, student of Boas, Edward Sapir (1884–1939), further reasoned that the linguistic forms of languages predispose speakers to certain perceptual interpretations, with their particular categories creating patterned speech-thought much like grooves in a road. Nevertheless, he reasoned that these grooves can be gotten out of, indicating that they were not chains determining and damning speakers to one code of representing the world (Sapir, 1921). By contrast, he shows that it is the connections between linguistic categories and associations between words that habituate linguistic thinking. Using translations from English into German and Yana,[4] Sapir demonstrates that certain relational concepts associated with grammatical forms (reference, tense, plurality and gender) may be more or less salient to particular speech communities:

> The Yana sentence has already illustrated the point that certain of our supposedly essential concepts may be ignored; both the Yana and the German sentence illustrate the further point that certain concepts may need an expression for which an English-speaking person, or rather an English-speaking habit, finds no need whatever.
>
> *Sapir, 1921, V, p. 13*

Where an essentialist position might interpret such linguistic oversights as evidence of inferior mental development or delayed social and cultural evolution, Boas and Sapir saw it as determined more externally, or ecologically—from a lack of necessity to conceptualize things in general terms (Boas, 1963, pp. 54–55). Importantly, it was recognized that there was some interplay between thought and culture, the individual and the social, for which language effected a crucial mediating role. For Sapir, the *instrument of language* makes the *product of thought* possible, while the product in turn refines the instrument (Sapir, 1921, I, pp. 14–15). Likewise, language was held to innovate social endeavour. In this way, early ethnolinguistic particularism voiced a scientific dissent to the rising fascism within Europe and racial disparity in America, where response to difference was otherwise reactionary (Boas, 1911, 1940; Sapir, 1921, X). It was clear that it was not that there was disparity in mental or civil development, but that there was just difference, which was habitualized through an inextricable tripartite of thought, language and culture (Lee, 2000).

A more fully articulated vision of this tripartite was given by Benjamin Lee Whorf (1897–1941), which might best be summed up in his notion of a "cultural mentality" (Whorf & Trager, 1938, in Lee, 1996). Whorf provided an initial procedure for shunting internally between language and mentality, and externally between language and culture, to reason about the implied metaphysics of a speech community. Incorporating field and process theory principles from the burgeoning discipline of quantum physics, Whorf conceptualized that the meaning associations that are activated by certain linguistic choices contribute to what is understood, and that such understandings are internalized through social sanction of norms for interpretation. Thus, what is explicit, or *overt*, takes its significance from what is implicit, or *covert*, in the whole of the language system. As examples, he offers the phonetically compatible names, *Alice* and *Ellis* (and a clever list of many others), as overt categories that activate the covert category of gender in English. To one less familiar with English, there is no outward indication of which name is assigned to which gender, whereas the native speaker/linguist is inducted into this custom through cultural priming, or reactances to mental associations with the deployment of the pronouns *she* and *he*, respectively (Whorf, 1937a, in *Language, Thought and Reality* [from hereon, LTR], Carroll, 1956, pp. 90–92). It is the interplay of, and rapport between, linguistic items of the language system as a whole that determines the potential of certain choices from that system to realize their significances to those who use them, revealing the essence of thought (Lee, 1996, p. 63; Whorf, 1936 [LTR], pp. 67–68). This insight into covert categories is not insignificant (Lee, 1996). If it is the covert categories that are most closely linked to unconscious behaviour, then this has consequences for the approach to the "calibration" of pictures of the universe between different language speakers (Whorf, 1940 [LTR], p. 214). This remains to be fully appreciated and extensively explored. In any case, to fuse field theory and Whorfian concepts, both the particles (overtly marked categories of the grammar) and the spaces that are perturbed between them (covert categories) contribute to *isolates of meaning* in the hidden workings of the mind of individuals, and by implication, the broader social consciousness of a speech community (Sapir, in Mandelbaum, 1949/1973).

Extending *isolates of meaning* out towards *isolates of experience* constitutes the "cultural" dimension of a cultural mentality. Working with a similar phenomenon to Kwakiutl for omitting generalized categories in Hopi, Whorf first describes the grammatical phenomenon: Hopi nouns have an individual sense, and so can take singular or plural forms ("a water", " two waters"). This is contrasted to what he calls SAE (Standard Average European), which requires a body-type or a container to individuate what the language segments as mass nouns (thus, "a droplet of water" and "a glass of water"). Relating these findings on the grammar externally to the matter of existence, he claims that "[t]he language has neither need for, nor analogies on which to build, the concept of existence as a duality of formless item and form" (Whorf, 1939 [LTR], pp. 140–142). In his own formulation, the tendency for matter to be pluralized in Hopi, rather than generalized or contained, constitutes a different segmentation of experience. Thus, in Hopi experience, all matters are individuated and are the main idea, while in English, it is the outline of the thing that is foregrounded, resulting in divergent interpretations of matter within the world (Whorf & Trager, 1938, p. 8). For both languages, the grammar is the initial framework for the analysis of everyday phenomena, through which its distinctive categories preference certain interpretations over others, which in turn influences personal and cultural activities (Whorf, 1939 [LTR], pp. 134–135). Taking from the early holographic model of the human brain, where the part *contains* the whole (Lee, 1996), consciousness was held to be an unfolding coherent whole, projected through semiotics such as language, onto the universe. Here again, elaborating isolates of experience as habitual patterning contributes to the depiction of generalized cultural mentalities, which might be calibrated in order to determine the factor of relativity active between distinct cultural mentalities. When Penny Lee states that "Linguistic Relativity [...] is predicated on the hypothesis that there are universal configurations of experience upon which different linguistic schemes of classification operate in a variable way" (Lee, 1996, p. 96), she highlights that relativity is far more sophisticated than a compromise between universalism and essentialism. At the very least, its beginning hypothesis incorporates ideas inherent in both.

Even to consider relativity as a hypothesis (as in the *Sapir–Whorf hypothesis*) is misleading, since it only points to the beginning assumptions rather than its evolution into a quite complex axiom (Hill & Mannheim, 1992, p. 383), having more evidence in its support than otherwise. Whorf at times refers to his formulations as a *Culture Complex*, which is apt in that it underscores both its elegance and the priority of culture. Yet this option does not emphasize the place of language and thinking, which are also at the heart of the theory. Even in his own use, Whorf is compelled to elaborate that it encompasses linguistic thinking (Whorf & Trager, 1938). Avoiding this shortcoming, Lee reformulates this as the *Whorf Theory Complex* (1996), yet this is perhaps even more of a nominal removal from the core concepts of the mind–language–culture tripartite, focusing only on one great thinker of the tradition. This focus on "Whorf" denies both the richness of the traditions that informed it, including ethnolinguistics, Theosophic Orientalism and also the startling epiphanies of relativists of the early 20th century (including Max Planck, David Hilbert, Albert Einstein, Niels Bohr, Paul Dirac and many more). It should not be disparaging at all to recall that Whorf incorporated a spectrum of world views in his thinking (Lee, 1996; Tymoczko, 2010), from Eastern philosophy and mathematical process theory to the biochemistry of his vocation. He interpreted the evidence for the language forms of thought by drawing together an interface of disciplines and epistemological orientations. This was a natural inclination of Boasian ethnolinguistics, given that its ethos was to encourage other cultures to speak of themselves, for themselves (Whorf, 1937b, [LTR]). More usefully, Lee also refers to it by an assignment Whorf himself gave: *the Linguistic Relativity Principle* (1940 [LTR], p. 214). In accordance with the fact that a principle encompasses a more considered system of ideas associated with an explanation for observed phenomena, the term is adopted here, along with *cultural mentality* to refer to the procedure of explanation.

An important critique of Whorf has been that the ways of talking exemplified in his contrastive studies between SAE and Hopi, and indeed in the fire incidents he investigated (see

Whorf, 1956 [LTR]), were not the only ways of talking about them. This is a valid point: Whorf's explanations were certainly empowered through his epistemological eclecticism, yet the extent of his engagement with data and informants and also the depth of his knowledge of the languages he was describing are not immediately apparent. It is not that he did *not* engage comprehensively with primary material: the reader of his work is only presented with the select cream of instances supporting any particular argument. Further, it could be argued that his fixation on morphosyntax was rather limited as an interpretation of human experience. However, it should be acknowledged that there was a place for anchoring contrastive grammatical descriptions in the real world, through texts, songs and other social artefacts and institutions, which he categorized under stylistics in his framework for language description in *Language: Plan and conception of arrangement* (1938 [LTR], pp. 126–133). The collation of texts and artefacts, as examples of habitual and symbolic ways of talking and representing the world, were fully consistent with the theory—see for example Sapir's emphasis on discourses of literature, art and aesthetic (1921, XI; see also Benedict, 1934; Hunt & Boas, 1921; Mandelbaum, 1949)—and anticipates contemporary approaches, such as Critical Discourse Analysis and ethnopoetics. Indeed, even Boas has been criticized as being excessively emic to the point of theoretical sterility, suspending any productive form of praxis (Harris, 1968, p. 251). Yet, his was an inductive approach: taking that mass of evidence together to reason by way of culminating waves of generalizable patterns in order to present the picture of the native as closely as possible from their own perspective (Benedict, 1943, p. 60; Boas, 1899, p. 95; Harris, 1999; Liron, 2003). Unsurprisingly, Boas held an important place for the comparison of languages in elucidating a broader vision of the universe and our experiences in the world of our immediate surroundings. The result of these principles was a more neutral approach to the significance of divergences observed between languages and cultures (Boas, 1940; Lucy, 2011). Even its strongest critics credit his form of particularism as a breaking away from the exoticism of essentialism towards a more relative rendition of humanity and lived experience. In, fact the entire project constitutes a pioneering effort towards interdisciplinarity, and issues a challenge for more systematic approaches in representing local contingencies towards explaining the Linguistic Relativity Principle (hereafter, the LRP).

5.2.2 Universal concerns: Systemic generalizations

In contrast to the essentialist programme, a practical consequence of universalism would be that translation was a simple process of linguistic transfer from one code to another. While even the most legendary of thinkers have held to this belief (see, for example, Penn on Aristotle, 1972, p. 42), anyone who has attempted the über perfectible process of translation would know that this is far from the case. This is partly due to the multivalency of meanings that can be imagined from a single utterance, where each of the translation options depends on the social situation, surrounding environment, rapport between interlocutors and so on (Catford, 1965; Jakobson, 1959). The refractile properties of a linguistic structure and its senses lends itself to a tension between acknowledging what was intended in the original and the spectrum of available selections to address norms in the alternative language (Toury, 1995). Although a universalist orientation to language (being uniform and unidirectional) has paved the way for innovations based on a principle of simplification, such as natural language processing (NLP), machine translation (MT) and translation memory (TM) applications, it also diminishes the complexity of real-world communication and cognitive reasoning, critical for navigating sameness and difference, generality and particularity within everyday contexts (Killman, 2015).

Rather, the linguistic behaviour of translation reveals something of the distinctiveness of different language speakers and alternate speech communities. If, in the infamous logic of

Wittgenstein in his *Philosophical Investigations*, language is the frame tracing our outline of reality rather than its essential nature (Wittgenstein, 1958, p. 148a point 114), then translation has to manage a double framing of reality, and often even a reality that is neither shared nor visible, being more like a picture frame than a window. The image of the world beyond may be only impressionistic. Nevertheless, it affords an image of realities that otherwise might be unimaginable. In this sense, it is a language activity that inherently assumes a role of calibration between cultures. In turn, reflecting systematically on translation can help identify patterns in divergent ways of understanding reality, and also reframe them in meaningful ways.

Polish anthropologist Bronislaw Malinowski (1884–1942) was among the first to recognize the contribution of translation to his work of describing the Kiriwinan language and culture of the Melanesian Trobriand Islanders. In his seminal monograph, *Coral gardens and their magic* (1935, Volume II, Part IV), he outlines procedures for translating "untranslatable words", such as *buyagu*, *bagula* and *baleko* (which have semantic overlaps), in his "Theory of Meaning" (1935, p. 231). Noting that an informant will refer to the contingent circumstances within which a word was used in order to elaborate its meaning, he highlights the interdependent existence of words (1935, p. 23). Each item in a language relates to another within the context of a certain situation (for any given culture), from which its comparative and contrastive salience emerges:

> Translation in the sense of defining a term by ethnographic analysis, that is, by placing it within its context of culture, by putting it within the set of kindred and cognate expressions, by contrasting it with its opposites, by grammatical analysis and above all by a number of well-chosen examples—such translation is feasible and is the only correct way of defining the linguistic and cultural character of a word.
>
> *Malinowski, 1935, p. 17*

In elaborating the meaning of the term *buyagu* (left undefined earlier), Malinowski's informant furnished the social, technical, physical, seasonal, economic and other associations implicit in its meaning as "a cleared plot of enclosed *and* fenced soil intended as a communal garden". The notion is further understood in comparison to *bagula* and *baleko* (referring to different stages of cultivation and ownership arrangements) as well as in opposition to *odila* and *yosewo* (uncut bushland outside the fenced/unfenced garden site). Thus, the meaning of *buyagu* could be conveyed comparatively, against cognates, and, also, in contrast to agnates. In translation, *buyagu* could no more be encapsulated by the commercial field of barley or rye than by the garden bed of pansies and peonies, since these have neither cognates nor agnates contributing to their particular meanings in the Trobriand context.

In his thesis on meaning, Malinowski's cognates and agnates were instruments to be indexed to their immediate and historical, situational and cultural contexts. For Malinowski to participate in and also relate the magic of coral gardens, he had to contemplate linguistic constellations situated within a different hemisphere and, indeed relate these very different interpretations of old and newly visible ones across hemispheric divides (1935, pp. 1–22). Within these contexts, the primary function of language was held to be pragmatic, that is, oriented toward some sort of social accomplishment (1946, in Jaworski & Coupland, 2006, p. 286). So, where the contrastive language descriptions of Boasian cultural anthropology led to representational insights, it was in the functional act of translation that social implications of language were foregrounded. In both cases, the anthropological task of having to translate across cultures provides a useful linguistic behaviour for reflecting on divergences between social groups and mentalities (Ingold, 1994; Rubel & Rosman, 2003).

Despite being criticized for a lack of training in linguistic structural analysis by Charles Voegelin and Zellig Harris (in Young, 2011, pp. 14–17), Malinowski's ideas made a lasting impression on European linguistics, including constructivist and functionalist thought. For example, his "sets of kindred expressions [...] and their opposites" (see earlier) can be seen in John Rupert Firth's (1890–1960) collocations, colligations and systems of alternatives in language use. This theoretical development engages Malinowski's context of situation and context of culture, to reason axially about the selections of collocation/colligations along a structural "horizontal" plane, with the systems of options that give such structures their value on a "vertical" plane (Firth, 1957, VII, p. 17). Malinowski's ideas also entered the Copenhagen linguistic circle, where Louis Hjelmslev (1899–1965), interestingly, brings his discussion of conjuncts and disjuncts into the territory of linguistic relativity:

> In a way, we can say that all functives of language enter into both a process and a system, contract both conjunction, or coexistence, and disjunction, or alternation, and that their definition in the particular instance as conjuncts and disjuncts, coexistents or alternants, depends on the point of view from which they are surveyed.
>
> *Hjelmslev, in his* Prolegomena, *1961, p. 35*

From this it is clear that the selection of constituents in a structure is relative not only to others within that same structure, but also to all alternatives that might be selected from the system of language. The notions of process and system, like covert and overt, may be equated to the particle and wave of field theory. The particle, as observable social and linguistic forms, emerges as a manifestation of the wave of potentiality carried in the socio-linguistic system. This way of thinking opens up a way for the description of languages that is based on inherent relevancies; that is, categories can be developed depending on the relationship between language features and their reflexivity to the situational context. Such description highlights the function of words and grammatical tendencies within their situations of use. In addition, the range of structures available from the paradigmatic resources of a language offers a descriptive template for ways that certain language speakers are habitualized to attend to different aspects of the world (Whorf, 1939 [LTR], 1956).

Translation in this discussion has been used to illustrate the implausibility of the universalist approach that takes its standards for categorizing phenomena from a limited collection of linguacultures. Yet, by the same token, the attentive description of languages enables reasoning towards more general principles that do make certain translation options more suitable than others for a given target audience context. It ought to be kept in mind that Malinowski reasoned *towards* generalizations, and that such an approach might continue to challenge linguistic theories, which is an important consideration, given the dynamic and ever-evolving nature of language. As with Boas, his logic began from extensive and intimate fieldwork with those cultures, from which he amassed copious data (including sacred ceremony, song and narratives) and worked with local participants in its interpretation towards outlining intrinsic linguistic categories.

To return to the theme of essentialism and universalism in reflections on languages, while there are arguments for either position, they fossilize the archaic idealisms of the arbitrary and non-arbitrary views of language and reality. Given the exclusivity of the universalist and essentialist relations to reality, it would be ideologically naïve to think that linguistic relativity is somehow a milder version of the two. The LRP has been conceptualized beyond either hypothesis towards a cultural mentality complex, where the particular informs the general. Furthermore, Malinowski's theory of meaning offers a complementary and principled approach to systematically representing what he identified as relative realities through translation and participant research, with translation as a means towards calibrating cultures. Indeed, where the Boasian approach sought for languages to speak of themselves, the Malinowskian one sought for them to speak to each other, and both revealed a relativity principle at work.

5.3 Contemporary approaches to translating cultures

Despite a brief dominance of the innate perspective through the early 20th century, aligning nicely as it did with the development of computerized technology, this has gradually been giving way to a renewed interest from the social sciences in the distinctiveness of diversity (Enfield, 2013; Hill & Mannheim, 1992). Among the avant-garde of the cultural turn was Clifford Geertz, with his manuscript *The interpretation of cultures* (1973). In this, he offered that cultures are intricate webs of symbolic meanings and enigmatical social expressions, which might only be interpreted through complementary *thick descriptions* (Geertz, 1973, pp. 5–6). Thick descriptive practice compels ecological validity through tracing intersecting networks between specific contextual manifestations, regular social activities and private realities, as well as reactances to external cultural contact (Hymes, 1996; Mühlhäusler, 2000). Crucially, the tension between difference and sameness is pre-eminent: where the calibration of cultures in the anthropological tradition validates diverse voices (Blommaert, 2005), and being heard, the theories themselves are enriched. The following discussion includes contemporary studies that have specifically focused on the LRP, and it draws upon ethnographic insights within Australian Indigenous contexts.

5.3.1 Conceptualizing space-time

The cognitive sciences were not untouched by the cultural turn, where the cognitive revolution was expressed through increased interdisciplinarity, involving psychology, linguistics, computer science, neuroscience, anthropology and philosophy (Miller, 2003). Rather than "pure cognitive sciences", thick descriptive practice scoped across disciplinary boundaries to cross-pollinate ideas on the nature of the interpenetrating biological brain, the conceptualizing mind, linguistics and society-culture (Enfield, 2000; Lee, 1996). Aligned with both a mental and a cultural orientation, there has been a surge in the number of cognition-based studies and monographs into the LRP (Darnell, 1998; Kockelman, 2010; Lee, 1996; Lucy, 1992; Penn, 1972; Seiferle-Valencia, 2017) as well as a number of broad scoping compendiums (Achard & Kemmer, 2004; Cook & Bassetti, 2011; Everett, 2013; Gentner & Goldin-Meadow, 2003; Gumperz & Levinson, 1996; Levinson & Wilkins, 2006; Niemeier & Dirven, 2000; Pütz & Vespoor, 2000). In particular, some seminal studies have debunked the primacy of previously assumed universals, such as left/right asymmetrical relations in the linguistic realities of the Central American Tenejapan world (Levinson & Brown, 1994) or an intrinsic experience of "coming" and "going" between Australian Arrernte and Solomon Islander Longgu speakers (Wilkins & Hill, 1995). With consideration of cultural and environmental agency in spatial reasoning through language, such studies concur with Sapir's suggestions on Yana, German and English (see Section 5.2.1).

Experiential conceptions of space have especially preoccupied investigations into linguistic relativity. Lucy, in his meta-studies on cognitive approaches to linguistic relativity, labels these domain-centred or behavioural-centred approaches (1996, 1997), depending on whether linguistic divergence with respect to a domain of shared reality results in variation in experiential framing or linguacultural behaviour. Where the two approaches conflate, these are distinguished from more structural, strictly linguistic approaches. As a physically accessible dimension of experience, space has been modelled along synesthetic transversals, where spatial behaviour is taken as indicative of less concrete experiences, such as temporal reasoning (Boroditsky & Gaby, 2010) or sound interval discrimination of pitch (Casasanto, 2016; Dolscheid et al., 2013). This resonates with Whorf's metaphysics of Hopi and his prediction that linguistic divergence would influence behaviour in the domains of space, time and matter (1936, 1941 [LTR]). So, it seems to be wholly congruent with canon.

However, many such studies are conducted in ways that can only be defined from an anthropological perspective as emaciated (thin treatments of all axes of the thought–language–culture tripartite) or even amputated (exploring only in part). For example, Boroditsky & Gaby's (2010) study of temporal reasoning through cardinal direction accuracy in the Pama-Nyungan language, Kuuk Thaayorre (Hale, 1964) is somewhat problematic. They conclude that "*there is no overlap between absolute spatial reference frames and time in the lexicon*" (2010, p. 1638, emphasis added; see also Gaby, 2003, pp. 14–15), yet this can only be true if the lexica of absolute reckoning is restricted to the Western terms given on a compass. Taylor's earlier ethnographic study of Thaayorre (1984) reveals the prominence of place in the conceptualization of time, tracing the common classifier for place, *raak*, through its abstraction to time-related events (such as *raak karrtham* and *raak wurripan* for wet and dry time, respectively) and larger stretches of time, as is imputed in phrases such as "the beginning of things/ place-time" (*raak kanpa minch*) and "the continuing of the present into the future/ the persistence of place in time" (*raak yuu-karr-ontam*) (1984, Ch. 2). Although it does not translate as a point on the Western compass, as an indicator of place, *raak* contains the possibility that the present physical world persists in the past and the future. If so, distant time need not be remembered, because the current place inhabits it and *east meets west*. Gaby had separately noted the integration of features of place in the cardinal orientation linguistic system of Thaayorre, such as the north and south banks of the river (2006/2017), and this would indicate the role of environmental features in cardinal fluency. The contribution of Boroditsky and Gaby's work is not in reckoning with Kuuk Thaayorre conceptualizations of time, but in highlighting the limitations of applying Western constructs of absolute/relative frames of reference. It points to a need to expand the account of space-time beyond specific lexical indicators of cardinal orientation to the complex web of geospatial and social discourses, which resonate with Thaayorre conceptualizations of time (Gaby, 2019, pers. comm.; Palmer et al., 2019).

It is critical to move beyond representational domains towards social relativity (Enfield, 2015). One means to do this is to move beyond morphosyntax towards more discursive and conversational explorations. Some innovative approaches to generating discourses for analysis employed in cognitive linguistics are exemplified by Ruth Berman and Dan Slobin (1994), Slobin (1996, 2000, 2003), as well as Christiane von Stutterheim (1999) and von Stutterheim et al. (2002). The premise here is contriving a "common" or "shared reality" to develop event-time narratives (Slobin, 1996, 2001) or text production tasks (von Stutterheim, 1999) in participants' respective languages. Using the wordless picture book, *Frog, where are you?* (by Mayer, 1969, cited in Slobin, 1996), Slobin and colleagues report that English, German, Spanish and Hebrew introduced linguistic-sensitive subjectivity to the "shared reality" (Slobin, 1996, p. 91), so that "[w]e were repeatedly surprised to discover how closely learners stick to the set of distinctions given to them by their language" (1994, p. 641), impacting the overall rhetorical style of even the youngest pre-schoolers (1996, pp. 77, 84). Likewise, findings from cross-linguistic text production stimulus tasks substantiate that the available linguistic devices preference different orientations to event-time conceptualization (von Stutterheim, 1999, p. 175; von Stutterheim et al., 2002). One thing to note about the Frog Story study is the scale of it. It has led to the general postulate of *thinking for speaking*, which lends cognitive processing priority to a linguistically determined model of language behaviour (see Slobin, 1996, 2000, 2003). The sheer number of participants, languages and cross-studies over time constitutes a collaborative tenor requisite to the discovery of new LRP generalizations. While the sequencing of events in the Frog Story might be questioned as privileging a Western narrative style, even critics incorporate the task in their toolkit of elicitation tasks (see later).

Innovating narrative through studying *interactions* affords a novel approach to behaviour during space-oriented tasks. An important example here, of course, is the series of studies on parent–child interactions in protolanguage (Bowerman & Choi, 2003; Choi & Bowerman, 1991). Perhaps the best example of rich comparative grammatical profiling of cultural mentalities in the spatial domain is that exemplified by the systematic work associated with the Max Planck Institute for Psycholinguistics (Levinson, 2003; Levinson & Wilkins, 2006). Native speaker interactions in negotiating meanings while performing orientation tasks, such as giving/taking directions, evidence how spatial linguistic items are spread variously across the clause. Based on the collaborative work in over 40 languages, they have proposed a typology for conceptual subdivisions of the spatial domain, including frames of reference (*Intrinsic*, *Relative* and *Absolute*) and spatial categories (*Topology* and *Motion*). Importantly, these categories are not claimed to be mutually exclusive systems, with a language being definitively located as a certain "type". Rather, the typology accounts for how different languages may incorporate aspects of each system to differing degrees in webs of semiotic resources (Levinson & Wilkins, 2006, p. 22). Wilkins, for example, finds that the use of Absolute frames of reference in Arrernte (a Central Australian language) is deployed for the description of motion, with other systems available for representing static location. It follows that cardinal directions are most prevalent in everyday language use and traditional narratives. In addition to remarking on the role of paralinguistic systems such as gesture, sign language and sand drawings, both indicate that the spatial orientation systems in Australian languages have fundamental social import (Levinson & Wilkins, 2006, pp. 58–62). The project is sophisticated and useful for comparison, validating findings such as: "There are robust correlations between frames of reference used in languages and frames of reference used in non-linguistic memory and reasoning, suggesting a major 'Whorfian' effect on language and cognition" (Levinson, 2003, p. xix). It would be timely for a complementary exploration beyond space and into the translation of time.

What is positive about the studies discussed so far is that they address the Boasian imperative of language contrast, which is what Enfield (2000) calls a *linguocentric* approach: centred not merely on language but on more than one language with both grammar and discourse in focus (in contrast to *linguaculture*, also used here, representing something more like a Cultural Complex). Yet, the complete picture needs to also entail comparative work, actively exploring points of similarity.

5.3.2 Translating the gravity of space-time relativity

Translation is a discipline that realizes some degree of calibration between languages, with written texts as units of comparative analysis. As such, it is a disciplinary space where Whorf can meet Malinowski. One translation scholar who has been particularly inspired by these two traditions is Juliane House (2000, 2011, 2016). House promotes culture through her assignment of a *Linguistic-Cultural Relativity* to refer to the recontextualization of full discourses (House, 2000). Now, while "culture" is intrinsic to the LRP, the focus she puts on it here reminds us of the importance of the context of situation in formulating meanings, since "translation is conceptualized not solely as a cognitive process, but as a sociocognitive act of recontextualization" (2011, p. 526). She posits that re-contextualizing the norms of discourse within one culture into those of another addresses the relativity paradox of translatability. In other words, translation is possible, despite the fact that a relativity of textured norms holds across linguacultures (House, 2000, p. 76; 2011, p. 521).

Yet, two things must be recognized before proceeding. The first is that the move from the individual to the collective in socio-cognitive norms of discourse demands a similar problematization of the "cultural mentality" beyond referential domains to functional events (see Enfield, 2015); and the second is the premise that there are conventions across all languages that might be considered "discourse", which necessitates *legitimating* discourses rather than reinventing them. To begin with, it is important to explore what a congruent cultural mentality beyond morphosyntax might entail (Lee, 2000). In her formulation, House recruits Whorf's notions of "overt" and "covert" morphosyntactic units up to the discourse unit, where a text as overt translation displays unfiltered discourse norms of the Source Culture (SC), whereas a text as covert translation is effectually disguised as a text generated within another linguaculture, and thus meets the norms of the Target Culture (TC). While she claims that the covert translation is more instructive on relativity, it is not clear whether the covertness of a translated text reveals more about SC or TC. The translation of these categories as explicitness and implicitness in text may assist in describing translation artefacts but may be only anaemic in translating the full implications of the LRP.

A thematic example may help to illustrate this, while also further tracing space-time realizations in Australian Indigenous cultures. Translation trends in conceptualizing cosmological narratives of Dreaming (Tonkinson, 1974; Walsh, 1997), such as Jukurrpa (of the Martu cultural groups of the Western Desert Pilbara region), reflect divergencies in temporal reasoning between those linguacultures involved. The early anthropological misnomer, *the Dreamtime*,[5] sets up a distinct period through its individuating abstraction as a unit of *time*. This evinces Western associations of a creative period as existing in a mythologically anodyne or temporally impervious past. Thus, the covert option reveals more about TC subtexts. But its rejection by Aboriginal groups in favour of the more overt preference, *the Dreaming* (Stanner, 1953/1979), is more revealing of SCs. This is consistent not only with the Kuuk Thaayorre space-time metaphors of "the beginning of place" and "the continuing of place into the future" (see Section 5.3.1) but also with the social significances of place in Dreaming re-enactment, as Dreaming places (*raak woochorrm*) and clan spirit centres (*raak parr'r woochorrm*) (Taylor, 1984, p. 249). The unexpectedness of the gerund form, as nominalized action, gives the reader reason to pause and thus stands as a flag to a distinction in the linguistic modelling of temporal experience, where "*Dreaming is dynamically manifest in the present and its spirituality traverses the temporal dimension such that it is neither solely of the past nor the present*" (Palmer, 2018, p. 111, emphasis added). It is unlikely that the subtext of a present ongoing is hidden from those groups who reject the more covert translation.[6]

Indeed, the coalescence of past–present–future in Indigenous linguacultures such as Thaayorre and Martu points to the contribution of cultural filters to the perception of the external world of space and time phenomena.[7] At the very least, this interpretation of space-time fusion commends itself to deeper contemplation on the gravity of cardinal direction competency in Indigenous groups. Michael Agar would identify these features of overtness as "rich points", since they are sites of cross-cultural learning in the face of incompetence (Agar, 1995). The point here is that *both* overt and covert translation can be revealing of relativity in different ways (Son, 2018), since overt and covert categories are operative both *within* a culture as well as *across* cultures. It was overt differences in everyday Trobriand Islander procedures that led Malinowski to many of his greatest insights. Likewise, Lee reminds us that the strong connection Whorf made between covert classes and the implicit metaphysics of a language "does not mean that he discounted the possibility of connections also between overt classes and world view" (Lee, 1996, p. 175). This is apposite in the case of translation, where the covertness of a translation may bury more deeply what is most different from the target readership. What insights might be gained with respect to the LRP depends on the set of linguistic-cultural sensitivities brought to the evaluation, and the particular orientation to the matter, since perspective and frames of reference are intrinsic to the relativity principle.[8]

It may be helpful to return to two important influences in Whorf's formulations: the hologram and field theory, which have increasingly been explored across both micro- and macro-domains of life. In a linguistic interpretation of field theory, texts and their subtexts might be related as the particle is to the wave. That is, a given linguistic form (including its overt and covert realizations) might be considered a particle that manifests from a wave of potential meanings (including all possible linguistic symbols and their meanings, explicit and implicit). The contribution of the hologram is one way to focus on the particle as it emerges (or is observed) from its waves of discrete settings. Somewhat analogous to the holographic model, the Peircian notion of *metaphor iconicity* maintains that there is a resemblance between semiotic forms and the phenomena they represent in the world. Similarly, the Hallidayan view conceptualizes text as standing in a metaphorical relation to the grammar of the clause (1981, 1997), and this idea is predicated on the clause having a semantic, non-arbitrary relation to its context.[9] The clause, in other words, maps onto text as an informational projection of multidimensional reality. Taking this approach to the LRP, Jim Martin, in his *Grammatical conspiracies in Tagalog* (1988), accumulated lexical, morphemic (as prototype) and grammatical (as cryptotype) evidence from clauses in Tagalog to suggest their collusion in cultural values of *family*, *face* and *fate*, with some contrast to Western emblems of *individualism*, *forthrightness* and *mastery of destiny*, traced in English (1988, pp. 282–286). This interpretation of grammar as covert is extended in Macdonald's thesis (2019), where she elaborates the contribution of disjunct at more general grammatical features (through system network modelling) to departures in the relative representation of music in English and Korean discourses, calibrating meanings through translation.[10] The idea of grammatical conspiracies as colluding towards cultural values hinges upon conceptualizing languages as socio-semiotic systems (Halliday & Hasan, 1985), with discourse interpreted as instances of socio-semiotic processes (Matthiessen, 2015). Thus, the isolate model of relativity is expanded to one of *systems of meaning* collaborating in *systems of experience* (see also *cryptosystems* in Halliday & Matthiessen, 1999). Furthermore, there is a shift in the conceptualization of "experience" from interaction with the environment to the enactment of relationships and embodiment of shared values as systems of meaning-making experience.

One obvious danger here is in mistranslating cultural values from grammatical and discourse evidence, due to cultural projections from the language of description, resulting in a "typing" of linguacultures somewhat synonymous with stereotyping. Is it realistic to say, for example, that *family* is not a cultural value of English-speaking cultures?—particularly as a global language spoken even by speakers of Tagalog (see Enfield's warning against making value claims; 2013, p. 162). One particular virtue of the use of Natural Semantic Metalanguage (NSM) (Goddard & Wierzbicka, 2002, 2004; Wierzbicka, 1992) to craft *cultural scripts* from translations (Wierzbicka, 2013) is its commitment to cultural neutrality. Cultural scripts deploy limited universally available linguistic currency to get a purchase on the associations activated in a particular cultural concept, without imposing associations from the language of description (Wierzbicka, 2013, p. 3—note that the binaries *left/right* and *come/go* are absent). By "articulating what is tacit knowledge about norms of evaluating, emoting, anticipating and so on" (Wierzbicka, 2013, p. 12) through the use of reduced scripts representing the logic of what *is* or *is not* implied in a domain of definition, these "fashions of speaking" may be verified by bilingual native speakers. In the

NSM model, these covert culturalized ways of thinking and emoting are formulated through overt universal primes.

Yet the depletion of language-specific descriptive resources may also be unnecessarily reductive. Given the reciprocity of the thought–language interrelationship to emancipate each other (Mandelbaum, 1949/1973), the stripping of language in scripting spare cultural assertions disaffects linguistic potential, compartmentalizing it too conclusively from value-laden thinking. How Western the logic of abductive validation is in the successive approximation method of developing cultural scripts might also be challenged. Arguing against a Bernsteinian tendency to view explicit forms as universal and implicit forms as more particulate and context specific, Hymes objects that "the actual fabric of relationships among kinds of meaning, communicative style and social consequences is more intricate" (Hymes, 1996, p. 48). A more ecologically expansive picture might better allow linguacultures to speak for themselves with respect to each other in an ultimately more cultivated yet comprehensible manner. One way in which anthropology contributes to this dilemma is through thick ecological description.

5.3.3 Calibrating the social implications of spatio-temporal frames of reference

While there are a number of neo-relativity theories that foreground both representational and pragmatic functions of language (see Hill & Mannheim, 1992; Leavitt, 2011), the second relativity model, or *Functional Relativity* (FR), proposed by Hymes (1964, 1996), is overviewed here because its capacity to calibrate cultures has some potential for translation and cognition. Emerging from *ethnopoetic* and *folklorist* traditions, this orientation is associated with spoken language and non-literate cultures. Yet, because of this focus on norms of interaction and the universal currency of speech, its techniques have been useful in reflexive introspection within literate cultures as well, as developed in *sociolinguistics* and *ethnomethodology* (Barnard, 2004; Gumperz, 1992; Ingold, 1994). Since the spoken word is far less contained than written text, it must be indexed more cohesively and creatively to the contexts complicit in meaning making (Gumperz, 1992), where "to describe the stratification of layers of significance is to describe increasingly thickly" (Rapport & Overing, 2000, p. 350). Given the emphasis on *functional* rather than *linguistic relativity*, it should be noted that there is not only an obvious bias towards the social dimension of the tripartite but even an assumption of social causality and priority. That said, this paradigm offers a comparative canvas where time and place themselves might become the medium for relating frames of reference and temporal reasoning, with social interaction as the centrepiece. Moreover, certain moves towards voice within this school provide useful means for hearing what is expressed as a whole, without the noise and interference of the artist or alternative projections.

Consider the following explanation of one mode of Dreaming discourse, Songlines, by Fitzroy Crossing man Joe Brown (2008, cited in Milroy & Revell, 2013, p. 3):

> every water got a song …
> people got to tell you story to make you happy and safe.
> Every place got a story

In FR, this constitutes a *communicative act*. Analysing it for its internal content and function builds somewhat on what has been revealed of Indigenous "cultural mentalities" in previous sections. That is, features of the landscape are prominent enough in Indigenous thinking to be connected to discourses of songs, also identified here as story. Such stories have a central social function, reflecting both more immediate concerns of navigating a sparse environment as well as the continuity between place and the cosmology of the creative era (Davenport,[11]

2019, pers. comm.). Under this analysis, a deeper function of cardinal direction fluency is its use in pointing to places of Dreaming, the locus of Martu-specific notions of time and of Law. Furthermore, linguistic patterning in such communicative acts corroborate in habitual 'ways of talking' that reveal highly valued concepts, such as balance (water-place), moral reciprocity (responsibility-response) and continuity (place-discourse-time) (see Marsh, 1977, p. 29; Stanner, 1979). For Martu man, Joshua Booth, the principle of 'rhyme' encapsulates these values and is essential to convey the meaning of any expression in translation into Martu Manyjilyjarra (Marsh, 1977, p. 25).

Yet this basic linguistic evidence only goes so far in terms of translatability and comprehensibility. In a number of interrelated *communicative events*, initiated elders from different Martu language groups have used Songline content to map waterhole locations of Dreaming stories, providing a semiotically tangible link between knowledge of country through stories of time. Ground truthing of the waterhole maps was verified in real space, which in turn validated their claims to stewardship of those lands since antiquity. The need for such semiotic shifts (from language to cartography) becomes apparent in contested discourse spaces, such as conflicting stories of the past (historical records of contact) and courtroom disputes over place (Native Title law courts of land stewardship), especially where communicative habits are almost incompatible. Although the enactment of Songlines may be broadcast in chant or dance, the message they imbibe of persistent connection to place is unlikely to offer a satisfactory testimony in the Native Title courtroom, where there is a strict dyadic protocol for engagement (see Walsh, 1997; Palmer, 2018). Indeed, access to sharing and listening to Songlines is highly restrictive, with Westerners almost entirely excluded from the Dreaming and matters of law (Stanner, 1979; Tonkinson, 1974). Rather than focusing on the exclusionary aspects of either language, an elaboration of the field of interaction can mediate divergent communicative styles and legitimate communicative events, through a dialectic of context. This has the potential to defuse the dangers of communicative divergence through highlighting the shared contextual spaces. Such elaborated *communicative situations*, whose properties include historical and politico-economical orders of indexicality, enables competing voices within these same contexts to be heard (Blommaert, 2005).

An important principle here is the operationalization of Sapir's notion of a *speech community* (1921) as one which is shifting, unstable and stratified, as opposed to the generic and unassailable terms of culture, society and language. The acknowledgement of Martu as comprised of multiple linguacultures with overlapping claims over land and Dreamtime stories facilitated their collaborative mapping project. Alignment of legal and anthropological definitions of "society" (see Section 5.1 and Palmer, 2018), allowed Songline maps to be admitted as evidence in Connection Reports (required by the states of Western Australia and Queensland) (Finlayson, 2001) in the successful Native Title bid, 1997–2002, of the Martu peoples (Rynne, 2002). Therefore, an important corollary of ecological validity is the validation of voices, or the "trusting that what people are saying is worth saying" (Davenport, 2019, pers. comm.), in order to penetrate why communicative acts are so fashioned. Indeed, articulating communication situations defines the field through which waves of communicative events are propagated. This has value towards the LRP, because the validation of voice equates to the calibration of communities.

Each of the approaches surveyed in this section has contributed variously in exploring physical, socio-cultural and conceptual experience between linguacultures. It would be useful to bring together both the field and the wave in a unified approach, exploring covertness/overtness on the micro-scale, implicitness/explicitness on the macro-scale and the interface between them.

5.4 Some suggested ways forward: Particle and wave as fully tripartite

David Bohm has adroitly argued for consciousness as holographic and reality as process (Bohm, 1980). Likewise, a process-oriented approach allows a reconfiguration of both the particle and the wave as fully tripartite. The veiled nature of cognitive processing, and the equally hidden social norms, might be seen as contributing more to covert wave-like potential (with overt expressions, nonetheless). By implication, language is perhaps a more salient overt particle-like feature that emerges from both fields (yet it, too, has covert features). Language and other forms of semiosis not only mediate the mind and culture in an observable way, but also radiate waves of implicature between individuals and society in all directions at once. The fully tripartite particle-phenomenon (revealed thought-speech-situation) is emergent from a wave that is also fully tripartite (implied ideology-semiotic valency-cosmos). Furthermore, translation as an endeavour mediates between these embodied tripartite complexes, mediating, in other words, between worlds.

One approach that aligns with this interpretation is exemplified in the work by Risku and colleagues at the University of Graz (Risku, 2014, 2017). Incorporating translation process research (TPR) and ethnographic description of the translator workplace environment, the focus of analysis is not the translation product in isolation, but the ecology of processes that influence translating as business performance (Risku, 2014, p. 334). To perhaps reformulate Risku, the research focus is on the waves of cognitive and social processes during a communicative event involving more than one speech community. Micro-behaviours measured through TPR technology (pause ratios, activity rhythm through eye gaze and keystroke logging) provide explicate points on the wave of mental processing. As a mirror to Functional Relativity, TPR conceptualizes these behaviours as cognitive events, while cognitive acts are the more implicit and immeasurable workings of the mind (Jakobsen, 2017). This inversion is not necessarily a contradiction but an enantiomorphism between the micro- and the macro-scale. Cognitive implicitness increases in the smallest of acts, while social implicitness becomes increasingly layered and more nuanced at the broadest scope. The ethnographic analysis emerges from an understanding that cognitive processes are variable according to material and socio-historical contexts.

Both cognitive and ethnographic dimensions allow for a degree of participant research to portray microcosms of translating within situated and dynamic performance boundaries. It should be rewarding to graft these microcosms into larger-scale collaborations, such as those comprehensive longitudinal studies that have yielded general frameworks and theoretical insights from the separate realities explored (Levinson & Wilkins, 2006; Slobin, 2003). The confluence of contrastive description and translation as comparison is paramount in managing the original paradox of sameness and difference. A further recommendation would be the use of *interpreting*. As a spoken form of speech community mediation, interpreting would lend itself nicely to ethnographic techniques. There may be quite distinct and revealing findings in a discourse space where meanings are negotiated online, with less room for revisions or embellishment. Furthermore, science and technology are advancing at such a rate that it should be possible to integrate thick descriptions of the psychology of interpreter behaviour, interpreting processes and outcomes of interaction within clearly depicted social ecologies. The recording and analysis of interpreting may offer a means of mitigating what has been identified by Risku as the potential for translator bias and positioning in retrospective interviews (2017). Interpreting deserves far more attention than it has received.

One final consideration for the future of explorations into the LRP is more comprehensive investigation into alternative semiotic systems which encode culture and cognition, such as social media, gaming, live streaming interaction, art and music. An outstanding example has been discussed in the Martu mapping of enduring connection to country (see Section 5.3.3). Similar multimodal maps, using artwork and embodied storytelling were used to retell the history of the Canning Stock Route, in effect retelling history from an alternative perspective, and allowing access by the majority of Australians to historical records kept in a very different manner. These 'cultural' maps formed part of a national exhibit combining colonial and Aboriginal recollections of colonisation in Western Australia (see Milroy & Revell, 2013). Importantly, the exhibition conveyed the distinct conceptualization of time and place in the Martu collective conscience in a way meaningful to others and true to themselves. Sapir advocated an early form of ethnopoetics in *Language: An introduction to the study of speech* (1921), which of necessity covered literature and art as a supplement to the indiscriminate and chaotic nature of more precise language. This was also something Malinowski addressed in his *Theory of magic* (Malinowski, 1935, part VI), with the language of magic having coefficients of "weirdness" and "intelligibility". This makes their accounts still vital and perhaps even more relevant now in an era of increasingly integrated multimodalities and eclectic intercultural spaces.

As a discourse of disparate cultural communities, anthropology has always been concerned with translating the minds of individuals and the behaviours of collectives within one culture across to those from another context. Anthropology is thus uniquely positioned as a meditation on humanity: bringing to mind the importance of ecological descriptive practice, the place of social interaction and the power of historical contemplations for a vigorous and expansive approach to the mind–language–culture tripartite.

Notes

1 Positions developed by Johann Georg Hamann (1730–1788), Johann Gottfried Herder (1744–1803) and later more comprehensively formulated by William von Humboldt (1767–1835).
2 The eventual position Immanuel Kant (1724–1804) held in his series of *Critiques*.
3 Book III of Locke's *Essay Concerning Human Understanding*, 1689.
4 Yana is a now extinct language of California.
5 According to W. E. H Stanner, the term "Dream Time" was first adopted by Spencer & Gillen for the Arrernte Dreaming concept, *Alchuringa* (Stanner, 1979, p. 23), as early as 1896.
6 Even the term "Dreaming" is criticized as insufficient for depicting entire epistemologies, some of which are quite varied in their function and content across language groups (Fletcher, 2003).
7 Philosopher Arthur Schopenhauer and physicist Erwin Schrödinger have both put forward that space and time may be perceived according to linguistic and cultural representations rather than the definitive nature of phenomena.
8 Compare, for example, Dirac's multiperspective measurement of wave function systems over time with Schrödinger's observer-only perspective, with the former leading to a unified theory of quantum mechanics and general relativity.
9 With a mostly arbitrary relationship being maintained at the lower phonemic and morphemic levels of language.
10 This study is unique in its attempt to incorporate both corpus-based contrastive description and translation as calibration.
11 Sue Davenport is an anthropologist associated with the Martu-run organization *Karnyirnninpa Jukkurrpa* (KJ).

Further reading

Carroll, J. C., Levinson, S. C., & Lee, P. (2012). *Language, thought and reality: Selected writings by Benjamin Lee Whorf* (2nd ed.). Cambridge, MA: MIT Press.
An essential compilation of the writings and musings of Benjamin Lee Whorf to appreciate the richness of the Linguistic Relativity Principle.

Leavitt, J. (2011). *Linguistic relativities: Language diversity and modern thought*. Cambridge: Cambridge University Press.

An illuminating, comprehensive and entertaining read covering the historical background feeding into contemporary interpretations of the Linguistic Relativity Principle.

Sapir, E. (1921). *Language: An introduction to the study of speech*. New York: Harcourt Brace and Company.

A classic that is even more incisive and relevant to studies on cognition, language and society today.

References

Achard, M., & Kemmer, S. (Eds.). (2004). *Language, culture and mind*. Stanford, CA: SCLI Publications.

Agar, M. (1995). *Language shock: Understanding the culture of conversation*. New York: William Morrow.

Barnard, A. (2004). *History and theory in anthropology*. Cambridge, New York: Cambridge University Press.

Barnard, A., & Spencer. J. (2002). *Encyclopedia of social and cultural anthropology*. New York: Routledge Taylor and Francis Group.

Bateson, G. (1972). *Steps to an ecology of mind: Essays in anthropology, psychiatry, evolution and epistemology*. San Francisco: Chandler.

Benedict, R. (1943). Obituary. Franz Boas. *Science, 97*(2507), 60–62.

Benedict, R. (1934/2005). *Patterns of culture* (First Mariner books ed.). Boston, New York: Houghton Mifflin Company.

Berman, R. A., & Slobin, D. I. (1994). *Relating events in narrative: A crosslinguistic developmental study*. New Jersey: L. Erlbaum.

Blommaert, J. (2005). *Discourse: A critical introduction*. Key Topics in Sociolinguistics. Cambridge, New York: Cambridge University Press.

Boas, F. (1899). Anthropology. *Science New Series, 9*(212), 93–96.

Boas, F. (1911). *Handbook of American Indian languages, part 1*. Washington: Government Print Office. https://archive.org/stream/handbookofameric00boas/handbookofameric00boas_djvu.txt

Boas, F. (1940). *Race, language and culture*. New York: The Free Press.

Boas, F. (1963). *Introduction to the handbook of American Indians*. Washington DC: Georgetown University Press.

Bohm, D. (1980). *Wholeness and the implicate order*. London and New York: Routledge.

Boroditsky, L., & Gaby, A. (2010). Remembrances of times east: Absolute spatial representations of time in an Australian Aboriginal community. *Psychological Science, 21*(11), 1635–1639.

Bowerman, M., & Choi, S. J. (2003). Space under construction: Language-specific spatial categorization in first language acquisition. In D. Genter, & S. Goldin-Meadow (Eds.), *Language in mind: Advances in the study of language and thought* (pp. 387–428). Cambridge, MA, London: MIT Press.

Carroll, J. B., Levinson, S. C., & Lee, P. (2012). *Language, thought and reality: Selected writings of Benjamin Lee Whorf* (2nd ed.). Cambridge, MA: MIT Press.

Casasanto, D. (2016). Linguistic relativity. In N. Riemer (Ed.), *Routledge handbook of semantics* (pp. 158–174). New York: Routledge.

Catford, J. C. (1965). *A linguistic theory of translation*. Oxford: Oxford University Press.

Choi, S., & Bowerman, M. (1991). Learning to express motion events in English and Korean: The influence of language-specific lexicalization patterns. *Cognition*, 4, 83–141.

Chomsky, N. (2000). *New horizons in the study of language and mind*. Cambridge: Cambridge University Press.

Cook, V., & Bassetti, B. (2011). *Language and bilingual cognition*. New York and Hove: Psychology Press, Taylor & Francis Group.

Darnell, R. (1998). *And along came Boas: Continuity and revolution in Americanist anthropology*. Amsterdam: John Benjamins.

Dolscheid, S., Shayan, S., Majid, A., & Casasanto, D. (2013). The thickness of musical pitch: Psychophysical evidence for linguistic relativity. *Psychological Science, 24*(5), 613–621.

Enfield, N. J. (2000). *On linguocentrism*. In M. Pütz, & M. Vespoor (Eds.), *Explorations in linguistic relativity*. Amsterdam studies in the theory and history of linguistic science. Series 4, Current issues in linguistic theory, Volume 199 (pp. 125–158). Amsterdam: John Benjamins.

Enfield, N. J. (2013). Language, culture, and mind: Trends and standards in the latest pendulum swing. *Journal of the Royal Anthropological Institute, 19*(1), 155–169.

Enfield, N. J. (2015). Linguistic relativity from reference to agency. *Annual Review of Anthropology, 44*, 207–224.

Everett, C. (2013). *Linguistic relativity: Evidence across languages and cognitive domains*. Berlin, Boston: De Gruyter Mouton.
Finlayson, J. (2001). Anthropology and connection reports in Native Title Claim applications. In G. Boeck (Ed.), *Land, rights, laws: Issues of Native Title, 2(9)*. Canberra: Native Title Research Unit, Australian Institute of Aboriginal and Torres Strait Islander Studies.
Firth, J. R. (1957). A synopsis of linguistic theory, 1930–1955. *Studies in Linguistic Analysis (Special volume of the Philological Society)*, pp. 1–32. Oxford: Blackwell.
Fletcher, M. (2003). "*Dreaming": Interpretation and representation* [unpublished PhD thesis]. Adelaide: Flinders University.
Gaby, A., & Kuuk Thaayorre language experts. (2017). *A grammar of Kuuk Thaayorre*. Berlin: Mouton DeGruyter [from the 2006 PhD thesis].
Geertz, C. (1973). *The interpretation of cultures: Selected essays by Clifford Geertz*. New York: Basic Books.
Gentner, D., & Goldin-Meadow, S. (Eds.). (2003). *Language in mind: Advances in the study of language and thought*. Cambridge, MA and London, England: MIT Press.
Goddard, C., & Wierzbicka, A. (Eds.). (2002). *Meaning and universal grammar: Theory and empirical findings* (Vols. 1 & 2). Amsterdam: John Benjamins.
Goddard, C., & Wierzbicka, A. (Eds.). (2004). Cultural scripts. *Intercultural Pragmatics, Special Issue, 1*(2), 153–166.
Gumperz, J. J. (1992). Contextualization and understanding. In A. Duranti, & C. Goodwin (Eds.), *Rethinking context* (pp. 229–252). Cambridge: Cambridge University Press.
Gumperz, J. J., & Levinson, S. C. (Eds.). (1996). *Rethinking linguistic relativity*. Cambridge: Cambridge University Press.
Hale, K. (1964). Classification of Northern Paman languages, Cape York Peninsula, Australia: A research report. *Oceanic Linguistics, 3*, 248–265.
Halliday, M. A. K. (1981). *How is a text like a clause?* Nobel Symposium on Text Processing, Topic 2: Aspects of text analysis and generation. Stockholm, 11–15 August 1980.
Halliday, M. A. K. (1997). Linguistics as metaphor. In J. J. Webster (Ed.). (2006). *On language and linguistics, Volume III in the collected works of M.A.K Halliday*. London and New York: Continuum.
Halliday, M. A. K., & Hasan, R. (1985). *Language, context and text: Aspects of language in a socio-semiotic perspective*. Melbourne: Deakin University Press.
Halliday, M. A. K., & Matthiessen, C. M. I. M. (1999). *Construing experience through meaning: A language-based approach to cognition*. London, New York: Cassell.
Harris, M. (1999). *Theories of culture in postmodern times*. Walnut Creek, London, New Delhi: Altamira Press.
Harris, Z. S. (1986). *Language and information: The Bampton lectures*. Columbia University. Retrieved December 2016 from Zellig S. Harris website: http://zelligharris.org/
Hill, J. H., & Mannheim, B. (1992). Language and world view. *Annual Review of Anthropology, 21*, 381–406.
Hjelmslev, L. (1961). *Prolegomena to theory of language*. Madison, Milwaukee and London: The University of Wisconsin Press.
House, J. (2000). Linguistic relativity and translation. In M. Pütz, & M. Vespoor (Eds.), *Explorations in linguistic relativity*. Amsterdam studies in the theory and history of linguistic science. Series 4, Current issues in linguistic theory, Volume 199 (pp. 125–158). Amsterdam: John Benjamins.
House, J. (2011). Translation and bilingual cognition. In V. Cook, & B. Bassetti (Eds.), *Language and bilingual cognition* (pp. 519–528). New York and Hove: Psychology Press, Taylor & Francis Group.
House, J. (2016). *Translation as communication across languages and cultures*. London and New York: Routledge Taylor & Francis Group.
Hunt, G., & Boas, F. (1921). *Ethnology of the Kwakiutl, based on data collected by George Hunt*. Washington DC: Government Printing Office.
Hymes, D. (1964). Two types of linguistic relativity (with examples from Amerindian ethnography). In W. Bright (Ed.), *Sociolinguistics: Proceedings of the UCLA Sociolinguistics Conference*, 1964, pp. 114–146.
Hymes, D. (1996). *Ethnography, linguistics narrative inequality: Toward an understanding of voice*. London, Bristol, PA: Taylor & Francis.
Ingold, T. (1994). *Companion encyclopedia of anthropology: Humanity, culture and social life*. London, New York: Routledge.
Jackendoff, R. (2002). *Foundations of language: Brain, meaning, grammar, evolution*. Oxford: Oxford University Press.
Jakobsen, A. L. (2008). Orientation, segmentation, and revision in translation. In G. Hansen (Ed.), *Empirical translation studies: Process and product* (pp. 191–204). Copenhagen Studies in Language Series, 27. Copenhagen: Samfundslitteratur.

Jakobsen, A. L. (2017). Translation process research. In J. W. Schwieter, & A. Ferreira (Eds.), *The handbook of translation and cognition* (pp. 21–49). Hoboken, NJ: Wiley-Blackwell.

Jakobson, R. (1959). On linguistic aspects of translation. In R. A. Brower (Ed.), *On translation* (pp. 232–239). Cambridge: Harvard University Press.

Jaworski, A., & Coupland, N. (2006). *The discourse reader* (2nd ed.). Abingdon and New York: Routledge.

Killman, J. (2015). Context as Achilles' heel of translation technologies: Major implications for end-users. *Translation and Interpreting Studies, 10(2)*, 203–222.

Kockelman, P. (2010). *Language, culture and mind: Natural constructions and social kinds*. Language, Culture and Cognition Series, 10. Cambridge: Cambridge University Press.

Koerner, E. F. K. (1999). *Linguistic historiography: Projects and prospects*. Amsterdam: John Benjamins.

Koerner, E. F. K. (2000). Towards a "full pedigree" of the "Sapir-Whorf hypothesis". In M. Pütz, & M. Vespoor (Eds.), *Explorations in linguistic relativity*. Amsterdam studies in the theory and history of linguistic science. Series 4, Current issues in linguistic theory, Volume 199 (pp. 1–24). Amsterdam: John Benjamins.

Lakoff, G. (1987). *Women, fire and dangerous things: What categories reveal about the mind*. Chicago and London: University of Chicago Press.

Langacker, R. W. (1987). *Foundations of cognitive grammar, Volume I, theoretical prerequisites*. Stanford: Stanford University Press.

Leavitt, J. (2011). *Linguistic relativities: Language diversity and modern thought*. Cambridge: Cambridge University Press.

Lee, P. (1996). *The Whorf Theory Complex: A critical reconstruction*. Studies in the History of the Language Sciences, Series III. Amsterdam: John Benjamins.

Lee, P. (2000). When is "linguistic relativity" Whorf's linguistic relativity? . In M. Pütz, & M. Vespoor (Eds.), *Explorations in linguistic relativity*. Amsterdam studies in the theory and history of linguistic science. Series 4, Current issues in linguistic theory, Volume 199 (pp. 45–68). Amsterdam: John Benjamins.

Leeman, J. (2015). *Cognitive testing of the American community survey language question in Spanish*. Washington, DC: Research and Methodology Directorate, Centre for Survey Measurement Study Series (Survey Methodology, 2015–2).

Levinson, S. C. (2003). *Space in language and cognition: Explorations in cognitive diversity*. Language, culture and cognition series 5. Cambridge: Cambridge University Press.

Levinson, S. C., & Brown, P. (1994). Immanuel Kant among the Tenejapans: Anthropology as empirical philosophy. *Ethos, 22*(1), 3–41.

Levinson, S. C., & Wilkins, D. P. (Eds.). (2006). *Language, culture, and cognition*, Vol. 6. Grammars of space: Explorations in cognitive diversity. Cambridge: Cambridge University Press.

Lucy, J. A. (1992). *Language diversity and thought: A reformulation of the linguistic relativity hypothesis*. Cambridge: Cambridge University Press.

Lucy, J. A. (1996). The scope of linguistic relativity: An analysis and review of empirical research. In J. J. Gumperz, & S. C. Levinson (Eds.), *Rethinking linguistic relativity* (pp. 37–69). Cambridge: Cambridge University Press.

Lucy, J. A. (1997). Linguistic relativity. *Annual Review of Anthropology, 26*, 291–312.

Lucy, J. A. (2011). Language and cognition: The view from anthropology. In V. Cook, & B. Bassetti (Eds.), *Language and bilingual cognition* (pp. 43–68). New York and Hove: Psychology Press, Taylor & Francis Group.

Macdonald, K. (2019). Construing musical discourses: Axial reasoning for a contrastive description of habitual ideational resources in English and Korean, with reflection on translation [unpublished PhD thesis]. Hong Kong Polytechnic University.

Malinowski, B. (1922). *Argonauts of the Western Pacific*. London: Routledge.

Malinowski, B. (1935). *Coral gardens and their magic: A study of the methods of tilling the soil and of agricultural rites in the Trobriand Islands, Volume II*. London: George Allen and Unwin, Ltd.

Malinowski, B. (1923/1946). The problem of meaning in primitive languages. In C. K Ogden, & I. A Richards (Eds.), *The meaning of meaning* (pp. 296–336). London: Routledge & Kegan Paul. Reproduced in A. Jaworski, & N. Coupland (Eds.), *The discourse reader* (3rd ed., 2014) (pp. 284–286). London and New York: Routledge.

Mandelbaum, D. G. (Ed.). (1949/1973). *Selected writings of Edward Sapir in language, culture and personality*. Berkeley: University of California Press.

Marsh, J. (1977). The notion of balance in Mantjiltjara grammar. In E. Brumby, & E. Vaszolyi (Eds.), *Language problems and Aboriginal education* (pp. 25–33). Mount Lawley College of Advanced Education.

Martin, J. R. (1988). Grammatical conspiracies in Tagalog: Family, face and fate—with regard to Benjamin Lee Whorf. In J. D. Benson, M. J. Cummings, & W. S, Greaves (Eds.), *Linguistics in a systemic perspective*. Current Issues in Linguistic Theory: IV. Amsterdam: John Benjamins.

Matthiessen, C. M. I. M. (2015). Register in the round: Registerial cartography. *Functional Linguistics, 2(9)*, 1–48.

Miller, G. A. (2003). The cognitive revolution: A historical perspective. *Trends in Cognitive Sciences, 7*(3), 141–144.

Milroy, J., & Revell, G. (2013). Aboriginal story systems: Remapping the West, knowing country, sharing space. *Occasion: Interdisciplinary Studies in the Humanities, 5*, 1–24.

Mühlhäusler, P. (2000). *Humboldt, Whorf and the roots of ecolinguistics*. In M. Pütz, & M. Vespoor (Eds.). *Explorations in linguistic relativity*. Amsterdam studies in the theory and history of linguistic science. Series 4, Current issues in linguistic theory, Volume 199 (pp. 89–100). Amsterdam: John Benjamins.

Niemeier, S., & Dirven, R. (Eds.). (2000). *Evidence for linguistic relativity*. Current Issues in Linguistic Theory Series, Volume 198. Amsterdam: John Benjamins.

Palmer, K. (2018). *Native Title anthropology: Strategic practice, the law and the state*. Canberra: Australian National University Press.

Palmer, W., Blythe, J., Gaby, A., Hoffman, D., & Ponsonnet, M. (2019). Geospatial natural language in Indigenous Australia: Research priorities. In K. Stock, C. Jones, & T. Tenbrick (Eds.), *Speaking of location* (11). Regensberg, Germany. CEUR-WS.org

Penn, J. M. (1972). *Linguistic relativity versus innate ideas: The origins of the Sapir-Whorf hypothesis in German thought*. Janua Linguarum—studia memoriae: Series Minor, 120. The Hague and Paris: Mouton.

Pinker, S. (1994). *The language instinct: The new science of language and the mind*. New York, NY: Harper Perennial Modern Classics.

Pullum, G. K. (1991). *The great Eskimo vocabulary hoax*. Chicago: Chicago University Press.

Pütz, M., & Vespoor, M. (Eds.). (2000). *Explorations in linguistic relativity*. Amsterdam studies in the theory and history of linguistic science. Series 4, Current issues in linguistic theory, Volume 199. Amsterdam: John Benjamins.

Rapport, N., & Overing, J. (2000). *Social and cultural anthropology: The key concepts*. London, New York: Routledge.

Risjord, M. (2007). Scientific change as political action: Franz Boas and the anthropology of race. *Philosophy of the Social Sciences, 37*(1), 24–45.

Risku, H. (2014). Translation Process Research as interaction research: From mental to socio-cognitive processes. In *MonTI Special Issue – Minding Translation*, 331–353.

Risku, H. (2017). Ethnographies of translation and situated cognition. In J. W. Schwieter, & A. Ferreira (Eds.), *The handbook of translation and cognition* (pp. 290–310). Hoboken, NJ and Malden, MA: Wiley Blackwell.

Rubel, P. G., & Rosman, A. (2003). *Translating cultures: Perspectives on translation and anthropology*. Oxford and New York: Berg.

Rynne, M. (2002). The Martu Native Title determination. *Native Title Newsletter, 5/2002*.

Sapir, E. (1921). *Language: An introduction to the study of speech*. New York: Harcourt Brace and Company.

Seiferle-Valencia, A. (2017). *Franz Boas's race, language and culture*. London: Routledge.

Slobin, D. (1996). From "thought and language" to "thinking to speaking". In J. J. Gumperz, & S. C. Levinson (Eds.), *Rethinking linguistic relativity* (pp. 70–96). Cambridge: Cambridge University Press.

Slobin, D. (2000). Verbalized events: A dynamic approach to linguistic relativity and determinism. In S. Niemeier, & R. Dirven (Eds.), *Evidence for linguistic relativity*. Current Issues in Linguistic Theory, Volume 198 (pp. 107–138). Amsterdam: John Benjamins.

Slobin, D. (2003). Language and thought online: Cognitive consequences of linguistic relativity. In D. Gentner, & S. Goldin-Meadow (Eds.), *Language in mind: Advances in the study of language and thought* (pp. 157–192). Cambridge, MA: MIT Press.

Son, J. Y. (2018). Back translation as a documentation tool. *International Journal for Translation & Interpreting Research, 10*(2), 89–100.

Stanner, W. E. H. (1953). *The Dreaming*.

Stanner, W. E. H. (1979). White man got no dreaming. *Essays, 1938–1973*.

Taylor, J. C. (1984). *Of acts and axes: An ethnography of socio-cultural change in an Aboriginal community, Cape York Peninsula* [PhD thesis]. James Cook University.

Tonkinson, R. (1974). *The Jigalong Mob: Aboriginal victors of the desert crusade*. Menlo Park, California: Cummings Publishing Company.

Toury, G. (1995). *Descriptive translation studies and beyond*. Amsterdam: John Benjamins.

Tymoczko, M. (2010). *Enlarging translation, empowering translators*. London and New York: Routledge.
von Stutterheim, C. (1999). How language-specific are processes in the conceptualizer? In R. Klabander, & C. von Stutterheim (Eds.), *Representations and processes in language production* (pp. 153–179). Studien zur Kognitionswissenschaft. Wiesbaden: Deutscher Universitätsverlag.
von Stutterheim, C., Nüse, R., & Murcia Serra, J. (2002). Crosslinguistic differences in the conceptualization of events. In H. Hasselgård, S. Johansson, C. Fabricius-Hansen, & B. Behrens (Eds.), *Information structures in a cross-linguistic perspective* (pp. 59–88). Amsterdam: John Benjamins.
Walsh, J. (1997). *Cross-cultural communication problems in Aboriginal Australia*. Discussion Paper, 7/97. The Australian National University North Australia Research Unit.
Whorf, B. L. (1936). Thinking in primitive communities. In J. B. Carroll (Ed.), *Language, thought and reality: Selected writings of Benjamin Lee Whorf* (22nd reprinting, pp. 65–86). Cambridge, MA: The MIT Press.
Whorf, B. L. (1937a). Grammatical categories. In J. B. Carroll (Ed.). (1956). *Language, thought and reality: Selected writings of Benjamin Lee Whorf* (pp. 87–101). Cambridge, MA: MIT Press.
Whorf, B. L. (1937b). Discussion of Hopi linguistics. In J. B. Carroll (Ed.). (1956). *Language, thought and reality: Selected writings of Benjamin Lee Whorf* (pp. 102–111). Cambridge, MA: MIT Press.
Whorf, B. L. (1938). Language: plan and conception of arrangement. In J. B. Carroll (Ed.). (1956). *Language, thought and reality: Selected writings of Benjamin Lee Whorf* (pp. 125–133). Cambridge, MA: MIT Press.
Whorf, B. L. (1939). The relation of habitual thought and behavior to language. In J. B. Carroll (Ed.). (1956). *Language, thought and reality: Selected writings of Benjamin Lee Whorf* (pp. 134–159). Cambridge, MA: MIT Press.
Whorf, B. L. (1940). Science and linguistics. In J. B. Carroll (Ed.). (1956). *Language, thought and reality: Selected writings of Benjamin Lee Whorf* (22nd reprinting, pp. 207–219). Cambridge, MA: MIT Press.
Whorf, B. L., & Trager, G. L. (1938). *The Yale report*. In P. Lee (Ed.). (1996), *The Whorf Theory Complex: A critical reconstruction* (pp. 251–280). Studies in the History of the Language Sciences, Series III. Amsterdam/Philadelphia: John Benjamins Publishing Company.
Wierzbicka, A. (1992). *Semantics, culture and cognition*. Oxford: Oxford University Press.
Wierzbicka, A. (2013). Translatability and the scripting of other peoples' souls. *The Australian Journal of Anthropology, 24*, 1–21.
Wilkins, D. P., & Hill, D. (1995). When "go" means "come": Questioning the basicness of basic motion verbs. *Cognitive Linguistics, 6*, 209–259.
Wittgenstein, L. (1958). *Philosophical investigations*. G. E. M. Anscombe (Trans.). Oxford: Blackwell.
Young, M. W. (2011). Malinowski's last word on the anthropological approach to language. *Pragmatics, 21*(1), 1–22.

6

Translation, contact linguistics and cognition

Haidee Kotze

6.1 Introduction

This chapter provides a sketch of the interdisciplinary interface between translation and contact linguistics, an area of research concerned with translation as a type of language contact. The chapter places particular emphasis on the (socio-)cognitive dimensions of translation and language contact. Kranich (2014, p. 97) explains the central focus of what she terms language contact through translation (LCTT) as follows:

> While translating a text from a source language (SL) to a target language (TL), the bilingual individual must activate his/her competence in both these languages. The product of this process can exhibit an impact of features of the SL on the target text (TT). This impact has been discussed under the label "interference" (Toury, 1995) as well as under the name of "shining-through" (Teich, 2003) in Translation Studies. If the same type of shining-through phenomenon occurs repeatedly in translations, it might spread to monolingual text production, that is to non-translated texts produced by TL authors.

This description clearly spells out one self-evident cognitive dimension of language contact: the bilingual language processing of the text producer (in this case, the translator), which demonstrates particular linguistic features as a consequence of the co-activation of two languages, most pertinently cross-linguistic influence (CLI). However, a second cognitive dimension of language contact is left implicit: The mechanisms by means of which contact-influenced features that occur in translations can "spread to" monolingual text production must, by definition, also involve cognitive processes. More specifically, these cognitive processes involve readers' exposure to and processing of LCTT-influenced linguistic features, under conditions that allow these features to become sufficiently cognitively entrenched in the linguistic repertoires of these readers that they themselves will re-use such features when they subsequently produce language. This iterative production–reception cycle is key to an understanding of (a) translation as a language-contact event that results in a contact-influenced language product and (b) the role of translations as a potential factor in contact-induced language change.

While both these processes clearly take place at the level of individual language processing, they also have an aggregate social dimension. It is only when contact-influenced forms are produced at a sufficiently high frequency to affect the linguistic representations of users who are exposed to these forms in the texts they read that contact-influenced forms gain a foothold to find their way into non-translational usage. The processes by means of which this may come about are constrained by a range of cognitive, linguistic and social factors (see also Muysken, 2013). These factors, and the way in which they influence both processes, are interwoven in complex ways, and form the focus of this chapter.

While there has been a growing interest in the interface between Translation Studies and contact linguistics, research in this area has been limited by a number of factors. First, the interdisciplinary relationship between Translation Studies and contact linguistics has been largely unidirectional: Translation Studies researchers have drawn more on ideas from contact linguistics[1] than vice versa (see Kranich, 2014). Second, a coherent theoretical framework that is consonant with what is known about contact as a factor in language variation and change more generally has been lacking in studies of translation as a form of language contact (see Malamatidou, 2016; Redelinghuys, 2019). Third, in the interdisciplinary conversation between Translation Studies and contact linguistics, the cognitive dimension of translation as a form of language contact and a factor in contact-induced linguistic change, has received little attention. Lastly, existing empirical research in this area has been restricted to a fairly limited number of language pairs, time periods and text types, with a focus on short-term rather than long-term change. This raises not only reservations about the generalizability of findings but also foundational questions about the meaning of the concept "language change" in this area of research.

Against this background, this chapter outlines the state of the art at the interface between Translation Studies and contact linguistics. The chapter is structured around the notion that any translation event is constituted by two cognitive processes (simultaneously embedded in, and cognitively representing, social processes and systems): The production process as well as the reception process (Kruger & Kruger, 2017). Section 6.2 focuses on translation production as an individual contact-influenced language processing event that creates the translated text. This section considers how cognitive and socio-cognitive factors shape the realization of the translated text as the product of a language-contact event, focusing first on the most obvious consequence of language contact, namely CLI, but also on other linguistic features that may be ascribed to bilingualism-influenced communication or contact effects. This section surveys existing research that has attempted to link the bilingual linguistic processing in translation to bilingual or contact-influenced text production more generally, including research that has argued that translation is a kind of contact variety that demonstrates similarities with other varieties of language characterized by language contact.

Section 6.3 takes as a starting point the cognitive processing of readers who are reading the text. It specifically considers how, and under what conditions, contact-influenced linguistic input from translation may find its way into multiple readers' cognitive representations of linguistic constructions, and how this exposure to contact-influenced constructions may, in favourable social conditions, potentially lead to language change in the target language.[2] This section outlines some theorizations of contact-induced language change through translation, and briefly surveys existing empirical work in the field. Section 6.4 concludes with an overview of the methodological and conceptual challenges in studying translation as a type of language contact.

6.2 Making the translated text: Individual language processing and the effects of language contact

Contact linguistics focuses on the linguistic outcomes that result from people using more than one language in a substantive, sustained and non-trivial way in a particular shared context, leading to the influence of one language on another (Thomason, 2001, pp. 1–2). These shared contexts may be of vastly different kinds, but they all presuppose some form of bi- or multilingualism. Contact linguistics highlights the complex interplay between individual and societal bi- and multilingualism (see Matras, 2009; Thomason & Kaufman, 1988). Against the background of the recognition of this complexity, Matras (2009), along with Weinreich (1953), argues that language contact should not, in the first instance, be considered as contact between language "systems". Instead, "the relevant locus of contact is the language processing apparatus of the individual multilingual speaker and the employment of this apparatus in communicative interaction" (Matras, 2009, p. 3).

This section therefore considers the language processing "apparatus" of the translator as an individual multilingual user, who is employing this apparatus in a very particular kind of communicative interaction. It raises the question of how translated texts reflect the traces of the translation process as a cognitive language-contact event. One important dimension of this is how the language processing apparatus of translators and their use of this apparatus in producing written texts are the same as or different from the language apparatus of other bilingual users using it to produce written texts. In other words: How are language producers socio-cognitively constrained when communicating under conditions of language contact? And are these constraints the same or different for translators and other writers producing texts under the influence of language contact?

This line of argumentation aligns with a strand of research in Translation Studies proposing that the recurrent linguistic features of translated language are not unique to translated language, but typify a broader set of varieties characterized by particular socio-cognitive constraints. Among these proposed constraints, language contact or bi- or multilingual discourse production is particularly prominent (see, for example, Bisiada, 2017; Chesterman, 2004; Gaspari & Bernardini, 2010; Granger, 2018; Kolehmainen et al., 2014; Kruger, 2012; Kruger & De Sutter, 2018; Kruger & Van Rooy, 2016a, 2016b; Lanstyák & Heltai, 2012; Ožbot, 2014; Shlesinger & Ordan, 2012; Steiner, 2008).[3] In brief, many of these scholars argue that the regularities of translated language are really regularities of contact-influenced language varieties more generally. Thus, the features that typify translated language are also evident in other written forms of contact-influenced varieties, such as contact-influenced first-language varieties of English, learner or second-language writing, indigenized second-language varieties of English (L2 varieties or New Englishes) and bilingualism-influenced communication more generally, since these varieties are constrained by similar cognitive and social factors associated with written communication in settings of language contact.

The following section focuses first on the most self-evident effect of contact-influenced language production, CLI. It considers the nature and proposed causes of CLI, and different types of CLI, integrating perspectives from contact linguistics and psycholinguistics. Subsequently it provides an overview of some studies of CLI in translation, specifically those emphasizing the cognitive dimension of language contact and the similarities between translations and other contact-influenced varieties.

6.2.1 Cross-linguistic influence (CLI)

6.2.1.1 The nature and causes of CLI

The most obvious consequence of language production under conditions of contact is CLI, defined as "the influence of a person's knowledge of one language on that person's knowledge or use of another language" (Jarvis & Pavlenko, 2008, p. 1), leading to a deviation from the conventions of either language in bilingual language production (Weinreich, 1953, p. 1).

It occurs when bilinguals move form-function units, word forms, patterns and schemas from one language to another during a communicative event (Matras, 2009, p. 99). CLI involves a complex set of interrelated phenomena occurring at various levels of language, discussed under rubrics such as borrowing, transfer, interference, calquing, code-switching, code-mixing, code-meshing, interlanguage, translanguaging and (at the most extreme end) creolization and pidginization. A full discussion of the relation among these phenomena falls outside the scope of this chapter; sufficient here is to note the existence of these phenomena and their general relation to CLI.

A key question in this respect is how bilinguals' languages are cognitively organized: whether they are separate or interlinked; at what levels linking exists; how they are controlled; and under what conditions one language may affect production in the other. Several models of bilingual language representation and processing have been proposed (see Kroll & Tokowicz, 2005; Matras, 2009 for overviews; Diamond & Shreve, 2010 for reflections on translation specifically), and are widely debated. The discussion that follows does not focus in detail on the nature of bilingual or multilingual language representation, as such. Instead, it focuses only on research demonstrating that, irrespective of how the cognitive organization of bilinguals' languages is modelled, the linguistic usage of bilinguals demonstrates the transfer of linguistic forms or patterns from one language to another.

From a psycholinguistic perspective, depending on the particular communicative situation, bilinguals' languages are cognitively activated to varying degrees, ranging from a fully monolingual mode (when the other language is largely inhibited or non-active) to a fully bilingual mode (when the two languages are both strongly activated) (Grosjean, 2008). When activation is more strongly towards the bilingual end of the scale, CLI becomes more likely. Taking a more functionalist approach, Matras (2009, p. 99) argues that bilinguals' linguistic knowledge is a complex repertoire of not only linguistic schemas, forms and constructions but also context-sensitive information about social conventions for the appropriate selection of these forms. This information is used, in part, for maintaining "demarcation boundaries" between languages, based on social expectations and communicative conventions. While there may be strong normative incentives in some communicative contexts to maintain the demarcation between languages very strictly, in some situations, control of these boundaries may be relaxed either inadvertently or on purpose.

CLI is a complex phenomenon that may occur for various (socio-cognitive) reasons. First, the effects of cross-linguistic cognitive priming at both lexical and structural levels are well documented in both corpus and psycholinguistic studies: when a second language is cognitively activated, regardless of whether this is in L2 communication, bilingual communication or translation, patterns from that language influence the patterns of the TL in often subtle ways that lead to the relative over- or under-representation of linguistic features in comparison to monolingual language production (see Bangalore et al., 2016; Gries & Kootstra, 2017; Hartsuiker et al., 2004; Hoey, 2011; Kootstra & Muysken, 2017; Loebell & Bock, 2003; Maier et al., 2017; Paradis, 1993; Pickering & Ferreira, 2008). At the psycholinguistic level, cross-linguistic priming is clear evidence of the co-activation of languages, and, as Kootstra and Muysken (2017, p. 215) point out, these psycholinguistic processes have real-world consequences for second-language acquisition, code-switching, code-mixing and contact-induced language change.

More generally, bilingualism research demonstrates that due to taxing processes of selection, switching and inhibitory control in a cognitive environment where languages are in competition, bilingual language processing is a more effortful cognitive environment, inducing higher processing costs (Costa & Sebastián-Gallés, 2014). This cognitively effortful processing environment may lead to a reduced level of control over the "demarcation boundaries" between different languages, leading to control lapses at particular communicatively sensitive junctures and allowing CLI (Matras, 2012, p. 41).

Another proposed reason for CLI effects is termed interlingual identification (Croft, 2000; Weinreich, 1953) or pivot matching (Matras & Sakel, 2007). This involves bilinguals' ability to create mental connections between linguistic units in their two languages that they perceive as corresponding, based on formal or functional properties (Croft, 2000, p. 145). This alleviates processing strain, since similar processing operations and selection procedures can be used for equivalent linguistic elements in both languages (Matras, 2009, p. 151; Matras & Sakel, 2007, p. 835). Interlingual identification is thus a kind of processing shortcut, but the consequence is that the cognitive representations of the two languages syncretize and converge, becoming less representationally distinct. In the context of Translation Studies, direct transfer or literal translation (based on parallel bilingual processing where direct links between cross-linguistic units exist) is often seen as a default strategy for reducing cognitive load and minimizing demands on working memory (see Carl & Dragsted, 2012; Schaeffer & Carl, 2013; Tirkkonen-Condit, 2005).

Lastly, CLI effects may be a consequence of a conscious, strategic, pragmatic choice by the language producer (in this case, the translator) to exploit some communicative advantage offered by transferring elements from the other language. In translation, this kind of deliberate socio-pragmatically motivated drawing on the resources of the SL may be intended to fill a lexical gap in the TL or, alternatively, to foreground the linguistic and cultural otherness of the source text in translation, as part of a foreignizing strategy or a conscious choice in favour of a source-text-oriented overt translation (see also Kranich, 2014).

While most research on CLI focuses on spoken language, the same principles are likely to apply to written translation, but with different constraints modulating the effects of CLI. As Kruger and Van Rooy (2016a, p. 29) point out, written translation operates at the extreme end of the bilingual activation mode, involving a rapid shuttling between two languages that is cognitively and pragmatically constrained by an existing text in the SL, thus theoretically creating a fertile ground for CLI. At the same time, other factors mitigate and constrain the effects of CLI in professional translation. The high degree of linguistic proficiency, biliteracy, task expertise and professional training of translators may lead them to be particularly aware of CLI and develop conscious strategies to avoid such effects (Kruger & Van Rooy 2016b, p. 121; see also Shreve & Lacruz, 2017 for an overview of research from psycholinguistics and cognitive science that demonstrates how factors like task schemas interact with bilingual language production in the context of translation). In written translation, it is likely that the incentive towards maintaining boundaries between languages and meeting the conventions of the target language is strong due to normative pressures and constraints on translation that have been internalized by translators, forming part of cognitive scripts for appropriate translational behaviour.

These socio-cognitive factors thus act as a powerful "brake" that keeps CLI in written translation in check. The degree to which translators choose to conform to the linguistic norms of the target culture depends on various factors, including whether they choose a source- or target-oriented approach to translation; a choice which is, in itself, shaped by translators' cognitive construal or framing of factors like text type, register, reader expectations and the function of the translation in the recipient system. Other factors may also play a role, such as the prestige of the languages involved and the degree of standardization of the TL. As argued by Toury (2012, p. 314), tolerance of CLI increases in situations where translations are done from a language with higher prestige into languages of lower prestige, or where the TL is not strongly standardized. Lastly, it should be emphasized that whether CLI effects appear in published translations also depends on other gatekeepers in the publishing industry, including editors and proofreaders.

6.2.1.2 Types of CLI

CLI may take various forms, usually designated as a distinction between lexical borrowing (also called matter replication or global code-copying) and structural or grammatical borrowing (also called pattern replication or selective code-copying) (see, for example, Fischer, 2013; Johanson, 2002; Matras, 2009; Sakel, 2007; Thomason, 2001).

Since lexical borrowing entails the direct replication of linguistic material, that is, morphemes and word forms, Matras (2009) uses the term "matter replication" rather than "borrowing". Matter replication appears to be strongly influenced by linguistic and socio-cultural factors. Gaps in the lexical inventory of the target language are an obvious motivation for matter replication; however, the prestige associated with the SL no doubt plays a crucial role: linguistic choices, including borrowings, are not just "determined by the ideas we want to get across, but also by the impression we want to convey on others, and by the kind of social identity that we want to be associated with" (Haspelmath, 2009, p. 48). This provides the socio-cognitive rationale for Toury's (2012) claim that tolerance of interference is typical of situations where translation is done from a language with higher prestige to one with lower prestige. Matras (2009, p. 149) argues that matter replication is subject to stringent selection and inhibition control, since it is strongly marked, and Backus and Verschik (2012) ascribe matter replication primarily to the semantic "attractiveness" of the word in question, which yields some communicative advantage. In the context of translation, matter replication is therefore more likely to be the result of a conscious decision-making process on the part of the translator.

Pattern replication, or structural transfer, does not involve the importation of linguistic matter from the SL to the TL. Instead, the use of linguistic material already present in the TL is changed in some way: "it is the patterns of distribution, of grammatical and semantic meaning, and of formal-syntactic arrangement at various levels [...] that are modelled on an external source" (Matras & Sakel, 2007, pp. 829–830). Backus and Verschik (2012) argue that whereas matter replication is governed by semantic factors that prompt conscious decision making, unconscious cognitive factors like frequency and entrenchment play a more important role in structural transfer. Linguistic patterns that belong to both languages are cognitively entrenched in the linguistic representation of bilinguals (Backus et al., 2011, p. 739). Where the degree of entrenchment of the source structure is stronger than that of the target structure, pattern replication may occur when the source pattern is transferred to the TL text (Backus & Verschik, 2012, p. 143), or where the influence of the SL pattern leads to a target-language structure undergoing an increase (or decrease) in frequency of use.

A further distinction that can be drawn is between overt transfer and covert transfer, adapted from Mougeon et al. (2005) but also raised in Heine and Kuteva (2005). In the case of overt transfer, the two languages in question differ in some respect in the inventories of linguistic constructions or patterns available to users. For a variety of reasons (strategic, or as a consequence of processing constraints), the SL construction is transferred across the two languages in the form of an innovative formal pattern or innovative functional use in the TL. This type of transfer is a qualitative change in usage, in the sense that it introduces a construction that did not exist prior to the contact event. It is comparatively rare, however (Backus, 2014, p. 96).

Much more common is covert transfer, which is a quantitative development only. In its most straightforward form, covert transfer involves a situation where contact causes individuals to start using a TL construction more frequently than they otherwise would have, under the influence of a similar SL construction which primes or activates the TL construction, and at the cost of other options in the TL. This may lead to the over-representation of particular word forms or syntactic constructions under the influence of the SL, or the replacement of more conventional syntactic-semantic constructions or collocations in the TL with less conventional constructions

(Mauranen, 2004). In the context of Translation Studies, a kind of "inverse" form of covert transfer is the unique items hypothesis (Tirkkonen-Condit, 2004), which posits that where there are mismatches between the lexical or syntactic inventories of the SL and TL, the activation of the SL may inhibit the selection of TL constructions that do not have a direct match or easy interlingual identification in the SL, leading to an under-representation of these linguistic items in translation.

Matter replication is, by definition, overt in nature. Pattern replication, however, may be overt or covert in nature—and is more often covert, taking complex shapes in the over- or under-representation of particular structural patterns under the influence of contact. If multiple linguistic features are affected by these patterns of over- and under-representation in translation, these frequency and distributional differences under the influence of contact may lead to overall stylistic, genre or register differences between translated and non-translated texts (see Baumgarten & Özçetin, 2008; Hansen-Schirra, 2011; Kruger & Van Rooy, 2018; Neumann, 2013, 2014).

6.2.1.3 Studies of CLI in translation: Towards a cognitive perspective

CLI has been widely investigated in the framework of corpus-based Translation Studies interested in the interplay between ST transfer "shining through" effects and target-oriented normalization effects in translation (see Hansen-Schirra, 2011; Lefer & Vogeleer, 2013; Teich, 2003). The majority of these studies focus on covert structural pattern replication (or structural transfer), demonstrating how the frequencies and distributional patterns of particular constructions in the TT are affected by CLI. The range of work in this area precludes a full discussion, and therefore the following brief discussion is not exhaustive but, rather, highlights some exemplary studies of CLI effects in translation.

Dai and Xiao (2011) demonstrate that passive constructions in Chinese translated from English are much more frequent than passive constructions in non-translated Chinese. The increased frequency of passives in translational Chinese is the consequence of CLI effects from English, which uses passive constructions more frequently than Chinese does. Cappelle and Loock (2013) focus on existential constructions in English and French, against the background that these constructions are much more frequent in the former than the latter language. They find that English translated from French contains many fewer existential constructions than original English; whereas French translated from English contains many more existential constructions than original French. They interpret these findings as strong evidence of CLI, clearly functioning at the covert, structural level. Cappelle (2012) and Cappelle and Loock (2017) investigate manner of motion verbs and phrasal verbs, respectively, in English translations from various SLs, and find that the typological differences between languages lead to different frequency patterns for the features in question in the English translations, yet again providing evidence in favour of CLI effects, also framed in terms of the unique items hypothesis. Baumgarten and Özçetin (2008) focus on speaker (first-person) pronouns in English and German translated and non-translated business communication. English and German have distinct stylistic preferences for personal pronoun use in this register, and through CLI, the English source texts introduce variation in the stylistic preferences of German.

These studies provide clear evidence of CLI effects, but as they investigate single features (and often registers), they provide a somewhat unidimensional picture. Other researchers, including Hansen-Schirra (2011) and Neumann (2013, 2014), have taken a broader approach, showing how sets of linguistic features demonstrate complex patters of over- and under-representation in translations compared with original texts, reflecting both CLI and (over-)adjustment to TL norms. This has the effect of altering the register profile of some translated registers compared with non-translated registers in the same language as a consequence of CLI. Van Oost et al. (2016) investigate another important dimension of CLI variability, namely the effect of SL status. Focusing on prepositional phrase placement, they show that there are strong shining-through effects in Dutch texts translated from German, and strong normalization in German texts translated from Dutch. They conclude: "These results confirm Toury's hypothesis that a less prestigious language such as Dutch is more tolerant towards higher frequencies of linguistic features which are typical of highly prestigious source languages such as German than the other way around" (Van Oost et al., 2016, p. 7).

These corpus-based studies clearly demonstrate the existence of CLI effects in translation, and its complexity. However, very few studies explore in any empirical depth how CLI works as a cognitive language-contact phenomenon. A particularly important kind of evidence in this respect is how CLI in translation is similar to or different from the kinds of CLI evident in other contact varieties, which illuminates how different socio-cognitive constraints shape different contact varieties. This notion has been explored theoretically by Kolehmainen et al. (2014), who focus on interlingual reduction (a general term for the under-representation of unique items) in translation, second-language acquisition and language-contact situations. Kruger and Van Rooy (2016b) is one of only a few studies that explicitly set out to empirically investigate translated English along with other contact-influenced forms of English, to determine how CLI plays out across different contact varieties and how different socio-cognitive factors promote and constrain the effects of CLI in varieties produced under different conditions of contact.

6.2.2 Other features of contact-influenced varieties: Increased explicitness, reduced complexity and conventionalization

While CLI effects are the most obvious consequence of communicative events occurring under conditions of language contact, a variety of other consequences have been identified, ascribed to various factors. The cognitive complexity of and effort involved in communicating under conditions of bilingual language activation may reduce working memory and increase the use of cognitive strategies to reduce cognitive load. Some of the proposed features of translated language, like increased explicitness of lexicogrammatical encoding, reduced lexicogrammatical complexity, and stylistic homogenization, have all been interpreted as collateral effects of the cognitive effort involved in translation, specifically—and are also evidenced across other forms of communication involving bilingual language activation.

For example, increased explicitness (also referred to as anti-deletion, hyperclarity or analyticity) is associated with translated language as well as various forms of bilingualism-influenced communication (such as L2 varieties of English) (see Hansen-Schirra et al., 2007; Kruger & De Sutter, 2018; Mesthrie, 2006; Szmrecsanyi & Kortmann, 2009). This increased explicitness is often explained by invoking findings that language processing contexts involving high cognitive demand (as is the case for language production under conditions of bilingual activation) are associated with the selection of more explicit and more analytical linguistic options (Mondorf, 2014; Rohdenburg, 1996). Following theories of processing efficiency (see Hawkins, 2003, 2004, 2014), explicit marking of dependency relationships of various kinds is regarded as a more efficient processing mechanism in contexts of high cognitive demand, since it reduces the processing effort of parsing syntactic and discourse relationships in linguistic production (see Kruger & De Sutter, 2018; Kruger & Van Rooy, 2016b).

However, increased explicitness, like reduced complexity and yet another familiar feature of translated language, conservatism or normalization, have also all been explained as a result of socio-cognitive factors: the translator's prioritization of the needs of the reader; her awareness of the "gap" between the source and the target text; and the risk-averse behaviour of translators. As opposed to the "pure" cognitive/psycholinguistic factors outlined earlier, these factors are socio-cognitive factors that are closely tied to two related concepts, those of cooperation and norms, which together form a script or a cognitive frame for translators' behaviour.

Pym (2015) argues that many of the features of translated language can be ascribed to translators' prioritization of principles of cooperation and their conscious or routinized avoidance of translation decisions and choices that increase the risk of miscommunication. Intersubjective coordination and cooperation is a key force that shapes language use and language structure (see Tomasello, 2008; Verhagen, 2005), and viewing translation as a special case of "talking to strangers" (Wray & Grace, 2007) across cultural and linguistic divides may well account for features like the avoidance of complexity and increased explicitness.

The concept of norms has been very influential in Translation Studies (see Schäffner, 1999; Toury, 2012); however, the cognitive aspect of norms has not been extensively theorized in this area of research. Linguistic norms are multidimensional. They have both an individual cognitive and a collective social dimension (Backus & Spotti, 2012; Harder, 2012). They can be both conscious (or explicit) and unconscious (or implicit) (Labov, 1972). They can be both overtly codified and sanctioned by top-down prescriptivist processes, and covertly emergent or bolstered by bottom-up processes (see Cameron, 1995; Curzan, 2014).

Backus and Spotti (2012) emphasize the relationship between the individual and social dimensions of a norm. They distinguish an individual internal norm, which refers to the way in which an individual habitually uses language. Where the overlaps in individuals' internal norms are considerable, it stimulates "their reification as a self-contained body of shared norms: a recognizable language or variety" (Backus & Spotti, 2012, p. 187). The aggregation of individual internal cognitive norms across groups of speakers leads to a cumulative internal norm, which is a view of internal norms as a social phenomenon (Backus & Spotti, 2012, p. 187). The relationship between individual and cumulative internal norms is complex and mutually reinforcing, driven by bottom-up and top-down accommodation processes, intersubjective alignment, frequency effects and social sanctioning. Norms are psychological constructs or socio-cognitive processing mechanisms, but they are operational in the sense that they "have the causal power to regulate community practices" (Harder, 2012, p. 297) while at the same time arising from and being shaped by such community practices. In addition to this "internal" view of norms, norms also, of course, have a codified, prescriptive dimension, which is "usually the result of explicit institutional agreement on how language either should or should not be used in a given linguistic interaction" (Backus & Spotti, 2012, p. 188). For many people, and particularly for some groups of people, such as professional translators, prescriptive guidelines on usage form a salient part of their internal norms.

In contexts of contact-influenced communication, cooperation and normativity may combine in unique ways to impose certain socio-cognitive constraints on language production, leading to the avoidance of complexity, increased explicitness, and hyperstandardization or conservatism, in order to ensure effective communication or to avoid normative sanction in conditions of communicative uncertainty. An important point is that features like increased explicitness, the avoidance of complexity and an over-adjustment to conventional norms do not just occur in translation but also in other varieties characterized by language contact. This argument is set out in some detail by Kruger and Van Rooy (2016a, 2018), who compare English translations and other varieties of English influenced by language contact in respect of the degree of explicitness, complexity and normativity. Their findings broadly find support for these shared effects of communication in contact situations, thus corroborating the argument set out in Lanstyák and Heltai (2012) that these features of translated language may well be seen as features of contact-influenced communication more generally, arising from a combination of psycholinguistic and socio-pragmatic factors.[4]

6.3 From the translated text to readers—and into new texts: Translation as a factor in language change?

As set out in Section 6.1, the translated text is constituted by two sets of cognitive processes: those of the people involved in the creation of the translated text, and those involved in receiving it. This latter group, of course, are also text producers, and the question is how their reception of contact-influenced features in translated texts influences their subsequent language production, if at all. This section explores this process, particularly to explain how the way in which people are exposed to translations may potentially play a role in processes of language variation and change more generally.

Numerous contributions in Translation Studies have used (diachronic) corpus methods to investigate the role of translation in language change (see, for example, Amouzadeh & House, 2010; Becher et al., 2009; Bisiada, 2013, 2016; Dai, 2016; House, 2011; Kolehmainen & Riionheimo, 2016; Kranich, 2014; Kranich et al., 2011; Kranich et al., 2012; Malamatidou, 2016, 2017; Neumann, 2011; Redelinghuys, 2019; Wurm, 2011). This research area has developed despite doubts about the role of translation in contact-induced language change. The reasons for these doubts are numerous. First, most language-contact studies assign primary importance to contact through direct, face-to-face conversational interaction between language users in close proximity (Kranich et al., 2011, pp. 11–12; Neumann, 2011, p. 236). In comparison, the role of written communication in contexts of language change (particularly through contact) has received much less attention, and major theoretical frameworks on language contact hardly reference translation (or, indeed, written communication) (Kranich, 2014). Also, the mechanisms and principles of translation-induced change are not well understood, and it is particularly difficult to distinguish translation effects from other types of contact-induced effects (Neumann, 2011). These concerns are at the root of the opinions of scholars like Hoey (2011, p. 164), who argues that "if we stay with accepted theories of language, we must conclude that [translation] is not responsible for more than a modicum of language change".

Despite these misgivings, several translation scholars have set out proposals for how translation may play a role in contact-induced language change. The following section briefly discusses some theoretical considerations outlined by Neumann (2011), Kranich (2014), Malamatidou (2016, 2017), and Redelinghuys (2019).

6.3.1 Theorizations of the role of translation in contact-induced language change

The discussion in Section 6.2 has clearly indicated that CLI occurs to a greater or lesser extent in translations, and that translations also may carry some other traces of language contact. However, it is more difficult to explain how it may come about that these contact-influenced features do not remain restricted to translated texts but might diffuse to monolingual text production. In other words, what requires explanation are the cognitive and social processes that would lead writers of original texts to adopt themselves the expressions they encounter in translation (Bisiada, 2013, p. 3) to such a degree that the linguistic conventions of the TL may be altered over time.

Earlier theorizations of LCTT focus on setting out the factors influencing LCTT, methodological requirements for testing the role of translation in language change, and hypotheses about language change through translation. For example, Neumann (2011) and Kranich (2014) set out some desiderata for theoretical and methodological approaches to language change through translation. Neumann (2011, p. 240) proposes that a methodology to test the hypothesis that translations play a role in language change should:

1. provide the means to determine a change in the properties of TL features (such as frequency) that could potentially be interpreted as adaptations to the properties of contrastively different SL features
2. allow the researcher to determine if evidence has been found for equivalent properties of the same features in translation into the TL, which mediates between properties of the SL and TL
3. be able to identify causal relationships between changes in the TL originals and the translations
4. assess and eliminate alternative explanations for changes in the TL originals.

Against the background of these stringent methodological demands, Kranich (2014, pp. 98–100) proposes ten hypotheses regarding LCTT, in line with what is generally known from studies of contact-induced linguistic change. Some of the hypotheses relate to the kind of borrowings associated with translation. She posits that while LCTT can involve all linguistic domains, lexical borrowing is more prominent than structural borrowing, and structural borrowing is restricted to syntactic borrowing. She also argues that structural borrowing relies on typological proximity, since it requires the establishment of functional analogies between the SL and the TL, which users can construe as equivalence relations. A further set of hypotheses relate to the kind of factors that influence the degree of CLI in translation and the uptake of CLI-influenced features by other users. In terms of the former, she highlights the importance of the degree of overtness of the translation strategy. In terms of the latter, she points out that the same factors that play a role in contact-induced change generally also play a role in LCTT (intensity and length of contact, socio-political dominance, prestige), but that LCTT influence is particularly strong in TL contexts where norms for written language and genre conventions are not well established.

While contributions like these connect well-known principles of contact-induced language change to translation, there is little systematic theorization of exactly how SL-influenced features propagate from translations to other texts. Malamatidou (2016) presents one possible theorization of language change through translation, adapting the Code-Copying Framework of Johanson (2002 and elsewhere) to translation as a case of language contact. In this model, CLI is referred to as "code-copying", and a distinction is made between global and selective code-copying. Global code-copying roughly corresponds to Matras's (2009) concept of matter replication, whereas selective code-copying corresponds to pattern replication (see Section 6.2.1). Various properties of the linguistic code can be copied, ranging across material, semantic, combinatorial and frequential properties (Malamatidou, 2016, p. 402).

The theory then attempts to explain how code-copying may diffuse from being single occurrences to being more widely used. Initially, a copy may be ephemeral (called a "momentary copy") but subsequently may become more widely used in a (bilingual) community, becoming a "habitualized copy". Once the copy becomes fully accepted in a speech community, it is a "conventionalized copy", and in a last step, it may start being used by monolingual speakers, at which point it is termed a "monolingual copy" (see Malamatidou, 2016, p. 404 for further discussion).

Malamatidou (2016) proceeds to apply these notions to translation, pointing out that in translation both kinds of copying may take place, although selective code-copying is more common, and frequential copying is particularly influential, which "results in a change in the frequency patterns of an existing lexical or morphosyntactic unit" (Malamatidou, 2016, p. 405). She suggests that these translation-influenced features may diffuse to non-translated texts—but still does not explain in detail how this might happen.

The gap in explaining how copies may "jump" the divide between translated texts and monolingual text production is a major shortcoming in existing theorizations of language change through translation, and is in part the consequence of limited consideration of the cognitive processes that are involved in language change through contact, and specifically translation. In response to this limitation, Redelinghuys (2019) outlines a broadly usage-based view (see Bybee, 2006, 2010) of language change through translation that integrates the cognitive and social dimensions of the contact-influenced translation event and the processes leading to the diffusion of translation-influenced forms to monolingual text production. Some principles and implications of a usage-based view of this kind are briefly elaborated in the remainder of this section.

There are two mechanisms that need accounting for in language change: the start of a linguistic change (innovation) and its spread through a linguistic community (propagation) (Brinton & Traugott, 2005). Innovation may refer to the creation of new linguistic forms or the development of new functions, but it may also refer to changes in distributional patterns and frequencies of existing forms, all of which generate variation in a linguistic system (Croft, 2000). Propagation refers to the diffusion of these innovative forms or usage patterns among individuals, groups and societies. Croft (2000) argues that innovation is individual and functional in nature because individuals produce innovative forms or unusual patterns in their attempts to bring about successful communication. This can happen in two ways: an individual may use an innovative form or pattern in a particular way on purpose, for some kind of communicative aim or effect; or the innovative form or pattern may be an "accidental" collateral effect of the general pursuit of communicative effectiveness (e.g. minimization of cognitive effort in communication; interpersonal alignment). Propagation, on the other hand, is social in nature because it arises from the structure of language communities.

Translations may first introduce innovative constructions, or changed frequency distributions for constructions, through overt or covert CLI effects. This may be done consciously, for strategic purposes, by translators, but is more likely to be an unconscious consequence of the bilingual language processing involved in translation. If a social environment exists in which a user is frequently exposed to contact-influenced constructions from translations (e.g. news translation; translation on the Internet; software localization, etc.), contact-influenced constructions may become increasingly cognitively entrenched in the linguistic repertoire of the user, either introducing a new linguistic construction or changing the strength of representation for competing constructions, making a construction that was previously less entrenched more strongly entrenched, and reducing the entrenchment of other, competing constructions. If a social environment exists in which enough users experience this kind of exposure, and these users subsequently interact with one another in spoken and written media, propagation and conventionalization of the contact-influenced construction are possible (see Backus, 2014). General processes of intersubjective alignment, coordination, cooperation and communication accommodation may drive propagation, interacting with frequency effects: As people accommodate their linguistic behaviour to that of other speakers, or to perceived norms for written text production, in the pursuit of communicative success, the frequency of particular forms or patterns of use increases. In this way, a feedback cycle of usage, frequency and entrenchment may be set in motion at both individual and social levels. These contact-influenced features or usage patterns may, in this way, be gradually integrated in the receiving language's repertoire or in a specific genre, register or domain (Kranich et al., 2011, p. 11).

Any potential for translation to play a role in language change is extremely tightly constrained by the social context of language contact. First, translation needs to be in widespread use. The kinds of innovative constructions or usage patterns cannot be normatively strongly proscribed, as this will short-circuit the feedback cycle (the same principles regarding norms discussed in Section 6.2.2 as applying to translators also apply to other language users). The social environment must be quite bi- or multilingual, with interactions between L1 and L2 users and interactions among L2 users too. Concomitant with this is the absence of language purism. The power relationships between the languages involved need to facilitate an openness to CLI effects.

Frequential changes under the influence of LCTT are by far the most common kind of change observed in studies of translation as a factor in language contact. Translation very infrequently introduces new structural patterns to a language—conventionalizing forces tend to resist such radical changes (see also Malamatidou, 2017). Language change associated with translation is thus most commonly associated with changes in communicative preferences in the TL under the influence of preferences from the SL, leading to stylistic convergence between the two languages, or register changes. These are the kinds of changes typically identified in existing studies of language contact through translation, briefly discussed in the following section.

6.3.2 Empirical studies of the role of translation in contact-induced language change

Kranich (2014) surveys a number of empirical studies of LCTT across various language pairs and across ancient, early modern and present-day contact situations. She finds that across all time periods, lexical items are introduced to the target language through translation. Derivational morphology likewise is affected across all periods, but inflectional morphology is not affected by contact-induced changes in early modern and present-day contact environments. Contact-influenced syntactic changes are common, but in modern-day contact settings, these are restricted to frequency effects (Kranich, 2014, p. 103), meaning that translations do not introduce new syntactic patterns into the TL but only contribute to a change in preferences for particular constructions under the influence of the SL. She highlights that this does not, strictly speaking, constitute syntactic change but, rather, reflects pragmatic or stylistic change—as also suggested earlier.

This is also the kind of change consistently found in one of the most extensive studies of LCTT to date, the Covert Translation project (1999–2011) (see, for example, Baumgarten & Őzçetin, 2009; Becher et al., 2009; House, 2006, 2011; Kranich, 2014; Kranich et al., 2011; Kranich et al., 2012). The project is based on the assumption that European languages, specifically German, are influenced to a greater or lesser extent by the omnipresence of Anglo-American linguistic and cultural norms (House, 2011, p. 189). The fundamental research question that guides the Covert Translation project is whether textual or cultural conventions of the target audience are disregarded in translation, with the result that source and target norms converge (Becher et al., 2009). The findings of the project are inconclusive, however, finding both convergence and divergence of communicative norms. Nevertheless, it does appear as though pragmatic features (like personal pronouns and modal markers) show CLI effects in English to German translations, particularly pronounced in more recent texts.

Beyond this project, others have extended the diachronic approach to investigate, for example, the role of translation in changes in preferences for hypotaxis versus parataxis, and the use of sentence-initial conjunctions in translated and non-translated German business articles (Bisiada, 2013, 2016); and the use of passive voice reporting verbs, and cleft- and pseudo-cleft constructions, in translated and non-translated Greek popular science articles (Malamatidou, 2016, 2018). Dai (2016) is an extensive study of the influence of English on Chinese, investigating the effects of translational "hybridity" or contact effects on changes in Chinese over a period of 70 years, focusing on a range of lexicogrammatical features. Redelinghuys (2019) investigates the role of translation in contact-influenced changes in South African English and Afrikaans over a period of a century, analysing the genitive alternation and modal auxiliary verbs. While she finds clear contact-related changes over time, she finds limited evidence that translation is responsible for either introducing these changes or propagating them. These studies investigate the relationship between language change, as evidenced in translation, and processes of change more generally evident in the target language, finding complex effects: translation may introduce CLI-influenced frequential changes that subsequently appear to disseminate to monolingual texts, but it may also exaggerate a change in progress (under the influence of more general contact effects), or it may hold it back, as a consequence of translators' normative awareness.

6.4 Concluding remarks: Methodological and theoretical challenges in investigating translation and language contact

Methodologically speaking, much of the research on contact effects arising from the individual translation event has relied on corpus methods. This method, while offering several advantages, also has limitations. A corpus allows researchers to statistically identify patterns based on "aggregate data that pools the productions of many speakers and writers" (Arppe et al., 2010, p. 3)—and translators. At the same time, it is also important not to lose sight of the effects of variation between individuals, and the mediated nature of the text: translations, like other written texts, are subject to editorial changes by a number of people other than the original writer or translator. Despite the fact that careful corpus designs and advanced multifactorial statistical methods can go some way towards separating out cognitive and social factors in language (and translation) production (see Kruger & De Sutter, 2018), a combination of corpus methods with experimental or quasi-experimental methods allowing us to understand language processing under more (or less) controlled conditions is essential in order to understand how communication under conditions of bilingual language activation affects language production in similar and different ways across different contact settings.

Current diachronic corpus-based research on the role of translation in contact-induced change remains inconclusive. All in all, it appears that under some (very limited) socio-linguistic conditions, in particular registers, the widespread use of translation may influence frequential patterns, leading to changes in pragmatic, stylistic or register preferences. Translations may also introduce lexical forms to a language. A major methodological challenge to this kind of research is to separate the effects of language contact through translation and language contact more generally (Neumann, 2011). Translation is, particularly in a contemporary globalized world, one among many gateways of language contact, and disentangling the effects of different sources of contact may well prove an intractable problem.

Other limitations of existing research include the relatively short timespans investigated (most studies of the modern period focus on spans of no more than 30 years), the focus on single registers, and limitations in the contact situations investigated (see Dai, 2016; Redelinghuys, 2019 for exceptions). More studies focusing on multiple registers, more diverse contact situations, and longer timeframes are needed to more definitively investigate the role of translation in contact-induced change.

Moving beyond corpora, a further crucial extension of this research is the application of process-type or experimental methods to understand both the production of the translated text as a contact-influenced event and the propagation of contact-influenced linguistic features from translations to monolingual text production. Taking the methods of iterative learning and agent-based modelling of language interaction, competition, change and evolution (e.g. Steels, 2011) into this area of research constitutes a frontier for studies of LCTT (see Fernández et al., 2017 for experimental evidence of the psycholinguistics of language change). The combination of corpus-based and experimental work will further assist in disentangling the complex web of social and cognitive factors that influence translation as a type of language contact.

Acknowledgements

This chapter draws on conversations and collaborative work with Karien Redelinghuys and Bertus van Rooy over a period of five years. Their contributions to the development of the arguments presented here are gratefully acknowledged.

Notes

1 As represented in, for example, wide-ranging work by Johanson (2002), Heine and Kuteva (2005), Thomason and Kaufmann (1988), Thomason (2001), Van Coetsem (2000), and Weinreich (1953).
2 In contact linguistics, a range of terms are used for the two languages involved in the contact situation. For ease of understanding, and to ensure consistency, this chapter avoids using these terms, and instead uses terms familiar to readers from Translation Studies: source language (SL) and target language (TL).
3 Bilingual language activation or discourse production is not the only constraint raised, however; the constraints introduced by reproducing or relaying an existing message are also often cited in this work.
4 It should be pointed out, however, that there are also differences between various contact-influenced varieties (e.g. L2 varieties) and translation, which can be ascribed to different conditions of contact, or different communication situations. For more detail, see Kruger and Van Rooy (2016a).

Further reading

Backus, A. (2014). Towards a usage-based account of language change: Implications of contact linguistics for linguistic theory. In R. Nicolai (Ed.), *Questioning language contact: Limits of contact, contact at its limits* (pp. 91–118). Leiden: Brill.
A theoretical proposal for the integration of usage-based theories with theories of language change, with a particular focus on language contact.

Kranich, S. (2014). Translations as a locus of language contact. In J. House (Ed.), *Translation: A multidisciplinary approach* (pp. 96–109). New York: Palgrave Macmillan..
An overview of theoretical and methodological issues in viewing translations as a possible factor in language change, and an outline of some key findings.

Kranich, S., Becher, V., & Höder, S. (2011). A tentative typology of translation-induced language change. In S. Kranich, V. Becher, S. Höder, & J. House (Eds.), *Multilingual discourse production: Diachronic and synchronic perspectives* (pp. 11–44). Amsterdam: John Benjamins.
An outline of relevant factors affecting translation-induced language change.

Kruger, H., & Van Rooy, B. (2016). Syntactic and pragmatic transfer effects in reported-speech constructions in three contact varieties of English influenced by Afrikaans. *Language Sciences, 56*, 118–131.
A case study of cross-linguistic influence in different contact varieties, proposing a set of constraints that play a role in different forms of language contact, including translation.

Malamatidou, S. (2016). Understanding translation as a site of language contact. *Target, 28*(3), 399–423.
A theoretical proposal for viewing translation as a form of language contact, accompanied by an empirical study.

Muysken, P. (2013). Language contact outcomes as the result of bilingual optimization strategies. *Bilingualism: Language and Cognition, 16*(4), 709–730.
A proposal for integrating the psycholinguistics of bilingualism with language-contact outcomes.

Neumann, S. (2011). Assessing the impact of translations on English-German language contact: Some methodological considerations. In S. Kranich, V. Becher, S. Höder, & J. House (Eds.), *Multilingual discourse production: Diachronic and synchronic perspectives* (pp. 233–256). Amsterdam: John Benjamins.
An outline of methodological considerations for studying language contact and change through translation.

References

Amouzadeh, M., & House, J. (2010). Translation as a language contact phenomenon: The case of English and Persian passives. *Languages in Contrast, 10*(1), 54–75.
Arppe, A., Gilquin, G., Glynn, D., Hilpert, M., & Zeschel, A. (2010). Cognitive corpus linguistics: Five points of debate on current theory and methodology. *Corpora, 5*(1), 1–27.
Backus, A. (2014). Towards a usage-based account of language change: Implications of contact linguistics for linguistic theory. In R. Nicolai (Ed.), *Questioning language contact: Limits of contact, contact at its limits* (pp. 91–118). Leiden: Brill.
Backus, A., Doğruöz, A. S., & Heine, B. (2011). Salient stages in contact-induced grammatical change: Evidence from synchronic vs. diachronic contact situations. *Language Sciences, 33*, 738–752.
Backus, A., & Spotti, M. (2012). Normativity and change: Introduction to the special issue on agency and power in multilingual discourse. *Sociolinguistic Studies, 6*(2), 185–212.
Backus, A., & Verschik, A. (2012). Copyability of (bound) morphology. In J. Johanson, & M. Robbeets (Eds.), *Copies versus cognates in bound morphology* (pp. 123–149). Leiden: Brill.
Bangalore, S., Behrens, B., Carl, M., Ghankot, M., Heilmann, A., Nitzke, J., Schaeffer, M., & Sturm, A. (2016). Syntactic variance and priming effects in translation. In M. Carl, S. Bangalore, & M. Schaeffer (Eds.), *New directions in empirical translation process research: Exploring the CRITTTPR-DB* (pp. 211–238). Cham: Springer.
Baumgarten, N., & Özçetin, D. (2008). Linguistic variation through language contact in translation. In P. Siemund, & N. Kintana (Eds.), *Language contact and contact languages* (pp. 293–316). Amsterdam: John Benjamins.
Becher, V., House, J., & Kranich, S. (2009). Convergence and divergence of communicative norms through language contact in translation. In K. Braunmüller, & J. House (Eds.), *Convergence and divergence in language contact situations* (pp. 125–151). Amsterdam: John Benjamins.
Bisiada, M. (2013). Changing conventions in German causal clause complexes: A diachronic corpus study of translated and non-translated business articles. *Languages in Contrast, 13*(1), 1–27.
Bisiada, M. (2016). Structural effects of English-German language contact in translation on concessive constructions in business articles. *Text & Talk, 36*(2), 133–154.
Bisiada, M. (2017). Universals of editing and translation. In S. Hansen-Schirra, O. Czulo, & S. Hofmann (Eds.), *Empirical modelling of translation and interpreting* (pp. 241–275). Berlin: Language Science Press.
Brinton, L. J., & Traugott, E. C. (2005). *Lexicalization and language change*. Cambridge: Cambridge University Press.
Bybee, J. (2006). From usage to grammar: The mind's response to repetition. *Language, 82*, 711–733.
Bybee, J. (2010). *Language, usage and cognition*. Cambridge: Cambridge University Press.
Cameron, D. (1995). *Verbal hygiene*. London: Routledge.
Cappelle, B. (2012). English is less rich in manner-of-motion verbs when translated from French. *Across Languages and Cultures, 13*(2), 173–195.
Cappelle, B., & Loock, R. (2013). Is there interference of usage constraints? A frequency study of existential *there is* and its French equivalent *il y a* in translated vs. non-translated texts. *Target: International Journal of Translation Studies, 25*(2), 252–275.
Cappelle, B., & Loock, R. (2017). Typological differences shining through: The case of phrasal verbs in translated English. In G. de Sutter, M.-A. Lefer, & I. Delaere (Eds.), *Empirical translation studies: New methodological and theoretical traditions* (pp. 235–263). Berlin: De Gruyter.
Carl, M., & Dragsted, B. (2012). Inside the monitor model: Processes of default and challenged translation production. *TC3: Translation: Computation, Corpora, Cognition, 2*(1), 127–145.
Chesterman, A. (2004). Hypotheses about translation universals. In G. Hansen, K. Malmkjær, & D. Gile (Eds.), *Claims, changes and challenges in translation studies: Selected contributions from the EST Congress, Copenhagen, 2001* (pp. 1–13). Amsterdam: John Benjamins.

Costa, A., & Sebastián-Gallés, N. (2014). How does the bilingual experience sculpt the brain? *Nature Reviews: Neuroscience, 15*(5), 336–345.
Croft, W. (2000). *Explaining language change: An evolutionary approach.* London: Longman.
Curzan, A. (2014). *Fixing English: Prescriptivism and language history.* Cambridge: University Press.
Dai, G. (2016). *Hybridity in translated Chinese: A corpus analytical framework.* Singapore: Springer.
Dai, G., & Xiao, R. (2011). SL "shining through" in translational language: A corpus-based study of Chinese translation of English passives. *Translation Quarterly, 62*, 85–108.
Diamond, B. J., & Shreve, G. M. (2010). Neural and physiological correlates of translation and interpreting in the bilingual brain: Recent perspectives. In G. M. Shreve, & E. Angelone (Eds.), *Translation and cognition* (pp. 289–322). Amsterdam: John Benjamins.
Fernández, E. M., De Souza, R. A., & Carando, A. (2017). Bilingual innovations: Experimental evidence offers clues regarding the psycholinguistics of language change. *Bilingualism: Language and Cognition,* 20, Special Issue 2 (Cross-linguistic priming in bilinguals: Multidisciplinary perspectives on language processing, acquisition and change), 251–268.
Fischer, O. (2013). The role of contact in English syntactic change in the Old and Middle English periods. In D. Schreier, & M. Hundt (Eds.), *English as a contact language* (pp. 18–40). Cambridge: University Press.
Gaspari, F., & Bernardini, S. (2010). Comparing non-native and translated language: Monolingual comparable corpora with a twist. In R. Xiao (Ed.), *Using corpora in contrastive and translation studies* (pp. 215–234). Newcastle: Cambridge Scholars Publishing.
Granger, S. (2018). Tracking the third code: A cross-linguistic corpus-driven approach to metadiscursive markers. In A. Čermáková, & M. Mahlberg (Eds.), *The corpus linguistics discourse: In honour of Wolfgang Teubert* (pp. 185–204). Amsterdam: John Benjamins.
Gries, S. Th., & Kootstra, G. J. (2017). Structural priming within and across languages: A corpus-based perspective. *Bilingualism: Language and Cognition,* 20, Special Issue 2 (Cross-linguistic priming in bilinguals: Multidisciplinary perspectives on language processing, acquisition and change), 235–250.
Grosjean, F. (2008). *Studying bilinguals.* Oxford: Oxford University Press.
Hansen-Schirra, S. (2011). Between normalization and shining-through: Specific properties of English-German translations and their influence on the target language. In S. Kranich, V. Becher, S. Höder, & J. House (Eds.), *Multilingual discourse production: Diachronic and synchronic perspectives* (pp. 135–162). Amsterdam: John Benjamins.
Hansen-Schirra, S., Neumann, S., & Steiner, E. (2007). Cohesive explicitness and explicitation in an English–German translation corpus. *Languages in Contrast,* 7, 241–265.
Harder, P. (2012). Variation, structure and norms. *Review of Cognitive Linguistics, 10*(2), 294–314.
Hartsuiker, R. J., Pickering, M. J., & Veltkamp, E. (2004). Is syntax separate or shared between languages? Cross-linguistic syntactic priming in Spanish-English bilinguals. *Psychological Science, 15*(6), 409–414.
Haspelmath, M. (2009). Lexical borrowing: Concepts and issues. In M. Haspelmath, & U. Tadmor (Eds.), *Loanwords in the world's languages: A comparative handbook* (pp. 35–54). Berlin: De Gruyter Mouton.
Hawkins, J. (2003). Why are zero-marked phrases close to their heads? In G. Rohdenburg, & B. Mondorf (Eds.), *Determinants of grammatical variation in English* (pp. 175–204). Berlin: Mouton De Gruyter.
Hawkins, J. (2004). *Efficiency and complexity in grammars.* Oxford: University Press.
Hawkins, J. (2014). *Cross-linguistic variation and efficiency.* Oxford: University Press.
Heine, B., & Kuteva, T. (2005). *Language contact and grammatical change.* New York: Cambridge University Press.
Hoey, M. (2011). Lexical priming and translation. In A. Kruger, K. Wallmach, & J. Munday (Eds.), *Corpus-based translation studies: Research and applications* (pp. 153–168). London: Continuum.
House, J. (2006). Covert translation, language contact, variation and change. *SYNAPS,* 25–47.
House, J. (2011). Using translation and parallel text corpora to investigate the influence of global English on textual norms in other languages. In A. Kruger, K. Wallmach, & J. Munday (Eds.), *Corpus-based translation studies* (pp. 189–208). New York: Continuum.
Jarvis, S., & Pavlenko, A. (2008). *Crosslinguistic influence in language and cognition.* London: Routledge.
Johanson, L. (2002). *Structural factors in Turkic language contacts.* London: Curzon.
Kolehmainen, L., Meriläinen, L., & Riionheimo, H. (2014). Interlingual reduction: Evidence from language contacts, translation and second-language acquisition. In H. Paulasto, L. Meriläinen, H. Riionheimo, & M. Kok (Eds.), *Language contacts at the crossroads of disciplines* (pp. 3–32). Newcastle: Cambridge Scholars Publishing.
Kolehmainen, L., & Riionheimo, H. (2016). Literary translation as language contact: A pilot study on the Finnish passive. *Literary Linguistics, 5*(3), 1–32.

Kootstra, G. J., & Muysken, P. (2017). Cross-linguistic priming in bilinguals: Multidisciplinary perspectives on language processing, acquisition, and change. *Bilingualism: Language and Cognition,* 20, Special Issue 2 (Cross-linguistic priming in bilinguals: Multidisciplinary perspectives on language processing, acquisition and change), 215–218.
Kranich, S. (2014). Translations as a locus of language contact. In J. House (Ed.), *Translation: A multidisciplinary approach* (pp. 96–109). New York: Palgrave Macmillan.
Kranich, S., Becher, V., & Höder, S. (2011). A tentative typology of translation-induced language change. In S. Kranich, V. Becher, S. Höder, & J. House (Eds.), *Multilingual discourse production: Diachronic and synchronic perspectives* (pp. 11–44). Amsterdam: John Benjamins.
Kranich, S., House, J., & Becher, V. (2012). Changing conventions in English-German translations of popular scientific texts. In K. Braunmüller, & C. Gabriel (Eds.), *Multilingual individuals and multilingual societies* (pp. 315–334). Amsterdam: John Benjamins.
Kroll, J. F., & Tokowicz, N. (2005). Models of bilingual representation and processing: Looking back and to the future. In J. F. Kroll, & A. M. B. de Groot (Eds.), *Handbook of bilingualism: Psycholinguistic approaches* (pp. 531–553). New York: Oxford University Press.
Kruger, H. (2012). A corpus-based study of the mediation effect in translated and edited language. *Target, 24*(2), 355–388.
Kruger, H., & De Sutter, G. (2018). Alternations in contact and non-contact varieties: Reconceptualising *that*-omission in translated and non-translated English using the MuPDAR approach. *Translation, Cognition and Behavior, 1*(2), 251–290.
Kruger, H., & Kruger, J. L. (2017). Cognition and reception. In J. Schwieter, & A. Ferreira (Eds.), *The handbook of translation and cognition* (pp. 71–89). London: Wiley-Blackwell.
Kruger, H., & Van Rooy, B. (2016a). Constrained language: A multidimensional analysis of translated English and a non-native indigenised variety of English. *English World-Wide, 37*(1), 26–57.
Kruger, H., & Van Rooy, B. (2016b). Syntactic and pragmatic transfer effects in reported-speech constructions in three contact varieties of English influenced by Afrikaans. *Language Sciences, 56*, 118–131.
Kruger, H., & Van Rooy, B. (2018). Register variation in written contact varieties of English: A multidimensional analysis. *English World-Wide, 39*(2), 214–242.
Labov, W. (1972). *Sociolinguistic patterns*. Philadelphia: University of Pennsylvania Press.
Lanstyák, I., & Heltai, P. (2012). Universals in language contact and translation. *Across Languages and Cultures, 13*(1), 99–121.
Lefer, M.-A., & Vogeleer, S. (Eds.). (2013). Interference and normalisation in genre-controlled multilingual corpora. *Belgian Journal of Linguistics,* 27.
Loebell, H., & Bock, K. (2003). Structural priming across languages. *Linguistics, 41*, 791–824.
Maier, R. M., Pickering, M. J., & Hartsuiker, R. J. (2017). Does translation involve structural priming? *The Quarterly Journal of Experimental Psychology, 70*(8), 1575–1589.
Malamatidou, S. (2016). Understanding translation as a site of language contact. *Target, 28*(3), 399–423.
Malamatidou, S. (2017). Why changes go unnoticed: The role of adaptation in translation-induced linguistic change. *Lingua, 200*, 22–32.
Malamatidou, S. (2018). *Corpus triangulation: Combining data and methods in corpus-based translation studies*. London: Routledge.
Matras, Y. (2009). *Language contact*. Cambridge: University Press.
Matras, Y. (2012). An activity-oriented approach to contact-induced language change. In I. Leglise, & C. Chamoreau (Eds.), *Dynamics of contact-induced change* (pp. 1–28). Berlin: Mouton de Gruyter.
Matras, Y., & Sakel, Y. (2007). Investigating the mechanisms of pattern replication in language convergence. *Studies in Language, 31*(4), 829–865.
Mauranen, A. (2004). Corpora, universals and interference. In A. Mauranen, & P. Kujamäki (Eds.), *Translation universals: Do they exist?* (pp. 143–164). Amsterdam: John Benjamins.
Mesthrie, R. (2006). Anti-deletions in an L2 grammar: A study of Black South African English mesolect. *English World-Wide, 27*, 111–145.
Mondorf, B. (2014). (Apparently) competing motivations in morpho-syntactic variation. In B. MacWhinney, A. Malchukov, & E. Moravcsik (Eds.), *Competing motivations in grammar and usage* (pp. 209–228). Oxford: University Press.
Mougeon, R., Nadasdi, T., & Rehner, K. (2005). Contact-induced linguistic innovations on the continuum of language use: The case of French in Ontario. *Bilingualism: Language and Cognition, 8*, 99–115.
Muysken, P. (2013). Language contact outcomes as the result of bilingual optimization strategies. *Bilingualism: Language and Cognition, 16*(4), 709–730.

Neumann, S. (2011). Assessing the impact of translations on English-German language contact: Some methodological considerations. In S. Kranich, V. Becher, S. Höder, & J. House (Eds.), *Multilingual discourse production: Diachronic and synchronic perspectives* (pp. 233–256). Amsterdam: John Benjamins.
Neumann, S. (2013). *Contrastive register variation: A quantitative approach to the comparison of English and German*. Berlin: de Gruyter.
Neumann, S. (2014). Cross-linguistic register studies: Theoretical and methodological considerations. *Languages in Contrast, 14*, 35–57.
Ožbot, M. (2014). The case for a common framework for transfer-related phenomena in the study of translation and language contact. In H. Paulasto, L. Meriläinen, H. Riionheimo, & M. Kok (Eds.), *Language contacts at the crossroads of disciplines* (pp. 131–160). Newcastle: Cambridge Scholars Publishing.
Paradis, M. (1993). Linguistic, psycholinguistic, and neurolinguistic aspects of "interference" in bilingual speakers: The Activation Threshold Hypothesis. *International Journal of Psycholinguistics, 9*, 133–145.
Pickering, M. J., & Ferreira, V. S. (2008). Structural priming: A critical review. *Psychological Bulletin, 134*(3), 427–459.
Pym, A. (2015). Translating as risk management. *Journal of Pragmatics, 85*, 67–80.
Redelinghuys, K. (2019). Language contact and change through translation in Afrikaans and South African English: A diachronic corpus-based study [unpublished PhD thesis]. Macquarie University / North-West University.
Rohdenburg, G. (1996). Cognitive complexity and increased grammatical explicitness in English. *Cognitive Linguistics, 7*, 149–182.
Sakel, J. (2007). Types of loan: Matter and pattern. In Y. Matras, & J. Sakel (Eds.), *Grammatical borrowing in cross-linguistic perspective* (pp. 15–30). Berlin: Mouton de Gruyter.
Schaeffer, M., & Carl, M. (2013). Shared representations and the translation process: A recursive model. *Translation and Interpreting Studies, 8*(2), 169–190.
Schäffner, C. (Ed.). (1999). *Translation and norms*. Clevedon: Multilingual Matters.
Shlesinger, M., & Ordan, N. (2012). More spoken or more translated? Exploring a known unknown of simultaneous interpreting. *Target, 24*(1), 43–60.
Shreve, G. M., & Lacruz, I. (2017). Aspects of a cognitive model of translation. In J. W. Schwieter, & A. Ferreira (Eds.), *The handbook of translation and cognition* (pp. 127–142). Hoboken: John Wiley.
Steels, L. (2011). Modeling the cultural evolution of language. *Physics of Life Reviews, 8*, 330–356.
Steiner, E. (2008). Empirical studies of translations as a mode of language contact: "Explicitness" of lexicogrammatical encoding as a relevant dimension. In P. Siemund, & N. Kintana (Eds.), *Language contact and contact languages* (pp. 317–346). Amsterdam: John Benjamins.
Szmrecsanyi, B., & Kortmann, B. (2009). Vernacular universals and Angloversals in a typological perspective. In M. Filppula, J. Klemola, & H. Paulasto (Eds.), *Vernacular universals and language contact: Evidence from varieties of English and beyond* (pp. 33–53). New York: Routledge.
Teich, E. (2003). *Cross-linguistic variation in system and text: A methodology for the investigation of translations and comparable texts*. Berlin: Mouton de Gruyter.
Thomason, S. G. (2001). *Language contact: An introduction*. Edinburgh: University Press.
Thomason, S. G., & Kaufman, T. (1988). *Language contact, creolization, and genetic linguistics*. Berkeley: University of California Press.
Tirkkonen-Condit, S. (2004). Unique items—over- or under-represented in translated language? In A. Mauranen, & P. Kujamäki (Eds.), *Translation universals: Do they exist?* (pp. 176–184). Amsterdam: John Benjamins.
Tirkkonen-Condit, S. (2005). The monitor model revisited: Evidence from process research. *Meta, 50*(2), 405–414.
Tomasello, M. (2008). *Origins of human communication*. Cambridge: MIT Press.
Toury, G. (1995). *Descriptive translation studies—and beyond*. Amsterdam: John Benjamins.
Toury, G. (2012). *Descriptive translation studies—and beyond* (revised ed.). Amsterdam: John Benjamins.
Van Coetsem, F. (2000). *A general and unified theory of the transmission process in language contact*. Heidelberg: Winter.
Van Oost, A., Willems, A., & De Sutter, G. (2016). Asymmetric syntactic patterns in German-Dutch translation: A corpus-based study of the interaction between normalisation and shining through. *International Journal of Translation, 28*(1–2), 7–25.
Verhagen, A. (2005). *Constructions of intersubjectivity: Discourse, syntax, and cognition*. Oxford: University Press.
Weinreich, U. (1953). *Languages in contact: Findings and problems*. The Hague: Mouton.

Wray, A., & Grace, G. W. (2007). The consequences of talking to strangers: Evolutionary corollaries of socio-cultural influences on linguistic form. *Lingua, 117*, 543–578.

Wurm, A. (2011). Translation-induced formulations of directives in Early Modern German cookbooks: An example of a translational effect. In S. Kranich, V. Becher, S. Höder, & J. House (Eds.), *Multilingual discourse production: Diachronic and synchronic perspectives* (pp. 87–108). Amsterdam: John Benjamins.

7

Translation, pragmatics and cognition

Fabio Alves

7.1 Introduction

In a very broad sense, pragmatics is the study of language use. Morris (1938) is widely acknowledged as the first author to provide a tentative definition of the goals and scope of pragmatics as "the science of the relationship of signs to their interpreters" (Morris, 1938, p. 30). Morris developed a typology of syntax, semantics and pragmatics within a general science of signs (semiotics) in the context of 1930s scholarship on philosophy of mind (see Tipton, 2019, p. 2) and focused on "the relation of signs to those who use the signs" (Mey, 2006, p. 51). In recent publications, both Spencer-Oatey and Žegarac (2010) and Tipton (2019) make such an acknowledgement to Morris (1938) in volumes dedicated to the interface between pragmatics and applied linguistics and between pragmatics and Translation Studies, respectively. However, a consensus about a unified definition of pragmatics seems hard to find. Spencer-Oatey and Žegarac (2010, p. 70) caution that "none of the many pragmatic theories and frameworks comes close to being a generally accepted paradigm, and in fact, there is no consensus as to the domain of pragmatics".

As Tipton (2019) puts it, historically, pragmatics has been dominated by an Anglo-American tradition, which is cognitive-philosophical oriented, and a Continental European tradition, which is socio-cultural-interactive oriented. The former builds on phenomena such as indexicality/deixis, speech acts, metaphor, implicit meaning, presuppositions, politeness and conversation analysis, whereas the latter draws on a broader perspective for studying language in general.

Focusing on the Anglo-American tradition, the first major breakthrough in this tradition can perhaps be traced back to the distinction between the linguistic form of the utterance and its communicative function (*illocutionary force*), as suggested in the works of Austin and Searle and their views that language should be seen as a form of action. Austin (1962) suggested that people use language to perform actions (such as requesting information, promising, offering, betting, etc.) that have an impact in some way on the world. He, and later Searle (1969), called these *speech acts* and tried to classify these actions into different categories of *felicity conditions* that enable *speech acts* to be performed successfully by language users.

Grice's (1957, 1975, 1989) theory of communication is built on the distinction between *what is said* (explicitly communicated) and *what is implicated* (implicitly communicated). For Grice, *speaker-meaning* is a way of expressing and recognizing intentions. Sperber and Wilson (1986/1995) point out that Grice (1975), in his William James Lectures, conveyed the fundamental idea that "once a certain piece of behaviour is identified as communicative, it is reasonable to assume that the communicator is trying to meet certain general standards" (Sperber & Wilson, 1986/1995, p. 33). This view led Grice to formulate a general principle for human communication: "Make your conversational contribution such as is required, at the stage at which it occurs, by the accepted purpose or direction of the talk exchange in which you are engaged" (Grice, 1975, p. 45). Grice called it the *co-operative principle*, which he then developed into nine maxims classified into four categories (maxims of quantity, quality, relation and manner) with the purpose of explaining how utterances convey not just explicit but also implicit meaning. To achieve this goal, Grice also coined the term *implicature* to provide an account of how implicit meaning is processed and generated.

Grice's ideas have been quite influential in the field of pragmatics. According to Sperber and Wilson (1986/1995, p. 38), his basic idea "offers a way of developing the analysis of inferential communication, as suggested by Grice himself in 'Meaning' (1957), into an explanatory model". Grice's views have been challenged (by Carston, 2002; Levinson, 2000; Sperber & Wilson, 1986/1995, among others) but have nevertheless remained influential.

Brown and Levinson's (1978, 1987) face model of politeness is also considered one of the influential models in the Anglo-American pragmatics tradition; it tries to explain the impact of social factors on people's use of language as "the public self-image that every member wants to claim for himself" (Brown & Levinson, 1987, p. 61).

Other relevant works within the Anglo-American tradition deal with cross-cultural pragmatics and the semantics of human interaction (Wierzbicka, 1991), pragmatic meaning and cognition (Marmaridou, 2000), and a specific cognitive approach to pragmatics with a focus on relevance, as shown in the works of Sperber and Wilson (1986/1995), Blakemore (1992) and Carston (2002), among others. This cognitive approach to pragmatics has been expanded to incorporate what is known as experimental pragmatics (Noveck & Sperber, 2004).

All these authors have provided input for establishing the goals and the scope of pragmatics within the Anglo-American tradition as seen in the comprehensive volumes published by Leech (1983), Davis (1991), Levinson (1993), Mey (1993), Yule (1996) and Verschueren (1999), among others.

In this chapter, we constrain the scope of pragmatics applied to translation within the Anglo-American tradition, and, while using Morris's (1938) definition as a sailing reference, we concentrate on translation-specific phenomena for our considerations applied to translation in general and to the translation–cognition interface in particular.

In Translation Studies, references to the impact of pragmatics on translation can be found in Hickey's (1998) edited volume, which, building on the Anglo-American tradition, focuses on topics such as speech acts and illocutionary function in translation, politeness, presuppositions, deictic features, discourse connectives, ellipsis, markedness and perlocutionary equivalence, expecting to shed light on "how pragmatics relate to translation and, in particular, [...] how pragmatic equivalence may be achieved in the process and product of the translator's art" (Hickey, 1998, p. 8).

Another trend in Translation Studies, as Tipton (2019) points out, has built on the works of Halliday (1978) and Halliday and Hasan (1985) on systemic functional linguistics. It has been particularly influential in developing communicative approaches to translation (see Baker, 1992, 1993; Hatim & Mason 1990, 1997; House, 1981, 1997, 2006, 2013, among others). However, these works build on a semantics-oriented semiotic theory of language (systemic functional linguistics), which does not include pragmatics as a specific theoretic component but deals with it as semiotic-semantic-related issues.

In Translation Studies, the German functionalist approach (Reiß & Vermeer, 1984) may be considered an original attempt to look at pragmatics-related phenomena. However, the relationship between pragmatics, translation and cognition per se was first explored by Gutt (1991/2000) and later expanded by Setton (1999), Doherty (2002), Alves (1995, 2007, 2009) and Gallai (2019), among others, leading to what Setton and Dawrant (2016, p. 473) call *cognitive pragmatics*, based on the fact that language is not a logical product but originates from the conventional practice of individuals, which depends on the particular context of the terms used by them.

In this chapter, we follow Gallai's (2019, p. 51) assertion that "even though other cognitive-pragmatic theories have been developed in the last three decades, relevance theory (Sperber & Wilson, 1986/1995) can be considered as the main theoretical framework in the area of cognitive pragmatics (Huang, 2007; Schmid, 2012) as well as the only cognitive pragmatic approach within translation studies". Thus, for the remaining sections of this chapter, we will build on the relevance-theoretic framework and on Gutt's seminal application of relevance theory to Translation Studies to ground our thinking.

7.2 Core topics

This section presents a relevance-theoretic approach to pragmatics in general and in its application to translation in particular. It starts by revisiting traditional axioms and looking at the relevance-comprehension heuristics and the notion of cognitive environment applied to translation. Next, considering translation as an act of interlingual interpretive language use, it explores the concept of interpretive resemblance (Gutt, 1991/2000) and looks at processing effort and cognitive effects in translation (Alves, 2007). It also deals with instances of metarepresentation and higher-order representations in translation (Gutt, 2004, 2005), the semantics–pragmatics interface in translation (Carston, 2002; Doherty, 2002) and issues related to explicitation and explicitness in translation (Alves, 2009; Englund-Dimitrova, 2005; Hansen-Schirra et al., 2007). Finally, it looks into experimental pragmatics (Noveck & Sperber, 2004) as a way to test hypotheses and provide empirical evidence to account for how translation functions as a cognitive activity.

7.2.1 Revisiting traditional axioms: The relevance-comprehension heuristics for translation

Sperber and Wilson (1986/1995) introduced relevance theory as an attempt to bring together Shannon and Weaver's (1949) code model and Grice's (1975) inferential model. The question they asked was "Should the code model and the inferential model be amalgamated?" (Sperber & Wilson, 1986/1995, p. 24). They claim that "there are at least two different modes of communication: the coding-decoding mode and the inferential mode" and add that "complex forms of communication can combine both modes" (Sperber & Wilson, 1986/1995, p. 27).

According to relevance theory, both communicator and audience must believe that the information effectively communicated is relevant enough to be worth the processing effort required to understand it. For Sperber & Wilson (1986/1995), every act of communication creates an expectation of relevance. This is achieved through two complementary forms of behaviour (ostensive and inferential), jointly generating what is called ostensive-inferential behaviour. On the one hand, the communicator is responsible for making ostensively manifest what he/she wants to communicate. On the other hand, the hearer must be prepared to dedicate the necessary effort to process ostensive stimuli inferentially. Ostensive-inferential comprehension requires mutual manifestness of a certain level for both communicator and audience. It typically involves several layers of metarepresentation. Relevance theory grounds these assumptions by formulating two general claims or "principles" about the role of relevance in cognition and in communication:

> *Cognitive principle of relevance.* Human cognition tends to be geared to the maximization of relevance.
> *Communicative principle of relevance.* Every act of communication conveys a presumption of its own optimal relevance.

Based on these two principles, Sperber & Wilson (1986/1995) assume that the presumption of optimal relevance conveyed by every utterance is strong enough to generate a specific relevance-oriented comprehension heuristic in which there is a supposedly positive relation between the effort needed to process the ostensive stimuli conveyed by the communicator and the cognitive effects generated in this process. There is no need for further effort if there are no extra gains, namely new cognitive effects. Relevance is thus defined as "a property of inputs to cognitive processes and analysed in terms of the notions of cognitive effect and processing effort" (Wilson, 2000, p. 423).

Gutt (1991/2000) builds on relevance-theoretic assumptions to apply them to the study of translation. He draws on the notions of cognitive environment and mutual manifestness to consider translation as a case of interlingual interpretive language use. It results from the relationship between processing effort and cognitive effects, mediated by higher-order representations (metarepresentations). We shall look at these topics in the following sections.

7.2.2 Rethinking context: The notions of cognitive environment and mutual manifestness

Work within social pragmatics has sometimes required the introduction of additional communicative norms to account for the unfolding of communicative processes (Spencer-Oatey & Žegarac, 2010, p. 76). Relevance theory sees it under a different light and assumes that the social factors that influence communication processes are best analysed as instances of a given context; namely, a set of assumptions which participants use in producing and interpreting acts of communication.

Sperber and Wilson (1986/1995) insist that the relevance-theoretic notion of context is different from the standard notion of the term. In relevance theory, context is defined as a mental construct, and it should be seen as effectively contributing to cognitive processes. Contextual input includes external stimuli, which can be perceived and attended to, and mental representations, which can be stored, recalled or used as premises in inference. For Sperber and Wilson, communication processes give rise to shared information, which implies that "some sharing of information is necessary if communication is to be achieved" (Sperber & Wilson, 1986/1995, p. 38). Because people speak different languages, entertain different representations of the world and vary in their perceptual abilities, a mental-oriented definition of context should entail those differentiations, which are defined as follows:

> A fact is *manifest* to an individual at a given time if and only if he is capable at that time of representing it mentally and accepting its representation as true or probably true.
> A *cognitive environment* of an individual is a set of facts that are manifest to him.
>
> *Sperber & Wilson, 1986/1995, p. 39*

It follows that a set of facts must be mutually manifested to a group of individuals engaged in human communication processes so that new cognitive effects can arise as modifications from previous states of an individual's cognitive environment in a set of facts that are mutually manifested to communicators and audience alike. Cognitive effects are changes in the individual's set of assumptions resulting from the processing of an input in a context of previously held assumptions. This type of processing may result in three types of cognitive effects: the derivation of new assumptions; the modification of the degree of strength of previously held assumptions; and the deletion of previously held assumptions. The possibility of achieving cognitive effects is what makes an input worth processing. Everything else being equal, inputs that yield greater cognitive effects are more relevant and more worthy of processing. As far as translation is concerned, the notions of cognitive environment and mutual manifestness (see Alves, 2007, 2009; Gutt, 1991/2000) are of paramount importance to the investigation of how translators and interpreters arrive at target products as well as to how audiences respond to the target texts (written or oral) they read or listen to.

7.2.3 Interpretive resemblance: Translation as an act of interlingual interpretive language use

Gutt (1991/2000) provided the first relevance-theoretic account of translation. His work caused controversies when it was published but grew to be considered a seminal contribution to Translation Studies. Gutt grounds his considerations on the inferential nature of human communication and on the notions of descriptive and interpretive language use (see also Gutt, 2001). Relevance theory postulates a distinction between descriptive and attributive (or interpretive) uses of language. Whereas a descriptive utterance is an interpretation of a thought, interpretive utterances are interpretations of a thought that is an interpretation of another thought or utterance, and consequently, a thought or an utterance attributed to another person or to a speaker at another time.

Gutt (1991/2000) first looks at the notions of overt and covert translation to make a distinction between "direct" and "indirect" translations. Building on House's (1981) theory of functional equivalence, in which translations should match the original text in function, Gutt states that for House only covert translations are capable of actually achieving functional equivalence. He disagrees with House's theoretical claims and suggests that instead one should focus on "direct" translations because of their (explicit or implicit) presumption of *interpretively resembling* original content. For Gutt, "a receptor language utterance is a direct translation of a source language utterance if and only if it purports to interpretively resemble the original completely in the context envisaged for the original" (Gutt, 1991/2000, p. 171).

The notion of interpretive resemblance derives from Gutt's view of translation as a case of interlingual interpretive language use. He looks at interpretive resemblances between propositional forms, between thoughts and utterances, and between utterances. Considering that the notion of interpretive resemblance between propositional forms cannot be applied to resemblances between utterances without some modification, Gutt insists that "the crucial point is the sharing of analytic and/or contextual implications" and goes on to claim that "since these implications are assumptions, we can say more generally that interpretive resemblance is characterized by the sharing of assumptions" (Gutt, 1991/2000, p. 46). He argues further that, ultimately, a relevance-theoretic approach could provide a unified account of translation. In other words, translators have the task of interpretively rendering the source text into the target language by providing the audience with communicative cues in the very same way as an original communicator provides the original audience with cues to enable inferential processing. This is the basis of a research programme based on the observation that "human beings have the remarkable ability to tell in one language what was first told in another language" Gutt (1991/2000, p. 205). Gutt calls it a competence-oriented research of translation (CORT) that "seeks to understand translation through understanding the communicative competence that makes it possible, for both the translator and his/her audience" Gutt (1991/2000, p. 205).

Several translation scholars have responded to Gutt's call and carried out empirical investigations based on CORT, focusing on aspects of translation process research (Alves, 1995, 2007, 2009; Alves & Gonçalves, 2003, 2013; Alves et al., 2010), translation competence (Alves & Gonçalves, 2007), conference interpretation (Setton & Dawrant, 2016), legal interpreting (Gallai, 2015, 2017), simultaneous interpreting (Setton, 1999, 2006; Vianna, 2005), audiovisual translation (Desilla, 2018, 2019), or other topics. We shall refer to these works in more detail in the following sections of this chapter.

7.2.4 Processing effort and cognitive effects in translation

Sperber and Wilson (1986/1995, p. 265) define a cognitive effect "as a contextual effect occurring in a cognitive system (e.g. an individual), and a *positive cognitive effect* as a cognitive effect that contributes positively to the fulfilment of cognitive functions or goals". They identify three types of cognitive effects: (1) cognitive effects that generate a conclusion drawn from old or new information together; (2) cognitive effects that strengthen an existing assumption; and (3) cognitive effects that contradict or eliminate an existing assumption. For Sperber and Wilson, the more cognitive effects one is able to generate, the more relevant is the information conveyed by the communicator.

7.2.4.1 The effort–effect relation in translation

In the relevance-theoretic framework, it follows that information is relevant for the audience to the extent that it generates cognitive effects by means of the necessary processing effort to modify existing assumptions about the world. In other words, human cognition is designed to maximize the generation of cognitive effects at the minimal cost of necessary processing effort. Thus, Sperber and Wilson (1986/1995) postulate a relation between processing effort and cognitive effects, which offers an alternative to empirically investigating the role of effort and effect in translation.

Using translation process data tracked with keylogging in conjunction with retrospective protocols, Alves (2007) carried out an empirical study to investigate the relation between processing effort and cognitive effects. The data analysis looked at segmentation patterns in terms of time taken on the task, including deletions and regressions, comparing keylogged data with metarepresentations of the task at hand, obtained through retrospective verbal protocols. By assessing verbally justified reasons for the output rendered by participants, the results showed that professional translators worked faster and purposefully, revealing meaningful correlations between the type of processing effort undertaken and the corresponding cognitive effects registered in the translation output and their corresponding metarepresentations conveyed through retrospection. Thus, Alves (2007) was in a position to corroborate the relevance-theoretic assumption that, for translation task execution carried out by professional translators, human cognition is designed to maximize the generation of cognitive effects with the least necessary processing effort. On the other hand, novice translators revealed a type of processing effort that showed a linear pattern of segmentation, with little meaningful processing effort and practically no justified retrospection for their translation output, showing weak cognitive effects and pointing to a type of cognitive behaviour that revealed a lack of ability to effectively carry out the translation task.

7.2.4.2 The role of conceptual and procedural encodings in translation

Blakemore (1987) introduced the conceptual–procedural distinction into relevance theory to account for differences between regular content words and discourse connectives. In relevance-theoretic terms, conceptually encoded information is encoded by open lexical categories, such as nouns, adjectives and verbs, to convey conceptual meaning, which is propositionally extendable. It can be enriched and contributes to the inferential processing of an utterance. Procedurally encoded information, on the other hand, is encoded via non-open morphological categories, such as negation, tenses, determiners, word order, etc. It cannot be extended in propositional terms but contributes decisively to the cognitive processing of an utterance by imposing procedural constraints on the construction of intended contexts and cognitive effects.

Wilson (2011, p. 9) argues that what distinguishes the conceptual–procedural distinction from the traditional semantic or pragmatic distinction is that "it carries definite cognitive commitments". For relevance theory, to say that a certain expression encodes conceptual or procedural meaning is to say that it has implications for the nature of the cognitive mechanisms involved. Additionally, Wilson (2011, p. 12) comments that "there is a fairly widespread view that the conceptual-procedural distinction is intended to be mutually exclusive, so that a single word cannot encode both types of meaning". She argues, however, that there is little textual evidence to support this interpretation of relevance theory and advocates that conceptual and procedural meaning should not be treated as mutually exclusive.

Alves and Gonçalves (2003, 2013) carried out empirical studies to investigate the conceptual–procedural distinction during translation task execution. Alves and Gonçalves (2003) built on the conceptual–procedural distinction to postulate that there would be fewer problems in the recognition of procedurally encoded information in the source text. Consequently, this would yield similar inferential processing among subjects, and translation decisions would be more structurally oriented. They also postulated that conceptually encoded information would be handled on the basis of individually available contextual assumptions. Therefore, solutions would be inferentially supported by contextual assumptions derived from the translators' cognitive environments and vary randomly among subjects. However, as Alves and Gonçalves (2003) have shown, procedurally encoded information needed contextual support to be processed effectively. Difficulties in retrieving the communicative clues conveyed by procedurally encoded information hindered the generation of positive cognitive effects, confirming relevance-theoretic assumptions about the hybrid nature of the conceptual–procedural distinction for translation task execution.

Alves and Gonçalves (2013) analysed keylogged data of professional translators when performing direct and inverse translation tasks. The analysis focused on the number and types of conceptual and procedural encodings found in micro /macro translation units (Alves & Vale, 2009, 2011). Results showed that processing effort in translation is greater in segments conveying procedurally encoded information than in segments rendering conceptually encoded information. Hybrid encodings, conveying conceptual information with a procedural function, also had an impact on translation task execution, requiring more processing effort in both direct and inverse translation tasks.

Alves et al. (2014) built on the conceptual–procedural distinction to investigate processing effort in translation task execution using keylogged and eye-tracking data. Their analysis showed that there are statistically significant differences when conceptual and procedural encodings are analysed in selected areas of interest, with instances related to procedural encoding requiring more processing effort to be translated. For Alves et al. (2014), as well as Alves and Gonçalves (2013), effort was not measured temporally in terms of time spent on a given segment but by the number of edits (replacement of unfinished segments, substitutions, deletions) and the distance of those editing actions from the first rendering of that given segment.

As far as interpreting is concerned, Setton (1999, p. 204) proposed the notion of "pragmatic clues to inference" by drawing on the distinction between conceptual and procedural information and their corresponding inferential computations. In other words, Setton (1999) showed that "the principle of relevance seems to apply particularly well to the real-time performance of simultaneous interpreters" and that "communication is successful when speakers' utterances are 'optimally relevant', i.e., when they give listeners access to maximum cognitive effects for minimum effort" (see Gallai, 2019, p. 63).

Altogether, these works provide empirical evidence that contextual assumptions play a role in handling procedurally and conceptually encoded information and point out that a relevance-theoretic view of translation can account for how implicatures and explicatures are expressed in different cognitive environments and, therefore, in different target texts.

7.2.5 Metarepresentation and higher-order representations in translation

The recognition of informative and communicative intentions depends on specific abilities, such as inferring and predicting the content of mental states. Metarepresentation is "the ability to represent how other people represent states of affairs in their minds" (Gutt, 2004, p. 80). Wilson (2012, p. 231) states that metarepresentation as "the ability to identify speaker meanings is nothing but the general mindreading ability applied to a specific communicative domain". Metarepresentation, as a mindreading ability, is often wrongly perceived as a misleading term which suggests the decoding of thoughts. According to relevance theory, texts or utterances are interpretive representations of an author's or a speaker's thoughts, which necessarily involve at least one level of metarepresentation to allow the recognition of informative and communicative intentions.

Drawing on the concept of metarepresentation, Gutt (2004) discusses translation as a *higher-order act of communication*. He argues that the primary concern of translators is to convey interpretive resemblance between source and target texts and not a representation of states of affairs. By metarepresenting bodies of thought, translators can achieve interpretive resemblance by drawing on cognitive environments of both original communicators and receptors of target texts. Clearly, the communicator, the translator and the audience have different cognitive environments. However, Gutt points out that "as soon as one recognizes the need to deal with different cognitive environments, it becomes clear that *metarepresentational skills* must be a core component of translation competence" (Gutt, 2004, p. 13).

Gutt (2005) reiterates the importance of CORT and argues that higher-order acts of communication can be applied to communicative situations in which the communicator and the audience do not share a mutual cognitive environment. In such cases—known as "secondary communication"—Gutt suggests that an additional cognitive level is needed for communication to succeed. Thus, Gutt (2005) claims, the capacity to metarepresent is a cognitive prerequisite for the ability to translate.

Drawing on Gutt (2004, 2005), Alves (2007) has shown that translators tend to regulate the relation between processing effort and cognitive effect on the basis of a multi-level process mediated by the metarepresentations they create. Pause analysis and retrospective data reveal that the relation between processing effort and cognitive effect is also conditioned by the translator's degree of metacognitive monitoring.

Alves and Gonçalves' (2007) cognitive model of translation competence emphasizes the central role played by metarepresentation and metacognition in the development of that competence.

Their model embeds translation competence in a comprehensive cognitive theory, claiming that the ability to translate requires highly complex metacognitive skills.

In the field of interpreting, Setton (1999, 2006), Setton and Dawrant (2016) and Vianna (2005) corroborate Gutt's view of translation and interpreting as attributed thought—stressing the fundamental underdeterminacy of linguistic encoding.

7.2.6 Explicitation and explicitness in translation

Explicitation has been often investigated as a translation-specific phenomenon. Blum-Kulka (1986) was perhaps the first scholar to propose a systematic study of explicitation in translation. For her, translations tend to be more explicit than their source-text counterparts, regardless of whether or not this increase in explicitation is imposed by differences in the linguistic systems of source and target texts. Baker (1993) drew on Blum-Kulka's (1986) proposal and used the tools of corpus linguistics to compare quantitative patterns between original texts and their translations, observing the type/token ratio. Olohan and Baker (2000) also drew on the explicitation hypothesis to analyse implicit and explicit occurrences of the conjunction "that" in original and translated texts. None of these works, however, were concerned with implications of explicitation for pragmatics.

Approaching explicitation from a cognitive perspective, Englund-Dimitrova (2005) proposed a distinction between explicitation processes that are norm-governed and those processes that have a strategic nature. The former type of explicitation, she claimed, would be determined by the constraints inherent in the linguistic systems in contrast and would not be cognitively relevant for investigations about the translation process. On the other hand, processes of strategic explicitation would arise from the difficulties found by the translator in solving translation problems that go beyond the scope of the constraints imposed by the linguistic systems in contrast. However, Englund-Dimitrova (2005) did not analyse instances of explicitation as renderings constrained by inferential processing.

Building on Steiner (2005), Hansen-Schirra et al. (2007) looked at explicitness, rather than explicitation, from a corpus linguistics perspective. They measured explicitness as a property of the encoding, not as a property of the communicative act as such, which, for them, is in the realm of explicitation. For Hansen-Schirra et al. (2007), although explicitation is of great significance for any attempt at understanding communication, their approach focuses on textual encoding as a means to provide a necessary prerequisite for investigations of translation as an act of communication. They observe that investigations of explicitation in translation from a cognitive pragmatic perspective fall more properly within the domain of relevance theory.

Sperber and Wilson (1986/1995) argue that the explicit side of communication is far more inferential than Grice envisaged. For them, both the explicit and the implicit side of communication involve making inferences from contextual assumptions on the basis of general pragmatic principles. Thus, Sperber and Wilson develop the notion of *explicature*, which is defined in terms of an inferential development of incomplete logical forms (verbally manifested) into propositional forms. In other words, explicatures serve to "flesh out" incomplete conceptual representations encoded by utterances and thus yield fully propositional content. Unlike explicit content, the *implicit* content or *implicature* within relevance theory is seen as an assumption that can only be derived pragmatically by means of pragmatic inferences. For Sperber and Wilson, the difference between explicatures and implicatures consists in the fact that the recovery of the former involves both decoding and inference, whereas the latter involves only inference. The inferential process integrates the semantic representation with *contextual assumptions* in order to reach an intended interpretation of the utterance, and is guided by the communicative principle of relevance. Further, as Carston (2002) argues, the semantic representation is not fully propositional but is just a template for utterance interpretation, which requires pragmatic inference in order to recover the proposition the speaker has intended. For Carston (2002), semantics is a relation between a linguistic form and the information it provides as input to the inference system, rather than a relation between a linguistic form and an entity in the world.

Applying the relevance-theoretic framework to cognitive-oriented empirical investigations of translation task execution, Alves (2009) used translation process data in combination with corpus linguistics annotations to assess the notions of "*norm-governed explicitations* and *strategic explicitations*" proposed by Englund-Dimitrova (2005, p. 236, italics in the original). Alves (2009) also looked at differences between processes of explicitness and explicitation proposed by Steiner (2005) and Hansen-Schirra et al. (2007). Drawing on the above-mentioned authors, Alves (2009) offered a process-oriented inferential account of explicitation in translation, which revealed that instances of explicitness in translation were predominantly related to norm-governed issues, whereas instances of explicitation in translation were predominantly strategic.

7.3 Recent developments and future directions

In this section, we look into some recent developments and point to some future directions concerning pragmatics at the interface with translation and cognition. We consider the possibilities offered by experimental pragmatics to investigate the impact of pragmatic factors on cognitive behaviour and the challenges that lie ahead as the field of cognitive pragmatics develops further within Cognitive Translation Studies.

7.3.1 Experimental pragmatics

We have seen in the previous sections that Gutt (1991/2000, 2001, 2004, 2005) has shown that elements of language can encode processing instructions which provide guidance to the audience as to how an expression is intended to be relevant. Gutt has also pointed out that these instructions are empirically grounded and can be used to make testable predictions about the success and failure of human communication. Gutt argues that the cause–effect framework provided by relevance theory can be used to predict problems when the stimuli and/or the audience lack ready access to certain pieces of information that are needed for consistency with the principle of relevance. Based on these assumptions, Gutt suggests that one could set up experiments to investigate cause–effect relations in translation, a suggestion that is well suited for grounding translation as a cognitive activity.

Psycholinguistics has developed sophisticated experimental methods in the study of verbal communication, but it has not used them to test pragmatic theories systematically. A combination of psycholinguistic methods with a pragmatic orientation lays down the bases for experimental pragmatics, a new disciplinary field that draws on pragmatics and psycholinguistics and also on the psychology of reasoning as proposed by Noveck and Sperber (2004), who state that "experimental pragmatics may contribute to linguistics and psychology, and to the cognitive sciences in general" (Noveck & Sperber, 2004, p. 19). They insist that the range of phenomena that pragmatics investigates is part of the much wider domain of psycholinguistics, which has paid very little attention to pragmatics. Noveck and Sperber (2004) define experimental pragmatics as the study of how linguistic properties and contextual factors interact in the interpretation of utterances. It is reasonable to expect that two fields of research dealing in part with the same material at the same level of abstraction would gain strength by joining forces, or at least by interacting actively. For pragmatics, the gain would be twofold. First, experimental evidence can be used, together with intuition and recordings, to confirm or disconfirm hypotheses. The high reliability and strong evidential value of experimental data put a premium on this sort of data, even though it is hard to collect and is generally more artificial than observational data (and therefore raises specific problems of interpretation).

Testing theories experimentally often leads one to revise and refine them in the light of new and precise evidence. Noveck and Sperber (2004, p. 9) suggest that, for experimental psycholinguistics, the gain from a greater involvement with pragmatics would be in taking advantage of the competencies, concepts and theories developed in this field in order to better describe and explain a range of phenomena that are clearly of a psycholinguistic nature, and to develop new experimental paradigms. Gibbs (2004), Van der Henst and Sperber (2004) and Noveck and Reboul (2008) are examples of works that test linguistic-pragmatic theories using experimental psychological methods. In Cognitive Translation Studies, Alves (2007, 2009), Alves and Gonçalves (2003, 2013) and Alves et al. (2014) present applications of experimental pragmatics drawing on relevance theory.

7.3.2 Emerging topics

As the cognitive pragmatics approach to translation becomes consolidated by applications of relevance theory, some authors have pointed out alternative and complementary possibilities of investigating the translation/pragmatics interface from a cognitive angle.

In contrast with the standard relevance-theoretic view of translation and interpreting as interlingual interpretive language use, Gallai's (2015, 2016, 2017) interdisciplinary study of procedural elements in interpreting draws a comparison between free indirect style or thought (FIT) representations in fiction and interpreter-mediated utterances in order to reassess the way in which attributed thoughts are represented in face-to-face interpreting. Gallai argues that "procedural elements have an important role to play in creating an illusion of being able to gain entry to the speaker's mind" (Gallai, 2019, p. 64).

Desilla (2018, 2019) looks at pragmatics and audiovisual translation and provides an overview of three salient pragmatic phenomena (speech acts, politeness and implicature), teasing out their significant role in the construal, translation and reception of audiovisual texts. Outlining Grice's pioneering study of implicature, Desilla (2018) treats implicatures as a sub-type of non-conventional indirectness and focuses on the cognitive psychological perspective of relevance theory, which, she argues, is equipped with the conceptual tools for understanding context selection. Desilla (2019) looks at how experimental pragmatics meets audiovisual translation by addressing methodological challenges in researching how film audiences understand implicatures.

Finally, Szpak (2017), Alves et al. (2019) and Szpak et al. (2019) look at translation in the brain from a combined perspective, which brings together the relevance-theoretic concept of metarepresentation and the concept of perspective taking used in theory of mind (ToM). Drawing on brain imaging studies, on tenets from relevance theory and on a ToM-oriented approach, the authors try to investigate how perspective taking interposes both metacommunicative and metapsychological processes (metarepresentation, mindreading and ToM) and locate it within a broad brain area, namely the left inferior parietal lobe. Szpak et al. (2019) provide a novel inferential account of neurophysiological data in relation to translation task execution, opening up new avenues for a metarepresentational view of translation. In their account, interpretive resemblance of attributed thoughts (metarepresentations) has proven to entail brain regions related to the identification of the target perspective (a metacommunicative process), which in turn involves brain regions related to the attribution of second-order mental representations to others (a metapsychological process).

The works mentioned here are only examples of new trends emerging in areas as diverse as legal interpreting, audiovisual translation and neuroimaging studies applied to translation task execution that have opened up new frontiers for the study of translation from a cognitive pragmatics perspective. As these new trends unfold, our understanding of the impact of pragmatics-related phenomena on the translation/cognition interface is bound to increase.

Further reading

Barron, A., Gu, Y., & Steen, G. (Eds.). (2017). *The Routledge handbook of pragmatics.* London: Routledge.
It provides an overview of pragmatics as an independent discipline within linguistics, including a historical account and the perspectives of further disciplinary development.

Gutt, E.-A. (2000). *Translation and relevance. Cognition and context* (2nd ed.). Manchester: St Jerome. (First edition published by Blackwell, 1991.)
It introduces applications of relevance theory to Translation Studies and provides an account of translation as interpretive language use.

Sperber, D., & Deirdre, W. (1995). *Relevance: Communication and cognition* (2nd ed.). Oxford: Blackwell. (First edition published by Blackwell, 1986.)
It presents the foundations of relevance theory and grounds it within the disciplinary framework of pragmatics.

Tipton, R., & Desilla, L. (Eds.). (2019). *The Routledge handbook of translation and pragmatics.* London: Routledge.
It outlines key theoretical concepts in pragmatics and presents an account of the disciplinary relationship between Translation Studies and pragmatics.

References

Alves, F. (1995). *Zwischen Schweigen und Sprechen: Wie bildet sich eine transkulturelle Brücke?* Hamburg: Dr. Kovac.

Alves, F. (2007). Cognitive effort and contextual effect in translation: A relevance-theoretic approach. *Journal of Translation Studies, 10*, 18–35.

Alves, F. (2009). Explicitation and explicitness in translation: A relevance-theoretic approach. In J. C. Costa, & F. J. Rauen (Eds.), *Topics on relevance theory* (pp. 100–117). Porto Alegre: Editora PUCRS.

Alves, F., & Gonçalves, J. L. (2003). A relevance theory approach to the investigation of inferential processes in translation. In F. Alves (Ed.), *Triangulating translation: Perspectives in process oriented research* (pp. 3–24). Amsterdam: John Benjamins.

Alves, F., & Gonçalves, J. L. (2007). Modelling translator's competence. Relevance and expertise under scrutiny. In Y. Gambier, M. Shlesinger, & R. Stolze (Eds.), *Doubts and directions in translation studies* (pp. 41–55) Amsterdam: John Benjamins.

Alves, F., & Gonçalves, J. L. (2013). Investigating the conceptual-procedural distinction in the translation process. A relevance-theoretic analysis of micro and macro translation units. *Target. International Journal of Translation Studies, 25*(1), 107–124. doi: 10.1075/target.25.1.09alv

Alves, F., Gonçalves, J. L., & Szpak, K. S. (2014). Some thoughts about the conceptual/procedural distinction in translation: A key-logging and eye-tracking study of processing effort. In *MonTI, 1,* 151–175.

Alves, F., Pagano, A., Neumann, S., Steiner, E., & Hansen-Schirra, S. (2010). Units of translation and grammatical shifts: Towards an integration of product- and process-based research in translation. In G. Shreve, & E. Angelone (Eds.), *Translation and cognition* (pp. 109–142). Amsterdam: John Benjamins.

Alves, F., Szpak, K. S., & Buchweitz, A. (2019). Translation in the brain: Preliminary thoughts about a brain-imaging study to investigate psychological processes involved in translation. In D. Li, V. Lei, & Y. He (Eds.), *Researching cognitive processes of translation* (pp. 121–138). Singapore: Springer.

Alves, F., & Vale, D. C. (2009). Probing the unit of translation in time: Aspects of the design and development of a web application for storing, annotating, and querying translation process data. *Across Languages and Cultures, 10*(2), 271–273.

Alves, F., & Vale, D. C. (2011). On drafting and revision in translation: A corpus linguistics oriented analysis of translation process data. *TC3: Translation, Corpora and Cognition, 1*, 105–122. Reprinted in S. Hansen-Schirra, S. Neumann, & O. Čulo (Eds.). (2017), *Annotation, exploitation and evaluation of parallel corpora: TC3 I* (pp. 89–109). Berlin: Language Science Press.

Austin, J. L. (1962). *How to do things with words.* Cambridge, MA: Harvard University Press.
Baker, M. (1992). *In other words: A coursebook on translation.* London: Routledge.
Baker, M. (1993). Corpus linguistics and translation studies. Implications and applications. In M. Baker, G. Francis, & T. Tognini-Bonelli (Eds.), *Text and technology: In honour of John Sinclair* (pp. 233–250). Amsterdam: John Benjamins.
Barron, A., Gu, Y., & Steen, G. (Eds.). (2017). *The Routledge handbook of pragmatics.* London: Routledge.
Blakemore, D. (1987). *Semantic constraints on relevance.* Oxford: Blackwell.
Blakemore, D. (1992). *Understanding utterances. An introduction to pragmatics.* Oxford: Blackwell.
Blum-Kulka, S. (1986). Shifts of cohesion and coherence in translation. In J. House, & S. Blum-Kulka (Eds.), *Interlingual and intercultural communication: Discourse and cognition in translation and second language acquisition* (pp.17–35). Tübingen: Narr.
Brown, P., & Levinson, S. C. (1978). Universals in language use: Politeness phenomena. In K. Scherer, & H. Giles (Eds.), *Social markers in speech* (pp. 291–347). Cambridge: Cambridge University Press.
Brown, P., & Levinson, S. C. (1987). *Politeness. Some universals in language usage.* Cambridge: Cambridge University Press.
Carston, R. (2002). *Thoughts and utterances. The pragmatics of explicit communication.* Oxford: Blackwell.
Davis, S. (Ed.) (1991). *Pragmatics: A reader.* Oxford: Oxford University Press.
Desilla, L. (2018). Pragmatics and audiovisual translation. In L. P. González (Ed.), *The Routledge handbook of audiovisual translation.* London: Routledge.
Desilla, L. (2019). Experimental pragmatics meets audiovisual translation: Tackling methodological challenges in researching how film audiences understand implicatures. In R. Tipton, & L. Desilla (Eds.), *The Routledge handbook of translation and pragmatics* (pp. 93–114). London: Routledge.
Doherty, M. (2002). *Language processing in discourse. A key to felicitous translation.* London: Routledge.
Englund-Dimitrova, B. (2005). *Expertise and explicitation in the translation process.* Amsterdam: John Benjamins.
Gallai, F. (2015). Legal interpreting and pragmatics: Are they compatible? In C. Zwischenberger, & M. Behr (Eds.), *Interpreting quality: A look around and ahead* (pp. 167–204). Berlin: Franke and Timme.
Gallai, F. (2016). Point of view in free indirect thought and in community interpreting. *Lingua, 175–176,* 97–121. Special issue edited by D. Wilson and R. Sasamoto. *Little words: Communication and procedural meaning.*
Gallai, F. (2017). Pragmatics and legal interpreters' codes of ethics. In J. Drugan, & R. Tipton (Eds.). Special Issue on Translation, Ethics and Social Responsibility. *The Translator, 23*(1), 177–196.
Gallai, F. (2019). Cognitive pragmatics and translation studies. In R. Tipton, & L. Desilla (Eds.), *The Routledge handbook of translation and pragmatics* (pp. 51–72). London: Routledge. doi: 10.4324/9781315205564–4
Gibbs, R. W. (2004). Psycholinguistic experiments and linguistic-pragmatics. In I. A. Noveck, & D. Sperber (Eds.), *Experimental pragmatics* (pp. 50–71). Basingstoke: Palgrave Macmillan.
Grice, H. P. (1957). Meaning. *The Philosophical Review, 66,* 377–388.
Grice, H. P. (1975). Logic and conversation. In P. Cole, & J. Morgan (Eds.), *Studies in syntax and semantics* (Vol. 3, Speech Acts, pp. 41–58). New York: Academic Press.
Grice, H. P. (1989) *Studies in the way of words.* Cambridge, MA: Harvard University Press.
Gutt, E.-A. (2000). *Translation and relevance. Cognition and context* (2nd ed.). Manchester: St Jerome. (First edition published by Blackwell, 1991.)
Gutt, E.-A. (2001). Pragmatic aspects of translation: Some relevance-theory observations. In L. Hickey (Ed.), *The pragmatics of translation.* Clevedon: Multilingual Matters.
Gutt, E.-A. (2004). Challenges of metarepresentation to translation competence. In E. Fleischmann, P. A. Schmitt, & G. Wotjak (Eds.), *Translationskompetenz: Proceedings of LICTRA 2001: VII. Leipziger Internationale Konferenz zu Grundfragen der Translatologie* (pp. 77–89). Tübingen: Stauffenburg.
Gutt, E.-A. (2005). On the significance of the cognitive core of translation. *The Translator, 11*(1), 25–49.
Halliday, M. A. K. (1978). *Language as social semiotic.* London: Edward Arnold.
Halliday, M. A. K., Hasan, R. (1985). *Language, context and text: Aspects of language in a social semiotic perspective.* Victoria: Deakin University Press.
Hansen-Schirra, S., Neumann, S., & Steiner, E. (2007). Cohesive explicitation in an English-German translation *corpus. Languages in Contrast,* 7(2), 241–265.
Hatim, B., & Mason, I. (1990). *Discourse and the translator.* London: Longman.
Hatim, B., & Mason, I. (1997). *The translator as communicator.* London: Routledge.
Hickey, L. (Ed.). (1998). *The pragmatics of translation.* Clevedon: Multilingual Matters.
House, J. (1981). *A model for translation quality assessment.* Tübingen: Narr.
House, J. (1997). *Translation quality assessment: A model revisited.* Tübingen: Narr.

House, J. (2006). Text and context in translation. *Journal of Pragmatics, 38*(3), 338–358.
House, J. (2013). Towards a new linguistic-cognitive orientation in translation studies. *Target. International Journal of Translation Studies, 25*(1), 46–60. doi: 10.1075/target.25.1.05hou
Huang, Y. (2007). *Pragmatics,* Oxford & New York: Oxford University Press.
Leech, G. (1983). *Principles of pragmatics.* London: Longman.
Levinson, S. (1993). *Pragmatics.* Cambridge: Cambridge University Press.
Levinson, S. (2000). *Presumptive meanings. The theory of generalized conversational implicature.* Cambridge, MA: MIT Press.
Marmaridou, S. (2000). *Pragmatic meaning and cognition.* Amsterdam: John Benjamins.
Mey, J. L. (1993). *Pragmatics: An introduction.* London: Blackwell.
Mey, J. L. (2006). Pragmatics: Overview. In K. Brown (Ed.) *Encyclopedia of language and linguistics* (2nd ed., vols. 1–14, pp. 51–62). Oxford: Elsevier.
Morris, C. H. (1938). Foundations of the theory of signs. In R. Carnap et al. (Eds.), *International encyclopedia of unified science.* Chicago: University of Chicago Press.
Noveck, I. A., & Reboul, A. (2008). Experimental pragmatics: A Gricean turn in the study of language. *Trends in Cognitive Sciences, 12*(11), 425–431.
Noveck, I. A., & Sperber, D. (Eds.). (2004). *Experimental pragmatics.* Basingstoke: Palgrave Macmillan.
Olohan, M., & Baker, M. (2000). Reporting that in translated English: Evidence for subconscious processes of explicitation? *Across Languages and Cultures, 1*(2), 141–158.
Reiß, K., & Vermeer, H. J. (1984). *Grundlegung einer allgemeinen Translationstheorie.* Berlin: Mouton de Gruyter.
Schmid, H.-J. (Ed.). (2012). *Cognitive pragmatics.* Berlin: De Gruyter.
Searle, J. R. (1969). *Speech acts: An essay in the philosophy of language.* Cambridge: Cambridge University Press.
Setton, R. (1999). *Simultaneous interpretation: A cognitive-pragmatic analysis.* Amsterdam: John Benjamins.
Setton, R. (2006). Context in simultaneous interpretation. *Journal of Pragmatics, 38*(3), 374–389.
Setton, R., & Dawrant, A. (2016). *Conference interpreting: A trainer's guide.* Amsterdam: John Benjamins.
Shannon, C., & Weaver, W. (1949). *The mathematical theory of communication.* Urbana, IL: University of Illinois Press.
Spencer-Oatey, H., & Žegarac, V. (2010). Pragmatics. In N. Schmitt (Ed.), *An introduction to applied linguistics* (2nd ed., pp. 70–88). London: Hodder & Stoughton.
Sperber, D., & Wilson, D. (1986/1995). *Relevance: Communication and cognition* (2nd ed.). Oxford: Blackwell. (First edition published by Blackwell, 1986.)
Steiner, E. (2005). Explicitation, its lexicogrammatical realization, and its determining (independent) variables—towards an empirical and corpus-based methodology. *SPRIKreports, 36,* 1–42.
Szpak, K. S. (2017). A atribuição de estados mentais em atividades de tradução: Um estudo conduzido com rastreamento ocular e ressonância magnética funcional [unpublished PhD thesis] (pp. 1–230). Federal University of Minas Gerais, Belo Horizonte.
Szpak, K. S., Alves, F., & Buchweitz, A. (2020). Perspective taking in translation: In search of neural correlates of representing and attributing mental states to others. In R. Muñoz Martín, & S. L. Halverson (Eds.), *The IATIS yearbook.* London: Routledge.
Tipton, R. (2019). Introduction. In R. Tipton, & L. Desilla (Eds.), *The Routledge handbook of translation and pragmatics* (pp. 1–9). London: Routledge.
Tipton, R., & Desilla, L. (Eds.). (2019). *The Routledge handbook of translation and pragmatics.* London: Routledge.
Van der Henst, J.-B., & Sperber, D. (2004). Testing the cognitive and communicative principles of relevance. In I. A. Noveck, & D. Sperber (Eds.), *Experimental pragmatics* (pp. 141–171). Basingstoke: Palgrave Macmillan.
Verschueren, J. (1999) *Understanding pragmatics.* London: Arnold.
Vianna, B. (2005). Simultaneous interpreting: A relevance-theoretic approach. *Intercultural Pragmatics, 2*(2), 169–190.
Wierzbicka, A. (1991). *Cross-cultural pragmatics. The semantics of human interaction.* Berlin: Mouton de Gruyter.
Wilson, D. (2000). Metarepresentation in linguistic communication. In D. Sperber (Ed.), *Metarepresentations* (pp. 411–448). Oxford: Oxford University Press.
Wilson, D. (2011). The conceptual–procedural distinction: Past, present, and future. In V. Escandell-Vidal, M. Leonetti, & A. Ahern (Eds.), *Procedural meaning: Problems and perspectives* (pp. 1–31). London: Emerald Group.
Wilson, D. (2012). Metarepresentation in linguistic communication. In D. Wilson, & D. Sperber (Eds.), *Meaning and relevance* (pp. 230–258). Cambridge: Cambridge University Press.
Yule, G. (1996). *Pragmatics.* Oxford: Oxford University Press.

8

Translation, ergonomics and cognition

Maureen Ehrensberger-Dow

8.1 Introduction

The relationship between translation, ergonomics and cognition might not be obvious upon first consideration, but a short explanation should suffice to illustrate how ergonomics is an interface discipline between the other two. The International Ergonomics Association (IEA)[1] defines that discipline as being "concerned with the understanding of interactions among humans and other elements of a system". Considering ergonomics as the interdisciplinary interface between translation and cognition is congruent with an understanding of translation both as an event embedded in a situation involving various actors and factors (see Chesterman, 2013; Risku, 2014; Toury, 2012) and as an individual cognitive activity (e.g. Hurtado & Alves, 2009). An ergonomics perspective of translation and cognition also resonates with Chesterman's (2009) proposal to add a branch he called "translator studies" to Holmes' (1972/2000) mapping of the then-emerging discipline of Translation Studies. In the following, a brief review of the history of ergonomics and human factors provides some context before its more recently recognized relevance to Translation Studies and cognition is discussed.

The first designation of "ergonomics" has been attributed to Jastrzebowski (1857/2006), who used the term to describe the "natural laws of work". Ergonomics has been used more or less interchangeably with "human factors" (Salvendy, 2012; Stramler, 1993) and is closely linked to "human engineering", all of which began to become established early in the 20th century. From an initial ambition to alter human behaviour in order for employees to operate new machines properly, the focus eventually changed to understanding human needs in order to adapt the technology they use to do their work. Although its early proponents were sometimes derided as doing little more than applying common sense, better ergonomics has been strongly associated with improved working conditions and safety for employees in many domains (e.g. GAO, 1997; Niu, 2010) as well as with added value for companies (e.g. Dul, 2003).

Much of the early research in the area of ergonomics and human factors was done in the areas of aviation and other military applications around the Second World War, but other industries also profited from (mostly occupational psychologists') insights into how processes and safety could be improved. According to the semi-centennial review of the Ergonomics and Human Factors Society,[2] the scope of the discipline is even broader: "to promote the discovery and exchange of knowledge concerning human behaviour that is relevant to the design of tools, devices, equipment, vehicles, vessels, inhabited spaces, procedures, processes, and systems composed of these and other elements" (Stuster, 2006, p. 1). In the words of the IEA, "ergonomics is a systems-oriented discipline which now extends across all aspects of human activity" and that "promotes a holistic approach in which considerations of physical, cognitive, social, organizational, environmental and other relevant factors are taken into account".[3] The number of medical, governmental and insurance websites and guidelines devoted to ergonomics (e.g. MedLinePlus,[4] US OSHA,[5] EU-OSHA[6] and Allianz[7]) attest to its importance, yet it has only become a topic as such in Translation Studies relatively recently. Nevertheless, the relevance of ergonomics to the study and practice of translation goes back at least to the beginning of the machine translation (MT) age.

The beginning of the technologization of human translation might be dated to the report produced by the Automatic Language Processing Advisory Committee (ALPAC, 1966), which recommended more support for research and development into machine-aided translation (while at the same time discouraging any further development of MT). The ALPAC report still referred to punched cards and magnetic tape as input media for the computer, but keyboards connected to visual display terminals soon supplanted those, and by the mid-1980s personal computers were becoming increasingly common at translation workplaces. The ergonomics of visual display units and other hardware associated with computer workstations was questioned early on and has continued to be an issue for heavy users (e.g. ANSI/HFES, 2007; Wahlström, 2005).

One of the first references to ergonomics in the context of (machine) translation work was by Bevan (1982), who rather presciently pointed out that

> The advent of cheap computer technology offers many exciting possibilities, but its full potential can only be realised if the user's psychological and ergonomic needs are fully understood. It is the responsibility of potential computer users to insist that machines are used to remove the drudgery from life and expand our horizons, rather than become our masters. The machines must serve our needs, and not we theirs.
>
> *Bevan, 1982, p. 78*

The rapid developments in machine-aided translation related to translation memory in the 1990s led to increased productivity and efficiency gains as well as to cost pressure and higher expectations. Translators were expected to make optimal use of the new tools at their disposal to produce equal quality in much less time. This inevitably meant longer, more intensive periods spent at their computers keyboarding and staring at the screen, with all the concomitant risks previously reserved for programmers and similar professions that involve screen-intensive tasks. The ergonomic benefits of looking away from the screen briefly to focus, for example, on a dictionary in one's lap or shifting one's position to retrieve a parallel text from a nearby bookshelf disappeared in the convenience of such information being mere clicks away. The increased risks for translators of computer vision syndrome and repetitive strain injury from excessive mouse use can be dated to this change.

These shifts in translation practice and in the language industry were not entirely unnoticed by the discipline of Translation Studies, especially by researchers interested in the translation process. An appreciation of translation as a situated activity (see Krüger, 2015; Risku, 2002, 2010) and as an enactment of embodied, distributed, situated cognition (see Muñoz Martín, 2010; Robbins & Aydede, 2008) has been growing over the last two decades. This view of cognition is also consistent with an ergonomics perspective, which recognizes that people work with systems that encompass actors in their professional environments and networks as well as factors such as tools, equipment and computer interfaces. Translators are expected to make the right decisions with the resources they have at their disposal in order to create high-quality texts that meet the needs of their clients and readers. However, they are subject to physical, temporal, economic, organizational and cultural constraints related to workflows, communication processes, project management, job security and status. Ergonomic issues in any of these areas can compromise the efficiency of the translation process and potentially impact on the quality of the target texts. This reality is recognized to a certain degree in models of translation competence that refer to contextual factors such as external information sources, working conditions, societal norms, psycho-physical disposition or psycho-physiological components (e.g. EMT, 2017; Göpferich, 2009; PACTE, 2003). Researchers with an ergonomics perspective attempt to understand the effect of these factors on translators who, despite high levels of competence or expertise, might not perform as well as expected or might have to exert extra effort to compensate for poor ergonomic conditions.

In widely available databases (e.g. Benjamins' *Translation Studies Bibliography*[8] or the University of Alicante's *Bitra. Bibliography of Translation and Interpreting*[9]), many of the academic publications that mention ergonomics concern sign language and other types of interpreting, which might be thought of as more physical than translation, but the interest in the latter is growing as well. The first academic conference devoted to the theme of translation and ergonomics was in Grenoble in 2010, hosted by Lavault-Olléon (2011a, 2011b). The next, held five years later, resulted in a second special issue on the topic in 2016 (Lavault-Olléon, 2016). The contributions at both conferences reflected the dramatic changes to translators' work that had been taking place over the previous decade. Computer workplaces, digital resources, online communication and emerging technologies had all become essential features of professional translation by the time of the first conference (see also O'Brien, 2012; Pym, 2011), and process researchers were beginning to carry out investigations at or about the translation workplace by the time of the second (e.g. Ehrensberger-Dow, 2014; LeBlanc, 2013; Marshman, 2014; Risku, 2014).

Although often primarily associated with physical factors, the discipline of ergonomics also covers two other major domains, which have been termed cognitive and organizational. The European Association of Cognitive Ergonomics dates its first conference to 1982, with a hand-typed program entitled "Cognitive engineering. A conference on problem solving with computers" (ECCE, 1982). The two main strands concerned the psychology of computer users (who at that time were mostly programmers) and human–computer interaction. The 2018 conference advertised multiple themes, including those devoted to motivational and emotional aspects of interaction with tools and how to study and support cognitive tasks. These are highly relevant to translation, of course, since it is a bilingual cognitive task with an increasing reliance on computer-aided translation (CAT) tools and other technology. As O'Brien (2012, p. 103) put it, "today translation is a form of human–computer interaction", and she suggested that "cognitive ergonomic studies of translation tools and the translation process itself" (O'Brien, 2012, p. 116) could contribute to improving the usability of translation technology so that translators can focus on the task at hand.

8.2 Core topics

Understanding ergonomics as being at the interface of translation and cognition involves appreciating not only what happens in the mind, as reflected by emerging target texts, but also how translators interact with, adapt to and shape their environments. An ergonomics perspective on translation can contribute to gaining insights into the sources of physical and cognitive overload related to processing language at a technologized workplace and the potential consequences for translator health (Section 8.2.1). It can also help us understand the influence of various physical, cognitive, environmental, social, organizational and other ergonomic issues on the translation process and the quality of the products as well as the complexity of their interactions (Section 8.2.2). These insights should be incorporated into various levels of translator training in order to enable students and professionals to benefit from taking control over their workplaces and tools pro-actively instead of having to suffer from negative experiences (Section 8.2.3).

8.2.1 Translator health, ergonomics and cognition

Translation is a complex cognitive activity that requires translators not only to understand a source text and mentally formulate or choose between target-text solutions but also to search for information, key in text, accept matches from translation memory and revise existing text. At a computer workplace, all this requires interacting with devices that involve the whole body and not just people's eyes and brains. Since translation is also primarily a sedentary activity done indoors, translators are subject to physical factors such as the design of desks, chairs and office equipment (see Salvendy, 2012, and Starrett & Cordoza, 2016, for general overviews). In addition, the position of the computer screen relative to the keyboard as well as the type and location of the mouse can affect posture and result in stiffness in the neck or back, fatigue and leg pain due to extended periods sitting in one position (see also Huysmans et al., 2015). Repetitive strain injury is also a risk, because entering text, clicking and scrolling can cause an overload on muscles of the upper extremities and back as well as the muscles and tendons of the hand, wrist and lower arm (see Lavault-Olléon, 2011a). Such musculoskeletal disorders are a major cause of absence from work in many countries and have been associated with awkward postures and repetitive activity (see da Costa & Vieira, 2010; Niu, 2010). A survey of UK translation sector professionals revealed that more than 25% of respondents reported health ailments related to the use of the computer (UK Translator Survey, 2016/2017).

In an interdisciplinary study (Meidert et al., 2016), professional translators' workplaces were assessed by occupational therapy and Translation Studies researchers. Although the dedicated workplaces in companies and institutions evinced a high ergonomic standard overall, most of the furniture and equipment had not been adjusted correctly for the individual translators using them. The ergonomics of most of the freelancers' workplaces was found to be sub-optimal with respect to furniture and equipment, and they reported the most severe health complaints of the three groups. Overall, the most commonly reported health problems related to eyes, neck, head and shoulder girdle, all of which have been linked with intense screen work. These workplace findings are consistent with those of an exploratory survey study with freelancers and commercial translators (Ehrensberger-Dow & O'Brien, 2015) and confirmed in a much larger international survey of over 1,800 professional translators (Ehrensberger-Dow et al., 2016). Many of the identified issues would actually be quite easy to remedy, with information on ergonomic posture and workplace setup freely available on the Internet from reliable sources.[10]

One of the largest employers of translators in the world, the EU Commission's Directorate-General for Translation, officially recognized the importance of ergonomics in 2011 by designating correspondents to offer advice and consulting about working conditions and well-being (Peters-Geiben, 2016). Considering that EU agencies have identified several risk factors for health that describe many translators' working conditions, this makes very good sense. For example, a recent publication from the European Agency for Safety and Health at Work (Elsler et al., 2017) highlights the risk associated with extended periods of sitting and poor posture, reporting that muscular-skeletal disorders are second only to cancer as the cause of work-related mortality and morbidity in the EU.

Translator health, and hence the cognitive capacity to produce high-quality work, can also be affected by environmental factors related to the ergonomics of the workplace. These include physical conditions such as temperature, lighting, office layout and ambient noise. Vischer (2007, p. 179) explains with an "environmental comfort model" that it is important for the physical and functional conditions of the workplace to be a good fit with the type of work that is being done. If a translator is cold or has to hold her head at an odd angle to avoid glare on her screen, for example, then her concentration is likely to suffer. A workplace case study of a professional translator (Ehrensberger-Dow & Hunziker Heeb, 2016) suggested that hunger and/or low-grade discomfort from extended sitting can also affect the frequency of errors in the translation process. That particular translator was allowed to work from home one day a week, an opportunity that many translators who work in open-plan offices might also appreciate. In an international survey of professional translators, most commercial staff translators reported that they shared their office with at least two other people, and almost 20% of them were in large offices with ten or more other people (Ehrensberger-Dow et al., 2016).

The potential for distractions and ambient noise as well as negative consequences for concentration, stress and health seems especially high for translators working in such shared workspaces (see Smith-Jackson & Klein, 2009; Vischer, 2008). Other conditions typical of professional translation work (i.e. "working at speed and to tight deadlines, not having enough time to do the job; frequent disruptive interruptions") are indicators of high work intensity, which in the updated report of the 6th European Working Conditions survey is associated with increased risk of serious ill-health (Eurofound, 2017, p. 51). Considering that the added value of human translation (i.e. over MT solutions) relates to uniquely human traits such as creativity, discourse awareness and understanding of the target audience, it seems obvious that the well-being of the translator should be a priority for employers and clients. An ergonomics perspective does justice to understanding human cognition as embodied and embedded and to the translation process as an example of an enacted, situated cognitive activity, as explained in the following.

8.2.2 Translation process, ergonomics and cognition

The translation process has been described at the micro-level as a cognitive activity (see Hurtado & Alves, 2009 for a good review of models) and as partially overlapping cycles of orientating to the task and/or problem, drafting and revising (e.g. Jakobsen, 2002). It has also been described at the macro-level of a workflow involving multiple actors and interactions (e.g. Gouadec, 2007/2010; Risku, 2014). This is very much in line with how the IEA defines cognitive ergonomics: "concerned with mental processes, such as perception, memory, reasoning, and motor response, as they affect interactions among humans and other elements of a system".[11] In many domains, cognitive ergonomics is primarily associated with features of human–computer interaction such as the design, organization and operation of user interfaces (see O'Brien, this volume), but it also includes mental load, decision making and stress that is related to work.

The international standard ISO 17100 (2015) for translation services makes it clear that proficiency using language technology and information sources is an integral part of the professional translation process. If such technology and information sources are in alignment with cognitive processes, then it is assumed that they will be easier to use and lead to more efficient performance, fewer errors and less stress than if they are not (e.g. Beale & Peter, 2008). For example, CAT tools are intended to make translators' work more efficient by providing external stores of previously translated segments, by relieving translators of repetitive tasks, and by ensuring consistent terminology. Pym (2011) pointed out that using tools essentially externalizes and extends human memory. This can free up cognitive resources during the drafting phase and let translators concentrate on higher-order problem solving and decision making. If functionalities and features of language technology are not intuitive, however, they can add a cognitive load to the already taxing bilingual task of translating new content or evaluating and editing TM matches.

Despite the reputed usefulness of language technology, there is still a lot of resistance to it among professional translators. O'Brien (2012) referred to the potential friction between translators and technology and drew on the notion of "cognitive friction" as defined by Cooper (2004, p. 19) in his discussion of the increasing technologization of society. Intuitive user interfaces and functionalities can allow translators to focus on the bilingual task at hand instead of being slowed down or distracted by technical problems. If the ergonomics of translation tools are less than ideal, though, translators could be expected to experience cognitive friction, which could in turn affect their flow of thinking and consequently the efficiency of the process. The findings reported by Bundgaard et al. (2016) are consistent with this: the translator they observed at her workplace seemed to be assisted by the MT-assisted translation memory she was using but at the same time demonstrated resistance to it. The analyses done by Teixeira and O'Brien (2017) in their workplace study revealed a considerable amount of attention switching between tasks and tools, leading them to suggest that this could interrupt cognitive flow and contribute to cognitive load.

In an international survey focused on the ergonomics of translation (Ehrensberger-Dow et al., 2016), about three-quarters of the respondents indeed reported that they used CAT tools, and virtually all the users said that they found them helpful at least some of the time. Rather worrying from an ergonomics perspective, however, was the finding that most of those respondents kept the default settings instead of customizing them to suit their needs. This suggests that they might be adjusting their cognitive processes to fit the machine instead of the converse. Similar conclusions have been drawn in other research with respect to post-editing MT (Mesa-Lao, 2014; Moorkens & O'Brien, 2013), integration of MT in CAT tools (Teixeira, 2014) and the usability of tools in general (Hansen-Schirra, 2012; Taravella & Villeneuve, 2013).

In their survey, Moorkens & O'Brien (2016) found that only about half of the respondents reported that they liked using TM technology. Many complained about performance issues and the default layout of the tool. In a follow-up study to the international ergonomics survey mentioned earlier, O'Brien et al. (2017) analysed the items about CAT tools in more detail. Over half of the users reported being irritated by their tools, and almost all of them took the time to explain why. The most common irritating feature that was mentioned was the complexity of the user interface, followed by segmentation, formatting issues, visual presentation and bugs. Irritation is not known to be conducive to decision making, let alone creativity or efficiency. Mitigating such irritations by making it easier for the translators concerned to choose their own tool and/or individualize their settings can improve cognitive ergonomics as well as contribute to the efficiency of their workflows.

The cognitive ergonomics of the translation process encompasses more than user interfaces and functionalities of CAT tools, however. Working conditions, time management and stress can all be associated with disturbances to the translation process (see Hansen, 2006). As explained in the previous section (8.2.1), ergonomic issues can be related to the physical conditions of the office, comfort of the furniture and usability of the equipment, but translators are also part of a complex network. Social factors in the translation process related to cognitive ergonomics include the ease with which collaboration and exchanges among translators can occur. In addition, possibilities for personal interactions between them and other agents in the chain of target-text production, such as project managers and revisers, can positively affect their sense of involvement and increase their job satisfaction. This is quite different from unwanted distractions from others, whether within or outside their networks, which can detrimentally affect concentration and cognitive processing (see Baethge & Rigotti, 2010).

Other social factors related to ergonomics concern job security, status and self-determination. Recent advances in neural machine translation (NMT) have created such a hype that many laypeople believe that high-quality, fully automatic translation is just around the corner, which is contributing to uncertainty among professionals. The ultimate usefulness of NMT in all domains and the future of human translation remain to be determined, but current CAT technology already allows the integration of MT with translation memory. The quality of the MT suggestions depends on the system, which is usually beyond the individual translator's control. This lack of control may be why some participants in a focus group study carried out at the European Commission's Directorate-General for Translation avoided using MT, expressing fear of its influence on translation performance as well as general discomfort with the technology (Cadwell et al., 2016). New expectations regarding productivity concomitant with developments in CAT and MT are best aligned with translators' competence and work patterns, and ideally accompanied by consultation and training in order to encourage involvement and empowerment.

The comparatively greater autonomy of freelance translators to determine the tools they use may account for why they seem less likely to find them irritating, a complaint more frequently expressed by their commercial and institutional counterparts (see O'Brien et al., 2017, or Ehrensberger-Dow & Massey, 2017). If the latter were given more of a voice in the choice of software and timing of CAT updates, they might be more inclined to take ownership of new technology and exploit it in their work. Even so, being under self-imposed or organizational pressure to spend extended periods engaged in very similar types of technologized translation work can be cognitively taxing. Taking regular mini-breaks might be a simple solution, but many people are unaware of the value of such techniques to improving cognitive performance. Despite a generally accepted recognition of the role of feedback in learning and in the development of expertise (e.g. Hoffman et al., 2014), models of the translation process rarely include the possibility of feedback on ergonomic issues. As outlined in the next section, heightening translators' awareness of the importance of ergonomics to cognitive processing in translation tasks would be a significant first step.

8.2.3 Translator training and ergonomics

One of the primary motivations for investigations into the ergonomics of translation has been to gain insights into physical, cognitive, social, environmental and organizational aspects in order to improve working conditions. The basic principle of good ergonomics is that workplaces and workflows should be translator centred and not technology centred. Simply by answering questions about their furniture, equipment, tools and working practices, translators can become more aware of the ergonomic conditions they are working under and the potential issues associated with these. In the Meidert et al. (2016) study mentioned in Section 8.2.1, health issues were more noticeable among younger professionals and freelancers, suggesting that older commercial and institutional translators had learned about the importance of good ergonomics on the job or from prior negative experience. To avoid such painful learning curves, information about good ergonomic practice can be included in university translation courses, discussed in professional development seminars for translators, and disseminated through professional associations (e.g. O'Brien & Ehrensberger-Dow, 2017).

Lavault-Olléon & Carré (2012) propose that metacognition, particularly with respect to cognitive ergonomics, should be fostered during translator training in order to prepare students for the current realities of, and possible changes to, professional translation. Including ergonomics in translator training can also contribute to empowering students and professionals to identify and change dysfunctional practices they might encounter in the workplace (see Robertson & O'Neill, 2003). If translators receive information about physical ergonomics early in their education and careers, they can procure suitable equipment and furniture and thereby minimize the risk of musculoskeletal complaints developing in their upper extremities, back, shoulders, arms and hands from extended periods of sitting in front of a computer. Good ergonomic practices for the use of CAT tools can also be conveyed during translator training programmes, for example in the context of considering their usability (see Brunette & O'Brien, 2011; Krüger, 2016). By including more explanation of the ergonomic benefits of individualizing settings, translators might be encouraged to take increased ownership of language technology.

Although ergonomics is usually understood to refer to working conditions, scholars have also applied ergonomic principles to the completion of any type of task, including those by learners in various educational settings. One of the first to take an ergonomic approach to education was Kao (1976), who included manual skills and scheduling as well as teaching materials, facilities, equipment and environment in his considerations. A more recent approach accounts more convincingly for the interaction of various ergonomic factors as learning progresses (see Benedyk et al., 2009) and could be used as a model in translator training. An ergonomic approach to translator training which also includes training in ergonomics could allow institutions and professional organizations to contribute to optimizing the deployment of human and technical resources at the workplace and ultimately ensuring the quality of translation and the future of the profession.

8.3 Recent developments and future directions

One of the most important lessons to be learned from putting the translator in focus and taking an ergonomics perspective is the potential for various aspects of the workplace to interact and impact on the human activity of translation. If cognitive resources are absorbed by trying to ignore distractions to finish a pressing job on time, for instance, this might impinge on creativity. This might result in lower quality, which in the worst case could lead to the loss of a client. The cognitive and economic benefits and costs of good and poor ergonomics, respectively, for human translation are under-researched topics that are especially deserving of attention as the pressures from technologies such as NMT increase. Although some ergonomic recommendations seem to be generally applicable, others may depend on factors such as geography (e.g. office temperature), culture (e.g. type of interactions) or employment status (e.g. freelance or staff). The following sections outline considerations and directions for research into cognitive load, the economics of translation workplace ergonomics, and comparisons of ergonomic conditions.

8.3.1 Cognitive load and ergonomics

Cognitive load, a construct originally from instructional psychology, is based on the assumption that humans have limited capacity to process information and that learners' working memory can be overloaded by different types of input (see Schnotz & Kürschner, 2007; Sweller, 2005). Translation is a multi-activity task, which can easily cause cognitive overload even when conditions are good. It requires translators to process input in one language and formulate output in the target language while thinking, retrieving and evaluating information from internal and external resources under tight temporal constraints. Translation never takes place in a vacuum, either: it is a situated activity that is influenced by societal expectations, information sources, technological aids, economic demands, organizational requirements and physical constraints (e.g.

Risku, 2010). Professional translators are subject to heavy demands on concentration, working memory and bilingual lexical retrieval processes, since they must also constantly keep in mind their client's requirements and target audience's needs. Just as models have been proposed to explain the effort or cognitive load involved in simultaneous interpreting (e.g. Gile, 1995/2009; Seeber, 2013), the theoretical construct of mental load has been used to explain how various factors such as time pressure, information content or input quality can affect translation performance (see Muñoz Martín, 2012, 2014).

The concept of cognitive load has been used in translation process research (e.g. Behrens, 2016; O'Brien, 2006), and has proven particularly useful in understanding that the load involved in accomplishing the intrinsic nature of the task itself (i.e. translation) is supplemented by extraneous load from dealing with task-external aspects (e.g. annoying features of CAT tools). The research reviewed in the preceding sections suggests that frustration, irritation and cognitive friction can all contribute to cognitive load. Emotional states and coping mechanisms also require the allocation of a certain amount of cognitive resources, which comes at the expense of other activities relying on the same resources, such as problem solving (Trémolière et al., 2016). There is still much potential for determining how increased cognitive load—from poor ergonomics, among other factors—can affect cognitive processing and ultimately, translation performance.

8.3.2 Workplace ergonomics and economic considerations

Calls for better ergonomic conditions can initially be met with scepticism by language service providers, translation project managers, and even translators working on their own account, since many people associate ergonomics with costly desks, adjustable chairs and strangely shaped keyboards. Such a limited focus on the cost of improving physical ergonomics ignores the economic benefits of increasing productivity by minimizing discomfort and sick leave related to musculoskeletal disorders. It is widely recognized that demanding cognitive work that requires intense concentration is best done in comfortable conditions (i.e. not too hot, cold or draughty). Office policies that give translators more control over basic aspects of their working environment, such as temperature, airflow and lighting, can foster a focus on the task and hence the quality of decision making as well as well-being and ultimately, job satisfaction.

Another ergonomic issue related to concentration that emerged in the research reported in Section 8.2.2 was the frequency of disturbances to the translation process by unrelated e-mails, chats and phone calls. If translators are often disturbed while working on a translation, they might have trouble entering or maintaining a state of flow (see Nakamura & Csikszentmihalyi, 2002). Productivity and quality can consequently suffer, since the translators have to search for the place they were before the interruption, and possibly re-read the source and/or target text in order to reconstruct in their minds the meaning that they are trying to convey.

Translators' working conditions also extend to organizational aspects such as team climate, diversity, innovativeness, respect, trust, collaboration and management support, all of which can contribute to productivity. Various scholars have expressed their concern that if an organization fails to address issues associated with such aspects, translators can feel disempowered and alienated, potentially reducing their commitment, agency and sense of responsibility for their decisions. For example, a lack of involvement in decision making at the workflow or organizational level may explain why so many translators have been resistant to taking new technology on board (see Cadwell et al., 2018). Lack of autonomy and self-determination can be detrimental to job satisfaction and company loyalty, which can result in costly staff turnover. Because of the potential for personal and societal costs, more research into the economic impact of translation workplace ergonomics is warranted.

8.3.3 *Comparisons of ergonomic conditions*

Ergonomic issues that emerged in the context of a translation workplace study carried out in Switzerland (see Ehrensberger-Dow & Massey, 2014) led to an exploratory study comparing professionals in Switzerland and Ireland (reported in Ehrensberger-Dow & O'Brien, 2015). The differences identified between the Swiss and Irish groups might have had less to do with the country than with the employment status of the respective groups (i.e. staff translators and freelancers, respectively). The international survey of professional translators launched at the FIT Congress in Berlin in 2014 took this into account, adding the category of institutional to staff and freelancer as well as leaving a space for any other self-declared employment status. By the time the online survey was closed four months later, translators from almost 50 countries had completed it, with about 100 or more responses from eight of those countries. The analyses of the total sample by employment status revealed, for example, that freelancers had the poorest physical ergonomic conditions but a greater degree of self-determination than the other groups, and that aspects of the workflow related to organizational ergonomics were better for the institutional translators than for the other two groups (Ehrensberger-Dow et al., 2016).

In a detailed comparison of the survey results from Finland and Switzerland, the translators from the two countries showed different ergonomic profiles, perhaps partly, but perhaps not only, because there were far more freelancers in the Finnish group (Ehrensberger-Dow & Jääskeläinen, 2019). Certain differences might relate to culture, such as the lower likelihood of Finnish translators using the phone to discuss their work and the higher use of e-mail than for the Swiss translators. The members of the Finnish group were also more likely to customize their CAT tools and much less likely to be irritated by them. Follow-up research would be needed to validate these and similar findings before generalizations about good ergonomic practice for the various forms of professional translation in different countries can be drawn, especially in light of rapid developments in technology and differences in working conditions.

Some caution is called for when interpreting results of any kind of survey research into ergonomics. The items in the online survey reported here were constructed based on the recommendations for good ergonomic practice for computer-related office work derived from the literature (e.g. Salvendy, 2012) and from guidelines published by insurance companies (SUVA, 2010) and governmental agencies (CCOHS, 2011), primarily from Western countries and not specifically for translation work. In addition to the usual problem of self-selection, awareness building about ergonomic issues might have taken place while the translators completed the survey. Another difficulty in generalizing from quantitative survey results is the challenge of reliably accessing information about soft issues related to organizational aspects of translation such as team climate, diversity, innovativeness, respect, trust, collaboration and management support. Other methodological approaches might be more suitable for investigating the organizational ergonomics of professional translation (e.g. Cadwell et al., 2016; Dam, 2013; Ehrensberger-Dow & Massey, 2017; Risku et al., 2017, this volume).

In any case, an ergonomics perspective can contribute to a greater understanding of the situated activity that translators engage in and the resources they draw on to deal with their complex bilingual work. Framing such research in terms of ergonomics might make it more accessible and generalizable to other domains and disciplines interested in human cognition.

Notes

1 www.iea.cc/whats/index.html
2 www.hfes.org/

3 www.iea.cc/whats/
4 https://medlineplus.gov/ergonomics.html
5 www.osha.gov/SLTC/ergonomics/
6 https://osha.europa.eu/en/publications/e-facts/efact13/view
7 www.allianzworldwidecare.com/v_1525074128583/en/docs/health-guides/Allianz_Health_Guide_Ergonomics_EN.pdf
8 https://benjamins.com/online/tsb/
9 https://aplicacionesua.cpd.ua.es/tra_int/usu/buscar.asp?idioma=en
10 For example, in German, French, Italian and English from the Swiss SUVA Accident Insurance Fund www.suva.ch/startseite-suva/service-suva/lernprogramme-suva/bildschirmarbeitsplatz-einrichten-suva.htm
11 www.iea.cc/whats/

Further reading

Ehrensberger-Dow, M. (2017). An ergonomic perspective of translation. In J. W. Schwieter, & A. Ferreira (Eds.), *The handbook of translation and cognition* (pp. 332–349). London: Wiley-Blackwell.
Physical and organizational issues, which are only touched on in this chapter, are dealt with in more depth in the 2017 volume.

Lavault-Olléon, E. (2016). Traducteurs à l'œuvre: une perspective ergonomique en traductologie appliquée / Translators at work: An ergonomic perspective in applied translation studies. *ILCEA 27 Approches ergonomiques des pratiques professionnelles et des formations des traducteurs*. https://journals.openedition.org/ilcea/4051
The first collection of articles specifically focused on the importance of various aspects of ergonomics for translation and interpreting practice and research.

van Egdom, G.-W., Kockaert, H., Segers W., & Cadwell, P. (forthcoming). Ergonomics in translator and interpreter training. Special Issue of *Interpreter and Translator Trainer*.
This is the first special issue devoted to ergonomics and the implications of an ergonomics perspective for T&I education.
See also the chapters on *Translation, human–computer interaction and cognition* (by Sharon O'Brien) and *Translation and situated, embodied, distributed, embedded and extended cognition* (by Hanna Risku and Regina Rogl) in the present Handbook.

References

ALPAC (1966). *Language and machines. Computers in translation and linguistics*. Washington, DC: National Research Council.
ANSI/HFES (2007). *Human factors engineering of computer workstations*. Santa Monica, CA: Human Factors and Ergonomics Society.
Baethge, A., & Rigotti, T. (2010). *Arbeitsunterbrechungen und Multitasking: Ein umfassender Überblick zu Theorien und Empirie unter besonderer Berücksichtigung von Altersdifferenzen*. Dortmund: Bundesanstalt für Arbeitsschutz und Arbeitsmedizin.
Beale, R., & Peter, C. (2008). The role of affect and emotion in HCI. In R. Beale, & C. Peter (Eds.), *Affect and emotion in HCI* (pp. 1–11). Berlin: Springer.
Behrens, B. (2016). The task of structuring information in translation. In M. Carl, S. Bangalore, & M. Schaeffer (Eds.), *New directions in empirical translation process research* (pp. 265–278). Cham: Springer.
Benedyk, R., Woodcock, A., & Harder, A. (2009). The hexagon-spindle model for educational ergonomics. *Work, 32*, 237–248.
Bevan, N. (1982). Psychological and ergonomic factors in machine translation. In V. Lawson (Ed.), *Practical experience of machine translation* (pp. 75–78). Amsterdam: North-Holland.
Brunette, L., & O'Brien, S. (2011). Quelle ergonomie pour la pratique postéditrice des textes traduits? *ILCEA* 14. https://journals.openedition.org/ilcea/1081
Bundgaard, K., Christensen, T. P., & Schjoldager, A. (2016). Translator-computer interaction in action—An observational process study of computer-aided translation. *The Journal of Specialised Translation, 25*, 106–130. www.jostrans.org/issue25/art_bundgaard.pdf
Cadwell, P., Castilho, S., O'Brien, S., & Mitchell, L. (2016). Human factors in machine translation and post-editing among institutional translators. *Translation Spaces, 5*(2), 222–243.

Cadwell, P., O'Brien, S., & Teixeira, C. S. C. (2018). Resistance and accommodation: Factors for the (non-) adoption of machine translation among professional translators. *Perspectives, 26*(3), 301–321.

CCOHS (2011). *Office Ergonomics Safety Guide* (6th ed.). Hamilton, ON: CCOHS.

Chesterman, A. (2009). The name and nature of translator studies. *Hermes–Journal of Language and Communication Studies, 42*, 13–22.

Chesterman, A. (2013). Models of what processes? *Translation and Interpreting Studies, 8*(2), 155–168.

Cooper, A. (2004). *The inmates are running the asylum: Why hi-tech products drive us crazy and how to restore the sanity*. Indianapolis, IN: Sams Publishing.

da Costa, B. R., & Vieira, E. R. (2010). Risk factors for work-related musculoskeletal disorders: A systematic review of recent longitudinal studies. *American Journal of Industrial Medicine, 53*(3), 285–323.

Dam, Helle V. (2013). The translator approach in translation studies—reflections based on a study of translators' weblogs. In M. Eronen, & M. Rodi-Risberg (Eds.), *Haasteena näkökulma, Perspektivet som utmaning, Point of view as challenge, Perspektivität als Herausforderung* (pp. 16–35). Vaasa: VAKKI Publications 2.

Dul, J. (2003). The strategic value of ergonomics for companies. In H. Luczak, & K. J. Zink (Eds.), *Human factors in organizational design and management—VII* (pp. 765–770). Santa Monica, CA: IEA Press.

ECCE (1982). *Cognitive engineering. A conference on the psychology of problem solving with computers*. Amsterdam: Vrije Universiteit Amsterdam. www.eace.net/proceedings/ECCE%201982.pdf

Ehrensberger-Dow, M. (2014). Challenges of translation process research at the workplace. *MonTI, Special Issue, 1*, 355–383.

Ehrensberger-Dow, M. (2017). An ergonomic perspective of translation. In J. W. Schwieter, & A. Ferreira (Eds.), *The handbook of translation and cognition* (pp. 332–349). London: Wiley-Blackwell.

Ehrensberger-Dow, M., & Hunziker Heeb, A. (2016). Investigating the ergonomics of a technologized translation workplace. In R. Muñoz Martín (Ed.), *Reembedding translation process research* (pp. 69–88). Amsterdam: John Benjamins.

Ehrensberger-Dow, M., Hunziker Heeb, A., Massey, G., Meidert, U., Neumann, S., & Becker, H. (2016). An international survey of the ergonomics of professional translation. *ILCEA, 27*. https://journals.openedition.org/ilcea/4004

Ehrensberger-Dow, M., & Jääskeläinen, R. (2019). Ergonomics of translation: Methodological, practical and educational implications. In H.V. Dam, M. N. Brøgger, & K. K. Zethsen (Eds.), *Moving boundaries in translation studies* (pp. 132–150). London: Routledge.

Ehrensberger-Dow, M., & Massey, G. (2014). Cognitive ergonomic issues in professional translation. In J. W. Schwieter, & A. Ferreira (Eds.), *The development of translation competence: Theories and methodologies from psycholinguistics and cognitive science* (pp. 58–86). Newcastle upon Tyne: Cambridge Scholars.

Ehrensberger-Dow, M., & Massey, G. (2017). Socio-technical issues in professional translation practice. *Translation Spaces, 6*(1), 104–121.

Ehrensberger-Dow, M., & O'Brien, S. (2015). Ergonomics of the translation workplace: Potential for cognitive friction. *Translation Spaces, 4*(1), 98–118.

Elsler, D., Takala, J., & Remes, J. (2017). *An international comparison of the cost of work-related accidents and illnesses.* Brussels: European Agency for Safety and Health at Work.

EMT (2017). *European master's in translation competence framework 2017*. Brussels: European Commission.

Eurofound (2017). *6th European working conditions survey—Overview report* (2017 update). Luxembourg: Publications Office of the European Union.

GAO (1997). *Worker protection. Private sector ergonomics programs yield positive results*. Washington, DC: United States General Accounting Office.

Gile, D. (1995/2009). *Basic concepts and models for interpreter and translator training*. Amsterdam: John Benjamins.

Göpferich, S. (2009). Towards a model of translation competence and its acquisition: The longitudinal study *TransComp*. In S. Göpferich, A. L. Jakobsen, & I. M. Mees (Eds.), *Behind the mind. Methods, models and results in translation process research* (pp. 17–43). Copenhagen: Samfundslitteratur.

Gouadec, D. (2007/2010). *Translation as a profession*. Amsterdam: John Benjamins.

Hansen, G. (2006). *Erfolgreich übersetzen. Entdecken und Beheben von Störquellen*. Tübingen: Narr Francke Attempto.

Hansen-Schirra, S. (2012). Nutzbarkeit von Sprachtechnologien für die Translation. *Trans-Kom, 5*(2), 211–226.

Hoffman, R. R., Ward, P., Feltovich, P. J., DiBello, L., Fiore, S. M., & Andrews, D. H. (2014). *Accelerated expertise: Training for high proficiency in a complex world*. New York: Psychology Press.

Holmes, J. S. (1972/2000). The name and nature of translation studies. In L. Venuti (Ed.), *The translation studies reader* (pp. 172–185). London: Routledge.

Hurtado Albir, A., & Alves, F. (2009). Translation as a cognitive activity. In J. Munday (Ed.), *The Routledge companion to translation studies* (pp. 54–73). London: Routledge.
Huysmans, M. A., van der Ploeg, H. P., Proper, K. I., Speklé, E. M., & van der Beek, A. J. (2015). Is sitting too much bad for your health? *Ergonomics in Design, July*, 4–8.
ISO 17100 (2015). *Translation services—Requirements for translation services.* Geneva: International Organization for Standardization.
Jakobsen, A. L. (2002). Translation drafting by professional translators and by translation students. In G. Hansen (Ed.), *Empirical translation studies: Process and product* (pp. 191–204). Copenhagen: Samfundslitteratur.
Jastrzebowski, W. (1857/2006). An outline of ergonomics, or the science of work based upon the truths drawn from the science of nature. In W. Karwowski (Ed.), *International encyclopedia of ergonomics and human factors* (Vol. 3, 2nd ed., pp. 129–141). Boca Raton, FL: CRC Press.
Kao, H. (1976). On educational ergonomics. *Ergonomics, 16*(6), 667–681.
Krüger, R. (2015). Fachübersetzen aus kognitionstranslatologischer Perspektive. Das Kölner Modell des situierten Fachübersetzers. *Trans-Kom, 8*(2), 273–313.
Krüger, R. (2016). Contextualising computer-assisted translation tools and modelling their usability. *Trans-Kom, 9*(1), 114–148.
Lavault-Olléon, E. (2011a). L'ergonomie, nouveau paradigme pour la traductologie. *ILCEA, 14*. https://journals.openedition.org/ilcea/1078?lang=en.html
Lavault-Olléon, E. (2011b). Une introduction à la problématique "Traduction et ergonomie". *ILCEA, 14*, Traduction et Ergonomie. https://journals.openedition.org/ilcea/1118
Lavault-Olléon, E. (2016). Traducteurs à l'œuvre: une perspective ergonomique en traductologie appliquée / Translators at work: an ergonomic perspective in applied translation studies. *ILCEA*, 27, Approches ergonomiques des pratiques professionnelles et des formations des traducteurs. https://journals.openedition.org/ilcea/4051
Lavault-Olléon, E., & Carré, A. (2012). Traduction spécialisée: L'ergonomie cognitive au service de la formation. *ASP, 62*, 67–77.
LeBlanc, M. (2013). Translators on translation memory (TM). Results of an ethnographic study in three translation services and agencies. *The International Journal for Translation & Interpreting, 5*(2), 1–13.
Marshman, E. (2014). Taking control: Language professionals and their perception of control when using language technologies. *Meta, 59*(2), 380–405.
Meidert, U., Neumann, S., Ehrensberger-Dow, M., & Becker, H. (2016). Physical ergonomics at translators' workplaces: Findings from ergonomic workplace assessments and interviews. *ILCEA, 27*. https://journals.openedition.org/ilcea/3996
Mesa-Lao, B. (2014). Gaze behaviour on source texts: An exploratory study comparing translation and post-editing. In S. O'Brien, L. Winther Balling, M. Carl, M. Simard, & L. Specia (Eds.), *Post-editing of machine translation: Processes and applications* (pp. 219–245). Newcastle upon Tyne: Cambridge Scholars.
Moorkens, J., & O'Brien, S. (2013). User attitudes to the post-editing interface. In *Proceedings of MT Summit XIV workshop on post-editing technology and practice* (pp. 19-25). Allschwil: The European Association for Machine Translation.
Moorkens, J., & O'Brien, S. (2016). Assessing user interface needs of post-editors of machine translation. In D. Kenny (Ed.), *IATIS yearbook 2016* (pp. 109–130). London: Routledge.
Muñoz Martín, R. (2010). On paradigms and cognitive translatology. In G. M. Shreve, & E. Angelone (Eds.), *Translation and cognition* (pp. 169–187). Amsterdam: John Benjamins.
Muñoz Martín, R. (2012). Just a matter of scope. Mental load in translation process research. *Translation Spaces, 1*, 169–178.
Muñoz Martín, R. (2014). Situating translation expertise. A review with a sketch of a construct. In J. W. Schwieter, & A. Ferreira (Eds.), *The development of translation competence. Theories and methodologies from psycholinguistics and cognitive science* (pp. 2–56). Newcastle upon Tyne: Cambridge Scholars.
Nakamura, J., & Csikszentmihalyi, M. (2002). Flow theory and research. In C. R. Snyder, & S. J. Lopez (Eds.), *Handbook of positive psychology* (pp. 89–105). Oxford: Oxford University Press.
Niu, S. (2010). Ergonomics and occupational safety and health: An ILO perspective. *Applied Ergonomics, 41*, 744–753.
O'Brien, S. (2006). Pauses as indicators of cognitive effort in post-editing machine translation output. *Across Languages and Cultures,* 7(1), 1–21.
O'Brien, S. (2012). Translation as human–computer interaction. *Translation Spaces, 1*, 101–122.
O'Brien, S., & Ehrensberger-Dow, M. (2017). Why ergonomics matters to translators. *ATA Chronicle, 46*(1), 12–14. www.atanet.org/chronicle-online/featured/why-ergonomics-matters-to-professional-translators/

O'Brien, S., Ehrensberger-Dow, M., Hasler, M., & Connolly, M. (2017). Irritating CAT tool features that matter to translators. *Hermes Journal of Language and Communication in Business, 56*, 145–162.

PACTE (2003). Building a translation competence model. In F. Alves (Ed.), *Triangulating translation: Perspectives in process-oriented research* (pp. 43–66). Amsterdam: John Benjamins.

Peters-Geiben, L. (2016). La prévention comportementale et contextuelle: Intégrer une approche ergonomique dans la formation des traducteurs. *ILCEA, 27*. http://journals.openedition.org/ilcea/4026

Pym, A. (2011). What technology does to translating. *Translation & Interpreting, 3*(1), 1–9. www.trans-int.org/index.php/transint/article/view/121/81

Risku, H. (2002). Situatedness in translation studies. *Cognitive Systems Research, 3*, 523–533.

Risku, H. (2010). A cognitive scientific view on technical communication and translation. Do embodiment and situatedness really make a difference? *Target, 22*(1), 94–111.

Risku, H. (2014). Translation process research as interaction research: From mental to socio-cognitive processes. *MonTI, Special Issue, 1*, 331–353.

Risku, H., Rogl, R., & Milosevic, J. (2017). Translation practice in the field: Current research on socio-cognitive processes. *Translation Spaces, 6*(1), 3–26.

Robbins, P., & Aydede, M. (Eds.) (2008). *The Cambridge handbook of situated cognition*. Cambridge: Cambridge University Press.

Robertson, M. M., & O'Neill, M. J. (2003). Reducing musculoskeletal discomfort: Effects of an office ergonomics workplace and training intervention. *International Journal of Occupational Safety and Ergonomics, 9*(4), 491–502.

Salvendy, G. (2012). *Handbook of human factors and ergonomics* (4th ed.). Hoboken, NJ: Wiley.

Schnotz, W., & Kürschner, C. (2007). A reconsideration of cognitive load theory. *Educational Psychology Review, 19*(4), 469–508.

Seeber, K. (2013). Cognitive load in simultaneous interpreting. Measures and methods. *Translation and Interpreting Studies, 8*(2), 18–32.

Smith-Jackson, T. L., & Klein, K. W. (2009). Open-plan offices: Task performance and mental workload. *Journal of Environmental Psychology, 29*, 279–289.

Starrett, K., & Cordoza, G. (2016). *Deskbound: Sitting is the new smoking*. London: Victory Belt Publishing.

Stramler, J. H. (1993). *The dictionary for human factors/ergonomics*. Boca Raton, FL: CRC Press.

Stuster, J. (2006). *The human factors and ergonomics society: Stories from the first 50 years*. Santa Monica, CA: IEA Press.

SUVA (2010). *Bildschirmarbeit: Wichtige Informationen für Ihr Wohlbefinden*. Lucerne: SUVA.

Sweller, J. (2005). Implications of cognitive load theory for multimedia learning. In R. E. Mayer (Ed.), *The Cambridge handbook of multimedia learning* (pp. 19–30). New York, NY: Cambridge University Press.

Taravella, A., & Villeneuve, A. O. (2013). Acknowledging the needs of computer-assisted translation tools users: The human perspective in human-machine translation. *The Journal of Specialised Translation, 19*, 62–74.

Teixeira, C. S. C. (2014). Perceived vs. measured performance in the post-editing of suggestions from machine translation and translation memories. In S. O'Brien, M. Simard, & L. Specia (Eds.), *Proceedings of the third workshop on post-editing technology and practice (WPTP-3)*. www.mt-archive.info/10/AMTA-2014-TOC.htm

Teixeira, C., & O'Brien, S. (2017). Investigating the cognitive ergonomic aspects of translation tools in a workplace setting. *Translation Spaces, 6*(1), 79–103.

Toury, G. (2012). *Descriptive translation studies—and beyond* (revised ed.). Amsterdam: John Benjamins.

Trémolière, B., Gagnon, M.-E., & Blanchette, I. (2016). Cognitive load mediates the effect of emotion on analytical thinking. *Experimental Psychology, 63*(6), 343–350.

UK Translator Survey (2016/2017). *UK translator survey final report*. European Commission Representation in the UK/ Chartered Institute of Linguists (CIOL)/ Institute of Translation and Interpreting (ITI).

van Egdom, G.-W., Kockaert, H., Segers, W., & Cadwell, P. (forthcoming). Ergonomics in translator and interpreter training. Special Issue of *Interpreter and Translator Trainer*.

Vischer, J. C. (2007). The effects of the physical environment on job performance: Towards a theoretical model of workspace stress. *Stress and Health, 23*, 175–184.

Vischer, J. C. (2008). Towards an environmental psychology of workspace: How people are affected by environments for work. *Architectural Science Review, 51*(2), 97–108.

Wahlström, J. (2005). Ergonomics, musculoskeletal disorders and computer work. *Occupational Medicine, 55*, 168–176.

9

Translation, ontologies and cognition

Adriana S. Pagano

9.1 Introduction

This chapter introduces ontologies with a view to showing their potential for approaching the concept of domain in models of translation competence and expertise. This represents an emergent perspective in Translation Studies, as few scholars have addressed ontologies in the discipline until recently (see, for instance, Budin, 2005), and those who have dealt with the topic have done so basically for terminology purposes. The chapter begins by problematizing how domain knowledge is at present accommodated in studies of translation competence and expertise. It then goes on to present the concept of ontology and provides a brief overview of different types of ontologies in order to make the case for a language-motivated ontology such as the generalized upper model (Bateman, 1990, 1992, 1995, 1997; Bateman et al., 2010) within a proposal for building domain databases to inform translation and other multilingual tasks. Drawing on systemic functional theory (Halliday & Matthiessen, 1999), a language-based approach to human cognition is taken, with language posited as central to human consciousness and actions. This supports a model whereby semantic representations are linked to forms that may share commonalities as well as differences across languages. Fine granularity in language description tuned to contextual variables ensures that text type-specific meanings are captured. Examples retrieved from questionnaires in the healthcare domain, originally written in English and translated and cross-culturally adapted into Brazilian Portuguese, will be used to illustrate the ontological approach. The chapter concludes with suggestions for future research on ontologies in translation and multilingual text production in different modes of human–machine interaction.

9.1.1 Domain knowledge in translation competence and expertise

The role of domain knowledge in translation tasks has been acknowledged by many (if not most) theoretical and empirical models of translation. Not only Translation Studies scholars but also computer scientists include domain in their modelling, the latter particularly for machine translation purposes. In most cases, domain is dealt with as an independent competence or module, disconnected from a language competence or component. This conceptualization builds on theories that establish a clear-cut separation between language, reality and thought, holding a "code view" of language that separates what is called "linguistic form" from "conceptual content".

This separation is evident in translator competence models (cf. Göpferich, 2009; PACTE, 2001, 2003), which posit translation as an activity in which "content" is decoded from and re-encoded into "linguistic form". These models draw on the distinction between declarative (know what) and procedural knowledge (know how). Declarative knowledge is seen as being implicated in domain knowledge, also called subject matter knowledge, while procedural knowledge is placed at the core of the strategic nature of translating proper. A similar picture can be seen in accounts of translation expertise (Shreve, 2006), which seems to point to the way most models accommodate the fact that translators acquire expertise in the so-called translation domain, i.e. they know how to solve translation problems, but, due to the task at hand, have to deal with domains they are not expert in, i.e. subject matter they have no knowledge about. As a result of this view, domain skills are deemed unnecessary within translation competence development curriculum design (cf. Hurtado, 2005, p. 43, endnote 19). To make up for the translator's lack of domain knowledge, developing documenting skills is proposed. In some models, domain is treated as a matter of accrued knowledge over time spent on deliberate practice within translation expertise approaches (Shreve, 2006, p. 30).

Interestingly, it is research in the field of expertise studies, particularly in expertise and disciplinary writing, that has provided insight into how domain knowledge in tasks involving text production, as in translation, is inextricably connected with linguistic knowledge. This challenges the view that dissociates language from reality, suggesting synergy between expertise in producing text and expertise in the subject matter of text (see esp. Scardamalia & Bereiter, 1991). The role of language in expertise is further supported by the notion of "interactional expertise" (Collins, 2004; Collins et al., 2006), a concept naming a particular type of expertise acquired through engagement in social activities within a particular domain, accounting for the fact that individuals can produce texts in that domain expertly even though they cannot become practitioners in the domain. In the case of translation, more specifically interpreting, Ribeiro (2007) has shown that interactional expertise can be developed by professional interpreters who work in a context of socialization in industrial settings. This allows those interpreters to talk with confidence about the workplace domain—steel-making in Ribeiro's research—although they are not able to be practitioners or contribute to steel-making practices. Interactional expertise seems to point to language's key role in domain construction, the issue of becoming a domain practitioner being possibly related to how domain-oriented socialization takes place.

In all cases, research pointing to the intrinsic relation between language and domain, such as the above-mentioned studies, could well benefit from a more comprehensive theory of language, one that does not model language as a code for representing things and relations in a pre-existing world of experience. One such theory is systemic functional linguistics (hereafter SFL) (Halliday & Matthiessen, 2014), which conceives of language as inextricably embedded in social context. From an SFL perspective, a domain is construed through language, and it is only through language that it can be construed in translation (Matthiessen, 2001).

Mapping a domain in such a way that information can be leveraged for translation purposes, other than for terminology queries, is challenging in the sense that information needs to be organized to cover how meanings are worded in a particular domain for particular text types in a language in a way that permits comparison across different languages. Ontologies are frameworks that allow this kind of organization, as the following discussion will attempt to show.

9.2 Core topics

9.2.1 Ontologies in domain modelling

An ontology is a way of abstracting and modelling some aspect of reality through the definition of concepts and their properties. This is formally specified in an unambiguous formal language with a natural language interface, so that it can be processed by machines and understood by human beings (see Gruber, 1993).

In order to devise a formal model of some domain of reality through an ontology, main concepts in that domain are defined and grouped into *classes*. One way of doing this is by mining a corpus of texts pertaining to the domain to extract likely candidates to terms and consulting domain experts (see Huang, 2010; Obrst, 2010; Uschold & King, 1995). For instance, to model a domain such as diabetes self-care, a first step is to mine texts on diabetes self-care management. Highly frequent words are likely candidates to terms, which can lead to formalization of concepts within the domain. For instance: person with diabetes; disease; healthcare provider; nurse; physician; dietitian; dentist; physiotherapist; diabetes type 1; diabetes type 2; blood sugar; insulin. Once concepts are established, they can be deployed in a conceptual map to define relationships between them. Figure 9.1 illustrates such a conceptual map.

In an ontology, concepts are posited as classes abstracting characteristics that apply to all members in the class. In Figure 9.1, "people" is part of "diabetes self-care" and abstracts characteristics of subclasses such as "person with diabetes" and "healthcare provider". "Nurse", "physician", "dentist", "dietitian" and "physiotherapist" are, in turn, subclasses of the class "healthcare provider". The relationships between classes are expressed by predicates such as "is a type of", "is a member of" or "has".

Individuals in an ontology are objects within a domain. Being instances of a class, individuals are bound to properties defining relations between them (object properties), which must be consistent with the relations holding between the classes they are instances of. If, for example, individual "patient 101" is an instance of the class "person with diabetes", that individual cannot be an instance of the class "healthcare provider". This is due to a restriction whereby "healthcare provider" and "person with diabetes" are two distinct subclasses individuals can be assigned to within an ontology. Even if in real life a person happens to be at the same time a healthcare provider and a patient with diabetes and, as a healthcare provider, takes care of his/her treatment him/herself, the two different roles that person takes—as a healthcare provider and as a person with diabetes—are modelled as two distinct individuals linked to two distinct classes in the ontology.

Explanations of how *classes* relate to one another, *individuals* to one another, and *classes* to *individuals* are *axioms*. *Axioms* are declarations about the elements of an ontology. *Axioms* are established by ontology designers; additionally, computer systems can use them to infer new *axioms*, thus expanding the power of an ontology to explain that experiential domain. For example, if individual "Dr. Lawrence" is an instance of the class "Physician" and "Dr. Wu" is an instance of the class "Nurse", based on the *axiom* "Dr. Lawrence works with Dr. Wu", a computer may create its own *axiom* "a physician works with a nurse". A computer can expand an ontology and its potential by inferring new *axioms* from those stated by the human developers who designed that ontology.

The methodology for building a domain ontology generally relies on categories posited by human experts or on the output of text mining for candidate terms that name conceptual categories, or a combination of the two procedures. In all cases, conceptual categories are arrived at after natural language terms are considered. In some ontologies, the boundary between

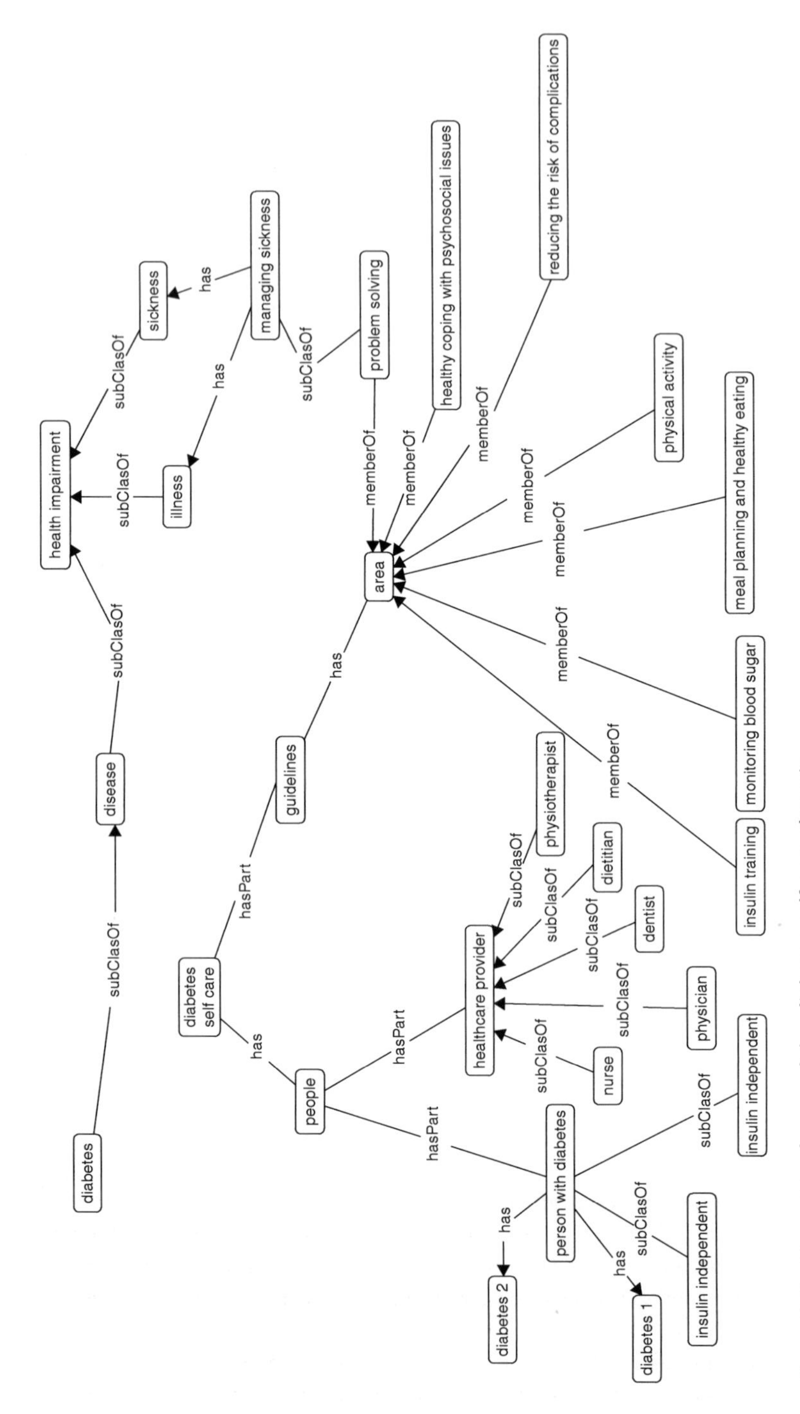

Figure 9.1 Conceptual map of the diabetes self-care domain

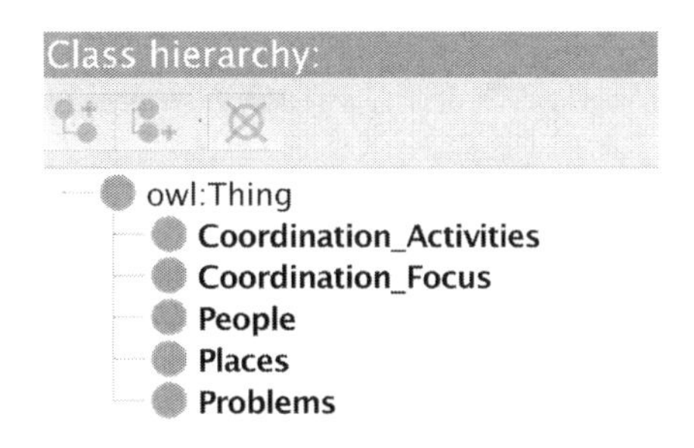

Figure 9.2 Protégé screenshot showing main classes in the nursing care coordination ontology

natural language terms and conceptual categories is nonexistent, as in the case of the nursing care coordination ontology,[1] developed by Sinclair School of Nursing, University of Missouri. This ontology seeks to represent activities and practices carried out by practising nurse care coordinators and was built drawing on notes written by practising nurses. The designers mined nurse notes for candidate terms that could point to concepts. These terms were organized by domain experts into concepts, which became the main classes in the ontology. At a more general level, the classes shown in Figure 9.2 were established.

Each class, in turn, has subclasses which further specify it. The class Coordination Activities, for example, has subclasses, which in turn are further subclassified, as shown in Figure 9.3.

This particular ontology relies on natural language to build a representation of a domain, and, in this sense, that representation reveals much about how nurses construe the domain of care they are in charge of. From an applied perspective, the ontology developers state that the nursing care coordination ontology is meant to match the language nurses use, so that the ontology could be used to automatically extract information from a larger corpus of nurses' notes. Making conceptual categories match natural language terms facilitates automatic instance recognition in texts. As an additional application, an ontology of this kind can be extended multilingually and used as a translation aid for texts within the specific domain. The potential of such a multilingual tool is further enhanced, however, if natural language is mapped within a linguistic ontology, as we will show in our discussion of the GUM ontology in Section 9.2.4.

The nursing care coordination ontology illustrates one of the various purposes domain ontologies are designed for, which range from fulfilling the basic need to better understand the way a domain is organized to automatized implementations in systems to allow discovery, execution and monitoring of services and resources. They can be used in decision support systems and be reused by other ontologies to further expand them (see Currás, 2010). An example of a domain ontology that reuses and is reused by other ontologies is the diabetes mellitus diagnosis ontology,[2] an ontology that represents diabetes diagnosis, treatment, manifestation, development, symptoms and laboratory tests (see Figure 9.4).

The diabetes mellitus diagnosis ontology indexes information that can be reused, for instance, in an ontology of hospital patients. If a particular patient has been assigned a number of symptoms, which are classified as diabetes symptoms in the diabetes mellitus diagnosis ontology, the reasoner mechanism within the patient ontology can infer that this particular patient has diabetes. This can support decision making in diagnosis and treatment. Ontologies can also be used to retrieve text dealing with particular topics from a large text corpus.

The diabetes mellitus diagnosis ontology was originally designed to be used in medical settings to assist accurate diagnosis and treatment based on patients' health condition and data. However, when reused in another ontology, it can potentially assist tasks of a different nature,

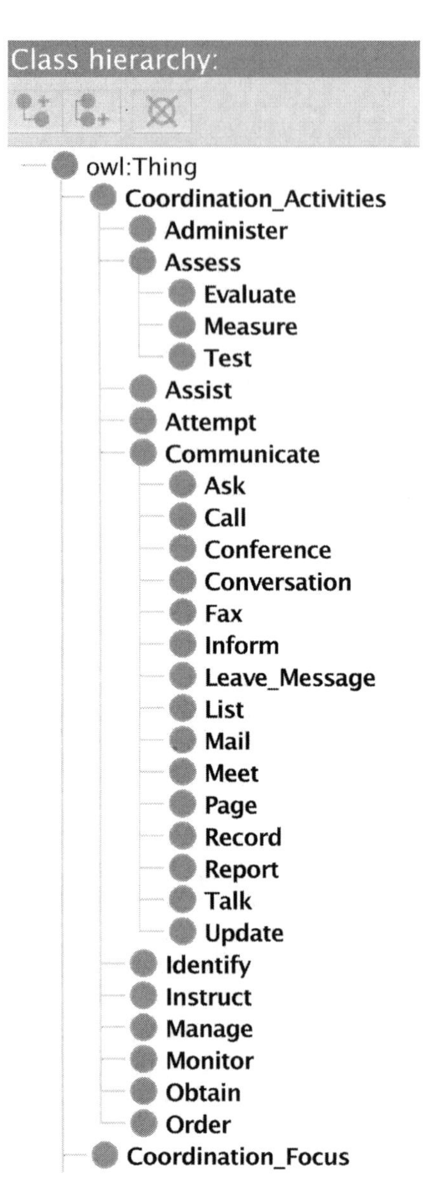

Figure 9.3 Protégé screenshot showing subclasses in the nursing care coordination ontology

such as text indexing. It can, for instance, be aligned with a similar ontology in languages other than English and used to retrieve and classify texts dealing with diabetes mellitus topics from a large multilingual text base.

Applied linguists and translators may regard domain ontologies as sources for terminology. However, ontologies offer many more resources to assist translation (both human and machine) and are far richer than dictionaries and thesauri. Through ontologies, analogous experiential domains in different contexts of culture can be compared. For example, a *diabetes self-care domain* ontology drawing on information from the North American and the Brazilian contexts can have analogous *classes* such as "nurse", "dentist", "physiotherapist" and 'physician". However,

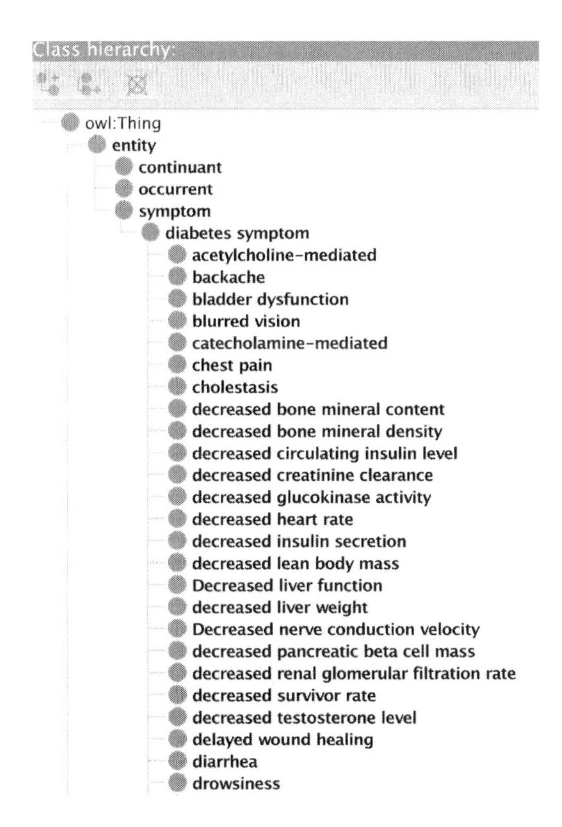

Figure 9.4 Protégé screenshot of classes in the diabetes mellitus diagnosis ontology

axioms applying in each context can be different. For instance, results of a diabetes patient's urine analysis test can be interpreted by professional healthcare providers, including nurses with a first degree in nursing, in the North American context. On the other hand, in the Brazilian context of culture, even though nurses are professional healthcare providers, they are not legally competent to interpret a urine analysis test result. Thus, in an ontology drawing on the North American context of culture, we could declare the *axiom* "a nurse can interpret urine analysis test results". That axiom would not hold in an ontology drawing on the Brazilian context of culture, where an *axiom* would need to state that "only a physician can interpret urine analysis test results".

A multilingual ontology mapping a domain common to different contexts of culture is particularly relevant in the case of medical translation and cross-cultural adaptation. Querying such an ontology would yield information that can be essential to cross-culturally adapt texts. While a bilingual dictionary or a thesaurus provides equivalents for words such as "nurse" in English and "enfermeira" [nurse] in Brazilian Portuguese, a domain ontology models both the conditions that support equivalence and the conditions under which such equivalence no longer holds, as is the case when urine analysis test results need to be assessed.

Modelling a domain in different contexts of culture is a challenge to automatic domain characterization and certainly requires manual design and validation by human experts. Nevertheless, the task is worth performing. If, on the one hand, such an endeavour demands a very thorough survey of concepts so that necessary *axioms* can account for specificities in each context of culture, its output, on the other hand, largely outperforms a terminology bank and provides a more informed perspective to translation and multilingual tasks. Still, as the following section will show, other types of ontology prove even more resourceful for domain modelling in translation and multilingual text production. This is the case of domain-independent ontologies or upper ontologies.

9.2.2 Upper ontologies

Regardless of their potential for a variety of uses, domain ontologies, such as the nursing care coordination ontology and the diabetes mellitus diagnosis ontology, are restricted to particular domains. This restriction can be overcome through the use of upper ontologies, or top ontologies, which are domain independent in the sense that they describe general concepts that can be used to model any domain. An example of an upper ontology is SUMO (suggested upper merged ontology)[3] (Niles & Pease, 2001), which has general categories for mapping material objects, abstractions and relations, as well as processes and attributes of human experience, as can be seen in Figure 9.5.

Both upper and domain ontologies are sometimes enhanced with links to lexical databases. In this case, vocabularies organized in hypernym, hyponym and synonym relationships are mapped onto ontological concepts (Gómez-Pérez et al., 2004) and allow more fine-grained applications, including text mining and document classification. SUMO, for example, has mappings to the lexical database WORDNET.[4] As an illustration, the lexical term "insulin" in WORDNET is mapped onto the term "insulin" in SUMO and can be located under the subclass "hormone", as can be seen in Figure 9.6.

The rationale underlying SUMO and other upper ontologies is that there are general categories to which all domains can ultimately relate. This assumes an abstract conceptual model of the world, the categories of which, as in the case of domain ontologies, need to be named by means of natural language. This assumption, however, is not without its weaknesses.

Labels for classes in an ontology, given on the assumption that they name abstract categories in the world, are no less natural language terms than words extracted from language in use in natural texts. This could be clearly observed in the nursing care coordination ontology, mentioned

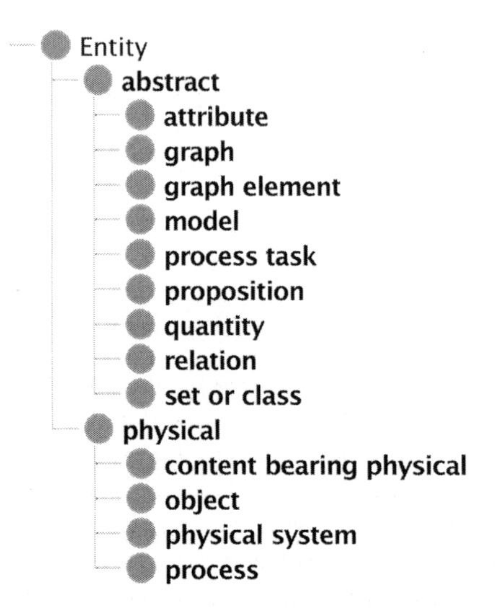

Figure 9.5 Protégé screenshot of primary classes in SUMO

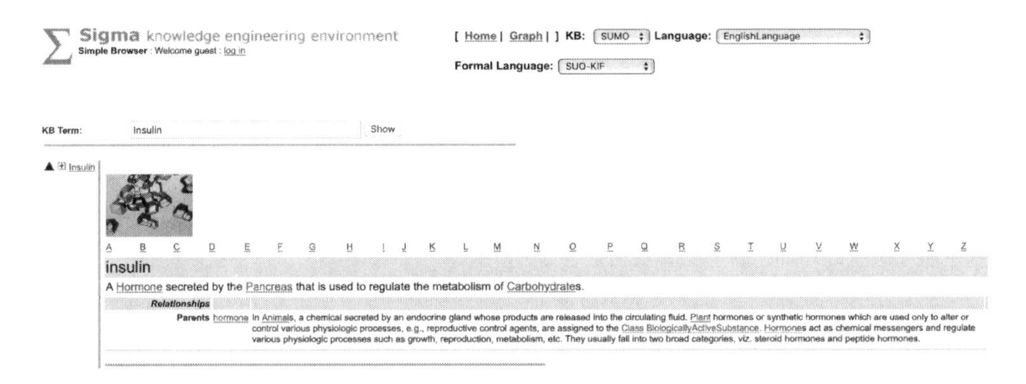

Figure 9.6 Screenshot of output for the word "insulin" by SUMO Search Tool

earlier, in which classes were named by the very words extracted from nurses' notes. This can also be concluded from the query performed in SUMO for the word "insulin", as seen in Figure 9.6, which was mapped onto the term "insulin".

Rather than choose arbitrary labels to name purportedly universal and consensual categories organizing human reality, adopting a theory of language, thought and reality as being inherently interconnected makes it possible to abstract ontological categories from the semantic resources of every language. This makes it possible, for example, to map "insulin" in a large text corpus and examine how different meanings are construed, not all of them being readily mapped onto particular terms as defined in a domain ontology. In other words, drawing on instances of "insulin" in natural text and analysing them as resources whereby grammar construes and represents reality offers us more insights into domain construal.

This is the proposal put forward by SFL, which views natural language as a "theory of human experience" (Halliday & Matthiessen, 2014, p. 30). In this sense, language motivates ontology classes, their relations between one another and axioms governing restrictions. This has been implemented in an upper ontology, the generalized upper model or GUM (Bateman et al., 1995), which makes it possible to model experiential domain drawing on SFL semantic and grammatical categories. GUM is a linguistically motivated ontology that models reality through meaning construal categories, relying on the SFL assumption that language is the starting point for construing our experience. A brief description of SFL is provided in the following section in order to show how categories in GUM are sufficiently abstract to describe meaning construed in distinct languages and to operationalize a shared representation of them which is relevant to multilingual tasks, including translation.

9.2.3 A linguistic approach to modelling domain

SFL conceives of language as a socio-semiotic system whereby we construe meanings that create our experience of the world (Halliday, 1978). Experience is not seen as a system external to language; rather, conceptualizations of the world are modelled as a level of representation inside the language system—called the semantic level—which is inherently linked to the grammatical level. Thus, what is commonly referred to as "knowledge" in expertise studies is approached as "meaning" in SFL (see Halliday & Matthiessen, 1999). SFL views domain ontologies as construed in and by language, and ontology classes ultimately as points of meaning condensation, as we will attempt to show in what follows.

Table 9.1 Description of context of situation variables for diabetes self-care questionnaires

Variable	*Description*
Field	social activity of surveying patient experience about an aspect of diabetes self-management, e.g. an insulin delivery system
Tenor	power relationships between respondent (patient) and insulin delivery system manufacturers; insulin user assumed to be familiarized with the basics of insulin use, though not expected to be a specialist
Mode	written text, with no dialogic turn taking (questions posed to respondents, but no exchange with enquirers)

In SFL, meaning entails an intrinsic relation between language and social context, which is modelled at two levels of abstraction: a more abstract level, called context of culture, encompassing the meaning potential available to language users to interpret the world and interact with each other; and a less abstract level, called context of situation, being a selection of meanings from the overall potential for a given situation.

Modelling language at the level of context of situation has a special vantage point: it allows us to focus on recurrent selected meanings from the large meaning potential and, at the same time, to make generalizations about particular instances of text that can be grouped in terms of similar selections of meanings (Halliday & Matthiessen, 1999). Context of situation can be characterized as the intersection of three sets of contextual variables. These are *field*, i.e. what social purpose language is being used for (*social activity*) and what experiences construed by language are about (*domain of experience*); *tenor*, i.e. what kinds of social relationship are being established between interlocutors; and *mode*, i.e. how meanings are organized in text, in terms of medium (spoken or written), channel (oral, graphic, electronic) and turn taking (monologue or dialogue).

The three sets of variables are interlinked but can clearly be made discrete, e.g. for the purpose of text analysis. If we take a domain and a particular text type as an example, such as patient questionnaires on diabetes self-care, we can roughly describe each variable as in Table 9.1.

Among the three contextual variables, *field* is the focus in domain modelling, and, in this sense, a domain of experience can be approached as the set of specific semantic selections made for a particular *field* (see Halliday & Matthiessen, 1999). Obtaining a profile of these selections allows us to construct a model based on the language used to construe an experiential domain. Thus, a model of questionnaires on diabetes self-care can be constructed by examining text samples and observing what kinds of phenomena are construed by language in those texts.

From an SFL perspective, language construes three types of phenomena: (i) phenomena perceived as doings and happenings taking place in an outer world, for instance, in such wordings as "a person with diabetes uses insulin", "insulin enters the bloodstream"; (ii) phenomena perceived as sensing-and-saying in an inner world of the speaker"s consciousness, as in "a person with diabetes feels embarrassed when injecting insulin" and "a person with diabetes worries about high blood glucose"; and (iii) phenomena perceived as relationships of being-and-having, established between objects and/or beings, as in "stomach, hips, thighs, buttocks and backs of the arms are common sites for injecting".

Each type of phenomena entails a specific configuration of semantic functions realized by grammatical functions. This configuration is called a semantic figure. Processes (goings-on) and participants (entities) are inherent functions to a semantic figure; circumstances (of time, space, manner, cause, etc.) are optional functions. The example in Table 9.2 illustrates these functions in a semantic figure of doing-and-happening.

Table 9.2 Example of a semantic figure with function labels

Clause	*The system*	*delivers*	*insulin*	*into the bloodstream.*
Function	participant	process	participant	circumstance

Table 9.3 Types of figures for phenomena in diabetes self-care domain illustrated with examples

Type of figure	*Example*
Figure of doing-and-happening	What type of *insulins* do *you* **use**? How many *injections* do *you* **take** per day? The *system* **delivers** *insulin* into the bloodstream. *Your blood sugar levels* will **go up** and **down** during the day.
Figure of sensing-and-saying	*Friends/family* **worry** about your blood glucose levels **Chat** privately with an experienced doctor online.
Figure of being-and-having	How *satisfied* **are** *you* with your current insulin delivery system? **Are** *your injections* ever *painful*? *The layer of fat on the stomach, hips, thighs, buttocks and backs of the arms* **are** *common sites for injecting insulin.*

For each type of figure, selections are made in the grammar for particular types of processes, participants and accompanying circumstances. In figures of doing-and-happening, material processes are selected, and participants can be either conscious or non-conscious beings. Figures of sensing-and-saying select mental and verbal processes and have conscious beings as participants. Figures of being-and-having select relational processes and have two inherent participants. Table 9.3 provides examples of each type, with processes in bold and participants in italic.

Drawing on Whorf (1956), SFL discriminates types of phenomena on the basis of grammatical reactances, that is, grammatical features that apply to types and distinguish them as a class, such as the number and nature of participants involved and their realization in the grammar of the language. As examples in Table 9.3 show, figures of doing-and-happening generally implicate one participant in happenings ("Your blood sugar levels will go up") and two or more participants in the case of doings ("How many injections do you take per day?"). Participants can be conscious ("you") or non-conscious beings ("injections", "blood sugar"). Figures of sensing-and-saying, in contrast, tend to select at least two participants and implicate a conscious being as main participant ("friends/family", "your blood glucose levels", "you", "an experienced doctor online"). Figures of being-and-having necessarily select two participants (conscious or non-conscious) to establish relations of identity, attribution or possession.

Patterns of semantic figures and their respective grammatical functions within a given domain can be deployed for domain characterization. Distinctive patterns reveal ways in which a domain is construed. In the case of diabetes self-care, patterns of agency upon agency, implicating an added participant role, are very frequent. There is thus a participant role of an agent inherent in a first process triggering a second participant role as agent to a second process. Table 9.4 shows examples of these agency patterns for different types of figure.

A distinctive feature in SFL modelling for the purpose of translation and multilingual applications is the principle of rank, whereby language units are organized hierarchically in terms of constituency, ranging from the highest level of the clause to the lowest of the morpheme, encompassing the levels of the group and the word in between. This is illustrated in Figure 9.7.

Table 9.4 Examples of agency patterns in the diabetes self-care domain

Figure	*Participant role 1*	*Process 1*	*Participant role 2*	*Process 2*	*Participant role 3*
doing-and happening	*The new delivery system*	*enables*	*insulin*	*to enter*	*the cells*
sensing-and saying	*Apps*	*help*	*people*	*remember*	*their insulin shot*
being-and-having	*Self-care tips*	*make*	*people*	*be*	*more active*

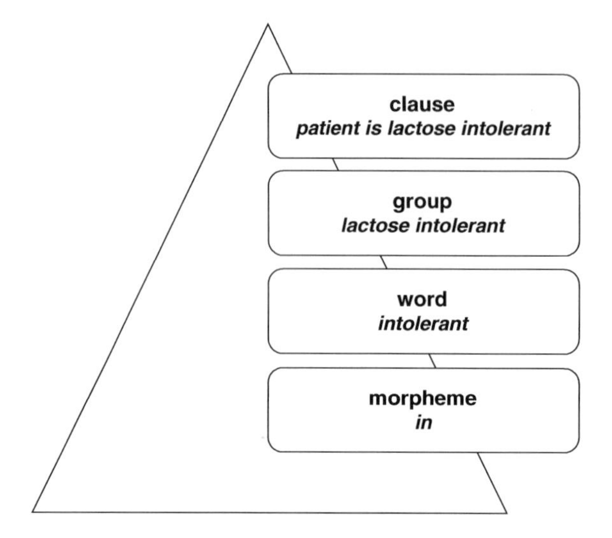

Figure 9.7 Levels of rank scale illustrated with an example

Table 9.5 Example illustrating one figure–one clause mapping

Clause	*Insulin*	*is replaced*
Function	participant	process

In Figure 9.7 we can follow the morpheme "in" construing negation in the examples provided at each level of the rank scale.

Taking rank into consideration is important, because phenomena within a domain can be construed by analogous forms at different ranks. There may be a one-to-one mapping between semantics and grammar, i.e. one figure in the semantics maps onto one clause in the grammar, as illustrated in Table 9.5.

A phenomenon, however, can also be construed and realized in the grammar by a unit at a lower level than the clause in the rank scale. The meaning illustrated in Table 9.5 can be construed and realized by a nominal group: "insulin replacement". This is usually done in order to "pack" meaning and use it in a larger construal. The clause "Insulin is replaced" is packed into the element "insulin replacement" in a new figure, such as that shown in Table 9.6.

Table 9.6 Example illustrating meaning packed into an element of a figure with participant function

Clause	*Insulin replacement*	*is known as*	*insulin therapy*
Function	Participant	process	participant

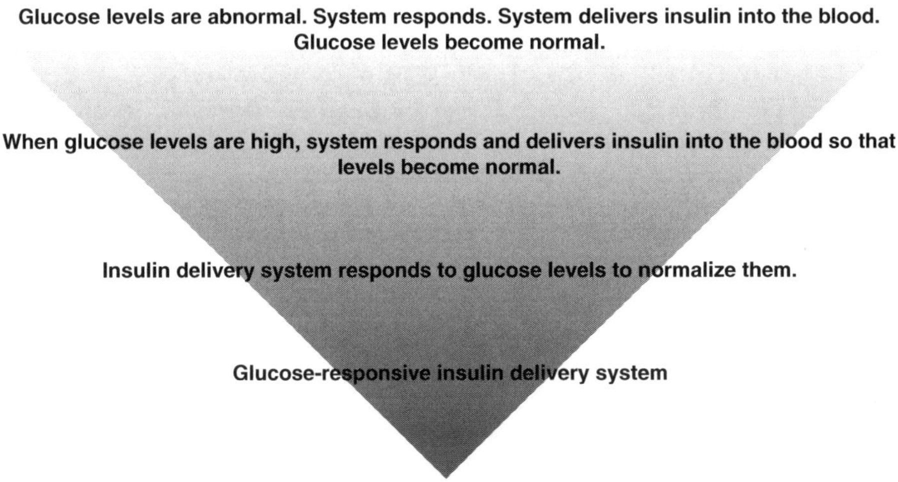

Glucose levels are abnormal. System responds. System delivers insulin into the blood. Glucose levels become normal.

When glucose levels are high, system responds and delivers insulin into the blood so that levels become normal.

Insulin delivery system responds to glucose levels to normalize them.

Glucose-responsive insulin delivery system

Figure 9.8 Meaning "packing" from several figures to a single element

Two or more figures can be densely "packed" and grammatically realized by units at a lower level than the clause in the rank scale, typically by nominal groups. Hence, meaning construed by several clauses is progressively "packed" into a nominal group, which condenses or distils grammatical meanings into lexical items, as shown in Figure 9.8.

In the example in Figure 9.8, meanings are progressively "packed" so that a multi-configuration of four figures logically concatenated and realized by four clauses ("Glucose levels are high. System responds. System delivers insulin into the blood. Glucose levels become normal") becomes a single nominal group ("Glucose-responsive insulin delivery system"). This process of "meaning packing" is approached by SFL through the notion of *grammatical metaphor* (Halliday & Matthiessen, 1999, p. 7) and explained as a case of related forms having "different mappings between the semantic and the grammatical categories". In this sense, the four forms in Figure 9.8 are related in meaning, but semantic functions are realized through different grammar configurations in each of them. As Halliday & Matthiessen (1999) state, related forms such as those in Figure 9.8 are not synonymous, since each representation is context specific and "aligns the experience in question with a different set of other experiences" (p. 523). Densely packed nominal groups tend to occur more in academic papers, whereas the less dense forms (realized as clauses) tend to occur more in news reports of science.

Meaning packing leaves out some of the once-explicit participants and processes. In the case of "glucose-responsive insulin delivery system", condensation leaves out the fact that glucose levels become abnormal, that the levels are detected by the system, and that the system responds by making insulin reach vital organs. At the same time, though, meaning packing allows phenomena to become things, which can be endowed with a reality of their own. Thus, a

Table 9.7 Example of a figure with densely packed meaning in a participant role

Clause	*we*	*developed*	*a novel and biocompatible glucose-responsive insulin delivery system*
Function	participant	process	Participant

Ashley recently started using a new form of **glucose responsive insulin** (GRI). This new insulin is given as a once-daily injection and it circulates in her body, only **activating when her blood glucose starts to rise. Once her glucose levels return to the normal range, the insulin stops working** and continues circulating until it's needed again.

Figure 9.9 Example of expert-to-lay explanation

"glucose-responsive insulin delivery system" becomes a thing and can be classified as "biocompatible" and assessed as being "novel", as seen in the example in Table 9.7.

An abstract entity, such as a system, can even become a concrete entity and physical object, as in "glucose-responsive insulin delivery device".

An element realized in the grammar by a nominal group, as in "glucose-responsive insulin delivery system", is the culminating point in a process whereby meanings have been distilled and selections in distinct grammatical systems have yielded a nominal group.

"Glucose-responsive insulin delivery system" can be approached from a lexical perspective as a lexical unit made up of five lexical items that form a collocation and can, in turn, enter into further collocational patterns. As a lexical unit, "glucose-responsive insulin delivery system" can predict its environment of occurrence, and a search in a large text corpus will yield academic papers, manuals, questionnaires and news reports in the healthcare domain.

When shifts in context variables take place, and communication is from expert to less expert, meaning may need to be unpacked for the purposes of explaining "glucose-responsive insulin delivery system" to a less specialized audience. This may be done as text unfolds, as seen in a short excerpt of an ad in Figure 9.9 (clauses in bold).

Figure 9.9 shows a single nominal group, "glucose responsive insulin", explained through a series of clauses. In this nominal group, "responsive" is explained by means of two conditions: "insulin activates when blood glucose starts to rise" and "once glucose levels return to normal range insulin stops working".

A comprehensive theory of language, such as SFL, accounts for the processes of "packing" and "unpacking" of meaning and thus offers a semantically motivated model for text analysis and text generation (see Bateman, 1990). Using an abstract semantic base linked to grammatical realizations, diverse structures can be generated for similar meanings. Different forms within a language system can be mapped to a shared semantic representation, as in Table 9.8.

Likewise, the model enables relating forms in two or more language systems that can be mapped onto a common semantic representation, even if configurations are different in each language and have different levels of metaphoricity. Table 9.9 shows examples retrieved from healthcare questionnaires in English and their translation counterparts in Brazilian Portuguese.

Table 9.8 Mapping of forms to a shared semantic representation

Rank	*Realization*
Clause	*System delivers insulin in response to glucose*
Nominal group	*Glucose-responsive insulin delivery system*

Table 9.9 Translation equivalents having a common semantic representation and different structural configuration (gloss underneath)

English	*Portuguese*
Age when you started using insulin	*Idade com a qual começou a usar insulina* gloss: Age with which you started using insulin
Age at start of insulin use	--------------------
Age when you were diagnosed as having diabetes	*Idade na qual foi diagnosticado com diabetes* gloss: Age at which you were diagnosed with diabetes
Age at diagnosis of diabetes	-------------------------

Table 9.10 Translation equivalents having a common semantic representation and different structural configuration (gloss underneath)

English	*Portuguese*
How easy it is to take insulin	------------------------
-----------------------------	Facilidade para aplicar insulina gloss: Easiness to take insulin

Cells in Table 9.9 with the actual equivalents found in texts are shaded, whereas unshaded cells show configurations that could have been used in the language. Dotted lines indicate no likely form in the language for an analogous structural realization.

Examples in Table 9.9 show a more metaphorical form in English—a circumstance of temporal location realized by a prepositional phrase ("at start"; "at diagnosis")—with implicit participant roles, and a less metaphorical one in Brazilian Portuguese—an embedded clause with explicit participant role. Unlike these examples, Table 9.10 shows a case of a less metaphorical form in English—a clause—and a more metaphorical form in Portuguese—nominalization.

Annotation of text retrieved from originals and translations illustrates deployment of semantic resources for domain construal in each language. Table 9.11 provides a further example of shared semantic representations for different structural configurations.

In Table 9.11, semantic configurations are related due to their occurrence as translation equivalents. Commonalities can be observed in terms of process and participants, but circumstantial meanings have different configurations. Whereas in English, meaning is circumscribed by a circumstance of temporal location, in Portuguese, a circumstance of contingency is selected. Further differences in configuration occur at a lower level in the rank scale, most precisely at the nominal group within the prepositional phrase realizing each circumstance. The nominal group "sick days" contains a packed figure: "sick" is not an attribute of "days" but of "person

Table 9.11 Translation equivalents having a common semantic representation and different structural configuration

English	Clause	*Take care of*	*your diabetes*	*during sick days*
	Grammatical function	process	participant	circumstance of temporal location
Brazilian Portuguese	Clause	*Cuide de*	*seu diabetes*	*em caso de gripe, resfriado, diarreia e infecções*
	Grammatical function	process	participant	circumstance of contingency
	gloss	*Take care of*	*your diabetes*	*in the event of the flu, the common cold, diarrhea and infections*

with diabetes". "Sick days", then, packs the meaning "days when a person with diabetes is sick or has a sickness". Grammatically, "sickness" is an attribute of "person with diabetes"; lexically, "sickness" partakes in a lexical set with "disease", with which it contrasts in that "sickness" is a state of being unwell and "disease" is a medical condition. In this sense, "diabetes" is a "disease", whereas "sickness" refers to states of being unwell, unrelated to diabetes.

Meaning packed in 'sick days' is to some extent made more explicit in the circumstance of contingency in Brazilian Portuguese. Since the meanings of "sickness" and "disease" are subsumed into a single lexical item in Brazilian Portuguese, namely "doença", which encompasses "disease", "sickness" and "illness", lexical items naming instances of "sickness", such as "flu", "common cold", "diarrhea" and "infections", construe the contrast between "disease" and "sickness" in Brazilian Portuguese, a contrast of fundamental importance, as "sick days" refer to health conditions not related to diabetes.

This example shows how mapping an experiential domain from a multilingual perspective, for instance in translation, can be enhanced if diversity of structural realizations both within and across languages are taken into account in the model (cf. Bateman, 1990) and mapped onto one another. This points to unpacking of meaning as a relevant methodological procedure, particularly for models that envision applications in assisted human translation as well as natural language processing and generation. Ontologies like GUM, as will be argued next, have great potential for building multilingually oriented representations of experiential domains and for explaining meaning with different degrees of packing.

9.2.4 Generalized upper model (GUM)

Ever since the first developments towards its construction, the GUM ontology has been devised for the purpose of supporting an architecture of natural language generation by allowing thorough descriptions of language. Instances of language can be annotated using GUM, the resulting annotations becoming instances of the classes in the ontology. The ontology is useful for building domain models as well as for providing a level of organization for interfacing domain models and natural language components.

Figure 9.10 shows GUM's main classes, built on the semantic configurations deployed by SFL to describe and explain human language. Screenshots are captured from the Protégé software.

As can be seen in Figure 9.10, GUM has three main *classes* (nodes): Configuration (corresponding to the concept of semantic figure previously introduced), Element (component

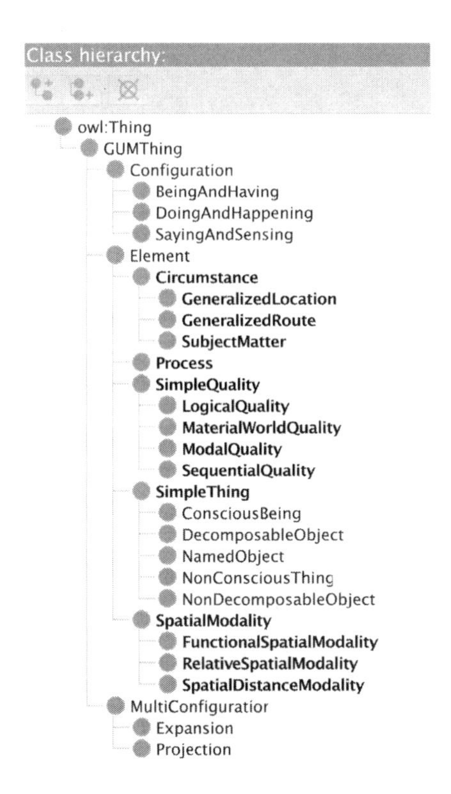

Figure 9.10 Protégé screenshot showing GUM's main classes

parts in a figure) and MultiConfiguration (combination of configurations). *Classes* are progressively subspecified from the more general to the more specific systems of semantics, realized by the grammar. The class identifier owl:Thing is predefined and encompasses the set of all individuals in an ontology.

Unlike domain ontologies, where categories are motivated by lexical words found in a given domain, GUM's categories are semantically motivated and pertain to functions, which enables mapping different structures within and across language systems onto a common base. Concepts in GUM have an underlying description aggregating the most relevant linguistic features for a particular grammatical function. Thus, differences in meaning construal between two wordings are accounted for in GUM through their distinctive patterns in the semantics and their realization in the grammar. For instance, clauses like "Patient denies history of diabetes" and "Patient has no history of diabetes" can be explained as two distinct instances of meaning construal within GUM. "Patient denies history of diabetes" is an instance of a configuration of sensing/saying, where the process is realized by the verb of saying "denies" and the main participant role, i.e. the sayer, is "Patient". The clause could be roughly reworded as "Patient says he/she has no history of diabetes." In contrast, "Patient has no history of diabetes" is an instance of a configuration of being/having, where there is a relational process construing possession, and "Patient" fulfils the main participant role of a carrier or possessor of an attribute "history of diabetes". The distinction between the two instances implicates the source for a particular statement—in other words, who the sayer is: a healthcare provider who is logging a report on a patient, or the patient him/herself.

Since GUM models any domain based on the language used to construe meanings, and its ontology framework is itself a comprehensive model of language, any particular domain can be modelled using GUM, and particular models can be integrated in the overall framework. GUM was conceived of to enable natural language processing applications and in that respect is particularly suitable for multilingual tasks including translation, since it constitutes an abstract framework organizing the grammar and lexis that generates meaning for particular domains. From the perspective of computational modelling, GUM operates as an interface between language categories and conceptual categories. From the perspective of translation competence skills, GUM can be used to map the language used to build a particular domain.

As will be argued in the following section, by drawing on a linguistically motivated ontology like GUM, text can be annotated regarding selections made in semantic systems for construing meanings in a given experiential domain. A mapping of selections in different languages constitutes an ideation base that can be queried by human translators and can be reused in a human-assisted translation or post-editing system.

9.2.5 Modelling domain through ontologies

The inextricable relation between language, thought and reality as conceived of by SFL posits a language-based approach to human cognition, whereby human capacity to construe meaning (semogenesis) is developed ontogenetically (as humans grow from child to adult); logogenetically (as text unfolds); and phylogenetically (as the language system evolves). The meaning potential developed throughout these evolving paths constitutes a "meaning base" (Halliday & Matthiessen, 1999) rather than a knowledge base. A meaning base is not an accumulation of concepts; nor is it a stock of stored ideas. Rather, it is a network of semantic resources that language offers to dynamically adapt to changes in the way we construe reality. As Halliday & Matthiessen (1999) put it,

> For any given domain the ideation base incorporates not only the known particulars of that domain but also the resources necessary for assimilating new information.
>
> *Halliday & Matthiessen, 1999, p. 14*

This dynamic movement accounts for the always-changing nature of domain and the suitability of a language ontology to map it, as language captures both actual instances of experience in a text and potential domain experience. In other words, the meaning resources deployed by language allow us to generate not only instantial individuals or relations within a domain, but also potential instances that language will generate to meet demands of human construal of a domain.

Building on a linguistically motivated upper ontology to model a domain implies conceiving of any and all domains as construed through language; hence, to model a domain, we examine the language used to construe it, and so language—clauses retrieved from text—is the source for annotations in the ontology.

Clause annotation is manually done by traversing a path of decisions through a system of choices from the more general to the more specific classes in the ontology. A sample clause like "the system delivers insulin into the bloodstream" has a path as shown in Figure 9.11.

Shaded rectangles in Figure 9.11 show the path traversed to obtain the semantic configuration of our sample clause. The class owl:Thing, which in all ontologies represents the class of all things, is the class within which the class GUM Thing is subsumed. From left to right, the path moves progressively along the subclasses selected to generate our sample clause: it is a configuration of Doing_And_Happening, more specifically Affecting_Action, more specifically Dispositive

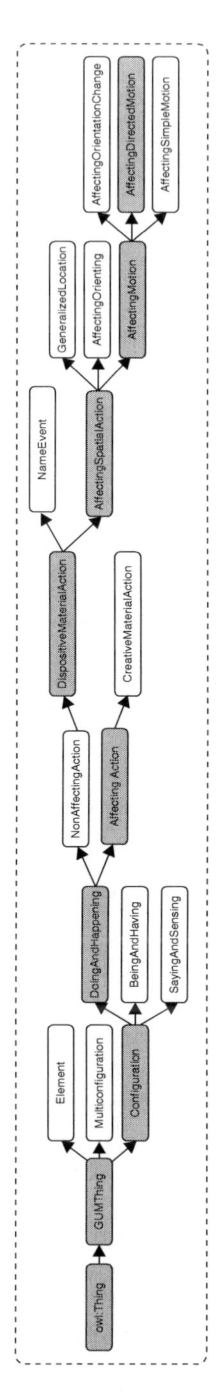

Figure 9.11 Protégé screenshot of ontology annotation fragment for clause "the system delivers insulin into the bloodstream"

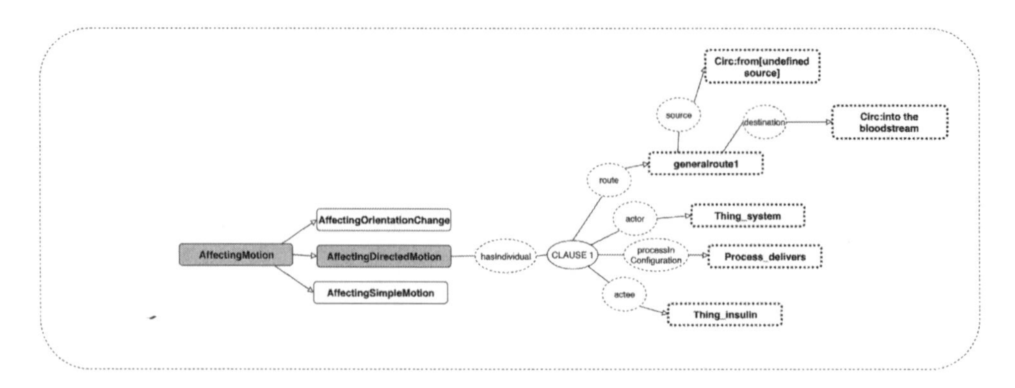

Figure 9.12 GUM representation of sample sentence

Material_Action, still more specifically Affecting_Spatial_Action, Affecting_Motion, reaching the highest granularity point of Affecting_Directed_Motion. Figure 9.12 shows a representation of categories annotated for our sample clause as an individual of the class Affecting_Directed_Motion with spatial modality.

Figure 9.12 shows that in our sample clause "deliver" can be classified as a process entailing directed motion, there being an actor ("system") having impact on an actee ("insulin"). There is spatial modality, with a route from an implicit source to an explicit destination ("the bloodstream"). Spatial meaning is thus construed, in this clause, as a movement of insulin into the bloodstream, and this is brokered by the insulin delivery system.

The counterpart clause in Brazilian Portuguese, "o sistema administra insulina" [the system administers insulin], has the path shown in Figure 9.13.

Shaded rectangles in Figure 9.13 show the path traversed for the semantic configuration of this sample clause. The path has the same starting point as that for the English sample clause and moves progressively along the classes Doing_And_Happening, Affecting_Action, Dispositive Material_Action. There is no spatial modality, though. Figure 9.14 shows a representation of categories annotated for our sample clause as an individual of the class Affecting_Directed_Motion.

Figure 9.14 shows that "administra" [administer] can be classified as a process having an actor ("system") with impact on an actee ("insulin"). No spatial modality is construed, as there is no route or destination to be reached by "insulin".

"The system delivers insulin into the bloodstream" and "o sistema administra insulina" [the system administers insulin] are mapped as forms in two different languages related to a shared representation. Both construe meaning regarding the performance of an insulin delivery system and share a similar path of choices along classes and subclasses in GUM, except for the fact that the English clause has spatial modality whereas its counterpart in Brazilian Portuguese does not. This is a significant difference, which can be further explored by examining other translation equivalents in order to see whether a pattern can be detected regarding spatial modality and its impact on how domain is construed in English and Brazilian Portuguese.

Once annotated in GUM, clauses aligned as translation equivalents can be analysed as potential meanings available in each language to construe this particular domain. They can be used as equivalents for translation purposes as well as for multilingual text generation. The fact that the two clauses can be assigned the status of translation equivalents, that is, they are generated by each language system for a common representation, prompts us to compare how domain is

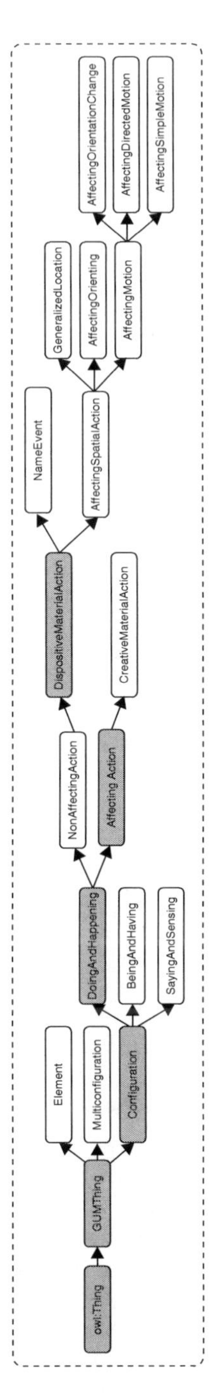

Figure 9.13 Protégé screenshot of ontology annotation fragment for "o sistema administra insulina"

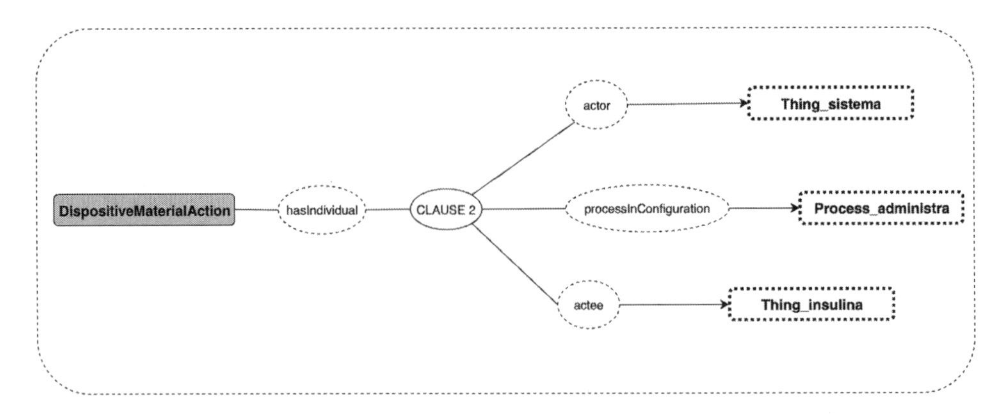

Figure 9.14 Classes and subclasses instantiating sample clause constituents

How satisfied are you with your current delivery system?				
	Completely satisfied	**Very satisfied**	**Somewhat satisfied**	**Not at all satisfied**
Difficulty in taking all insulin prescribed				
Uncertainty about getting the amount of insulin intended				

Figure 9.15 Excerpt from healthcare questionnaire

construed in each language and to use this comparison to generate further clauses by leveraging the annotations in GUM.

Annotating text through an ontology involves treating text units as instances of classes in the ontology, enabling connections between individual instances as members of a set making up a class. These connections are actually the multiple networks of meanings that make up a domain. A linguistically motivated ontology captures agency of processes as well as temporal and spatial relationships and explicates meaning packing. This is particularly elucidating in multilingual applications, since, as shown earlier, the level of meaning packing may be different in different languages, and a common semantic representation enables mapping different structural configurations to a common representation.

An ontological map of a domain also provides insight into how meanings are interrelated in each language system, which is another source of insight for translation and multilingual generation. The example in Figure 9.15, retrieved from a healthcare questionnaire, illustrates this case.

The two items in the questionnaire that the respondent is requested to rate are grammatically realized by nominal groups ("difficulty" and "uncertainty") that pack meanings regarding participant roles and processes. Using the grammar to unpack those two nominals makes it possible to render explicit the participant roles implicated in the processes, i.e. who does what to whom; who sets actions going. Table 9.12 shows unpacked meanings.

Both nominalizations involve a figure of sensing realized in the grammar by the processes "thinks" and "feels", having "insulin delivery system user" as the main participant role. This is in

Table 9.12 Realization of packed and unpacked meanings

Packed meaning	*Meaning unpacked*
Difficulty in taking all insulin prescribed	The insulin delivery system user thinks it is difficult for him/her to take the complete insulin dose prescribed by the physician
Uncertainty about getting the amount of insulin intended	The insulin delivery system user feels uncertain about whether (s) he gets into his/her bloodstream the amount of insulin that (s)he intends to get

Sobre seu sistema atual de administração de insulina, você está satisfeito em relação a ...? **GLOSS: About your current system of insulin administration, you are satisfied regarding ...?**				
	Totalmente satisfeito **Gloss: Completely satisfied**	**Muito satisfeito** **Gloss: Very satisfied**	**Mais ou menos satisfeito** **Gloss: More or less satisfied**	**Nada satisfeito** **Gloss: Not satisfied at all**
Dificuldade de aplicar a dose toda de insulina prescrita **GLOSS: Difficulty to apply the whole dose of insulin prescribed**				
Dúvida se o sistema administra a quantidade de insulina que precisa usar **GLOSS: Doubt whether the system administers the quantity of insulin that you need to use**				

Figure 9.16 Excerpt from translated and cross-culturally adapted healthcare questionnaire

accordance with the particular text type under analysis: a questionnaire to survey users' opinion about their current insulin delivery system. Figures of doing are also implicated. In the first nominalization, "take" construes the meaning of consume as medicine, as "take a medication". In the second, "get" construes the meaning of acquire, there being spatial modality, that is, motion and location ("into her bloodstream").

The examples in Figure 9.16 translated and cross-culturally adapted in Brazilian Portuguese are displayed with an English gloss underneath.

The choices in Brazilian Portuguese reveal a pattern of connectivity. Meaning in "aplicar a dose toda de insulina prescrita" [to apply the whole dose of insulin prescribed] and "o sistema administra a quantidade de insulina que precisa usar" [the system administers the quantity of insulin that you need to use] is construed by figures of doing, with an actor and an actee, following the pattern observed for the clause analysed in Figures 9.13 and 9.14: "sistema administra insulina" [system administers insulin].

Unlike English, which, as seen in Figures 9.11 and 9.12, follows a pattern of spatial modality, Brazilian Portuguese does not. This difference between the two languages accounts for the main participant role in the clause "Dúvida se o sistema administra a quantidade de insulina que precisa usar" [Doubt whether the system administers the quantity of insulin that you need to use] being "system" instead of "system user" as in English. This particular difference can only be captured by a language-motivated ontology, which maps instances on a text onto semantic resources in the ideation base. This amounts to saying that new instances to be annotated in GUM will follow patterns of instances already annotated, as GUM is meant to enable mapping the general potential for meaning making in a language regardless of particular domains. In this sense, GUM, as will be argued next, is a resource with great potential to deal with domain in translation tasks.

9.2.6 Why ontologies in Translation Studies?

Two questions posed by Chandrasekaran et al. (1999) regarding artificial intelligence, at a time when ontologies were being introduced in that field, can be asked at this time apropos of Translation Studies: what are ontologies and why do we need them?

Ontologies are an emergent theme of Translation and Cognitive Translation Studies. Few Translation Studies scholars have explored ontologies for translation purposes, and most of them have done so as terminological aids. The full potential of the ontological approach to translation, however, as we have argued, lies in their theoretical and methodological power to model domain, drawing on the language used in that domain. Resuming our initial discussion on how domain knowledge impacts translation tasks, it can be argued that there is no separation between domain concepts and domain language. Domain is construed through language along a path of expertise by novices who become experts and by experts who further specialize in the field. Not being experts in the domain of the text they deal with, translators have to rely on representations of domain with sufficient linguistic granularity and powerful generation potential to allow them to attain renditions as if they were domain experts. This can be achieved with the assistance of resources such as ontologies.

Unlike domain ontologies, GUM is a linguistically motivated ontology and draws on a language-based approach to human cognition. Any domain can be mapped by using GUM, and the patterns of choice emerging in domain mapping are useful to translators, as they show how a domain is organized, which particular configurations are selected, and which participants and processes make up that domain. Meaning resources potentially offered by the language system and particular selections among them are captured in GUM though text-based annotation. Not having "domain knowledge", i.e. not having developed semantic resources to construe meanings to operate within a particular domain, a translator can rely on a "meaning base" as captured in GUM to make translation decisions, which clearly outperforms conventional tools and does not reduce domain to a matter of mere terminology.

The following section concludes this chapter by summarizing some arguments for the introduction of ontologies into the discipline of Translation Studies and offering suggestions for future research on ontologies in translation and multilingual text production in different modes of human–machine interaction.

9.3 Future directions

As seen in this chapter, a mapping of selections in different languages by means of an ontology organizing meaning construed by those selections constitutes an ideation base that can be queried by human translators and can be reused in a human-assisted translation or language generation system.

The choices of types of figures and semantic functions characterize a domain, and an ontology is able to capture those functions and reveal connectivity between them. Because language is the means to construe a domain, every linguistic choice pertaining to a domain is connected to other choices in that domain. This can be clearly seen when we compare the way a similar domain is construed in two languages. Through incrementally annotating clauses pertaining to a domain in an ontology, patterns characteristic of that domain can be obtained; these make up the ideation base that can inform translation and multilingual tasks.

A single ontology enables annotations of meanings, not only regarding different contexts of culture and language systems, English and Brazilian Portuguese for instance, but also related to lay and expert construals of a domain. By mapping packed and unpacked meaning, an ontology retraces the ontogenetic development whereby novices become experts, as well as the phylogenetic and semogenetic progression of a domain, whereby figures realized by clauses become nominalizations. An ontology can also model the language used by lay and expert domain practitioners, as in the case of healthcare. The latter perspective is particularly relevant for tuning text production to particular configurations of tenor variables involving power relationships, expertise and social distance between interlocutors (healthcare providers—patients).

As argued in this chapter, the potential of ontology in Translation Studies is further enhanced when domains are modelled upon linguistically motivated ontologies, such as GUM, and translation is viewed from a wider angle within the scope of multilingual text processing. Matthiessen (2001, pp. 42–43) advocates such a perspective when he proposes to explore "where the outer limits of translation as a phenomenon lie—where translation ceases to be re-construal of meaning and shades into first-time construal of meaning". Ontological modelling of domain implies thinking translation as one among other modes of multilingual text production and within a continuum of human–machine interaction.

Translation is produced within a context with variables pertaining to what a translated text is expected to be in a given context of situation in a given context of culture. These translation context variables, nonetheless, are inextricably bound to the context variables whereby a non-translated text is generated. Source and target texts used to feed ontology classes with instances are able to capture the conflation of these two sets of contextual variables and point to patterns of meaning that can be leveraged to inform text production for translation purposes and natural language generation.

Domain models obtained by building ontologies are also important as task-oriented descriptions of language in use for the purposes of translator training and education. This fills the gap observed in translator competence and expertise models, which make no provision for ways in which to deal with development of domain knowledge other than developing search skills (PACTE, 2001) or engaging in extensive practice (Shreve, 2006). If, as expertise studies have suggested, discourse skills are inextricably connected to domain skills (Scardamalia & Bereiter, 1991), introducing an ontology approach in translator education can be a promising avenue, and one that can be an object of experimental studies that might offer valuable contributions to existing models of translation competence and expertise.

Last, but certainly not least, an ontology approach to translation, in particular a language-based one, can bring insights into our theoretical understanding of the nature of human cognition.

Notes

1 Available at https://bioportal.bioontology.org/ontologies/NCCO
2 Available at http://bioportal.bioontology.org/ontologies/DDO
3 www.adampease.org/OP/
4 https://wordnet.princeton.edu

Further reading

Bateman, J. A., Hois, J., Ross, R., & Tenbrink, T. (2010) A linguistic ontology of space for natural language processing. *Artificial Intelligence, 174*(14), 1027–1071.

This article describes the rationale underlying the generalized upper model (GUM) and its recent expansion to cover the area of spatial semantics.

Halliday, M. A. K., & Matthiessen, C. M. I. M. (1999). *Construing experience through meaning: A language-based approach to cognition.* London: Continuum.

Halliday, M. A. K., & Matthiessen, C. M. I. M. (2014). *Halliday's introduction to functional grammar* (4th ed.). London: Routledge.

These two titles are must-reads for those who want to learn about systemic functional linguistics and the way domain is modelled within this theory.

Matthiessen, C. M. I. M. (2001). The environments of translation. In E. Steiner, & C. Yallop (Eds.), *Exploring translation and multilingual text production: Beyond content* (pp. 41–124). Berlin / New York: Mouton de Gruyter.

This is a comprehensive discussion of translation from a systemic functional perspective on language with many insights into modelling translation at different levels of granularity.

References

Bateman, J. A. (1990). Finding translation equivalents: An application of grammatical metaphor. *Proceedings of the 13th international conference on computational linguistics, COLING* (pp. 13–18). Finland: University of Helsinki.

Bateman, J. A. (1992). The theoretical status of ontologies in natural language processing. In S. Preub, & B. Schmitz (Eds.), *Papers from the workshop on text representation and domain modelling: Technical University Berlin,* October 9th to 11th, 1991, KIT-Report 97 (pp. 50–100). Technische Universität Berlin, Fachbereich 20 Informatik.

Bateman, J. A. (1995). On the relationship between ontology construction and natural language: A socio-semiotic view. *International Journal of Human-Computer Studies, 43*(5–6), 929–944.

Bateman, J. A. (1997). Enabling technology for multilingual natural language generation: The KPML development environment. *Natural Language Engineering, 3*(1), 15–55.

Bateman, J. A., Henschel, R., & Rinaldi, F. (1995). Generalized upper model 2.0: documentation (Technical report). GMD/Institut für Integrierte Publikations- und Informationssysteme, Darmstadt, Germany. http://purl.org/net/gum2

Bateman, J. A., Hois, J., Ross, R., & Tenbrink, T. (2010). A linguistic ontology of space for natural language processing. *Artificial Intelligence, 174*(14), 1027–1071.

Budin, G. (2005). Ontology-driven translation management: Knowledge systems and translation. In H. V. Dam, J. Engberg, & H. Gerzymisch-Arbogast (Eds.), *Knowledge systems and translation* (pp. 103–124). Boston: De Gruyter.

Chandrasekaran, B., Josephson, J. R., & Benjamins, V. R. (1999). What are ontologies, and why do we need them? *IEEE Intelligent Systems, 14*(1), 20–26.

Collins, H. M. (2004). Interactional expertise as third kind of knowledge. *Phenomenology and the Cognitive Sciences, 3*, 125–143.

Collins, H. M., Evans, R., Ribeiro, R., & Hall, M. (2006). Experiments with interactional expertise. *Studies in History and Philosophy of Science, 37*(4), 656–674.

Currás, E. (2010). *Ontologies, taxonomies and thesauri in systems science and systematics.* Witney: Chandos.

Gómez-Pérez, A., Fernandez-Lopez, M., & Corcho, O. (2004). *Ontological engineering: With examples from the areas of knowledge management, e-commerce and the semantic web.* London: Springer.

Göpferich, S. (2009). Towards a model of translation competence and its acquisition: The longitudinal study TransComp. In S. Göpferich, A. L. Jakobsen, & I. M. Mees (Eds.), *Behind the mind: Methods, models and results in translation process research* (pp. 12–38). Copenhagen: Samfundslitteratur.

Gruber, T. R. (1993). Toward principles for the design of ontologies used for knowledge sharing? *International Journal of Human-Computer Studies, 43*(5–6), 907–928.

Halliday, M. A. K. (1978). *Language as social semiotic.* London: Edward Arnold.

Halliday, M. A. K., & Matthiessen, C. M. I. M. (1999). *Construing experience through meaning: A language-based approach to cognition*. London: Continuum.
Halliday, M. A. K., & Matthiessen, C. M. I. M. (2014). *Halliday's introduction to functional grammar* (4th ed.). London: Routledge.
Huang, C.-R. (Ed.). (2010). *Ontology and the lexicon: A natural language processing perspective*. Cambridge: Cambridge University Press.
Hurtado Albir, A. (2005). A aquisição da competência tradutória. In A. Pagano, C. M. Magalhães, & F. Alves (Eds.), *Competência em tradução: Cognição e discurso* (pp. 19–38). Belo Horizonte: Editora UFMG.
Matthiessen, C. M. I. M. (2001). The environments of translation. In E. Steiner, & C. Yallop (Eds.), *Exploring translation and multilingual text production: Beyond content* (pp. 41–124). Berlin/New York: De Gruyter.
Niles, I., & Pease, A. (2001). Towards a standard upper ontology. *International Conference on Formal Ontology in Information Systems* (FOIS 2001) (pp. 2–9). New York: ACM.
Obrst, L. (2010). Ontological architectures. In R. Poli, M. Healy, & A. Kameas (Eds.), *Theory and applications of ontology: Computer applications* (pp. 27–66). Dordrecht: Springer.
PACTE (2001). La competencia traductora y su adquisición. *Quaderns. Revista de Traducció, 6*, 39–45.
PACTE (2003). Building a translation competence model. In F. Alves (Ed.), *Triangulating translation: Perspectives in process oriented research* (pp. 43–66). Amsterdam: John Benjamins.
Ribeiro, R. (2007). The role of interactional expertise in interpreting: The case of technology transfer in the steel industry. *Studies in History and Philosophy of Science, 38*, 713–721.
Scardamalia, M., & Bereiter, C. (1991). Literate expertise. In K. A. Ericsson, & J. Smith (Eds.), *Toward a general theory of expertise* (pp. 172–194). Cambridge: CUP.
Shreve, G. M. (2006). The deliberate practice: Translation and expertise. *Journal of Translation Studies, 9*(1), 27–42.
Uschold, M., & King, M. (1995). *Towards a methodology for building ontologies*. Edinburgh: University of Edinburgh.
Whorf, B. L. (1956). *Language, thought, and reality. Selected writings of Benjamin Lee Whorf*. Cambridge: MIT.

10

Translation, corpus linguistics and cognition

Stella Neumann and Tatiana Serbina

10.1 Historical background: The corpus approach in Translation Studies

Empirical approaches to translation started to gain ground in the mid-1980s. A first wave of process-based studies (e.g. Königs, 1987; Krings, 1986) appeared around that time and were soon followed by first papers proposing a corpus-based approach to translation (e.g. Johansson & Hofland, 1994; Toury, 1995/2012/1995). Particularly inspired by Baker's (1993) paper advocating the creation and analysis of translation corpora, a corpus-oriented approach quickly gained ground in Translation Studies, yielding a flurry of reports of typical properties of translated texts (Hansen, 2003; Kenny, 1998; Laviosa-Braithwaite, 1996; Mauranen & Kujamäki, 2004b; Olohan & Baker, 2000; Teich, 2003, etc.), giving rise to what has been referred to as corpus-based Translation Studies (CBTS). Until recently, process- and product-based approaches were carried out more or less independently of each other, although both approaches were obviously concerned with translators' behaviour.

In earlier CBTS, cognition was not explicitly addressed as a source of potential explanations of corpus findings, not least because corpus linguists were traditionally at least agnostic towards the role of cognition in language. As Anderman and Rogers (2008, p. 13) point out, the underlying idea of corpus linguistics was initially to study language by observation rather than through introspection or experimentation. Discussing the notion of universals in translation, Mauranen and Kujamäki (2004a, p. 2) refer to a distinction in general linguistics between "universals which can be traced back to general cognitive capacities in humans, and those which relate linguistic structures and the functional uses of languages" and then go on to link them to different characteristics of translated language, namely cognitive translation processes, social and historical determinants of translation, and the typical linguistic features of translations. Claims like these suggest that corpus-based translation scholars were perhaps not primarily interested in but at least aware of cognitive aspects of translation, even if they did not refer to cognition explicitly. Instead, they variously referred to translators' behaviour during the translation process, subconscious processes or understanding. Steiner (2002, p. 216), for example, discusses three sources of explanation for the specific properties of translations observed in corpora, namely contrastive differences in the language pair, differences between registers, and individual differences in the process of understanding during translating. He also refers to "internal 'fatigue'" (Steiner,

2002, p. 219). One way of operationalizing such references to understanding is explained by Hansen-Schirra:

> Cognitively speaking, the decoding phase during translation, which includes comprehension, understanding, and conceptualization of the source text, triggers the phenomena introduced above [translation properties]. Translators transfer their own understanding and interpretation into the target texts, which make them more explicit and easier to understand. In case of ambiguities or complex structures, this results in resolving, unpacking, and deconstruing [sic] the source text structures (including defective and unexpected text passages) and translating them with unambiguous, clearer, and optimized ones.
>
> *2017, pp. 235–236*

Chesterman (2010, p. 43) explicitly argues that the most obvious cause for typical features of translations found in corpora is a cognitive one, "something in the mind of translators that affects the way they process texts simultaneously in two languages". This is not altogether implausible, as corpora provide a window on systematic patterns of translators' behaviour.

The goal of this chapter is not to give a comprehensive overview of CBTS in general but to focus on the interplay between corpus linguistic and cognitive approaches in Translation Studies (see also Rodríguez-Inés, 2017), assuming that both approaches are novel in their interest in the empirical facts of translation. The chapter is organized as follows. First, we very briefly introduce some sources of corpus-based approaches in Translation Studies and some central notions of corpus research. Section 10.2 will be concerned with a detailed discussion of various aspects of corpus-based research in Translation Studies and its interaction with cognitive approaches. Finally, in Section 10.3, we will summarize more recent developments in CBTS and conclude with some recommendations for further reading in Section 10.4.

A precursor of the corpus-based approach in Translation Studies can be found in Even-Zohar's (1979) polysystem theory (see Baker, 1993; Kruger, 2012; Øveras, 1998). It emphasizes a matter-of-fact perspective on the empirical reality of texts and consequently strongly rejects a focus on particularly valued texts. Even-Zohar's theory characterizes the individual literary work as part of systems that capture traditions, genres, etc. Translations are seen as being part of, and potentially playing a central role in, the literary system. Linking the approach to Holmes' outline of Translation Studies, Gideon Toury (1995/2012) went on to work out what an empirical or—in his words—descriptive approach to translation would have to cover. Specifically, he discussed potential laws that should be empirically investigated.[1] This was, in turn, picked up by Baker (1993), who linked Toury's programmatic claims as well as small-scale studies of traditional literary corpora, such as Vanderauwera (1985), to Sinclairian corpus linguistics (Sinclair, 1991).

Initially, the term "corpus" referred to a collection of texts used for some kind of text analysis, typically in literary studies. With the advent of corpus linguistics (see Johansson, 2008 for a historical overview), the concept was narrowed down to large, electronic collections of texts compiled in a principled way to achieve balance and, ideally, representativeness of some population of texts, which can be used for systematic empirical analysis. Translation corpora, in a more specific sense, are corpora containing translated texts, i.e. products of the translation process (see Anderman & Rogers, 2008 for the sources of corpus linguistics in Translation Studies). For the purposes of comparison, quite often they also include non-translated texts, either the source texts of the included translations or originals in the same language as the translated texts.

In a parallel development, Stig Johansson, an early proponent of corpus methods in linguistics, initiated the compilation of some of the best-known translation corpora (see later). His main interest was linguistic rather than translation-related, focusing primarily on contrastive aspects for which he used translations as a *tertium comparationis*. Similarly, Fabricius-Hansen (1999) demonstrated how using corpora could help shed light on translation-related linguistic questions. Although Fabricius-Hansen in particular based her study on assumptions about the (incremental) processing of complex sentences, these approaches are not directly linked to cognitive aspects of translation.

One main reason for collecting translation corpora is to investigate the particular linguistic traits making translations distinguishable from non-translated texts (for collections of CBTS see, e.g., Fantinuoli & Zanettin, 2015; Ji, 2016; Kruger et al., 2011; Laviosa, 1998; Mauranen & Kujamäki, 2004b; Oakes & Ji, 2012). However, the compilation of a sufficiently large corpus is costly and subject to copyright issues. As a consequence, corpora compiled for individual studies are usually fairly small. In addition to corpora collected for the purpose of individual corpus investigations, a number of well-known corpora have been used in more than one study. These include the Translational English Corpus[2] (TEC; Baker, 1996), the English-Norwegian Parallel Corpus (ENPC; Johansson & Hofland, 1994), the German-English CroCo Corpus (Hansen-Schirra et al., 2012), the Dutch Parallel Corpus (Macken et al., 2011), the MeLLANGE Learner Translator Corpus[3] (Kunz et al., 2010), the interpreting corpus EPIC (Monti et al., 2005) and the ZJU Corpus of Translational Chinese[4] (ZCTC; Xiao & Hu, 2015).

As corpora are by definition samples of the overall population of all texts, carefully considering design criteria in terms of balance and representativeness (coverage of different authors and publishers, spoken and written language, a range of genres/registers, and all this in relation to translation) is of great importance. The actual compilation according to the specified criteria may involve compromising on certain criteria due to unavailability of texts. The optimal size of a corpus is not an entirely straightforward question and largely depends on how vigorously aspects of balance have been taken into consideration in the previous steps. A well-balanced corpus may reflect the distribution of linguistic features, even if it is comparatively small (see Biber, 1990). Likewise, a very large corpus may turn out to be uninformative if texts are sampled mainly from the same sources or if the distribution between parts of the corpus is imbalanced. Usually, corpus size is given in number of words; however, this may be inappropriate, especially for translation-related research questions, as translations have been shown to deviate from their source texts in length (Frankenberg-Garcia, 2009). Moreover, they display differences in length depending on the translation direction (Neumann & Hansen-Schirra, 2012). This is likely to be a more general phenomenon in translation, resulting in difficulties in determining the appropriate size in number of words.

Compilation of modern corpora usually involves a range of manual or automatic pre-processing steps to ensure computer readability of the data, including encoding of specific features such as formatting, labelling errors, tokenization of the stream of symbols into words, punctuation marks, etc., automatic enrichment with linguistic annotation of word classes (so-called part-of-speech tagging) and occasionally also other types of linguistic information. If the corpus includes matched originals and translations, it will usually also include sentence alignment, and occasionally alignment of smaller units (on the importance of analysing aligned units see Steiner, 2012). Various types of comparison used in Translation Studies (see Hansen-Schirra & Teich, 2009) have their advantages and disadvantages. Comparing original texts and translations in the same language (in so-called comparable corpora) ensures the comparability of linguistic features. However, this makes it difficult to specify the exact factors that could explain the observed phenomena in translations: they could be triggered by the specific linguistic facts of the corresponding source-text unit, they could be due to some general feature of the source or target language (including their registers), or they could be brought in by the translator without any traceable reason in the texts or languages. This problem is resolved by parallel corpora, i.e.

corpora of matching source and target texts, especially if the texts are aligned at least at sentence level. However, parallel corpora introduce the challenge of having to determine comparable features that capture phenomena which are formally and/or functionally equivalent. Corpora also differ as to whether both translation directions in a language pair are covered. Coverage of both directions is very useful as a first indication of how pervasive a certain translation phenomenon is: if some assumed universal feature of translation can only be found in one translation direction, this already calls into question the universal character of the feature. However, bi-directional corpora stratified into registers raise the question of comparability of registers. It is quite likely that corresponding registers in a language pair will display some incomparability. In such cases, claims about directionality effects can be called into question.

Although the definition of corpus in corpus linguistics is quite specific, the ways such corpora are used still cover a fairly wide methodological space. Tummers et al. (2005) point out that there is a range from what they call corpus-illustrated research through descriptive operationalization to explanatory research using bivariate or multivariate statistical analysis. Some earlier translation-related studies can arguably be classified as corpus-illustrated research, in which the corpus is used as a resource for examples to underpin the (theoretical) linguistic discussion. Most translation-related studies fall into the descriptive category; while some of them include statistical significance testing, others focus on the comparison of frequency counts, often simply represented by bar charts. Only a few recent studies use multivariate statistics (e.g. Delaere et al., 2012; Hu et al., 2019; Ji, 2016; Lapshinova-Koltunski, 2017; Oakes & Ji, 2012).

In a stricter methodological sense, corpus-based approaches can be categorized as observational studies. Experimental studies involve control over unwanted factors and deliberate manipulation of some variable assumed to have an effect on some phenomenon that is to be explained (Butler, 1985, p. 65). Observational studies, by contrast, draw on data outside the researcher's control, where it is impossible for the researcher to manipulate a variable in order to measure its effect on the dependent variable. This has implications particularly for control over confounding factors, which would allow robust explanations of phenomena in terms of the manipulated independent variables. Corpus linguists cannot manipulate independent variables because they are not able to control the generation of the data they are analysing. Rather than investigating the effect the independent variable has on the dependent variable, the corpus linguist investigates correlations between the different variables. What may be of help in some cases is metadata: the more information we have about the author(s), the text production circumstances, etc., the more possible it will be to filter the texts accordingly, thus simulating the manipulation of independent variables. But in many cases, the corpus linguist will simply face a different task than the experimental researcher in comparing two measurements of the same linguistic variable in different data sets and interpreting the results in light of the characterization of the data sets.

Laboratory experiments are sometimes criticized for the extent to which potentially confounding factors are excluded. In extreme cases this may lead to laboratory artefacts which cannot be replicated under authentic conditions. In contrast, less controlled experiments allowing more authentic conditions will yield results that are difficult to link to one particular explanation, because various factors could lead to the observed behaviour. A major advantage of corpus-based investigations is the authenticity of the data, including translational data. However, this is not to say that corpora are necessarily neutral in this respect. Establishing what is authentic or, more generally, appropriate data is at the corpus compilers' discretion. They may choose to consider certain types of language use as inadequate, incorrect or not accepted in a given culture (see Mauranen, 2004, p. 73) and consequently only to include esteemed texts or texts judged to display a certain quality. In the context of CBTS, for example, a relevant question is whether to include non-professional translations, translations of texts written by non-native speakers, etc., as all these factors potentially affect the translation product. The less restrictive corpus compilers are, the more likely the corpus will be to reflect the empirical reality of the translations the consumer is actually exposed to. However, variation in the collected texts will be reflected by unexplained variance in the corpus study if the corpus does not contain relevant information on the translator and details on the translation process. For instance, providing information about the experience and training of the translators of corpus texts will give access to a comparison of translation products based on notions such as expertise. If such information is not available, corpus texts will still reflect differences in the translators' expertise, but the variance in the data cannot be linked to this potential explanation. Regardless of these design considerations, a main characteristic of corpus studies is that they involve a certain amount of idealization (McEnery & Wilson, 2001, p. 77): The search will retrieve exactly those patterns that match the query. If the linguistic phenomenon is characterized by high variability, it is likely that the corpus analysis will not retrieve all relevant cases.

10.2 Core topics

Existing (published) translations are rarely used as material to investigate the cognitive processes the translator engages in, since they reflect the *result* or final product of cognitive processing taking place during translation task execution. The actions that lead up to this result, i.e. the translator's behaviour reflecting the processing, are no longer observable in the product. Although it is possible that a translation contains particular linguistic traits that are identifiable as a result of, say, increased cognitive demand during translating, it is also possible that this demand is not apparent in the product because the final translation is the most likely one of a given source-text segment despite having been produced in more than one attempt, i.e. in a prolonged process (see Section 10.2.3). Alternatively, corpus findings might hint at increased cognitive demand during task execution, e.g. in the form of simplified grammatical structures, but process-based experiments do not retrieve behavioural correlates of increased demand. Instead, they suggest automatized translation behaviour (Heilmann et al., in press). Halverson (2015, p. 323) points out that "[t]he basic consensus emerging from this methodological discussion is that while corpus data provides evidence of aggregate patterns that must be explained, this type of data alone does not provide direct evidence of online cognitive processes".

It therefore appears plausible that cognition during translating is usually investigated with the help of behavioural measures such as eye movements and keystroke logging, which could be used as a window to cognitive activity. Nevertheless, as will be shown later, corpus studies are still useful for generating or testing assumptions about cognitive processes during translating. The discussion of the interaction between CBTS and cognition will be structured into three areas. The first area discusses the use of corpora as a means of making and testing assumptions about cognition in the above sense, that is, without necessarily committing to theoretical claims about cognition. The second area examines the use of corpora against the theoretical background of cognitive linguistics. Approaches in this framework do not necessarily adopt a corpus method as the default method to test claims about mental processing. The third area deliberately bridges the gap between experimental and corpus-based methodologies by triangulation or by developing integrated research designs. In the next sub-sections, we will discuss these three areas in more detail, concentrating on the potentials and limitations of each approach as a way of gaining insight into translation as a cognitive activity.

10.2.1 Corpus-based inferences about cognition based on eliminating other sources of explanation

The first of the three areas, i.e. corpus-based inferences about cognition, includes studies that draw on cognition or some related concept such as the (unconscious) process of understanding to explain instances of language use which cannot be explained by other factors like contrastive differences or register characteristics. Since corpora do not give access to the actual behaviour during task execution, inferences are thus made by eliminating the other potential explanations.

The question of whether the specific phenomena observed in translations are actually universal, i.e. whether they occur irrespectively of the translator, the languages involved, the specific linguistic feature, etc., has triggered much debate in CBTS (see among others Becher, 2010; Bernardini & Zanettin, 2004; Chesterman, 2010; Zufferey & Cartoni, 2014). However, the sheer fact that machine learning approaches achieve high accuracy in classifying translated versus non-translated texts (see, e.g., Baroni & Bernardini, 2006; Volansky et al., 2015) suggests that there must be something peculiar about translations that computers can pick up on with the help of certain—usually fairly shallow—features (for more detail see Evert & Neumann, 2017). To some extent, this could be due to the fact that features of translations are probabilistic in nature and more specifically conditional rather than deterministic; therefore, simple frequency counts are likely to be too concrete (Toury, 2004). At the same time, as Toury points out, claims that are too general run the risk of not being informative and therefore have to be made as specific as possible. This, in turn, cannot be operationalized by investigating individual features. It does not do justice to the complexities of translation, as "there is probably no single variable affecting translation which cannot be enhanced, mitigated, maybe even offset by the presence of another" (Toury, 2004, p. 25). Probabilistic concepts that have "the possibility of *exception*" built into them (Toury, 2004, p. 29, emphasis in original) are to be preferred to the notion of universals. Therefore, it appears advisable that one should be careful when labelling the observed peculiarities as translation universals for as long as we do not have compelling evidence of the universal character of those peculiarities. In view of such considerations, we will refer to phenomena such as explicitation of information implicit in the source text, adaptation of features to the norms of the target language (normalization), simplification of linguistic structures (Baker, 1996) and shining through or interference of the source language (Mauranen, 2004; Teich, 2003) as properties or features of translations rather than as universals.

The main task of corpus-based studies is, then, to examine the corpus findings and link them to plausible sources of explanation. It seems that, among the potential explanations, the ones that are most immediately related to the translation process are the ones translation scholars are actually interested in. Baker (1993, p. 243), for example, makes a distinction between features linked to "the nature of the translation process itself" and features resulting from the "confrontation of specific linguistic systems". She explicitly characterizes the assumed features of translation as "a product of constraints which are inherent in the translation process itself" (ibid.). While this is a far cry from making claims about cognitive processing, it at least closes in on the translation process during which the translator displays a particular behaviour. In corpus analyses, isolating explanations linked to the translation process itself can happen in different ways. The most indirect way is to address the suspected feature right away and design the study in such a way that it precludes any other explanation. A more direct way involves the explicit discussion of various sources of explanation and closing in on cognitive aspects by eliminating other explanations. Lastly, studies might also work the other way around, i.e. rather than eliminating other explanations, rule out cognitive aspects as an explanation.

In an early corpus study, Olohan & Baker (2000) link the assumed translation property of explicitation to what they call subconscious processes[5] in the title of the paper. The concept of unconscious processes or unintentional language use is not explained, but we might reasonably assume that the authors are interested in indications of mental processes. In a sense, they even strengthen their assumptions about mental processes by specifically referring to unconscious phenomena. Unfortunately, Olohan & Baker (2000) do not go into more detail about the extent to which their findings are actually evidence of such mental processes. This evidence can be arrived at indirectly by eliminating other potential explanations. Contrastive differences can be ruled out, since the Translational English Corpus contains English translations from a wide range of source languages. They report statistically significantly more occurrences of the optional complementizer *that* in translations than in comparable original English texts and come to the conclusion that "inherent, subliminal processes" must lead translators to spell out the implicit grammatical information in their target texts. In a follow-up study, Olohan (2002, p. 166) corroborates Olohan and Baker's (2000) findings with further features and comes to the conclusion that explicitating optional syntactic elements "may be considered subliminal or subconscious, rather than a result of deliberate decision-making of which the translator is aware".

In a study of interference in translation, Mauranen (2004) explicitly addresses the simultaneous activation of source- and target-language systems in the brain during translating as a form of bilingual processing. However, she does not link the results of her analysis of the Corpus of Translated Finnish, which contains Finnish translations where the source language is specified, to conclusions about processing of the two languages. A special case of elimination of factors not linked to the translation process is a priori ruling out certain factors as irrelevant or straightforward.

Addressing different possible explanations in a more direct way, Hansen-Schirra et al. (2007) examine various combinations of subcorpora in the CroCo Corpus and features allowing them to eliminate language contrasts, register differences and understanding as sources of explanation, respectively. Their paper also covers features for which contrasts or register differences are identified as the explanation (findings on pronominal reference are explained by contrastive differences), thus allowing the authors to link findings on lexical density to what they call the translation process, i.e. understanding (Hansen-Schirra et al., 2007, p. 225). It should be noted that, like the other papers cited earlier, they also remain vague about whether they actually refer to cognitive processing. Zufferey and Cartoni (2014) examine a set of different factors in their study of the potential explicitation of connectives, but this set does not include cognition as a factor.

A more complex research design in combination with multivariate statistical techniques can add reliability to the interpretation of corpus findings. The computation of occurrences of 27 lexicogrammatical features in the bi-directional CroCo corpus allows Evert and Neumann (2017) to question the assumed parallel activation of both languages during translation as suggested by Mauranen (2004). They show that translations into German systematically display more similarities with their English source texts than translations and source texts in the opposite direction. This leads to the conclusion that the cognitive assumption of parallel activation cannot be the sole explanation for shining through because of the detected directionality effect of this phenomenon. If the cognitive explanation of parallel activation of both language systems were the main—or only—explanation of shining through, the effect should be virtually identical in both translation directions. While parallel activation appears to be a plausible explanation for translators' behaviour in general (see empirical evidence reported in Balling et al., 2014), other explanations such as social factors have to be considered, too. This shows that assumptions about cognition can be tested with a corpus-based approach. In principle, this result could also be achieved in an experimental study; however, the necessary size in terms of participants and their specification (coverage of both translation directions) virtually precludes experimental testing.

10.2.2 Corpus-based investigations of cognitive linguistic concepts

According to Halverson (2013, p. 66), who has been one of the main proponents of introducing cognitive linguistic explanations (e.g. Goldberg, 2006; Langacker, 2008) into Translation Studies, "the most striking consequence of adopting a cognitive perspective is that it places human cognition and human agency at the centre of the causal picture". Halverson (2013) and Tabakowska (2013) both suggest that the explanatory power of a cognitive linguistic theory can provide a better understanding of the cognitive factors playing a role in the translation process.

Over the last few decades, cognitive linguistics has been increasingly applied within Translation Studies to provide theoretical explanations for various translation phenomena (e.g. Halverson, 2017; Szymańska, 2011). In cognitive linguistic approaches, linguistic phenomena are explained by means of general cognitive processes, namely categorization, analogy, chunking, schematization and association (Langacker, 2008, p. 16). Entrenchment, one of the central concepts of cognitive linguistics, is primarily related to the process of chunking. It is assumed to take place when "through repetition or rehearsal, a complex structure is thoroughly mastered to the point that using it is virtually automatic and requires little conscious monitoring" (Langacker, 2008, p. 16). The link between cognitive entrenchment of a linguistic construction and its use as observed in corpora is established through the so-called "corpus-to-cognition principle" (Schmid, 2010, pp. 101–102): frequency of use identified by means of corpora is assumed to give insight into the level of entrenchment of the analysed linguistic patterns in the speakers' minds. Stefanowitsch and Flach (2017, pp. 102–103) argue that there are two possible perspectives on the relationship between corpora and cognition: corpus-as-output and corpus-as-input. According to the former perspective, the corpus is seen as a sample of linguistic usage, which reflects the cognitive representations of linguistic phenomena, while the latter perspective assumes that the corpus is a sample of linguistic exposure, which forms the cognitive representations over the course of life-long language acquisition. On a more general note, the authors point out that entrenchment is a theoretical construct and as such, cannot be measured directly by any method. Instead, the measurement of entrenchment depends on its operationalization. Since this cognitive concept is closely related to the frequency of usage and the frequency of exposure, corpus linguistic quantitative measures appear to be possible means of operationalization (Stefanowitsch & Flach, 2017).

Halverson (2003, 200) applies the concept of entrenchment to the study of translations, arguing that the repetition of certain translation routines leads to their entrenchment and to easier activation during the translation process. For her model of the translation process, Halverson combines the theoretical framework of cognitive grammar (Langacker, 2008) with models of semantic representation of lexical material in bilinguals (e.g. de Groot, 1992). According to Halverson's model, source-text (ST) material—the lexical/phonological representation of a linguistic item—activates the relevant parts of the cognitive semantic network available to the speaker, which is shared for languages she/he is familiar with. Depending on the number of shared semantic elements between the corresponding structures in the source and target languages (SL and TL), the phonological representation in the TL, which is required to produce a translation, is assumed to be more or less readily available. Halverson argues that, even if the SL items do not directly activate any particular TL elements, the more often the translator actively searches for the appropriate translation solution, the more entrenched the corresponding connection between the two nodes in the semantic network becomes (Halverson, 2003, p. 215), making the selected translation strategy more automatic.

Halverson further suggests that some of the nodes in the cognitive semantic network are more central or salient than others. Halverson (2003) refers to the gravitational pull hypothesis to provide a cognitive explanation for the corpus linguistic findings discussed in the literature, such as translation properties. In its latest version, the gravitational pull hypothesis is divided into three possible sources of translational features, namely "source language salience (gravitational pull), target language salience (magnetism), and link strength effects (connectivity)" (Halverson, 2017, p. 15), the last referring to entrenched translation solutions.

From the perspective of conceptual structures, Halverson (2007) links the concept of translation shifts to the construal operations related to general cognitive processes. Depending on the construal operation, the same situation is conceptualized in a different way, resulting in different linguistic structures. Halverson (2007, pp. 118–119) identifies three general reasons for translation shifts, based on cognition, convention and context. Szymańska (2011) uses the notion of construction to study translation shifts: since constructions represent multiple formal and functional levels of analysis and display a certain range in size and abstractness, they are believed to be suitable for explaining such multi-layer translation concepts as equivalence and translation shifts. While it may be possible to identify corresponding constructions that share some formal and functional features (Leino, 2010, p. 131), translators may encounter "constructional resistance", that is, "the difficulty (sometimes impossibility) of finding a TL construct representing all the functional properties of a ST construct (semantic, pragmatic, discourse, stylistic) while keeping the formal similarity as well as producing a 'natural' TL expression" (Szymańska, 2011, p. 127). Constructional resistance, which occurs because constructions—complex bundles of formal and functional features—are language specific, may therefore lead to translation shifts. Tabakowska (1993, 2013) views constructions as scene conceptualizations: during translation, potential differences in the conceptualization of the same situation and its linguistic realization in different languages may lead to translation shifts (Tabakowska, 1993, pp. 73–77).

In these accounts, constructions thus serve as translation units: assuming that a complete inventory of constructions is available, constructions may be used in corpus-based studies to systematically classify various translation shifts at different levels of analysis (Szymańska, 2011, p. 216). Recent corpus research on translation shifts drawing on the notion of construction, though bound by the range of identified constructions, has shown the potential of combining this theoretical framework with multivariate statistical approaches. Using the comparable and parallel parts of the CroCo corpus (Hansen-Schirra et al., 2012), Serbina (2015) investigated shifts to and from argument structure constructions on several levels of abstractness. The results indicate that one of the reasons for translation shifts may be different frequency distributions of the corresponding constructions in English and German, which, according to the corpus-to-cognition principle, reflect different levels of entrenchment in the speakers' minds. To add the process perspective on the construction shifts, the corpus-based study was complemented by a small-scale translation experiment (see Heilmann et al., in press, for more comprehensive experimental testing).

In another corpus-based study adopting the framework of construction grammar, Kruger and van Rooy (2016) focus on reported-speech constructions in different contact varieties. The study draws on cognitive assumptions to explain the over- or under-representation of the features of the investigated constructions. For instance, the frequent presence of the complementizer *that* in indirect speech (see Olohan & Baker, 2000, discussed in Section 10.2.1) in translated English and second-language writing is interpreted in terms of the assumed reduction in processing effort and the potentially higher level of entrenchment of this variant in the minds of bilingual speakers of English and Afrikaans.

Kruger and De Sutter (2018) also investigate the use of the complementizer *that*. They focus on the predictors for the use of explicit *that*, namely cognition, risk avoidance and contrastive differences. The study takes into account multiple factors considered as operationalizations of cognition and risk-avoidance-related sources of explanations. Kruger and De Sutter (2018) apply a multifactorial statistical analysis to compare the effect of these factors on the presence of the complementizer in different varieties. The influence of contrastive effects is included in the research design indirectly by analysing two pairwise comparisons between the standard variety and the two language-contact varieties. While corpus-based work applying this statistical approach appears promising for assessing the relative importance of various groups of factors on the omission of *that*, the authors observe that to clearly separate cognition and risk-avoidance factors, it is necessary to carry out more controlled corpus-based and experimental studies (Kruger & De Sutter, 2018, pp. 279–280).

A constructional account of translations with a focus on frame semantics (see, e.g., Fillmore, 1982) is presented in Čulo (2013). The study investigates two cases of constructional mismatch for the language pair English-German. One of these constructions is the unagentive subject construction, that is, an extreme case of non-agentive subjects (see Section 10.2.3). Čulo (2013) reports an analysis of the CroCo corpus according to which this construction could be translated into German using another topicalization construction. This constructional shift may be accompanied by a frame-semantic shift. Building on this analysis, Czulo (2017) proposes a frame-semantic and constructional model of translation.

The overview of research presented in this sub-section has shown that cognitive linguistics allows assumptions to be made about the cognitive reasons for various phenomena observed in translated language, thus taking into account the important role of cognitive processes during the process of translation. As discussed in the cognitive linguistic literature, corpus data—conceptualized as usage data both reflecting cognition and (re-)forming it—could be used as a link to cognition, provided the appropriate operationalization of the cognitive theoretical constructs is adopted. However, this overview has also indicated that the triangulation of corpus linguistic methods with experimental work can provide further insights into the analysed research questions.

10.2.3 Combined corpus-based and experimental studies

The third broad area in the interaction between CBTS and cognition entails complementary process- and product-based analyses of translation data (see also Hansen-Schirra, this volume). One way of combining product and process analyses is by sequentially (or cyclically) investigating the same linguistic phenomenon using a corpus linguistic and an experimental research design. Both approaches can also inform each other by using the one to generate hypotheses, which are then tested with the help of the other. This strand of research often uses corpus data to study the final products of translations, drawing on available large parallel or comparable corpora (see Section 10.1). Thus, corpus-based research is complemented by experimental data, which provides an additional process-based perspective not present in such corpora. Typically, process data is used to test hypotheses about cognition. Therefore, in this approach, corpus data is only indirectly linked to the study of cognition, either by preparing the ground for experiments that aim at testing hypotheses related to cognition, or by testing experimental evidence on a larger quantitative basis using corpora.

Vandepitte's research group has been focusing on the analysis of non-prototypical agents, i.e. non-agentive subjects, combined with proto-agents requiring predicates. A translation experiment performed using the software Morae indicated slower translation speed for sentences containing this type of agent (Vandepitte & Hartsuiker, 2011). In addition, a corpus study has shown that a combination of non-prototypical agents and the verb *give* is likely to be changed in translations from English to Dutch (Doms et al., 2016). The experimental research preceding the corpus analysis is considered to be evidence for existing restrictions on the combination of non-human agents and action verbs in Dutch—this evidence is taken as a starting point for the subsequent corpus study.

S. Hansen (2003) is an early example of a more direct triangulation of the product- and process-based perspectives. She combines a detailed analysis of comparable and parallel corpora with an experimental pilot study. The corpus analyses study the translation properties of normalization and anti-normalization (i.e. avoidance of typical features) and establish the influence of the source language on the features of the translated language. The experiment tests the process of translating the linguistic features that displayed specific patterns in the corpus analysis.

Alves et al. (2010) combine corpus and experimental research to study the phenomenon of grammatical metaphor, more specifically the process of de-metaphorization, from both product and process perspectives. The authors specifically address the advantages of triangulating corpus-based and experimental methods by noting, on the one hand, that corpora provide authentic data but make it difficult to test assumptions about cognition, while, on the other hand, experiments provide a more direct perspective on the process, making the connection between the observed behavioural data and cognitive processes somewhat more straightforward. A combined process-product analysis could, therefore, benefit from the advantages of both methods and help overcome their individual disadvantages.

In Serbina et al. (2017), analyses of experimental data on the shifts between nouns and verbs are preceded by a study of the proportions of these word classes in the CroCo Corpus. Data from a translation experiment in the English-German translation direction shows that shifts from verbs to nouns are most frequent: these results are in line with contrastive tendencies observed in the corpus analysis, suggesting that translators follow the norms of the target language. Keystroke logging analyses additionally show that these shifts are typically performed in one step without additional intermediate solutions. Moreover, cognitive effort, particularly when operationalized through the measure of fixation count, increases when a shift either to nouns or to verbs is present compared with cases where the word class is not changed.

Halverson (2017) tests the latest version of the gravitational pull hypothesis (see Section 10.2.2) in a cyclic approach drawing on frequency distributions in the English-Norwegian Parallel Corpus (ENPC) and the British National Corpus (BNC) as well as on results reported in the previous literature, corroborating these in a sentence generation task. The results of these initial analyses are then reviewed again using a combination of corpus and keystroke logging data.

An approach that further integrates product and process analyses treats translation process data collected with the help of keystroke logging as a corpus. While the studies described earlier use two different types of data to combine product- and process-based perspectives, treating keystroke logs as a corpus allows researchers to analyse the same data using different methods. The idea of creating a corpus consisting of translation process data was introduced in Pagano et al. (2004) using data stored in the Corpus on Process for the Analysis of Translations (CORPRAT), a corpus which can be manually annotated and semi-automatically queried using the software LITTERAE, designed to analyse translation units stored as translation process data (Alves & Vale, 2009, 2011).

A recent study by Kajzer-Wietrzny et al. (2016) also uses data collected in an experiment for analysing the source and the final target texts from a process perspective by focusing on different phases of translation. However, it is of less interest in the present context, as the study does not have a strong focus on cognition.

The Keystroke Logged Translation Corpus (KLTC), introduced in Serbina et al. (2015), is also based on data collected during translation experiments. This corpus can be used to analyse intermediate versions of individual words and sentences, defined as "variants of the unfolding texts produced at certain points in time during the translation process" (Serbina et al., 2015, p. 102). Further pushing the boundaries between corpus-based and experimental methods, this corpus is enriched with part-of-speech annotation of these intermediate versions even if they do not make it to the final translation product. Similarly to the corpora described in Section 10.1, querying for this type of annotation in a corpus consisting of the keystroke log files (Alves & Vale, 2009, 2011) allows researchers to generalize beyond individual observations. Appropriate querying strategies for this new type of corpus still require further development.

10.3 Recent developments and future directions

Until now, the large number of individual papers in CBTS somehow does not seem to have brought a major breakthrough in terms of fully illustrating the empirical facts of translation. To some extent this is due to the inconclusiveness of the results, which, in turn, can be partly explained by limitations of some of the studies. So, one might come to the conclusion that perhaps the approach has not lived up to its promise after all, were it not for the machine learning findings mentioned in Section 10.2.1, which suggest that there must be something about translations that makes them so easy to spot.

So, possibly, the methodological approach needs to be refined. Often, studies investigate individual linguistic features in one language pair and usually in a fairly limited data set in one genre/register. However, translations are pushed into different directions by different features, and different contexts will involve different expectations as to the extent to which translations are allowed to be identifiable as such (House, 1997). Moreover, excluding the (aligned) source texts will make it utterly impossible to decide whether a feature of the target text is specific to translation without checking whether the feature was present in the source text (see Steiner, 2012). Also, it may be necessary to use more strictly quantitative, explanatory designs, as suggested by Tummers et al. (2005).

In recent years, a number of CBTS studies have worked towards overcoming these limitations. A number of corpus studies have employed more complex designs that include different registers, different translation directions and sometimes also different features, using advanced statistical techniques, and thus yielding a more differentiated picture of the multifactorial make-up of translations (e.g. Delaere, 2015; Evert & Neumann, 2017; Kunz et al., 2017; Serbina, 2015; Vandevoorde, forthcoming, etc.).

An important recent development in CBTS that calls into question the conventional conceptualization of the translation process in empirical translation research is concerned with a closer investigation of the influence of editing/proofreading on the translation workflow. Bisiada (e.g. 2013) reports findings from a corpus of translations submitted to a journal and the final versions of the translations after revision by the journal editor. It turns out that many of the phenomena described by CBTS scholars as translation properties appear to be introduced by the editor rather than by translators (for indirect evidence in newspapers see also Delaere, 2015). These findings are further strengthened by Kruger (2017), who reports statistically significant linguistic differences in a corpus of originals before and after editing, which suggest that editors introduce the features that have also been described as properties of translation, and plausibly conjectures that edited translations should show similar effects. Although these approaches are, strictly speaking, not concerned with cognitive aspects of translation, they are of great relevance to cognitive approaches to translation because they relativize the role of the translator's cognition in relation to the published translation, in which those aspects that might require the highest cognitive demand on the part of the translator are not present because they are changed during editing.

A potential development towards the integration of process- and product-based translation research could emerge from the CRITT TPR database (Carl et al., 2016). At this stage, this is a database containing translation and post-editing experiments. However, the way it is structured and analysed bears some striking similarities to corpus linguistic methods. While at this stage it is impossible to search the database in the same way a corpus can be searched, it contains the relevant information and could be exploited in a similar way.

It seems appropriate to conclude that CBTS has become mature in recent years: there is a trend towards more complex research designs involving statistical analysis. In line with a general trend in corpus linguistics, this also involves a tendency to take into consideration cognitive explanations. Halverson (2015, p. 311) even claims that "the adoption of some cognitively oriented explanatory models is driving the integration of process and product perspectives on translational phenomena". Also, the analysis of process data as a corpus will probably see some additional expansion.

Depending on the details of the research design, it is also possible that a very large corpus with rich metadata can be used to discover behavioural patterns that might be interpretable with respect to cognitive explanations. Generally speaking, the use of larger, but still balanced, corpora should level out idiosyncratic aspects of the individual translators and the specific situations in which they produced their translations, so that patterns of translators' behaviour become visible. Only such more complex designs—ideally with more meta-information on the details of the production context—can reduce the gap between high-level theoretical assumptions and limits of the empirical research design.

Hansen-Schirra (2017, p. 238) discusses links between psychological mechanisms and properties of translation: she suggests that priming (of the source language) could lead to shining through, whereas monitoring or inhibition (of the source language) could lead to normalization. This appears to be a very promising avenue for future research, as it could provide an explanation for the mechanisms that lead to these features. However, as Evert and Neumann (2017) show, the linguistic characteristics of translations are not exhausted by cognitive explanations.

Notes

1 Although Toury never carried out any corpus analyses, he discussed quantitative analyses and specifically the role of probabilities for understanding specific characteristics of translations far beyond what was typically done in Translation Studies in Toury (2004).
2 Retrieved 16 March 2018 from www.alc.manchester.ac.uk/translation-and-intercultural-studies/research/projects/translational-english-corpus-tec/
3 Retrieved 16 March 2018 from http://corpus.leeds.ac.uk/mellange/ltc.html
4 Retrieved 3 April 2018 from www.lancaster.ac.uk/fass/projects/corpus/ZCTC/
5 Olohan and Baker indeed use the term "subconscious" despite its psychoanalytical connotations. In what follows, we will use the more neutral term "unconscious".

Further reading

De Sutter, G., Lefer, M.-A., & Delaere, I. (Eds.). (2017). *Empirical translation studies. New methodological and theoretical traditions*. Berlin: Mouton de Gruyter.

This volume presents empirical work in the area of corpus-based Translation Studies with particular emphasis on cognitive questions. The papers collected in this publication use cutting-edge methodology drawing on data triangulation and advanced statistical analyses.

Halverson, S. L. (2015). Cognitive Translation Studies and the merging of empirical paradigms: The case of "literal translation". *Translation Spaces, 4*(2), 310–340.
Using the concept of "literal translation", Halverson demonstrates the combination of product- and process-based methods for theorizing translation grounded in cognitive linguistics.

Hansen-Schirra, S., Czulo, O., & Hofmann, S. (Eds.). (2017). *Empirical modelling of translation and interpreting.* Berlin: Language Science Press.
This publication provides a recent overview of both empirical translation and interpreting studies. The papers apply a broad range of methodologies, often combining corpus-based analyses with process-based methods including eye-tracking, keystroke logging and functional MRI experiments as well as surveys.

Oakes, M. P., & Ji, M. (Eds.). (2012). *Quantitative methods in corpus-based translation studies. A practical guide to descriptive translation research.* Amsterdam: John Benjamins.
The papers in this volume apply state-of-the-art quantitative methods, some of which are frequently used in cognitively oriented corpus linguistics, to translation corpora.

Rojo, A., & Ibarretxe-Antuñano, I. (Eds.). (2013). *Cognitive linguistics and translation: Advances in some theoretical models and applications.* Berlin: Mouton de Gruyter.
This volume contains a range of papers taking a cognitive linguistic perspective on Translation Studies. It includes both theoretical and empirical work in the area.

References

Alves, F., Pagano, A., Neumann, S., Steiner, E., & Hansen-Schirra, S. (2010). Units of translation and grammatical shifts: Towards an integration of product- and process-based research in translation. In G. Shreve, & E. Angelone (Eds.), *Translation and cognition* (pp. 109–142). Amsterdam: John Benjamins.

Alves, F., & Vale, D. (2009). Probing the unit of translation in time: Aspects of the design and development of a web application for storing, annotating, and querying translation process data. *Across Languages and Cultures, 10*(2), 251–273.

Alves, F., & Vale, D. C. (2011). On drafting and revision in translation: A corpus linguistics oriented analysis of translation process data. *Translation: Computation, Corpora, Cognition, 1*(1), 105–122.

Anderman, G. M., & Rogers, M. (2008). The linguist and the translator. In G. M. Anderman, & M. Rogers (Eds.), *Incorporating corpora. The linguist and the translator* (pp. 5–17). Clevedon: Multilingual Matters.

Baker, M. (1993). Corpus linguistics and translation studies. Implications and applications. In M. Baker, G. Francis, & E. Tognini-Bonelli (Eds.), *Text and technology. In honour of John Sinclair* (pp. 233–250). Amsterdam: John Benjamins.

Baker, M. (1996). Corpus-based translation studies: The challenges that lie ahead. In H. Somers (Ed.), *Terminology, LSP and translation. Studies in language engineering in honour of Juan C. Sager* (pp. 175–186). Amsterdam: John Benjamins.

Balling, L. W., Hvelplund, K. T., & Sjørup, A. C. (2014). Evidence of parallel processing during translation. *Meta: Journal des Traducteurs, 59*(2), 234–259. https://doi.org/10.7202/1027474ar

Baroni, M., & Bernardini, S. (2006). A new approach to the study of translationese: Machine-learning the difference between original and translated text. *Literary and Linguistic Computing, 21*(3), 259–274. https://doi.org/10.1093/llc/fqi039

Becher, V. (2010). Abandoning the notion of "translation-inherent" explicitation: Against a dogma of translation studies. *Across Languages and Cultures, 11*(1), 1–28. https://doi.org/10.1556/Acr.11.2010.1.1

Bernardini, S., & Zanettin, F. (2004). When is a universal not a universal? Some limits of current corpus-based methodologies for the investigation of translation universals. In A. Mauranen, & P. Kujamäki (Eds.), *Translation universals. Do they exist?* (pp. 51–62). Amsterdam: John Benjamins.

Biber, D. (1990). Methodological issues regarding corpus-based analyses of linguistic variation. *Literary and Linguistic Computing,* 5(4), 257–269.

Bisiada, M. (2013). From hypotaxis to parataxis: An investigation of English–German syntactic convergence in translation [unpublished PhD dissertation]. The University of Manchester. www.research.manchester.ac.uk/portal/en/theses/from-hypotaxis-to-parataxis-an-investigation-of-englishgerman-syntactic-convergence-in-translation(c404ecc5-80a3-46fb-93a6-9e8ad7dc5792).html

Butler, C. (1985). *Statistics in linguistics.* Oxford: Blackwell.

Carl, M., Bangalore, S., & Schaeffer, M. (2016). *New directions in empirical translation process research: Exploring the CRITT TPR-DB.* Cham: Springer.

Chesterman, A. (2010). Why study translation universals? In R. Hartama-Heinonen, & P. Kukkonen (Eds.), *Kiasm* (pp. 38–48). https://helda.helsinki.fi/handle/10138/24313
Čulo, O. (2013). Constructions-and-frames analysis of translations: The interplay of syntax and semantics in translations between English and German. *Constructions and Frames, 5*(2), 143–167.
Czulo, O. (2017). Aspects of a primacy of frame model of translation. In S. Hansen-Schirra, O. Czulo, & S. Hofmann (Eds.), *Empirical modelling of translation and interpreting* (pp. 465–490). Berlin: Language Science Press.
de Groot, A. M. B. (1992). Bilingual lexical representation: A closer look at conceptual representations. In R. Frost, & L. Katz (Eds.), *Orthography, phonology, morphology, and meaning* (pp. 389–412). Elsevier.
Delaere, I. (2015). Do translators walk the line? Visually exploring translated and non-translated texts in search of norm conformity [unpublished PhD dissertation]. University of Ghent, Ghent. http://hdl.handle.net/1854/LU-5888594
Delaere, I., De Sutter, G., & Plevoets, K. (2012). Is translated language more standardized than non-translated language?: Using profile-based correspondence analysis for measuring linguistic distances between language varieties. *Target, 24*(2), 203–224. https://doi.org/10.1075/target.24.2.01del
De Sutter, G., Lefer, M.-A., & Delaere, I. (Eds.). (2017). *Empirical translation studies. New methodological and theoretical traditions*. Berlin: Mouton de Gruyter.
Doms, S., De Clerck, B., & Vandepitte, S. (2016). Non-human agents as subjects in English and Dutch: A corpus-based translation study. In T. Ruchot, & P. Van Praet (Eds.), *Atypical predicate-argument relations* (pp. 87–112). Amsterdam: John Benjamins. https://doi.org/10.1075/lis.33.04dom
Even-Zohar, I. (1979). Polysystem theory. *Poetics Today, 1*(1/2), 287–310. https://doi.org/10.2307/1772051
Evert, S., & Neumann, S. (2017). The impact of translation direction on characteristics of translated texts. A multivariate analysis for English and German. In G. De Sutter, M.-A. Lefer, & I. Delaere (Eds.), *Empirical translation studies. New theoretical and methodological traditions* (pp. 47–80). Berlin: Mouton de Gruyter.
Fabricius-Hansen, C. (1999). Information packaging and translation: Aspects of translational sentence splitting (German-English/Norwegian). In M. Doherty (Ed.), *Sprachspezifische Aspekte der Informationsverteilung* (pp. 175–214). Berlin: Akademie Verlag.
Fantinuoli, C., & Zanettin, F. (Eds.). (2015). *New directions in corpus-based translation studies*. Berlin: Language Science Press.
Fillmore, C. J. (1982). Frame semantics. In Linguistics Society of Korea (Ed.), *Linguistics in the morning calm* (pp. 111–137). Seoul: Hanshin.
Frankenberg-Garcia, A. (2009). Are translations longer than source texts?: A corpus-based study of explicitation. In A. Beeby, P. Rodríguez Inés, & P. Sánchez-Gijón (Eds.), *Corpus use and translating. Corpus use for learning to translate and learning corpus use to translate* (Vol. 82, pp. 47–58). https://doi.org/10.1075/btl.82.05fra
Goldberg, A. E. (2006). *Constructions at work: The nature of generalization in language*. Oxford: Oxford University Press.
Halverson, S. L. (2003). The cognitive basis of translation universals. *Target, 15*(2), 197–241. https://doi.org/10.1075/target.15.2.02hal
Halverson, S. L. (2007). A cognitive linguistic approach to translation shifts. In W. Vandeweghe, S. Vandepitte, & M. van de Velde (Eds.), *The study of language and translation* (pp. 105–121). Amsterdam: John Benjamins.
Halverson, S. L. (2013). Implications of cognitive linguistics for translation studies. In A. Rojo, & I. Ibarretxe-Antuñano (Eds.), *Cognitive linguistics and translation advances in some theoretical models and applications* (pp. 33–74). Berlin: Mouton de Gruyter.
Halverson, S. L. (2015). Cognitive translation studies and the merging of empirical paradigms: The case of "literal translation". *Translation Spaces, 4*(2), 310–340. https://doi.org/10.1075/ts.4.2.07hal
Halverson, S. L. (2017). Gravitational pull in translation: Testing a revised model. In G. de Sutter, M.-A. Lefer, & I. Delaere (Eds.), *Empirical translation studies* (pp. 9–45). Berlin: Mouton de Gruyter.
Hansen, S. (2003). *The nature of translated text. An interdisciplinary methodology for the investigation of the specific properties of translations*. Saarbrücken: DFKI/Universität des Saarlandes.
Hansen-Schirra, S. (2017). EEG and universal language processing in translation. In J. W. Schwieter, & A. Ferreira (Eds.), *The handbook of translation and cognition* (pp. 232–247). Hoboken, NJ: John Wiley & Sons, Inc. https://doi.org/10.1002/9781119241485.ch13
Hansen-Schirra, S., Czulo, O., & Hofmann, S. (Eds.). (2017). *Empirical modelling of translation and interpreting*. Berlin: Language Science Press.

Hansen-Schirra, S., Neumann, S., & Steiner, E. (2007). Cohesive explicitness and explicitation in an English-German translation corpus. *Languages in Contrast, 7*(2), 241–265.
Hansen-Schirra, S., Neumann, S., & Steiner, E. (2012). *Cross-linguistic corpora for the study of translations—Insights from the language pair English-German*. Berlin: Mouton de Gruyter.
Hansen-Schirra, S., & Teich, E. (2009). Corpora in human translation. In A. Lüdeling, & M. Kytö (Eds.), *Corpus linguistics. An international handbook* (pp. 1159–1175). Berlin: de Gruyter.
Heilmann, A., Freiwald, J., Serbina, T., & Neumann, S. (in press/online first). Animacy and agentivity of subject themes in English-German translation. Lingua. https://doi.org/10.1016/j.lingua.2020.102813.
House, J. (1997). *Translation quality assessment. A model revisited*. Tübingen: Gunter Narr Verlag.
Hu, X., Xiao, R., & Hardie, A. (2019). How do English translations differ from non-translated English writings? A multi-feature statistical model for linguistic variation analysis. *Corpus Linguistics and Linguistic Theory, 15*(2), 347–382.
Ji, M. (Ed.). (2016). *Empirical translation studies: Interdisciplinary methodologies explored*. Bristol, CT: Equinox.
Johansson, S. (2008). Some aspects of the development of corpus linguistics in the 1970s and 1980s. In A. Lüdeling, & M. Kytö (Eds.), *Corpus linguistics. An international handbook* (pp. 33–53). Berlin: De Gruyter.
Johansson, S., & Hofland, K. (1994). Towards an English-Norwegian parallel corpus. In U. Fries, G. Tottie, & P. Schneider (Eds.), *Creating and using English language corpora* (pp. 25–37). Amsterdam: Rodopi.
Kajzer-Wietrzny, M., Whyatt, B., & Stachowiak, K. (2016). Simplification in inter- and intralinguistic translation: Combining corpus linguistics, key logging and eye-tracking. *Poznan Studies in Contemporary Linguistics, 52*(2), 235–267.
Kenny, D. (1998). Corpora in translation studies. In M. Baker (Ed.), *Routledge encyclopedia of translation studies* (pp. 50–53). London: Routledge.
Königs, F. G. (1987). Was beim Übersetzen passiert. Theoretische Aspekte, empirische Befunde und praktische Konsequenzen. *Die Neueren Sprachen, 86*(2), 162–185.
Krings, H. P. (1986). *Was in den Köpfen von Übersetzern vorgeht. Eine empirische Untersuchung zur Struktur des Übersetzungsprozesses an fortgeschrittenen Französischlernern*. Tübingen: Gunter Narr Verlag.
Kruger, A., Wallmach, K., & Munday, J. (Eds.). (2011). *Corpus-based translation studies: Research and applications*. London: Continuum.
Kruger, H. (2012). *Postcolonial polysystems: The production and reception of translated children's literature in South Africa*. Amsterdam: John Benjamins. https://doi.org/10.1075/btl.105
Kruger, H. (2017). The effects of editorial intervention. Implications for studies of the features of translated language. In G. De Sutter, M.-A. Lefer, & I. Delaere (Eds.), *Empirical translation studies. New methodological and theoretical traditions* (pp. 113–156). Berlin: De Gruyter.
Kruger, H., & De Sutter, G. (2018). Alternations in contact and non-contact varieties. *Translation, Cognition & Behavior, 1*(2), 251–290. https://doi.org/10.1075/tcb.00011.kru
Kruger, H., & van Rooy, B. (2016). Syntactic and pragmatic transfer effects in reported-speech constructions in three contact varieties of English influenced by Afrikaans. *Language Sciences, 56*, 118–131.
Kunz, K., Castagnoli, S., & Kübler, N. (2010). Corpora in translator training. A program for an eLearning course. In D. Gile, G. Hansen, & N. K. Pokorn (Eds.), *Why translation studies matters* (pp. 195–208). Amsterdam: John Benjamins.
Kunz, K., Degaetano-Ortlieb, S., Lapshinova-Koltunski, E., Menzel, K., & Steiner, E. (2017). English-German contrasts in cohesion and implications for translation. In G. De Sutter, M.-A. Lefer, & I. Delaere (Eds.), *Empirical translation studies. New methodological and theoretical traditions* (pp. 265–311). Berlin: De Gruyter.
Langacker, R. W. (2008). *Cognitive grammar: A basic introduction*. Oxford: Oxford University Press.
Lapshinova-Koltunski, E. (2017). Exploratory analysis of dimensions influencing variation in translation. In G. De Sutter, M.-A. Lefer, & I. Delaere (Eds.), *Empirical translation studies. New methodological and theoretical traditions* (pp. 207–234). Berlin: De Gruyter.
Laviosa, S. (Ed.). (1998). L'approche basée sur le corpus/The corpus-based approach. *Meta: Journal des Traducteurs, 43*.
Laviosa-Braithwaite, S. (1996). Comparable corpora: Towards a corpus linguistic methodology for the empirical study of translation. In M. Thelen, & B. Lewandowska-Tomaszczyk (Eds.), *Translation and meaning*. Part 3 (pp. 153–163). Maastricht: Euroterm.
Leino, J. (2010). Results, cases, and constructions: Argument structure constructions in English and Finnish. In H. C. Boas (Ed.), *Contrastive studies in construction grammar* (pp. 103–135). Amsterdam: John Benjamins.
McEnery, T., & Wilson, A. (2001). *Corpus linguistics. An introduction* (2nd ed.). Edinburgh: Edinburgh University Press.

Macken, L., De Clercq, O., & Paulussen, H. (2011). Dutch parallel corpus: A balanced copyright-cleared parallel corpus. *Meta: Journal des Traducteurs, 56*(2), 374–390. https://doi.org/10.7202/1006182ar
Mauranen, A. (2004). Corpora, universals and interference. In A. Mauranen, & P. Kujamäki (Eds.), *Translation universals. Do they exist?* (pp. 65–82). Amsterdam: John Benjamins.
Mauranen, A., & Kujamäki, P. (2004a). Introduction. In A. Mauranen, & P. Kujamäki (Eds.), *Translation universals. Do they exist?* (pp. 1–11). Amsterdam: John Benjamins.
Mauranen, A., & Kujamäki, P. (Eds.). (2004b). *Translation universals. Do they exist?* Amsterdam: John Benjamins.
Monti, C., Bendazzoli, C., Sandrelli, A., & Russo, M. (2005). Studying directionality in simultaneous interpreting through an electronic corpus: EPIC (European Parliament Interpreting Corpus). *Meta: Journal des Traducteurs, 50*(4). https://doi.org/10.7202/019850ar
Neumann, S., & Hansen-Schirra, S. (2012). Corpus methodology and design. In S. Hansen-Schirra, S. Neumann, & E. Steiner (Eds.), *Cross-linguistic corpora for the study of translations: Insights from the language pair English-German* (pp. 21–34). Berlin: Mouton de Gruyter. https://doi.org/10.1515/9783110260328
Oakes, M. P., & Ji, M. (Eds.). (2012). *Quantitative methods in corpus-based translation studies. A practical guide to descriptive translation research.* Amsterdam: John Benjamins.
Olohan, M. (2002). Leave it out! Using a comparable corpus to investigate aspects of explicitation in translation. *Cadernos de Tradução, 1*(9), 153–169.
Olohan, M., & Baker, M. (2000). Reporting "that" in translated English. Evidence for subconscious processes of explicitation? *Across Languages and Cultures, 1*(2), 141–158.
Øveras, L. (1998). In search of the third code. An investigation of norms in literary translation. *Meta: Journal des Traducteurs, 43*(4), 557–570.
Pagano, A., Magalhães, C., & Alves, F. (2004). Towards the construction of a multilingual, multifunctional corpus: Factors in the design and application of CORDIALL. *TradTerm, 10*, 143–161.
Rodríguez-Inés, P. (2017). Corpus-based insights into cognition. In J. W. Schwieter, & A. Ferreira (Eds.), *The handbook of translation and cognition* (pp. 265–289). Hoboken, NJ: John Wiley & Sons, Inc.
Rojo, A., & Ibarretxe-Antuñano, I. (Eds.). (2013). *Cognitive linguistics and translation: Advances in some theoretical models and applications.* Berlin: Mouton de Gruyter.
Schmid, H.-J. (2010). Does frequency in text really instantiate entrenchment in the cognitive system? In D. Glynn, & K. Fischer (Eds.), *Quantitative methods in cognitive semantics* (pp. 101–133). Berlin: de Gruyter.
Serbina, T. (2015). A construction grammar approach to the analysis of translation shifts: a corpus-based study [PhD dissertation]. Publikationsserver der RWTH Aachen University). https://publications.rwth-aachen.de/record/538325
Serbina, T., Hintzen, S., Niemietz, P., & Neumann, S. (2017). Changes of word class during translation—Insights from a combined analysis of corpus, keystroke logging and eye-tracking data. In S. Hansen-Schirra, O. Czulo, & S. Hofmann (Eds.), *Empirical modelling of translation and interpreting* (pp. 177–208). Berlin: Language Science Press.
Serbina, T., Niemietz, P., & Neumann, S. (2015). Development of a keystroke logged translation corpus. In C. Fantinuoli, & F. Zanettin (Eds.), *New directions in corpus-based translation studies* (pp. 11–34). Berlin: Language Science Press. http://langsci-press.org/catalog/book/76
Sinclair, J. M. (1991). *Corpus, concordance, collocation.* Oxford: Oxford University Press.
Stefanowitsch, A., & Flach, S. (2017). The corpus-based perspective on entrenchment. In H.-J. Schmid (Ed.), *Entrenchment and the psychology of language learning* (pp. 101–127). Washington: American Psychological Association.
Steiner, E. (2002). Grammatical metaphor in translation. Some methods for corpus-based investigations. In H. Hasselgård, S. Johansson, B. Behrens, & C. Fabricius-Hansen (Eds.), *Information structure in a cross-linguistic perspective* (pp. 213–228). Amsterdam: Rodopi.
Steiner, E. (2012). A characterization of the resource based on shallow statistics. In S. Hansen-Schirra, S. Neumann, & E. Steiner (Eds.), *Cross-linguistic corpora for the study of translations: Insights from the language pair English-German* (pp. 71–90). Berlin: Mouton de Gruyter.
Szymańska, I. (2011). Construction grammar as a framework for describing translation: A prolegomenon. In M. Pawlak, & J. Bielak (Eds.), *New perspectives in language, discourse and translation studies* (pp. 215–225). Berlin: Springer.
Tabakowska, E. (1993). *Cognitive linguistics and poetics of translation.* Tübingen: Narr.
Tabakowska, E. (2013). (Cognitive) grammar in translation: Form as meaning. In A. Rojo, & I. Ibarretxe-Antuñano (Eds.), *Cognitive linguistics and translation* (pp. 229–250). Berlin: Mouton de Gruyter.
Teich, E. (2003). *Cross-linguistic variation in system and text. A methodology for the investigation of translations and comparable texts.* Berlin: Mouton de Gruyter.

Toury, G. (2004). Probabilistic explanations in translation studies. Welcome as they are, would they qualify as universals? In A. Mauranen, & P. Kujamäki (Eds.), *Translation universals. Do they exist?* (pp. 15–32). Amsterdam: John Benjamins.

Toury, G. (1995/2012). *Descriptive translation studies—and beyond* (revised (2nd) ed.). Amsterdam: John Benjamins.

Tummers, J., Heylen, K., & Geeraerts, D. (2005). Usage-based approaches in cognitive linguistics: A technical state of the art. *Corpus Linguistics and Linguistic Theory, 1*(2), 225–261.

Vandepitte, S., & Hartsuiker, R. (2011). Metonymic language use as a student translation problem: Towards a controlled psycholinguistic investigation. In C. Alvstad, A. Hild, & E. Tiselius, (Eds.), *Methods and strategies of process research: Integrative approaches in translation studies* (pp. 67–92). Amsterdam: John Benjamins.

Vanderauwera, R. (1985). *Dutch novels translated into English: The transformation of a "minority" literature.* Amsterdam: Rodopi.

Vandevoorde, L. (forthcoming). *Semantic differences in translation: Exploring the field of inchoativity.* Berlin: Language Science Press.

Volansky, V., Ordan, N., & Wintner, S. (2015). On the features of translationese. *Digital Scholarship in the Humanities, 30*(1), 98–118. https://doi.org/10.1093/llc/fqt031

Xiao, R., & Hu, X. (2015). *Corpus-based studies of translational Chinese in English-Chinese translation.* Berlin: Springer.

Zufferey, S., & Cartoni, B. (2014). A multifactorial analysis of explicitation in translation. *Target, 26*(3), 361–384. https://doi.org/10.1075/target.26.3.02zuf

11

Translation, linguistics and cognition

Kirsten Malmkjær

11.1 Introduction

Linguistics is the academic discipline that investigates the nature and uses of spoken, written and signed language. It has a number of sub-disciplines, which cover the sounds of language (phonemics, phonetics and phonology), the structure of language (morphology and grammar), meaning (semantics), language in society (socio-linguistics), language in the individual (psycholinguistics) and language in text (text linguistics, discourse and conversational analysis, and genre analysis). In addition, historical linguistics deals with the history of languages, and comparative or contrastive linguistics identifies, describes and compares major types of languages. In this volume, a separate chapter discusses the relationships between translation, psycholinguistics and cognition, so I shall not discuss psycholinguistics here. Like many other disciplines, linguistics is the home to a number of different "schools", each with a set of basic assumptions that will influence how each of the sub-disciplines mentioned is approached or whether it is approached at all. For example, approaches that are especially focused on cognition may pay relatively little attention to socio-linguistics. Similarly, not each sub-discipline will pay the same amount of attention to cognition.

11.2 Core topics

11.2.1 Cognition and universalism in linguistics

Issues of cognition are not central in historical linguistics, although practitioners readily acknowledge that it is likely that human cognition constrains the form of possible human languages (Anderson et al., 2010, p. 229, and see later); similarly, within typological linguistics, Tomlin (1986, p. 3) has suggested that "[t]ypological explanation, to the extent that it can be separated from areal, genetic, and historical influences, derives from universal principles of linguistic organization and processing", which, in turn, "derive from fundamental cognitive constraints on the processing of linguistic information during discourse production and comprehension".

The predominance among languages of the Subject-Verb-Object (SVO) and Subject-Object-Verb (SOV) types may be explained with reference to three principles, of which two may be considered to have a cognitive issue at their root. These are the Theme First Principle, which says that information that is central to moving the discourse forward (thematic) tends to be placed in initial position in simple main clauses; and the Animated First Principle, according to which noun phrases that denote the most animated discourse participants will be placed early in the discourse (the third principle is the Verb-Object Bonding principle, according to which the object of a transitive clause is more tightly bound to the verb than to the subject) (Tomlin, 1986, p. 4). Arguably, these principles provide a functional account of the consequences of the cognitive constraints on human language. These consequences are available to direct observation, whereas the constraints themselves are inferred from the consequences. Thus, one of the best-known observers of languages, Joseph Harold Greenberg (1915–2001), used observations of 30 languages as the basis for his list of 45 linguistic universals (Greenberg, 1963, on the basis of a paper delivered at a conference in 1961), which, in turn, he sought to account for in terms of "some general principles [...] which seem to underlie a number of different universals and from which they may be deduced" (1963, p. 96). He writes, for example (1963, p. 76–77; italics original),

> The vast majority of languages have several variant orders but a single dominant one. Logically, there are six possible orders: SVO, SOV, VSO, VOS, OSV, and OVS. Of these six, however, only three normally occur as dominant orders. The three which do not occur at all, or at least are excessively rare, are VOS, OSV, and OVS. These all have in common that the object precedes the subject. This gives us our first universal:
>
> *Universal 1.* In declarative sentences with nominal subject and object, the dominant order is almost always one in which the subject precedes the object.

Greenberg thus focuses on how universals manifest in languages, and he allows for variation among languages with respect to the presence within them of one or other feature (as evidenced in the expression "almost always" in universal 1, quoted in the extract). A rather different, though, at least initially, equally observation-based account of the nature of human language was provided by Avram Noam Chomsky (b. 1928), whose observation of the complexity of human languages led him to conclude that unless something like universals resided in the human mind, it would simply not be possible for small children to learn any language. He writes (Chomsky, 1965 p. 27):

> A theory of linguistic structure that aims for explanatory adequacy incorporates an account of linguistic universals, and it attributes tacit knowledge of these universals to the child. It proposes, then, that the child approaches the data with the presumption that they are drawn from a language of a certain antecedently well-defined type, his problem being to determine which of the (humanly) possible languages is that of the community in which he is placed. Language learning would be impossible unless this were the case. The important question is: What are the initial assumptions concerning the nature of languages that the child brings to language learning, and how detailed and specific is the innate schema (the general definition of "grammar") that gradually becomes more explicit and differentiated as the child learns the language?

In a sense, these two approaches to commonality between languages (Greenberg's focus on the verbal data and Chomsky's focus on the mind) reflect two broad approaches to linguistics in the West, which relate to two major modes of understanding human reason in general, namely, rationalism and empiricism; Chomsky even titles one of his early publications *Cartesian Linguistics* (1966) in acknowledgement of his debt to the French philosopher, René Descartes (1596–1650), from whose work, *Discours de la méthode* (1637, Part IV; English translation by Donald A. Cress,

1988), stems the famous saying (which Descartes gives in Latin, although the *Discours* is written in French), "*cogito ergo sum*"; "I think, therefore I am."

Broadly speaking, rationalism ascribes considerable innateness of cognitive phenomena to people and may hence encourage universalist understandings of humanity, while empiricism considers socialization and cognitive development to be mainly a question of acquiring and learning from experience. Note, though, that, as Quine (1969, pp. 95–96) puts it, even a behaviourist (that is, someone who believes that everything is learned from experience—an empiricist, in other words) "is knowingly and cheerfully up to his neck in innate mechanisms". However, the behaviourist's mechanisms are understood as means for learning; they are not the nascent concepts imagined by rationalists, which merely require a triggering by experience to come into full bloom. Translation is likely to be of more interest to people looking at differences between languages than it is for linguists of the opposite persuasion; and, in any case, Chomsky (1965, pp. 30 and 202) believes that nothing follows about the possibility of translating between languages from the existence of universals. It is interesting to note in passing that Saussure (of whom more later), though he believes that "universal kinship between languages is not probable", makes a somewhat similar point to Chomsky's, namely, that even if there was a universal kinship between languages "it could not be proved because of the excessive number of changes that have intervened" (1959, p. 192). Nida (1964, pp. 60–69) has a flirtation with the use of early, Chomsky-style transformations in his chapter on Linguistic Meaning, but this does not by any means amount to a full-scale application of the theory to the study of translation.

11.2.2 Descriptivist linguistics, translation and cognition

As indicated earlier, movements in linguistics tend to reflect movements of thought more generally. Chomsky's approach is basically rationalist, whereas that of Franz Boas (1858–1942), Edward Sapir (1884–1939) and Benjamin Lee Whorf (1897–1941), the so-called descriptivists (Sampson, 1980: Ch. 3), is basically empiricist. Their urge to describe was influenced by a sense that the Amerindian languages would soon become lost as their speakers became increasingly integrated into the culture and societies of non-native Americans; and they used translation both as a discovery procedure and as a way of illustrating the differences between languages. For example, Sapir (1921, pp. 86–98) uses versions of the sentence "the farmer kills the duckling" in English, German, Yana, Chinese and Kwakiutl Indian to show that some or all of the 13 concepts that our sentence happens to embody may not only be expressed in different form but may also be differently grouped among themselves; that some among them may be dispensed with; and that other concepts, not considered worth expressing in English idiom, may be treated as absolutely indispensable to the intelligible rendering of the proposition (1921, pp. 94–95).

In German, for example, the definite reference cannot be expressed in isolation from gender, so that the first "the" is translated into the masculine definite article, *der*, because the German word for "farmer", *Bauer*, is masculine; whereas the second "the" has to be translated as *das*, which is the neuter definite article, because *Entelein*, the German word for "duckling", is neuter: *Der Bauer tötet das Entelein*. English does not have grammatical gender, so the concept is not expressed in the English sentence. Nor does English "the" indicate number; it would remain unchanged had there been several farmers killing several ducklings; in German, again, such pluralities would have been shown in the article, which would have been expressed with *die*, as well as in the verb. The early Translation Studies scholar John Cunnison (Ian) Catford (1917–2009) similarly advocates the use of translations provided by "a competent bilingual informant or translator" as a way of discovering "textual equivalents" between languages (1965, p. 27).

Sapir's emphasis is on describing languages, and he stresses that what he calls "the mere content of language" (1921, p. 234) is "intimately related to culture. A society that has no knowledge of theosophy has no name for it." However, Sapir also warns that "the linguistic student should never make the mistake of identifying a language with its dictionary" and emphasizes that "when it comes to linguistic form, Plato walks with the Macedonian swineherd". These pronouncements testify to Sapir's faith that all languages relate to the same human cognitive scheme, although he refers to it as "a sliding scale"; and even Whorf, well known for his "principle of relativity, which holds that all observers are not led by the same physical evidence to the same picture of the universe, unless their linguistic backgrounds are similar, or can in some way be calibrated" (1940, p. 214), speaks of a "higher mind" (1942, p. 257) possessed by every human, which "can function in every linguistic system" (1942, p. 258). Saussure (1959, p. 23), too, writes of the need for the linguist to "acquaint himself with the greatest possible number of languages in order to determine what is universal in them by observing and comparing them".

In Europe in the late 19th and early 20th century, a third "school" of linguistics arose as the manifestation within linguistics of the general movement known as structuralism (although the oft-acclaimed "father" of this movement, Ferdinand de Saussure (1857–1913), was in fact a linguist). A further school, the Prague School, was functionalist in nature in the sense that its adherents "analysed a given language with a view to showing the respective functions played by the various structural components in the use of the entire language" (Sampson, 1980, p. 103). From this school stems the so-called functional sentence perspective adopted in contemporary functionalist linguistics, as represented prominently by, for example, Michael Halliday (see later). Contemporary rationalist, or, more accurately expressed, cognitive, approaches to linguistics include cognitive grammar, pioneered by Ronald Langacker (b. 1942) and Rene Dirven (1932–2016), and, most recently, the relevance-theoretic approach developed by Dan Sperber (b. 1942) and Deirdre Wilson (b. 1941), although the latter locate themselves within the philosophy of language in the Gricean tradition rather than within linguistics. In spite of that, relevance theory has enjoyed a more systematic application to translation than any of the schools of linguistics mentioned earlier, in the work of Ernst August Gutt (e.g. 1991, 2000a, 2000b, 2004, 2005). However, Translation Studies scholars have also adopted systemic-functionalist linguistics and Langacker's cognitive approach.

11.2.3 Structuralist linguistics (Geneva School)

The first major publication in what has become known as the modern academic discipline of linguistics is the Swiss linguist Ferdinand de Saussure's posthumously published *Cours de linguistique générale* (*Course in General Linguistics*) (1916). This publication was based on notes taken by students who had attended a course which Saussure delivered at the university of Geneva between 1907 and 1911; the notes were collected and edited by Charles Bally (1865–1947) and Albert Sechehaye (1870–1946), fellow members of the so-called Geneva School of Linguistics. Saussure's course and the subsequent publication were remarkable in that they heralded a turn in European linguistics away from the historical study of the development of languages over the course of time (diachronic linguistics) towards studying the composition of individual languages at a moment in time (synchronic linguistics) (see Sanders, 2006, pp. 770). The history of publication of the *Cours* raises issues concerning the degree to which the understanding of linguistics that it presents accurately reflects Saussure's views (see Cobley, 2006, p. 757 and Sanders, 2006, p. 769); the fact remains, however, that generations of linguistics students have been presented with the views published in the *Cours/Course*,[1] and this has profoundly affected the development of structural linguistics. The name "structural linguistics" reflects Saussure's alleged view of language "as a system of interdependent terms in which the value of each term results solely from the simultaneous presence of the others" (1959, p. 114). Such a view could be considered problematic in the context of translation, given that different languages tend to reflect different classifications and groupings of *realia* and the relationships between them. However, given that translators deal in clauses and states of affairs rather than in words and things, the systems-feature of languages rarely causes insurmountable difficulties for them. As Jakobson explains (1959/1987, p. 431),

> All cognitive experience and its classification is conveyable in any existing language. Whenever there is a deficiency, terminology can be qualified and amplified by loanwords or loan translations, by neologisms or semantic shifts, and, finally, by circumlocutions. [...] No lack of grammatical devices in the language translated into makes impossible a literal translation of the entire conceptual information contained in the original.

Saussure's *Cours/Course* presents language as

> a social product [...] a storehouse filled by members of a given community through their active use of speaking, a grammatical system that has a potential existence in each brain, or, more specifically, in the brains of a group of individuals. For language is not complete in any speaker; it exists perfectly only within a collectivity.
>
> *1959, pp. 13–14*

So, each speaker holds a proportion of language in their brain, and the *Cours/Course* stresses that rather than simply linking a thing and a sound, the linguistic sign links "a concept and a sound-image" (1959, pp. 66). Of the three Saussurean terms that are all to an extent covered by the English word "language" (*langage*, *langue* and *parole*), "*langue* refers to both the underlying linguistic structure and to the speaker's mental representation of it" (Sanders, 2006, p. 770; *langage* is the term for language as a human characteristic, and *parole* means language in use). *Langue* becomes "embedded in the brain of everyone who has learned a given language"; it is "a network of relationships" (Koerner, 2006, p. 756). Saussure is thus clear that the sound patterns that constitute utterances relate to a cognitive phenomenon (a mental representation) rather than a physical referent out there in the world; of the link between the latter and the mental representation, he has less to say; nor does he address translation.

11.2.4 Systemic functional linguistics, cognition and translation

According to Saussure (1959: 130), when speaking, interlocutors make choices from ranges of possible meanings, a view that has been inherited by systemic functional linguistics, a major contemporary school of linguistics which has enjoyed immense popularity among Translation Studies scholars, in particular in mainland and greater China (see e.g. Webster & Peng, 2017; Zhang, 2015). This approach to language was developed by the British linguist Michael Alexander Kirkwood Halliday, who took forward the work of John Rupert Firth (1890–1960), the first professor of general linguistics in Britain. Inspired by the anthropologist Bronislav Malinowski (1884–1942), who saw meaning as totally dependent on context, Firth was especially focused on theorizing the relationship between language and its surroundings. These are of two kinds: The non-linguistic context of people, things and events, and the linguistic context within which each linguistic unit finds itself. He is known for stressing that "[y]ou shall know a word by the company it keeps" (Firth, 1957, p. 11), an aspect of the inspiration for the corpus work instigated at the University of Birmingham by John McHardy Sinclair (1942–2007). The relationship between language and its physical context has remained a particular focus of the work of Halliday. A major difference between this work and that of the Relevance Theorists (see later) is that in the work of the latter, context is considered from a more overtly cognitive point of view than it is in the Hallidayan approach, even though Halliday has achieved a considerable level of theorization of the mind–language–context relationship.

According to Halliday, language is realized through the activation of three functional components, or "metafunctions", of speakers' "meaning potential", namely, the ideational, the interpersonal and the textual, which correspond to three abstractions from physical context: (i) the field (what is perceived by the interlocutors as taking place), (ii) the tenor (the interactants with their characteristics and the relationships between them) and (iii) the mode (the part that the language is playing). The field tends to influence choices among nouns and their numbers, verbs and their tenses, and adjectives and adverbs; the tenor tends to influence choices of pronouns, modes of address and mood; and the mode tends to influence style (ranging from formal to casual) and the choice between speech, writing and signing (Halliday, 1978). Speakers develop expectations about the kinds of language that will occur in certain contexts, and these "register expectations" are evidenced in the register (configuration of linguistic choices) of texts.

Whereas Halliday himself has said little about translation, his approach has, as mentioned earlier, been enthusiastically embraced by Translation Studies scholars (see, for example, the papers collected in the journal *Target*, 27(3), 2015). The first of these enthusiasts was Catford (1965), according to whom, famously (1965, p. 1),

> Translation is an operation performed on languages: a process of substituting a text in one language for a text in another. Clearly, then, any theory of translation must draw upon a theory of language—a general linguistic theory.

The general linguistic theory that Catford fastens onto is "essentially that developed [...] by M. A. K. Halliday", as expressed, at that time, in Halliday (1961); and his "starting-point is a consideration of how language is related to the human social situations in which it operates" (Catford, 1965, p. 1). Cognition does not figure with any prominence in Catford's application of Hallidayan linguistics to the description and explanation of translation. However, as Gideon Toury (1942–2016) shows, it takes only a minor modification of Catford's definition of translation equivalence to illustrate the cognitive aspect of that phenomenon. Where Catford (1965, p. 50) has "*translation equivalence occurs when an SL and a TL text or item are relatable to (at least some of) the same features of substance*" (italics in the original), Toury (1980, p. 37) has "Translation equivalence occurs when a SL and a TL text (or item) are relatable to (at least some of) the same relevant features." The notion of relevance, which Toury adds to Catford's definition, brings cognition into play, because "it is to be regarded as an abbreviation for 'relevant *for* something' or 'relevant *from* a certain point of view,' or the like" (Toury, 1980, p. 38); and even though Toury here writes of this point of view as being that of either the source text or the target text, it is clear that this is metaphorical for, or an abbreviation of, a description of all the people involved in the production of a translation. Furthermore, as Toury (1999, p. 18) also reminds us, we must distinguish between translation as a textual, socially embedded phenomenon and translation as a mental act, which "is indeed cognitive". Nevertheless, cognition is not a central notion in the tradition of descriptive Translation Studies pioneered by Toury, and I will now turn to a school of linguistics that places cognition at its centre, and which has captured the imagination of a number of Translation Studies scholars.

11.2.5 Cognitive linguistics and translation

In 2012, the Gutenberg Research Award was awarded to Leonard Talmy in recognition of his pioneering work in cognitive linguistics.[2] Other prominent proponents include George Lakoff (b. 1941), Ronald Langacker (b. 1942) and Charles Fillmore (1929–2014). According to Talmy (2006, p. 543) "cognitive linguistics is concerned with the patterns in which and processes by which conceptual content is organized in language", particular attention being paid to "space and time, scenes and events, entities and processes, motion and location, and force and causation". Cognitive linguistics attributes ideational and affective categories to cognitive agents, "such as attention and perspective, volition and intention, and expectation and affect", and many of its proponents are especially interested in metaphor (see in particular Lakoff and Johnson, 1980). A distinction is drawn between conceptual content and conceptual structure, with the former being represented by open-class forms such as nouns and verbs and the latter by closed-class forms like number and inflections. Deixis is another category of particular interest in cognitive linguistics, because this system positions speakers with reference to aspects of the speech situation. For example, in English, the contrast between "here" and "there" relates to where the speaker is relative to the topic of speech, and tenses and temporal adverbs position the topic in time relative to the speaker's now. Languages differ with respect to such features, and it is not unlikely that this influences cognition. Catford (1965: 37) gives a particularly interesting example, because it occurs within what is broadly one language, in its different varieties. In standard English, he points out, "this" indicates singularity and closeness, and "these" indicates plurality and closeness. "That" indicates singularity and distance, and "those" indicates plurality and distance. In contrast, North East Scots has three terms in its place-deictic system, "this", "that" and "yon", but makes no distinction according to number. Catford uses this example to show that when translating between the two languages, with respect to these sets of terms, "we cannot 'transfer meaning'. There is no way in which, for example, Scots *that* can be said to 'mean the same' as English *that* or *this* or *these* or *those*." Catford does not go on to discuss what this may mean for the ways in which the two speaker groups conceptualize place and space, because his focus is on the relationship between language and physical context rather than on the relationship between language and cognition. A cognitive linguist would likely use the example to argue that speakers of North East Scots express their conception of spatial divisions by means of this triad of terms, just as speakers of English do with their dyad, but would add that the conceptions of space and its divisions by the two speaker groups must therefore also differ. Thus, Slobin (1996, p. 195) draws on Talmy's distinction (1985, 1991) between satellite-framed and verb-framed languages to argue that speaker groups are predisposed by their language typologies to view events in particular ways. In verb-framed languages, the verb carries the core information or "schema" of an utterance; in satellite-framed languages, this is carried by the satellite. He illustrates this with the example of a bottle floating out of a cave. In English, this event might be described by saying "The bottle floated out", with the satellite, "out", conveying the motion involved. In Spanish, the natural expression would be *La botella salió flotando*, that is, "the bottle exited floating", with the verb, *salió*, carrying the directional information (Slobin, 1996, p. 196).

Slobin uses translation into Spanish of 80 motion events taken from four English novels, and of 60 motion events from three Spanish novels into English, in order to explore what he refers to as the "rhetorical slants" or "rhetorical sets" of the two languages (1996, pp. 209–210, 217). He shows that "English loses more in translation than does Spanish", because the English translators are able to follow the original, whereas the Spanish translators tend to reduce the depiction of the motion (the "path") (1996, p. 210). He concludes that (1996, p. 218) typologies of grammar have consequences for "typologies of rhetoric." The effects of such typologies on usage may be strong enough to influence speakers' narrative attention to particular conceptual domains.

Within Translation Studies, cognitive linguistics has been applied to explain the kinds of regularities identified by Blum-Kulka (1986) and Blum-Kulka and Levinston (1983). Blum-Kulka and Levinston (1983, p. 119) find lexical simplification operating in diverse linguistic activities, including the speech of second-language learners, children, adults speaking to children, simplified reading texts, pidgins and translation. This ubiquity suggests to them that lexical simplification "operates according to universal principles [which] [...] derive from certain aspect of *semantic competence*" (1983, pp. 118–119; italics in the original), including awareness of semantic relations like hyponomy, antonymy and converseness and of the possibility of paraphrase. However, Blum-Kulka (1986, p. 18) attributes "shifts in levels of explicitness; i.e. the general level of the target text's explicitness is higher or lower than that of the source text" to the translation process itself: such shifts are, she contends (1986, p. 19), "inherent in the process of translation" and possibly in any process of language mediation (1986, p. 21). This notion has sparked a long debate about the existence and nature of so-called translation universals, which can be tracked in Mauranen and Kujamäki (2004). Halverson (2003) explains the existence of these phenomena "with reference to general characteristics of human cognition" (2003, pp. 197–198—see also Chapter 1, Section 3, this volume), building on "the theory of cognitive grammar, primarily as elaborated by Langacker (1987)". She lists Langacker's fundamental assumptions as follows: (i) "language is symbolic in nature" (Langacker, 1987, p. 11; Halverson, 2003, p. 198); (ii) "language is an integral part of human cognition. An account of linguistic structure should therefore articulate with what is known about cognitive processing in general" (Langacker, 1987, p. 12; Halverson, 2003, p. 199); (iii) cognitive grammar is a "usage-based approach to language" (Langacker, 1987, pp. 45–47, 494; Halverson, 2003, p. 199), so it considers language to occur during events, and events are of various types. In this respect, cognitive grammar resembles Halliday's systemic functional account, in which, as mentioned earlier, speakers are considered to be aware of various registers, i.e. of the kinds of language that are appropriate to kinds of events. In Langacker's account, event types that are especially deeply entrenched in a speaker's cognition are called routines (1987, p. 100; Halverson, 2003, p. 199), and cognitive grammar shares with connectivist psycholinguistic theories the notion that repeated activation of such routines strengthens their entrenchment. Although "cognitive salience [...] will not be an absolute predictor of translation choices" (Halverson, 2003, p. 233), "translated language at an aggregate level will show an overall over- or under-representation of specific structures" compared with text that has not been translated, because of the influence of the language being translated from and previous associations made in previous translation activities (Halverson, 2003, p. 233). Halverson (2003, *passim*) theorizes this effect in terms of her notion of gravitational pull, a metaphor borrowed from space science, in which it refers to the attraction between planets. Halverson (2014), not unreasonably, argues that a cognitive approach to translation will help to place the translator centrally in the discipline, a goal shared by many 21st-century orientations in Translation Studies (see e.g. Berneking, 2017).

The final cognitive approach to language and translation to be discussed in this chapter is one that returns to the notion of relevance that we saw Toury (1980, p. 37) introduce in the previous section, although in the approach to be discussed next, that of relevance theory (Sperber & Wilson, 1986a), the notion of relevance is exposed to detailed scrutiny and defined from a decidedly cognitive perspective.

11.2.6 Relevance theory, cognition and translation

Relevance theory is an inferential approach to communication developed by Dan Sperber and Deirdre Wilson on the basis of the work of Grice (1957, 1968, 1975, 1978). Sperber and Wilson (1990, p. 40) explain that "[i]nferential communication involves the formation and evaluation of hypotheses about the communicator's intentions" and that (ibid., p. 41) human cognition predisposes humans to pay particular attention to what is most relevant to them. Information is relevant to a person if it interacts with information, also called "assumptions", that the person already possesses in such a way that the existing information is strengthened by the new information, or in such a way that an existing assumption is contradicted and eliminated. Strengthening, adding to and eliminating existing assumptions are called "contextual effects" ("cognitive effects" in the second edition of 1995) of new information, and "new information is relevant in any context in which it has contextual effects, and the greater its contextual effects, the more relevant it will be" (ibid., p. 43). However, the effort a hearer has to make in order to process new information also has to be taken into consideration, so relevance is defined as follows (ibid., p. 44):

(a) Other things being equal, the greater the contextual effects, the greater the relevance.
(b) Other things being equal, the smaller the processing effort, the greater the relevance.

If an act of inferential communication achieves a range of contextual effects without putting the hearer to unjustifiable processing effort, then it is optimally relevant, and "every act of inferential communication creates an assumption of optimal relevance" (ibid., p. 45). That is, a hearer can assume that what a speaker says to them is going to be relevant enough for them to invest in an effort to process it; and the more effort is required, the more contextual effects will accrue to them. What is said need not be identical to the full interpretation it causes or is meant to cause a hearer to arrive at; often, what is said only interpretively resembles what the intended communication is. Wilson and Sperber (1988 p. 138) define interpretive resemblance as follows: "two propositional forms P and Q [...] *interpretively resemble* one another in a context C to the extent that they share their analytic and contextual implications in the context C" (emphasis in original). Analytic implications follow from an utterance alone without the use of context; for example, "Nils is Tom's brother" analytically implies "Nils is a male sibling of Tom." In various contexts, however, the utterance of these words may have various contextual implications. For example, if Jonas asks Amy whether Tom will help Nils out of a difficulty, Amy may respond to Jonas, "Nils is Tom's brother." The implications of her utterance will depend on the interactants' background assumptions about many things but probably prominently including, for example, family relationship. When two propositions share all their implications, one is a literal interpretation of the other.

The theory has been met with considerable scepticism in linguistics. For example, Levinson (1989, p. 456) contends that Sperber and Wilson (1986a) and the theory in general

> relies on improbable presuppositions about human cognition; it underplays the role of usage in pragmatic theory; it ignores many current developments in semantics, pragmatics and the study of inference; it is too ambitious and globally reductive; and anyway the theory is obscure and it is not clear how it could be made to have clear empirical application.

Within Translation Studies, too, there have been criticisms (see e.g. Malmkjær, 1992; Tirkkonen-Condit, 1992). However, some translation scholars have embraced the theory enthusiastically;

for example, Boase-Beier makes frequent references to Sperber and Wilson (1995, the second edition of 1986a) in her discussion of cognitive stylistics and translation (Boase-Beier, 2006, Ch. 4), and Gutt (1991, p. 188) goes so far as to claim that "the range of phenomena commonly considered as translation" can be accounted for by relevance theory (Sperber and Wilson, 1986a, 1986b, 1987; Wilson and Sperber, 1985; 1988).

According to Gutt (1991, pp. 101–102), the intended interpretation of a translation should resemble the original "in respects that make it adequately relevant to the audience—that is, that offer adequate contextual effects" and "it should be expressed in such a manner that it yields the intended interpretation without putting the audience to unnecessary processing effort". The principle of relevance thus constrains a translation both in terms of what it should convey and in terms of how this should be expressed. Both constraints are context determined, since the principle of relevance is context dependent. And (1991, p. 102):

> These conditions seem to provide exactly the guidance that translators and translation theorists have been looking for: they determine in what respects the translation should resemble the original—only in those respects that can be expected to make it adequately relevant to the receptor language audience. They determine also that the translation should be clear and natural in expression in the sense that it should not be unnecessarily difficult to understand.

It is the responsibility of the target audience for a translation to familiarize themselves with the context assumed by the original communicator, even if this may be difficult (1991, p. 166). The translator's task is to form a communicative intention and then decide whether this communicative intention can in fact be communicated (1991, pp. 180–181):

> Thus, the translator is confronted not only with the question of *how* he should communicate, but *what* he can reasonably expect to convey by means of his translation. The answer to this question will be determined by his view of the cognitive environment of the target audience, and it will affect some basic decisions. It will, for example, have a bearing on whether he should engage in interpretive use [translation] at all or whether descriptive use [non-translation] would be more appropriate.

If the translator judges that it is "relevant to the audience to recognize that the receptor language text is presented in virtue of its resemblance to an original in another language", i.e. if the translator decides to translate, then he will have to consider further what degree of resemblance he could aim for, being aware that communicability requires that the receptor language text resemble the original "closely enough in relevant respects" (Wilson & Sperber, 1988, p. 137). To determine what is close enough resemblance in relevant respects, the translator needs to look at both the likely benefits, that is, the contextual effects, and also at the processing effort involved for the audience. Thus, he will have to choose between indirect and direct translation, and also decide whether resemblance in linguistic properties should be included.

The jury is still out on relevance theory, and it has proved difficult to establish a method by which it might be tested. A potential strength of the theory is that it does not require separate accounts of processing for literal and figurative language; but there are other theories that share the same advantage (e.g. Davidson, 1978). In terms of the focus of the present chapter, however, relevance theory presents an interesting attempt at providing a cognitive account of the relationship between language and context, something that has exercised linguists since at least the time of the descriptivist movement.

11.3 Concluding remarks

Early in this chapter, I mentioned the opposition between empiricism and rationalism and its relationship to descriptive and cognitive linguistic theories, respectively. It is clear, though, that there can be no language use without cognitive engagement, just as there cannot be cognitive activity of the type that comes anywhere near what we think of as thought that is not expressible in language. As the strict empiricist philosopher, John Locke (1632–1704), has it,

> *Words are the sensible Signs of his Ideas who uses them* [...] The comfort and advantage of society not being to be had without communication of thoughts, it was necessary that man should find out some external sensible signs, whereof those invisible ideas, which his thoughts are made up of, might be known to others.
>
> *Locke 1690, Book Three, Chapter II, paragraph 2; italics in the original*

Perhaps the best way of viewing the various ways of approaching the question of the relationship between the linguistic approaches reviewed in this chapter is to see them as different concentrations on a phenomenon through which we do indeed achieve "the comfort and advantage of society"—language, which enables us to share much of what we have in mind.

Notes

1 All references in this article are to the Fontana/Collins edition initially published in 1959 and translated by Wade Baskin.
2 Retrieved 24 March 2018 from www.uni-mainz.de/eng/15383.php.

Further reading

Carl, M., Bangalore, S., & Schaeffer, M. (Eds.) (2016). *New directions in empirical translation process research: Exploring the CRITT TPR-DBV*. Heidelberg: Springer.

A comprehensive introduction to the Translation Process Research Database (TPR-DB), which features more than 500 hours of annotated recorded translation process data, and to some of the studies that have explored it.

Dancygier, B. (Ed.) (2017). *The Cambridge handbook of cognitive linguistics*. Cambridge: Cambridge University Press.

A survey of cognitive linguistics that provides in-depth explanations of its key concepts and major research foci, and of the theoretical approaches adopted.

Ehrensberger-Dow, M., Englund Dimitrova, B., Hübscher-Davidson, S., & Norberg, U. (Eds.) (2015). *Describing cognitive processes in translation: Acts and events*. Amsterdam: John Benjamins.

A selection of articles focusing on the cognitive and mental processes involved in translating and interpreting and encouraging research at the interface between sociological and cognitive approaches to the translating/interpreting act.

References

Anderson, J. M., Dawson, H. C. D., & Joseph, B. D. (2010). Historical linguistics. In K. Malmkjær (Ed.), *The Routledge linguistics encyclopedia* (3rd ed., pp. 225–251). London: Routledge.

Berneking, S. (2017). A sociology of translation and the central role of the translator. *The Bible Translator, 67*(3), 265–281.

Blum-Kulka, S. (1986). Shifts of cohesion and coherence in translation. In J. House, & S. Blum-Kulka (Eds.), *Interlingual and intercultural communication: Discourse and cognition in translation and second language acquisition studies* (pp. 17–35). Tübingen: Gunter Narr Verlag.

Blum-Kulka, S., & Levinston, E. A. (1983). Universals of lexical simplification. In C. Færch, & G. Kasper (Eds.), *Strategies in interlanguage communication* (pp. 119–139). London: Longman.
Boase-Beier, J. (2006). *Stylistic approaches to translation*. Manchester: St. Jerome Publishing.
Carl, M., Bangalore, S., & Schaeffer, M. (Eds.) (2016). *New directions in empirical translation process research: Exploring the CRITT TPR-DBV*. Heidelberg: Springer.
Catford, J. C. (1965). *A linguistic theory of translation: An essay in applied linguistics*. London: Oxford University Press.
Chomsky, A. N. (1965). *Aspects of the theory of syntax*. Cambridge, MA: The MIT Press.
Cobley, P. (2006). Saussure: Theory of the sign. In K. Brown (Ed.), *Encyclopedia of language and linguistics*. (2nd ed., Vol. 10, pp. 757–768). Amsterdam: Elsevier.
Dancygier, B. (Ed.) (2017). *The Cambridge handbook of cognitive linguistics*. Cambridge: Cambridge University Press.
Davidson, D. (1978). What metaphors mean. *Critical Inquiry, 5*(1), 31–47.
de Saussure, F. (1916). *Cours de linguistique générale (Course in General Linguistics)*. Lausanne/Paris: Payot.
de Saussure, F. (1959). *Course in general linguistics* (W. Baskin, Trans.). London: Fontana.
Ehrensberger-Dow, M. Englund Dimitrova, B. Hübscher-Davidson, S., & Norberg, U. (Eds.) (2015). *Describing cognitive processes in translation: Acts and events*. Amsterdam: John Benjamins.
Firth, J. R. (1957). *Papers in linguistics 1934–1951*. London: Oxford University Press.
Greenberg, J. H. (1963). Some universals of grammar with particular reference to the order of meaningful elements. In J. H. Greenberg (Ed.), *Universals of language*. (2nd ed., pp 73–113). Cambridge, MA: The MIT Press.
Grice, H. P. (1957). Meaning. *Philosophical Review, 66*, 377–388.
Grice, H. P. (1968). Utterer's meaning, sentence meaning and word meaning. *Foundations of Language, 4*, 225–242.
Grice, H. P. (1975). Logic and conversation. In P. Cole, & J. Morgan (Eds.), *Syntax and semantics 3: Speech acts* (pp. 41–58). New York: Academic Press.
Grice, H. P. (1978). Further notes on logic and conversation. In P. Cole, & J. Morgan (Eds.), *Syntax and semantics 9: Pragmatics* (pp. 113–128). New York: Academic Press.
Gutt, E.-A. (1991). *Translation and relevance: Cognition and context*. Oxford: Basil Blackwell.
Gutt, E.-A. (2000a) Translation as interlingual interpretive use. In L. Venuti (Ed.), *The translation studies reader* (pp. 376–396). London: Routledge.
Gutt, E.-A. (2000b). Issues of translation research in the inferential paradigm of communication. In M. Olohan (Ed.), *Intercultural faultlines: Research models in translation studies 1. Textual and cognitive aspects*. Manchester: St Jerome.
Gutt, E.-A. (2004). Challenges of metarepresentation to translation competence. In E. Fleischmann, P. A. Schmitt, & G. Wotjak (Eds.), *Translationskompetenz. Tagungsberichte der LICTRA* [Leipzig International Conference on Translation Studies] (pp. 77–89). Tübingen: Stauffenberg.
Gutt, E.-A. (2005). On the significance of the cognitive core of translation. *The Translator, 11*(1), 25–49.
Halliday, M. A. K. (1961). Categories of the theory of grammar. *Word, 17*, 241–292.
Halliday, M. A. K. (1978). *Language as social semiotic*. London: Edward Arnold.
Halverson, S. L. (2003). The cognitive basis of translation universals. *Target, 15*(2), 197–292.
Halverson, S. L. (2014). Reorienting translation studies: Cognitive approaches and the centrality of the translator. In J. House (Ed.), *Translation: A multidisciplinary approach* (pp. 116–139). Basingstoke: Palgrave Macmillan.
Jakobson, R. (1959/1987). On linguistic aspects of translation. In R. A. Brower. (Ed.), *On translation* (pp. 232–239). Oxford: Oxford University Press. Reprinted in K. Pomorska, & S. Rudy (Eds.) (1987). *Language and literature* (pp. 428–435). Cambridge, MA: The Belknap Press of Harvard University Press.
Koerner, E. F. K. (2006). Saussure, Ferdinand (-Mongin) de (1857–1913). In K. Brown (Ed.), *Encyclopedia of language and linguistics* (2nd ed., Vol. 10, pp. 754–757). Amsterdam: Elsevier.
Lakoff, G., & Johnson, M. (1980). *Metaphors we live by*. Chicago: University of Chicago Press.
Langacker, R. (1987). *Foundations of cognitive grammar 1*. Stanford: Stanford University Press.
Levinson, S. C. (1989). A review of *Relevance. Journal of Linguistics, 25*, 455–472.
Locke, J. (1690). *An essay concerning humane understanding*. London: Thomas Bassett.
Malmkjær, K. (1992) Review of Gutt, E.-A. (1991), Translation and relevance: Cognition and context, Oxford, Basil Blackwell. *Mind and Language*, 7(3), 298–309.
Mauranen, A., & Kujamäki, P. (Eds.). (2004). *Translation universals: Do they exist?* Amsterdam: John Benjamins.
Nida, E. A. (1964). *Toward a science of translating*. Leiden: E. J. Brill.

Quine, W. van O. (1969). Linguistics and philosophy. In S. Hook (Ed.), *Language and philosophy: A symposium* (pp. 95–98). New York: New York University Press.
Sampson, G. (1980). *Schools of linguistics: Competition and evolution*. London: Hutchinson.
Sanders, C. (2006). Saussurean tradition in 20th-century linguistics. In K. Brown (Ed.), *Encyclopedia of language and linguistics* (2nd ed., Vol. 10, pp. 769–773). Amsterdam: Elsevier.
Sapir, E. (1921). *Language: An introduction to the study of speech*. New York: Harcourt, Brace and Company.
Slobin, D. (1996). Two ways to travel: Verbs of motion in English and Spanish. In M. Shibatani, & S.A. Thompson (Eds.), *Grammatical constructions: Their form and meaning* (pp. 195–233). Oxford: Clarendon Press.
Sperber, D., & Wilson, D. (1986a). *Relevance: Communication and cognition* (2nd ed. 1995). Oxford: Basil Blackwell.
Sperber, D., & Wilson, D. (1986b). Loose talk. *Proceedings of the Aristotelian Society*, 86, 153–171.
Sperber, D., & Wilson, D. (1987). Précis of *Relevance: Communication and cognition*. *Behavioural and Brain Sciences*, 10, 697–754.
Sperber, D., & Wilson, D. (1990). Rhetoric and relevance. In D. Wellbery, & J. Bender (Eds.), *The ends of rhetoric: History, theory, practice* (pp. 140–155). Redwood City: Stanford University Press.
Talmy, L. (1985). Lexicalization patterns: Semantic structure in lexical forms. In T. Shopen (Ed.), *Language typology and syntactic description III: Grammatical categories and the lexicon* (pp. 36–149). Cambridge: Cambridge University Press.
Talmy, L. (1991). Paths to realization: A typology of event conflation. *Proceedings of the seventeenth annual meeting of the Berkeley Linguistics Society*, 480–519.
Talmy, L. (2006). Cognitive linguistics. In K. Brown (Ed.), *Encyclopedia of language and linguistics* (2nd ed., Vol. 2, pp. 542–546). Amsterdam: Elsevier.
Tirkkonen-Condit, S. (1992). A theoretical account of translation—without translation theory? *Target*, 4(2), 237–245.
Tomlin, R. S. (1986) *Basic word order: Functional principles*. London: Croom Helm.
Toury, G. (1980). Translated literature: System, norms performance: Toward a TT-oriented approach to literary translation. In *In search of a theory of translation* (pp. 35–50). Tel Aviv: Porter Institute.
Toury, G. (1999). A handful of paragraphs on "translation" and "norms". In C. Schäffner (Ed.), *Translation and norms* (pp. 9–31). Clevedon: Multilingual Matters Ltd.
Webster, J. J., & Peng, X. (Eds.) (2017). *Applying systemic functional linguistics: The state of the art in China today*. London: Bloomsbury.
Whorf, B. L. (1940). Science and linguistics. Reprinted from *Technological Review, 42*(6), 247–231 and 247–248 in J. B. Caroll (Ed.), *Language, thought and reality: Selected writings of Benjamin Lee Whorf* (pp. 207–219). Cambridge, MA: The MIT Press.
Whorf, B. L. (1942). Language, mind and reality. Reprinted from *Theosophist*, January and April issues, in J. B. Caroll (Ed.), *Language, thought and reality: Selected writings of Benjamin Lee Whorf* (pp. 246–270). Cambridge, MA: The MIT Press.
Wilson, D., & Sperber, D. (1988). Representation and relevance. In R. M. Kempson (Ed.), *Mental representations: The interface between language and reality* (pp. 133–153). Cambridge: Cambridge University Press.
Zhang, M. F. (2015). *Functional approaches to English-Chinese translation*. Beijing: Foreign Languages Press.

12

Translation, psycholinguistics and cognition

Agnieszka Chmiel

12.1 Introduction

In order to study cognitive processes involved in translation, it is natural to draw upon psychology and cognitive science. In fact, according to Munday (2001), these two play a leading role in the study of the process of translation and interpreting. Psycholinguistics, an area at the intersection of psychology and language, is featured on the level of areas of linguistics relevant to translation, according to Snell-Hornby's integrated approach to translation (1988).

Psycholinguistics has developed since the mid-1960s (Traxler & Gernsbacher, 2011) and focuses on psychological mechanisms of linguistic processing. It tries to explain how language is processed by the human mind and includes such areas as spoken and written language comprehension and production. Traditionally, psycholinguists have been interested in discovering the psychology of language in a monolingual individual. However, since the majority of people are currently bilingual, a lot of psycholinguistic research is conducted in the area of bilingualism and multilingualism (Bialystok et al., 2012). Another interesting population for psycholinguists to study is code-switchers, i.e. bilinguals who frequently mix their two languages (Verreyt et al., 2015). It is thus a natural extension to also look into the population of translators and interpreters, who use their languages in very specific circumstances, which might impact the way their minds process languages. As Dijkstra et al. (2018, p. 17) claim when discussing translation: "although every bilingual from a young child to a professional interpreter can to some extent perform this task, it is one of the most complex language activities that a human speaker can engage in".

Historically, Translation Studies embraced psycholinguistics as one of the sources of interdisciplinarity in the mid-1980s with the start of translation process research (TPR) (Alves, 2015) and the empirical turn (Snell-Hornby, 2006), when experimental research on translation and interpreting gained impetus. In fact, studies by Krings and Lörscher with the application of think-aloud protocols (TAPs) were deemed "the first highly visible contribution of psychology to research into written translation" (Ferreira et al., 2015, p. 5).

Krings (1986) used the method of think-aloud protocols and asked his participants to think aloud and verbalize their thoughts while translating. The idea, borrowed from psychology, was to get an insight into the black box of the translator's mind and examine word choice decision processes. The premise was that such data could be interpreted as observable indicators of unobservable mental processes (Lörscher, 1996). The same method was later used by Jääskeläinen (1989), Tirkkonen-Condit (1989) and Lörscher (1991), bringing consistent accounts of different approaches to translation used by professionals and non-professionals. While the former used their background knowledge, analysed the text holistically and focused on the sense, the latter focused more narrowly on words and phrases. Concurrent TAPs were later criticized for their limited value, as verbalizations could interfere with the cognitive process under scrutiny (Jakobsen, 2003).

The application of TAPs in the study of the translation process gave way to TPR, which is a very dynamic trend in Translation Studies. It focuses on unveiling the complexities and intricacies of the translation process. It has now an almost 30-year-long history (Alves, 2015), and psycholinguistics, alongside cognitive studies, expertise studies and neuroscience, is one of the major neighbouring disciplines influencing it. TPR uses "a behavioural-cognitive experimental methodological paradigm" (Jakobsen, 2017, p. 21). Jakobsen (2017) claims that TPR and psycholinguistics share the same assumption of the mind–brain–behaviour correlation: in other words, by observing certain behaviours (such as keylogging, eye movements or brain activity) we can make inferences about processes in the translator's black box.

Although TPR has been "greatly inspired by the methodological rigor of cognitive psychology and experimental psycholinguistics" (Jakobsen, 2017, p. 23) with highly controlled variables, translation process researchers still focus on ecological validity and making experimental conditions as similar to the authentic translation task as possible. According to Alves (2015, p. 30), "[a]ssuring a reasonable level of ecological validity for the translation task under investigation is a major problem in terms of developing a rigorous experimental design with controlled variables". The current strategy is to sacrifice some experimental control for the sake of ecological validity and to offset less strictly controlled conditions and more numerous confounding variables with more advanced statistical methods that can to a certain extent control these confounds (such as linear mixed models) (Jakobsen, 2017).

There are two potential solutions to the problem: either benefit from Big Data advantages by using corpus-based data and confirm the results by running a psycholinguistic experiment (e.g. Hansen-Schirra's studies on nominalizations and cognates: Hansen-Schirra, 2011; Hansen-Schirra et al., 2017—see also Hansen-Schirra & Nitzke, this volume) or use an approach that seems a fair compromise between experimental control and ecological validity. This approach involves using constructed rather than authentic texts in experiments so that the texts include appropriately controlled and distributed experimental items. Such a method of creating experimental material has been applied, for instance by Jensen et al. (2009) and Balling et al. (2014) in a study of written translation and by Chmiel and Mazur (2013) in their eye-tracking experiment on sight translation. In such a setup, experimental participants work with a text, so that the task seems authentic to the participant (except for other experiment-related circumstances, such as location, presence of the experimenter, or the eye calibration procedure necessary to record eye movements). However, the text used may be manipulated so that it includes neatly distributed and strictly controlled, matched and normed experimental stimuli (words or sentences). Thus, what appears to the participant to be a single text to be interpreted is in fact a series of experimental trials combined within one text. Obviously, certain confounding variables are still present. For example, spill-over effects due to a troublesome sentence located prior to the experimental stimulus might still skew the data. However, experimental stimuli in such a design are much more controlled than in an authentic text.

TPR scholars have frequently adopted the method of data triangulation (Alves, 2003), that is, combining data obtained via different tools from the same translation or interpreting event to shed more light on the translation process. It is now standard practice to combine keylogging with eye tracking and retrospective verbal reports. TPR provides useful information about reading and typing patterns in translation, lexical and syntactic processing and directionality. These core topics will be reviewed in the following sections.

It was interpreting scholars who welcomed influences from psycholinguistics in their research earlier than researchers on written translation. Psychologists became "intrigued by simultaneous interpreting as a challenge to prevailing theories on the limits of human processing capacity", and psycholinguists "seized upon simultaneous interpreting as a means of testing their hypotheses concerning the role of input segmentation as well as hesitations and pauses in speech production" (Pöchhacker & Shlesinger, 2002, p. 25). Eva Paneth, a practising interpreter, wrote the first study (her MA thesis) combining interpreting and psycholinguistics (1957/2002). It was based on empirical data collected while observing interpreters at work during conferences.

Many scholars following Paneth focused on temporal aspects of processing in simultaneous interpreting, such as the delay between the source text and the target text, later known as ear-voice span (EVS) (Goldman-Eisler, 1972; Oléron & Nanpon, 1965). Gerver's 1971 doctoral dissertation was deemed "the most comprehensive and influential piece of psychological investigation into simultaneous interpreting" (Pöchhacker & Shlesinger, 2002, p. 26), triggering many further experimental studies. Chernov's (1979) work on semantic aspects of interpreting also contributed to this early cross-fertilization between interpreting and psycholinguistics. According to Chernov, information processing and comprehension in simultaneous interpreting are supported by anticipating sound patterns, semantic and syntactic structures, and the overall sense.

The relation between interpreting and psycholinguistics has not always been very close, and certain models of interpreting were based on intuition rather than insights from psychology (e.g. Gile's effort models (2009)). In fact, as Gile (2015a, p. 46) himself points out, his models "were developed intuitively first and findings from cognitive psychology were gradually integrated into them as he discovered relevant work in that discipline".

More recently, the need for combining psycholinguistics with interpreting studies has been re-emphasized (Chmiel, 2010; Gile, 2015b). Chmiel (2010) points to a potential synergistic effect and argues that the interaction between the two disciplines can be a two-way street mutually beneficial to both fields. Interpreting scholars have drawn upon memory and mental lexicon models and have applied experimental methodology developed by psycholinguists in their research. The main contribution of interpreting studies to psycholinguistics is the population of interpreters who, due to the extreme temporal constraints and extreme cognitive load involved in interpreting, have to exercise extreme language control (Hervais-Adelman et al., 2015). For instance, "interpreters may prove extremely useful in studies on the structure and flexibility of the mental lexicon" (Chmiel, 2010, p. 229). Moreover, by examining trainees throughout their training in a longitudinal study (in which their performance in experimental tasks is compared pre- and post-training), useful insights might be generated about how training influences linguistic processing and memory. By comparing interpreters working only into their first language (L1) with bi-directional interpreters (working both into L1 and into L2), we can learn about how directionality shapes the structure of the mental lexicon.

However, there are numerous problems related to conducting highly controlled experimental studies. Gile (2015a) mentions small experimental samples and difficult access to professional interpreters, who are often unwilling to participate in research (which may lead to underpowered studies, i.e. experiments too weak to find an expected effect), high dropout rates when recruiting trainees in longitudinal studies, the choice of dependent variables (for instance, it might be difficult to establish one accurate translation for a given word) and ecological validity (translating single words is hardly comparable to an authentic communicative act at a conference). Some of these problems may be overcome; for instance, experimental samples may be increased by remunerating interpreters for their time in the laboratory and collecting longitudinal data from groups in two or three consecutive years to make sure the number of participants is high.

12.2 Core topics

The following sections present some core issues at the interface of translation, cognition and psycholinguistics. As the reader will notice, some of them have been examined more in a particular type of translation, while some have been tackled equally across written translation and interpreting. We also include (where relevant) a brief overview of studies in audiovisual translation and media accessibility (audio description and respeaking). This overview is by no means exhaustive but serves as a comprehensive survey of the interaction of translation, cognition and psycholinguistics.

12.2.1 Lexical processing

Words are the main tool used in translation, so lexical processing has been an object of study in psycholinguistic and process-oriented Translation Studies. Obviously, translation is much more than replacing words of the source language with those of the target language. However, such studies are part of incremental research and may tell us something about how words are processed in the translator's mind at a basic level.

Because translation equivalents co-occur frequently in translation practice, direct memory associations may be formed between them, and "the more often the same two terms (words or longer phrases) co-occur in a translation act, the stronger the memory connection between them will be" (de Groot & Christoffels, 2006, p. 198). Translators have to actively engage in language control: activation and inhibition (Paradis, 1984). However, such language control differs from that exercised by regular bilinguals. Non-switching bilinguals may simply inhibit one language, while translators have to activate comprehension in the source language and production in the target language (Paradis, 1994). According to the monitor model, literal translation equivalents are selected unless there is a problem and other solutions have to be adopted (Tirkkonen-Condit, 2005). In the following, we present selected lexical processing studies pertaining first to comprehension and then to production. Many of these studies involve interpreters and not written translators, but, according to García et al. (2014), linguistic effects of translation experience may be independent of whether the experience pertains to interpreting or written translation.

Comprehension studies involve various research methods, such as self-paced reading (where participants read each word of a sentence separately and control when to see the next word), error detection, lexical decision (participants decide whether a string of letters is a word or not) and semantic decision (participants decide whether two words are semantically similar). Interpreters generally do not outperform trainees and controls in basic single-word-level comprehension, as has been shown by Bajo et al. (2000), but interpreting experience influences more complex processing such as detection of semantic errors (Morales, Yudes, et al., 2015), semantic categorization (Bajo et al., 2000) and recognition of semantic congruence (Elmer et al., 2010).

As regards production studies, García et al. (2014) found an effect of training but no effect of experience in a word translation task performed by translators and trainees. These findings were interpreted as follows: early training, when the focus is on interlinguistic associations, strengthens links between translation equivalents. These become so strong that further translation practice has no effect on them. In a study by Christoffels et al. (2006), interpreters outperformed non-interpreting students but not English teachers in a picture naming task, suggesting that not only interpreting experience, but also other types of practice involving proficient language use, might boost lexical retrieval. In another study employing a sentence repetition task, translators outperformed bilingual controls (Ibáñez et al.,) and showed no switching costs between languages, unlike bilinguals, suggesting more efficient bilingual language control.

Cognates are frequently used as experimental items in lexical processing studies. These are words sharing the same meaning and form across languages and have thus been "identified as a valuable means to study lexical access and representation in the bilingual lexicon" (Tercedor, 2011, p. 178). Numerous studies point to cognate facilitation effect, i.e. faster processing due to cross-linguistic similarity (Kroll et al., 2002). Translation scholars have also been interested in how cognates are processed and whether cognate or non-cognate translation equivalents are used. Hansen-Schirra et al. (2017) conducted a series of studies and corpus analyses on cognate translations with translation trainees and found that whether cognate or non-cognate translation equivalents are selected depends on the language status (e.g. German accepts more cognates than Slovene), context (more non-cognate solutions in text as opposed to isolated words) and translation training (more experienced trainees use fewer cognates). The authors interpret the findings by claiming that the mental lexicon is modulated with experience and that experienced students have more attentional resources available for monitoring. Tercedor (2011) explained the frequent occurrence of cognates in translation via the priming effect, which means that due to cross-linguistic similarity, cognate equivalents are activated first and selected as translation equivalents. It seems that participants with more translation experience tend to avoid cognate translations more than controls (Lijewska & Chmiel, 2015; Tercedor, 2011; but see Lijewska & Chmiel, 2015), while there are more cognate translation equivalents in interpreting than in written translation due to greater cognitive effort and temporal constraints in the former (Oster, 2017; Shlesinger & Malkiel, 2005).

12.2.2 Syntactic processing

In general, syntactic processing in translation might be facilitated by similarity of syntactic structures (Maier et al., 2017). Similar structures have shared representations and prime equivalent translation structures. When working between languages from different language families, translators may be more frequently forced to engage in syntactic restructuring (Gernsbacher & Shlesinger, 1997).

Numerous studies have focused on word order that requires reordering in the target language and syntactic structures that necessitate restructuring in translation. For instance, Jensen et al. (2009) found that source-text items in L1 (Danish) that required a changed word order in L2 (English) generated longer gaze times, interpreted as increased processing effort. This finding was corroborated by Balling et al. (2014) in translation involving the same language pair. Additionally, they conducted two further experiments with L1 and L2 reading to confirm that longer gaze times for non-congruent experimental items (i.e. those that required a word order change) were related to parallel processing in translation rather than the language-specific difficulty of such structures. These findings were also in line with Schaeffer et al. (2016), who found that areas of interest that required syntactic source–target-language reordering increased first fixation durations (interpreted as a representation of automatic processes) and total reading time (understood as a representation of later and more conscious processing).

An interesting approach to syntactic processing was offered by Bangalore et al. (2016), who used empirical data from the CRITT-TPR database (Carl et al., 2016) to examine how syntactic variation influences cognitive effort, operationalized by reading and typing patterns. They found that syntactic structures that could be easily rendered into the target language without any restructuring requirements were translated literally and that structures with higher syntactic entropy (i.e. with several potential equivalent structures in the target language) induced longer reading times and less smooth typing. They interpreted their findings in favour of cross-linguistic priming of shared representations, which means that if a structure is shared between source and target language, it is easier to translate. These results corroborated the findings of earlier experimental studies: by Vandepitte and Hartsuiker (2011) with constructions that required restructuring the agent and the predicate in the target language and by Ruiz et al. (2008).

In another study, Serbina et al. (2017) looked into English–German restructuring based on word-class shifts by using keylogging and eye-tracking data obtained from professional translators and domain experts. They found numerous shifts from verbs to other parts of speech, explained by the feature of German being a more nominal language than English. They also reported increased cognitive effort (operationalized as total fixation duration and fixation count) for more complex structures and changes in grammatical complexity.

Studies on syntactic processing in interpreting and sight translation are less numerous. Timarová et al. (2015) found that syntactic processing in interpreting correlated positively with experience but not with working memory scores; i.e. more experienced interpreters, regardless of their memory capacity, performed better when interpreting sentences with double subject-relative clauses. As in studies on written translation, sentences requiring restructuring in the target language triggered greater cognitive effort in interpreting, as evidenced by pupil dilation (Seeber & Kerzel, 2011) and information retention (Viezzi, 1989). Shreve et al. (2011) manipulated syntax by including either simple sentences or a complex sentence with embedded relative clauses into the text. On the basis of reading and disfluency measures, they concluded that paragraphs containing less complex sentences were less effortful to process. Interestingly, more difficult syntactic structures triggered longer translation durations but shorter reading times in sight translation (Chmiel & Lijewska, 2019), suggesting that the participants looked away from the text to avoid visual interference from source-language structures. Additionally, an experience effect was found in syntactic processing in simultaneous interpreting, with professionals performing better than trainees (Riccardi, 1996). However, in another study, trainees were found to restructure more than professionals (Setton & Motta, 2007), while Chmiel and Mazur (2013) found no difference in the processing of simple and complex sentences in sight translation performed by more and less advanced interpreting trainees.

Syntactic processing has also been examined in audiovisual translation. Gerber-Morón and Szarkowska (2018) and Gerber-Morón et al. (2018) examined reading of syntactically segmented and non-syntactically segmented subtitles, and although no differences were found in comprehension, higher cognitive load and different reading patterns were reported for non-syntactically segmented subtitles. Additionally, longer fixations (Perego et al., 2010) and more difficult processing (Rajendran et al., 2013) were reported for such subtitles in other studies. Chmiel, Lijewska, et al. (2017) examined structural paraphrasing in respeaking, i.e. live subtitling in which the translator repeats the spoken content to be transformed into subtitling text in an automatic speech recognition system. They found that the translation and interpreting experience of respeakers did not modulate paraphrasing quality or speed. Syntax was also examined in audio description: Mazur and Chmiel (2016) manipulated sentence complexity and found that simpler syntax generated higher recall, comprehension and visualization scores among blind participants.

12.2.3 Temporal aspects of processing

Processing in translation can also be studied by focusing on its temporal aspects. According to Timarová et al. (2011, p. 121), "[t]ime lag provides insight into the temporal characteristics of simultaneity in interpreting, speed of translation and also into the cognitive load and cognitive processing involved in the translation/interpreting process."

In simultaneous interpreting, the most important temporal measure is EVS, also known as décalage, defined as the interval between the interpreter's hearing the source-text unit and its production in the target language (Lee, 2002; Pöchhacker, 2004). EVS have been found to last 2–6 s (Defrancq, 2015; Oléron & Nanpon, 1965; Timarová et al., 2011). It is modulated by the syntactic complexity of the source text (Adamowicz, 1989), sentence length (Lee, 2002), source-text speed (Lee, 2002), working with or without text (Lamberger-Felber, 2001), and previous preparation (Díaz-Galaz et al., 2015).

Pauses have also been studied as a temporal aspect of interpreting. They have been linked to problems with both comprehension and production (Goldman-Eisler, 1972; Piccaluga et al., 2005). They can also be used strategically to gain time while processing the input (Tissi, 2000). Pausing patterns can be modulated by experience (Tóth, 2013), directionality and source-text speed (Piccaluga et al., 2005). In sight translation, speech disfluencies (unfilled and filled pauses, repetitions and revisions) were also found to provide information about cognitive processes, such as visual interference, since more difficult sentences generated less fluent output (Shreve et al., 2011). Both EVS and pauses have also been examined as temporal aspects of respeaking: Chmiel, Szarkowska, et al. (2017) found that interlingual respeaking was more difficult than intralingual, as the former generated longer EVS and longer pauses, and because both EVS and pauses were longest when respeaking a text spoken with a fast delivery rate.

The equivalent of EVS in written translation is EKS, or eye-key span, the interval between reading the source-text unit and typing its translation in the target language. Timarová et al. (2011) compared professional translators and trainees and found more stable EKS patterns in the group of professionals. The authors interpreted the higher EKS values for trainees as reflecting the construction of meaning of the source text and switching to the target-language production. Dragsted (2010) also focused on temporal aspects of translation performed by translators and trainees. She noticed integrated coordination, i.e. almost simultaneous target-text production with source-text reading, in the group of professionals, and sequential coordination, i.e. processing of larger chunks by comprehending them first and producing later, in the group of students. In line with research on interpreters, professional translators had a shorter EKS (approx. 2.8 s) as compared with trainees (approx. 7.2 s). They also had a much lower number of pauses (both production related and source-text reading related), which suggests more cognitive effort on the part of translation trainees.

With the application of keylogging software, it became possible to analyse temporal aspects of typing in translation. Pauses in translation production have been interpreted as manifestations of mental organization and problem solving (Immonen, 2006) and reflections of cognitive processing (O'Brien, 2006). Immonen and Mäkisalo (2010) found that pause length correlates with phrase boundary types and functions: pauses before main and subordinate clauses are similar (unlike in monolingual text production), suggesting separate processing of both, despite the fact that the subordinate clause is not a stand-alone segment but functions within the whole sentence. Dragsted (2005) reported that the pausing/writing ratio increases with text difficulty. She suggested two different translation modes based on pausing patterns: the analytic mode with short segment size, slow production and long pauses, and the integrated processing mode with long segment size, fast production and short pauses. Her experiment showed that professionals used the integrated mode when translating an easy text and the analytic mode when translating a difficult text, while trainees used the analytic mode for both texts.

The studies on temporal aspects of translation have also popularized the concept of cognitive rhythm as proposed by Schilperoord (1996). According to Jakobsen (2003, p. 89), "human language processing (comprehension as well as production) proceeds by chunking of the information stream, creating a cognitive rhythm" that is marked by translation units, or segments of the source text processed by the translator (be it a word, a phrase or a sentence) (Buchweitz & Alves, 2006). It might also be defined as "'bursts' of creativity in between pauses" (Saldanha & O'Brien, 2014) and is linked to the analysis of translation stages and other activities, such as typing, insertions, deletions, revisions and use of information sources (Saldanha & O'Brien, 2014). Buchweitz and Alves (2006) have shown, for instance, that cognitive rhythm slows down if translators operate on small translation units consisting of single words only.

Finally, on the macro-level, the process of translation has been divided into the following stages: orientation (involving reading the source text), drafting (production of the first draft of the translation) and revision (final editing and correcting of the draft) (Jakobsen, 2002). Recent studies show that depending on the translator's personality or style, translators may tend to combine revision with drafting or turn to revision after completing the drafting stage (Carl, Dragsted, & Jakobsen, 2011; Lehka-Paul & Whyatt, 2016).

12.2.4 Memory and executive functions

Memory and cognitive control performed via executive functions, i.e. mental processes that regulate cognition (Nour et al., forthcoming), have sparked interest among translation scholars, especially among interpreting researchers, because translation is a complex activity that taxes cognitive processing. In fact, functional brain changes due to interpreting experience have been found in areas supporting multilingual executive control (Hervais-Adelman et al., 2015). Many scholars have set out to examine how memory and executive control functions interact with translation.

Memory studies involving simple span tasks (such as digit and letter span, in which participants recall previously presented stimuli) have shown inconclusive results. Some studies report interpreter advantage over controls (Babcock & Vallesi, 2015; Christoffels et al., 2006), while others do not (Köpke & Nespoulous, 2006; Padilla et al., 1995). Complex span tasks (such as reading span, where participants recall final words of previously presented sentences) show interpreter advantage depending on the task: interpreters outperform controls in the reading and operational span task (Babcock & Vallesi, 2015; Christoffels et al., 2006) and the speaking span (Christoffels et al., 2006) but not in the listening span (Köpke & Nespoulous, 2006) or the symmetry span task (Babcock & Vallesi, 2015). The majority of studies found no difference between professional interpreters and trainees. This pertains to both simple span tasks (Köpke & Nespoulous, 2006; Padilla et al., 1995) and complex span tasks (Liu et al., 2004; Yudes et al., 2012) except for one study by Díaz-Galaz et al. (2015), who found better reading span scores for interpreters than for trainees. Longitudinal studies focusing on how interpreter training affects memory found a strengthening effect in a simple span task in a group of sign language interpreting trainees (Macnamara & Conway, 2015) as well as in reading span (Chmiel, 2018a) and short-term verbal memory measures in a group of interpreting trainees (Babcock et al., 2017). It is interesting to note that the last study used students of written translation as a control group and found no improvement in the short-term memory score as a result of training. Higher memory scores have been associated with better performance in sign language interpreting (Macnamara & Conway, 2015) and simultaneous interpreting (Chmiel, 2018a; Timarová et al., 2015), but see Timarová et al. (2015).

Following Nour et al. (forthcoming), we classify executive functions into inhibition, shifting and updating. Inhibition is the ability to suppress a dominant and automatic response to a stimulus. Shifting means switching between multiple tasks, and updating involves monitoring and applying incoming information when necessary (Nour et al., forthcoming). As Nour et al. (forthcoming) conclude in their systematic review, interpreters have shown advantage over controls in shifting and updating but not in inhibition (Babcock & Vallesi, 2015; Dong & Liu, 2016; Morales, Padilla, et al., 2015). Similarly, longitudinal studies that compare trainees at various stages of training show that interpreter training improves shifting and updating but not inhibition (Babcock et al., 2017; Dong & Liu, 2016; Dong et al., 2018; Macnamara & Conway, 2014).

12.2.5 Directionality

Translators and interpreters are usually required to work into their mother tongue. Such L2–L1 translation is sometimes referred to as direct translation (da Silva et al., 2017) or backward translation (Christoffels et al., 2003; García & Muñoz, this volume). Translation out of one's mother tongue into the foreign language, i.e. L1–L2 translation, is known as L2 translation (Whyatt, 2018), forward translation (Christoffels et al., 2003; García & Muñoz, this volume), inverse translation (da Silva et al., 2017) or retour in the case of conference interpreting (Pavlović, 2007). For the sake of clarity, we will use the most self-explanatory labels in this section, i.e. L1–L2 and L2–L1 translation.

The common approach is that translators should work only into their L1 due to lower proficiency in their non-native L2. However, L1–L2 translation is frequently used around the world, especially when L1 is a language of limited diffusion (Ferreira & Schwieter, 2017). Moreover, as Whyatt (2018, p. 90) argues, L1–L2 translation, especially if L1 is a minor language and L2 is English, "has been vilified without solid empirical evidence". Fortunately, many researchers have applied a descriptive, rather than a prescriptive, approach to understand the advantages and disadvantages of translation into both directions.

Many studies focusing on directionality in translation focused on cognitive effort operationalized by eye movements or keystrokes. The patterns of results are mixed. Ferreira et al. (2016) found that translators spent more time in L1–L2 translation; their fixation durations were also longer in L1–L2 translation, but the fixation count was higher in L2–L1 translation. Pavlović and Jensen (2009) asked trainees and translators to perform translation in both directions. Their results show that L1–L2 translation turned out to generate greater cognitive effort only as measured by pupil dilation, while task length, gaze time and fixation durations did not differ statistically. The same pattern of results was obtained for target-text production in L1–L2 translation. In a keylogging study, Jakobsen (2003) found fewer text production keystrokes in L1–L2, a task speed decrease by 16% as compared with L2–L1 translation, and no difference in the revision stage between directions. These results are not uniform, and small sample sizes might have been one of the reasons. In such studies, it is not uncommon to conduct a study on four or eight participants in a group, which might compromise the power of these experiments.

Additionally, many such studies focus on reading measures globally for L1–L2 and L2–L1 translation, while more focus should be put on a separate analysis of source-text and target-text processing as a function of directionality. A good example is a study by Whyatt (2018), who compared translations performed by professional translators into L1 and L2. She measured fixation duration when reading the source text in the orientation phase, typing speed and pauses during target-text production, and the percentage of total task time devoted to revision, and found only slight differences as a result of translation directionality. In another study, da Silva et al. (2017) examined translators with one year of experience performing translation both from and into L2. They found that directionality influenced fixation count on the source text and target text and the total reading time of the target text: they reported more effort when reading target text in L1–L2 translation and more numerous fixations on L2 text when it was a source text in

L2–L1 translation. These results show an L1 processing advantage, in line with many studies on directionality in interpreting reviewed below.

The majority of directionality studies on interpreting show L2–L1 advantage, manifested in greater efficiency and more automaticity in L1 production (Gran & Fabbro, 1988), greater coherence and production ease (Donovan, 2005), greater fluency (Mead, 2005) and faster word translation (Chmiel, 2016). Opposing results were found in interpreting linguistically complex texts (Tommola & Helevä, 1998), as well as in anticipation and quality (Kurz & Färber, 2003). Chmiel (2018b) found that for both professional interpreters and trainees, words in L1 primed (i.e. accelerated recognition of) words in L2, but not vice versa, suggesting language asymmetry and L1 advantage.

12.2.6 Reading patterns

Translation scholars, especially those involved in TPR, have embraced eye-tracking as a method that can shed more light on the translation process. Obviously, the link between eye movements and cognitive processing is not as direct as originally posited by Just and Carpenter (1980) in their eye-mind hypothesis. Although many eye-tracking studies of translation interpret prolonged or more numerous fixations as an index of increased cognitive effort, one has to bear in mind a potential temporal misalignment—the eyes might fixate on one word, but the mind might be processing the neighbouring one (Jakobsen, 2017). Although reading research has developed since the 1970s (Rayner, 1998) and has established both early and late reading measures as indices of lexical and semantic processing and integration, these measures are of limited use to translation scholars because reading for translation differs so much from regular reading. One of the most obvious reasons is that the translator alternatively or concurrently reads and processes two texts.

Numerous studies have used eye trackers to determine unique reading patterns in translation. One of the now classic studies was by Jakobsen and Jensen (2008), who found that reading patterns differ depending on the purpose of reading. They compared reading while typing the target text, sight translation, reading for comprehension and reading for orientation before translation. Not surprisingly, reading while producing translation generated more and longer fixations than other tasks, while reading for comprehension generated fewer fixations than the remaining and cognitively more demanding tasks. These results were not corroborated by Alves et al. (2011), who found the longest fixations in sight translation. These conflicting results might be explained by different participant profiles in the Danish and Brazilian experimental groups, their different levels of experience with sight translation, and distinct task construals used in the studies.

A more recent study by Schaeffer et al. (2017) is an example of a well-controlled experiment, in which professional translators read single sentences either for comprehension or for translation. The authors manipulated the number of words constructing the translation equivalent (the source-language word could be translated by using either one target-language word or a phrase). The stimuli were matched for frequency, length and predictability (variables known to affect reading times). The study showed task effects in all reading measures, suggesting that the reading purpose changes reading patterns due to co-activation of both linguistic systems when reading for translation. It also offered further evidence for the recursive model of translation (Schaeffer & Carl, 2013) and parallel operation of vertical and horizontal translation processes.

Hvelplund (2017) conducted an even more fine-grained analysis of various reading types involved in translation and analysed fixation durations and pupil dilation separately in source-text reading, source-text reading while typing, reading of the existing target text, and reading of the emerging target text (while typing). He found longer fixation durations and larger pupil sizes for the target text as compared with the source text and in the emerging target text as compared with the existing target text. The latter finding was explained as reflecting such processes as verification of spelling and punctuation in the produced text and, more importantly, reformulation, i.e. construction of a pre-verbal version of the source text in the target language, comparison of meaning, and subsequent lexical and syntactic encoding of the target text for typing purposes.

In other studies, reading patterns in translation have been found to be modulated by syntactic complexity (Balling et al., 2014; Bangalore et al., 2016; Jensen et al., 2009), the number of potential translation equivalents of a fixated source-language word (Schaeffer et al., 2016), directionality (da Silva et al., 2017; Ferreira et al., 2016; Pavlović & Jensen, 2009; Whyatt, 2018), time pressure (Sharmin et al., 2008), source-text difficulty (Hvelplund, 2011), translation experience (Carl et al., 2011; Dragsted, 2010) and translation style (Carl et al., 2011).

New applications of eye-tracking measures are being proposed as the field is developing. For instance, Hvelplund (2016) used eye-tracking measures combined with keylogging measures to operationalize cognitive efficiency in translation. Chmiel and Lijewska (2019) suggested a novel measure for analysing sight translation: percentage of dwell time, understood as the percentage of total sight translation time spent viewing the sentence. A low value of the percentage of dwell time means more looking away from the text while sight translating to avoid source-text interference. Finally, Kruger and Steyn (2014) introduced a Reading Index for Dynamic Texts, used to analyse subtitle reading and to determine to what degree a subtitle was read in the context of a dynamic audiovisual material.

12.3 Recent developments and future directions

Certain models of translation and interpreting can be considered psycholinguistically motivated, and these are briefly reviewed below. Authors such as de Groot and Christoffels (2006), Macizo and Bajo (2006), and Ruiz et al. (2008) distinguish between vertical and horizontal translation strategies. In vertical or conceptually mediated translation, the source-language message is first decoded, and its conceptual representation is activated through phonological, morphological and semantic analysis; then its lexical representation in the target language is activated for production. In horizontal or structurally mediated translation, source-language utterances are directly transcoded into their target-language equivalents through memory associations (de Groot & Christoffels, 2006). Translators and interpreters might use either strategy depending on context.

Additionally, words in two languages might have semantic fields overlapping to a greater or smaller extent. In a study involving non-translating bilinguals, words with one-to-one mappings, i.e. words that have a single translation equivalent, have been found to be translated faster, while words with many translation equivalents are translated more slowly because activation reaches more candidates and inhibition must take place to choose the right equivalent (Tokowicz & Kroll, 2007). This modulation of word translation by the number of its translation equivalents has been studied in written translation and is known as the word translation entropy effect (Schaeffer et al., 2016).

The division between vertical and horizontal translation has been used as a framework for defining other models in translation. Tirkkonen-Condit (2005) proposed the monitor model of translation, according to which literal or horizontal translation is the default used by professionals and trainees alike. Production is constantly monitored, and a non-literal solution is proposed when problems occur. This model was empirically tested by Carl and Dragsted (2017), who compared eye-tracking and keylogging data from copying and translating and found more interventions from the monitor mechanism in translation, as manifested by less linear typing and reading patterns.

The monitor model and the horizontal/vertical translation dichotomy have served as a basis for another interesting model that interfaces with psycholinguistics, i.e. the recursive model of translation (Schaeffer & Carl, 2013). According to this model, most of the translation is done horizontally thanks to shared bilingual representations (both lexical and syntactic). Such horizontal priming (i.e. activation of parallel target-text structures due to exposure to source-text structures) is the default translation mode. To test the model empirically, Schaeffer and Carl (2013, p. 184) compared verbatim recall (as a measure of priming) after comprehension and translation and concluded that "for the translator, every sentence acts as a prime—when cross-linguistic similarity allows". In the recursive model, translation proceeds mainly horizontally (through early activation of shared representations), and vertical processes serve as the monitor. When the horizontally produced target text is not acceptable, vertical processes adapt the text to the target-language norms. These processes also control equivalents, so both languages have to be active during production. "Vertical processes access the output from the automatic default procedure recursively in both the source and the target language and monitor consistency as the context during translation production increases" (Schaeffer & Carl, 2013, p. 186). The recursive model thus integrates the monitor model with the horizontal/vertical translation strategies and is backed by experimental data and psycholinguistic findings from priming studies.

Another interesting attempt to combine translation process models with psycholinguistics is Halverson's gravitational pull hypothesis (Halverson, 2004), which uses the premises of the distributed feature model (DMF) of the bilingual mental lexicon to explain translation universals. According to DMF (de Groot, 1992), word meanings or senses are represented as sets of features. The overlap in the meaning of translation equivalents is larger if more features are shared. Following psycholinguistic research, Halverson assumes one knowledge store for both languages and various patterns of activation between concepts and words. For instance, more concrete and more frequent words are more easily activated and translated faster, as they exert "gravitational pull". As a result, highly salient features (not only words but also structures) will be more easily activated and over-represented in the translated text. Interestingly, Gile (2009) uses a similar metaphor of gravitational forces in his gravitational model of linguistic availability, in which more frequently used language constituents become more available. Although this model does not interface with psycholinguistic models in any way, it uses a similar concept of a dynamic structure of language that undergoes change due to linguistic experience.

Gile is the author of probably the most influential models of interpreting, the so-called effort models (Gile, 2009). The models posit that the interpreter distributes limited cognitive resources among various efforts, such as listening and analysis, production, memory and coordination. Even the most experienced interpreters make errors due to insufficient resources as they work near the saturation level, called by Gile the tightrope hypothesis (Gile, 2009). According to de Groot and Christoffels (2007), clear similarities between Gile's approach to interpreting as a divided-attention task and psycholinguistic theories can be found. However, the models have been criticized by Seeber (2011) for being based on an obsolete single-resource theory of attention (Kahneman, 1973) and for assuming possible shifts of attention between efforts. Seeber proposed his own cognitive load model based on the multiple resource model (Wickens, 1984), according to which tasks may interfere with each other depending on whether they are supported by similar mechanisms. Comprehension and production in interpreting are divided into demand vectors, which have various conflict coefficients that reflect interference. The model was tested empirically based on pupil dilation data from a task involving simultaneous interpreting of symmetrical and asymmetrical structures. Cognitive load increased in asymmetrical structures, thus lending support to local cognitive load variation posited by the model. However, the model has not been developed further so far to account for other types of interpreting, and no further studies have been conducted to test its assumptions.

The intersection of translation, cognition and psycholinguistics brings an important benefit in the form of research designs and methodology. Group comparisons are especially interesting as being directly related to the interpreter advantage hypothesis put forward by García (2014, p. 219), according to which "task-specific cognitive skills developed by professional interpreters [...] generalize to more efficient linguistic and executive abilities in non-interpreting tasks". In fact, many studies have compared professional interpreters, interpreting trainees and non-interpreting bilinguals in a variety of both interpreting and interpreting-related tasks (Babcock & Vallesi, 2015; Bajo et al., 2000; Chmiel, 2018a; Dong & Liu, 2016). The findings of these studies may be cautiously interpreted as a certain superiority of interpreters over trainees and trainees over controls in such aspects as information prioritizing, cognitive capacity management, fluent delivery, semantic processing, syntactic restructuring, verb anticipation and various aspects of working memory and executive functions, such as shifting and updating. It is possible that interpreting experience and training may trigger interpreter advantage. However, further studies and replications are needed to make a more definitive statement.

Group comparisons have also been used in study designs involving written translation. Translation tasks performed by professional translators have been found to last for a shorter time and trigger less cognitive effort than in the case of trainees, as measured with eye-tracking (Dragsted, 2010; Hvelplund, 2016; Pavlović & Jensen, 2009) and keylogging data (Jakobsen, 2003). Professional translators further outperform trainees in flexibility and adaptability (Hvelplund, 2016), shorter or more stable EKS and a lower number of pauses (Dragsted, 2010; Timarová et al., 2011), but not in the number of revisions (Jakobsen, 2003).

The range of methods used by translation scholars is growing. Experiments include translation and interpreting tasks with high ecological validity (e.g. written translation, sight translation, simultaneous interpreting with and without text, consecutive interpreting and paraphrasing) and translation-related tasks (such as single word translation or sentence translation), reading for translation, and more controlled experimental tasks such as lexical decision or picture naming. Research methods are mostly behavioural, including eye tracking and keylogging, priming, screen capturing, and audio and video recording. The application of neurolinguistic methods (such as electroencephalography (EEG) and functional magnetic resonance imaging (fMRI)) is on the rise. The dependent variables, i.e. what is measured in empirical studies due to experimental manipulation, usually include reaction times, accuracy and quality of output, reading measures (such as, for instance, first fixation duration and total reading time), keylogging measures (such as keystroke count and revisions) and temporal measures (EVS, EKS and pauses). In line with Alves' call for data triangulation in Translation Studies (Alves, 2003), more and more studies are using a mixed-methods approach (for instance, they combine eye tracking and keylogging or reaction times and event-related potentials in EEG studies) to arrive at results well grounded in data from various sources.

As Translation Studies further interacts with psycholinguistics and taps into the richness of psycholinguistic research methods with rigid control of variables and experimental conditions, it should also embrace most recent developments in statistics. The use of linear mixed models (Baayen et al., 2008) is becoming more and more predominant in psycholinguistics and has found its first applications in Translation Studies as well. Linear mixed models are gaining popularity over the more traditional analysis of variance (ANOVA) because they handle unbalanced designs better (e.g. where the number of participants or items is not equal in each experimental condition) and take into consideration random factors of both participants and items (e.g. the fact that some participants may be slower than others and that some stimuli might be easier than others) (Baayen et al., 2008; Barr et al., 2013). Additionally, it should now be standard to report effect size for statistically significant tests (i.e. not only whether the observed difference is significant but also how big it is) and to take care of experimental power. This means having sufficient numbers of participants and items in the experimental studies, so that the expected effect is not missed due to an underpowered experiment, or so that the significant effect is not due to type M error (an exaggerated effect found by chance) (Vasishth et al., 2018). It is important to report statistical analyses in full in order to ensure replicability of studies. If the same effect is reported repeatedly by many scholars, it becomes a much stronger foundation for building our knowledge about psycholinguistic aspects of translation than a single study.

Another interesting development in psycholinguistics that may have direct influence on translation is the latest Multilink model of the bilingual mental lexicon (Dijkstra et al., 2018), which posits language non-selective lexical access and a multiplicity of cross-level activations. This model, as the authors themselves admit, is especially suited to explain word translation mechanisms but has not been tested on translators or interpreters so far. In general, future studies should focus on more integration between translation and psycholinguistics, as scholars attempt to explain their experimental results with psycholinguistic models and integrate existing translation models with the current state of knowledge based on psycholinguistic studies. The ensuing synergy will help translation scholars discover more about the nature of processing in translation and will help psycholinguists find out more about linguistic processing under difficult conditions.

The overview of the psycholinguistically oriented studies in translation included in the present chapter shows the great potential of studies that aim at gaining insights into the cognition of translation, be it lexical or syntactic processing, memory and executive functions, reading or temporal patterns of processing. With the application of rigid methodologies and study designs, statistical modelling, and the integration of cognitive processing models in psycholinguistics and translation, we can hope to discover more about the cognitive nature of the fascinating task of translation in the future.

Further reading

Dong, Y., & Cai, R. (2015). Working memory and interpreting: A commentary on theoretical models. In Z. Wen, M. B. Mota, & A. McNeill (Eds.), *Working memory in second language acquisition and processing* (pp. 63–84). Bristol: Multilingual Matters.

An overview of studies on working memory in interpreting and a discussion of current models of memory and interpreting.

Kruger, J.-L. (2016). Psycholinguistics and audiovisual translation. *Target: International Journal of Translation Studies, 28*(2), 276–287.

A systematic overview of psycholinguistic studies in audiovisual translation.

Santilli, M., Vilas, M. G., Mikulan, E., Martorell Caro, M., Muñoz, E., Sedeño, L., Ibáñez, A., & García, A. M. (2018). Bilingual memory, to the extreme: Lexical processing in simultaneous interpreters. *Bilingualism: Language and Cognition, 22*(2), 331–348. doi:10.1017/S1366728918000378

One of the latest studies focusing on interpreter advantage and on the effect of extreme processing demands of simultaneous interpreting on lexical processing.

References

Adamowicz, A. (1989). The role of anticipation in discourse: Text processing in simultaneous interpreting. *Polish Psychological Bulletin, 20*(2), 153–160.

Alves, F. (2003). Triangulation in process oriented research in translation. In F. Alves (Ed.), *Triangulating translation. Perspectives in process oriented research* (pp. vii–x). Amsterdam: John Benjamins.

Alves, F. (2015). Translation process research at the interface. In A. Ferreira, & J. W. Schwieter (Eds.), *Psycholinguistic and cognitive inquiries into translation and interpreting* (pp. 17–40). Amsterdam: John Benjamins.

Alves, F., Pagano, A., & da Silva, I. (2011). Towards an investigation of reading modalities in/for translation: An exploratory study using eye-tracking data. In S. O'Brien (Ed.), *Cognitive explorations of translation* (pp. 175–196). London: Continuum.

Baayen, R. H., Davidson, D. J., & Bates, D. M. (2008). Mixed-effects modeling with crossed random effects for subjects and items. *Journal of Memory and Language, 59*(4), 390–412. doi:10.1016/j.jml.2007.12.005

Babcock, L., Capizzi, M., Arbula, S., & Vallesi, A. (2017). Short-term memory improvement after simultaneous interpretation training. *Journal of Cognitive Enhancement, 1*(3), 254–267. doi:10.1007/s41465-017-0011-x

Babcock, L., & Vallesi, A. (2015). Are simultaneous interpreters expert bilinguals, unique bilinguals, or both? *Bilingualism: Language and Cognition, 22*(2), 403–417. doi:10.1017/s1366728915000735

Bajo, M. T., Padilla, F., & Padilla, P. (2000). Comprehension processes in simultaneous interpreting. In A. Chesterman, & N. G. Gallardo-San Salvador, Y. (Eds.), *Translation in context* (pp. 127–142). Amsterdam: John Benjamins.

Balling, L. W., Hvelplund, K. T., & Sjørup, A. C. (2014). Evidence of parallel processing during translation. *Meta: Journal des traducteurs, 59*(2), 234–259. doi:10.7202/1027474ar

Bangalore, S., Behrens, B., Carl, M., Ghankot, M., Heilmann, A., Nitzke, J., Schaeffer, M., & Sturm, A. (2016). Syntactic variance and priming effects in translation. In M. Carl, S. Bangalore, & M. Schaeffer (Eds.), *New directions in empirical translation process research* (pp. 211–238). Cham: Springer.

Barr, D. J., Levy, R., Scheepers, C., & Tily, H. J. (2013). Random effects structure for confirmatory hypothesis testing: Keep it maximal. *Journal of Memory and Language, 68*(3), 255–278. doi:10.1016/j.jml.2012.11.001

Bialystok, E., Craik, F. I., & Luk, G. (2012). Bilingualism: Consequences for mind and brain. *Trends in Cognitive Science, 16*(4), 240–250. doi:10.1016/j.tics.2012.03.001

Buchweitz, A., & Alves, F. (2006). Cognitive adaptation in translation: An interface between language direction, time, and recursiveness in target text production. *Letras de Hoje, 41*(2), 241–272.

Carl, M., & Dragsted, B. (2017). Inside the monitor model: Processes of default and challenged translation production. In O. Czulo, & S. Hansen-Schirra (Eds.), *Crossroads between contrastive linguistics, translation studies and machine translation*. Berlin: Language Science Press.

Carl, M., Dragsted, B., & Jakobsen, A. L. (2011). A taxonomy of human translation styles. *Translation Journal, 16*(2), 155–168.

Carl, M., Schaeffer, M., & Bangalore, S. (2016). The CRITT Translation Process Research Database. In M. Carl, M. Schaeffer, & S. Bangalore (Eds.), *New directions in empirical translation process research* (pp. 13–54). Cham: Springer.

Chernov, G.V. (1979). Semantic aspects of psychological research in simultaneous interpretation. *Language and Speech, 22*, 277–295.

Chmiel, A. (2010). Interpreting Studies and psycholinguistics: A possible synergy effect. *Why Translation Studies Matters, 88*, 223–236.

Chmiel, A. (2016). Directionality and context effects in word translation tasks performed by conference interpreters. *Poznań Studies in Contemporary Linguistics, 52*(2), 269–295.

Chmiel, A. (2018a). In search of the working memory advantage in conference interpreting—Training, experience and task effects. *International Journal of Bilingualism, 22*(3), 371–384. doi:10.1177/1367006916681082

Chmiel, A. (2018b). Meaning and words in the conference interpreter's mind—effects of interpreter training and experience in a semantic priming study. *Translation, Cognition & Behaviour, 1*(1), 21–41.

Chmiel, A., & Lijewska, A. (2019). Syntactic processing in sight translation by professional and trainee interpreters. Professionals are more time-efficient while trainees view the source text less. *Target, 31*(3), 378–397. doi: 10.1075/target.18091.chm

Chmiel, A., Lijewska, A., Szarkowska, A., & Dutka, Ł. (2017). Paraphrasing in respeaking—comparing linguistic competence of interpreters, translators and bilinguals. *Perspectives. Studies in Translation Theory and Practice, 26*(5), 725–744. doi:10.1080/0907676x.2017.1394331

Chmiel, A., & Mazur, I. (2013). Eye tracking sight translation performed by trainee interpreters. In C. Way, S. Vandepitte, R. Meylaerts, & M. Bartłomiejczyk (Eds.), *Tracks and treks in translation studies* (pp. 189–205). Amsterdam: John Benjamins.

Chmiel, A., Szarkowska, A., Koržinek, D., Lijewska, A., Dutka, Ł., Brocki, Ł., & Marasek, K. (2017). Ear–voice span and pauses in intra- and interlingual respeaking: An exploratory study into temporal aspects of the respeaking process. *Applied Psycholinguistics, 38*(5), 1201–1227. doi:10.1017/s0142716417000108

Christoffels, I. K., de Groot, A. M. B., & Kroll, J. F. (2006). Memory and language skills in simultaneous interpreters: The role of expertise and language proficiency. *Journal of Memory and Language, 54*(3), 324–345. doi:10.1016/j.jml.2005.12.004

Christoffels, I. K., de Groot, A. M. B., & Waldorp, L. J. (2003). Basic skills in a complex task: A graphical model relating memory and lexical retrieval to simultaneous interpreting. *Bilingualism: Language and Cognition, 6*(3), 201–211. doi:10.1017/s1366728903001135

da Silva, I. A. L., Alves, F., Schmaltz, M., Pagano, A., Wong, D., Chao, L., Leal, A. L. V., Quaresma, P., Garcia, C., & da Silva, G. E. (2017). Translation, post-editing and directionality. A study of effort in the Chinese-Portuguese language pair. In A. L. Jakobsen, & B. Mesa-Lao (Eds.), *Translation in transition. Between cognition, computing and technology* (pp. 107–133). Amsterdam: John Benjamins.

Defrancq, B. (2015). Corpus-based research into the presumed effects of short EVS. *Interpreting, 17*(1), 26–45. doi:10.1075/intp.17.1.02def

de Groot, A. M. B. (1992). Determinants of word translation. *Journal of Experimental Psychology: Learning, Memory, and Cognition, 18*, 1001–1018.

de Groot, A. M. B., & Christoffels, I. K. (2006). Language control in bilinguals: Monolingual tasks and simultaneous interpreting. *Bilingualism: Language and Cognition, 9*(2), 189–201. doi:10.1017/s1366728906002537

de Groot, A. M. B., & Christoffels, I. K. (2007). Processes and mechanisms of bilingual control: Insights from monolingual task performance extended to simultaneous interpretation. *Journal of Translation Studies, 10*(1), 17–41.

Díaz-Galaz, S., Padilla, P., & Bajo, M. T. (2015). The role of advance preparation in simultaneous interpreting: A comparison of professional interpreters and interpreting students. *Interpreting, 17*(1), 1–25. doi:10.1075/intp.17.1.01dia

Dijkstra, T. O. N., Wahl, A., Buytenhuijs, F., Van Halem, N., Al-Jibouri, Z., De Korte, M., & Rekké, S. (2018). Multilink: A computational model for bilingual word recognition and word translation. *Bilingualism: Language and Cognition*, 1–23. doi:10.1017/s1366728918000287

Dong, Y., & Cai, R. (2015). Working memory and interpreting: A commentary on theoretical models. In Z. Wen, M. B. Mota, & A. McNeill (Eds.), *Working memory in second language acquisition and processing* (pp. 63–84). Bristol: Multilingual Matters.

Dong, Y., & Liu, Y. (2016). Classes in translating and interpreting produce differential gains in switching and updating. *Frontiers in Psychology, 7*, 1297. doi:10.3389/fpsyg.2016.01297

Dong, Y., Liu, Y., & Cai, R. (2018). How does consecutive interpreting training influence working memory: A longitudinal study of potential links between the two. *Frontiers in Psychology, 9*, 1–12. doi:10.3389/fpsyg.2018.00875

Donovan, C. (2005). Teaching simultaneous interpretation into B: A challenge for responsible interpreter training. *Communication and Cognition. Monographies, 38*(1–2), 147–166.

Dragsted, B. (2005). Segmentation in translation. Differences across levels of expertise and difficulty. *Target, 17*(1), 49–70.

Dragsted, B. (2010). Coordination of reading and writing processes in translation: An eye on uncharted territory. In G. M. Shreve, & E. Angelone (Eds.), *Translation and cognition* (pp. 41–62). Amsterdam: John Benjamins.

Elmer, S., Meyer, M., & Jancke, L. (2010). Simultaneous interpreters as a model for neuronal adaptation in the domain of language processing. *Brain Res, 1317*, 147–156. doi:10.1016/j.brainres.2009.12.052

Ferreira, A., & Schwieter, J. W. (2017). Directionality in translation. In J. W. Schwieter, & A. Ferreira (Eds.), *The handbook of translation and cognition* (pp. 90–105). Hoboken: John Wiley & Sons.

Ferreira, A., Schwieter, J. W., & Gile, D. (2015). The position of psycholinguistic and cognitive science in translation and interpreting. In A. Ferreira, & J. W. Schwieter (Eds.), *Psycholinguistic and cognitive inquiries into translation and interpreting* (pp. 3–15). Amsterdam: John Benjamins.

Ferreira, A., Schwieter, J. W., Gottardo, A., & Jones, J. (2016). Cognitive effort in direct and inverse translation performance: Insight from eye-tracking technology. *Cadernos de Tradução, 36*(3), 60–80. doi:10.5007/2175-7968.2016v36n3p60

García, A. M. (2014). The interpreter advantage hypothesis: Preliminary data patterns and empirically motivated questions. *Translation and Interpreting Studies, 9*(2), 219–238. doi:10.1075/tis.9.2.04gar

García, A. M., Ibáñez, A., Huepe, D., Houck, A. L., Michon, M., Lezama, C. G., & Rivera-Rei, A. (2014). Word reading and translation in bilinguals: The impact of formal and informal translation expertise. *Frontiers in Psychology, 5*, 1302. doi:10.3389/fpsyg.2014.01302

Gerber-Morón, O., & Szarkowska, A. (2018). Line breaks in subtitling: An eye tracking study on viewer preferences. *Journal of Eye Movement Research, 11*(3), 1–22. doi:10.16910/jemr.11.3.2

Gerber-Morón, O., Szarkowska, A., & Woll, B. (2018). The impact of text segmentation on subtitle reading. *Journal of Eye Movement Research, 11*(4), 1–18.

Gernsbacher, M. A., & Shlesinger, M. (1997). The proposed role of suppression in simultaneous interpretation. *Interpreting, 2*(1–2), 119–140. doi:10.1075/intp.2.1–2.05ger

Gerver, D. (1971). *Simultaneous interpretation and human information processing* (unpublished doctoral dissertation). Department of Psychology, University of Durham.

Gile, D. (2009). *Basic concepts and models for interpreter and translator training*. Amsterdam: John Benjamins.

Gile, D. (2015a). The contributions of cognitive psychology and psycholinguistics to conference interpreting. In A. Ferreira, & J. W. Schwieter (Eds.), *Psycholinguistic and cognitive inquiries into translation and interpreting* (pp. 41–64). Amsterdam: John Benjamins.

Gile, D. (2015b). The contributions of cognitive psychology and psycholinguistics to conference interpreting. A critical analysis. In A. Ferreira, & J. W. Schwieter (Eds.), *Psycholinguistic and cognitive inquiries into translation and interpreting* (pp. 41–66). Amsterdam: John Benjamins.

Goldman-Eisler, F. (1972). Segmentation of input in simultaneous translation. *Journal of Psycholinguistic Research, 1*(2), 127–140.

Gran, L., & Fabbro, F. (1988). The role of neuroscience in the teaching of interpretation. *The Interpreters' Newsletter, 1*, 23–41.

Halverson, S. (2004). The cognitive basis of translation universals. *Target, 15*(2), 197–241.

Hansen-Schirra, S. (2011). Between normalization and shining-through: Specific properties of English–German translations and their influence on the target language. In S. Kranich, V. Becher, S. Höder, & J. House (Eds.), *Multilingual discourse production: Diachronic and synchronic perspectives* (pp. 133–162). Amsterdam: John Benjamins.

Hansen-Schirra, S., Nitzke, J., & Oster, K. (2017). Predicting cognate translation. In S. Hansen-Schirra, O. Czulo, & S. Hofmann (Eds.), *Empirical modelling of translation and interpreting* (pp. 3–22). Berlin: Language Science Press.

Hervais-Adelman, A., Moser-Mercer, B., & Golestani, N. (2015). Brain functional plasticity associated with the emergence of expertise in extreme language control. *Neuroimage, 114*, 264–274. doi:10.1016/j.neuroimage.2015.03.072

Hvelplund, K. T. (2011). Allocation of cognitive resources in translation: An eye-tracking and key-logging study [PhD thesis]. Frederiksberg: Samfundslitteratur.

Hvelplund, K. T. (2016). Cognitive efficiency in translation. In R. Muñoz Martín (Ed.), *Reembedding translation process research* (pp. 149–169). Amsterdam: John Benjamins.

Hvelplund, K. T. (2017). Four fundamental types of reading during translation. In A. L. Jakobsen , & B. Mesa-Lao (Eds.), *Translation in transition. Between cognition, computing and technology* (pp. 55–80). Amsterdam: John Benjamins.

Ibáñez, A. J., Macizo, P., & Bajo, M. T. (2010). Language access and language selection in professional translators. *Acta Psychologica, 135*(2), 257–266. doi:10.1016/j.actpsy.2010.07.009

Immonen, S. (2006). Translation as a writing process: Pauses in translation versus monolingual text production. *Target. International Journal of Translation Studies, 18*(2), 313–336.

Immonen, S., & Mäkisalo, J. (2010). Pauses reflecting the processing of syntactic units in monolingual text production and translation. *Hermes. Journal of Language and Communication in Business, 23*(44), 45–61.

Jääskeläinen, R. (1989). Translation assignment in professional vs. non-professional translation: A think-aloud protocol study. In C. Séguinot (Ed.), *The translation process* (pp. 87–98). Toronto: H. G. Publications.

Jakobsen, A. L. (2002). Orientation, segmentation and revision in translation. In G. Hansen (Ed.), *Empirical translation studies: Process and product* (pp. 191–204). Copenhagen: Samfundslitteratur.

Jakobsen, A. L. (2003). Effects of think aloud on translation speed, revision, and segmentation. In F. Alves (Ed.), *Triangulating translation* (pp. 69–96). Amsterdam: John Benjamins.

Jakobsen, A. L. (2017). Translation process research. In A. Ferreira, & J. W. Schwieter (Eds.), *The handbook of translation and cognition* (pp. 19–49). Hoboken: Wiley and Sons.

Jakobsen, A. L., & Jensen, K. (2008). Eye movement behaviour across four different types of reading task. In S. Göpferich, A. L. Jakobsen, & I. Mees (Eds.), *Looking at eyes—Eye tracking studies of reading and translation processing* (pp. 103–124). Copenhagen: Samfundslitteratur.

Jensen, K. T. H., Sjørup, A. C., & Balling, L. W. (2009). Effects of L1 syntax on L2 translation. In I. M. Mees, F. Alves, & S. Göpferich (Eds.), *Methodology, technology and innovation in translation process research* (pp. 319–336). Copenhagen: Samfundslitteratur.

Just, M. A., & Carpenter, P. A. (1980). A theory of reading: From eye fixations to comprehension. *Psychological Review,* 87(4), 329–354.
Kahneman, D. (1973). *Attention and effort*. Englewood Cliffs, NJ: Prentice-Hall.
Köpke, B., & Nespoulous, J.-L. (2006). Working memory performance in expert and novice interpreters. *Interpreting, 8*(1), 1–23.
Krings, H. P. (1986). *Was in den Köpfen von Übersetzern vorgeht: Eine empirische Untersuchung zur Struktur des Übersetzungsprozesses an fortgeschrittenen Französischlernern.* Tübingen: Narr.
Kroll, J. F., Michael, E., Tokowicz, N., & Dufour, R. (2002). The development of lexical fluency in a second language. *Second Language Research, 18*(2), 137–171.
Kruger, J.-L. (2016). Psycholinguistics and audiovisual translation. *Target: International Journal of Translation Studies, 28*(2), 276–287.
Kruger, J.-L., & Steyn, F. (2014). Subtitles and eye tracking: Reading and performance. *Reading Research Quarterly, 49*(1), 105–120. doi:10.1002/rrq.59
Kurz, I., & Färber, B. (2003). Anticipation in German-English simultaneous interpreting. *Forum, 1*(2), 123–150.
Lamberger-Felber, H. (2001). Text-oriented research into interpreting: Examples from a case-study. *Hermes, 26*, 39–63.
Lee, T.-H. (2002). Ear voice span in English into Korean simultaneous interpretation. *Meta, 47*(4), 596–606.
Lehka-Paul, O., & Whyatt, B. (2016). Does personality matter in translation? Interdisciplinary research into the translation process and product. *Poznań Studies in Contemporary Linguistics, 52*(2), 317–349. https://doi.org/10.1515/psicl-2016-0012
Lijewska, A., & Chmiel, A. (2015). Cognate facilitation in sentence context—Translation production by interpreting trainees and non-interpreting trilinguals. *International Journal of Multilingualism, 12*(3), 358–375.
Liu, M., Schallert, D. L., & Carroll, P. J. (2004). Working memory and expertise in simultaneous interpreting. *Interpreting, 6*(1), 19–42.
Lörscher, W. (1991). *Translation performance, translation process, and translation strategies: A psycholinguistic investigation.* Tübingen: Narr.
Lörscher, W. (1996). A psycholinguistic analysis of translation processes. *Meta: Journal des traducteurs, 41*(1), 26. doi:10.7202/003518ar
Macizo, P., & Bajo, M. T. (2006). Reading for repetition and reading for translation: Do they involve the same processes? *Cognition, 99*(1), 1–34. doi:10.1016/j.cognition.2004.09.012
Macnamara, B. N., & Conway, A. R. (2014). Novel evidence in support of the bilingual advantage: Influences of task demands and experience on cognitive control and working memory. *Psychonomic Bulletin and Review, 21*(2), 520–525. doi:10.3758/s13423-013-0524-y
Macnamara, B. N., & Conway, A. R. A. (2015). Working memory capacity as a predictor of simultaneous language interpreting performance. *Journal of Applied Research in Memory and Cognition*, 5(4), 434–444. doi:10.1016/j.jarmac.2015.12.001
Maier, R. M., Pickering, M. J., & Hartsuiker, R. J. (2017). Does translation involve structural priming? *Quarterly Journal of Experimental Psychology, 70*(8), 1575–1589. doi:10.1080/17470218.2016.1194439
Mazur, I., & Chmiel, A. (2016). Should audio description reflect the way sighted viewers look at films? Combining eye-tracking and reception study data. In A. Matamala, & P. Orero (Eds.), *Researching audio description. New approaches* (pp. 97–122). Basingstoke: Palgrave Macmillan.
Mead, P. (2005). Directionality and fluency: An experimental study of pausing in consecutive interpretation into English and Italian. *Communication and Cognition. Monographies, 38*(1–2), 127–146.
Morales, J., Padilla, F., Gomez-Ariza, C. J., & Bajo, M. T. (2015). Simultaneous interpretation selectively influences working memory and attentional networks. *Acta Psychologica (Amst), 155*, 82–91. doi:10.1016/j.actpsy.2014.12.004
Morales, J., Yudes, C., Gomez-Ariza, C. J., & Bajo, M. T. (2015). Bilingualism modulates dual mechanisms of cognitive control: Evidence from ERPs. *Neuropsychologia, 66*, 157–169. doi:10.1016/j.neuropsycho logia.2014.11.014
Munday, J. (2001). *Introducing translation studies: Theories and applications*. London: Routledge.
Nour, S., Struys, E., Woumans, E., Hollebeke, I., & Stengers, H. (forthcoming). An interpreter advantage in executive functions? A systematic review. *Interpreting.*
O'Brien, S. (2006). Pauses as indicators of cognitive effort in post-editing machine translation output. *Across Languages and Cultures,* 7(1), 1–21. doi:10.1556/Acr.7.2006.1.1
Oléron, P., & Nanpon, H. (1965). Recherches sur la traduction simultanée. *Journal de Psychologie Normale et Pathologique, 62*(1), 73–94.

Oster, K. (2017). The influence of self-monitoring on the translation of cognates. In S. Hansen-Schirra, O. Czulo, & S. Hofmann (Eds.), *Empirical modelling of translation and interpreting* (pp. 23–39). Berlin: Language Science Press.

Padilla, P., Cañas, J. J., & Padilla, F. (1995). Cognitive processes of memory in simultaneous interpretation. In J. Tommola (Ed.), *Topics in interpreting research*. Turku: University of Turku, Centre for Translation and Interpreting.

Paneth, E. (1957/2002). An investigation into conference interpreting. In F. Pöchhacker, & M. Shlesinger (Eds.), *The interpreting studies reader* (pp. 30–40). London/New York: Routledge.

Paradis, M. (1984). Aphasie et traduction. *Meta, 29*, 57–67.

Paradis, M. (1994). Toward a neurolinguistic theory of simultaneous translation: The framework. *International Journal of Psycholinguistics, 10*(3), 319–335.

Pavlović, N. (2007). Directionality in translation and interpreting practice. Report on a questionnaire survey in Croatia. *Forum, 5*(2), 79–99.

Pavlović, N., & Jensen, K. T. H. (2009). Eye tracking translation directionality. In A. Pym, & A. Perekrestenko (Eds.), *Translation research projects 2* (pp. 93–109). Tarragona: Intercultural Studies Group.

Perego, E., Del Missier, F., Porta, M., & Mosconi, M. (2010). The cognitive effectiveness of subtitle processing. *Media Psychology, 13*(3), 243–272.

Piccaluga, M., Nespoulous, J.-L., & Harmegnies, B. (2005). Disfluencies as a window on cognitive processing. An analysis of silent pauses in simultaneous interpreting. *Proceedings of DiSS'05, Disfluency in Spontaneous Speech Workshop. 10–12 September 2005, Aix-en-Provence, France* (pp. 151–155).

Pöchhacker, F. (2004). *Introducing interpreting studies*. London/New York: Routledge.

Pöchhacker, F., & Shlesinger, M. (Eds.). (2002). *The interpreting studies reader*. London and New York: Routledge.

Rajendran, D. J., Duchowski, A. T., Orero, P., Martínez, J., & Romero-Fresco, P. (2013). Effects of text chunking on subtitling: A quantitative and qualitative examination. *Perspectives, 21*(1), 5–21.

Rayner, K. (1998). Eye movements in reading and information processing: 20 years of research. *Psychological Bulletin, 124*(3), 372–422.

Riccardi, A. (1996). Language-specific strategies in simultaneous interpreting. In C. Dollerup, & V. Appel (Eds.), *Teaching translation and interpreting 3*. Amsterdam: John Benjamins.

Ruiz, C., Paredes, N., Macizo, P., & Bajo, M. T. (2008). Activation of lexical and syntactic target language properties in translation. *Acta Psychologica, 128*(3), 490–500. doi:10.1016/j.actpsy.2007.08.004

Saldanha, G., & O'Brien, S. (2014). *Research methodologies in translation studies*. London: Routledge.

Santilli, M., Vilas, M. G., Mikulan, E., Martorell Caro, M., Muñoz, E., Sedeño, L., Ibáñez, A., & García, A. M. (2018). Bilingual memory, to the extreme: Lexical processing in simultaneous interpreters. *Bilingualism: Language and Cognition, 22*(2), 331–348. doi:10.1017/S1366728918000378

Schaeffer, M., & Carl, M. (2013). Shared representations and the translation process: A recursive model. *Translation and Interpreting Studies, 8*(2), 169–190. doi:10.1075/tis.8.2.03sch

Schaeffer, M., Dragsted, B., Hvelplund, K. T., Balling, L. W., & Carl, M. (2016). Word translation entropy: Evidence of early target language activation during reading for translation. In M. Carl, S. Bangalore, & M. Schaeffer (Eds.), *New directions in empirical translation process research: Exploring the CRITT TPR-DB* (pp. 183–210). Cham: Springer.

Schaeffer, M., Paterson, K. B., McGowan, V. A., White, S. J., & Malmkjær, K. (2017). Reading for translation. In A. L. Jakobsen, & B. Mesa-Lao (Eds.), *Translation in transition. Between cognition, computing and technology* (pp. 17–53). Amsterdam: John Benjamins.

Schilperoord, J. (1996). *It's about time: Temporal aspects of cognitive processes in text production*. Amsterdam: Rodopi.

Seeber, K. G. (2011). Cognitive load in simultaneous interpreting: Existing theories—new models. *Interpreting, 13*(2), 176–204. doi:10.1075/intp.13.2.02see

Seeber, K. G., & Kerzel, D. (2011). Cognitive load in simultaneous interpreting: Model meets data. *International Journal of Bilingualism, 16*(2), 228–242. doi:10.1177/1367006911402982

Serbina, T., Hintzen, S., Niemietz, P., & Neumann, S. (2017). Changes of word class during translation—Insights from a combined analysis of corpus, keystroke logging and eye-tracking data. In S. Hansen-Schirra, O. Czulo, & S. Hofmann (Eds.), *Empirical modelling of translation and interpreting* (pp. 177–208). Berlin: Language Science Press.

Setton, R., & Motta, M. (2007). Syntacrobatics: Quality and reformulation in simultaneous-with-text. *Interpreting, 9*(2), 199–230.

Sharmin, S., Špakov, O., Räihä, K.-J., & Jakobsen, A. L. (2008). Effects of time pressure and text complexity on translator's fixations. *Proceedings of eye tracking research and applications symposium (ETRA 2008)* (pp. 123–126). Savannah, GA: Association for Computing Machinery.

Shlesinger, M., & Malkiel, B. (2005). Comparing modalities: Cognates as a case in point. *Across Languages and Cultures, 6*(2), 173–193.
Shreve, G. M., Lacruz, I., & Angelone, E. (2011). Sight translation and speech disfluency. Performance analysis as a window to cognitive translation processes. In C. Alvstad, A. Hild, & E. Tiselius (Eds.), *Methods and strategies of process research: Integrative approaches in translation studies* (pp. 93–120). Amsterdam: John Benjamins.
Snell-Hornby, M. (1988). *Translation studies: An integrated approach.* Amsterdam: John Benjamins.
Snell-Hornby, M. (2006). *The turns of translation studies: New paradigms or shifting viewpoints?* Amsterdam: John Benjamins.
Tercedor, M. (2011). Cognates as lexical choices in translation: Interference in space-constrained environments. *Target, 22*(2), 177–193. doi:10.1075/target.22.2.01ter
Timarová, Š., Čeňková, I., & Meylaerts, R. (2015). Simultaneous interpreting and working memory capacity. In A. Ferreira, & J. W. Schwieter (Eds.), *Psycholinguistic and cognitive inquiries into translation and interpreting* (pp. 101–126). Amsterdam: John Benjamins.
Timarová, Š., Dragsted, B., & Hansen, I. G. (2011). Time lag in translation and interpreting: A methodological exploration. In C. Alvstad, A. Hild, & E. Tiselius (Eds.), *Methods and strategies of process research: Integrative approaches in translation studies* (pp. 121–146). Amsterdam: John Benjamins.
Tirkkonen-Condit, S. (1989). Professional vs. non-professional translation: A think-aloud protocol study. In C. Séguinot (Ed.), *The translation process* (pp. 73–85). Toronto: H. G. Publications.
Tirkkonen-Condit, S. (2005). The monitor model revisited: Evidence from process research. *Meta: Journal des traducteurs, 50*(2), 405. doi:10.7202/010990ar
Tissi, B. (2000). Silent pauses and disfluencies in simultaneous interpretation: A descriptive analysis. *The Interpreters' Newsletter, 10*, 103–127.
Tokowicz, N., & Kroll, J. F. (2007). Number of meanings and concreteness: Consequences of ambiguity within and across languages. *Language and Cognitive Processes, 22*(5), 727–779. doi:10.1080/01690960601057068
Tommola, J., & Helevä, M. (1998). Language direction and source text complexity: Effects on trainee performance in simultaneous interpreting. In L. Bowker, M. Cronin, D. Kenny, & J. Pearson (Eds.), *Unity in diversity. Current trends in translation studies* (pp. 177–186). Manchester: St. Jerome.
Tóth, A. (2013). The study of pauses and hesitations in conference interpreters' target language output [unpublished doctoral thesis]. Eötvös Loránd University, Budapest.
Traxler, M., & Gernsbacher, M. A. (Eds.). (2011). *Handbook of psycholinguistics.* Amsterdam: Elsevier.
Vandepitte, S., & Hartsuiker, R. J. (2011). Metonymic language use as a student translation problem. In C. Alvstad, A. Hild, & E. Tiselius (Eds.), *Methods and strategies of process research: Integrative approaches in translation studies* (pp. 67–92). Amsterdam: John Benjamins.
Vasishth, S., Mertzen, D., Jager, L. A., & Gelman, A. (2018). The statistical significance filter leads to overconfident expectations of replicability. *Journal of Memory and Language, 103*, 151–175.
Verreyt, N., Woumans, E. V. Y., Vandelanotte, D., Szmalec, A., & Duyck, W. (2015). The influence of language-switching experience on the bilingual executive control advantage. *Bilingualism: Language and Cognition, 19*(01), 181–190. doi:10.1017/s1366728914000352
Viezzi, M. (1989). Information retention as a parameter for the comparison of sight translation and simultaneous interpretation: An experimental study. *The Interpreters' Newsletter, 2*, 65–69.
Whyatt, B. (2018). Old habits die hard. Towards understanding L2 translation. *Między Oryginałem a Przekładem, 24*(41), 89–112. doi:10.12797/MOaP.24.2018.41.05
Wickens, C. D. (1984). Processing resources in attention. In R. Parasuraman, & D. R. Davies (Eds.), *Varieties of attention* (pp. 63–102). New York: Academic Press.
Yudes, C., Macizo, P., & Bajo, T. (2012). Coordinating comprehension and production in simultaneous interpreters: Evidence from the Articulatory Suppression Effect. *Bilingualism: Language and Cognition, 15*(02), 329–339. doi:10.1017/s1366728911000150

13
Translation, neuroscience and cognition

Adolfo M. García and Edinson Muñoz

13.1 Introduction

Like any other form of human cognition, the processes targeted in this book are associated with, if not dependent on, specific patterns of brain activity. Of course, as shown throughout the volume, mental aspects of translation and interpreting can be inferred in the absence of neurological evidence. Yet, revealing findings from neuroscience and related disciplines have accumulated over the decades, offering more direct insights on the psychobiological foundations of interlingual reformulation (IR)—namely, translation and interpreting in their various modalities (García, 2019; García, Mikulan, et al., 2016). The present chapter provides a succinct yet comprehensive survey of this interdisciplinary interface, addressing its main methods, topics, findings, implications and challenges.

A few years ago, the neural basis of IR was deemed a "known unknown" for Translation Studies (Tymoczko, 2012). Arguably, however, this is not because critical research is scant but, rather, because it has not been incorporated by mainstream circles in the field. In fact, scientific knowledge on the translating and interpreting brain has been produced since the late 1920s, and it has expanded considerably in the last 25 years (Elmer, 2012; García, 2013, 2015a). Broadly speaking, works on the topic can be periodized in three stages.

First, in the early and mid-20th century, a series of single-case studies on bilingual aphasics showed that translation processes could become variously disturbed following brain lesions, even when other abilities in the native and non-native language (L1 and L2, respectively) were relatively spared. Specifically, Kauders (1929), Veyrac (1931) and Gastaldi (1951) documented three hitherto unknown disorders that compromised verbal and cognitive control mechanisms mediating backward translation (BT, from L2 to L1) and/or forward translation (FT, from L1 to L2)—see Section 13.2.1. In retrospect, the evidence gleaned throughout this period was far from robust, as it was scarce, mostly anecdotal and devoid of apt theoretical frameworks. Moreover, it lacked both systematic translation assessments and detailed neurological and neuropsychological descriptions. However, it paved the way for more detailed examinations of normal and pathological translation processes, and it was eventually resurfaced to inform suggestive hypotheses and models (Fabbro, 1999; García, 2012, 2015a)—see Section 13.2.2.

Then, in the last decades of the 20th century, methodological and technological advances led to significant breakthroughs. The incipient clinical corpus described earlier was considerably broadened thanks to the development of a standardized test including translation tasks (Paradis, 1979),[1] the accumulation of more meticulous case reports (García, 2015a; Paradis, 1989) and the detection of a fourth neurological disorder affecting translation skills (Paradis et al., 1982)—for details, see Section 13.2.1. Also, joint efforts by neurolinguists and translation scholars gave rise to pioneering experimental reports on brain lateralization (Fabbro et al., 1990; Fabbro et al., 1991), functional connectivity patterns (Kurz, 1994, 1995) and evoked hemodynamic activity (Klein et al., 1995; Price et al., 1999) during IR—see Sections 13.2.1 and 13.2.2. At the same time, neurocognitive accounts of translation and interpreting began to emerge in the literature, including compendiums of theoretical reflections (Bouton, 1984; Paradis, 1984), position papers (Fabbro & Gran, 1997) and even full-fledged models of critical mechanisms (Fabbro, 1999; Paradis, 1994).

Finally, since the year 2000, research on the neurocognitive basis of IR has extended in several directions. Behavioural evidence on processing of cross-language equivalents (García, 2015b) and executive skills in simultaneous interpreters (SIs) (García, 2014) has grown in parallel with relevant neuroscientific studies. The latter include assessments of verbal and non-verbal mechanisms through techniques as varied as positron emission tomography (PET) (e.g. Rinne et al., 2000), functional near-infrared spectroscopy (fNIRS) (e.g. Quaresima et al., 2002), event-related potentials (e.g. Proverbio et al., 2004), functional magnetic resonance imaging (fMRI) (e.g. Lehtonen et al., 2005), transcranial direct current stimulation (tDCS) (e.g. Liuzzi et al., 2010), direct cortical stimulation (e.g. Borius et al., 2012), structural MRI (e.g. Elmer, Hänggi, et al., 2014), and scalp-level and intracranial electroencephalographic (EEG) recordings (e.g. Dottori et al., 2020; García, Mikulan, et al., 2016). Indeed, this empirical corpus has become sufficiently large to warrant systematic reviews focused on the neurocognitive correlates of translation directionality, the mechanisms engaged by different translation units, and the anatomo-functional impact of interpreting expertise (Diamond & Shreve, 2010, 2017; Elmer, 2012; García, 2013; Moser-Mercer, 2010). Moreover, a substantial part of the evidence has been jointly interpreted to inform a neuroanatomical model of putative translation routes (García, 2012).

As this recapitulation suggests, research on the neurocognitive basis of IR has been gaining momentum within and outside Translation Studies—so much so that it may be ready to assert itself as a full-blown sub-discipline in the field (García, 2019). The remainder of this chapter provides a comprehensive outlook on this thriving interdisciplinary arena. In Section 13.2.1, an overview is offered of the methods used in existing studies, ranging from behavioural to lesion-based and neuroscientific approaches. Next, Section 13.2.2 considers key findings on the neurocognitive basis of IR, including insights into anatomical, functional and plastic properties of its fundamental systems as well as integrative theoretical models proposed therefrom. Finally, Section 13.2.3 addresses the challenges and opportunities facing the field's short- and long-term development. All in all, this chapter illustrates the myriad contributions that neuroscience can bring to cognitive translatology.

13.2 Core topics

13.2.1 Methods

Research on neural aspects of IR profits from numerous behavioural tasks, which may be used on their own, administered to brain-damaged patients, or combined with neuroscientific recordings from healthy participants. This methodological repertoire is briefly introduced in the following sub-sections.

13.2.1.1 Behavioural measures

Behavioural measures involve verbal or non-verbal stimuli to which participants must respond following predefined instructions. Outward performance is assumed to reflect the operation of specific target mechanisms, such as those supporting reading, translation or cognitive control—namely, the collection of processes involved in the coordination, selection, inhibition and anticipation of relevant mental operations (Diamond, 2013). Typically, experimental tasks comprise a number of contrastive conditions (e.g. BT and FT) conceived in terms of three types of variables: controlled variables (i.e. factors that are established as similar between conditions), independent variables (i.e. factors manipulated by design in the experiments, such as translation directionality) and dependent variables. The latter are indexed by the participants' responses, which may consist in oral productions (e.g. translating words out loud), manual actions (e.g. pressing a key to indicate that two words are similar in meaning) or other types of overt decisions (e.g. grabbing a card and choosing which other card it should be paired with). The two most important types of dependent variable are accuracy or number of hits (measures of performance efficacy, typically calculated as the sum or ratio of correct responses) and response time (a measure of processing efficiency, represented by the speed with which the subject responds).

Several behavioural instruments have proven informative for the field. For example, Fabbro et al. (1991) assessed hemispheric dominance for cross-linguistic processing in SIs via a dichotic listening test. Participants listened to a source-/target-language sentence pair, with L1 and L2 sentences delivered to different ears, and decided whether they were similar in meaning. Since initial auditory processes are contralateralized, an advantage for right-ear responses in such a study can be interpreted as a left-hemisphere preference for the process in question, and vice versa.[2]

Other tasks tap into particular linguistic operations. For instance, overt translation tasks have been used to examine processing differences between BT and FT, as compared with single-language tasks, in professional translators and lay bilinguals[3] (e.g. García et al., 2014). Additional aspects of cross-linguistic processing, such as the recruitment of semantic vs. form-level mechanisms, have been assessed through equivalent recognition tasks, in which participants must decide whether pairs of words constitute feasible translation equivalents (e.g. Ferré et al., 2006).[4]

Behavioural tests can also be used for investigating executive functions, such as inhibitory control (the ability to suppress prepotent responses for successful task performance), mental set-shifting (the capacity to change the dominant cognitive scheme in response to dynamic circumstances) or working memory (the system mediating transient storage of verbal or visual information) (Diamond, 2013). For example, to assess whether information storage skills were enhanced in SIs relative to lay bilinguals, Christoffels, de Groot, and Kroll (2006) asked participants to memorize and repeat increasingly longer lists of words. As in any other word-span task, the number of recalled items was taken to reflect the storage capacity of working memory.[5]

Although measures of accuracy, number of hits and response time yield reliable information about the final outcome of a process, they are blind to its associated neurological activity. Thus, when considered on their own, behavioural tasks only warrant limited, indirect inferences of neurocognitive mechanisms—for example, faster responses for one task than another can be interpreted as greater strength of the underlying connections (García, 2015b). However, they are essential to understand the contributions of specific neuroanatomical and neurophysiological mechanisms in lesion studies and neuroscientific experiments, as described in the following sub-section.

13.2.1.2 Lesion studies

A powerful approach to identify which brain regions prove critical for specific IR processes consists in administering relevant tasks to brain-damaged bilinguals. Single-case studies allow single dissociations to be established, that is, empirical demonstrations that circumscribed cerebral damage can disturb a given process of interest (e.g. BT) while others (e.g. FT) remain fully or relatively spared. Moreover, joint analyses of two or more cases may warrant the postulation of double dissociations, namely, contrastive patterns in which a lesion to area A compromises function X but not Y, while a lesion to area B yields opposite results (Dunn & Kirsner, 2003).

Both types of dissociation have been observed in individual and integrative analyses of patients featuring translation neuropathologies (Fabbro, 1999, 2001), that is, neurocognitive impairments characterized by selective or differential deficits in IR skills relative to other linguistic processes. As shown in Table 13.1, reports on bilinguals with aphasia and other neurological conditions have revealed four such disorders (spontaneous translation, inability to translate, paradoxical translation behaviour and translation without comprehension), each affording specific insights into the neural basis of IR—see Section 13.2.2.

Lesion studies feature a number of methodological and conceptual caveats (Dunn & Kirsner, 2003). First, as they are "natural experiments", the specific regions affected differ widely among patients, which limits comparability between studies. Also, the ensuing results may not be directly informative of a region's functional role in a healthy subject, due to possible disease-related compensatory or otherwise plastic mechanisms. Moreover, since none of the available lesion studies on IR has incorporated *in vivo* brain recordings, they are blind to the full range of circuits engaged by the tasks and to their interactive temporal dynamics. In addition, single-case studies vary in terms of the subjects' overall cognitive profile, language combinations, L2 proficiency and age of acquisition, which means that fine-grained generalizations can rarely be derived therefrom. Notwithstanding, dissociations warrant the inferences that (i) certain processes are critically and differentially dependent on specific brain areas and that (ii) two or more cognitive mechanisms (e.g. those subserving BT and FT) are functionally autonomous, in the sense that one can become impaired while the other(s) remain(s) partially or fully functional. In cognitive translatology, interpretations along these lines have informed several lines of inquiry, motivating the rejection of long-standing theoretical postulates (García, 2012, 2015a) and the formulation of explicit neuroanatomical models—see Section 13.2.2.

13.2.2 Neuroscientific methods

Neuroscientific methods allow aspects of brain anatomy and function to be quantified. Moreover, some of them, like tDCS, can modulate neurophysiological activity and influence cognitive performance. In most cases, such measures are combined with analyses of behavioural outcomes in online or offline tasks. If the experiment's design is adequate, neural differences between two target conditions (e.g. BT and FT) or groups (e.g. SIs and lay bilinguals) can be inferred to constitute key biological correlates of the independent variable at hand (e.g. translation directionality or interpreting expertise). This section describes the generalities of the major neuroscientific methods employed so far to study IR—for a more in-depth treatment, see García, Mikulan, et al. (2016).

First, structural imaging methods (e.g. structural MRI) illuminate features of brain anatomy, such as grey matter density or cortical thickness in a region of interest. In particular, they are useful to assess whether certain experiences or profiles involve plastic adaptations in critical areas. This, for example, has been shown to be the case in studies comparing the volume of executive control regions in SIs and lay multilinguals (Becker et al., 2016).

Second, haemodynamic methods (e.g., fMRI, PET and fNIRS) measure regional blood flow and associated metabolic demands, on the premise that these will increase in regions subserving an ongoing cognitive process. Such tools offer excellent spatial resolution but poor temporal resolution. So, they allow identification of which brain areas are engaged by a given process, but they are not useful to ascertain *when* that process occurred. Among other things, they have been used to reveal differential neurocognitive mechanisms engaged by BT and FT (Klein et al., 1995; Quaresima et al., 2002; Rinne et al., 2000)—see Section 2.2.

Third, electrophysiological methods (e.g. ERPs and EEG connectivity) tap into fine-grained modulations of electrical brain activity as recorded through electrodes on the scalp. The peak time and amplitude of the signals, as well as their synchronization or desynchronization, warrant inferences on various aspects of an ongoing cognitive process, such as its overall demands, the type of mechanisms it recruits (e.g. syntactic, semantic) or its reliance on integrated or segregated information flows across cortical sites. Although they offer low spatial resolution, these methods are optimal to examine the temporal dynamics of a target process, even before behavioural responses are made. They have been employed, for instance, to identify distinctive demands placed by specific translation units (Christoffels et al., 2013) and to assess the coupling of sensory and articulatory mechanisms in SIs (Elmer & Kühnis, 2016)—see Section 13.2.2.

While these methods are non-invasive, others, like intracranial EEG or direct cortical stimulation, involve incisions, perforations or electrical perturbations in predetermined brain regions to assess the latter's direct role in a process of interest. Thus, they offer invaluable evidence on the putative basis of specific neurocognitive functions. Since they are only applied on neurological patients during presurgical assessment, they typically offer small time windows to test hypotheses of marginal clinical interest. Despite these caveats, they have been applied to assess the relative independence of translation relative to single-language processes (Borius et al., 2012) and the spatio-temporal dynamics of BT vs. FT (García, Mikulan, et al., 2016) –see Section 13.2.3.

In sum, neuroscientific methods offer critical (and, in many cases, real-time) evidence about the contributions of particular regions or neural mechanisms to specific cognitive processes. Importantly, they can be combined with multiple behavioural tasks designed to target all sorts of mental operations. Relative to other approaches, their application in experiments assessing IR has been limited so far. However, they have been growingly used to such an end in recent years, illuminating numerous topics of relevance for Translation Studies, as shown in Section 13.2.3.

13.2.3 Insights into the translating and interpreting brain

The above-mentioned methods have afforded multiple findings on the neural basis of IR. In particular, convergent evidence warrants preliminary conclusions on four major topics, namely: (i) the contributions of each hemisphere to translation skills, (ii) the neurofunctional organization of putative routes, (iii) the temporal dynamics of task-specific activity and (iv) the impact of interpreting expertise on neuroplasticity. A concise account of these issues is offered here.

13.2.3.1 Hemispheric specializations for interlingual reformulation processes

The two cerebral hemispheres constantly exchange information during any cognitive process. However, some higher-order functions rely more critically on the contributions of one or the other (Stephan et al., 2007). Such forms of functional specialization are well established in the domain of language, with some 90% of the population exhibiting left dominance for basic language skills (e.g. phonological and syntactic processing) (Mazoyer et al., 2016), alongside major contributions of the right hemisphere for multiple verbal and otherwise communicative functions (Lindell, 2006). A similar pattern seems to characterize neurocognitive mechanisms supporting IR.

The most critical pathways subserving translation processes appear to be left-lateralized. A recent review of translation neuropathologies shows that, in 18 out of 21 cases, the disorders resulted exclusively from left-sided lesions (García, 2015a). Compatibly, Borius et al. (2012) showed that direct electrostimulation of left frontal regions directly inhibited translation processes. Moreover, separate neuroimaging studies offering whole-brain recordings have shown that, relative to other language functions, IR yields activation increases only in the left hemisphere (Klein et al., 1995; Lehtonen et al., 2005; Rinne et al., 2000). This aligns with the view that the neural networks implicated in linguistic processes during translation and interpreting are largely embedded within gross areas supporting more general verbal domains (García, Mikulan, et al., 2016).

However, this does not mean that the right hemisphere is inactive during IR. In a PET study (Price et al., 1999), predominant left-sided activation increases for translation relative to single-language reading were accompanied by *bilateral* engagement of the anterior cingulate and the basal ganglia. Also, functional connectivity experiments with professional translators (García, Mikulan, et al., 2016) and interpreters (Kurz, 1994, 1995) have revealed significant signal exchanges across both hemispheres during BT and FT. In addition, concreteness effects during translation of cognates have been reported to involve maximal event-related potential (ERP) (in particular, N400) modulations over right centrotemporal sites (Janyan et al., 2009b). Finally, behavioural (including dichotic listening) experiments (Fabbro et al., 1990, 1991; Proverbio & Adorni, 2011) and electrophysiological research (Proverbio et al., 2004) indicate that linguistic processes are less left-lateralized in SIs relative to other groups, with additional imaging evidence pointing to neuroplastic adaptations of right hemisphere regions triggered by interpreting training (Hervais-Adelman et al., 2015; Hervais-Adelman et al., 2017)—for details, see "Neuroplasticity in simultaneous interpreters" below.

In sum, despite an arguable dominance of left-sided regions for IR, relevant neurocognitive mechanisms seem to be widely distributed across both hemispheres, especially in the case of subjects with considerable interpreting expertise. However, these conclusions must be taken with reserve, as the impact of right hemisphere lesions on translation skills remains poorly studied (García, 2015a), available neuroimaging results are not entirely consistent (García, 2013) and behavioural methods may prove sub-optimal to estimate lateralization of functions (Paradis, 1992, 1995, 2003, 2008). More research is thus necessary to fully understand the differential contributions of each hemisphere during IR.

13.2.3.2 Functional organization of the systems subserving interlingual reformulation

The available evidence also sheds light on the neurofunctional organization of the systems subserving IR. Broadly speaking, critical hubs include perisylvian (superior and inferior temporal gyri, Broca's area), extrasylvian (parieto-occipital, superior parietal and supplementary motor cortices) and subcortical (cerebellum, putamen, globus pallidus and insula) regions involved in linguistic and executive functions at large (García, 2013, 2015a). These areas, especially in the left hemisphere, exhibit differential activation increases for translation as compared with other verbal processes (García, 2013), and damage to them systematically disturbs IR skills (García, 2015a).

Some of the mechanisms supporting IR within such regions are partially dissociable from single-language processing skills. Reports on inability to translate and paradoxical translation behaviour (Table 13.1) indicate that the systems supporting BT can remain functional when those involved in L1 production are impaired (and vice versa), and that FT abilities can be spared even when L2 processing is disturbed (and vice versa) (de Vreese et al., 1988; Eviatar et al., 1999;

Table 13.1 Translation neuropathologies

*Disorder**	*Principal manifestation*	*Lesion site*
Spontaneous translation (G.-Caballero et al., 2007)	Involuntary and immediate translation of utterances, often accompanied by an inability to translate willingly.	Left hemisphere: frontal lobe, basal ganglia, temporal lobe, parietal lobe
Inability to translate (Aglioti & Fabbro, 1993)	Severe or complete incapacity to translate in one or both directions, even when single-language processes are (relatively) spared.	
Paradoxical translation behaviour (Paradis et al., 1982)	Incapacity to translate into a language available for spontaneous production, with spared abilities to translate into a language unavailable for spontaneous production.	Left hemisphere: temporal lobe, parietal lobe
Translation without comprehension (Paradis et al., 1982)	Preserved ability to translate utterances despite unawareness of the meaning of the source unit.	

* Citations in the first column refer to representative cases of each disorder.
Source: based on García (2015a).

Fabbro & Paradis, 1995; Paradis et al., 1982). Compatibly, direct electrostimulation evidence has shown that inhibition of specific left frontal and temporal sites can interfere with single-language production without affecting translation skills (Borius et al., 2012). Of note, cases of translation without comprehension (Table 13.1) suggest that IR systems include a semantically unmediated lexical route, as they show that adequate cross-linguistic equivalents can be found even when relevant conceptual representations cannot be accessed (de Vreese et al., 1988; Paradis et al., 1982).

Moreover, partially autonomous mechanisms seem to be recruited depending on translation directionality. Cases of inability to translate and paradoxical translation behaviour indicate that BT and FT can become differentially or even selectively compromised following brain lesions, and that such patterns can be either transient or stable (Aglioti & Fabbro, 1993; Byng et al., 1984; de Vreese et al., 1988; Fabbro & Paradis, 1995; Nilipour & Ashayeri, 1989; Paradis et al., 1982)—see Figure 13.1A, which displays results from a patient exhibiting inability to translate. Also, evidence from hemodynamic methods shows that FT involves wider activity patterns than BT in frontostriatal regions, such as the putamen (Klein et al., 1995) and Broca's area (Quaresima et al., 2002; Rinne et al., 2000), suggesting that at least some brain structures play a more critical role in the former direction—Figure 13.1B. In addition, electrophysiological findings indicate that, as compared with BT, FT involves greater participation of the right hemisphere (Kurz, 1994, 1995) and enhanced bilateral frontotemporal connectivity (García, Mikulan, et al., 2016), reinforcing the view that each translation direction recruits partially distinct neurocognitive resources.

Finally, different regions seem to be implicated during IR depending on the translation unit. A comparison of activity patterns during translation of words and sentences suggests that the former rely mainly on temporal regions associated with declarative memory, whereas the latter differentially implicate frontostriatal circuits subserving procedural functions (García, 2013). Indeed, damage to frontostriatal networks disrupts sentence translation more markedly than word translation (Fabbro & Paradis, 1995)—Figure 13.1C. These two broad unit types also differ in their engagement of basal ganglia structures during IR, with words yielding activation peaks in the putamen and the caudate nucleus (Price et al., 1999), and sentences involving hemodynamic

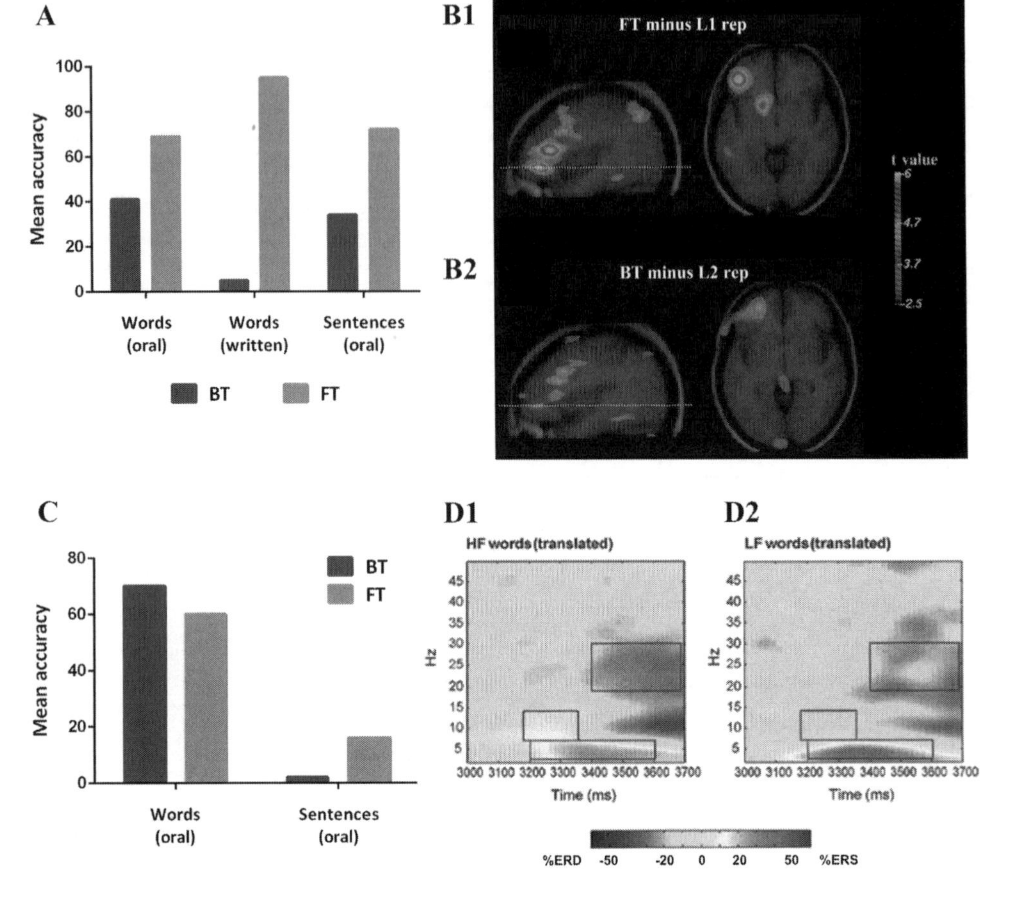

Figure 13.1 Functional organization of the systems subserving interlingual reformulation. **A.** Translation performance of patient E.M. in her first study, including assessments of word and sentence translation in both directions. **B.** Averaged PET subtraction image of cerebrospinal fluid increases in the left inferior frontal cortex for 12 subjects, superimposed upon the averaged MRI scans. **B1.** Activity pattern upon subtraction of L1 repetition from FT. **B2.** Activity pattern upon subtraction of L2 repetition from BT. **C.** Translation performance of patient El.M., including assessments of word and sentence translation in both directions. **D.** Time-frequency representations of event-related synchronization and desynchronization (ERS and ERD, respectively) for high-frequency words (**D1**) and low-frequency words (**D2**) at one representative electrode position (Cz).

Notes: Panel A: BT: backward translation; FT: forward translation. Panel D: Charts are plotted for the first 700 ms after stimulus onset (3,000 ms) and for the frequency range of 2–50 Hz. The red colour indicates event-related bandpower increases (ERS), and the blue colour represents event-related bandpower decreases (ERD). The frequency bands and time intervals that show ERS/ERD differences between the experimental conditions are highlighted with black rectangles. Note that fully coloured images are available in the digital version of the chapter.

Source: Panel A: data from Aglioti & Fabbro, 1993. Panel B: reprinted with permission from D. Klein, B. Milner, R. J. Zatorre, E. Meyer, and A. C. Evans, The neural substrates underlying word generation: a bilingual functional-imaging study, Proceedings of the National Academy of Sciences, 92(7), 2899–2903, Copyright (1995) National Academy of Sciences, USA Panel C: data from Fabbro & Paradis, 1995. Panel D: reprinted from Brain Research Bulletin, 72(1), R. H. Grabner, C. Brunner, R. Leeb, C. Neuper, and G. Pfurtscheller, Event-related EEG theta and alpha band oscillatory responses during language translation, pages 57–65, Copyright (2007), with permission from Elsevier.

increases in the globus pallidus (Lehtonen et al., 2005). Also, which specific neurocognitive mechanisms are recruited during translation depends on more fine-grained linguistic properties of the units at hand. For example, electrophysiological signals over parietal and frontal regions are differentially synchronized and desynchronized for high- relative to low-frequency source words (Grabner et al., 2007)—Figure 13.1D.[6] In the same vein, distinct neural modulations have been observed depending on the level of sublexical (Christoffels et al., 2013; Janyan et al., 2009) or semantic (Moldovan et al., 2016) overlap between source and target items.

In short, IR implicates widely distributed networks spanning cortical and subcortical areas involved in general language functions. Yet, far from constituting an undifferentiated, all-purpose system, those circuits include specific hubs which are distinctively associated with particular directions (BT vs. FT), processing levels (lexical vs. conceptual) and translation units (e.g., words vs. sentences). Beyond anatomical considerations, these mechanisms exhibit particular functional dynamics, as described in the following section.

13.2.3.3 Temporal dynamics of interlingual reformulation

The neurocognitive systems described earlier may involve different temporal dynamics depending on the demands that each instance of IR places on them. In particular, both directionality and the linguistic properties of the translation unit are key modulators of the underlying processes. For instance, behavioural studies in healthy subjects show that BT is typically performed faster than FT—although this effect is typically attenuated as translation competence increases (Christoffels et al., 2006; García et al., 2014; Kroll & Stewart, 1994; McElree et al., 2000)—and that cognate and concrete words are translated faster than non-cognate (de Groot, 1992; de Groot et al., 1994; van Hell & de Groot, 1998; García et al., 2014) and abstract (Christoffels et al., 2003; García et al., 2014; Sánchez-Casas et al., 1992) words, respectively. While these findings, derived from RT measures, reflect differences in the overall period from process onset to behavioural manifestation, EEG methods allow the inner timecourse of IR to be examined.

A number of neuroscientific studies have inquired into the temporal specificities of directionality effects. In an ERP study on word translation, Christoffels et al. (2013) found that FT yielded more positive amplitudes than BT in the P2 component over central and parietal sites, suggesting that each translation direction involves differential neurocognitive mechanisms as early as 200 ms after source-word onset—Figure 13.2A. The same pattern was reported by García, Mikulan, et al. (2016) through direct intracranial ERP recordings in the posterior fusiform gyrus, a region implicated in lexico-semantic processing—Figure 13.2B. Also, the two translation directions differ in their later temporal dynamics (from 400 to 700 ms after stimulus onset). As compared with BT, FT involves less negative-going amplitudes in the N400 component over centro-parietal sites (Christoffels et al., 2013), a modulation that was replicated through direct recordings of activity in the posterior fusiform and the anterior middle temporal gyri (García, Mikulan, et al., 2016). Such findings have been proposed to reflect increased implicit attentional demands by FT alongside more effortful comprehension of source words in BT (Christoffels et al., 2013; García, Mikulan, et al., 2016).

Also, distinct temporal dynamics during IR have been observed depending on specific linguistic features of the translation unit. For example, word frequency modulates synchronization and desynchronization patterns of EEG signals in a period from 200 to 600 ms after source-item presentation (Grabner et al., 2007)—see Figure 13.1D. In addition, relative to control words, interlingual homographs (lexical pairs with similar form but different meanings across languages) are translated more slowly and elicit more negative amplitudes around 400 ms post-stimulus presentation, suggesting additional processing effort (Christoffels et al., 2013). Also, abstract cognates, as compared with concrete cognates, yield N400-like deflections mainly over centro-temporal

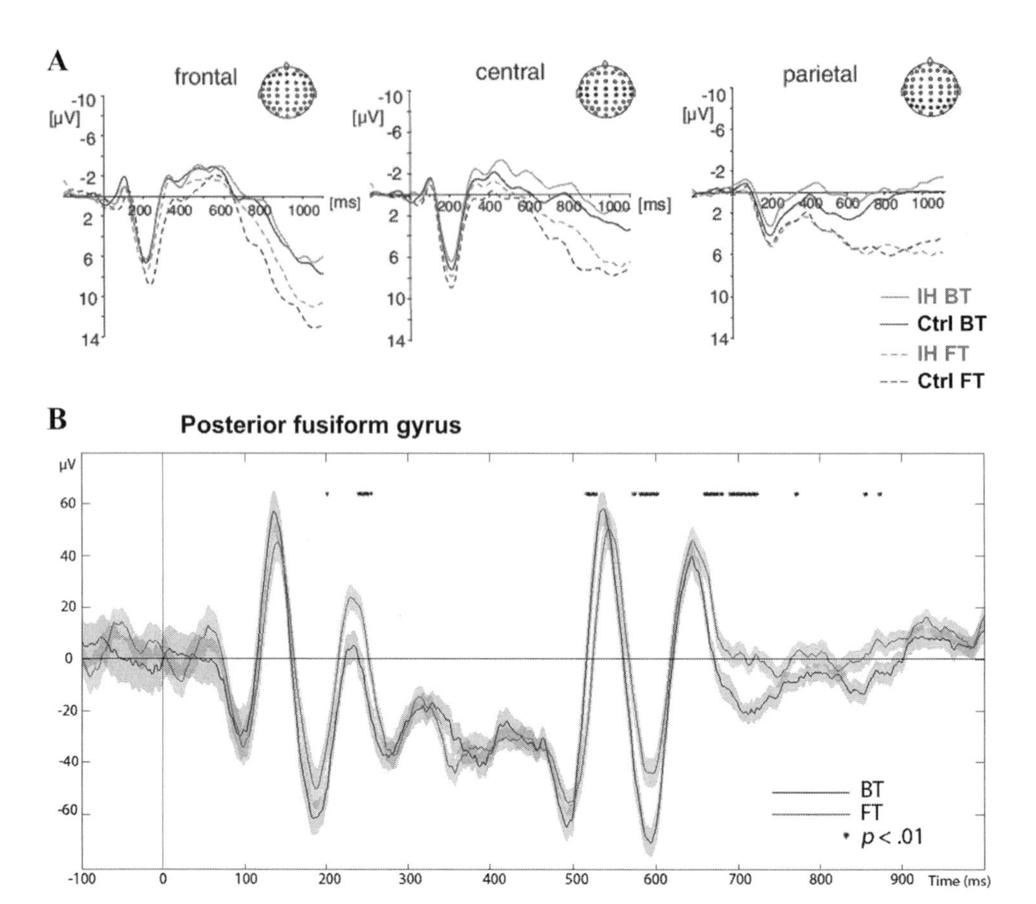

Figure 13.2 Temporal dynamics of interlingual reformulation. **A.** ERP waveforms for interlingual homographs (IHs, grey) and control words (Ctrl, black) for forward translation (FT, dotted line) and backward translation (BT, solid line). **B.** Intracranial ERP recordings during FT and BT from a proficient bilingual.

Notes: Panel A: In the 150–270 ms window, FT yielded more positive amplitudes than BT at central and parietal sites. In the 300–600 ms window, BT yielded more negative amplitudes than FT over central and parietal sites. The same was true for IHs relative to control words over central sites. For presentation purposes only, the signal was low-pass filtered (10 Hz) and the waveforms were averaged across electrodes of three regions: frontal, central and parietal. Panel B: Activity patterns in the posterior fusiform gyrus showed that, relative to BT, FT yielded more positive modulations in an early (220–250 ms) window and less negative modulations in a late (500–750) window.

Sources: Panel A: reprinted from Journal of Cognitive Psychology, 25(5), I. K. Christoffels, L. Ganushchak, and D. Koester. Language conflict in translation: An ERP study of translation production, pages 646–664, Copyright (2013), with permission from Taylor & Francis. Panel B: reprinted with permission from A. M. García, E. Mikulan, and A. Ibáñez (2016), A neuroscientific toolkit for translation studies, in R. Muñoz Martín (Ed.), Reembedding Translation Process Research (pp. 21–46), Amsterdam: John Benjamins.

sites (Janyan et al., 2009), showing that semantic access during source-word processing operates in a window comparable to that observed during single-language tasks. Moreover, additional ERP evidence indicates that word meaning can be accessed before activation of the target word (Moldovan et al., 2016), although the latter can also be activated automatically and unconsciously during processing of its translation equivalent in a single-language context (Thierry & Wu, 2007).

Overall, high-temporal-resolution methods reveal that well-established behavioural effects are linked to underlying neurophysiological factors. In particular, translation directions and various types of translation unit entail distinctive neural dynamics covering very early and considerably late time windows. The fact that most of these findings could hardly have been obtained without EEG tools highlights the relevance of neuroscience in the development of cognitive translatology.

13.2.3.4 Neuroplasticity in simultaneous interpreters

While most of the evidence reviewed here comes from subjects who are not translators or interpreters, many of the ensuing conclusions are probably generalizable across professionals and non-professionals. However, another line of research has revealed a number of neuroplastic changes associated with expertise in a most demanding form of IR, namely, simultaneous interpreting (Chernov, 2004). These studies complement and extend behavioural research showing that, relative to non-interpreter bilinguals, SIs exhibit advantages in tasks taxing particular linguistic or executive skills (García, 2014; García et al., 2019; Santilli et al, 2018).

First, expertise in simultaneous interpreting entails structural changes in task-relevant areas. As compared with monolinguals, SIs have been reported to exhibit reduced grey matter volume in the left cingulate gyrus, the pars opercularis, the bilateral pars triangularis and the middle part of the insula, arguably reflecting cortical pruning linked to more efficient language control (Elmer, Hänggi, et al., 2014)—Figure 13.3A. Notably, however, SIs seem to possess *more* grey matter volume than multilinguals in the left frontal pole, a region supporting executive domains (Becker et al., 2016). Though at odds with the previous finding (probably because of methodological factors), this result aligns with the overall view that expertise in SI can induce neuroanatomical adaptations. Indeed, a longitudinal MRI study showed that, after roughly 14 months of intensive practice, interpreting trainees exhibited greater cortical thickness in regions supporting phonetic, tactic and/or executive functions, such as the left planum temporale, the superior temporal and anterior supramarginal gyri, and the right parietal, angular and dorsal premotor cortices (Hervais-Adelman et al., 2017)—Figure 13.3B.

Moreover, SIs seem to be characterized by functional adaptations in critical neurocognitive mechanisms. For example, in comparison to non-interpreters, they exhibit a differential bias towards linguistic stimuli in speech-to-noise tasks, accompanied by distinct modulations of the N400 and P600 components over distributed scalp sites (Elmer, Klein, et al., 2014). Also, during intra- and interlinguistic semantic association tasks, they exhibit greater functional coupling between dorsal stream regions subserving sensory-to-articulation mapping (Elmer & Kühnis, 2016) as well as enhanced N400 modulations for all language combinations but the one corresponding to their professionally trained interpreting direction (Elmer et al., 2010). Furthermore, SIs' advantages in switching and dual-task performance are accompanied by greater functional connectivity in the left inferior frontal gyrus, a region implicated in verbal and cognitive control functions (Becker et al., 2016). In addition, sustained interpreting practice seems to reduce the recruitment of the right caudate nucleus (a key hub for executive control) during actual simultaneous interpretation sessions (Hervais-Adelman et al., 2015).

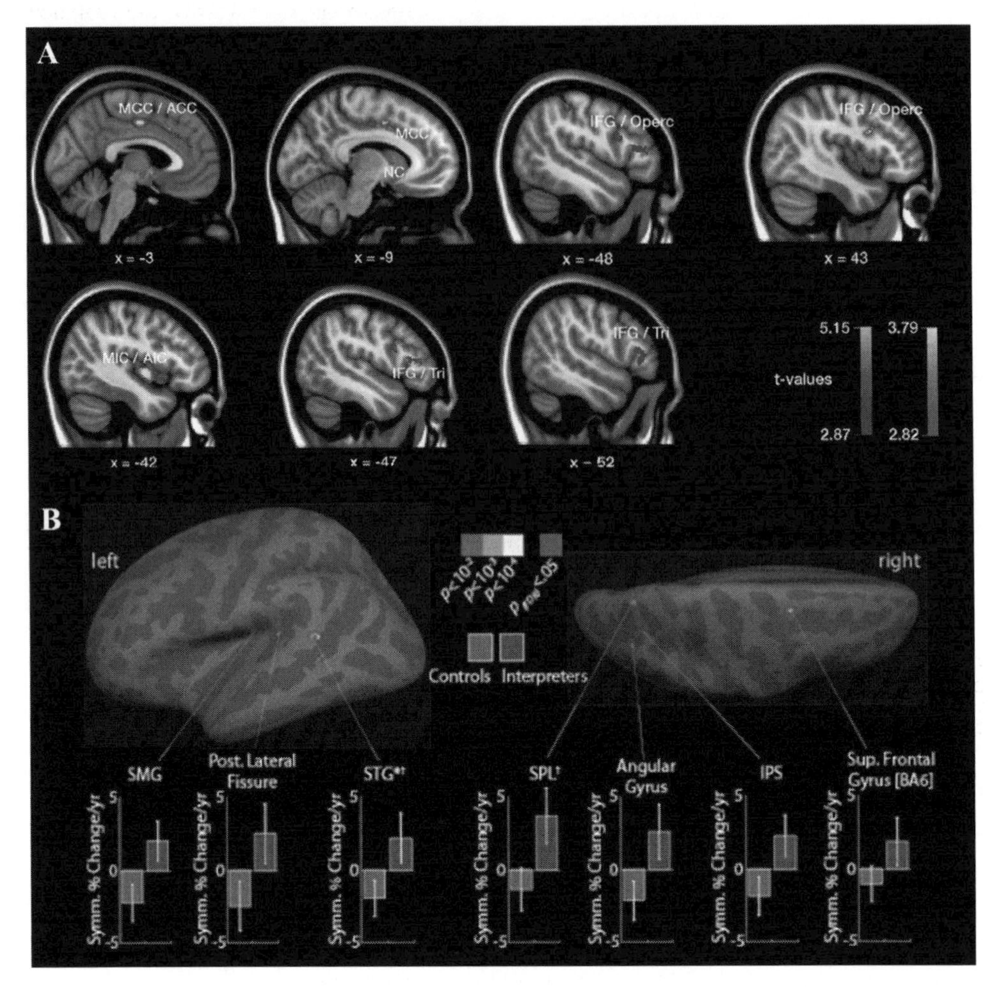

Figure 13.3 Neuroplasticity in simultaneous interpreters. **A.** Grey matter differences between professional simultaneous interpreters and bilingual control subjects (blue, left part of the figure). **B.** Training-induced cortical thickness changes in simultaneous interpreting students. Regions showing a significant main effect of group in mean per cent annualized cortical thickness change, projected on a canonical inflated white-matter surface. Dark patches represent sulci; light patches represent gyri.

Notes: Panel A: MCC: middle cingulate cortex; ACC: anterior cingulate cortex; MIC: middle insular cortex; AIC: anterior insular cortex; IFG: inferior frontal gyrus; Operc: pars opercularis; Tri: pars triangularis; NC: nucleus caudatus. Panel B: For clarity of display, the clusters having reached significance (at $p < .0001$) are displayed at a threshold of $p < .01$. Colour coding indicates significance level. The asterisk (*) denotes clusters in which the peak reaches whole-brain FDR-corrected significance at $p < .05$. The dagger (†) denotes clusters that reach whole-brain cluster-corrected significance at $p < .05$, with a cluster-forming threshold of $p < .0005$. Bar plots show symmetrized per cent change for both groups at the peak co-ordinates of the cluster. Error bars represent 95% confidence intervals. SMG: supramarginal gyrus; STG: superior temporal gyrus; SPL: superior parietal lobule; IPS: intraparietal sulcus. Note that fully coloured images are available in the digital version of the chapter.

Sources: Panel A: reprinted from Cortex, 54, S. Elmer, J. Hänggi, and L. Jäncke, Processing demands upon cognitive, linguistic, and articulatory functions promote grey matter plasticity in the adult multilingual brain: Insights from simultaneous interpreters, pages 179–189, Copyright (2014), with permission from Elsevier. Panel B: reprinted from Neuropsychologia, 98, A. Hervais-Adelman, B. Moser-Mercer, M.M. Murray, and N. Golestani, Cortical thickness increases after simultaneous interpretation training, pages 212–219, Copyright (2017), with permission from Elsevier.

Taken together, these results strongly suggest that the extreme cognitive demands inherent in sustained interpreting practice trigger structural and functional neuroplastic adaptations. These changes are multidimensional, as they are reflected in the grey matter density, cortical thickness, electrophysiological modulations and connectivity patterns of key neural systems supporting relevant verbal and non-verbal functions. Future research along the same lines is essential to further illuminate the neurocognitive impact of expert performance in this and other forms of IR.

13.2.3.5 Neurocognitive models of interlingual reformulation

Despite the accumulation of evidence, neurocognitive models of IR have been advanced very sparsely. The most explicit—and, thus, testable—proposals have been set forth by García (2012), Fabbro (1999) and Paradis (1994, 2009), as described next.

Building on lesion studies and neuroscientific findings, García (2012) has formulated a neurocognitive account of the linguistic systems mediating IR. The model is broadly consistent with the evidence presented in a preceding section, "Functional organization of the systems subserving interlingual reformulation". In particular, it posits the existence of neurofunctionally independent routes for translation, as opposed to monolingual speech production; BT, as opposed to FT; and form-based, as opposed to conceptually mediated, translation. These routes are claimed to rely mainly on perisylvian and frontostriatal regions of the left hemisphere. Moreover, the model postulates that word and sentence translation are differentially related to posterior networks implicated in declarative memory, and by frontostriatal pathways subserving procedural memory, respectively—clarifying that no translation unit is exclusively reliant on any such circuits. Moreover, while these broad organizational features are assumed to hold for any bilingual, the role of procedural and declarative mechanisms is proposed to depend on L2 competence. This model is characterized by its systematic construction, its inclusion of key constraints regarding the overall organization of bilingual memory, and its reliance on multidimensional evidence. However, it is limited in that it overlooks the role of relevant non-verbal mechanisms and dynamic aspects of online processing.

While the above model is presumed valid for any and all forms of IR, the one construed by Fabbro (1999) is specifically concerned with systems supporting simultaneous interpreting. Still, it aligns with the former in postulating separate routes for BT and FT and in acknowledging the partial functional autonomy of other linguistic systems involved in L1 and L2 processing. In addition, it is more explicit regarding the role of right hemisphere regions (e.g. right premotor and temporo-parietal areas) and subcortical structures (e.g. the cingulate gyrus) in prosodic, emotional, attentional and pragmatic functions. Notwithstanding, this model is blind to the distinction between form-level and conceptually mediated routes, and it fails to recognize the differential engagement of specific brain regions depending on the translation unit. Be that as it may, Fabbro's model constitutes the first and hitherto most comprehensive neurocognitive account of the verbal and non-verbal systems involved in simultaneous interpreting.

Finally, Paradis (1994, 2009) has also relied on neurolinguistic insights to describe cognitive aspects of simultaneous interpreting. Though less explicit in neuroanatomical terms, this proposal also assumes a distinction between conceptually mediated and form-level routes, further positing that each of them supports specific processing strategies. Conceptually mediated mechanisms would be mainly recruited under Strategy 1, which involves decoding the source unit across hierarchically organized linguistic levels (phonology, morphology, syntax and lexical semantics) prior to the construal of a non-linguistic representation. Thereon, the corresponding processing levels in the target language would be traversed in reverse order, leading to the production of the translation proper. On the other hand, form-level mechanisms would be more critical for Strategy 2, which consists in the direct transcoding of units at any hierarchical level of linguistic processing, in the absence of (non-verbal) conceptual processing. Paradis (2009) adds that Strategy 1 would operate automatically and implicitly, whereas Strategy 2 would be conscious and explicit. Thus, the former would rely mainly on procedural memory, while the latter would more directly depend on declarative memory functions. Still, these hypotheses require experimental testing to be judged for plausibility.

All in all, existing models of the neurocognitive basis of IR are decidedly scant. Also, despite their broad compatibility, they target relatively different phenomena, which precludes fine-grained comparisons among them. Furthermore, since their original formulation, they have not been updated to incorporate the most recent findings enumerated throughout the preceding sections, which leaves open multiple avenues for future works. That being said, they all succeed in advancing falsifiable hypotheses (a cornerstone for any empirical discipline) and in showing that theoretical breakthroughs in Translation Studies can be forged through an interdisciplinary dialogue with neuroscience.

13.3 Recent developments and future directions

In light of the previous pages, rather than a "known unknown" (Tymoczko, 2012), the neuroscience of IR might be better described as an "unknown known" within Translation Studies. Though poorly disseminated throughout the discipline, copious findings have been reported concerning the lateralization, functional organization, temporal dynamicity and plastic adaptability of putative neurocognitive systems. Promisingly, such areas of inquiry have been gaining momentum in recent years, as attested by the very existence of this chapter in the present handbook. Now, as research continues to expand, a number of challenges and opportunities should be prioritized in the field's agenda.

First, a substantial part of the evidence comes from lay bilinguals, that is, subjects lacking formal experience in translation and interpreting. While some findings (e.g. the partial independence of mechanisms for BT and FT, or the predominance of left-sided regions for verbal processes in both directions) are likely generalizable to all bilinguals, others could be specific to those who have developed expert IR skills. In this sense, neuroscientific studies targeting professional translators and interpreters (e.g., García, Mikulan, et al., 2016; Proverbio et al., 2009), or comparing such populations with lay bilinguals (e.g. Becker et al., 2016; Elmer, Klein, et al., 2014; Hervais-Adelman et al., 2017), are crucial to illuminate questions of more direct relevance to Translation Studies. The challenge for the neuroscience of IR, as it were, is to move from studies on translation and interpreting to studies on translators and interpreters.

Second, it would also be useful to establish modality-specific profiles via direct comparisons between the latter two groups. No neuroscientific study has hitherto examined this issue, so that the literature is moot on which findings are general to expertise in IR as opposed to only one of its subforms. In fact, given the broad differences in the verbal and (more particularly) non-verbal mechanisms involved in translation and interpreting (Schäffner, 2004; Schwieter & Ferreira, 2017), research in this direction might even be highly relevant to understand brain plasticity beyond the particular interests of translation scholars.

Third, efforts should be made to develop more naturalistic designs. As is the case with most experiments in neuroscience, available findings stem from atomistic tasks in which subjects respond to (pseudo-)randomized lists of decontextualized words or sentences. While direct unit-by-unit correspondences certainly play a role in translation (Christoffels et al., 2003; Darò & Fabbro, 1994; Paradis, 1994, 2009), token-based paradigms are blind to the contextual nature of *skopos*-driven textual processes. Promisingly, new approaches based on behavioural interventions (Trevisan et al., 2017) and hemodynamic recordings (Desai et al., 2016; Huth et al., 2016), as well as analysis of narrative comprehension (García et al., 2017) and production (García, Carrillo, et al., 2016) in brain-lesioned individuals, allow the psychobiological aspects of discourse-level processing to be explored. Though their application in cross-linguistic paradigms abounds in methodological challenges, these and other strategies may inaugurate a tighter rapprochement between neuroscience and Translation Studies at large.

Fourth, multidimensional approaches should be favoured to gain more comprehensive insights into targeted phenomena. Available research has mostly relied on data from individual neuroscientific methods, thus shedding light on anatomical *or* hemodynamic *or* electrophysiological aspects—for a notable exception, see Becker et al. (2016). However, any aspect of brain function involves concerted interactions among all such dimensions, as well as several others. To capture such complexities, future studies should integrate measures from two or more of those levels, as is done in other areas of cognitive neuroscience (García-Cordero et al., 2016; Melloni et al., 2015, 2016). Moreover, correlations or regressions could be made between ensuing neurocognitive findings and offline assessments of situated performance (e.g. quantitative estimations of translation or interpreting quality). Major breakthroughs could be attained through these yet unattempted forms of empirical triangulation.

Fifth, the majority of the evidence so far is correlational in nature, which limits the possibility of postulating justifiable mechanistic models. Promisingly, some reports in the literature have circumvented this caveat by either conducting longitudinal studies (Hervais-Adelman et al., 2015, 2017) or employing invasive (Borius et al., 2012) and non-invasive (Liuzzi et al., 2010) stimulation techniques. Both approaches offer a much firmer ground to identify causal links between performance or experience in IR and specific neural phenomena, such as activation patterns or plastic changes. More evidence along these lines would afford an invaluable empirical basis to construct explanatory and predictive models in the field.

Sixth, neuroscientists and translation scholars should work more closely together to promote theoretical cross-fertilization. Current studies have rarely been informed by constructs from the core of Translation Studies. Moreover, most cognitive models of IR have overlooked neurocognitive evidence, an omission that has sometimes resulted in incomplete or inaccurate proposals (see García, Mikulan, et al., 2016). As shown by recent works in other fields, such as systemic functional linguistics (García & Ibáñez, 2016), progress in theory building can be made by overcoming the divide between biological and non-biological approaches to mental phenomena.

Finally, all these developments could contribute to bridging the long-standing gap between theory and practice in Translation Studies. Indeed, neuroscientific works have paved the way for appliable innovations in other language-related arenas, such as the teaching of reading (Dehaene, 2010) or the design of interventions for boosting discourse-processing skills (Trevisan et al., 2017). Though speculative at this juncture, the integration of neuroscience into the central agenda of Translation Studies could foster similar advancements for the teaching and practice of translation and interpreting.

In summary, cognitive aspects of IR can be understood in concrete biological terms. Through a combination of behavioural measures, lesion studies and neuroscientific methods, researchers from within and outside Translation Studies have examined several relevant topics. In particular, results to date indicate that the key involvement of left perisylvian and frontostriatal regions during IR is accompanied by distinct contributions from the right hemisphere. Also, different neurocognitive mechanisms seem to be recruited depending on translation directionality and the linguistic properties of the source segment. Moreover, sustained practice of at least one specific modality (namely, simultaneous interpreting) triggers anatomical and functional changes in critical neural systems. Importantly, some of these findings have been incorporated in descriptive theoretical models, which can inspire new testable hypotheses.

Above and beyond these milestones, possibilities abound for further neurocognitively oriented research with more direct involvement of translation scholars. A profitable path lies open for those ready to tackle the challenge. Hopefully, by disseminating both existing and prospective breakthroughs, this chapter will represent a stepping stone in that direction.

Notes

1 In particular, Part C of the Bilingual Aphasia Test (Paradis, 1979, 2011) includes tasks tapping into cross-linguistic equivalent recognition, word translation and sentence translation, all in both backward and forward directions. All versions of this instrument can be freely downloaded at www.mcgill.ca/linguistics/research/bat
2 However, see Paradis (1992, 1995, 2003, 2008) for a critique of this approach.
3 The term "lay bilingual" refers to an individual who is competent in (at least) two languages but lacks formal experience in translation and interpreting.
4 For a review of behavioral tasks tapping into cross-linguistic equivalent processing, see García (2015b).
5 For a review of research on executive functions in SIs, see García (2014).
6 Event-related synchronization and desynchronization are two functional phenomena whereby distributed neuronal populations become spatially linked, during a cognitive process, to form transient functional networks. They can be tapped via time-frequency analyses, which enable the detection of spectral power changes across time points and frequency bands in an EEG signal—for details, see Singer (1993).

Further reading

To delve more deeply into the topics of this chapter, the reader may wish to consult the following works.

For detailed accounts of translation disorders following brain lesions:

Aglioti, S., & Fabbro, F. (1993). Paradoxical selective recovery in a bilingual aphasic following subcortical lesions. *Neuroreport, 4*(12), 1359–1362.
Fabbro, F., & Paradis, M. (1995). Differential impairments in four multilingual patients with subcortical lesions. In M. Paradis (Ed.), *Aspects of bilingual aphasia* (pp. 139–176). Oxford: Pergamon.
García, A. (2015). Translating with an injured brain: neurolinguistic aspects of translation as revealed by bilinguals with cerebral lesions. *Meta: Translators' Journal, 60*(1), 112–134.
Paradis, M., Goldblum, M.-C., & Abidi, R. (1982). Alternate antagonism with paradoxical translation behavior in two bilingual aphasic patients. *Brain and Language, 15*(1), 55–69.

For more details on neuroscientific methods and their relevance to studying IR:

García, A. M., Mikulan, E., & Ibáñez, A. (2016). A neuroscientific toolkit for translation studies. In R. Muñoz Martín (Ed.), *Reembedding translation process research* (pp. 21–46). Amsterdam: John Benjamins.

For examples and discussions of neuroscientific experiments on IR:

Christoffels, I. K., Ganushchak, L., & Koester, D. (2013). Language conflict in translation: An ERP study of translation production. *Journal of Cognitive Psychology, 25*(5), 646–664. doi: 10.1080/20445911.2013.821127
García, A. M. (2013). Brain activity during translation: A review of the neuroimaging evidence as a testing ground for clinically-based hypotheses. *Journal of Neurolinguistics, 26*(3), 370–383.
Hervais-Adelman, A., Moser-Mercer, B., Murray, M. M., & Golestani, N. (2017). Cortical thickness increases after simultaneous interpretation training. *Neuropsychologia, 98*, 212–219. doi: 10.1016/j.neuropsychologia.2017.01.008
Klein, D., Milner, B., Zatorre, R. J., Meyer, E., & Evans, A. C. (1995). The neural substrates underlying word generation: A bilingual functional-imaging study. *Proceedings of the National Academy of Sciences, 92*(7), 2899–2903.

Acknowledgements

This work was partially supported by CONICET; Programa Interdisciplinario de Investigación Experimental en Comunicación y Cognición (PIIECC), Facultad de Humanidades, USACH; and the INECO Foundation.

References

Aglioti, S., & Fabbro, F. (1993). Paradoxical selective recovery in a bilingual aphasic following subcortical lesions. *Neuroreport, 4*(12), 1359–1362.

Becker, M., Schubert, T., Strobach, T., Gallinat, J., & Kühn, S. (2016). Simultaneous interpreters vs. professional multilingual controls: Group differences in cognitive control as well as brain structure and function. *NeuroImage, 134*, 250–260.

Borius, P.Y., Giussani, C., Draper, L., & Roux, F. E. (2012). Sentence translation in proficient bilinguals: A direct electrostimulation brain mapping. *Cortex, 48*(5), 614–622.

Bouton, C. (1984). Le cerveau du traducteur: de quelques propositions sur ce thème. *Meta: Translators' Journal, 29*(1), 44–56.

Byng, S., Coltheart, M., Masterson, J., Prior, M., & Riddoch, J. (1984). Bilingual biscriptal deep dyslexia. *The Quarterly Journal of Experimental Psychology, 36*(3), 417–433.

Chernov, G. V. (2004). *Inference and anticipation in simultaneous interpreting: A probability-prediction model.* Amsterdam: John Benjamins.

Christoffels, I. K., de Groot, A. M. B., & Kroll, J. F. (2006). Memory and language skills in simultaneous interpreters: The role of expertise and language proficiency. *Journal of Memory and Language, 54*(3), 324–345.

Christoffels, I. K., de Groot, A. M. B., & Waldorp, L. J. (2003). Basic skills in a complex task: A graphical model relating memory and lexical retrieval to simultaneous interpreting. *Bilingualism: Language and Cognition, 6*(3), 201–211.

Christoffels, I. K., Ganushchak, L., & Koester, D. (2013). Language conflict in translation: An ERP study of translation production. *Journal of Cognitive Psychology, 25*(5), 646–664.

Darò, V., & Fabbro, F. (1994). Verbal memory during simultaneous interpretation: Effects of phonological interference. *Applied Linguistics, 15*, 365–381.

de Groot, A. M. B. (1992). Determinants of word translation. *Journal of Experimental Psychology: Learning, Memory, and Cognition, 18*, 1001–1018.

de Groot, A. M. B., Dannenburg, L., & van Hell, J. G. (1994). Forward and backward word translation by bilinguals. *Journal of Memory and Language, 33*(5), 600–629. https://doi.org/10.1006/jmla.1994.1029

Dehaene, S. (2010). *Reading in the brain: The new science of how we read.* London: Penguin.

Desai, R. H., Choi, W., Lai, V. T., & Henderson, J. M. (2016). Toward semantics in the wild: Activation to manipulable nouns in naturalistic reading. *Journal of Neuroscience, 36*(14), 4050–4055. doi: 10.1523/jneurosci.1480–15.2016

de Vreese, L. P., Motta, M., & Toschi, A. (1988). Compulsive and paradoxical translation behaviour in a case of presenile dementia of the Alzheimer type. *Journal of Neurolinguistics, 3*(2), 233–259.

Diamond, A. (2013). Executive functions. *Annual Review of Psychology, 64*, 135–168.

Diamond, B., & Shreve, G. (2010). Neural and physiological correlates of translation and interpreting in the bilingual brain: Recent perspectives. In G. Shreve, & E. Angelone (Eds.), *Translation and cognition* (pp. 289–322). Amsterdam & Philadelphia: John Benjamins.

Diamond, B. J., & Shreve, G. M. (2017). Deliberate practice and neurocognitive optimization of translation expertise. In J. W. Schwieter, & A. Ferreira (Eds.), *The handbook of translation and cognition* (pp. 476–495). New York: John Wiley & Sons.

Dottori, M., Hesse, E., Santilli, M., Vilas, M., Martorell Caro, M., Fraiman, D., Sedeño, L., Ibáñez, A., & García, A. M. (2020). Task-specific signatures in the expert brain: Differential correlates of translation and reading in professional interpreters. *NeuroImage, 209*, 116519.

Dunn, J. C., & Kirsner, K. (2003). What can we infer from double dissociations? *Cortex, 39*(1), 1–7.

Elmer, S. (2012). The investigation of simultaneous interpreters as an alternative approach to address the signature of multilingual speech processing. *Zeitschrift für Neuropsychologie, 23*(2), 105–116. doi: 10.1024/1016–264X/a000068

Elmer, S., Hänggi, J., & Jäncke, L. (2014). Processing demands upon cognitive, linguistic, and articulatory functions promote grey matter plasticity in the adult multilingual brain: Insights from simultaneous interpreters. *Cortex, 54*, 179–189. http://dx.doi.org/10.1016/j.cortex.2014.02.014

Elmer, S., Klein, C., Kühnis, J., Liem, F., Meyer, M., & Jäncke, L. (2014). Music and language expertise influence the categorization of speech and musical sounds: Behavioral and electrophysiological measurements. *Journal of Cognitive Neuroscience, 26*(10), 2356–2369.

Elmer, S., & Kühnis, J. (2016). Functional connectivity in the left dorsal stream facilitates simultaneous language translation: An EEG study. *Frontiers in Human Neuroscience, 10*, 60. https://doi.org/10.3389/fnhum.2016.00060

Elmer, S., Meyer, M., & Jancke, L. (2010). Simultaneous interpreters as a model for neuronal adaptation in the domain of language processing. *Brain Research, 1317*, 147–156.
Eviatar, Z., Leikin, M., & Ibrahim, R. (1999). Phonological processing of second language phonemes: A selective deficit in a bilingual aphasic. *Language Learning, 49*(1), 121–141. doi: 10.1111/1467–9922.00072
Fabbro, F. (1999). *The neurolinguistics of bilingualism: An introduction*. Hove: Psychology Press.
Fabbro, F. (2001). The bilingual brain: Cerebral representation of languages. *Brain and Language, 79*(2), 211–222. doi: 10.1006/brln.2001.2481
Fabbro, F., & Gran, L. (1997). Neurolinguistic research in simultaneous interpretation. In Y. Gambier, D. Gile, & C. Taylor (Eds.), *Conference Interpreting: Current trends in research. Proceedings of the international conference on interpreting: What do we know and how?* (pp. 9–27). Amsterdam: John Benjamins.
Fabbro, F., Gran, L., Basso, G., & Bava, A. (1990). Cerebral lateralization in simultaneous interpretation. *Brain and Language, 39*(1), 69–89.
Fabbro, F., Gran, B., & Gran, L. (1991). Hemispheric specialization for semantic and syntactic components of language in simultaneous interpreters. *Brain and Language, 41*(1), 1–42.
Fabbro, F., & Paradis, M. (1995). Differential impairments in four multilingual patients with subcortical lesions. In M. Paradis (Ed.), *Aspects of bilingual aphasia* (pp. 139–176). Oxford: Pergamon.
Ferré, P., Sánchez-Casas, R., & Guasch, M. (2006). Can a horse be a donkey? Semantic and form interference effects in translation recognition in early and late proficient and nonproficient Spanish-Catalan bilinguals. *Language Learning, 56*(4), 571–608. doi: 10.1111/j.1467–9922.2006.00389.x
García, A. M. (2012). *Traductología y neurocognición: Cómo se organiza el sistema lingüístico del traductor.* Córdoba: Facultad de Lenguas de la UNC.
García, A. M. (2013). Brain activity during translation: A review of the neuroimaging evidence as a testing ground for clinically-based hypotheses. *Journal of Neurolinguistics, 26*(3), 370–383.
García, A. M. (2014). The interpreter advantage hypothesis: Preliminary data patterns and empirically motivated questions. *Translation and Interpreting Studies, 9*(2), 219–238. doi: 10.1075/tis.9.2.04gar
García, A. M. (2015a). Translating with an injured brain: Neurolinguistic aspects of translation as revealed by bilinguals with cerebral lesions. *Meta: Translators' Journal, 60*(1), 112–134.
García, A. M. (2015b). Psycholinguistic explorations of lexical translation equivalents: Thirty years of research and their implications for cognitive translatology. *Translation Spaces, 4*(1), 9–28.
García, A. M. (2019). *The neurocognition of translation and interpreting*. Amsterdam: John Benjamins.
García, A. M., Bocanegra, Y., Herrera, E., Moreno, L., Carmona, J., Baena, A., Lopera, F., Pineda, D., Melloni, M., Legaz, A., Muñoz, E., Sedeño, L., Baez, S., & Ibáñez, A. M. (2018). Parkinson's disease compromises the appraisal of action meanings evoked by naturalistic texts. *Cortex, 100*, 111–126.
García, A. M., Carrillo, F., Orozco-Arroyave, J. R., Trujillo, N., Vargas Bonilla, J. F., Fittipaldi, S., Adolfi, F., Nöth, E., Sigman, M., Fernández Slezak, D., Ibáñez, A., & Cecchi, G. A. (2016). How language flows when movements don't: An automated analysis of spontaneous discourse in Parkinson's disease. *Brain and Language,* 162, 19–28.
García, A. M., & Ibáñez, A. (2016). Processes and verbs of doing, in the brain: Theoretical implications for systemic functional linguistics. *Functions of Language, 23*(3), 305–335.
García, A. M., Ibáñez, A., Huepe, D., Houck, A., Michon, M., Gelormini Lezama, C., Chadha, S., & Rivera-Rei, Á. (2014). Word reading and translation in bilinguals: The impact of formal and informal translation expertise. *Frontiers in Psychology, 5*, 1302. doi: 10.3389/fpsyg.2014.01302
García, A. M., Mikulan, E., & Ibáñez, A. (2016). A neuroscientific toolkit for translation studies. In R. Muñoz Martín (Ed.), *Reembedding translation process research* (pp. 21–46). Amsterdam: John Benjamins.
Garcia, A. M., Muñoz, E., & Kogan, B. (2019). Taxing the bilingual mind: Effects of simultaneous interpreting experience on verbal and executive mechanisms. *Bilingualism: Language and Cognition*. Retrieved 26 March 2020 from www.cambridge.org/core/journals/bilingualism-language-and-cognition/article/taxing-the-bilingual-mind-effects-of-simultaneous-interpreting-experience-on-verbal-and-executive-mechanisms/8C4C7D2FB95FD880B89C4FCC19AEA477
García-Caballero, A., García-Lado, I., Gonzalez-Hermida, J., Area, R., Recimil, M. J., Juncos Rabadan, O., Lamas, S., Ozaita, G., & Jorge, F. J. (2007). Paradoxical recovery in a bilingual patient with aphasia after right capsuloputaminal infarction. *Journal of Neurology, Neurosurgery, and Psychiatry, 78*(1), 89–91. doi: 10.1136/jnnp.2006.095406
García-Cordero, I., Sedeño, L., de la Fuente, L., Slachevsky, A., Forno, G., Klein, F., Lillo, P., Ferrari, J., Rodriguez, C., Bustin, J., Torralva, T., Baez, S., Yoris, A., Esteves, S., Melloni, M., Salamone, P., Huepe, D., Manes, F., García, A. M., & Ibañez, A. (2016). Feeling, learning from and being aware of inner

states: interoceptive dimensions in neurodegeneration and stroke. *Philosophical Transactions of the Royal Society B: Biological Sciences, 371*(1708). doi: 10.1098/rstb.2016.0006

Gastaldi, G. (1951). Osservazioni su un afasico bilingüe. *Sistema Nervoso, 2*, 175–180.

Grabner, R. H., Brunner, C., Leeb, R., Neuper, C., & Pfurtscheller, G. (2007). Event-related EEG theta and alpha band oscillatory responses during language translation. *Brain Research Bulletin, 72*(1), 57–65.

Hervais-Adelman, A., Moser-Mercer, B., & Golestani, N. (2015). Brain functional plasticity associated with the emergence of expertise in extreme language control. *Neuroimage, 114*, 264–274.

Hervais-Adelman, A., Moser-Mercer, B., Murray, M. M., & Golestani, N. (2017). Cortical thickness increases after simultaneous interpretation training. *Neuropsychologia, 98*, 212–219. doi: 10.1016/j.neuropsychologia.2017.01.008

Huth, A. G., de Heer, W. A., Griffiths, T. L., Theunissen, F. E., & Gallant, J. L. (2016). Natural speech reveals the semantic maps that tile human cerebral cortex. *Nature, 532*(7600), 453–458. doi: 10.1038/nature17637

Janyan, A., Popivanov, I., & Andonova, E. (2009). Concreteness effect and word cognate status: ERPs in single word translation. In K. Alter, M. Horne, M. Lindgren, M. Roll, & J. von Koss Torkildsen (Eds.), *Brain talk: Discourse with and in the brain* (pp. 21–30). Lund: Lunds Universitet.

Kauders, O. (1929). Über polyglotte Reaktionen bei einer sensorischen Aphasie. *Zeitschrift für die gesamte Neurologie und Psychiatrie, 149*, 291–301.

Klein, D., Milner, B., Zatorre, R. J., Meyer, E., & Evans, A. C. (1995). The neural substrates underlying word generation: A bilingual functional-imaging study. *Proceedings of the National Academy of Sciences, 92*(7), 2899–2903.

Kroll, J. F., & Stewart, E. (1994). Category interference in translation and picture naming: Evidence for asymmetric connections between bilingual memory representations. *Journal of Memory and Language, 33*, 149–174.

Kurz, I. (1994). A look into the "black box"—EEG probability mapping during mental simultaneous interpreting. In M. Snell-Hornby, F. Pöchhacker, & K. Kaindl (Eds.), *Translation studies. An interdiscipline* (pp. 199–207). Amsterdam: John Benjamins.

Kurz, I. (1995). Watching the brain at work—An exploratory study of EEG changes during simultaneous interpreting (SI). *The Interpreters' Newsletter, 6*, 3–16.

Lehtonen, M. H., Laine, M., Niemi, J., Thomsen, T., Vorobyev, V. A., & Hugdahl, K. (2005). Brain correlates of sentence translation in Finnish–Norwegian bilinguals. *Neuroreport, 16*(6), 607–610.

Lindell, A. K. (2006). In your right mind: Right hemisphere contributions to language processing and production. *Neuropsychology Review, 16*(3), 131–148. doi: 10.1007/s11065–006–9011–9

Liuzzi, G., Freundlieb, N., Ridder, V., Hoppe, J., Heise, K., Zimerman, M., Dobel, C., Enriquez-Geppert, S., Gerloff, C., Zwitserlood, P., & Hummel, F. C. (2010). The involvement of the left motor cortex in learning of a novel action word lexicon. *Current Biology, 20*(19), 1745–1751. doi: 10.1016/j.cub.2010.08.034

McElree, B., Jia, G., & Litvak, A. (2000). The time course of conceptual processing in three bilingual populations. *Journal of Memory and Language, 42*, 229–254.

Mazoyer, B., Mellet, E., Perchey, G., Zago, L., Crivello, F., Jobard, G., Delcroix, N., Vigneau, M., Leroux, G., Petit, L., Joliot, M., & Tzourio-Mazoyer, N. (2016). BIL&GIN: A neuroimaging, cognitive, behavioral, and genetic database for the study of human brain lateralization. *Neuroimage, 124*, 1225–1231.

Melloni, M., Billeke, P., Baez, S., Hesse, E., de la Fuente, L., Forno, G., Birba, A., García-Cordero, I., Serrano, C., Plastino, A., Slachevsky, A., Huepe, D., Sigman, M., Manes, F., García, A. M., Sedeño, L., & Ibáñez, A. (2016). Your perspective and my benefit: Multiple lesion models of self-other integration strategies during social bargaining. *Brain, 139*(11), 1–19. doi: 10.1093/brain/aww231

Melloni, M., Sedeño, L., Hesse, E., García-Cordero, I., Mikulan, E., Plastino, A., Marcotti, A., López, J. D., Bustamante, C., Lopera, F., Pineda, D., García, A. M., Manes, F., Trujillo, N., & Ibáñez, A. (2015). Cortical dynamics and subcortical signatures of motor-language coupling in Parkinson's disease. *Scientific Reports, 5*, 11899.

Moldovan, C. D., Demestre, J., Ferré, P., & Sánchez-Casas, R. (2016). The role of meaning and form similarity in translation recognition in highly proficient balanced bilinguals: A behavioral and ERP study. *Journal of Neurolinguistics, 37*, 1–11.

Moser-Mercer, B. (2010). The search for neuro-physiological correlates of expertise in interpreting. In G. Shreve, & E. Angelone (Eds.), *Translation and cognition* (pp. 263–288). Amsterdam & Philadelphia: John Benjamins.

Nilipour, R., & Ashayeri, H. (1989). Alternating antagonism between two languages with successive recovery of a third in a trilingual aphasic patient. *Brain and Language, 36*(1), 23–48. doi: 10.1016/0093–934x(89)90050–3

Paradis, M. (1979). L'aphasie chez les bilingues et les polyglottes. In A. R. Lecours, F. L'hermitte, et al. (Eds.), *L'aphasie* (pp. 605–616). Paris: Flammarion.
Paradis, M. (1984). Aphasie et traduction. *Meta: International Translator's Journal, 29*, 57–67.
Paradis, M. (1989). Bilingual and polyglot aphasia. In F. Boller , & J. Grafman (Eds.), *Handbook of neuropsychology*,Vol. 2 (pp. 117–140). Amsterdam: Elsevier.
Paradis, M. (1992). The Loch Ness Monster approach to bilingual language lateralization: A response to Berquier and Ashton. *Brain and Language, 43*(3), 534–537.
Paradis, M. (1994). Towards a neurolinguistic theory of simultaneous translation: The framework. *International Journal of Psycholinguistics, 10*, 319–335.
Paradis, M. (1995). Another sighting of differential language laterality in multilinguals, this time in Loch Tok Pisin: Comments on Wuillemin, Richardson, and Lynch (1994). *Brain and Language, 49*(2), 173–186. http://dx.doi.org/10.1006/brln.1995.1027
Paradis, M. (2003). The bilingual Loch Ness Monster raises its non-asymmetric head again—or, why bother with such cumbersome notions as validity and reliability? Comments on Evans et al. (2000). *Brain and Language, 87*(3), 441–448.
Paradis, M. (2008). Bilingual laterality: Unfounded claim of validity: A comment on. *Neuropsychologia, 46*(5), 1588–1590; author reply 1591–1583. doi: 10.1016/j.neuropsychologia.2008.01.029
Paradis, M. (2009). *Declarative and procedural determinants of second languages*. Amsterdam: John Benjamins.
Paradis, M. (2011). Principles underlying the Bilingual Aphasia Test (BAT) and its uses. *Clinical Linguistics and Phonetics, 25*(6–7), 427–443. doi: 10.3109/02699206.2011.560326
Paradis, M., Goldblum, M.-C., & Abidi, R. (1982). Alternate antagonism with paradoxical translation behavior in two bilingual aphasic patients. *Brain and Language, 15*(1), 55–69.
Price, C. J., Green, D. W., & von Studnitz, R. (1999). A functional imaging study of translation and language switching. *Brain, 122*, 2221–2235.
Proverbio, A. M., & Adorni, R. (2011). Hemispheric asymmetry for language processing and lateral preference in simultaneous interpreters. *Psychology, 2*(01), 12–17.
Proverbio, A. M., Adorni, R., & Zani, A. (2009). Inferring native language from early bio-electrical activity. *Biological Psychology, 80*(1), 52–63.
Proverbio, A. M., Leoni, G., & Zani, A. (2004). Language switching mechanisms in simultaneous interpreters: An ERP study. *Neuropsychologia, 42*(12), 1636–1656.
Quaresima, V., Ferrari, M., van der Sluijs, M. C., Menssen, J., & Colier, W. N. (2002). Lateral frontal cortex oxygenation changes during translation and language switching revealed by non-invasive near-infrared multi-point measurements. *Brain Research Bulletin, 59*(3), 235–243.
Rinne, J. O., Tommola, J., Laine, M., Krause, B. J., Schmidt, D., Kaasinen, V., Teräs, M., Sipilä, H., & Sunnari, M. (2000). The translating brain: Cerebral activation patterns during simultaneous interpreting. *Neuroscience Letters, 294*(2), 85–88.
Sánchez-Casas, R. M., García-Albea, J. E., & Davis, C. W. (1992). Bilingual lexical processing: Exploring the cognate/non-cognate distinction. *European Journal of Cognitive Psychology, 4*(4), 293–310. doi: 10.1080/09541449208406189
Santilli, M., Vilas, M., Mikulan, E., Martorell Caro, M., Muñoz, E., Sedeño, L., Ibáñez, A., & García, A. M. (2018). Bilingual memory, to the extreme: Lexical processing in simultaneous interpreters. *Bilingualism: Language and Cognition*. doi: https://doi.org/10.1017/S1366728918000378
Schäffner, C. (2004). Translation research versus interpreting research: Kinship, differences and prospects for partnership. In C. Schäffner (Ed.), *Translation research and interpreting research: Traditions, gaps and synergies* (pp. 1–9): Clevedon (UK): Multilingual Matters.
Schwieter, J. W., & Ferreira, A. (2017). Bilingualism in cognitive translation and interpreting studies. In J. W. Schwieter, & A. Ferreira (Eds.), *The handbook of translation and cognition* (pp. 144–164): John Wiley & Sons, Inc. Hoboken, NJ.
Singer, W. (1993). Synchronization of cortical activity and its putative role in information processing and learning. *Annual Review of Physiology, 55*, 349–374.
Stephan, K. E., Fink, G. R., & Marshall, J. C. (2007). Mechanisms of hemispheric specialization: Insights from analyses of connectivity. *Neuropsychologia, 45*(2–4), 209–228. doi: 10.1016/j.neuropsycholo gia.2006.07.002
Thierry, G., & Wu, Y. J. (2007). Brain potentials reveal unconscious translation during foreign-language comprehension. *Proceedings of the National Academy of Sciences, 104*(30), 12530–12535. doi: 10.1073/pnas.0609927104

Trevisan, P., Sedeño, L., Birba, A., Ibáñez, A., & García, A. M. (2017). A moving story: Ecological motor training selectively improves the appraisal of action meaning evoked by naturalistic narratives. *Scientific Reports, 7*, 12538. doi: 10.1038/s41598–017–12928-w

Tymoczko, M. (2012). The neuroscience of translation. *Target, 24*(1), 83–102.

van Hell, J., & de Groot, A. M. B. (1998). Disentangling context availability and concreteness in lexical decision and word translation. *The Quarterly Journal of Experimental Psychology Section A, 51*(1), 41–63. doi: 10.1080/713755752

Veyrac, G.-J. (1931). *Étude de l'aphasie chez les sujets polyglottes*. Paris: L. Arnette.

Part III

Translation and types of cognitive processing

14

Translation, effort and cognition

Daniel Gile and Victoria Lei

14.1 Introduction: The emergence of the "effort" topic in the history of Translation and Interpreting Studies (TIS)

In translation and interpreting (henceforth "Translation"), some direct source-Text to target-Text transfers seem effortless and immediate. Such is the case with proper names, well-established technical terms and standard phrases when they are known to the translator or interpreter. However, these short Text segments (Text stands for written texts and spoken and signed speeches) are generally embedded in Texts that require reflection, problem solving and decision making as to what the author or speaker actually means and intends to achieve, and how the Translator should reformulate the message in the target language so as to meet applicable professional and social requirements. Such analyses and decision-making processes involve effort. In particular, translation generally requires the search for ad hoc information, both linguistic and extra-linguistic, to identify target-appropriate lexical and phraseological usage, as well as reflection and decisions on how to address the tension arising from differences in information-explicitness requirements between the source and the target language ("linguistically and culturally induced information"—see Gile, 2009, Ch. 3). In interpreting, considerable added pressure is associated with stringent time constraints.

Interestingly, at the beginning of scientific investigation into translation, starting in the 1950s, this aspect of the translator's work was given little attention. The initial focus was on lexical and stylistic contrasts between languages (e.g. Mounin and Vinay and Darbelnet's work) and on the related issue of equivalence (e.g. Nida's innovative distinction between formal and dynamic equivalence), and then moved on to philosophical, cultural and sociological issues (e.g. Toury's norms). One exception was Jiří Levý's (1967) formulation of the game-theory-based minimax principle, according to which translators invest as little effort as possible to achieve maximum effect.

Investigators of translation became more interested in the topic with the advent of translation process research, starting in the mid-1980s, and the integration of concepts from cognitive science. Interestingly, Gutt (1991/2000) introduced the notion of (cognitive) processing effort into translation theory via Sperber and Wilson's relevance theory (1986), which basically also postulates optimization of human cognitive investment as a function of expected gain. Gutt quotes Levý, but only refers to the minimax principle once (in a footnote, on page 20), interpreting it and criticizing it in terms of pragmatics. Levý's use of the term seems to refer not to intense cognitive effort but to effort in the more mundane sense of the word, similar to Pym's (2004) effort measurable in "time, hardship, technology costs, interpersonal exchanges".

In cognitive psychology, a key reference discipline for cognitive research into Translation, cognitive effort is viewed as the engaged proportion of limited-capacity central processing (see Tyler et al., 1979). The link between the abstract concept of cognitive effort and the more mundane meaning of effort in everyday life becomes clear when considering the distinction between so-called automatic operations, which are very fast and require little attention, and non-automatic operations, which are far slower and do require attentional resources—and are associated with a feeling of effort (Kahneman, 2011). According to Kahneman, effort is basically unpleasant, and "System 2", the symbolic name he gives to a set of mechanisms that govern non-automatic operations—those which require significant attention—is "lazy", feels uncomfortable when exerting effort and would like nothing more than to hand the control of cognition back to "System 1", which relies on heuristics and performs automatic operations. When humans are forced to deploy effortful activities, as a result of a phenomenon called "ego depletion", they are less inclined to take on new effortful activities later. This intrinsic "laziness" of human cognition, which ties in with Zipf's principle of least effort in human behaviour (1949), is not unrelated to relevance theory. It also draws attention to the potentially important role of motivation, be it under a personal ethical philosophy, under a code of ethics or under the threat of sanctions if sufficient effort is not devoted to achieving quality.

To pioneer psychologists and psycholinguists who showed interest in the cognition of simultaneous interpreting (SI) in the 1960s and early 1970s, the relevance of limitations of cognitive resources in interpreting was clear. Actually, to a large extent, their interest in SI was triggered by curiosity as to how interpreters managed to conduct multiple operations with limited attentional resources under cognitive pressure, for instance under high-speed delivery conditions (see Gerver, 1969). Gerver talks about an information-handling system, which is subject to overload if required to carry out complex processes at too fast a rate, and of evidence that within this system, attention is shared between the input message, processes involved in translating a previous message, and the monitoring of current output. He adds that while under normal conditions attention can be shared between these processes, when the total capacity of the system is exceeded, less attention can be paid to either input or output if interpretation is to proceed at all. Hence, less material is available for recall for reformulation, and more omissions and errors in the output will occur. A similar analysis, initially based on introspection, was developed by Gile a little over a decade later with his Effort models. Kirchhoff (1976), a translation and interpreting teacher with an interest in bilingualism, drew on psycholinguistics to attribute the fundamental challenge that arises when interpreting between structurally dissimilar languages to the need to process larger segments, presumably because of the need to store more information before being able to reformulate the source-speech message. She talks about "cognitive load" and the need to choose "appropriate strategies" as a function of available capacity.

Investigation of interpreting by psychologists and psycholinguists in this early history of interpreting studies was short-lived. Seleskovitch, Lederer and their followers, who developed interpretive theory in the 1970s and 1980s and gained a virtually exclusive influence in the field for close to two decades, were conference interpreters. As practitioners, they were aware of the existence of such cognitive pressure (as evidenced by tactical advice given to students in the classroom on how to handle various difficulties and on the need to focus one's attention on listening rather than note-taking in consecutive interpretation, and on listening rather than reading in simultaneous interpretation with text) but did not incorporate it into their theory, probably in order to avoid giving salience to language issues and what they considered the excessive influence of theories and methods from cognitive psychology. Chernov, who developed his own theory in the USSR in parallel, did so, albeit indirectly, through his idea that anticipation was what made SI possible (Chernov, 1994). Starting in the early 1980s, Gile brought the focus back to cognitive effort with the Effort models in an attempt to account for a large proportion of errors, omissions and infelicities observed and measured in interpreting performance. Other interpreting studies researchers followed, including Tommola (Tommola & Hyönä, 1990; Tommola & Niemi, 1986) and Setton (1998), who adopted relevance theory as one of the pillars of his own model of interpreting, and many others in the following years.

Over the past two decades or so, with the emergence of cognitive translation and interpreting studies (CTIS), the concept of effort, especially cognitive effort,[1] with its implications and measurements, has gained much visibility in research into written translation as well, making it one of the topics where investigators of translation and interpreting meet.

14.2 Effort and Translation performance

14.2.1 The relevance and importance of effort in Translation performance

The most obvious reason for which the topic of effort deserves attention in TIS is the link between the amount of effort invested by the Translator and the quality of his/her performance. The relation is not straightforward. The correlation can be assumed to be strong for low effort investment: a translation done carelessly will most probably exhibit weaknesses, linguistic infelicities if not inaccuracies, and outright errors. As the amount of effort invested in ad hoc information collection, in analytical thinking about the source Text and implications of the Translation brief, and in revision increases, infelicities and errors are likely to recede. What is more difficult to determine is how much quality improvement will ensue from incremental additional efforts. For instance, what is the return of an increasing number of iterations of self-revision and revision by others or of consulting more documentary sources when retrieving thematic, terminological and phraseological information? Beyond a certain point, further effort may not contribute much or may even become counterproductive, though where this point lies can vary greatly.

Another reason for the importance of the topic is that effort, in the sense of the use of one's attentional resources to perform demanding tasks, has become and continues to be an important tool in the investigation of various translation and interpreting phenomena, as illustrated later in this chapter.

Finally, in the light of recent interest in the place of risk and risk analysis in Translation behaviour, and in particular in the links between risk and effort as discussed by Pym (2015), it will be argued here that effort is also an ethical issue (see also Pym, this volume, for a discussion of translation, risk management and cognition).

14.2.2 How much effort investment?

Cognitive psychologists tend to focus on "effort" as it arises in the form of non-automatic operations drawing upon limited available attentional resources. When engaged in such operations, people may only become aware of associated discomfort when it reaches high intensity, as happens regularly in SI. A feeling of discomfort also arises when low-intensity effort is prolonged, as in the preparation of glossaries, in repeated self-revision in translation, or in repeated search for terminological information in external sources.

Both high-intensity and low-intensity effort are found in translation and interpreting, but in TIS, authors do not generally make the distinction. And while "general theoreticians", in particular Levý and Pym, seem to focus on low-intensity effort, investigators interested in Translation cognition focus more on high-intensity effort, i.e. on intensive use of attentional resources (see also Hvelplund, this volume, for a discussion of translation, attention and cognition).

According to what could be termed universals in professional ethics and codes of conduct, in view of the link between effort and quality, it would seem natural for translators and interpreters to invest considerable effort in their work. Levý, however, believes that translators seek to optimize the effort/gain ratio. Taken at face value and without qualifications, such behaviour is ethically debatable. Pym (2015) advocates a principle whereby so-called high-risk translation problems deserve more effort than low-risk translation problems. His use of the term "risk" is problematic: in Pym (2011), he defines it as the probability of an undesired outcome of an action, while in Pym (2015), the term seems to denote stakes. Setting aside this terminological ambiguity, the idea that more effort should be invested in high-stake Text segments makes sense—provided as much effort as is required to achieve acceptable quality is invested in lower-stake segments as well.

Gile's sequential model of translation (Gile, 2009) includes iterative tests of source-text comprehension and of fidelity and acceptability on translation units and aggregates until satisfactory solutions are found. According to this model, which specifically refers to risk and recommends the adoption of solutions with the best combination of loss/gain associated with their respective probabilities of occurrence, effort is only discontinued when satisfactory solutions are found. It was developed for didactic purposes and cannot be taken as a good reflection of reality, not only because in real life, operations it describes as a sequence are often performed with overlapping and backtracking, but also because entirely satisfactory solutions to translation problems are not always found.

When, and how much, do translators and interpreters invest effort as they address Translation challenges in real life? Risk management has the potential to account for some effort-related behaviour, especially in a business context (for instance as regards investment in equipment and translation software, or in human resources management), and to some extent in individual Translation behaviour—in interpreting, minimizing the risk of cognitive interference can be a determinant of coping tactic selection (see Section 14.3.1). However, other factors may be far more important. Professionalism, personal motivation and resistance to fatigue-induced ego depletion as well as socially induced emulation between peers could be stronger determinants, as suggested, for instance, by a recent survey among conference interpreters by Zwischenberger (2017), who found that the strongest reason for satisfaction/dissatisfaction was succeeding/failing to fulfil the respondents' own standards. This point is taken up again later in this chapter.

14.3 TIS research on cognitive effort

14.3.1 Cognitive effort in TIS literature

With the spectacular growth of cognitive investigations into translation over the past two decades, from general considerations of "effort" in the everyday sense of the word, a new focus has been crystallizing around cognitive effort, which is associated with longer information processing time, more pauses, filled or unfilled (e.g. Plevoets & Defrancq, 2016; Shreve et al., 2011), slower target-text production, which includes longer gaze time and longer reading time (e.g. Dragsted,

2012; Schaeffer & Carl, 2014), and physiological changes. In particular, Jakobsen et al. (2007) used task time as a metric to compare the cognitive effort required to render idioms and non-idioms in written translation and sight translation, and found that idiomatic expressions slowed down production in both modes.

Patterns of segmentation have also been correlated with cognitive effort in written translation. In a study using eye tracking and keyloggings, Jakobsen (2011) observes that comprehension, reformulation, and the actual typing and monitoring of the translation product overlap at times, as some of the chunks recorded are too long to be possibly stored in short-term memory, and proposes a six-step "micro-cycle" of motor and cognitive efforts, stressing that some of the steps can be skipped, while some may be repeated again and again.

Relevance theory was integrated into experimental translation process research, as it "postulates an optimal cognitive relation between the minimum processing effort necessary for the generation of the maximum cognitive effects" and "offers a productive way to investigate the role of effort in translation task resolution" (Alves & Gonçalves, 2013, p.108). The cross-linguistic priming effect in written translation has also been taken up in research into translation cognition. Various studies have sought to find out whether translation involving larger cross-linguistic differences takes up more cognitive effort (e.g. Bangalore et al., 2016; Schaeffer et al., 2016; Vandepitte & Hartsuiker, 2011). Schaeffer & Carl (2014) put forward a recursive model of translation consisting of an early priming process based on shared semantic and syntactic representations between the source text (ST) and the target text (TT) ("horizontal translation" based on translinguistic equivalences, which is automatic, and "vertical translation", which is conceptually mediated and requires conscious effort).

One phenomenon that clearly distinguishes interpreting from translation (except for the case of sight translation, a hybrid) is that while in translation, the intensity of expended cognitive effort is relatively low most of the time, and the feeling of discomfort associated with effort mostly only arises if it is prolonged, in interpreting, it can be very high, e.g. when the speaker is fast or speaks with a strong, unfamiliar accent, or when the speech is informationally dense (a list of "problem triggers" is drawn up in Gile, 2009 and corroborated empirically inter alia in Mankauskienė, 2018).

When discussing high-intensity effort in interpreting, it is convenient to use the cognitive psychological construct of working memory (WM), as defined by Baddeley and Hitch (1974) and later elaborated on through different avatars, e.g. by Miyake and Shah, Ericsson, Kintsch, Cowan, and Barrouillet and Camos (for a clear and thorough presentation, see Gieshoff, 2018). This concept refers to some virtual cognitive space where information is stored for up to a few seconds and processed. WM is postulated to have very limited storage space of just a few "chunks" of information, which disappear rapidly if not "refreshed" with some effort, and is associated with "executive functions": inter alia, attention allocation and attention switching.

Intense cognitive effort arises naturally from the simultaneous processing of information coming from the speaker and production of the interpreter's target speech (in the simultaneous mode) or notes (in the consecutive mode). When interpreters feel that they are close to saturation or anticipate the imminent arrival of such a state (when under high cognitive load), they may resort to various coping tactics to relieve the pressure, e.g. buying time to better understand an utterance or decide how to reformulate the message (stalling) or making sure the recipients understand a reference or idea (paraphrasing, explicating and explaining). Many have a cognitive cost insofar as they take up time and further effort, with the associated risk of overloading WM and jeopardizing the fragile balance of attention allocation between Efforts to the detriment of the processing of neighbouring speech segments through a cascade effect (Gile, 2009, Ch. 8; Pointurier-Pournin & Gile, 2012 for signed language interpreting). The specific dynamics of such phenomena, including measurements of the level of cognitive effort expended and reaction chains, are clearly a key to better insight into interpreting cognition.

Gile (2009) considers that while interpreting, the combined attentional requirements of all functional Efforts (reception, production, short-term memory operations and attention management) tend to be close to maximum available capacity, which makes interpreters vulnerable to sudden increases of requirements (through problem triggers) and to attention management errors, as well as to attention fluctuations. This assumption is referred to as the "tightrope hypothesis".

Interpreters are aware of this and, as explained earlier, use various coping tactics (actions with an immediate objective, as opposed to strategies, actions with a wider, longer-term objective) to alleviate cognitive pressure and optimize their target speech. Ideally, they seek to reformulate the maximum amount of information meant by the speaker to be transmitted to the addressees subject to compatibility with the speaker's assumed intention and interests (in the default value of speaker-loyalty—see Gile, 2009), including compliance with applicable social norms.

The selection and implementation of such coping tactics can require much deliberate effort on the part of the interpreter, who may be tempted to opt for the (effortless) omission tactic or for another effort-sparing but sub-optimal tactic in terms of communication efficiency, such as simply repeating a term as it was pronounced by the speaker without translating it.

In other words, in the interpreter's tactical behaviour, conscientiousness (or "professionalism" in the everyday sense of the word) and personal motivation can be determinants of effort investment.

Actually, this applies to low-intensity effort as well, and thus to interpreting strategies such as conference preparation: by acquiring new relevant information about the future meeting to which they have been assigned (e.g. the participants, the stakes, relevant concepts and ideas, and relevant terminological and phraseological information) or by refreshing it in their long-term memory, interpreters expect to prime the information and therefore to need less cognitive effort to understand the speakers' utterances and to reformulate them in the target language. Preparatory investment in low-intensity effort is thus assumed to reduce high-intensity effort requirements when the meeting starts.

Besides these low- and high-intensity efforts related to information, socio-affective effort can also be an important part of interpreting work in public-service settings (e.g. healthcare, courts or asylums), where, at least on one side, stakes are high and principals may be in emotional states that require special attention. Interpreters need to get attuned to the specific needs in the situation at hand and make the right decisions with respect to interventionism, sometimes beyond the standard boundaries defined by professional codes of conduct (this has become a central topic in community interpreting and signed language interpreting circles in the past two decades or so—see e.g. Janzen, 2005).

Since the 1990s, interpreting researchers have been following more closely developments in cognitive science in their investigation of cognitive load and cognitive effort. In particular, referring to Gile's Effort models, Seeber (2011) sought to fine-tune the analysis by drawing on Wickens' multiple-resource model (2002). According to Wickens, contrary to a single-resource model postulated by Kahneman (1973), cognitive operations draw on a shared pool of resources to a small extent but rely mostly on different resource pools depending on the processing stage (i.e. central processing and response processes), the type of process (manual/spatial and vocal/verbal), the processing modality (aural or visual), and within visual processing, on focal vs. ambient vision. In his view, such a model gives a better account of the relative amount of interference between tasks involving WM to store or transform information. Seeber uses this model to calculate the relative (theoretical) amount of cognitive interference from various interpreting-related tasks, namely shadowing, sight translation and SI. Note that Barrouillet & Camos (2012) argue for a different time-based resource-sharing model (TBRS), according to which maintenance of memory traces depends on attentional focusing within a single resource-limited central system.

14.3.2 Quantifying cognitive effort

Being a construct, cognitive effort can only be measured indirectly through indicators. This section lists the main indicators and measurement methods found in research into TIS over the years, with a special focus on recent research.

14.3.2.1 Rating scales

They involve post-hoc questionnaire surveys asking participants to rate the level of cognitive load they experienced when performing a task, under the assumption that increased capacity invested in a task is associated with "subjective feelings of effort or exertion" (O'Donnell & Eggemeier, 1986, 42/7), on which people can report retrospectively (Paas et al., 2003). A study of translation difficulty by Sun and Shreve (2014) is an attempt to apply subjective rating to TIS. In the study, a four-category rating scale (Mental Demand, Effort, Frustration level and Performance) adapted from the NASA-TLX (Task Load Index) developed by Hart and Staveland (1988) was used for evaluation of post-hoc subjective workload. Sun (2018) believes the method can also be employed to measure cognitive load in post-editing of machine translation. Chen (2017a) points out that the rating scale techniques have been widely used in research related to Cognitive Load Theory, and their potential in TIS is still to be tapped.

14.3.2.2 Behavioural indicators

14.3.2.2.1 Hesitation pauses

Lacruz and Shreve (2014), which explores "the promise of simple pause metrics as tools for measuring cognitive effort during post-editing" (p. 266), puts forward the average pause ratio (average time per pause ÷ average time per word) as a potential measure for cognitive effort in post-editing, suggesting that high pause densities during post-editing, which tend to produce low average pause times, indicate high levels of cognitive effort. The measure is tested out in a few case studies, which demonstrate a significant negative correlation between the average pause ratio and the event-to-word ratio.

In a large corpus of interpreted and non-interpreted texts, Plevoets & Defrancq (2016) operationalize informational load, which generates cognitive load, in terms of delivery rate, lexical density, percentage of numerals and average sentence length. Interpreted texts were analysed based on the interpreter's output and compared with the input of non-interpreted texts, and the effect of source-Text features was measured by counting the occurrence rate of the speech disfluency u(h)m. Interpreters produced significantly more uh(m)s than non-interpreters in original speeches. The authors attribute this difference mainly to the effect of lexical density on the output side. The main source predictor of uh(m)s in the target Text was shown to be the delivery rate of the source Text.

14.3.2.2.2 Dependency distance and lexical simplification

Liang et al. (2017) used reduction of dependency distance as an indicator of cognitive effort. Dependency distance (DD), the distance (the number of words) between two syntactically related words in a sentence, is assumed to be correlated with cognitive load during speech production, because to produce a correct sentence, traces of the first word in the relevant syntactic relation have to be kept in WM until the last syntactically dependent word is produced. By virtue of the law of least effort, speech producers are assumed to tend to minimize DD. Liang et al.

compared mean DDs from SI, consecutive interpreting (CI) and written translations and found them to be largest for translated texts, second largest for SI and smallest for CI. They conclude that CI may entail heavier cognitive load than SI. This inference is problematic, since under the heavy cognitive pressure of simultaneous interpretation, interpreters are not as free to reorganize their information into sentences as they are in consecutive interpretation, because such reorganization would entail much waiting and possibly saturate WM. A similar comment applies to a related study by Lv and Liang (2018) using lexical simplification (choosing more frequent words, shorter words, etc.) as an indicator of cognitive effort, which concludes that CI may be associated with higher cognitive effort based on the finding that lexical (and associated) simplification was larger for CI than for SI.

14.3.2.2.3 Ear-voice span and eye-key span

Ear-voice span, the time lag between source-speech segments and corresponding target-speech segments, was a focus of interest of early investigators working on SI (e.g. Oléron & Nanpon, 1965; Gerver, 1969; Goldman-Eisler, 1967), perhaps partly because it actually measured the "simultaneousness" of SI, but also because it was related to questions about the storage capacity of WM. Also note that in cognitive psychology and psycholinguistics, reaction times are a central dependent variable in experiments, and time lag was a likely candidate to be used as an indicator of cognitive effort. Lee (2003) studied a corpus of authentic English to Korean simultaneously interpreted speeches and found that longer source-language sentences induced longer lags, reduced accuracy and longer intra-sentence pauses in the target speech, all of which suggest that the lag indicates more processing effort. However, longer lags can also generate higher cognitive load insofar as they force the interpreter to store more information from the incoming speech before reformulation becomes possible. This ambiguity is also found in Chen's use of ear-pen span (Chen, 2017b), the lag between the source speech and the notes being taken by the interpreter.

Combining keylogging and eye-tracking technologies has facilitated accurate measurement of time lag in translation. Dragsted and Hansen (2008) introduced the concept of eye-key span, the time lag between the first fixation on source-text words and the beginning of the typing of the corresponding target-text words. Translation and interpreting students were recruited to translate from English into Danish. The authors claim that longer eye-key span is associated with "problem words", such as metaphorical expressions, that require more cognitive effort, as these words also elicit more regressive eye movements and longer accumulated fixation durations.

14.3.2.2.4 Eye movement patterns

When reading is involved, the number of fixation counts, the duration of fixations and the number of regressive saccades are considered as indicating cognitive effort (e.g. Just & Carpenter, 1980). Sjørup (2008) found that the processing of metaphors was associated with longer fixations than the processing of literary expressions. Jakobsen and Jensen (2008) investigated cognitive processing by professional translators and translation students in four different types of reading tasks—reading for comprehension, reading in preparation for translation, sight translating, and reading while typing a translation. For both groups of subjects, sight translation and written translation were associated with significantly longer task time, more fixation counts and longer total gaze time (total duration of all fixations) than the two reading tasks that did not involve actual translation. Eye movement patterns and task time also suggest that all four tasks were more effortful for the students than for the professionals, and in written translation, that students allocated more effort to reading the source text, while the professionals invested more visual attention in the target text. A study by Pavlović and Jensen (2009), with professional translators and translation students with native Danish and English as their L2, investigated directionality in translation using eye movement patterns as well as pupillometry to measure cognitive effort. The results corroborated the hypothesis that in both translation directions, processing the target text demands more cognitive effort than processing the source text: it was associated with longer gaze time, longer average fixation duration and larger pupil dilation. However, only pupil dilation values and task time supported the hypothesis that translating from L1 to L2 demands more cognitive effort than translating from L2 to L1. In both groups, L2-to-L1 translation was associated with longer gaze time. Chang (2009) used similar eye-tracking metrics to compare cognitive effort in L1-to-L2 and L2-to-L1 translation. The study, which includes two language combinations, recruited two groups of translation students, one with Mandarin and English, and one with Spanish and English. They were asked to type-copy a text in L1, type-copy a text in L2, type-translate from L2 to L1 and type-translate from L1 to L2. The results suggest that for both language combinations, L1-to-L2 translation requires more cognitive effort, as it is associated with significantly larger pupil dilation, a significantly larger number of fixations, higher fixation frequency, higher blink frequency and longer task time. Eye movement patterns also indicate that for both translation directions, more visual attention was allocated to the TT.

14.3.2.3 *Brain imaging and associated techniques*

14.3.2.3.1 Electroencephalography (EEG)

Electroencephalography (EEG) records electrical cerebral activities from the scalp and has been identified as a physiological index for measuring online cognitive effort (e.g. Antonenko et al., 2010). Compared with neuroimaging techniques based on blood-flow measurement as listed later, EEG provides high temporal resolution but low spatial resolution. It is therefore suitable for assessing online task-evoked cognitive effort, but not for drawing inferences on the locations of brain activation.

Petsche et al. (1993) investigated directionality in SI by comparing EEG changes in three professional interpreters interpreting mentally (to prevent physiological artefacts from speech utterance) into their L1 and L2 and compared coherence of the EEG signals as an indicator of cognitive effort, which turned out to be larger when interpreting into L2. Lachaud (2011) combined EEG (for measuring time-frequency power), eye tracking (for measuring fixation duration and pupillary responses) and keylogging (for measuring response time) to look at the cognitive effort expended when transcoding between English and Norwegian at the word level when Deceptive Cognates (false friends), True Cognates and Non-cognates are involved. Translation involving Deceptive Cognates was found to be the most effortful: it was associated with significantly larger time-frequency power, longer total fixation duration, bigger pupil size amplitude variation and longer reaction times. EEG and keylogging data also suggest that transcoding Non-cognates requires significantly more cognitive effort than transcoding True Cognates. Eye-tracking data did not demonstrate significant differences between the two.

An ERP study by Koshkin et al. (2018) is the first published work testing Gile's Effort models in a "naturalistic setting requiring the participants to interpret continuous prose overtly" using neuroimaging techniques. Event-related potentials (ERPs), which reflect brain responses to events, are calculated by averaging a continuous EEG signal over a large number of trials (e.g. Antonenko et al., 2010). The study tried to find out whether more resources given to WM would result in reduced availability of attention for listening. Based on previous functional magnetic resonance imaging (fMRI) studies (e.g. Corbetta & Shulman, 2002; Mayer et al., 2007) claiming that attention and WM are subserved by overlapping brain regions, it was assumed that changes in listening effort and WM load would be reflected by changes in the amplitude of ERPs (N1 and P1). Interpreters were asked to interpret speeches from English into Russian and vice versa. The source and target speeches were transcribed and manually time-stamped. The WM load of each unit was defined by ear-voice span and estimated using the number of content words in the unit, the number of content words in the unit weighted by their frequency, and all the words in the unit weighted by their respective syllabic lengths. A larger negativity in the P1 and N1 amplitude was found when the WM load was lower and vice versa, suggesting that the listening effort and the WM effort competed for cognitive resources.

14.3.2.3.2 Positron emission tomography (PET)

PET looks at blood flow and oxygen consumption in different parts of the body, including the brain. Increased blood flow and metabolism imply increased activities and effort. It is a relatively invasive method compared with EEG, functional near-infrared spectroscopy (fNIRS) and fMRI, if only because it involves injecting a tracer into a vein.

Rinne et al. (2000) used PET to measure brain activation in professional interpreters during SI between Finnish (L1) and English (L2) in both directions. Interpreting into L1 elicited left frontal activation, while interpreting into L2 elicited much more extensive left fronto-temporal activation. The results indicate that SI activates predominantly left-hemispheric structures and imply that translating into L2 demands more cognitive effort.

14.3.2.3.3 Functional magnetic resonance imaging (fMRI)

fMRI is a non-invasive neuroimaging technology using magnetic resonance to measure brain activity by detecting changes associated with blood flow. The technology provides high spatial resolution and allows precise localization of activated areas of the brain. Larger amplitude of activations may imply enhanced cognitive effort. Ahrens et al. (2010) conducted an fMRI experiment with student interpreters to compare brain activity between two conditions—SI from Spanish (L2) to German (L1) and free speech production in German. SI and free speech production elicited different activations in the brain. Chang (2009) used fMRI to investigate novice interpreters' cognitive effort during L1-to-L2 and L2-to-L1 interpreting. Interpreters with Mandarin as L1 and English as L2 were recruited to perform sight interpreting from L1 to L2 and vice versa during the scan. The results suggest that L1-to-L2 translation was more effortful, as it involved more cortical areas than L2-to-L1 translation.

14.3.2.3.4 Functional near-infrared spectroscopy (fNIRS)

fNIRS is a portable neuroimaging technique that makes non-invasive optical measurements of blood flow in the brain. As fMRI does, it detects brain activation by measuring blood oxygenation changes. However, compared with fMRI and PET, which require the subject to lie still during the scan, fNIRS has the potential to allow more ecologically valid experiments for translation and interpreting studies. Lin et al. (2018) combined behavioural measures and fNIRS technology to examine the cognitive effort associated with pairing, transphrasing and non-translation. Students of interpreting were asked to perform three tasks—a pairing task, a transphrasing task and a non-translation task. The stimuli were visually presented two-character cultural-specific items in Chinese (L1), which subjects were required to interpret into English (L2). The area of interest was the left prefrontal cortex (PFC), which includes Broca's area and has been associated with lexical search, semantic processing, bilingual processing, production of speech and cognitive control. Pairing elicited the most intense activation, which was localized in Broca's area, while transphrasing induced the longest, most extensive activation overall in the left PFC, suggesting that the latter is most likely to lead to cognitive overload in SI. However, the ecological validity of the study is limited, as the stimuli used in the study were at word level. And as the study only investigated one part of the brain, the results do not give a full picture of the cognitive effort involved in the task.

14.3.2.4 Pupillometry

It is widely accepted that the pupil reacts to cognitive activity (e.g. Kahneman et al., 1969). Pupil dilation, which is usually measured with an eye tracker, can therefore be seen as a response to increased cognitive load (e.g. Holmqvist et al., 2011; Rayner et al., 2006). Pupillometry is less invasive and more convenient than some of the other physiological measures presented here. Tommola and Niemi (1986) used pupillometry to investigate the impact of source-Text syntactical complexity on cognitive load. Tommola and Hyönä (1990) and Hyönä et al. (1995) further tested the method by comparing pupil dilation levels associated with three language processing tasks, namely listening, shadowing and SI, which clearly require different levels of mental effort. Relative pupillary responses to the three tasks were consistent with the hypothesized differences in task difficulty. O'Brien (2006) recruited professional translators to translate an English text into their L1s using a translation memory tool. Their pupillary responses were recorded and analysed in combination with other measures, including processing speeds and retrospective protocols to compare the cognitive effort required to process four types of translation memory matches into German or French, namely "No Match", "Fuzzy Match", "Machine Translation Match" and "Exact Match". The results indicate that No Match, which means the translator needs to translate from scratch, required the largest cognitive effort, as suggested by the largest average pupil dilation and the lowest processing speed. Exact Match required the least effort, with the smallest average pupil dilation and the fastest processing speed. Pavlović and Jensen (2009) and Chang (2009) applied pupillometry as one of the indicators of cognitive effort in their studies of translation directionality. Seeber and Kerzel (2011) used pupillometry to measure cognitive effort during SI from German into English. Pupil dilation elicited by interpreting German verb-final structures into English was significantly larger than that induced by interpreting verb-initial structures, suggesting that translating between non-parallel structures takes up more cognitive effort than translating between parallel structures. Hvelplund's (2011) study, which combined keylogging and eye-tracking technologies to investigate the allocation of cognitive resources in the translation process, applied an array of measures, including task time, fixation duration and retrospection, but used pupillometry as the sole indicator to measure cognitive effort. Professional translators and student translators participated in a series of experiments involving translating English (L2) texts into Danish (L1) under two conditions—with or without time constraints. The results confirmed that under all conditions, translation was more effortful for the students than for the professionals, and for both groups, translating under time pressure generated more cognitive effort than translating without time pressure. However, no significant difference in pupil dilation was found between translating easier texts and more difficult texts, though the subjects did confirm that one text was more difficult than the other in post-hoc interviews. This suggests that pupillary responses might not always be sensitive enough to reflect cognitive load differences and the effort incurred by translating texts of different levels of complexity. As pointed out by Gieshoff (2018), pupil dilation also responds to sources of arousal other than cognitive effort. Hvelplund also found that more cognitive effort was involved in target-text processing than in source-text processing or source-text/target-text parallel processing.

14.4 Methodological and other challenges in the use of the concept of effort in TIS

Scientific research is effective when it discovers/uncovers facts, when it produces theories that help explain them and predict what is yet uncovered, and when it develops concepts and methods that help uncover more facts and develop and test theories. The feeling of effortfulness in translation and interpreting is part of reality. The concept of cognitive effort has demonstrated its usefulness in developing theories that explain phenomena observed in translation and interpreting. How predictive these theories are depends on the level of accuracy and detail sought. For instance, the Effort models and the associated tightrope hypothesis have been shown to successfully predict holistic phenomena such as increases in the rate of errors, omissions and infelicities in interpreting under certain working conditions (in particular high delivery speed) and in the presence of well-identified problem triggers. However, the precision of their predictions is limited because of their very nature as a holistic conceptual framework. The tightrope hypothesis did not quantify the interpreters' "closeness" to saturation. It only indicated functionally that they tended to work close enough to saturation to become vulnerable to unexpected hikes in attention requirements and to attention management errors. More specific cognitive theories could do better, provided researchers find a way to offset the high variability arising from the large number of highly influential parameters, which are determinants of performance and are difficult to control in ecologically valid experiments.

Measurements of cognitive effort per se are also of potential use in uncovering interesting correlations; for example, if it could be shown, beyond general principles as postulated by existing theories, that certain tactical behaviour patterns and certain types of Translation quality deteriorations tend to occur at specific thresholds of cognitive load or cognitive effort. Many indicators provide relative data but not interval scales against which particular Translational phenomena can be identified. What is the level of effort at which errors of specific types occur, and what is the level of effort at which Translators give up trying hard and adopt less effortful but also less efficient strategies and tactics as regards the quality of the final product? Answers may come from triangulation studies measuring Translation behaviour and cognitive effort while eliciting retrospective reports (as in Gumul, 2018, in which many retrospective comments on 240 interpretations by students turned out to point to processes described by the Effort models).

Yet another limitation of many quantitative cognitive effort indicators such as pupillometry, gaze duration and hesitation pauses in their use to test ideas and theories about interpreting cognition (the problem may be less conspicuous in translation, where source-text reading is easier to separate from target-text writing) is that while they may be able to detect variations in cognitive effort intensity, they are not specific enough to point to the particular sub-processes or process components that caused them. How does one know whether increased cognitive effort is caused by a comprehension difficulty, a production difficulty or by the retrieval of information from memory, and what is the triggering speech segment? When considering ear-voice span, or ear-pen span in the case of CI, as in Chen (2017b), lag time can be taken as an indicator of cognitive effort possibly arising from processing difficulty associated with a previously heard source-speech segment. However, it can also indicate a tactical pause, which is not linked to any such difficulty, for instance when the interpreter reviews previous notes to make sure s/he will be able to read them when reformulating the speech.

Brain imaging techniques may provide one solution to the problem, if they show that distinct processes tend to mobilize specific areas of the brain or generate specific activity. However, in many cases, a far simpler solution can also help, again in the form of retrospective reports. Cued retrospection, as already used with keylogging techniques a long time ago, inaccurate and incomplete as it is, could well help interpret more specifically causes of increased cognitive effort as detected by objective indicators.

14.5 Concluding remarks

In human Translation, effort cannot be reduced to one parameter in a productivity or profitability equation. On the professional side and on the didactic side, it is a determinant of quality, and as such, an ethical obligation—up to a certain point. It is also, for at least a decade or so in every Translator's early career, an investment in future expertise (Ericsson, 1996).

On the research side, it is also a useful concept for the investigation of Translation behaviour, including tactics, strategies and sub-optimal performance (e.g. errors, omissions and infelicities) and also socio-affective behaviour determinants.

It is particularly useful in research into Translation cognition, especially with the help of theories from cognitive sciences and with relevant technology that provides physiological evidence of mental activity. Pupillometry and brain imaging techniques, which are becoming less and less invasive, hold much promise, but when aiming for the best results, it is important to take on board the specific technical and social environment of Translation and to consider triangulating these "objective" techniques with more qualitative techniques, in particular retrospection, which can be of much help when seeking to interpret quantitative data correctly.

Note

1 In this chapter, "cognitive load" is used to denote the cognitive pressure that a process imposes by virtue of environmental and task-specific factors, while "cognitive effort" refers to the effort actually expended by the Translator when performing the task.

Further reading

Chen, S. (2017). The construct of cognitive load in interpreting and its measurement. *Perspectives, 25*(4), 640–657.
A good overview and discussion of cognitive load in interpreting and beyond.

Gieshoff, A. C. (2018). The impact of audio-visual speech input on work-load in simultaneous interpreting. [Doctoral Dissertation, Universität Mainz]. Germersheim: Johannes Gutenberg-Universität Mainz.
This doctoral dissertation includes very well-written discussions of cognitive effort and associated theories.

Gile, D. (2009). *Basic concepts and models for interpreter and translator training* (revised ed.). Amsterdam: John Benjamins.
This book, which is largely devoted to interpreting and translation cognition, offers further simple-language explanations on points mentioned in this chapter.

Kahneman, D. (2011). *Thinking, fast and slow*. New York: Farrar, Straus and Giroux.
A very fundamental book, which explains many facets of less than rational decisions in humans, but as regards this chapter more specifically, it is an excellent introduction to effortful and less effortful thinking and acting in humans.

Li, D., Lei, V. L. C., & He, Y. (Eds.). (2019). *Researching cognitive processes of translation*. Cham, Heidelberg, New York, Dordrecht & London: Springer.
A good overview of existing research and technologies available, including technical details, and reflections about future directions.

References

Ahrens, B., Kalderon, E., Krick, C. M., & Reith, W. (2010). fMRI for exploring simultaneous interpreting. In D. Gile, G. Hansen, & N. K. Pokorn (Eds.). *Why translation studies matters* (pp. 237–248). Amsterdam & Philadelphia: John Benjamins.
Alves, F., & Gonçalves, J. L. (2013). Investigating the conceptual-procedural distinction in the translation process: A relevance-theoretic analysis of micro and macro translation units. *Target. International Journal of Translation Studies, 25*(1), 107–124.
Antonenko, P., Paas, F., Grabner, R., & Van Gog, T. (2010). Using electroencephalography to measure cognitive load. *Educational Psychology Review, 22*(4), 425–438.

Baddeley, A. D., & Hitch, G. (1974). Working memory. *Psychology of Learning and Motivation, 8*, 47–89.

Bangalore, S., Behrens, B., Carl, M., Ghankot, M., Heilmann, A., Nitzke, J., & Sturm, A. (2016). Syntactic variance and priming effects in translation. In M. Carl, S. Bangalore, & M. Schaeffer (Eds.), *New directions in empirical translation process research* (pp. 183–210). Cham, Heidelberg, New York, Dordrecht & London: Springer.

Barrouillet, P., & Camos, V. (2012). As time goes by: Temporal constraints in working memory. *Current Directions in Psychological Science, 21*(6), 413–419.

Chang, C.Y. (2009). Testing applicability of eye-tracking and fMRI to translation and interpreting studies: An investigation into directionality [Unpublished doctoral dissertation]. Imperial College London, United Kingdom.

Chen, S. (2017a). The construct of cognitive load in interpreting and its measurement. *Perspectives, 25*(4), 640–657.

Chen, S. (2017b). Note-taking in consecutive interpreting: New data from pen recording. *Translation & Interpreting, 9*(1), 4–23.

Chernov, G. V. (1994). Message redundancy and message anticipation in simultaneous interpretation. In S. Lambert, & B. Moser-Mercer (Eds.), *Bridging the gap: Empirical research in simultaneous interpretation* (pp.139–153). Amsterdam & Philadelphia: John Benjamins.

Corbetta, M., & Shulman, G. L. (2002). Control of goal-directed and stimulus-driven attention in the brain. *Nature Reviews Neuroscience, 3*(3), 201–214.

Dragsted, B. (2012). Indicators of difficulty in translation—correlating product and process data. *Across Languages and Cultures, 13*(1), 81–98.

Dragsted, B., & Hansen, I. G. (2008). Comprehension and production in translation: A pilot study on segmentation and the coordination of reading and writing processes. In S. Göpferich, A. L. Jakobsen, & I. M. Mees (Eds.), *Looking at eyes: Eye-tracking studies of reading and translation processing* (Copenhagen Studies in Language, 36, 9–29). Copenhagen: Samfundslitteratur.

Ericsson, K. A. (Ed.). (1996). *The road to excellence: The acquisition of expert performance in the arts and sciences, sports and games*. Mahwah, NJ: Lawrence Erlbaum Associates.

Gerver, D. (1969). The effects of source language presentation rates on the performance of simultaneous conference interpreting. In E. Foulke (Ed.), *Proceedings of the second Louisville Conference on rate and/or frequency-controlled speech* (pp.162–184). Louisville, Kentucky: Center for Rate-controlled Recordings, University of Louisville.

Gieshoff, A. C. (2018). The impact of audio-visual speech input on work-load in simultaneous interpreting [Doctoral dissertation, Universität Mainz]. Germersheim: Johannes Gutenberg-Universität Mainz.

Gile, D. (2009). *Basic concepts and models for interpreter and translator training* (revised ed.). Amsterdam: John Benjamins.

Goldman-Eisler, F. (1967). Sequential temporal patterns and cognitive processes in speech. *Language and Speech, 10*(3), 122–132.

Gumul, E. (2018). Searching for evidence of Gile's Effort Models in retrospective protocols of trainee simultaneous interpreters. *Między Oryginałem a Przekładem, 24*(42), 17–39. (Special issue: *Points of view on translator and interpreter education* edited by A. Jankowska, O. Mastela, & Ł. Wiraszka).

Gutt, E. (1991/2000). *Translation and relevance: Cognition and context* (2nd ed.). Manchester: St. Jerome.

Hart, S. G., & Staveland, L. E. (1988). Development of NASA-TLX (Task Load Index): Results of empirical and theoretical research. *Advances in Psychology, 52*, 139–183.

Holmqvist, K., Nyström, M., Andersson, R., Dewhurst, R., Jarodzka, H., & Van de Weijer, J. (2011). *Eye tracking: A comprehensive guide to methods and measures*. Oxford: Oxford University Press.

Hvelplund, K. T. (2011). Allocation of cognitive resources in translation: An eye-tracking and key-logging study. [Doctoral dissertation]. Copenhagen: Samfundslitteratur. https://static-curis.ku.dk/portal/files/131448126/2011_Allocation_of_cognitive_resources_in_translation_Hvelplund.pdf

Hyönä, J., Tommola, J., & Alaja, A. M. (1995). Pupil dilation as a measure of processing load in simultaneous interpretation and other language tasks. *The Quarterly Journal of Experimental Psychology Section A, 48*(3), 598–612.

Jakobsen, A. L. (2011). Tracking translators' keystrokes and eye movements with Translog. In C. Alvstad, A. Hild, & T. Elisabet (Eds.). *Methods and strategies of process research: Integrative approaches in translation studies* (pp. 37–55). Amsterdam: John Benjamins.

Jakobsen, A. L., & Jensen, K. T. H. (2008). Eye movement behaviour across four different types of reading task. In S. Göpferich, A. L. Jakobsen, & I. M. Mees (Eds.), *Looking at eyes: Eye-tracking studies of reading and translation processing* (Copenhagen Studies in Language, 36, 103–124). Copenhagen: Samfundslitteratur.

Jakobsen, A. L., Jensen, K. T. H., & Mees, I. M. (2007). Comparing modalities: Idioms as a case in point. In F. Pöchhacker, A. L. Jakobsen, & I. M. Mees (Eds.), *Interpreting studies and beyond: A tribute to Miriam Shlesinger* (Copenhagen Studies in Language, 35, 217–249). Copenhagen: Samfundslitteratur.
Janzen, T. (Ed.). (2005). *Topics in signed language interpreting*. Amsterdam: John Benjamins.
Just, M. A., & Carpenter, P. A. (1980). A theory of reading: From eye fixations to comprehension. *Psychological Review, 87*(4), 329.
Kahneman, D. (1973). *Attention and effort*. Englewood Cliffs, NJ: Prentice-Hall.
Kahneman, D. (2011). *Thinking, fast and slow*. New York: Farrar, Straus and Giroux.
Kahneman, D., Tursky, B., Shapiro, D., & Crider, A. (1969). Pupillary, heart rate, and skin resistance changes during a mental task. *Journal of Experimental Psychology, 79*(1), 164–167.
Kirchhoff, H. (1976). Das Simultandolmetschen: Interdependenz der Variablen im Dolmetschprozess, Dolmetschmodelle und Dolmetschstrategien. In H. W. Drescher, & S. Scheffzek (Eds.), *Theorie und Praxis des Übersetzens und Dolmetschen* (pp. 59–71). Frankfurt am Main: Peter Lang.
Koshkin, R., Shtyrov, Y., Myachykov, A., & Ossadtchi, A. (2018). Testing the efforts model of simultaneous interpreting: An ERP study. *PLOS ONE, 13*(10), e0206129.
Lacruz, I., & Shreve, G. M. (2014). Pauses and cognitive effort in post-editing. In S. O'Brien, L. W. Balling, M. Carl, & M. Simard (Eds.), *Post-editing of machine translation: Processes and applications* (pp. 246–273). Newcastle upon Tyne, UK: Cambridge Scholar.
Lachaud, C. M. (2011). EEG, EYE and KEY: Three simultaneous streams of data for investigating the cognitive mechanisms of translation. In S. O'Brien (Ed.), *Cognitive explorations of translation* (pp. 131–153). London & New York: Continuum.
Lee, T. H. (2003). Tail-to-tail span: A new variable in conference interpreting research. *Forum, 1*(1), 41–62.
Levý, J. (1967). Translation as a decision process. In *To honor Roman Jakobson. Janua Linguarum Series Major 33*, 1171–1182. Berlin: Mouton de Gruyter.
Li, D., Lei, V. L. C., & He, Y. (Eds.). (2019). *Researching cognitive processes of translation*. Cham, Heidelberg, New York, Dordrecht & London: Springer.
Liang, J., Fang, Y., Lv, Q., & Liu, H. (2017). Dependency distance differences across interpreting types: Implications for cognitive demand. *Frontiers in Psychology, 8*, 2132. www.ncbi.nlm.nih.gov/pmc/articles/PMC5733006/
Lin, X., Lei, V. L. C., Li, D., & Yuan, Z. (2018). Which is more costly in Chinese to English simultaneous interpreting, "pairing" or "transphrasing"? Evidence from an fNIRS neuroimaging study. *Neurophotonics, 5*(2), 025010.
Lv, Q., & Liang, J. (2018). Is consecutive interpreting easier than simultaneous interpreting?—A corpus-based study of lexical simplification in interpretation. *Perspectives, 27*(1). doi: 10.1080/0907676X.2018.1498531
Mankauskienė, D. (2018). Sinchroniniam vertimui iš anglų kalbos į lietuvių kalbą būdingi kliuviniai (Problem triggers in simultaneous interpreting from English into Lithuanian) [unpublished doctoral dissertation]. Vilnius University, Lithuania.
Mayer, J. S., Bittner, R. A., Nikolić, D., Bledowski, C., Goebel, R., & Linden, D. E. (2007). Common neural substrates for visual working memory and attention. *Neuroimage, 36*(2), 441–453.
O'Brien, S. (2006). Eye-tracking and translation memory matches. *Perspectives. Studies in translatology, 14*(3), 185–205.
O'Donnell, R. D., & Eggemeier, F. T. (1986). Workload assessment methodology. In K. R. Boff, L. Kaufman, & J. P. Thomas (Eds.). *Handbook of perception and human performance*. Vol. 2. Cognitive processes and performance (42/1–42/49). Hoboken, NJ: Wiley.
Oléron, P., & Nanpon, H. (1965). Recherches sur la traduction simultanée. *Journal de psychologie normale et pathologique, 62*(1), 73–94.
Paas, F., Tuovinen, J. E., Tabbers, H., & Van Gerven, P. W. (2003). Cognitive load measurement as a means to advance cognitive load theory. *Educational Psychologist, 38*(1), 63–71.
Pavlović, N., & Jensen, K. (2009). Eye tracking translation directionality. *Translation Research Projects, 2*, 93–109.
Petsche, H., Etlinger, S. C., & Filz, O. (1993). Brain electrical mechanisms of bilingual speech management: An initial investigation. *Electroencephalography and Clinical Neurophysiology, 86*(6), 385–394.
Plevoets, K., & Defrancq, B. (2016). The effect of informational load on disfluencies in interpreting. *Translation and Interpreting Studies. The Journal of the American Translation and Interpreting Studies Association, 11*(2), 202–224.
Pointurier-Pournin, S., & Gile, D. (2012). Les tactiques de l'interprète en langue des signes face au vide lexical: une étude de cas. *Jostrans*, 17, 164–183.

Pym, A. (2004). Text and risk in translation. In M. Sidiropoulou, & A. Papaconstantinou (Eds.), *Choice and difference in translation—The specifics of transfer* (pp. 27–42). Athens: University of Athens.

Pym, A. (2011). Translation research terms: A tentative glossary for moments of perplexity and dispute. In A. Pym (Ed.), *Translation research projects 3* (pp. 75–110). Tarragona: Intercultural Studies Group.

Pym, A. (2015). Translating as risk management. *Journal of Pragmatics, 85*, 67–80.

Rayner, K., Chace, K. H., Slattery, T. J., & Ashby, J. (2006). Eye movements as reflections of comprehension processes in reading. *Scientific Studies of Reading, 10*(3), 241–255.

Rinne, J. O., Tommola, J., Laine, M., Krause, B. J., Schmidt, D., Kaasinen, V., & Sunnari, M. (2000). The translating brain: cerebral activation patterns during simultaneous interpreting. *Neuroscience Letters, 294*(2), 85–88.

Schaeffer, M., & Carl, M. (2014). Measuring the cognitive effort of literal translation processes. In U. Germann, M. Carl, P. Koehn, G. Sanchis-Trilles, F. Casacuberta, R. Hill, & S. O'Brien (Eds.), *Proceedings of the workshop on humans and computer-assisted translation* (pp. 29–37). Stroudsburg, PA: Association for Computational Linguistics.

Schaeffer, M., Dragsted, B., Hvelplund, K. T., Balling, L. W., & Carl, M. (2016). Word translation entropy: Evidence of early target language activation during reading for translation. In M. Carl, S. Bangalore, & M. Schaeffer (Eds.), *New directions in empirical translation process research* (pp. 183–210). Cham, Heidelberg, New York, Dordrecht & London: Springer.

Seeber, K. (2011). Cognitive load in simultaneous interpreting: Existing theories—new models. *Interpreting, 13*(2), 176–204.

Seeber, K. G., & Kerzel, D. (2011). Cognitive load in simultaneous interpreting: Model meets data. *International Journal of Bilingualism, 16*(2), 228–242.

Setton, R. (1998). *Simultaneous interpretation: A cognitive-pragmatic analysis*. Amsterdam & Philadelphia: John Benjamins.

Shreve, G. M., Lacruz, I., & Angelone, E. (2011). Sight translation and speech disfluency: Performance analysis as a window to cognitive translation process. In C. Alvstad, A. Hild, & E. Tiselius (Eds.), *Methods and strategies of process research: Integrative approaches in translation studies* (pp. 93–120). Amsterdam & Philadelphia: John Benjamins.

Sjørup, A. C. (2008). Metaphor comprehension in translation. In S. Göpferich, A. L. Jakobsen, & I. M. Mees (Eds.), *Looking at eyes: Eye-tracking studies of reading and translation processing* (Copenhagen Studies in Language, 36, 53–78). Copenhagen: Samfundslitteratur.

Sperber, D., & Wilson, D. (1986). *Relevance: Communication and cognition*. Oxford: Blackwell.

Sun, S. (2018). Measuring difficulty in translation and post-editing: A review. In D. Li, V. L. C. Lei, & Y. He (Eds.), *Researching cognitive processes of translation* (pp. 139–168). Singapore: Springer.

Sun, S., & Shreve, G. M. (2014). Measuring translation difficulty: An empirical study. *Target. International Journal of Translation Studies, 26*(1), 98–127.

Tommola, J., & Hyönä, J. (1990). Mental load in listening, speech shadowing and simultaneous interpreting: A pupillometric study. In J. Tommola (Ed.), *Foreign language comprehension and production* (pp. 179–188). Turku: AFinLA.

Tommola, J., & Niemi, P. (1986). Mental load in simultaneous interpreting: An on-line pilot study. In L. S. Evenson (Ed.), *Nordic research in text linguistics and discourse analysis* (pp.171–184). Trondheim: Tapir.

Tyler, S. W., Hertel, P. T., McCallum, M. C., & Ellis, H. C. (1979). Cognitive effort and memory. *Journal of Experimental Psychology: Human Learning and Memory, 5*(6), 607–617.

Vandepitte, S., & Hartsuiker, R. J. (2011). Metonymic language use as a student translation problem: Towards a controlled psycholinguistic investigation. In C. Alvstad, A. Hild, & E. Tiselius (Eds.), *Methods and strategies of process research: Integrative approaches in translation studies* (pp. 67–92). Amsterdam: John Benjamins.

Wickens, C. D. (2002). Multiple resources and performance prediction. *Theoretical Issues in Ergonomics Science, 3*(2), 159–177.

Zipf, G. K. (1949). *Human behavior and the principle of least effort*. Cambridge, MA: Addison-Wesley Press Inc.

Zwischenberger, C. (2017). Professional self-perception of the social role of conference interpreters. In M. Biagini, M. S. Boyd, & C. Monacelli (Eds.), *The changing role of the interpreter: Contextualizing norms, ethics and quality standards* (pp. 52–73). New York & Oxon: Routledge.

15

Translation, attention and cognition

Kristian Hvelplund

15.1 Introduction: Attention as a cognitive process

When we attend to something, we focus our conscious awareness on a certain object while we "filter out" information by suppressing and ignoring non-relevant information. In psychology, the study of attention dates back to the 1890s, when American psychologist William James proposed a definition of attention:

> [Attention] is the taking possession by the mind in clear and vivid form, of one out of what seem several simultaneously possible objects or trains of thought. Focalization, concentration, of consciousness are of its essence. It implies withdrawal from some things in order to deal effectively with others.
>
> *William James, 1890, p. 403*

Attention involves focusing on specific environmental stimuli while ignoring other stimuli. Attention to specific aspects of a person's environment is necessary to perform tasks: when we read, we pay attention to the book's letters and words, possibly ignoring or suppressing sounds that could grab our attention, and when we engage in spoken conversation, we listen to the person's words while we ignore other non-relevant sounds. While there is a relationship between attention and cognition, they are different: attention is a type of cognitive process by which the mind engages cognitive resources on a specific object. Cognition concerns the mental processing of sensory information and information already held in a person's memory:

> cognition refers to all processes by which the sensory input is transformed, reduced, elaborated, stored, recovered, and used. [Cognition] is concerned with these processes even when they operate in the absence of relevant stimulation.
>
> *Ulric Neisser, 1967, p. 4*

Thus, attention can be construed as a behavioural and sometimes observable cognitive process that is intimately linked with underlying cognitive processes. In empirical translation research, experimental studies of attention and cognitive processing assume such a link between manifestations of attention (for example eye movements, typing events, verbalizations from think-aloud protocols and EEG signals) and the "invisible" cognitive processes that are involved in the manipulation of written and spoken source-language input that is to be written or spoken in the target language.

Few translation and interpreting studies have explicitly concerned *attention* as an isolated cognitive process, but a large number of studies have had an indirect interest in attention as a gateway to studying the underlying cognitive processes. Concerned specifically with *attention*, empirical translation studies have covered a range of issues and topics. For instance, the term *attention unit* has been used as an indicator of a cognitive translation unit or processing unit based on verbalized think-aloud data (e.g. Bernardini, 2001; Jääskeläinen, 1990, 1996) and combined eye-tracking and keylogged data (e.g. Hvelplund, 2011, 2016). The related *activity unit* (Carl & Schaeffer, 2017) similarly uses behavioural eye-tracking and keylogged data to describe underlying cognitive processes. The *coordination of attention* to source text and target text has been studied from a combination of eye-tracking and keylogged data (Dragsted, 2010; Dragsted & Hansen, 2008), and the role of attentional division in translation has been discussed in both translation research (Hansen, 2005; Hvelplund, 2011, 2016; Jensen, 2011) and interpreting research (Darò & Fabbro, 1994; Lambert, 2004; Mizuno, 2005). Related to attention, *working memory* has been considered central to understanding the processing mechanisms associated with translation (e.g. Dragsted, 2004; Englund Dimitrova, 2005; Hvelplund, 2011, 2016; Kosma, 2007; Rothe-Neves, 2003) and even more extensively in interpreting (e.g. Darò & Fabbro, 1994; Gile, 1995; Jin, 2010; Köpke & Signorelli, 2012; Mizuno, 2005; Moser-Mercer, 2010; Padilla et al., 1999; Padilla et al., 2005; Timarová et al., 2015).

In addition to these studies, which have explicitly concerned attention or working memory theory in relation to translating or interpreting, a large body of translation and interpreting studies has been indirectly interested in attention as a proxy to understanding the underlying cognitive processes. This chapter will not concern these many studies, since that would involve virtually all experimental translation and interpreting research that uses behavioural data as an indication of cognitive processes. Instead, the chapter will consider examples of research that in some way are helpful to the characterization and discussion of attention in relation to translation and interpreting. Readers are referred to overviews of translation process research using think-aloud (Bernardini, 2001, Jääskeläinen, 2002), keylogging (Jakobsen, 2006) and eye tracking (Hvelplund, 2017a).

The following sections will introduce attention as a fundamental cognitive process that is necessary for translating. While the overall focus is mainly on written translation, interpreting will also be given some consideration in comparison with translation. Section 15.2 presents three experimental research methodologies, which are often applied in translation process research to study attention and cognition: think-aloud protocols, keylogging and eye tracking. Drawing on models and concepts from psychology and cognitive psychology, Section 15.3 will outline and discuss working memory and attention theory in relation to translation and interpreting and review relevant translation and interpreting research. Finally, Section 15.4 will take a look to the future and outline other methods that could be used to examine the translator's object of attention.

15.2 Core topics

15.2.1 Measuring attention in translation

Various methods are used to gain access to the translator's object of attention during translation. Three important methods are eye tracking, which captures the location of a person's eye movements, keystroke logging, which registers typing events during a writing task, and think-aloud protocol, which is a method to collect verbal data during the execution of a task. These three methods have had a defining impact on our current understanding of translation as a cognitive process. These three research methods are presented in the following, with brief consideration given to their advantages and applicability in translation (and interpreting) research, but the emphasis will be placed on eye-tracking methodology, which is currently very much in vogue as a way to record attention (and thus cognitive processes) during the execution of translation.

Think-aloud protocol (TAP) analysis was brought over from psychology (e.g. Ericsson & Simon, 1984) to Translation Studies in the 1980s (e.g. Jääskeläinen, 1990; Jääskeläinen & Tirkkonen-Condit, 1991; Krings, 1986) as interest in translation as a cognitive activity accelerated (Jakobsen, 2017, p. 24). The principle behind the method of TAP is that a person verbalizes thoughts related to a specific task while she carries out that task. The assumption is that a person's verbalized thoughts reflect attention to the item being verbalized and thus, the invisible ongoing cognitive processes associated with that task. In the late 1990s, keylogging was proposed as a method to complement TAP in order to "gain better knowledge about cognitive processes by adding a technological, behavioural, and therefore less subjective supplement to introspective 'tapping'" (Jakobsen, 2017, p. 28). Keylogging, also known as keystroke logging, is the process of registering keystroke typing events on a computer using a keylogging program, which records the key that was pressed and the time that key was pressed. Popular keylogging programs for translation research are Translog (Jakobsen & Schou, 1999) and its successor Translog II (which also captures eye movements) (Carl, 2012; Jakobsen, 2011). Analyses of keylogging data rest on the assumption that writing events reflect ongoing cognitive events:

> the idea [is] that the process of writing a translation constitutes behaviour that can be studied quantitatively—across time—and interpreted as a correlate of mental processing. The assumption is further that it will be possible to triangulate qualitative and quantitative data and test hypotheses derived from analyses of qualitative data against quantitative data, and vice versa.
>
> *Jakobsen, 1998, p. 74*

Keystroke logging has been used in multiple studies as an indicator of cognitive processing (e.g. Dragsted, 2004; Immonen, 2006; Jakobsen, 1998, 2003, 2005; Jakobsen & Schou, 1999; Jensen, 2001) using typing events as an attention proxy. With respect to the cognitive process *attention*, registration of keystroke output can provide an estimation of the translation item that was at the translator's focus of attention just prior to typing it. Keylogging has the advantage that typing events are registered without interfering with the translation process, and it is therefore a non-intrusive alternative to TAP. A disadvantage of keylogging, however, is that the focus of the translator's attention is not available during writing pauses. Measures often reported in translation process research include pause duration and typing speed (e.g. Dragsted, 2004; Immonen, 2006; Jakobsen & Schou, 1999). However, other combinatory measures, such as eye-key span (measuring the time lag from source-text reading to target-text typing) (Dragsted, 2010; Dragsted & Hansen, 2008; Timarová et al., 2011), are emerging. In fact, an increasing number of studies combining keylogged and eye-tracking data are being carried out (e.g. Carl & Schaeffer, 2017; Hvelplund, 2011, 2016; Jakobsen, 2011; Martínez-Gómez et al., 2014) to achieve a higher degree of *completeness* in the description of the translation process.

Eye tracking is the most recent method among the three main methodological trends in translation process research, dating back to O'Brien's (2006) study on translation memory matches using pupillary data. For several decades, eye tracking has been a key method in psychology and the cognitive sciences to explore attention and cognitive processing. In the last decade, eye tracking has become a very popular research method to get insight into translators' focus of visual attention and the underlying cognitive processes that are associated with translation. Eye movement data are collected with a device known as an eye tracker, which is a piece of hardware that can register the relative position of eye movements with reasonable precision. Video-based eye trackers, which are the preferred type of equipment in empirical translation research, use near-infrared illumination produced by diodes located on the eye tracker to identify the location of the eyes relative to the content on the computer monitor. The infrared light is reflected on the participant's cornea, and this reflection is captured by a high-resolution camera that is located in the middle of the eye-tracking device. Different types of trackers exist, including head-mounted systems (such as eye-tracking glasses) and head-fixating trackers (where the head is strapped to the eye tracker for better precision). However, in translation process research, *remote* video-based trackers are by far the most popular eye-tracking system. Remote eye trackers come in two different designs: older designs, where the diodes and camera are built into a computer monitor, and newer designs, where diodes and camera are built into a device that is mounted onto a regular computer monitor. There are three main reasons why remote systems are preferred over other systems. Firstly, remote trackers are less intrusive, as they allow free head movement. Translators often need to monitor typing activity and consult offline dictionaries, and this is complicated by head-fixation trackers. Secondly, temporal and spatial precision is considerably higher for remote eye trackers than for eye-tracking glasses. And finally, data recorded in a two-dimensional static display environment with the remote setup is much easier to analyse than data collected in a three-dimensional dynamic environment, such as with the eye-tracking glasses. See Hvelplund (2014) for a presentation of different kinds of eye trackers and their advantages and drawbacks, and Alves et al. (2009), O'Brien (2009) and Hvelplund (2014) for useful guidelines concerning the use of eye tracking in translation experiments.

15.2.1.1 Eye movements and assumptions

In psychology, cognitive sciences, psycholinguistics, translation process research and other fields using eye tracking, eye movements are assumed to reflect ongoing mental processing. In other words, manifested visual attention is assumed to correlate with underlying cognitive attention. This relationship between visual focus and cognitive focus rests on two basic assumptions formulated by psychologists Just and Carpenter (1980). The primary assumption is the eye-mind assumption:

> the eye-mind assumption posits that there is no appreciable lag between what is being fixated and what is being processed.
>
> *Just & Carpenter, 1980, p. 331*

A traditional interpretation of this assumption is that the eyes will focus on the item that is currently the object of cognitive processing. For instance, during reading, the eyes will fixate on words for precisely as long as it takes to comprehend those words. The second important assumption proposed by Just and Carpenter is the immediacy assumption:

> interpretations at all levels of processing are not deferred; they occur as soon as possible.
>
> *ibid.*

The typical interpretation of the immediacy assumption is that cognitive processing of an item starts right when that item comes into visual focus and processing continues until the item leaves visual focus, at which point cognitive processing ends.

These basic assumptions seem reasonable, but reservation has been expressed with regard to a straightforward interpretation (e.g. von der Malsburg & Vasishth, 2011). We are not always mentally engaged in the object our eyes are looking at, and thus, disagreement between observed focus of attention and actual cognitive focus of attention can make it difficult to rely blindly on the eye-mind assumption. Posner (1980, p. 5) points out that "it is important to distinguish between overt changes in orienting that can be observed in head and eye movements, and the purely covert orienting that may be achieved by the central mechanism alone". *Overt* orienting reflects the behavioural, manifested focus of visual attention, and not necessarily *covert* orienting, which concerns the actual object of attention. With respect to the case of translation, eye-tracking analysis is complicated by these potentially covert changes in attention: although the translator is looking at the source text, he may well be considering possible target language equivalents of that specific source text word, and when looking at the target text, the translator may well be constructing meaning hypotheses based on source text content (Hvelplund, 2014, p. 210).

Another weakness of a strict interpretation of the immediacy assumption concerns asynchrony between manifested orientation of the eye and the mind's object of orientation. Holmqvist et al. (2011, p. 379) point out that the eye is roughly 250 ms ahead of the mind, so for translation, this means that the translator will not cognitively process an item until a quarter of a second after it has entered into visual focus. Another kind of asynchrony is of a mechanical nature: drift (ibid.). Drift occurs when the registered eye position and the reader's actual eye position become gradually asynchronous during the course of the data collection session process. Despite these concerns, eye tracking is still considered a viable method for observing manifested attention in translation: while covert attention is a factor to consider, we cannot ignore the many instances during the translation process where ST [source-text] words have been read for the purpose of translating them into the TL [target language]. During those instances, visual focus will have been overt manifestation of cognitive focus (*Hvelplund, 2014, p. 211*).

15.2.1.2 Eye-tracking measures

In translation process research, *eye fixation* is a very popular indicator of visual attention and the underlying cognitive processes. A fixation is a kind of eye movement where the eye is kept relatively still in order to extract visual information from a specific item or object (Duchowski, 2007, p. 46). A range of quantifiable eye-tracking measures are used to register the object of visual attention, including fixation count, fixation duration, time-to-first fixation, first and second fixation durations, and other less frequently used measures such as pupil size (Hvelplund, 2014, p. 212) and blink rate (Chang, 2009). Eye movement measures are typically interpreted as correlates of mental processing (in line with Just and Carpenter's eye-mind and immediacy assumptions); more specifically, changes in fixation duration, count, pupil size, etc. indicate changes in the workload that is placed on working memory. Longer fixations and larger pupils indicate more cognitive effort, while shorter fixations and smaller pupils indicate the opposite.

15.2.2 Attention, cognition and translation

As introduced at the beginning of this chapter, behavioural indicators of attention such as eye movements, typing and thinking aloud can be interpreted as correlates of ongoing cognitive processing in a person's mind. But how do these attentional manifestations tie in with the underlying cognitive processing system? What are the cognitive mechanisms that guide attentional selection, focus and coordination, and how can the study of attention and memory improve our knowledge of the translation and interpreting processes? This section will consider these questions against the backdrop of psychology research, and cognitive psychology in particular.

15.2.2.1 The human memory system

The human memory system is a network of specialized memories that are responsible for the manipulation, storage and retrieval of information. As an information processing system, the human memory system is proposed to consist of a sensory memory (or *registers*), working memory and long-term memory (Baddeley, 2007). Sensory memory is responsible for the initial automatic filtering of environmental input, such as visual, auditory and haptic information, before actual cognitive processing takes place. In psychology, there has been an ongoing debate about whether and how sensory information is filtered by the perceptual system before reaching working memory. An early selection filter theory proposed by Broadbent (1958) suggests that sensory information is selected based on shared physical properties, such as colour, shape or tone, rather than semantic properties. According to this theory, the human memory system is simply not equipped to process all sensory information registered by the sensory organs, and pre-attentive selection ensures that only relevant information is forwarded to working memory for semantic analysis. The late selection theory proposed by Deutsch and Deutsch (1963) suggests that information is selected at a higher stage and that there is no limited capacity when processing sensory information. According to this theory, all incoming sensory information is evaluated for its semantic properties by the response system (e.g. working memory).

Working memory (WM) is a cognitive system that is involved in the conscious retrieval, temporary storage and manipulation of information. Where sensory memory is considered to be outside cognitive control, working memory content may be consciously manipulated in accordance with specific tasks and goals (Baddeley, 2007). Working memory is a key part of attentional processing in translation and interpreting, and we shall return to working memory and attention in Section 15.2.2.2.

Long-term memory (LTM) is a permanent storage system, which can retain seemingly unlimited amounts of information for years (Anderson, 2000, p. 205). In terms of the relationship between attention and LTM, Baddeley (1999, p. 294) points out that: "nothing is likely to get into long-term memory unless you attend to it". LTM is often said to consist of two types of memory: procedural memory and declarative memory (Eysenck & Keane, 2020). Procedural memory (also known as implicit memory) retains knowledge of how to perform automatically specific tasks, for instance walking, reading, typing, etc. Declarative memory (also known as explicit memory) contains factual knowledge and memory of previous experiences. The procedural and declarative memory systems are closely connected, and as in other cognitive information processing tasks, both memories are involved in the translation process (Alves, 2005). Procedural knowledge involves the "ability to carry out the transfer process from the comprehension of the source text to the re-expression of the target text, taking into account the purpose of the translation and the characteristics of the target text readers" (PACTE, 2005, p. 58). Thus, procedural knowledge involves knowledge of how to carry out translation as well as the specific task-dependent sub-competences associated with translation. During the translation process, declarative knowledge is activated when the translator identifies possible meanings of lexical units during language comprehension and evaluates potential target-language equivalents during language reformulation.

15.2.2.2 Working memory and attention in translation and interpreting

As noted earlier, working memory is a cognitive system involved in the retrieval, temporary storage and manipulation of information to perform complex tasks (Baddeley, 2007, p. 1). Working memory theory has been used in translation process research (e.g. Dragsted, 2004; Englund Dimitrova, 2005; Hvelplund, 2011, 2016; Kosma, 2007) and in interpreting research (e.g. Darò & Fabbro, 1994; Jin, 2010; Köpke & Signorelli, 2012; Mizuno, 2005; Padilla et al., 1999; Padilla et al., 2005; Timarová et al., 2015) as a framework to evaluate attention and the underlying cognitive processes associated with translation and interpreting activity.

A central aspect of working memory is capacity. Working memory capacity is limited in terms of storage and information decay: working memory can only retain a small amount of information, and that information will disappear from working memory within seconds. Early working memory span tests demonstrated that between five and nine memory items can be held in working memory (Miller, 1956, p. 81), and information can only be held in working memory for a short while (Peterson & Peterson, 1959, p. 193), such that 50% of items can be recalled after 3 seconds and only 5% after 18 seconds. Working memory storage and processing capacity have not been studied in isolation in translation process research; however, there is indication from TAP studies and keylogging research that professional translators tend to work on larger translation units than non-professionals (e.g. Dragsted, 2004; Lörscher, 1991). In interpreting, the overall indication from empirical studies is that interpreters' basic storage capacity is not different from those of interpreting students and non-interpreters (Timarová et al., 2014, p. 140). However, in terms of processing capacity, there is indication that interpreters outperform the latter two groups in so-called free recall tasks with articulatory suppression (Köpke & Signorelli, 2012, p. 183). In this type of test, the participant is asked to remember and recall items after a few moments while speaking at the same time. This observation indicates that interpreters are better able to focus attention while inhibiting irrelevant information than non-interpreters are.

Attention and working memory are closely connected (Fougnie, 2009), and according to Deutsch and Deutsch's (1963) late selection theory, working memory is, to some extent, involved in focusing attentional resources on a specific task or object. In their multi-component model introduced earlier, Baddeley & Hitch (1974) outline three key components of working memory: the central executive and two slave-systems, the phonological loop and the visuospatial sketchpad. The central executive is responsible for directing attentional resources to the phonological loop and the visuospatial sketchpad (Baddeley, 2007, p. 7). These two slave-components temporarily store and process aural information (i.e. sound) and visual and spatial information, respectively. A third slave-system was suggested by Baddeley (2000, p. 421), *the episodic buffer*, which is responsible for the integration of sensory information held in the phonological and visuospatial components with information from LTM to create "a form of temporary representation" (ibid.).

In translation and interpreting, working memory is responsible for efficiently directing attention to tasks such as language comprehension and language production, and sub-tasks such as reading and writing (Hvelplund, 2011). Taking a closer look at working memory from a translation- and interpreting-oriented perspective, the visuospatial sketchpad plays a central role in text-based translation due to the mainly visual input of this modality (ibid.), while the phonological loop is critical to the activity of interpreting, which involves the processing of aural information (Timarová, 2012). This means that translation skills and interpreting skills are not necessarily transferrable, since the modalities tap into different working memory components. It should be noted that the visuospatial sketchpad is likely to play some role in interpreting, since the interpreting situation most often involves visual attention to the speakers and possibly also to the interpreter's notes.

An efficient translation and interpreting process is related to the central executive's ability to process attentional input in the slave-systems and integrate that information with information from LTM in temporary episodes in the episodic buffer—quickly and with few attentional and cognitive resources. According to Baddeley (2007, p. 124), there are four executive processes that are governed by the central executive: the ability to focus attention (*attentional focus*), the ability to divide attention between concurrent tasks (*attentional division*), the ability to switch attention between tasks (*attentional switching*), and the ability to integrate information from working memory and from LTM (*memory integration*). The three attention processes outlined here are discussed in the following in relation to translation and interpreting.

15.2.2.3 Attentional focus

Attentional focus is the ability to maintain cognitive resources directed to one specific task for as long or as short a time as necessary to complete the task. During attentional focus, task-irrelevant and potentially interfering information is inhibited or ignored. This ability to keep attention focused is related to expertise (Baddeley, 2007, p. 124), and activity exposure is likely to improve the ability to focus attention.

During translation, source-text interpretation, target-language reformulation and related tasks such as dictionary consultation and using parallel texts compete for attentional resources: the translator has to sustain attention to the source text for as long as it takes to arrive at a plausible meaning hypothesis (see Gile, 1995). Similarly, attention to target-language reformulation must be sustained for as long as it takes to formulate a qualified rendition of that meaning hypothesis in the target language (Hvelplund, 2011, p. 45). During comprehension and reformulation, attention may be directed to relevant aids that are necessary to either comprehend source text or produce target text. Premature attentional disengagement from these activities may result in translation error. While it is the privilege of the translator to decide how much time is spent on the individual translation sub-tasks, the interpreter has to follow the pace of the speakers and arrive at a plausible meaning hypothesis during or shortly after the speaker's utterance. In simultaneous interpreting, this urgency is even more acute, and the interpreter has to focus attention on source-language comprehension and target-language reformulation optimally by focusing only on the relevant task and ignoring irrelevant information. In addition, a source-language item is presented only once during simultaneous interpreting, and therefore the load on working memory is likely to be high, since the aural content held in the phonological loops needs to be refreshed *and* this item must be recreated in the target language in a short time. In fact, without this urgency, source-text items held in the interpreter's working memory would probably disappear within seconds (see Mizuno, 2005 for an elaborate model of simultaneous interpreting and working memory).

15.2.2.4 Attentional division

Attentional division refers to the ability to attend to multiple tasks more or less simultaneously (Baddeley, 2007, p. 133). Divided attention is involved in many activities that require the mind to attend to two or more different stimuli at the same time. In a learning environment, the student needs to attend to the teacher's lecture while writing down notes, and when driving, the driver has to attend to other traffic while operating the car. Attentional division is possible, as "tasks could [...] be run using highly practised existing schemata" (Baddeley, 2007, p. 124). In other words, some tasks are *automated* and rely on habitual processing (ibid.), while *controlled processing* occurs when habitual processing is insufficient or unavailable. In accordance with this framework, experience has an impact on a person's ability to divide attention to multiple stimuli, and successful attentional division is the result of activity exposure, by which a certain level of automaticity has been achieved for one or more tasks.

In translation and interpreting, attention is divided between multiple tasks more or less simultaneously. The translator allocates attention to source-text reading while at the same time writing the target text, and the (simultaneous) interpreter splits attention between source and target languages when listening to the speaker's output and verbally producing a translation at the same time. In translation process research, manifestation of parallel attention has been described in a number of studies. In a comparison of experienced professional translators and less experienced non-professional translators, Hvelplund (2011, p. 129) observed that the former group were engaged in parallel source-text reading and target-text writing during 12.1% of the task. The corresponding rate for the student group was 7.3%. This finding points to two conclusions: 1) split attention to simultaneous tasks occurs in translation and 2) the ability to divide attention to multiple tasks is a function of experience. This kind of manifested split attention in translation is possible because one activity occurs automatically: "the activity of typing can become partly automatised [...] and will not demand many attentional or cognitive resources" and "[reading] [...] is an inherently automated activity. Automatic identification of meaning occurs as soon as words enter visual focus, and this process can be interrupted only when looking away from the words" (Hvelplund, 2016, p. 152). A later study revealed that cognitive effort is higher when the translator attends to the source text and the target text at the same time: fixations are significantly longer during simultaneous reading and writing, and pupils are significantly larger (Hvelplund, 2017a). This means that, although one process runs automatically, the load on working memory is higher compared with translation activity where only one task is attended to.

15.2.2.5 Attentional switching

Attentional switching concerns the activity of intentionally reallocating working memory capacity from one task to another. Switching of attention between tasks is not cost free: cognitive effort increases during attentional switching tasks, as indicated by slowing of performance, and this slow-down effect is associated with task complexity and expertise, such that attention switching costs are higher during difficult tasks and higher for less experienced individuals (Baddeley, 2007, p. 130ff).

Translation is an activity that consists of at least two tasks: source-text comprehension and target-text reformulation, and often also auxiliary tasks such as dictionary consultation and parallel text consultation, as noted earlier. It could be argued that there are two kinds of attention switching between source text and target text: switching where attention shifts overtly (see Posner, 1980)—the eyes move from source text to target text and vice versa—and that where attention shifts covertly—e.g. the eyes fixate on source-text content, but the translator is in fact considering target-text rendition. While overt attention shifting between source text and target text occurs every 0.8 seconds on average (Hvelplund, 2011, p. 143), covert attention shifts are probably more frequent. Research has demonstrated that lexical access of the target language occurs automatically during source-text reading (Balling et al., 2014; Ruiz et al., 2008; Schaeffer et al., 2015) and thus not only when the translator looks at the target text. A pertinent question is whether it makes sense at all to distinguish between source-text processing and target-text processing as separate categories of attentional focus, or whether attention is directed only to one ongoing activity, namely mapping of target-text content based on a source-text template. The role of working memory and the attention switching mechanisms associated with translation processing is still unclear, and future research may examine this relationship more closely. There is some indication that overt attentional switching may incur some cognitive cost during translation, as indicated by fixation count increase (Jakobsen & Jensen, 2008, p. 120) and larger pupils (Hvelplund, 2011, p. 192). In translation process research, attentional switching and switching costs have not been the object of systematic investigation, and in interpreting research, where covert attentional switching is likely to occur frequently throughout the interpreting task, there is no research so far concerned with this aspect of attention.

15.2.2.6 Distribution and coordination of attention in translation

Distribution of attention during translation has been examined with eye-tracking data (e.g. Sharmin et al., 2008) and eye-tracking data in combination with keylogged data (e.g. Hvelplund, 2011; Jensen, 2011). Sharmin et al. (2008) were interested in differences between source-text and target-text reading during translation, and they observed significantly larger fixations during target-text reading. Hvelplund (2011) similarly observed different pupil sizes, as pupils were significantly larger for target-text fixations than for those on the source text. The higher complexity of target-text processing is proposed as an explanation for this difference, as it involves integration of source-text and target-text information (see Baddeley's memory integration construct presented earlier) as well as concurrent typing. Concerning the amount of visual attention during translation, Jakobsen and Jensen (2008, p. 114) noticed that professional translators looked at the target text nearly twice as much as they looked at the source text. In addition, in a study of 24 translators (12 professionals and 12 students), data from Jensen (2011, p. 223) show that around 71% of the professionals' gaze time and 63% of the students' gaze time were allocated to the target-text area of the screen. The remainder (29% and 37%, respectively) was allocated to the source text (with and without concurrent typing). The same distribution pattern emerges irrespective of translation direction (Pavlović & Jensen, 2009): source-text / target-text processing in L1 translation, 36% / 64%, respectively, and in L2 translation, 31.5% / 68.5%, respectively. Going beyond the binary division of the translation task into source text and target text processing, dubbing translation has been examined as an example of a translation with multiple areas of attention (Hvelplund, 2017b): the source-text area attracted 26% of visual attention, the target text 47.5%, the film area 21.6% and the dictionary area 4.9%.

Concerning the coordination of attention to both source and target texts, Dragsted and Hansen (2008) propose the eye-key span to measure the time lag from first or last fixation on a source-text word to the first typing event associated with the translation of that source-text word. Dragsted and Hansen examined the eye-key span in 16 MA translation and interpreting students and detected eye-key spans (from last fixation) ranging from 0.5 seconds to 32 seconds. The authors draw a comparison with the ear-voice span in simultaneous interpreting (ibid., p. 14) and point out that the ear-voice span of 2–3 seconds indicates "closer integration of the comprehension and the production phase, and thus a more condensed coordination effort of shorter duration" (ibid.).

15.2.2.7 Attention units

Originating in psychology (Newell & Simon, 1972, p. 313), the concept of *attention unit* has been the object of interest in translation process research at least since the 1990s (e.g. Bernardini, 2001; Jääskeläinen, 1990, 1996). The attention unit may be considered the manifestation of attention to a specific item or object. In think-aloud studies, an attention unit has been defined as the *marked* processing during translation:

> Those instances in the translation process in which the translator's "unmarked processing" is interrupted by shifting the focus of attention onto particular task-relevant aspects.
>
> *Jääskeläinen, 1990, p. 173; cited in Bernardini, 2001, p. 249*

Concerning eye-tracking and keylogged data, Hvelplund (2017a, p. 251) defines an attention unit as "a unit of uninterrupted source and/or target text attention indicated by successive fixations, saccades and typing events", and the variation in attention unit duration can be used as an indicator of cognitive management (Hvelplund, 2011). Carl and Schaeffer's (2017) *activity unit* similarly uses eye-tracking and keylogged data to record the focus of the translator's attention into units. The duration of attention units varies according to the research methodology used. In translation process research using eye tracking, attention units have an overall average duration of around 0.8 seconds: more specifically, source-text (reading) attention units are 846 milliseconds, target-text (reading or writing) attention units are 1141 milliseconds, and parallel source-text (reading)/ target (writing) units are 429 milliseconds (Hvelplund, 2011, p. 143). In addition to variation according to activity, attention unit duration is also sensitive to expertise and time pressure.

15.2.2.8 Attention and cognitive efficiency

Attention and working memory are important for efficient cognitive performance. The ability to process selectively relevant sensory information and to retain information relevant to a specific task is central to efficient task performance (Fougnie, 2009). In translation process research, the term *cognitive efficiency* has been used to describe this ability to focus attentional resources and to select, implement and manipulate relevant information (Hvelplund, 2016). Cognitive efficiency is suggested to be a composition of *cognitive flexibility* and *cognitive automaticity*, cognitive constructs which are both thought to be controlled by working memory and the attentional processes governed by the central executive. Cognitive flexibility is important for the efficient interpreting and translating process: "the efficiency of the interpreting process rests, in part, on the interpreter's ability to flexibly focus attention on, switch attention to and divide attention between those efforts"[1] (Hvelplund, 2016, p. 153) and "the translator's ability to adjust allocation of [attentional] resources codetermines overall processing efficiency" (ibid.). Eye-tracking data indicate that experienced translators are better than less experienced translators at allocating attention to relevant parts of the translation process: attention shifting between source and target texts is performed more efficiently by the former group. While that study does not concern the oral modality, a similar outcome might be hypothesized for interpreting, as interpreters—even more acutely—are compelled to perform cognitively in an efficient way due to the temporal confines of the interpreting situation. Within the context of cognitive efficiency, cognitive automaticity is the execution of a task with little attention and few cognitive resources allocated to its completion (Anderson, 2000, p. 98). Automated task execution requiring little or no attention and cognitive processing has been considered a correlate of expertise in translation (Dragsted, 2004; Hvelplund, 2016; Jääskeläinen & Tirkkonen-Condit, 1991) as well as in interpreting (Lambert, 2004).

15.3 Looking ahead: Translation Studies, attention and beyond

In the years to come, we are likely to see an increasing number of studies that embrace other technologies to collect data about the translator's and interpreter's focus of attention. MRI (magnetic resonance imaging) is one promising technology, which has already received some attention in translation research (Chang, 2009) and interpreting research (e.g. Ahrens et al., 2010; Hervais-Adelman et al., 2015; Hervais-Adelman et al., 2017). Using functional MRI, measurements of changes in blood flows in the brain may be helpful to determine the object of overt as well as covert attention during translation and interpreting. Studies using this imaging technique may help us to better understand the relationship between a) attention and executive control in translation and interpreting and b) those areas of the brain that are typically associated with attention, such as the prefrontal cortex. Other methods that might also be helpful to the exploration of attention as a key cognitive mechanism in translation and interpreting include electroencephalography (EEG), electrodermal activity (EDA) and heart rate monitoring.

Unlike the methods of eye tracking and keylogging, discussed in the sections above, the behavioural and psycho-physiological methods outlined here are far more intrusive, and this intrusiveness may impact the reliability of the process data, as it is very likely that the observing process will attract some attention away from the translation and interpreting process. A method yet to be introduced in translation process research is mouse tracking, which is used to collect information about the mouse cursor position on the computer monitor. Mouse tracking is an inexpensive and non-intrusive tracking alternative compared with eye tracking and MRI. It is substantially less intrusive compared with other methods that would require the participant to lie still in an MRI scanner, work in front of an eye tracker or produce text with a heart rate monitor attached. Mouse tracking is a popular research method in usability testing and web designing. In translation research specifically, mouse tracking could be used to enrich keylogged data and eye-tracking data, and thus provide an even better resolution in terms of the location of the translator's attention, without compromising the reliability of the recorded data as a reflection of the translator's focus of attention.

Note

1 Gile's four 'Efforts' include listening and analysis, memory, production, and coordination (Gile, 1995, p. 186).

Further reading

Baddeley, A. D. (2007). *Working memory, thought, and action*. Oxford: Oxford University Press.
Concerns human memory and specifically discusses Baddeley and Hitch's multi-component model of working memory. This model has been a key theoretical framework in a number of translation process studies.

Duchowski, A. T. (2007), *Eye tracking methodology: Theory and practice*. London: Springer.
A highly useful and comprehensive introduction to eye-tracking systems. The book outlines the human visual system and key issues in visual perception, and also surveys a variety of different eye-tracking technologies.

Eysenck, M. W., & Keane, M. T. (2020). *Cognitive psychology: A student's handbook* (8th ed.). New York: Psychology Press.
A thoroughly researched and easily accessible handbook in cognitive psychology. With its strong focus on human cognition, the handbook is a highly useful companion that outlines the fundamentals of cognitive psychology as well as recent developments within the field.

Hvelplund, K. T. (2014). Eye tracking and the translation process: Reflections on the analysis and interpretation of eye-tracking data. In R. Muñoz Martín (Ed.), *MonTI Special Issue: Minding translation [Con la traducción en mente*], 201–223. Alicante: Publicaciones de la Universidad de Alicante.
Discusses major methodological issues involved in the use of eye tracking in translation research, focusing specifically on challenges in the analysis and interpretation of eye-tracking data as reflections of cognitive processes during translation.

References

Ahrens, B., Kalderon, E., & Krick, C. M. (2010). fMRI for exploring simultaneous interpreting. In D. Gile, G. Hansen, & N. K. Pokorn (Eds.), *Why translation studies matters* (pp. 237–250). Amsterdam: John Benjamins.

Alves, F. (2005). Bridging the gap between declarative and procedural knowledge in the training of translators: Meta-reflection under scrutiny. *Meta, 50*(4), 1–25.

Alves, F., Pagano, A., & da Silva, I. (2009). A new window on translators' cognitive activity: Methodological issues in the combined use of eye tracking, key logging and retrospective protocols. In I. M. Mees, F.

Alves, & S. Göpferich (Eds.), *Methodology, technology and innovation in translation process research* (pp. 267–291). Copenhagen: Samfundslitteratur.
Anderson, J. R. (2000). *Cognitive psychology and its implications* (5th ed.). New York: Worth.
Baddeley, A. D. (1999). *Essentials of human memory*. Hove: Psychology Press.
Baddeley, A. D. (2000). The episodic buffer: a new component of working memory? Trends in Cognitive Sciences, 4(11), 417–423.
Baddeley, A. D. (2007). *Working memory, thought, and action*. Oxford: Oxford University Press.
Baddeley, A. D., & Hitch, G. J. (1974). Working memory. In G. A. Bower (Ed.), *The psychology of learning and motivation: Advances in research and theory* (Vol. 8, pp. 47–89). New York: Academic Press.
Balling, L. W., Hvelplund, K. T., & Sjørup, A. C. (2014). Evidence of parallel processing during translation. *Meta, 59*(2), 234–259.
Bernardini, S. (2001). Think-aloud protocols in translation research: Achievements, limits, future prospects. *Target, 13*(2), 241–263.
Broadbent, D. E. (1958). *Perception and communication*. Oxford: Pergamon.
Carl, M. (2012). Translog-II: A program for recording user activity data for empirical reading and writing research. *Proceedings of the eighth international conference on language resources and evaluation, European Language Resources Association (ELRA)*, 4108–4112. Istanbul: European Language Resources Association (ELRA).
Carl, M., & Schaeffer, M. (2017). Sketch of a noisy channel model for the translation process. In S. Hansen-Schirra, S. Hofmann, & B. Meyer (Eds.), *Empirically modelling translation and interpreting* (pp. 71–116). Berlin: Language Science Press.
Chang, V. C. (2009). Testing applicability of eye-tracking and fMRI to translation and interpreting studies: An investigation into directionality [PhD dissertation, Imperial College London]. London: Imperial College London.
Darò, V., & Fabbro, F. (1994). Verbal memory during simultaneous interpretation: Effects of phonological interference. *Applied Linguistics, 15*(4), 365–381.
Deutsch, J. A., & Deutsch, D. (1963). Attention: Some theoretical considerations. *Psychological Review, 93*, 283–321.
Dragsted, B. (2004). Segmentation in translation and translation memory systems: An empirical investigation of cognitive segmentation and effects of integrating a TM-System into the translation process [PhD thesis, Copenhagen Business School]. Copenhagen: Samfundslitteratur.
Dragsted, B. (2010). Coordination of reading and writing processes in translation. In G. M. Shreve, & E. Angelone (Eds.), *Translation and cognition* (pp. 41–62). Amsterdam: John Benjamins.
Dragsted, B., & Hansen, I. G. (2008). Comprehension and production in translation: A pilot study on segmentation and the coordination of reading and writing processes. In S. Göpferich, A. L. Jakobsen, & I. M. Mees (Eds.), *Looking at eyes. Eye-tracking studies of reading and translation processing*. (Copenhagen Studies in Language, 36, pp. 9–30.) Copenhagen: Samfundslitteratur.
Duchowski, A. T. (2007), *Eye tracking methodology: Theory and practice*. London: Springer.
Englund Dimitrova, B. (2005). *Expertise and explicitation in the translation process*. Amsterdam: John Benjamins.
Ericsson, K. A., Simon, H. A. (1984). *Protocol analysis: Verbal reports as data*. Cambridge, MA: MIT Press.
Eysenck, M. W., & Keane, M. T. (2020). *Cognitive psychology: A student's handbook* (8th ed.). New York: Psychology Press.
Fougnie, D. (2009). The relationship between attention and working memory. In N. B. Johansen (Ed.), *New research on short-term memory*. New York: Nova Science Publishers.
Gile, D. (1995). *Basic concepts and models for interpreter and translator training*. Amsterdam: John Benjamins.
Hansen, G. (2005). Experience and emotion in empirical translation research with think-aloud and retrospection. *Meta, 50*(2), 511–521.
Hervais-Adelman, A., Moser-Mercer, B., Michel, C. M., & Golestani, N. (2015). fMRI of simultaneous interpretation reveals the neural basis of extreme language control. *Cerebral Cortex, 25*(12), 4727–4739.
Hervais-Adelman, A., Moser-Mercer, B., Murray, M. M., & Golestani, N. (2017). Cortical thickness increases after simultaneous interpretation training. *Neuropsychologia, 98*, 212–219.
Holmqvist, K., Nystrom, M., Andersson, R., Dewhurst, R., Jarodzka, H., & van de Weijer, J. (2011). *Eye tracking: A comprehensive guide to methods and measures*. New York: Oxford University Press.
Hvelplund, K. T. (2011). Allocation of cognitive resources in translation: An eye-tracking and keylogging study [PhD dissertation, Copenhagen Business School]. Copenhagen: Samfundslitteratur.

Hvelplund, K.T. (2014). Eye tracking and the translation process: Reflections on the analysis and interpretation of eye-tracking data. In R. Muñoz Martín (Ed.), *MonTI Special Issue: Minding Translation [Con la traducción en mente]*, 201–223. Alicante: Publicaciones de la Universidad de Alicante.
Hvelplund, K. T. (2016). Cognitive efficiency in translation. In R. Muñoz Martín (Ed.), *Reembedding translation process research* (pp. 149–170). Amsterdam: John Benjamins.
Hvelplund, K.T. (2017a). Eye tracking in translation process research. In J.W. Schwieter, & A. Ferreira (Eds.), *The handbook of translation and cognition* (pp. 248–264). New Jersey: Wiley-Blackwell.
Hvelplund, K. T. (2017b). Eye tracking and the process of dubbing translation. In J. Díaz-Cintas, & K. Nikolić (Eds.), *Audiovisual translation: Expanding borders* (pp. 110–125). Bristol: Multilingual Matters.
Immonen, S. (2006). Translation as a writing process: Pauses in translation versus monolingual text production. *Target, 18*(2), 313–336.
Jääskeläinen, R. (1990). Features of successful translation processes: A think-aloud protocol study [Licentiate thesis, University of Joensuu]. Joensuu: University of Joensuu.
Jääskeläinen, R. (1996). Hard work will bear beautiful fruit. A comparison of two think-aloud protocol studies. *Meta, 41*(1), 61–74.
Jääskeläinen, R. (2002). Think-aloud protocol studies into translation: An annotated bibliography. *Target, 14*(1), 107–136.
Jääskeläinen, R., & Tirkkonen-Condit, S. (1991). Automated processes in professional vs. non-professional translation: A think-aloud protocol study. In S. Tirkkonen-Condit (Ed.), *Empirical research in translation and intercultural studies* (pp. 89–109). Tübingen: Gunter Narr.
Jakobsen, A. L. (1998). Logging time delay in translation. In G. Hansen (Ed.), *LSP texts and the process of translation.* (Copenhagen Working Papers 1, pp. 71–101.) Copenhagen: Copenhagen Business School.
Jakobsen, A. L. (2003). Effects of think aloud on translation speed, revision and segmentation. In F. Alves (Ed.), *Triangulating translation. Perspectives in process oriented research* (pp. 69–95). Amsterdam: John Benjamins.
Jakobsen, A. L. (2005). Investigating expert translators' processing knowledge. In H. V. Dam, J. Engberg, & H. Gerzymisch-Arbogast (Eds.), *Knowledge systems and translation* (pp. 173–189). Berlin and New York: Mouton de Gruyter.
Jakobsen, A. L. (2006). Research methods in translation: Translog. In E. Lindgren, & K. P. H. Sullivan (Eds.), *Computer keystroke logging and writing: Methods and applications* (Studies in Writing, Vol. 18, pp. 95–105). Oxford: Pergamon Press.
Jakobsen, A. L. (2011). Tracking translators' keystrokes and eye movements with Translog. In C. Alvstad, A. Hild, & E. Tiselius (Eds.), *Methods and strategies of process research: Integrative approaches in translation studies* (pp. 37–55). Amsterdam: John Benjamins.
Jakobsen, A. L. (2017). Translation process research. In J. W. Schwieter, & A. Ferreira (Eds.), *The handbook of translation and cognition* (pp. 21–49). New Jersey: Wiley-Blackwell.
Jakobsen, A. L., & Jensen, K. T. H. (2008). Eye movement behaviour across four different types of reading task. In S. Göpferich, A. L. Jakobsen, & I. M. Mees (Eds.), *Looking at eyes: Eye-tracking studies of reading and translation processing.* (Copenhagen Studies in Language 36, pp. 103–124). Copenhagen: Samfundslitteratur.
Jakobsen, A. L., & Schou, L. (1999). Translog documentation. In G. Hansen (Ed.), *Probing the process in translation: Methods and results* (pp. 1–36). Frederiksberg: Samfundslitteratur.
James, W. (1890). *Principles of psychology*. New York: Holt.
Jensen, A. (2001). The effects of time on cognitive processes and strategies in translation [PhD dissertation, Copenhagen Business School]. Working Papers in LSP, No. 2001-2. Frederiksberg.
Jensen, K.T. H. (2011). Distribution of attention between source text and target text during translation. In S. O'Brien (Ed.), *Continuum studies in translation: Cognitive explorations of translation* (pp. 215–236). London and New York: Continuum.
Jin, Y. (2010). Is working memory working in consecutive interpreting? [PhD thesis, University of Edinburgh]. Edinburgh: The University of Edinburgh.
Just, M. A., & Carpenter, P. A. (1980). A theory of reading: From eye fixations to comprehension. *Psychological Review, 87*(4), 329–354.
Kosma, A. (2007). The specific functioning of working memory in translation. *Meta, 52*(1), 22–28.
Köpke, B., & Signorelli, T. M. (2012). Methodological aspects of working memory assessment in simultaneous interpreters. *International Journal of Bilingualism, 16*(2), 183–197.
Krings, H. P. (1986). *Was in den Köpfen von Übersetzern vorgeht.* Tübingen: Gunter Narr.
Lambert, S. (2004). Shared attention during sight translation, sight interpretation and simultaneous interpretation. *Meta, 49*(2), 294–306.

Lörscher, W. (1991). *Translation performance, translation process, and translation strategies: A psycholinguistic investigation.* Tübingen: Günter Narr.
Martínez-Gómez, P., Minocha, A., Huang, J., Carl, M., Bangalore, S., & Aizawa, A. (2014). Recognition of translator expertise using sequences of fixations and keystrokes. *Proceedings of the symposium on eye tracking research and applications*, 299–302. New York: Association for Computing Machinery.
Miller, G. A. (1956). The magical number seven, plus or minus two: some limits on our capacity for processing information. *Psychological Review, 63*, 81–97.
Mizuno, A. (2005). Process model for simultaneous interpreting and working memory. *Meta, 50*(2), 739–752.
Moser-Mercer, B. (2010). The search for neuro-physiological correlates of expertise in interpreting. In G. M. Shreve, & E. Angelone (Eds.), *Translation and cognition* (pp. 263–287). Amsterdam: John Benjamins.
Neisser, U. (1967). *Cognitive psychology*. Englewood Cliffs, NJ: Prentice-Hall.
Newell, A., & Simon, H. A. (1972). *Human problem solving*. Englewood Cliffs, NJ: Prentice-Hall.
O'Brien, S. (2006). Eye-tracking and translation memory matches. *Perspectives: Studies in Translatology, 14*, 185–203.
O'Brien, S. (2009). Eye tracking in translation-process research: Methodological challenges and solution. In I. M. Mees, F. Alves, & S. Göpferich (Eds.), *Methodology, technology and innovation in translation process research* (pp. 251–266). Copenhagen: Samfundslitteratur.
PACTE (2005). Investigating translation competence: Conceptual and methodological issues. *Meta, 50*(2), 609–619.
Padilla, F., Bajo, M. T., & Macizo, P. (2005). Articulatory suppression in language interpretation: Working memory capacity, dual tasking and word knowledge. *Bilingualism: Language and Cognition, 8*, 207–219.
Padilla, P., Bajo, M. T., & Padilla, F. (1999). Proposal for a cognitive theory of translation and interpreting: A methodology for future empirical research. *The Interpreter's Newsletter, 9*, 61–78.
Pavlović, N., & Jensen, K. T. H. (2009). Eye tracking translation directionality. In A. Pym, & A. Perekrestenko (Eds.), *Translation research projects* (pp. 101–119). Tarragona: Universitat Rovira i Virgili.
Peterson, L. P., & Peterson, M. J. (1959). Short-term retention of individual verbal items. *Journal of Experimental Psychology, 58*(3), 193–198.
Posner, M. I. (1980). Orienting of attention. *Quarterly Journal of Experimental Psychology, 32*(1), 3–25.
Rothe-Neves, R. (2003). The influence of working memory features on some formal aspects of translation performance. In F. Alves (Ed.), *Triangulating translation: Perspectives in process oriented research* (pp. 97–119). Amsterdam: John Benjamins.
Ruiz, C., Paredes, N., Macizo, P., & Bajo, M. T. (2008). Activation of lexical and syntactic target language properties in translation. *Acta Psychologica,* 128, 490–500.
Schaeffer, M., Dragsted, B., Hvelplund, K. T., Balling, L. W., & Carl, M. (2015). Word translation entropy in translation: Evidence of early target language activation during reading for translation. In M. Carl, S. Bangalore, & M. Schaeffer (Eds.), *New directions in empirical translation process research: Exploring the CRITT TPR-DB* (pp. 183–210). Berlin: Springer.
Sharmin, S., Špakov, O., Räihä, K., & Jakobsen, A. L. (2008). Where on the screen do translation students look while translating, and for how long? In S. Göpferich, A. L. Jakobsen, & I. M. Mees (Eds.), *Looking at eyes: Eye-tracking studies of reading and translation processing*. (Copenhagen Studies in Language 36, pp. 31–51). Copenhagen: Samfundslitteratur.
Timarová, Š. (2012). Working memory in conference simultaneous interpreting [PhD thesis, University of Leuven]. Leuven: University of Leuven.
Timarová, Š., Čeňková, I., Meylaerts, R., Hertog, E., Szmalec, A., & Duyck, W. (2014). Simultaneous interpreting and working memory executive control. *Interpreting, 16*(2), 139–168.
Timarová, Š., Čeňková, I., Meylaerts, R., Hertog, E., Szmalec, A., & Duyck, W. (2015). Simultaneous interpreting and working memory capacity. In A. Ferreira, & J. W. Schwieter (Eds.), *Psycholinguistic and cognitive inquiries into translation and interpreting* (pp. 101–126). Amsterdam: John Benjamins.
Timarová, Š., Dragsted, B., & Hansen, I. G. (2011). Time lag in translation and interpreting. In C. Alvstad, A. Hild, & E. Tiselius (Eds.), *Methods and strategies of process research: Integrative approaches in translation studies* (pp. 121–146). Amsterdam: John Benjamins.
von der Malsburg, T., & Vasishth, S. (2011). What is the scanpath signature of syntactic reanalysis? *Journal of Memory and Language, 65*(2), 109–127.

16

Translation, emotion and cognition

Caroline Lehr

16.1 Introduction

In an increasingly interdependent world, the indispensability of translation for overcoming language barriers and its essential contribution to supporting linguistic heterogeneity add to the need for a deeper understanding of the complex communicative events that involve authors, translators and readers and the use of two languages. Research into the translation process is central to this understanding, not least because it may allow us to give recommendations for practice and to refine teaching methods.

In its beginnings, translation process research has mainly investigated the translation process from a classical cognitive perspective that emphasized the role of information processing (e.g., Shreve & Koby, 1997). However, this perspective does not capture the entire cognitive architecture that underlies human behaviour, especially because it does not emphasize the role of emotion sufficiently. This neglect is all the more striking as, in the past few decades, psychological research has increasingly acknowledged that emotion is central to the organization of human cognition and that "few thoughts are entirely free of feelings and emotions influence thinking" (Ellsworth & Scherer, 2003, p. 572). Moreover, in the neurosciences, an understanding of the neural circuits that underlie emotional experience has emerged (LeDoux, 1995).

Over the years, several translation scholars have insisted that emotion deserves increased attention, and that the phenomenon of translation should not be reduced to its rational dimensions only (Jääskeläinen, 1996; Lee-Jahnke, 2011; Risku, 2004; Robinson, 1991). However, the significance of emotion for the translation process as well as for the resulting product has for a long time not been systematically integrated into translation theory, and focused empirical investigations have been lacking. More recently, and in line with research in neighbouring disciplines, translation scholars have acknowledged the relevance of emotion for studying the translation process and have begun to examine the role of emotion in the activity of translating (Hubscher-Davidson, 2018; Lehr, 2014). The following seeks to provide an overview of how emotions have been addressed and studied by translation scholars to date.

16.1.1 Emotions during the early stages of translation research

In Translation Studies, emotions have for a long time rarely been at the centre of interest. This is certainly linked to the theoretical influence of what can now be considered the classical cognitive paradigm, which dominated cognitive psychology for several decades and placed its emphasis on rational processes in cognition. It may also be due to an unwillingness of this fairly young scientific discipline to link its object of study to what is perceived as "irrational behaviour", as this may appear to run contrary to scientific principles and may also be considered particularly inappropriate in a discipline with an applied orientation towards professional life. Nevertheless, even if emotions were seldom the focus of interest, several translation scholars showed awareness of the importance of emotions. Already at a very early stage of translation theory, Nida and Taber (1969) put particular emphasis on the importance of rendering the emotional impact of a text in its translation. In the context of Bible translation, where the emotional reaction of the reader takes centre stage, they granted emotions an important role in reader response, as illustrated by their definition of message as "the total meaning or content of a discourse, the concepts and feelings which the author intends the reader to understand and perceive" (1969, p. 205). In their theory, where the focus was on the translation product and on assisting the translator by providing suggestions of translation procedures, Nida and Taber addressed a central aspect of the relationship between translation and emotion: the translation of a text's emotionality. In the 1980s, important works in translation and interpreting theory (e.g., Seleskovitch & Lederer, 1984) moved the translator and cognitive processes into the centre of interest, and, at the same time, shifted attention away from the translation product. Scholars started to investigate the translation process empirically to obtain insight into the mental processes that underlie this complex cognitive activity, and began to develop models of the translation process.

16.1.2 Modelling the translation process within the classical cognitive paradigm

Research studying the translation process examines cognitive processes and therefore has to be coherent with generally prevailing theories on cognition. This has important implications for theory inside the field of translation process research, as it must seek compatibility with the paradigms of cognitive psychology, that is, the generally accepted perspectives which determine the set of practices of a discipline in a given period (Kuhn, 1962). The first dominating paradigm in translation process research became what is now considered the "classical" cognitive paradigm, which conceived of the human brain as a symbol-processing system and used computational functions to describe cognitive processes, with the aim of developing a detailed process-based understanding of cognitive functionalities (e.g. Barber, 1988). Influenced by the prior experience of people working in psycholinguistics (e.g. Shreve & Koby, 1997), the integration of this paradigm into translation process research helped to answer some fundamental questions. At an early stage of research, it allowed an elucidation of the cognitive basis of the translation process, its basic entities and their interaction with each other. Models of the translation process were framed within this paradigm and divided the process generally into three core processes that rely on encoding and decoding of information in memory: comprehension, transfer between the two languages, and production (Danks & Griffin, 1997; Hönig, 1995; Krings, 1986; Padilla et al., 1999). Language comprehension processes and the reading of the source text formed the onset of the translation process and were modelled in accordance with Kintsch's (1988) construction-integration model (Padilla et al., 1999). This model assumes that the information in the text is integrated and combined with the readers' knowledge, leading to the construction of a mental model of the text. Based on the mental model and the interpretation of the source text as well as expectations towards the prospective target text, the translator then develops a macro-strategy. The macro-strategy constitutes the frame of reference for the translation and guides the associative competence (Hönig, 1995) of the translator. The associative competence is the competence that enables proper language transfer and, whenever necessary, is complemented by the translator's competence in information mining and research. The translated text then emerges from an interplay of macro- and micro-level interpretation, judgements and decisions (Levy, 1967), processes that were seen as primarily rational in the classical cognitive paradigm.

16.1.3 The relevance of emotion for understanding translators' decision making

The perspective of rational information processing outlined earlier allowed an understanding of the basic elements and regularities of the translation process. Yet, it was limited in its explanatory power. In a meta-analytical approach to translation research, Robinson (1991) stated as a main point of criticism that translation research had, until then, regarded translation as a fundamentally cognitive process, not taking into account the importance of emotions for human beings. In line with this criticism, further translation scholars began to include emotions in their theoretical considerations and to acknowledge the general importance of emotion for decision making during the translation process. Describing potentially influential factors in the translation process, Chesterman (2002) and Krings (2005) emphasized that Translation Studies must also take into account the effects of internal, subjective factors on translational choices and listed, among others, the translator's emotional state, motivation, attitude towards a text and personality as potential internal causes of translator behaviour.

In a similar vein, in her model of decision making in translation, Durieux (2007) assigned a central role to emotion and to subjective evaluations of objects or events which elicit emotions and then exert an influence on information processing and translators' decision making. More specifically, Lederer (2003) and Hansen (2006) focused on the influence of emotions elicited through the translator's interaction with the text. Hansen explains that

> be it in connection with [...] some themes or words, impulses in the form of images, experiences, associations, and emotions immediately emerge and influence the process and the decision during the process. During the act of translation emotions and earlier experiences [...] are activated and these have an impact on the actual decisions.
>
> *2006, p. 76*

In addition, in her influential interpretive model, based on the idea that the identification of sense and its re-expression underlies all translating, Lederer (2003) considers the translator's emotional reaction to the text to be a prerequisite for the translation of the text's emotionality and regards emotions as a necessary factor in translators' decision making. She states that the emotionality of the text can only be translated if the translator feels the emotion of the text and experiences the text's "affective components" (2003, p. 50).

These theoretical considerations were complemented by empirical studies, which identified several variables influencing the translation process and its components, but which could not be fully explained without considering affective aspects. For example, risk taking in decision making (Pym, 2005), the translator's way of managing uncertainty (Angelone, 2010), as well as time pressure (Jensen & Jakobsen, 2000) and resulting stress (Bayer-Hohenwarter, 2009) were found to influence translation. The classical cognitive paradigm, however, was limited in its ability to explain the role and influence of most of the aforementioned variables by insufficiently recognizing that humans differ from mere processing devices by being emotional creatures. As a consequence, without entirely re-examining the assumptions underlying the classical cognitive paradigm, the view of the translation process had to be broadened to increase the understanding of variables inaccessible in the classical cognitive view, and to do more justice to the translator as a human being. In the following, we will take a look at relevant neighbouring disciplines and interface research on emotion and translation process research to respond to this need.

16.2 Core topics

16.2.1 What are emotions?

The question of the origins and nature of emotion already attracted the interest of Aristotle and Descartes, who reflected upon the types of situations that produce special reactions and behaviours in humans and give rise to a conscious feeling of a particular quality over a period of time. William James (1884) is claimed to be the first psychologist who attempted to establish a comprehensive explanation of emotion. He postulated that emotional experience arises from bodily responses and the conscious experience of these responses. In the decades following James's theory, the study of emotions remained to a large extent outside the mainstream of psychology. Based on the early conceptions of human intelligence likening the mind to a rational information processing system (e.g., Reed, 1982), cognition and "irrational" emotion were considered distinct concepts and human faculties, with emotion seen, rather, as a source of intrusion or interruption. A few psychologists, however, did not lose sight of emotional phenomena and provided increasing evidence of the interdependence of cognition and emotion. They showed that emotions arise from our thinking and perceptions of situational meaning (Arnold, 1960; Leventhal & Scherer, 1987; Zajonc, 1980) and revealed that emotions play a fundamental role in decision making, "the purpose of reasoning" (Damasio, 1994, p. 165), and that people make decisions sometimes primarily at an emotional level (Bechara, 2004). In the past few decades, the field of affective sciences, devoted to all aspects of affect and emotion, has been steadily growing (Sander & Scherer, 2009) and has contributed significantly to our understanding of the crucial role emotion plays in human cognition and behaviour. Today, psychological research largely recognizes emotion as a fundamental principle of human behaviour that derives from evolution and neurobiological development as well as culture and organizes perception, thought and action tendencies (Izard, 1991). Emotions are thought to very largely determine the contents and focus of human consciousness (Izard, 2009) by activating relevant associative networks in memory, altering attention and shifting certain behaviours upwards in response hierarchies. Physiologically, emotions rapidly organize the responses of biological systems such as facial expression, voice and autonomic nervous activity (Levenson, 1999). Through these means, they allow human beings to rapidly make sense of complex environments and to efficiently adapt and respond to changing environmental demands and stimuli that have more direct relevance for their well-being and survival than others (Smith & Lazarus, 1990).

Definitions of affective phenomena, and in particular the term *emotion*, have given rise to many controversies, and terms are still used in an inconsistent fashion (Scherer, 2005). According to the definition by Scherer (2005, p. 697), the term *emotion* refers to "an episode of interrelated, synchronized changes in the states of all or most of the five organismic subsystems[1] in response to the evaluation of an external or internal stimulus event as relevant to major concerns of the organism". As outlined by this definition and by Shuman and Scherer (2014), emotions are generally viewed as episodes that can be evoked by a variety of stimuli

(Ekman, 1992; Russell, 2003; Scherer, 2009). The emotion-eliciting stimuli can be stimuli that are indeed occurring, such as being anxious about a feedback one receives on a translation, but also remembered or imagined stimuli, such as memories of past failures or worrying about potential outcomes. As emotions are triggered by these internal or external stimuli that are of major significance to the individual, they have also been called relevance detectors. In contrast to emotions, *moods* refer to "affective states that are of long duration, low intensity and a certain diffuseness" (Fridja in Sander & Scherer, 2009, p. 258). Moods are not perceived as a response to a distinct event; they can, however, result from an experienced emotion. As an overarching term, including both emotions and moods, *affect* is used in a "generic sense for a mental state that is characterized by emotional feeling as compared with rational thinking" (Fridja & Scherer, 2009, p. 10).

16.2.2 The componential view of emotions

In addition to viewing emotions as episodes, researchers generally agree that emotions have a componential nature. That is, they involve different components which are associated with different functions and influence each other in complex ways. It is commonly assumed that five components interact during an emotion episode: the appraisal component, the action tendency component, the physiological component, the expression component and the subjective feeling component. The multi-componential nature of emotion becomes evident in the definition of emotion that we have adopted (Scherer, 2005). It can also be illustrated by self-reports of emotion (Shuman & Scherer, 2014): for example, "I didn't have enough time to properly finalize the translation", "I don't want to get feedback from the client", "I feel jittery", and "I am afraid" refer to the different components of the emotion.

The first expression describes cognitive appraisals of the situation, including goal frustration ("I couldn't finalize") and lack of coping potential ("I didn't have enough time"). Appraisals are at the onset of an emotional episode and evaluate an object or event for its emotional significance. These evaluations can be processed unconsciously or consciously (Leventhal & Scherer, 1987). They include evaluations of novelty, pleasantness, goal compatibility, coping potential, congruency with an individual's ideal self, and conformity to an individual's norms and values.

The second expression indicates an avoidance action tendency associated with an emotional state. The function of the action tendency component of emotion is to prioritize actions that are needed in a given situation and to ensure the preparation of appropriate action. For example, feeling afraid is associated with an urge to avoid the situation (Fridja, 1988). Moreover, action tendencies are associated with specific cognitive and motivational processes. Positive emotions, for example, are thought to broaden thought-action repertoires and to enhance creativity (Fredrickson, 1998), whereas negative emotions promote systematic and detailed information processing (Schwarz, 2002).

The third expression refers to the physiological component of emotion. This component regulates and supports the bodily reactions during an emotional response. For example, when an event is appraised as relevant to an individual, this is associated with increased activity in such brain areas as the amygdala and the frontal cortex (Sander et al., 2005). Also, action tendencies are supported by changes in the autonomic nervous system. The experience of anger, for example, goes together with an increase in heart rate and an increase of blood flow to the hands and arms (Levenson et al., 1990).

Further, observable motor activities are linked to emotion. In emotional situations, facial expressions, such as smiling or frowning, or changes in the voice, such as raised pitch, can be apparent. The expression component of emotion serves a communicative function, for example, when we express our feelings of happiness after receiving positive feedback by smiling at the feedback giver.

The noticeable responses and changes in the appraisal, the action tendency, the physiological and the expression component, then, give rise to the particular quality of the subjective emotion experience (Scherer, 2009), which may be labelled as, for example, "I am afraid." As a fifth component, the subjective feeling component integrates the information of the previous four components and is considered to have a monitoring function, as feeling fear enables the individual to take regulatory efforts to reduce the emotion, and feeling less fear signals that the regulation was successful. Emotion-regulatory processes, attempts to influence the quality, intensity and expression of emotions (McRae & Gross, 2009), may therefore be part of an emotion episode, especially when we try to upregulate and change negative emotional states. Individuals may undertake different strategies, such as situation selection or attentional deployment, to shape the trajectory of an emotional response (Gross & Thompson, 2007). These strategies are dependent on inter-individual differences, related to broad domains of personality and emotional competences which influence and moderate emotion processes.

16.2.3 Translation as an emotion episode—A perspective on the translation process beyond the classical cognitive approach

Based on our understanding of an emotion episode and the modelling of the translation process within the classical cognitive paradigm, the aim of this section is to integrate translation process research and recent research on emotion. We will provide a perspective on the translation process which attempts to broaden the classical cognitive approach to translation by framing the translation process within the componential view of emotion. As outlined before, this componential view of emotion enables an understanding of an emotion episode during translation by dividing it into different components. In the following, we will therefore discuss the role of the five components that form an emotion episode in the translation process.

16.2.3.1 The appraisal component during the translation process

As explained earlier, emotions are triggered by internal or external stimuli that are of major significance to the individual. During the translation process, cognitive evaluations of objects or events thus form the onset of an emotion episode. These appraisals are based on the translator's subjective perception of the circumstances, and the result of this cognitive evaluation is assumed to produce changes in the other four components of emotion, leading to different emotional states and an adaptation to the current situation. A translator may, for example, be happy about a particular translation solution or be worried about the functioning of a translation memory.

In the translator's workplace, potentially emotion-eliciting stimuli can be divided into four major groups: the text that is translated, performance assessments, the translator's working conditions and the translator's personal well-being. Emotion-eliciting stimuli during the translation process can be not only actually occurring stimuli but also remembered or imagined stimuli, such as past experiences or worry about potential positive or negative outcomes of one's actions. Moreover, two types of emotions can be distinguished: integral and incidental emotions. Integral emotions, such as text-related emotions, refer to emotional responses that are directly linked to the object of judgement or decision and are experienced through direct exposure to the object itself or in response to some representation of it. In the translator's work, the text that is translated constitutes an important potential stimulus, as emotions are a vital part of the mental representation of texts (Hansen, 2006; Lederer, 2003). During the translation process, emotional responses to text content and aesthetic responses co-exist. A translator can, for example, get annoyed because of the poor style of a text or be moved by the fate of a person described in a narrative. Incidental emotions, on the other hand, are unrelated to the judgements or decisions being made during translation and are elicited through other stimuli that are present in the translation situation (Loewenstein & Lerner, 2003). Performance assessments, for example feedback during training or feedback from revisors or clients, are an important group of appraisals associated with incidental emotions. In addition, the translator's working conditions involve a range of emotional stimuli. For example, time pressure, submission deadlines, job uncertainty, adaptation to working with new translation technologies, contacts with clients or other team members, and workload have the potential to elicit a range of positive and negative incidental emotions as well as stress. The last term refers to a particular emotional response to "either acute or chronic strains" (Uchino et al., 2009, p. 383). The last group of appraisals can be subsumed under the category that relates to the translator's personal well-being. These appraisals depend on stimuli entirely external to the translation situation. For example, anger elicited by a traffic jam in the morning can shape the emotional state of the translator at work.

16.2.3.2 The action tendency component of emotion and its influence on translators' decisions

The action tendency component of emotion ensures the preparation and direction of appropriate action during an emotion episode. It is of crucial importance in the context of translation, as action tendencies are associated with specific cognitive and motivational processes, which can influence the translator's information processing, perception and mental representation of the translation situation. Three core influences of emotion have been identified: (1) emotion-congruence effects, (2) the processing consequences of affect and (3) inferential mechanisms (Forgas, 1995; Fredrickson & Branigan, 2005; Schwarz & Clore, 1983). These different influences of emotion facilitate the processing of emotion-congruent information, influence the translator's cognitive processing style or lead to generally more positive or negative judgements.

(1) Emotion-congruence effects are based on the assumption that during an emotional state, information that is associated with the emotion is primed, its processing is facilitated, and it is more likely to be used in information encoding and retrieval. This leads to attentional selectivity to emotion-congruent features in the translation task and a potential mediation of perception by the emotional state (Forgas, 1995).
(2) Moreover, emotion may influence not only what people think but also how people think, through its effects on the way information is processed. Research studying the processing consequences of affect (Fredrickson, 1998; Schwarz, 2002) assumes that negative emotions function as a warning signal, indicating that the environment is threatening and that these concerns must be addressed. Individuals therefore become more motivated to identify, alleviate or eliminate the problem, resulting in increased attention to the details at hand and a more analytic, systematic processing strategy. Conversely, positive emotions signal that the environment is safe and are associated with reliance on less demanding heuristic processing and prior general knowledge. Also, they imply a tendency to explore, as well as a general openness to the unusual, and have the ability to widen attention and to broaden people's momentary thought-action repertoires. Through this, positive emotions stimulate diverse thoughts and actions, enhance creativity and allow a building up of intellectual resources.
(3) Lastly, influences of emotion on inferential mechanisms start from the assumption that people attend to their feelings as a source of information (Schwarz & Clore, 1983). As affective states convey information about the positive or negative aspect of things, they lead

> to generally more positive or negative judgements. This influence of affective states increases with its perceived relevance for the judgement, but it may also be misattributed, so that emotions aroused by one event may affect judgements in an entirely different situation.

As a consequence, through their influences on both lower-level and higher-level cognition, emotions can modulate attention, interpretation, judgement, problem solving and decision-making processes during the translation process. Influences of emotion on translational choices may then become visible in the translated text and may have an influence on the perception of the text by the reader, but may also be retraceable during the process itself, through their influence on the use of auxiliary devices, time spent translating, pauses, production speed, segmentation or revision behaviour.

16.2.3.3 The physiological and the expression component of emotion during the translation process

When a translator is experiencing an emotion, the physiological component of emotion ensures an appropriate response from the body system. Apart from increased activity in brain areas implicated in emotion processing, such as the amygdala and the frontal cortex (Sander et al., 2005), cardiovascular and electrodermal responses (Mauss & Robinson, 2010) or changes in eye properties (Bradley et al., 2008) will indicate activation in the autonomic nervous system as a function of emotional responding. Moreover, higher stress levels have been shown to affect salivation during interpreting (Moser-Mercer, 2003) as well as translators' levels of adrenaline (Bayer-Hohenwarter, 2009).

Further, the expression component of emotion can be observable during the translation process. It includes the verbal and non-verbal communication of emotion and is particularly important at an inter-individual level, for example during a translator's interactions with colleagues or clients. A translator may smile at a client to express happiness, or the acoustic properties of the translator's voice can be affected by his anger in social interactions with colleagues. Moreover, variations in body movements, such as the adoption of expansive and diminutive body postures, can be behavioural indicators of emotional reactions (Tracy & Robins, 2004) and can convey specific information about the translator's emotional state while translating.

16.2.3.4 The translator's subjective feeling

During an emotion episode, activation in the different emotion components gives rise to the translator's feeling, i.e. the conscious experience of emotion. The subjective feeling component underlies the experience of feeling good or bad, the experience of a particular discrete emotion, and the extent to which translators enjoy the translation task. This emotion component is continuously updated and has an important monitoring and regulatory function. A negative feeling, for example, allows the translator to make regulatory efforts and to employ an emotion regulation strategy, such as situation selection, where a translator may choose not to translate a particular text to avoid frustration.

16.2.3.5 The dynamic relations of the emotion components

While zooming in on single emotion components allows us to describe an entire emotional episode as well as the processes within the five emotion components, it is important to note that these components are not independent of each other and interact in complex and dynamic ways during the translation process. As appraisals and the changes they produce form the context for the following subjective evaluations of a situation, recursive effects of an elicited emotion can occur, and emotions may not only exert a momentary influence but impact on whole sequences of action during the translation process. Recursive effects of emotions happen through feedback from the pattern of emotional reaction in the different components on the ongoing process of cognitive evaluation, which then again influences the other emotion components, namely subsequent action tendencies, physiological response, expression and subjective feeling.

16.3 Recent developments and future directions

16.3.1 Empirical investigations into influences of emotion on the translation process and translation performance

16.3.1.1 Exploratory empirical evidence from think-aloud protocols

Against this theoretical backdrop, it is now time to turn our attention towards the more recent empirical evidence that has been provided for influences of emotion on the translation process and on the resulting product. At earlier stages of translation process research, several studies based on think-aloud protocols provided some exploratory empirical evidence for the relevance of emotion for the translation process and insights into how emotions may influence translation performance. In many cases, this evidence emerged accidentally as a by-product of research that had an essentially exploratory nature. In Kussmaul's (1991) study, which he describes as an attempt to "find out more about [...] translation problems which involve creative thinking" (1991, p. 91), two translators were asked to discuss their activities with the experimenter while translating a text from English into German. From his study, Kussmaul concluded that creative solutions during the translation process are associated with moments of positive affect. A few years later, Tirkkonen-Condit's and Laukkanen's (1996) explicit aim was to "shed light on the affective side of translators' decisions" (Tirkkonen-Condit & Laukkanen, 1996, p. 45). They analysed evaluative statements of professional translators in think-aloud protocols and compared differences between routine tasks and non-routine tasks. Tirkkonen-Condit and Laukkanen concluded from their observations that the "affective differences" (1996, p. 48) between the tasks may have been related to the better quality of the translation in the routine task. In particular, they suggested a positive relation between a translator's confidence and translation quality, for "in a feeling of security" (1996, p. 50), the subject was more likely to assume the "role of a communicator" (1996, p. 56) and detach herself from the source text. Even if Tirkkonen-Condit's and Laukkanen's study involved only two subjects, they suggested a new relation between a translator's affective state and translation performance. In particular, their findings indicated that detachment from the source text may be emotion sensitive. Moreover, Jääskeläinen (1996) compared think-aloud protocol data and translation evaluations from a sample that was composed of subjects with three different proficiency levels: bilingual laymen, students and professionals. Jääskeläinen's comparison revealed that contrary to the expectations underlying the study, neither a translator's degree of proficiency nor specific translation procedures accounted for translation quality. Rather, Jääskeläinen pointed out, "the effort invested in the process bears fruit as higher translation quality" (1996, p. 66). Her results illustrated the relevance of motivation, which is largely guided by the affective system (Higgins, 2009), for translation performance. In a similar vein, Fraser (1996) observed, in a study relying on think-aloud protocols and retrospective data, that in order to achieve high standards in their work, professional translators invest emotional commitment. In conclusion, the studies mentioned provided indications that affective factors can influence translation performance in different ways: either emotions promote translation quality by enhancing creativity or through motivational processes, or emotions have a negative influence by inhibiting detachment from the source text. It was a methodological issue, however, that all the studies relied on very small samples. Building on this exploratory evidence, further translation scholars set out to examine the role of affect and emotion with larger samples.

16.3.1.2 Larger-scale studies focusing on influences of emotion on translators' decision making and translation performance

Acknowledging the potential significance of emotion for the translation process as well as the resulting product, Lehr (2014) conducted a focused empirical investigation aiming at providing insight into how translators' emotional state influences decision making and performance. The two-phase study relied on reader responses and expert evaluations of translation quality and involved 42 professional translators, who completed two translations in their usual work environments. Starting from the notion that the text is an important emotional stimulus when translating (Hansen, 2006; Lederer, 2003), but that also other situational aspects can elicit emotional reactions, such as, for example, feedback (Kussmaul, 1991), the study focused on the relation between two different types of emotions, integral and incidental emotions, and performance in translation. The first phase of the study examined the influence of integral text-related emotions on the emotionality of a translated text, that is, the text's potential to prompt an emotional response. Based on the assumption that this relation could be explained through emotion-congruence effects and primed emotional information (Niedenthal et al., 1997), translations by translators who had themselves experienced a more or a less intense emotional response to the text were rated for emotionality by readers and compared. No evidence in support of the assumption that the emotional response of the translator influences the emotionality of the translation was found, suggesting that emotion-congruence effects at lower levels of processing only have a limited impact on more controlled processes and the translation product. Indeed, the study's results indicate, rather, that emotion-congruence effects in more controlled language processing, such as translation, may be subject to other processes, for example intentions to overcome a bias.

In the second part of the study, Lehr (2014) examined the influence of incidental emotions, which the literature has associated most notably with the processing consequences of affective states and differences in accuracy and creativity (Bohner & Schwarz, 1993; Kussmaul, 1991). With this aim, translations from translators who had received either positive or negative feedback on a previous translation task were compared. The feedback induced emotions that were clearly separated on the positive–negative axis. Moreover, the comparison of translation evaluations between the two groups showed higher ratings for idiomatic expression and stylistic appropriateness after positive feedback, criteria that can be attributed to the creativity category, and higher ratings for terminology after negative feedback, a criterion that can be attributed to accuracy in translation. The study thus found that translators' emotional state can influence the translated text through its influences on cognitive processing style. More specifically, emotions aroused by positive or negative performance feedback seem to have an influence on particular aspects of accuracy, fundamental to all translation activity, and creativity, which is necessary at certain points in the text to varying degrees. Having framed the translation process within the componential view of emotion, we would assume that influences of emotion on accuracy and creativity in translation occur through the recursive effects of emotion on subsequent appraisals during the translation process. The results outlined here seem to indicate that the influence of affective states may be manifested most clearly in instances when there is a need for something "on top" of the basic routine processes, such as, for example, very careful scrutiny for terminology or finding a particularly idiomatic formulation. Effects of affective states may thus be especially impactful when the task becomes increasingly difficult and translators cannot draw on routinized solutions. Similar tendencies, namely that positive affect seems to promote creativity and negative affect accuracy, were observed by Rojo & Ramos Caro (2016), although their results lacked statistical significance. In line with the results from the first phase of Lehr's (2014) study, their observations indicate that decision making in professional translation may only to a certain extent be susceptible to emotion effects, and that emotion effects in the translation process may be subject to routine procedures and other controlled processes, for example motivations to be accurate. How emotions influence translation performance, and how strong the effects are, no doubt depends on numerous variables that remain to be further investigated and have to be integrated with other factors that influence emotion processes, such as inter-individual differences between translators.

16.3.2 Differences between individual translators—emotions as traits and competences

Research suggesting that emotional states may influence translation performance and therefore the actual quality of a translated text does not merely imply that emotions are a significant variable in translation performance; it allows us to draw an additional important conclusion. If translators' emotional states influence their performance, then not only are translators supposed to identify and accurately re-express emotions, but regulating and using emotions is also part of performing well as a translator. In line with this conclusion, Hubscher-Davidson (2014; 2018) emphasizes that emotion processes can be related to or moderated by inter-individual differences in emotional competences. Emotional competences can be collectively defined as the ability "to optimally use the emotion mechanism as it has been shaped by evolution" (Scherer, 2009, p. 92), and they can be divided into four core competences: understanding, identification, utilization and regulation of emotion (Mayer & Salovey, 1997). Psychological research has studied emotional competences from different perspectives. One perspective sees emotional competences as a set of abilities and a form of intelligence (Mayer & Salovey, 1997); others perceive them as conceptually related to dimensions of personality, and therefore a set of personality traits related to emotion (Petrides & Furnham, 2003). Debates on the status of emotional competences as intelligence or trait have given birth to a tripartite model, which tries to capture all aspects of emotional competence and posits three levels: knowledge, abilities and traits (Mikolajczak et al., 2009). The knowledge level refers to the complexity and breadth of knowledge about emotions; for example, our knowledge about which emotion regulation strategies exist and how efficient they are. The ability level of emotional competence refers to the ability to apply such knowledge in an emotional situation and to implement a given strategy. It focuses on what people are able to do. Finally, the trait level of the tripartite model refers to people's tendencies to behave in a certain way in emotional situations. The focus of this third level is not on what people are able to do but on what people consistently do. It addresses their dispositions; whether and how they use their emotion-related knowledge and abilities in their everyday life.

In particular, the last trait level of emotional competence has attracted the interest of translation scholars. Regarding emotional competences as personality-based and emotion-related dispositions, Hubscher-Davidson (2018) conducted an extensive study comprising 155 professional translators. Using the Trait Emotional Intelligence Questionnaire (Petrides, 2009) as an assessment instrument, her aim was to explore professional translators' individual differences in emotional competences, and how they are related to various aspects of behaviour and professional success. In her study, Hubscher-Davidson found several interesting correlations. For example, she reports that emotion regulation skills are positively associated with the acquisition of literary translation experience. Although the direction of the effect remains unclear, as translators with good competences in emotion regulation may also benefit from good interpersonal relations and therefore be able to acquire more clients and professional experience, Hubscher-Davidson suggests that translators may develop their emotion regulation skills through literary translation work and the emotional aspects it involves. She argues that one could assume that the more they translate literature, the more translators engage in multifaceted emotional experiences and have opportunities to improve the way they handle these. With regard to emotion regulation, Hubscher-Davidson also finds a trend towards a positive association between emotion regulation competences and translators' job satisfaction and success. She explains that inadaptive emotion regulation strategies, such as suppression, may impair translators' performance. They may lead translators to experience more stress or fewer good interpersonal contacts in teams or with clients, and they may also interfere with the self-control translators must have in order to know the limits of their task and not to interfere with the author's original work. Further, Hubscher-Davidson finds a trend for translators' job success to be positively associated with trait emotional competences in emotion expression. In an attempt to integrate situational and personality-related aspects of translation performance, Hubscher-Davidson suggests that one could explain this result through the relation between positive affect and creativity, addressed by other studies in translation process research (Lehr, 2014; Rojo & Ramos Caro, 2016). As individuals high in emotion expression tend to experience more positive affect, which is conducive to a more creative processing style, these translators may have more instances of creative expression and be able to produce translations of a higher quality.

In the preceding sections, we have studied how both translators' situational affect and emotional competences can influence translation performance. The existing evidence allows us to conclude that the ability to deal with emotions and, consequently, emotional competences should be viewed as an integral part of translation competence. Hence, emotional competences should also be integrated into translator training in order to enhance translators' employability and to prepare them to handle emotions they may encounter in the workplace. This may involve training translators to apply specific emotional competences, such as emotion regulation strategies, in their professional life and may also include the integration of some theoretical knowledge about emotion in translator education.

16.4 Concluding remarks

In this chapter, we have reviewed how emotion, which, for a long time, was neglected as the unwanted irrational side of human thinking and behaviour, has made its way into translation process research. Emotion and affect emerged accidentally in think-aloud protocols as an influential variable, and their relevance was acknowledged through the integration of theory from neighbouring disciplines, arguing that emotion cannot be dissociated from human thinking processes. Based on targeted empirical studies during the past years, we can now conclude, on empirical grounds, that emotion is an important variable in translation performance and that it should be integrated into our conception of translation competence. As emotions have only fairly recently become a topic of focused investigation in translation process research, numerous questions remain to be addressed by future research to increase our understanding of how and when emotions impact on translators' work, and how situational and inter-individual factors interact in this influence. It follows, also, that emotion should become a topic of learning in the training of translators in order to enhance translators' emotional competences through specific exercises. In addition, we will have to address topics raised by new technological developments, such as the role of emotion in the adaptation to and interaction with technology, the capacity of machine translation engines to render emotional content, or the influence of emotion on new forms of translation activity, such as post-editing or transcreation. Addressing these issues should allow us to continuously develop our models of the translation process and to get a more comprehensive understanding of performance in translation, now and in the future.

Note

1 According to Scherer's (2005) emotion theory, the following five organismic subsystems are involved in an emotion episode and underlie the different emotion components:

Organismic subsystem	Emotion component
Information processing (evaluation of objects and events)	Appraisal component
Support (system regulation)	Physiological component
Executive (preparation and direction of action)	Action tendency component
Action (communication of reaction and behavioural intention)	Expression component
Monitor (monitoring of internal state and organism–environment interaction)	Subjective feeling component

Further reading

Fredrickson, B. L., & Branigan, C. A. (2005). Positive emotions broaden the scope of attention and thought-action repertoires. *Cognition and Emotion, 19*(3), 313–332.
A central study providing evidence for the broaden-and-build theory of positive emotions.

Hubscher-Davidson, S. (2018). *Translation and emotion. A psychological perspective*. New York: Routledge.
This recent book provides a comprehensive overview of the role of emotion in the translation process.

Loewenstein, G., & Lerner, J. S. (2003). The role of affect in decision-making. In R. Davidson, K. R. Scherer, & H. Goldsmith (Eds.), *Handbook of affective sciences* (pp. 619–642). New York: Oxford University Press.
A key paper that well explains how affect and emotion may influence decision making.

Rojo, A., & Ramos Caro, M. (2016). Can emotion stir translation skill? In R. Muñoz Martín (Ed.), *Reembedding translation process research* (pp. 107–130). Amsterdam: John Benjamins.
One of the rare studies in translation process research examining how emotion influences creativity in translation.

Scherer, K. R. (2005). What are emotions? And how can they be measured? *Social Science Information, 44*, 693–727.
One of the fundamental theoretical papers in emotion theory, outlining the component process model of emotion.

References

Angelone, E. (2010). Uncertainty, uncertainty management, and metacognitive problem solving in the translation task. In E. Angelone, & G. M. Shreve (Eds.), *Translation and cognition* (pp. 17–40). Amsterdam: John Benjamins.
Arnold, M. (1960). *Emotion and personality*. New York: Columbia University Press.
Barber, P. (1988). *Applied cognitive psychology: An information processing framework*. London, New York: Methuen.
Bayer-Hohenwarter, G. (2009). Methodological reflections on the experimental design of time-pressure studies. *Across Languages and Cultures, 10*(2), 193–206.
Bechara, A. (2004). The role of emotion in decision-making: Evidence from neurological patients with orbitofrontal damage. *Brain & Cognition, 55*, 30–40.
Bohner, G., & Schwarz, N. (1993). Mood states influence the production of persuasive arguments. *Communication Research, 20*(5), 696–722.
Bradley, M. M., Miccoli, L., Escrig, M. A., & Lang, P. J. (2008). The pupil as a measure of emotional arousal and autonomic activation. *Psychophysiology, 45*, 602–607.
Chesterman, A. (2002). Semiotic modalities in translation causality. *Across Languages and Cultures, 3*(2), 145–158.

Damasio, A. R. (1994). *Descartes' error. Emotion, reason and the human brain*. New York: Harper Collins.
Danks, H. J., & Griffin, J. (1997). Reading and translation: A psycholinguistic perspective. In H. J. Danks, G. M. Shreve, S. B. Fountain, & M. K. McBeath (Eds.), *Cognitive processing in translation and interpreting* (pp. 161–175). Thousand Oaks: Sage.
Durieux, C. (2007). L'opération traduisante entre raison et émotion. *META Translators' Journal, 52*(1), 48–55.
Ekman, P. (1992). An argument for basic emotions. *Cognition and Emotion, 6*, 196–200.
Ellsworth, P., & Scherer, K. R. (2003). Appraisal processes in emotion. In R. J. Davidson, K. R. Scherer, & H. Goldsmith Hill (Eds.), *Handbook of affective sciences* (pp. 572–595). Oxford: Oxford University Press.
Forgas, J. P. (1995). Mood and judgment: The affect infusion model. *Psychological Bulletin, 117*(1), 39–66.
Fraser, J. (1996). Mapping the process of translation. *META Translators' Journal, 41*(1), 84–96.
Fredrickson, B. L. (1998). What good are positive emotions? *Review of General Psychology, 2*, 300–319.
Fredrickson, B. L., & Branigan, C. A. (2005). Positive emotions broaden the scope of attention and thought-action repertoires. *Cognition and Emotion, 19*(3), 313–332.
Fridja, N. H. (1988). The laws of emotion. *American Psychologist, 43*(5), 349–358.
Fridja, N. H. (2009). Mood. In D. Sander, & K. R. Scherer (Eds.), *The Oxford companion to emotion and the affective sciences* (p. 258). Oxford: Oxford University Press.
Fridja, N. H., & Scherer, K. R. (2009). Affect. In D. Sander, & K. R. Scherer (Eds.), *The Oxford companion to emotion and the affective sciences* (p. 10). Oxford: Oxford University Press.
Gross, J. J., & Thompson, R. A. (2007). Emotion regulation: Conceptual foundations. In J. J. Gross (Ed.), *Handbook of emotion regulation* (pp. 3–24). New York: Guilford Press.
Hansen, G. (2006). *Erfolgreich Übersetzen, Entdecken und Beheben von Störquellen*. Tübingen: Gunter Narr.
Higgins, E. T. (2009). Motivation. In D. Sander, & K. R. Scherer (Eds.), *The Oxford companion to emotion and the affective sciences* (pp. 266–267). Oxford: Oxford University Press.
Hönig, H. G. (1995). *Konstruktives Übersetzen*. Tübingen: Narr.
Hubscher-Davidson, S. (2014). Emotional intelligence and translation studies: A new bridge. *META Translators' Journal, 58*(2), 324–346.
Hubscher-Davidson, S. (2018). *Translation and emotion. A psychological perspective*. New York: Routledge.
Izard, C. E. (1991). *The psychology of emotions*. New York, London: Plenum Press.
Izard, C. E. (2009). Emotion theory and research: Highlights, unanswered questions, and emerging issues. *Annual Review of Psychology, 60*, 1–25.
Jääskeläinen, R. (1996). Hard work will bear beautiful fruit. A comparison of two think-aloud protocols. In F. G. Königs (Ed.), special issue of *Les processus de la traduction—META Translators' Journal, 41*(1), 60–74.
James, W. (1884). What is an emotion? *Mind*, 9(34), 188–205.
Jensen, A., & Jakobsen, A. L. (2000). Translating under time pressure: An empirical investigation of problem-solving activity and translation strategies by non-professional and professional translators. In A. Chesterman, N. G. San Salvador, & Y. Gambier (Eds.), *Translation in context* (pp. 105–116). Amsterdam: John Benjamins.
Kintsch, W. (1988). The role of knowledge in discourse comprehension: A construction integration model. *Psychological Review, 95*, 163–182.
Krings, H. P. (1986). *Was in den Köpfen von Übersetzern vorgeht. Eine empirische Untersuchung zur Struktur des Übersetzungsprozesses an fortgeschrittenen Französischlernern*. Tübingen: Narr.
Krings, H. P. (2005). Wege ins Labyrinth—Fragestellungen und Methoden der Übersetzungsprozessforschung im Überblick. In H. Lee-Jahnke (Ed.), special issue of *Processus et Cheminements en Traduction et Interprétation—META Translators' Journal, 50*(2), 342–358.
Kuhn, T. S. (1962). *The structure of scientific revolutions*. Chicago: University of Chicago Press.
Kussmaul, P. (1991). Creativity in the translation process: Empirical approaches. In K.V. Leuven-Zwart, & T. Naaijkens (Eds.), *Translation studies: The state of the art* (pp. 91–101). Amsterdam: Rodopi.
Lederer, M. (2003). *The interpretive model*. Manchester, Northampton: St. Jerome.
LeDoux, J. E. (1995). Emotion: Clues from the brain. *Annual Review of Psychology, 46*, 209–235.
Lee-Jahnke, H. (2011). Trendsetters and milestones in interdisciplinary process-oriented translation: Cognition, emotion, motivation. In M. Forstner, & H. Lee-Jahnke (Eds.), *Translation and interpreting in a new geopolitical setting - CIUTI Forum 2010* (pp. 109–151). Bern: Peter Lang.
Lehr, C. (2014). The influence of emotion on language performance [PhD dissertation, University of Geneva]. https://archive-ouverte.unige.ch/unige:42306
Levenson, R. W. (1999). The intrapersonal functions of emotions. *Cognition & Emotion, 13*(5), 481–504.

Levenson, R.W., Ekman, P., & Friesen, W.V. (1990). Voluntary facial action generates emotion-specific autonomic nervous system activity. *Psychophysiology, 27*, 363–384.
Leventhal, H., & Scherer, K. R. (1987). The relationship of emotion to cognition: A functional approach to a semantic controversy. *Cognition and Emotion, 1*, 3–28.
Levy, J. (1967). Translation as a decision process. In *To honor Roman Jakobson: Essays on the occasion of his seventieth birthday* (Vol. II, pp. 1171–1182). The Hague: Mouton.
Loewenstein, G., & Lerner, J. S. (2003). The role of affect in decision-making. In R. Davidson, K. R. Scherer, & H. Goldsmith (Eds.), *Handbook of affective sciences* (pp. 619–642). New York: Oxford University Press.
McRae, K., & Gross, J. J. (2009). Emotion regulation. In D. Sander, & K. R. Scherer (Eds.), *The Oxford companion to emotion and the affective sciences* (pp. 337–339). Oxford: Oxford University Press.
Mauss, I. B., & Robinson, M. D. (2010). Measures of emotion: A review. In J. De Houwer, & D. Hermans (Eds.), *Cognition and emotion: Review of current theories and research* (pp. 99–127). Hove, New York: Psychology Press.
Mayer, J. D., & Salovey, P. (1997). What is emotional intelligence? In P. Salovey, & D. Sluyter (Eds.), *Emotional development and emotional intelligence: Educational implications* (pp. 3–31). New York: Basic Books.
Mikolajczak, M., Petrides, K.V., Coumans, N., & Luminet, O. (2009). An experimental investigation of the moderating effects of trait emotional intelligence on laboratory-induced stress. *International Journal of Clinical and Health Psychology, 9*, 455–477.
Moser-Mercer, B. (2003). Remote interpreting: Assessment of human factors and performance parameters. Retrieved December 2018 from https://aiic.net/page/1125/remote-interpreting-assessment-of-human-factors-and-pe/lang/1
Nida, E., & Taber, C. (1969). *The theory and practice of translation*. Brill: Leiden.
Niedenthal, P. M., Halberstadt, J. B., & Setterlund, M. B. (1997). Being happy and seeing "Happy": Emotional state mediates visual word recognition. *Cognition & Emotion, 11*(4), 403–432.
Padilla, B., Bajo, M. J., & Padilla, F. (1999). Proposal for a cognitive theory of translation and interpreting: A methodology for future empirical research. *The Interpreter's Newsletter, 9*, 61–78.
Petrides, K. V. (2009). *Technical manual for the Trait Emotional Intelligence Questionnaires (TEIQue)*. London: London Psychometric Laboratory.
Petrides, K.V., & Furnham, A. (2003). Trait emotional intelligence: Behavioural validation in two studies of emotion recognition and reactivity to mood induction. *European Journal of Personality, 17*, 39–57.
Pym, A. (2005). Text and risk in translation. In K. Aijmer, & C. Alvstad (Eds.), *New tendencies in translation studies* (pp. 69–82). Göteborg: Göteborg University.
Reed, S. K. (1982). *Cognition: Theory and applications*. Monterey, CA: Brooks Cole.
Risku, H. (2004). *Translationsmanagement—Interkulturelle Fachkommunikation im Informationszeitalter*. Tübingen: Gunter Narr.
Robinson, D. (1991). *The translator's turn*. London: John Hopkins University Press.
Rojo, A., & Ramos Caro, M. (2016). Can emotion stir translation skill? In R. Muñoz Martín (Ed.), *Reembedding translation process research* (pp. 107–130). Amsterdam: John Benjamins.
Russell, J. A. (2003). Core affect and the psychological construction of emotion. *Psychological Review, 110*(1), 145–172.
Sander, D., Grandjean, D., & Scherer, K. R. (2005). A systems approach to appraisal mechanisms in emotion. *Neural Networks, 18*, 317–352.
Sander, D., & Scherer, K. R. (2009). *The Oxford companion to emotion and the affective sciences*. Oxford: Oxford University Press.
Scherer, K. R. (2005). What are emotions? And how can they be measured? *Social Science Information, 44*, 693–727.
Scherer, K. R. (2009). The dynamic architecture of emotion: Evidence for the Component Process Model. *Cognition & Emotion, 23*(7), 1307–1351.
Schwarz, N. (2002). Situated cognition and the wisdom of feelings: Cognitive tuning. In L. Feldman Barret, & P. Salovey (Eds.), *The wisdom of feeling* (pp. 144–166). New York: Guilford Press.
Schwarz, N., & Clore, G. L. (1983). Mood, misattribution and judgments of well-being: Informative and directive functions of affective states. *Journal of Personality and Social Psychology, 45*, 513–523.
Seleskovitch, D., & Lederer, M. (1984). *Interpréter pour traduire*. Paris: Didier Erudition.
Shreve, G. M., & Koby, G. S. (1997). *What's in the "black box"? Cognitive science and translation studies*. Thousand Oaks: Sage.
Shuman, V., & Scherer, K. R. (2014). Concepts and structures of emotions. In R. Pekrun, & L. Linnenbrink-Garcia (Eds.), *International handbook of emotions in education* (pp. 13–35). Oxford: Oxford University Press.

Smith, C. A., & Lazarus, R. S. (1990). Emotion and adaptation. In L. A. Pervin (Ed.), *Handbook of personality: Theory and research* (pp. 609–637). New York: Guilford.

Tirkkonen-Condit, S., & Laukkanen, J. (1996). Evaluations—A key towards understanding the affective dimension of translational decisions. In F. G. Königs (Ed.), special issue of *Les processus de la traduction—META Translators' Journal, 41*, 45–59.

Tracy, J. L., & Robins, R. W. (2004). Show your pride: Evidence for a discrete emotion expression. *Psychological Science, 15*(3), 194–197.

Uchino, B. N., Ruiz, J. M., & Holt-Lunstad, J. (2009). Stress. In D. Sander, & K. R. Scherer (Eds.), *The Oxford companion to emotion and the affective sciences* (p. 383). Oxford: Oxford University Press.

Zajonc, R. B. (1980). Feeling and thinking: Preferences need no inferences. *American Psychologist, 35*, 151–175.

17

Translation, creativity and cognition

Gerrit Bayer-Hohenwarter and Paul Kußmaul

17.1 Introduction

17.1.1 Why does creativity matter in Translation Studies?

Creativity is the pivotal term in the title of this chapter, and it is explored from a cognitive perspective here. According to conventional wisdom, someone who writes creatively produces texts that did not exist before; translators, however, are bound to source texts and have to be in some sense faithful to them. In Translation Studies this widespread notion has traditionally been reflected in terms such as equivalence, adequacy and invariance (e.g. Koller, 1979/2011; Neubert, 2004).

Nevertheless, for quite some time it has been popular in Translation Studies, as in many areas of life, to use the term *creativity* as an attractive phrase suggesting prestige. But this does not mean that one has come to grips with what creativity really means. In the old tradition of faithful translation, it was not felt that there was such a need.

It was Christiane Nord (1989), who opened the door to more freedom when she suggested replacing "faithfulness" with "loyalty". Loyalty, as seen within a functionalist paradigm, notably the *skopos* theory of Hans J. Vermeer (see Nord, 1997, pp. 123–128; Vermeer, 1989), takes account of the fact that the translator has a double responsibility: to the partners on the source-text side and also to the partners on the target-text side. The translator, when taking account of the partners on the target-text side, will quite often find it suitable to produce shifts, that is, use words that are not literal reproductions of the source text but new expressions, which, as will be seen, can be called creative solutions. *Skopos* theory can indeed be seen as the overall framework for creative translation.

Creative translation has become an important topic in Translation Studies in recent years. In times when translation memories and machine translation play an ever-increasing role in the translation business, and in times when the "digital revolution" is causing a disruption of traditional business models in many professions, it is important to point out those qualities of translation services that only human translators are able to deliver.

17.1.2 Translational creativity research in a nutshell

Whereas the psychological discipline of creativity research saw its birth in 1950, creativity in translation research was discussed predominantly within the literal-versus-free debate until the

1990s. According to Wilss, the reason was that translational creativity could "neither be clearly conceptualized, nor measured, nor weighted nor described precisely" (Wilss, 1988, p. 111, our translation). The first empirical study seems to be Wilde's (1994) type/token analysis. It investigated the relation between Language for Specific Purposes (LSP) elements and creative elements in promotional/advertising texts and found that even LSP-specific syntactic structures can work as creative elements (Wilde, 1994, p. 25). Also in the 1990s, Kußmaul ventured to undertake a large-scale series of investigations into translational creativity. His work was based on observation and empirical data from the translation classroom and demonstrates the value of cognitive and psychological insights for translation research, translation practice and translator training (Kußmaul, 1991, 1993, 1998, 2000a, 2000b, 2000c, 2000d, 2005, 2007).

Even today, much of the research appears to be limited to the perception that creativity in translation is what goes beyond any literal rendition and is studied in the context of literary translation (e.g. several articles in Cercel et al., 2017; Mariaule, 2017). Other research into translational creativity has been of limited scope. Studies were carried out frequently with regard to specific text types, e.g. promotional texts (Jettmarová, 1998; Quillard, 1998, 2001), technical texts (Byrne, 2006; Durieux, 1991; Schmitt, 2005), legal texts (Nida, 1998; Pommer, 2006; Šarčević, 2000), religious texts (Nida, 1998; Nord, 2005), audiovisual translation (Chaume Varela, 1998; Fontcuberta i Gel, 1997) and popular music (Kaindl, 2005). Other investigations have dealt with pedagogic aspects (Bastin, 2000, 2003; Forstner, 2005; Lee-Jahnke, 2005; Mackenzie, 1998). Dancette et al. (2007) developed criteria for translational creativity; Hubscher-Davidson (2005, 2006) and Hague (2009) place the focus of their research on the creative personality of translators. Thomä (2003) created a translation-specific creativity test and analysed the creativity of 30 students in their first or second semester of studies in English Language and Literature with that of 16 professional translators in her PhD thesis. Kenny (2001, 2006) and Laviosa (1998) compared the degree of creativity inherent in source texts and target texts and observed a trend towards less original and more conventional English target texts, which Stewart (2000) attributes to the use of corpora by translators. A conference in 2005 in Portsmouth (Kemble & O'Sullivan, 2006) brought practising translators, researchers and translation teachers together. In translation process research, Heiden (2005) carried out an investigation from a keylogging study. She found that most creative translations are first created in the main phase and that long revision phases are a strong indicator of creative translation processes. Fontanet (2005) conducted self-experiments with technical texts and found that the problem-solving processes use divergent thinking differently for comprehension problems and production problems. Audet (2008a, 2008b) reports on a framework for the analysis of translational creativity that was developed following the analysis of think-aloud data of translation processes. It resembles text-analytic approaches and allows qualitative creativity assessment. Other process-oriented studies were carried out by Kußmaul (2007); Hubscher-Davidson (2005, 2006); and Cho (2006).

The large number of studies on aspects of limited scope and conceptualization issues has paved the way for quantitative research. The focus of interest is now on large-scale empirical investigations that extend beyond sample text analyses and the discussion of conceptualization issues, e.g. research such as Adamczuk (2005) or the corpus linguistic studies by Kenny (2001, 2006); Laviosa (1998); and Stewart (2000). Studies using psycholinguistic methods such as key-logging or think-aloud were carried out by Kußmaul (2007); Fontanet (2005); Heiden (2005); Hubscher-Davidson (2005, 2006); Cho (2006); Audet and Dancette (2005); Dancette et al. (2007); and Audet (2008a, 2008b). On a cognitive level of analysis, Kußmaul (2000d, p. 31) introduced the concept of obligatory shifts (which also entails the existence of "optional shifts") and developed his types of creative translation. The term *shift* has long been used in translation theory from the point of view of structural semantics (e.g. Catford, 1965), but Kußmaul's typology of cognitive shifts was based on scenes-and-frames theory (Kußmaul, 2000b, 2000c; see

Bayer-Hohenwarter, 2009, p. 42, for a more detailed discussion). In contrast to shifts, literal translation is, according to Ballard (1997, p. 90), a sort of conscious or unconscious ideal of equivalence. Observations by Englund Dimitrova (2005) and empirical evidence by Zhong (2005) and Tirkkonen-Condit et al. (2008) also support this view. Here, all authors seem to adhere to the general consensus that "the literal translation reflex" is presumably a cognitive universal, but certainly not a universally valid, acceptable and accepted quality standard.

The study by Bastin and Betancourt (2005) was the first to measure creativity according to strict criteria and analyse the development of translational creativity. From 2008 onwards, a large-scale empirical study on translational creativity was carried out within the TransComp project (Bayer-Hohenwarter, 2012; Göpferich, 2009; Göpferich et al. 2008; Göpferich et al., 2011). In this context, a comprehensive framework and a sophisticated creativity assessment procedure (see e.g. Bayer-Hohenwarter, 2010, p. 98) were developed for the analysis of translational creativity. The creative performance of different translators, such as students in different semesters (first to sixth), was compared with that of professionals, and the development of their performance in time was traced. This study comprised an analysis of 652 source text (ST)–target text (TT) pairs from 163 experimental trials.

Regarding the results of substantial empirical studies on translational creativity, the following results seem to be based on sound evidence:

- **Divergent thinking**: It functions optimally with sufficiently developed evaluation competence (Bastin, 2003, p. 353; Kußmaul, 2000d, pp. 76–80).
- **Normalization**: Strong normalization trends in translations into English occur across various text types (Kenny, 2001, 2006; Laviosa, 1998; Stewart, 2000). The opposite trend was observed for translations of promotional texts into French (Quillard, 1998, 2001).
- **Metaphorization**: It is an important creative translation strategy (Adamczuk, 2005); ST metaphors in non-literary texts that are highly conventionalized are usually preserved by translators in the TT (Pisarska, 1989).
- **Learnability**: Translational creativity is competence dependent and thus potentially learnable and teachable (Bayer-Hohenwarter, 2012, p. 299).
- **Switch competence**: The more experienced translators are better able to switch more economically between a cognitively less demanding routine mode and a cognitively more demanding creativity mode (Bayer-Hohenwarter, 2012, p. 302); see also Section 17.2.4.
- **Routinized creativity**: The use of certain creative strategies has become routine for professional translators (Bayer-Hohenwarter, 2012, p. 235, Göpferich et al., 2011, p. 75).
- **Creativity in non-literary texts**: Creative cognitive processes are also useful in the translation of non-literary (instructive) texts (Bayer-Hohenwarter, 2011a, 2012, p. 314; Popescu & Cohen-Vida, 2015).

17.2 Core topics

17.2.1 The creative product

In creativity research, the creative product is defined as both new and appropriate to the task (see, for instance, the survey by Preiser, 1976, pp. 1–7). Accordingly, in Translation Studies, a creative translation is a translation that often involves changes (as a result of shifts) when compared with the source-text, thereby bringing in something that is new and also appropriate to the task that was set, i.e. to the translation assignment (or purpose). There seems to be agreement on this double quality of novelty and of appropriateness or adequacy. This is mentioned in many studies, for instance in Kußmaul (2000d, *passim*), Siever (2010, p. 196), Bayer-Hohenwarter (2009, *passim*) (see also Section 17.2.4).

Changes are a matter of degree; there are small changes and big changes. Sometimes the translator can follow pre-existing linguistic patterns. For instance, Ballard quotes as an example of a creative translation *derrière Winston* for *behind Winston's back* (1997, p. 93). The translation certainly involves a change, but it is a small one.

It is when there are no pre-existing patterns in the target language that a larger amount of creativity is involved. In the following example no shifts occur, but a new word is created. The quote is taken from August Wilhelm Schlegel's German translation of the well-known lines from Hamlet's monologue (act III, scene 1):

> And thus the native hue of resolution
> Is sicklied o'er with the pale cast of thought

While preserving rhyme and metre (which was the task he set himself), Schlegel created a new word for *sicklied over* and wrote:

> Der angebornen Farbe der Entschließung
> Wird des Gedankens Blässe angekränkelt
>
> *Shakespeare,* Dramatische Werke*, übersetzt von August Wilhelm Schlegel und Ludwig Tieck, Berlin: Lambert Schneider o. Jg.: 522*

Angekränkelt is a word that did not exist before. Grimms *Deutsches Wörterbuch* (Erster Band, Leipzig: S. Hirzel, 1854) quotes Schlegel as the source. He created it, and its meaning very precisely renders that of *sicklied over.* The fact that it is quoted by Grimm can be taken as a sign of its appropriateness (see Kußmaul, 2000d, pp. 30–31).

We quoted from literature, but creative translation is by no means restricted to literature. It can be observed in all kinds of texts; for instance, also in specific-purpose and in technical texts (Bayer-Hohenwarter, 2011a, 2012, pp. 53–54).

17.2.2 The creative process: The phase model

Many psychologists agree that the creative process can be divided into four phases: 1. preparation, 2. incubation, 3. illumination and 4. evaluation (for an overview see Preiser, 1976, pp. 42–49; Shorthouse & Maycroft, 2017, pp. 142–149). The four-phase model has, with some modifications, been taken over by translation researchers on creativity (e.g. Balacescu & Stefanink, 2006; Fontanet, 2005; Kußmaul, 1991, 2007).

The preparation phase focuses on the stage of comprehension of the source text, where text analysis and interpretation play a major part and where the function of the target text is being established. Moreover, in this phase problems are recognized, since problem solution is closely connected with creativity (see Kußmaul, 2000d, *passim*; Wilss, 1996, pp. 47–48). Comprehension itself can indeed be creative. Bayer-Hohenwarter presents an extensive list of researchers who share this opinion (Bayer-Hohenwarter, 2012, p. 115). She also noticed in her case studies that paraphrases, among other phenomena, can be a sign of "deeper" (and thus creative) understanding of a text passage (Bayer-Hohenwarter, 2012, pp. 115–117).

In the incubation phase, according to psychologists, thinking seems to be mainly associative and subconscious. According to Guilford (1975, p. 40), often set in motion by brainstorming, two kinds of thinking take place during this phase: "fluency of thinking", i.e. producing a large number of thoughts and ideas in a short space of time, and "divergent thinking", i.e. not thinking along strict and logical lines but finding several possible solutions to a problem (cf. Shorthouse & Maycroft, 2017, pp. 157–159). Observing these thinking processes in translators has proved difficult. In think-aloud protocols and in programs used to record and study human reading and writing processes on a computer (such as Translog), mental activities are mostly reflected in pauses. If these pauses have been creative, then they are followed by suggestions for a translation. This means that, after all, we can only observe the final products but not the actual processes (Kußmaul, 1991, p. 94, 2000d, *passim*).

The illumination phase appears to be closely connected with evaluation (Kußmaul, 1991, 2000d, pp. 76–80). Therefore, these two phases are not treated separately here. Preiser (1976) in his survey draws attention to the fact that phases do not normally simply follow each other in a sequence, but there are moves backward and forward (see Shorthouse & Maycroft, 2017, p. 154); in other words, the phase model is a theoretical construct. For translation this can mean, for instance, that (normally) silent verbalizations of what is comprehended in the preparation phase may, in fact, represent a creative translation, and evaluating solutions during the incubation phase can prevent good ideas from getting lost (Kußmaul, 2000d, pp. 79–80). Nevertheless, the phase model can be regarded as a means to structure observation of the process.

17.2.3 Visualization

In his book *The act of creation / Der göttliche Funke* (1966), the author and journalist Arthur Koestler points out that creative thinking is predominantly visual thinking, which is a specific kind of imagination. He illustrates this by a number of reports about famous poets and scientists. For instance, he mentions a dream of the chemist August Kekulé about the Uroboros, in which he saw a snake that bit itself in its tail. This revealed to him the shape of the benzene ring (see Koestler, 1966, pp. 174–182).

In the field of translation, the interpretive theory developed by the "Paris School" has recommended visualization as a method to build sense (Seleskovitch, 1978, p. 55; Seleskovitch & Lederer, 1989, pp. 24–26).

In order to explain what goes on in the minds of translators when they are creative, Kußmaul (2000d) has made use of cognitive semantics, especially of scenes-and-frames semantics, as developed by Charles Fillmore (1977), and prototype semantics, with its notions of core and fuzzy edges as presented by Eleanor Rosch (1973). When comprehending a text, translators produce mental representations, and these representations lead to translations. It seems that especially by visualizing (imagining) a scene with core elements, translators have good chances of producing creative solutions (Kußmaul, 2005). For instance, students in one of Kußmaul's classes were asked to translate the following text:

> 8 p.m. Off to dinner party. All the Smug Marrieds keep inviting me on Saturday nights now I am alone again, seating me opposite an increasingly horrifying selection of single men.
>
> *Fielding, 1996, p. 212*

The students were tape recorded while discussing the translation of "Smug Marrieds". They explicitly visualized a scene, a cliché: "the people that are settled in their [...] nice house with a garden, and then there is the dog, the cat", which, obviously, they retrieved from their memory. The details of this scene, being part of a cliché, are thus core elements by definition; in other words, they are prototypical. The visualization of a scene with prototypical elements (in Fillmore's sense) led to a translation. The students suggested: "der Club der ganz ekelhaft glücklich Verheirateten" (the club of the disgustingly happily married ones), which for the final version was slightly modified into "der Club der ganz ekelhaft glücklichen Ehepaare" (the club of the disgustingly happy married couples). This translation can be called creative, since it contains words expressing a negative attitude not explicitly expressed in the source text but certainly implied. The phrase thus highlights the feelings of the narrator (see Kußmaul, 2005, pp. 388–389). Visualizations also play a central role in the case study we present towards the end of our chapter (see later).

Bayer-Hohenwarter (2012, p. 111) adopted the concept of visualization and extended it to include other forms of imagination such as acoustic, arithmetic, graphologic, kinaesthetic and process-related imagination. Agnetta and Cercel (2017) approach the issue of how comprehension takes place in a similar way by conceptualizing it as an "act of hearing".

Martín de León (2017) took up Kußmaul's concept of visualization and looked at the mental imaging patterns of five Spanish students when they translated English texts into Spanish. The participants were asked to report about the images that occurred to them while translating, and to a certain extent a correlation between their ability to imagine and the quality of their translations could be observed. Due to the small scale of the study, the findings cannot be generalized but may still be seen as indicative (Martín de León, 2017, p. 16).

17.2.4 Defining, measuring and evaluating creative translating

In psychology, creativity has been assumed to be an elusive concept that seems to defy precise definition and measurement because of its multicomponential nature. According to Wittgenstein's idea of family resemblances (Wittgenstein, 1958; see Lakoff, 1987, p. 16), there are many concepts which cannot be defined by common properties with clear boundaries. Translational creativity is such a concept. Creative translation products and processes can be characterized by qualities such as rareness, outstanding quality, high cognitive effort, fluency or non-literalness, but none of these individual qualities are mandatory. Consequently, it is impossible to set up an exhaustive list of criteria that can reasonably be regarded as necessary and sufficient for a definition of translational creativity (cf. e.g. Amelang et al., 1981, p. 46). Two criteria, however, that any creative process or product must meet, as mentioned, are novelty and adequacy (e.g. Amabile, 1996; Csikszentmihalyi, 1997; Torrance, 1988). The factorial approach suggested by Guilford (1950) provides a very comprehensive framework. It comprises nine dimensions, or basic abilities, which are a prerequisite for creativity: novelty, fluency, flexibility, ability to synthesize, ability to analyse, ability to reorganize/redefine, complexity/span of ideational structure, and evaluation. Generally, it seems possible to attribute all manifestations of translational creativity, e.g. non-literalness, generativity as measured by Krings' variant factor (1988, 2001) or Kußmaul's types of creative translation (2000a, 2000c), to one of these dimensions, whereby novelty, fluency and flexibility are commonly perceived as the prototypical creativity dimensions.

In Bayer-Hohenwarter (2012, p. 92), several criteria were devised to measure creativity quantitatively across different units of analysis (i.e. chunks of text) and across different experimental texts, regardless of their text type. This measurement was carried out on a product level and on a process level of analysis using think-aloud data and Translog data. In this study, translational novelty was defined as a manifestation of (1) exceptional performance that considerably exceeded translational routine, (2) uniqueness or rareness within the TransComp data corpus (= originality) and (3) a non-obligatory translational shift (cf. Bayer-Hohenwarter, 2012, pp. 108–109; Kußmaul, 2000d, pp. 23–24), not all of which aspects must be present. Flexibility was defined as the ability to transgress fixedness (e.g. literalness in translation) and fluency as the ability to produce a large number of translation variants and/or adequate translation solutions spontaneously or even automatically. Acceptability was defined as *skopos* adequacy. In addition, translator profiles were described based on specific strengths and weaknesses in different areas of Guilford's framework, e.g. high fluency but little evaluation competence or high flexibility but little fluency.

One particular aspect of flexibility was developed starting from Kußmaul's types of creative translation (2000c) and the levels of categorization in the theory of basic-level primacy suggested by Brown (1958). With reference to this, it can be argued that target-text (TT) versions that belong to the same level of categorization as the corresponding source-text (ST) element can generally be considered "natural" and less creative than TT versions that belong to a different level of categorization. This explains why "literal" translations that are on the same level of categorization as the corresponding ST element are commonly (and reasonably) regarded as less creative than non-literal translations. Following this, the three basic creative procedures—abstraction, modification and concretization—were developed (Bayer-Hohenwarter, 2009, 2011b, 2012). This basic cognitive typology refers to "directions of thought" as in the tree structure used in terminology management, i.e. upward (searching for a more general way of expression), sideways (searching for a different way of expression) and downward (searching for a more specific way of expression) with reference to the ST element. These strategies are opposed to mere reproduction.

Example:
German ST: hund
TT abstraction: animal, mammal
TT concretization: poodle
TT modification: cur, puppy, lap dog, doggy
TT reproduction: dog

Abstraction refers to using TT solutions that are more vague, general or abstract TT as compared with the ST ("upward"). Concretization refers to the evoking of a more explicit, more detailed and more precise TT idea compared with the ST ("downward"). Modification includes strategies such as re-metaphorization or changes of perspective ("sideways"). Reproduction is a non-creative strategy involving the same cognitive level (not upward, downward or sideways). It is assumed that changes such as paraphrase, addition and deletion cannot be directly attributed to any one of these three procedures, because they refer to the linguistic form but are not necessarily cognitive concepts: An addition, for instance, means that a linguistic element has been added, but this can lead to a more abstract or a more precise idea in the TT according to Bayer-Hohenwarter's cognitive strategies. The basic creative procedures suggested may therefore have very different manifestations at the form level. It is assumed that all translation products can reasonably be assigned to either abstraction, modification, concretization or reproduction and that all procedures except reproduction can be considered creative because they deviate from the initial level of categorization, i.e. the level represented in the ST.

In the large-scale empirical study on translational creativity carried out within the TransComp project (Bayer-Hohenwarter, 2012; Göpferich, 2009; Göpferich et al., 2011), creativity was conceptualized using a number of creativity indicators, including the above-mentioned procedures abstraction, modification and concretization. The creativity assessment relies on a scoring system in which acceptability and several indicators reflecting the dimensions of novelty, flexibility and fluency, as measured in the translation process and in the translation product, are rewarded by bonus points. The scores for process and product creativity are transformed to percentages and added up to form an overall creativity score (for details see Bayer-Hohenwarter, 2010, 2012). In the study of 652 ST-TT pairs from 163 experimental trials, no direct correlation was found between high creativity and high formal levels of competence. Instead, in line with the notion of cognitive economy, the results were seen as an indication that the more experienced translators were able to switch more economically between a cognitively less demanding routine mode and a cognitively more demanding creativity mode (hence named "switch competence"). In line with this model of cognitive economy underlying translational creativity, the notion of "cognitive brake" was introduced (Bayer-Hohenwarter, 2012, pp. 306, 307). It was suggested that factors such as lack of concentration, lack of competence, fixation on irrelevant details, and laziness can prevent the activation of this cognitive brake.

On the basis of her experience with translations from English into German by non-native speakers of German, Hagemann suggested a principle for evaluating creativity in student translation tests that may also be applicable to translations into the mother tongue. She asked how far-reaching the positive effect of a creative translation was. She distinguished between two kinds of creative translation: 1) Translations that are just as adequate in the target context and culture and for a given translation assignment as the ST item and 2) translations that optimize the text (Hagemann, 2007, p. 108). Furthermore, she distinguished how far-reaching the optimizing creative translation was: a) It affected the text on a micro-level and two sentences succeeding each other, or b) it affected the text on a macro-level and thus more than two sentences (Hagemann, 2007, p. 108). A positive evaluation might be of considerable psychological value to students in a translation curriculum.

17.3 Some recent developments

As we have said, one of the main topics investigated within creative translating is creative shifts, i.e. what happens to the source text when it is translated. But some work has also been done on observing the comprehension process of the source text as a first stage of creative translating. Recently, Bayer-Hohenwarter (2017) has looked at comprehension problems in a real-life sample of translations from the point of view of complex analogical reasoning. She found that unsuccessful translations often result from a fixation on the surface structures of the source text instead of considering the source concept level and remaining open to various possibilities of interpretation. She advocates joint translations, in which both translator and reviewer should have the same language combinations, but one of them should have the language of the source text as his/her mother tongue. This would help minimize miscomprehension due to fixation.

Taking up the concept of switch competence and the notion of cognitive economy, Martínez Martínez and Teich (2017) operationalize the notion of routine in terms of entropy and surprisal, both measured in terms of the probability of a target-text unit in a given context. The more routinized the translators' behaviour, the lower the entropy and the lower the surprisal. They analysed 72 translations of 69 difficulty units in one source text. The professionals produced "translation spaces (i.e. target text solutions) with lower entropy than learners", and the learners produced target-text solutions with more variation. Moreover, it turned out that professionals tend to produce preferred solutions with a lower surprisal than trainees and that professionals' behaviour seems to be more consistent. The entropy of translation spaces appeared to be almost ten times higher for learners than for professionals. This seems to corroborate the assumption that the switch competence of professionals is more highly developed than that of learners, which allows them to work more economically.

Recently, a new term, "transcreation", has been added to translation, first suggested by Nina Sattler-Hovdar (2016) and taken up by Michael Schreiber (2017). It seems to widen the concept of creative translation and is used to highlight processes that go beyond what is traditionally called translation. It takes account, for instance, of newly invented slogans in marketing texts (e.g. BMW: Freude am Fahren. English: The ultimate driving machine). In Translation Studies, this kind of activity has been covered by the terms "localization", "rewriting" and "adaptation". Moreover, it has long found its place within Vermeer's (1989) *skopos* theory and Holz-Mänttäri's (1984) theory of translatorial action. But it can be regarded as a useful marketing label, and if translation is seen from a business point of view, translators may well be able to increase their chances of receiving higher fees by using this label.

An even wider view is taken by Risku, Milošević, et al. (2017a) in an empirical study based on the concept of situated cognition. They show that creativity in the translation departments of organizations is not necessarily restricted to the actual translations but also takes place in the interactions between people, and they identify a number of factors that contribute to creativity in the translation workplace. In an exploratory empirical study, Risku, Pichler, et al. (2017b) investigate the expectations of regular clients of an Austrian translation agency regarding the translation of marketing materials, thus placing the focus of research on the client perspective.

17.3.1 Case study

The following case study is an example from a translator's professional practice and may help to illustrate the relevance of a number of features of creative translating mentioned earlier and to understand current developments. The data is from a virtual social media platform for translators that was set up for professional exchange. An estimated number of several hundred members who rarely know each other in person, and frequently do not even register by their regular name but under an alias, use this platform for a variety of purposes. They discuss specific translation problems and general challenges of a translator's everyday life, such as defaulting clients, inappropriate inquiries and prices. The inquiry discussed in the following was made on 19 August 2017 in German and is here translated into English:

> I need to translate a somewhat delicate passage and simply don't know what to do.
>
> Context: A woman invited to an event would like a man to join her. The man usually wears jeans and T-shirts, but all of a sudden appears dressed in a suit. She is speechless and then explains to him that all women at the event will admire him because he is so handsome in this suit, because women like well-dressed men. And she says:
>
> "…now you're walking, talking suit porn."
>
> And later again: "… you're suit porn in the flesh."
>
> Any ideas how to translate "suit porn" [into German]? I've been trying to figure out a good translation since yesterday, but no adequate solutions yet. Many thanks!

Within 30 hours, 20 persons had made 117 comments or interactions (some people only posted links) and made 55 target-text suggestions, some of which were the following:

> Example 1: Purer Sex auf zwei Beinen.
> Literal translation: Pure sex on two legs.
> Example 2: Der Anzug ist der reine Schenkelöffner.
> Literal translation: This suit is the perfect leg-opener.
> Example 3: In dem Anzug schubst dich garantiert keine von der Bettkante.
> Literal translation: In this suit no woman will push you away from the edge of the bed.
> Example 4: Du bist ein wandelnder Porno in dem Anzug.
> Literal translation: In this suit you are a walking porn film.

The challenge for the translator seems to have been to correctly judge which level of sexual connotation was appropriate for the translation of "suit porn" in the given context. The following observations are worth noticing:

1) The 23rd comment suggests that the target-text suggestions with strong sexual connotations could be far off the mark. It is argued that "suit porn" is a social media term corresponding to "food porn", and that over the course of time a shift in meaning has taken place with the effect that the sexual connotation has lost importance or may even have disappeared altogether. According to Wikipedia, "food porn":

   > is a glamourized spectacular visual presentation of cooking or eating in advertisements, infomercials, blogs, cooking shows or other visual media, foods boasting a high fat and calorie content, exotic dishes that arouse a desire to eat or the glorification of food as a substitute for sex. Food porn often takes the form of food photography and styling that presents food provocatively, in a similar way to glamour photography or pornographic photography.

2) These comments were made by one male translator whose native tongue is probably (judging from his name and place of education) not German, and by a female translator whose mother tongue is clearly English. All translators who, judging by their names, are native speakers of German are mainly driven by primary associations and visualizations with a strong sexual connotation. The very few proponents of the theory that this sexual connotation might not necessarily be so prominent must vehemently defend their point of view. The female translator eventually does so by underlining that her mother tongue is English and that she is convinced of her judgement.
3) After 32 comments, the first source (*Urban Dictionary*) is cited. According to this, "talking suit porn" is an idiom with the following meaning:

   > Graphs, charts, spreadsheets, and power-point presentations illustrating vague descriptions of corporate growth, fiscal decline, projected expenditures, etc. Typically, those involved in authoring such empty and meaningless propaganda claim six-figure salaries yet produce no tangible profits, capital, or services for their organization which offsets the aforementioned salary.

4) For a long time, no person leaving a comment or link to this thread sets out to discuss the wider context or type of publication where the text is going to be published. It is assumed that it is a widespread phenomenon that members of such groups tend to disclose as little context as possible to avoid possible competitors getting in touch with their clients. Only the 41st posting mentions for the first time that the translator has googled the corresponding passage to identify its source. Comment no. 70 finally includes a link to the source text.
5) A look at the publication from where the example is taken shows that it is a romantic novel and that "suit porn" appears three times in the same chapter on two subsequent pages. The sexual connotation seems to be intended, but starting out with a vague allusion, gradually building up to a clear statement.
6) The translations of "suit porn" suggested so far and the discussion on the social media platform clearly show that creative translating must find a balance between novelty and adequacy. The translation examples 1) to 4) were obviously inspired by visualizations triggered by the seemingly strong sexual connotations of the phrase "suit porn". But were they adequate? What becomes clear in the discussion on the social media platform is that the phrase "suit porn" is lexicalized, and the meaning of "porn" has no sexual connotations any more. Nevertheless, in the context of the novel, the sexual connotations are, as it were, revived. The example thus very clearly shows two basic features of the comprehension process in general: Comprehension

is determined by (1) the role of the meaning as preserved in the lexicon, here in the *Urban Dictionary* and in the memory of the native speakers of English, and (2) by the role of the context, here the scene described earlier, where the man suddenly appears dressed in a suit.

The following target-text variants could be considered if, for the first vague allusion, a relatively unexcited, sober translation for "walking, talking suit porn" is needed corresponding to the model "food porn":

- Das ist doch das reinste Kopfkino. (Literal: This makes women's imaginations go wild.)
- Reißerischer Aufzug, reißerischer Anzug. (Literal: Lurid act, lurid suit.)
- Heiße Luft, heißes Outfit. (Literal: Hot air, hot outfit.)

17.3.2 Results

From the perspective of translational creativity, these examples from the collaborative working practice of translators are interesting for the following reasons:

1) Visualization and other forms of imagination can be regarded as cognitive default patterns just like literalness and equivalence. In order to fulfil the criterion of acceptability/adequacy, which is commonly regarded as one of the necessary prerequisites for a truly creative translation, the translator must be able to use his or her "cognitive brake". The examples given in this case study clearly show how dangerous it can be to make decisions solely on the grounds of the most flowery visualizations and linguistically and metaphorically most expressive primary associations. However tempting and creative primary associations may be, the focus must always consider the traditional parameters such as the translation task, context and medium, target group and text function.
2) Visualizations relying on frames and scenes may well take different forms depending on certain translators' characteristics, such as their mother tongue. For source-text comprehension, the spontaneous imaginations of native speakers of the source text can generally be assumed to be more reliable than those of native speakers of the target language or other languages.
3) The interactions taking place on the virtual translator platform as described in this case study may serve as a model of collaboration as practised in a contemporary translator's everyday life. It shows how translations can be the product of the cognitive resources of a large number of individuals with no commonalities except their profession and interest in a particular topic, which they voluntarily comment on. One might call this "Facebook brainstorming".
4) Even if translations in collaborative settings are often the product of the cognitive resources of a large number of individuals, it is the translator who is the contractor paid for the specific project and bound to clients' expectations. It is s/he who must take responsibility for selecting one of the many proposed target texts. The process of generating target-text suggestions may well be "outsourced", but judging which is the most suitable solution and deciding on the final target text always remains the task of the person commissioned to undertake the translation, the person bound by economic, institutional and social factors.

17.4 Concluding remarks

When we observe how the Internet, social media and the digital revolution influence the work of translators at present, we realize that they are bringing about vast changes. The Internet has long been used by translators to assist in finding translations with the lowest entropy and least surprisal, and hence the "mainstream solutions" that are least likely to be disputed by colleagues, clients and technical tools alike: Just google different target-text solutions to find out which one has more hits. This can make translations acceptable, but sometimes it can make them rather flat. Social media translator platforms can be used as more refined tools for gathering different opinions on how well suited a particular target-text solution is in a given context and for a particular target group. The case study has shown how creativity can even be "outsourced" and how crowd intelligence can speed up the translation process. For research, social media platforms can possibly be used as sources of empirical data in future studies.

The digital revolution that is under way is likely to affect all trades and professions, from factory workers to doctors and lawyers. What production robots, autonomous vehicles, surgery robots, telemedicine and legal technology are for these professions (e.g. Brünjes, 2016), Google translate, post-editing of machine translation, and translation vendor platforms are for the translation business.

In the context of these recent developments, creative translators seem to be those who understand (1) where high-quality human translation is needed, (2) how it can best be achieved, and (3) how it can best be sold to meet customers' expectations, for instance by using new labels (e.g. "transcreation") for old concepts.

Technological disruption has been affecting the translation business in a radical way and will most likely do so even more radically in the future. But the need for the highest possible translation quality is greater than ever in the case of the few types of translation projects for which computer-assisted translation (CAT) tools are of no avail. For these, translational creativity with a high degree of evaluation competence will still be needed.

Training translators to be creative will give them the chance to do a really good job. If they aim at the best, they will be good even when the best is not needed. And they will be able to recognize, evaluate and appreciate human peak performance in times when the creative cognitive abilities of translators are challenged by the digital revolution.

Further reading

Bayer-Hohenwarter, G. (2012). *Translatorische Kreativität. Definition, Messung, Entwicklung*. Tübingen: Narr.
An empirical study that measures and compares creativity and routine and their development on a product and process level in professionals and students on the basis of 652 source-text/target-text pairs from 163 experimental trials over three years.

Kußmaul, P. (2000). *Kreatives Übersetzen*. (Studien zur Translation 10). Tübingen: Stauffenburg.
A study that applies the results of cognitive linguistics and creativity research to translation.

Lakoff, G. (1987). *Women, fire, and dangerous things. What categories reveal about the mind*. Chicago/London: University of Chicago Press.
Lakoff explores the effects of cognitive metaphors on mental categories. His central concept of chaining of categories seems to have a close affinity with the notion of divergent thinking in creativity research.

Shorthouse, J., & Maycroft, N. (2017). *Where is creativity: A multidisciplinary approach*. London: Routledge.
A book that explores the place and role of creativity in our modern world from various points of view and is firmly based on creativity research.

References

Adamczuk, M. (2005). Task-specific creativity in simultaneous interpreting. [Unpublished dissertation.] University of Vienna.

Agnetta, M., & Cercel, L. (2017). Was heißt es, den (richtigen) Ton in der Übersetzung zu treffen? In L. Cercel, M. Agnetta, & M. T. Lozano (Eds.), *Kreativität und Hermeneutik in der Translation. Translationswissenschaft, Band 12* (pp. 185–213). Tübingen: Narr.

Amabile, T. M. (1996). *Creativity in context. Update to the social psychology of creativity.* Boulder/ Oxford: Westview Press.
Amelang, M., Bartussek, D., Stemmler, G., & Hagemann, D. (1981). *Differentielle Psychologie und Persönlichkeitsforschung.* 6. Stuttgart: Kohlhammer.
Audet, L. (2008a). La création dans le processus traductif. Analyse théorique et empirique de la littérarité dans les traductions en français d'une nouvelle hongroise. [Unpublished dissertation.] University of Montreal.
Audet, L. (2008b). Évaluation de la traduction littéraire: de la 'sensibilité à la littérarité' à la 'littérarité en traduction'. *TTR, 21*(1), 127–172.
Audet, L., & Dancette, J. (2005). Le mouvement de la création dans la traduction littéraire. *Bulletin suisse de linguistique appliquée, 81*, 5–24.
Balacescu, I., & Stefanink, B. (2006). Kognitivismus und übersetzerische Kreativität. *Lebende Sprachen, 51*(2), 50–61.
Ballard, M. (1997). Créativité et traduction. *Target, 9*(1), 85–110.
Bastin, G. (2000). Evaluating beginners' re-expression and creativity: A positive approach. *The Translator, 6*(2), 231–245.
Bastin, G. (2003). Aventures et mésaventures de la créativité chez les débutants. *Meta, 48*(3), 347–360.
Bastin, G., & Betancourt, M. (2005). Les avatars de la créativité chez les traducteurs débutants. In J. Peeters (Ed.), *On the relationships between translation theory and translation practice* (Studien zur romanischen Sprachwissenschaft und interkulturellen Kommunikation 19) (pp. 213–224). Frankfurt: Lang.
Bayer-Hohenwarter, G. (2009). Translational creativity: How to measure the unmeasurable. In S. Göpferich, A. L. Jakobsen, & I. M. Mees (Eds.), *Behind the mind: Methods, models and results in translation process research* (Copenhagen Studies in Language 37) (pp. 39–59). Copenhagen: Samfundslitteratur.
Bayer-Hohenwarter, G. (2010). Comparing translational creativity scores of students and professionals: Flexible problem-solving and/or fluent routine behaviour? In S. Göpferich, F. Alves, & I. M. Mees (Eds.), *New approaches in translation process research* (Copenhagen Studies in Language 39) (pp. 83–111). Copenhagen: Samfundslitteratur.
Bayer-Hohenwarter, G. (2011a). Kreativität in populärwissenschaftlichen und instruktiven Texten im Vergleich: Kein Raum für Kreativität beim Übersetzen von instruktiven Texten? *Trans-Kom, 4*(1), 49–75.
Bayer-Hohenwarter, G. (2011b). "Creative shifts" as a means of measuring and promoting translational creativity. *Meta, 56*(3), 663–692.
Bayer-Hohenwarter, G. (2012). *Translatorische Kreativität. Definition, Messung, Entwicklung.* Tübingen: Narr.
Bayer-Hohenwarter, G. (2017). Denken in Analogien—kreatives Lösen von Verstehensproblemen im Übersetzungsprozess. In L. Cercel, M. Agnetta, & M. T. Lozano (Eds.), *Kreativität und Hermeneutik in der Translation* (Translationswissenschaft Band 12) (pp. 427–454). Tübingen: Narr.
Brown, R. (1958). How shall a thing be called? *Psychological Review, 65*, 14–21.
Brünjes, J. (2016). Digitalization could soon make every second junior lawyer redundant. Retrieved 8 April 2019 from www.bucerius-education.de/article/digitalization-could-soon-make-every-second-junior-lawyer-redundant/
Byrne, J. (2006). Suppression as a form of creativity in technical translation. In I. Kemble, & C. O'Sullivan (Eds.), *Proceedings of the Conference held on 12th November 2005 in Portsmouth* (pp. 6–13). Portsmouth: University of Portsmouth.
Catford, J. C. (1965). *A linguistic theory of translation.* London: Oxford University Press.
Cercel, L., Agnetta, M., & Lozano, M. T. (Eds.) (2017). *Kreativität und Hermeneutik in der Translation* (Translationswissenschaft Band 12). Tübingen: Narr.
Chaume Varela, F. (1998). Textual constraints and the translator's creativity in dubbing. In A. Beylard-Ozeroff, J. Králová, & B. Moser Mercer (Eds.), *Translator's strategies and creativity* (pp. 15–22). Amsterdam: John Benjamins.
Cho, S. E. (2006). Translator's creativity found in the process of Japanese-Korean translation. *Meta, 51*(2), 378–388.
Csikszentmihalyi, M. (1997). *Creativity. Flow and the psychology of discovery and invention.* New York: Harper.
Dancette, J., Audet, L., & Jay-Rayon, L. (2007). Axes et critères de la créativité en traduction. *Meta, 52*(1), 108–122.
Durieux, C. (1991). Liberté et créativité en traduction technique. In M. Lederer, & F. Israël (Eds.), *La liberté en traduction. Actes du colloque international tenu à l'E.S.I.T. les 7, 8 et 9 juin 1990 réunis par Marianne Lederer* (Collection „traductologie" 7) (pp. 169–179). Paris: Didier Erudition.
Englund Dimitrova, B. (2005). *Expertise and explicitation in the translation process.* Amsterdam: John Benjamins.
Fielding, H. (1996). *Bridget Jones's Diary.* London: Picador.

Fillmore, C. J. (1977). Scenes-and-frames semantics. In A. Zampolli (Ed.), *Linguistic structures processing* (pp. 55–88). Amsterdam: North Holland.
Fontanet, M. (2005). Temps de créativité en traduction. *Meta, 50*(2), 432–447.
Fontcuberta i Gel, J. (1997). Creatividad en la traducción audiovisual. In P. Fernandéz Nistal, & J. M. Bravo Gozalo (Eds.), *Aproximaciones a los estudios de traducción* (pp. 217–230). Valladolid: Servicio de Apoyo a la Enseñanza.
Forstner, M. (2005). Bemerkungen zu Kreativität und Expertise. *Lebende Sprachen, 50*(3), 98–104.
Göpferich, S. (2009). Adding value to data in translation process research: The TransComp asset management system. In I. M. Mees, F. Alves, & S. Göpferich (Eds.), *Methodology, technology and innovation in translation process research: A tribute to Arnt Lykke Jakobsen* (Copenhagen Studies in Language 38) (pp. 159–182). Copenhagen: Samfundslitteratur.
Göpferich, S., Bayer-Hohenwarter, G., Prassl, F., & Stadlober, J. (2011). Exploring translation competence acquisition: Criteria of analysis put to the test. In S. O'Brien (Ed.), *Cognitive explorations of translation.* (Continuum studies in translation) (pp. 57–85). New York/London: Continuum.
Göpferich, S., Bayer-Hohenwarter, G., & Stigler, H. (Eds.) (2008 and updated in following years until the end of the research project). *TransComp—The development of translation competence. [Corpus and asset management system for the longitudinal study TransComp.]* Graz: University of Graz. Retrieved 18 March 2020 from https://gams.uni-graz.at/context:tc
Grimms Deutsches Wörterbuch, Erster Band (1854). Leipzig: S. Hirzel.
Guilford, J. P. (1950). Creativity. *American Psychologist, 5*, 444–454.
Guilford, J. P. (1975). Creativity: A quarter century of progress. In I. A. Taylor, & J. W. Geztels (Eds.), *Perspectives in creativity* (pp. 37–59). Chicago: Aldine.
Hagemann, S. (2007). Zur Evaluierung kreativer Übersetzungsleistungen. *Lebende Sprachen, 3*, 102–109.
Hague, D. R. (2009). Prophets and pandemonium: Creativity in the translating self. *New Voices in Translation Studies, 5*, 16–28.
Heiden, T. (2005). Blick in die Black Box: Kreative Momente im Übersetzungsprozess: eine experimentelle Studie mit Translog. *Meta, 50*(2), 448–472.
Holz-Mänttäri, J. (1984). *Translatorisches Handeln. Theorie und Methode.* Helsinki: Suomalainen Tiedeakatemia.
Hubscher-Davidson, S. (2005). Psycholinguistic similarities and differences between subjects as seen through TAPS. *International Journal of Translation, 17*(1–2), 97–107.
Hubscher-Davidson, S. (2006). Using TAPs to analyse creativity in translation. In I. Kemble, & C. O'Sullivan (Eds.), *Proceedings of the conference held on 12th November 2005 in Portsmouth* (pp. 63–71). Portsmouth: University of Portsmouth.
Jettmarová, Z. (1998). Literalness as an overall strategy for translating advertisements in the Czech Republic. In A. Beylard-Ozeroff, J. Králová, & B. Moser-Mercer (Eds.), *Translator's strategies and creativity. Selected papers from the 9th international conference on translation and interpreting, Prague, September 1995* (pp. 97–105). Amsterdam: John Benjamins.
Kaindl, K. (2005). Kreativität in der Übersetzung von Popularmusik. *Lebende Sprachen, 50*(3), 119–124.
Kemble, I., & O'Sullivan, C. (Eds.) (2006). *Proceedings of the conference held on 12th November 2005 in Portsmouth.* Portsmouth: University of Portsmouth.
Kenny, D. (2001). *Lexis and creativity in translation. A corpus-based study.* Manchester: St. Jerome.
Kenny, D. (2006). Creativity in translation: Opening up the corpus-based approach. In I. Kemble, & C. O'Sullivan (Eds.), *Proceedings of the conference held on 12th November 2005 in Portsmouth* (pp. 72–81). Portsmouth: University of Portsmouth.
Koestler, A. (1966). *Der göttliche Funke. Der schöpferische Akt in Kunst und Wissenschaft.* Bern/München/Wien: Scherz Verlag.
Koller, W. (2011). *Einführung in die Übersetzungswissenschaft.* 8. neubearbeitete Auflage 2011, 1. Auflage 1979. Tübingen: Narr Francke Attempto.
Krings, H. P. (1988). Blick in die "Black Box"—eine Fallstudie zum Übersetzungsprozess bei Berufsübersetzern. In R. Arntz (Ed.), *Textlinguistik und Fachsprache. Akten des Internationalen übersetzun gswissenschaftlichen AILA-Symposions Hildesheim, 13.-16. April 1987* (pp. 393–411). Hildesheim: Olms.
Krings, H. P. (2001). *Repairing texts: Empirical investigations of machine translation post-editing processes.* [Texte reparieren.] Translated by G. S. Koby et al. Translation of the author's Habilitationsschrift (1994). Kent: Kent State University Press.
Kußmaul, P. (1991). Creativity in the translation process: Empirical approaches. In K. M. Van Leuven-Zwart, & T. Naaijkens (Eds.), *Translation studies: The state of the art. Proceedings of the first James S. Holmes symposium on translation studies* (pp. 91–101). Amsterdam/Atlanta: Rodopi.

Kußmaul, P. (1993). Empirische Grundlagen einer Übersetzungsdidaktik: Kreativität im Übersetzungsprozeß. In J. Holz-Mänttäri, & Nord, C. (Eds.), *Traducere Navem. Festschrift für Katharina Reiß zum 70. Geburtstag* (pp. 275–286). Tampere: Tampereen yliopisto.

Kußmaul, P. (1998). Kreativität. In M. Snell-Hornby, H. G. Hönig, P. Kußmaul, & P. A. Schmitt (Eds.), *Handbuch Translation* (pp. 178–180). Tübingen: Stauffenburg.

Kußmaul, P. (2000a). Gedankensprünge beim Übersetzen. In M. Kadric, K. Kaindl, & F. Pöchhacker (Eds), *Translationswissenschaft. Festschrift für Mary Snell-Hornby zum 60. Geburtstag* (pp. 305–317). Tübingen: Stauffenburg.

Kußmaul, P. (2000b). A cognitive framework for looking at creative mental processes. In M. Olohan (Ed.), *Intercultural faultlines: Textual and cognitive aspects* (Research models in translation studies 1) (pp. 57–71). Manchester: St. Jerome.

Kußmaul, P. (2000c). Types of creative translating. In A. Chesterman, N. Gallardo San Salvador, & Y. Gambier (Eds.), *Translation in context. Selected contributions from the EST congress, Granada 1998* (Benjamins translation library 39) (pp. 117–126). Amsterdam: John Benjamins.

Kußmaul, P. (2000d). *Kreatives Übersetzen* (Studien zur Translation 10). Tübingen: Stauffenburg.

Kußmaul, P. (2005). Translation through visualization. *Meta, 50*(2), 378–391.

Kußmaul, P. (2007). *Verstehen und Übersetzen. Ein Lehr- und Arbeitsbuch.* Tübingen: Narr.

Lakoff, G. (1987). *Women, fire, and dangerous things. What categories reveal about the mind.* Chicago/London: University of Chicago Press.

Laviosa, S. (1998). Core patterns of lexical use in comparable corpus of English narrative prose. *Meta, 43*(4), 557–570.

Lee-Jahnke, H. (2005). Unterrichts- und Evaluierungsmethoden zur Förderung des kreativen Übersetzens. *Lebende Sprachen, 50*(3), 125–132.

Mackenzie, R. (1998). Creative problem-solving and translator training. In A. Beylard-Ozeroff, J. Králová, & B. Moser-Mercer (Eds.), *Translator's strategies and creativity. Selected papers from the 9th international conference on translation and interpreting Prague, September 1995* (pp. 201–206). Amsterdam: John Benjamins.

Mariaule, M. (2017). Créativité, création et recréation en traduction: un flou conceptuel. *Parallèles, 27*(2), 83–96.

Martín de León, C. (2017). Mental imagery in translation processes. *Hermes—Journal of Language and Communication in Business, 56*, 201–220.

Martínez Martínez, J. M., & Teich, E. (2017). Modeling routine in translation with entropy and surprisal. A comparison of learner and professional translations. In L. Cercel, M. Agnetta, & M. T. Lozano (Eds.), *Kreativität und Hermeneutik in der Translation.* Translationswissenschaft Band 12 (pp. 403–426). Tübingen: Narr.

Neubert, A. (2004). Equivalence in translation. In H. Kittel, et al. (Eds.), *Übersetzung—Translation—Traduction. Ein internationales Handbuch zur Übersetzungsforschung* [An international encyclopedia of translation studies], Vol. 1 (pp. 329–342). Berlin/New York: Walter de Gruyter.

Nida, E. A. (1998). Translators' creativity versus sociolinguistic constraints. In A. Beylard-Ozeroff, J. Králová, & B. Moser-Mercer (Eds.), *Translators' strategies and creativity. Selected papers from the 9th international conference on translation and interpreting Prague, September 1995* (pp. 127–136). Amsterdam: John Benjamins.

Nord, C. (1989). Loyalität statt Treue. Vorschläge zu einer funktionalen Übersetzungstypologie. *Lebende Sprachen, 34*(3), 100–105.

Nord, C. (1997). *Translation as a purposeful activity. Functionalist approaches explained.* Manchester: St. Jerome.

Nord, C. (2005). Kreativität und Methode—spannende oder gespannte Beziehung? *Lebende Sprachen, 50*(3), 137–141.

Pisarska, A. (1989). *Creativity of translators—The translation of metaphorical expressions in non-literary texts.* Poznań: UAM.

Pommer, S. (2006). *Rechtsübersetzung und Rechtsvergleichung. Translatologische Fragen zur Interdisziplinarität* (Europäische Hochschulschriften Reihe XXI, Linguistik, 290). Frankfurt: Lang.

Popescu, A.-V., & Cohen-Vida, M.-I. (2015). Can the specialized translator be creative? In *Proceedings of the 7th World Conference on Educational Sciences (WCES-2015), 05–07* (pp. 1195–1202). Retrieved 30 April 2019 from www.sciencedirect.com/

Preiser, S. (1976). *Kreativitätsforschung.* Darmstadt: Wissenschaftliche Buchgesellschaft.

Quillard, G. (1998). Translating advertisements and creativity. In A. Beylard-Ozeroff, J. Králová, & B. Moser-Mercer (Eds.), *Translator's strategies and creativity. Selected papers from the 9th international conference on translation and interpreting, Prague, September 1995* (pp. 23–31). Amsterdam: John Benjamins.

Quillard, G. (2001). La traduction des jeux de mots dans les annonces publicitaires. *TTR: traduction, terminologie, rédaction. Traductologie et diversité*, 14(1), 117–157.
Risku, H., Milošević, J., & Rogl, R. (2017a). Creativity in the translation workplace. In L. Cercel, M. Agnetta, & M. T. Amido Lozano (Eds.), *Kreativität und Hermeneutik in der Translation* (pp. 455–459) Tübingen: Narr.
Risku, H., Pichler, T., & Wieser, V. (2017b). Transcreation as a translation service: Process requirements and client expectations. *Across Languages and Cultures, 18*(1), 53–77.
Rosch, E. (1973). Natural categories. *Cognitive Psychology, 4*, 328–350.
Šarčević, S. (2000). Creativity in legal translation: How much is too much? In A. Chesterman, N. Gallardo San Salvador, & Y. Gambier (Eds.), *Translation in context. Selected papers from the EST congress, Granada 1998* (pp. 281–291). Amsterdam: John Benjamins.
Sattler-Hovdar, N. (2016). *Translation—Transkreation. Vom Über-Setzen zum Über-Texten*. Berlin: BDÜ.
Schmitt, P. A. (2005). Grenzen der Kreativität. *Lebende Sprachen, 50*(3), 104–111.
Schreiber, M. (2017). Kreativität in Translation und Translationswissenschaft: Zwei Fallbeispiele und ein Vorschlag. In L. Cercel, M. Agnetta, & M. T. Lozano (Eds.), *Kreativität und Hermeneutik in der Translation*. Translationswissenschaft Band 12 (pp. 349–358). Tübingen: Narr.
Seleskovitch, D. (1978). *Interpreting for international conferences*. Washington: Pen and booth.
Seleskovitch, D., & Lederer, M. (1989). *Pédagogie raisonnée de l'interprétation* (Collection Traductologie n° 4.). Brussels: Didier Érudition.
Shakespeare, W. (no date). *Dramatische Werke, übersetzt von August Wilhelm Schlegel und Ludwig Tieck*. Berlin: Lambert Schneider.
Shorthouse, J., & Maycroft, N. (2017). *Where is creativity: A multidisciplinary approach*. London: Routledge.
Siever, H. (2010). *Übersetzen und Interpretation*. Frankfurt am Main: Peter Lang.
Stewart, D. (2000). Conventionality, creativity and translated text: The implications of electronic corpora in translation. In M. Olohan (Ed.), *Intercultural faultlines: Textual and cognitive aspects. Research models in translation studies, 1* (pp. 73–91). Manchester: St. Jerome.
Thomä, S. (2003). Creativity in translation. An interdisciplinary approach [Unpublished dissertation, Universität Salzburg].
Tirkkonen-Condit, S., Mäkisalo, J., & Immonen, S. (2008). The translation process—interplay between literal rendering and a search for sense. *Across Languages and Cultures, 9*(1), 1–15.
Torrance, E. P. (1988). The nature of creativity as manifest in its testing. In R. J. Sternberg (Ed.), *The nature of creativity. Contemporary psychological perspectives* (pp. 43–75). Cambridge: Cambridge University Press.
Urban Dictionary (2010). www.urbandictionary.com/define.php?term=suit-porn.
Vermeer, H. J. (1989): Skopos and commission in translational action. In L. Venuti (Ed.), *The translation studies reader* (pp. 221–232). London/New York: Routledge.
Wikipedia (2017). https://en.wikipedia.org/wiki/Food_porn, last edited on 3 October 2017.
Wilde, U. (1994). Werbesprache—zwischen Kreativität und Fachsprachlichkeit. Analyse einiger Beispiele aus dem Französischen. *Fachsprache, 15*(1–2), 18–26.
Wilss, W. (1988). *Kognition und Übersetzen: zu Theorie und Praxis des menschlichen und des maschinellen Übersetzens*. Tübingen: Niemeyer.
Wilss, W. (1996). *Knowledge and skills in translator behaviour*. Amsterdam: John Benjamins.
Wittgenstein, L. (1958). *Philosophische Untersuchungen*. Frankfurt: Suhrkamp.
Zhong, Y. (2005). A matter of principles: Empirical treatments of translation principles—a case study. *Meta, 50*(2), 495–510.

18

Translation, metaphor and cognition

Christina Schäffner and Paul Chilton

18.1 Introduction

In characterizing translation as socially contexted behaviour, Toury (1995) differentiated between the translation act and the translation event. The act of translation refers to the cognitive aspects of translating as a decision-making process, and the translation event is the situational, socio-cultural, historical and ideological context in which the act is embedded. In other words, the translator's decision making is influenced by the concrete situation, assignment, environment, purpose, etc. After an initial focus on translation as a linguistic process in the 1960s/1970s, in which decision making was described as a choice between various target-language options (e.g. Levý, 1967/2000), cultural as well as cognitive aspects received more attention. In respect of cognition, the question was asked: what happens in a translator's mind in the act of translating?

The development from a linguistics-based translation theory to functionalist, cultural, sociological and cognitive approaches is reflected in changing definitions of translation and changing research methods. For investigating cognitive processes of translation, cognitive linguistics has been seen as beneficial. Cognitive linguistics (CL) studies the relationships between language structures and elements of conceptualization. The basic premises of CL include the following: language is an integral part of cognition and grounded in human conceptualization; natural language structures reflect cognitive features and mechanisms influenced by our bodily, physical, social and cultural experience, i.e. language is embodied; and meaning construction is a complex, dynamic and situated process; thus, meaning can be said to be emergent (for overviews of CL see e.g. Croft & Cruse, 2004; Lee, 2001).

CL can make a significant contribution to the description and explanation of translation phenomena (see e.g. Rojo & Ibarretxe-Antuñano, 2013: Schwieter & Ferreira, 2017). Cognitive Translation Studies, or cognitive translatology (Muñoz Martín, 2013a), has in fact already entered the discipline of Translation Studies (TS), under the influence of both cognitive linguistics and cognitive science (CS) more generally. Cognition was initially thought of as an information processing activity, with the mind viewed as a computer-like machine. In what is now labelled second-generation CS, "language and behaviour are [...] seen as being deeply rooted in sensorimotor or bodily processes that condition the way we perceive and construct the world" (Muñoz Martín, 2013b, p. 241). This approach emphasizes the situated, embodied and emergent nature of human cognition. Investigation of the situated and embedded nature of translation takes into account that translators use a variety of tools, including dictionaries, glossaries and computer-assisted translation tools (CAT). From the perspective of CS, translation is thus understood as a form of distributed cognition (e.g. Teixeira & O'Brien, 2017), in which the processes in a translator's mind (internal representations) interact with external representations provided by the tools.

A development similar to that outlined earlier for TS, that is, a development from a linguistic to a socio-cultural perspective, can be seen clearly in one strand of CL that is of particular interest to TS—the investigation of the phenomenon of metaphor. In classical rhetoric, metaphor tended to be thought of as a persuasive device, in literary studies as an ornamental or emotive device, and in structural linguistics as a deviation from a more literal meaning. With the emergence of CL, however, conceptual metaphor theory (CMT), initially developed by Lakoff and Johnson (1980/2003), introduced a new perspective. CMT explains metaphor as a mapping across conceptual domains. Metaphors are not the preserve of orators or poets but a pervasive and fundamental aspect of cognition that serves to organize and structure the way we think about and act in and on the world around us. CMT thus belongs to the second generation of CS. Metaphorical expressions in a particular language are manifestations of underlying mappings that are not linguistic but cognitive (Lakoff, 1993, pp. 202–203). Here, "mapping" is used in the logico-mathematical sense (moving one element of a set to another): in metaphor, an element already known in one conceptual domain is transferred to a more abstract, less well-understood conceptual domain. For example, the expression "I'm at a crossroads in my life" is a linguistic realization of the conceptual metaphor LIFE IS A JOURNEY.

It is important to note that metaphor is not the only concern of CL, and certainly not the only aspect of language that concerns TS and the practice of translation itself. All linguistic communications involve non-linguistic brain activity; there is no isolated language module generating meanings without being integrated with knowledge and experience. This is the case at the levels of syntax, lexis and morphology (taking phonetics and phonology as separate systems). Thus, TS and translators need to consider not only whether the different grammatical patterns of different languages may carry different mental representations (e.g. via the different word order of different languages) but also, in the process of practical translation, whether, for example, a passive sentence in the source text (ST) should or can be rendered by an active sentence in the target text (TT), since the mental representations (i.e. meanings) could very well be different. The same question arises, even more obviously, in the case of lexical structures. There are no one-to-one correspondences; the vocabulary structure of one language is not the same as that of another. This means not only that the lexical networks of languages differ but also that the individual lexemes from those networks are meaningfully different, i.e. are involved in different mental representations. One example is spatial prepositions, which differ conceptually across languages. English speakers use the same preposition, *on*, for location on horizontal and vertical surfaces, while German speakers use different ones: *an* for a vertical surface and *auf* for a horizontal one. In particular phrases, two languages may refer to the same type of object, but the choice of preposition may reflect different conceptualizations. For example, French has *dans le train* as a translation equivalent for English "on the train", French thus conceptualizing trains as containers and English as platforms.

Lakoff and Johnson (1980/2003) identify three fundamental types of metaphor: orientational, ontological and structural. Orientational metaphors organize "a whole system of concepts with respect to one another [and] most of them have to do with spatial orientation" (Lakoff & Johnson, 1980/2003, p. 14), as in HAPPY IS UP. Ontological metaphors (e.g. THE BODY IS A CONTAINER) are "ways of viewing events, activities, emotions, ideas, etc., as entities and substances" (ibid., p. 25) derived from bodily interactions with physical objects. Structural metaphors relate experience of different kinds of things in such a way that one concept, usually an abstract one, is metaphorically structured in terms of another, more concrete one, e.g. ARGUMENT IS WAR. Incidentally, translation itself has served as both source domain (e.g. DREAM ANALYSIS IS TRANSLATION) and target domain (e.g. TRANSLATION IS BRIDGE BUILDING; see also St André, 2010; Guldin, 2016).

This chapter focuses on how metaphor has been analysed in TS. After presenting some key arguments of CMT in respect of cognition, language and culture, we summarize some relevant research in TS, in particular process research.

18.2 Core issues

18.2.1 Core issues for metaphor studies

Core issues for metaphor research concern aspects such as the differentiation between linguistic and conceptual metaphors, universality or culture specificity of metaphor, cross-cultural variation, conceptualization in metaphor production versus comprehension, pragmatic and discursive functions of metaphors, and the value of Lakoff and Johnson's conceptual metaphor theory in comparison to other cognitive theories, e.g. blending theory (Turner & Fauconnier, 2002). Chilton (2019) notes that conceptual blending theory

> builds on and is compatible with CMT, though different from it in the following respects. It is essentially a model of online cognitive processes, rather than a model of relatively stable input conceptual frames and mappings. Its input spaces can be multiple, the mappings may flow both ways. In addition to a minimum of two input spaces, the theory postulates a generic space that contains the abstract conceptual commonality of the inputs, and, crucially, a blend space in which conceptual structure emerges selectively from the inputs.
>
> *Chilton, 2019, p. 253*

For CMT, metaphor is first and foremost a matter of cognition, with metaphor in language being secondary. As Semino and Demjén (2017a) argue, Lakoff and Johnson "did not regard the different linguistic forms metaphor can take and the different functions it can perform in discourse as worthy of attention in their own right" (p. 3). Conceptual metaphors are manifested linguistically, and metaphor can thus be observed most directly in language, i.e. in texts. It is these linguistic manifestations (the metaphorical expressions) that discourse participants (as well as translators) encounter (on recent developments in metaphor research in relation to language, see Semino & Demjén, 2017b). More recent research that has considered the textual or discursive manifestations of metaphors has revealed how metaphors shape a text, thus giving insights into the strategic use of language. It has also been observed that metaphor varies across registers and genres (e.g. Semino, 2008, and particularly for political discourse Charteris-Black, 2004; Chilton, 1996; Musolff, 2016).

Metaphor research also asks whether conceptual metaphors are universal or culture specific. Lakoff and Johnson's location of the roots of metaphorical cross-domain mappings in bodily experience suggests a fundamentally universalist view. However, in the Afterword to the 2003 edition of their book, they make a distinction between primary metaphors, which are "grounded in the everyday experience that links our sensory-motor experience to the domain of our subjective judgements" (Lakoff & Johnson, 2003, p. 255), and complex metaphors, which are "composed of primary metaphors and [...] make use of culturally based conceptual frames"

(p. 257). Primary metaphors thus seem to be universal, whereas complex metaphors may differ between cultures.

A significant amount of research into cross-cultural variation of metaphor has been conducted by Kövecses (e.g. 2005, 2014). Kövecses (2005) argues that "complex metaphors are more important to cultural situations", since it is complex metaphors "with which people actually engage in their thought in real cultural contexts" (p. 11). He investigated, for example, cultural differences in the conceptualization of emotionally and cognitively central body parts (as seats of emotional and/or cognitive activity). He, too, argues for investigating metaphor in language, since the study of linguistic metaphors "may provide a good clue to finding the systematic conceptual correspondences between domains (i.e. to conceptual metaphors)" (p. 32). Based on a comparison of the linguistic expressions of particular conceptual metaphors in different languages, Kövecses (2014, p. 33) summarizes possibilities for translation as follows:

(i) same literal meaning—same figurative meaning—same conceptual metaphor
(ii) different literal meaning—same figurative meaning—same conceptual metaphor
(iii) different literal meaning—same figurative meaning—different conceptual metaphor

Some of the cross-linguistic comparisons of metaphor were based on translation. However, translations tended to be sentence based and/or elicited in experiments. Such tests made Kövecses (2014) conclude that differences in metaphorical conceptual systems and the context in which they emerge cause problems in metaphor translation. Similarly, Brdar and Brdar-Szabó (2017) tested whether equivalent metaphors are available in the target language and which one test participants would opt for, illustrated with cross-linguistic asymmetries in the use of time metaphors. That is, differences and similarities in metaphors across languages were not established on the basis of authentic source and target texts. These issues concerning metaphor, in particular the differentiation between linguistic and conceptual metaphor and the cross-cultural similarity and variation, make the question of what happens to metaphors in the process of translation a relevant one.

18.2.2 Core issues for Translation Studies

Core issues for TS were initially translatability, that is, *whether* metaphor can be translated (e.g. Dagut, 1976), and methods for translating metaphors, that is, *how* metaphors can, or should, be translated. Research was predominantly informed by more traditional views of metaphor as a linguistic phenomenon and mainly devoted to metaphorical expressions, with little consideration of the conceptual level of which the linguistic expressions were representations. Investigations mainly dealt with products; that is, source texts (ST) and target texts (TT) were compared to identify how metaphors in the ST had been dealt with. Such descriptive studies often resulted in the production of lists of translation methods, or translation procedures, that supplemented or replaced similar lists which had been set up earlier with a view to guiding translators in how to translate metaphors (e.g. Newmark, 1981).

In such earlier publications, three translation procedures are recurring: metaphor into same metaphor, metaphor into different metaphor, and metaphor into represented sense. They correspond to van den Broeck's (1981) three modes of *sensu stricto*, substitution and paraphrase. From a TT perspective, Toury (1995) added to these complete omission ("metaphor into zero") as well as "non-metaphor into metaphor" and "zero into metaphor" (pp. 82–83; for an overview, see e.g. Schäffner, 2004, 2017a). It was realized, however, that dealing with metaphor in translation is not simply a matter of identifying potential linguistic correspondences in source and target languages, but also of reflecting correspondences between their conceptual systems. An increasing number of such product-oriented studies have therefore been based on CMT (e.g. Al-Harrasi, 2001; Shuttleworth, 2011, 2017; Tcaciuc, 2014). For example, based on a descriptive analysis of English translations of Arabic political speeches, Al-Harrasi (2001, pp. 277–288) presented a list of translation procedures that is significantly different from previously produced ones, which were based on linguistic categories. His list includes procedures such as Instantiating the Same Conceptual Metaphor (with sub-procedures such as Concretising an Image Schematic Metaphor, or Same Mapping but a Different Perspective), Using a Different Conceptual Metaphor, and Deletion of the Expression of the Metaphor. Only a few recent publications that are based on CMT will briefly be mentioned in the following section. They share a concern with identifying translation procedures for metaphors by reflecting on the relationship between linguistic expressions and conceptual metaphors. Moreover, all of them are empirical studies working with authentic translations.

18.2.2.1 Metaphor analysis in translations as products

Shuttleworth (2011) uses a multilingual corpus to study the translation of metaphor in original English popular-scientific texts, exploring how metaphorical expressions have been dealt with both at the micro-level (translation procedures for individual metaphorical expressions) and at the macro-level (clusters of mapping) for a variety of target languages. Retention of the metaphorical expression was identified as the default procedure across all languages. Shuttleworth also concludes that what is "lost" in translation is individual metaphorical expressions rather than entire mappings. Such a cross-lingual analysis of scientific metaphors in translation is expanded further in Shuttleworth (2017).

Some other product-oriented research has analysed the treatment of specific types of metaphors in translation. For example, Safarnejad et al. (2014) have studied how metaphors of the emotions happiness and sadness have been rendered from Persian into English. Their corpus, however, was rather small, consisting of a Persian novel and two of its translations into English. They used the metaphor identification procedure (MIP) proposed by the Pragglejaz group (2007), a step-by-step method for identifying metaphor in language, that is, for deciding whether a lexical unit in a text can be marked as metaphorical (for more information on MIP and its more refined and extended variant MIPVU, see Steen, 2017). Their interest was in identifying the emotive metaphorical expressions in the ST and their underlying conceptual metaphors. A comparison revealed differences in the cognitive mappings in Persian and English, reflecting different models of conceptualizing experiences in each culture. They list their identified translation strategies as 1) translation of the source metaphorical expression to the equivalent target-language metaphor, 2) translation of the source metaphorical expression to non-metaphor in the target language, 3) mistranslation and 4) literal translation (Safarnejad et al., 2014, p. 110). It is, however, strange to see that mistranslation is presented as a strategy, in particular since these strategies are suggested to translators as means "to overcome the ordinary barriers while conveying Persian metaphor into English" (p. 110).

Burmakova and Marugina (2014) analysed the anthropomorphous metaphor, in particular the conceptual mapping between MAN and NATURE concepts in a contrastive study of Russian short stories and their translations into English. Their theoretical framework is Mandelblit's (1995) cognitive translation hypothesis. Mandelblit suggested two possible scenarios in the translation of metaphors. In the case of similar mapping conditions, no conceptual shift occurs between the metaphors in the source and target languages, whereas for different mapping conditions, a conceptual shift does take place from source language (SL) to target language (TL). Their results show that in the majority of cases (more than 50%), the anthropocentric perspective and the source domain were preserved in the TT. For metaphors with similar mapping conditions, the most commonly used translation procedure was reproducing the same image in the TL. Metaphors with different mapping conditions were paraphrased.

Yan et al. (2010) used larger corpora to investigate how cross-cultural variation in conceptual metaphor affects the translation of metaphorical expressions (i.e. the linguistic level) and how the translation of these metaphorical expressions affects the conceptual metaphors they express, illustrated with FEAR metaphors in translations from English to Chinese. One of the results of their study is that "expressions of English metaphors are not necessarily translated as expressions of the same metaphors in Chinese even in cases when the metaphor is shared by the two languages" (Yan et al., 2010, p. 49). Their findings reflect that conceptual shifts do not occur only in the case of different mapping conditions, as initially suggested by Mandelblit's cognitive translation hypothesis.

Most of the research into metaphor in translation is product oriented (e.g. the chapters in Miller & Monti, 2014; Trim & Śliwa, 2019). However, such product-oriented studies can make only a limited contribution to the study of the actual processing of metaphor in translation. Analysing the different translation strategies and linking metaphorical expressions to the conceptual metaphors by which they are sanctioned can only amount to speculating about aspects of cognition and reasoning involved. The (albeit fewer) process-oriented investigations of metaphor in translation are therefore particularly valuable for exploring aspects of translation, metaphor and cognition (see also Schäffner & Shuttleworth, 2013).

18.2.2.2 Metaphor analysis in the translation process

In general, translation process research is concerned with cognitive aspects of task performance (for an overview, see e.g. Ehrensberger-Dow, 2018; Hansen, 2013; also Lacruz & Jääskeläinen, 2018). In respect of metaphor, core issues of process-oriented studies are the processing of metaphor in translation (are metaphors processed differently than non-metaphorical language?), difficulties (are metaphors more difficult to translate than non-metaphorical language?), and cognitive load (does translating metaphors demand more cognitive effort?). Empirical and experimental studies have made use of methods often employed in process research, especially think-aloud protocols (TAPs), keystroke logging and/or eye-tracking methods, sometimes combined with retrospective interviews. In these studies, too, the majority of the researchers have referred to CMT.

Thinking aloud is an introspective method. Translators are asked to verbalize what they are thinking while carrying out a translation task. These verbalizations are recorded and transcribed, with the resulting TAPs serving as input for the researcher's analysis (for an overview, see e.g. Jääskeläinen, 2009). In respect of metaphors, TAPs have been used by Tirkkonen-Condit (e.g. 2001) to test Mandelblit's (1995) cognitive translation hypothesis. Mandelblit had hypothesized that "metaphorical expressions take more time and are more difficult to translate if they exploit a different cognitive domain than the target language equivalent expression" (Tirkkonen-Condit, 2001, p. 11). Mandelblit herself (1995, p. 493) had tested her hypothesis with reference to time, and found that metaphorical expressions took more time and were more difficult to translate if they exploited a different cognitive domain, which required the translator to make a conceptual shift. Using TAPs of professional translators in simulated translation tasks, Tirkkonen-Condit (2001) measured time and the length of TAP segments, counted the lines of TT produced, and also asked the translators to comment on their own satisfaction with their translations. The research confirmed Mandelblit's hypothesis. The translators required more time for translating metaphorical expressions for which they had to search for another conceptual domain. Metaphorical expressions with different domains also resulted in more verbalization in the TAPs and in more translation solutions. The overall conclusion is that delay in the translation process, and the related uncertainty in the translation of different domain metaphors, is evidence of concept mediation. In other words, "the fact that translation difficulty is increased by domain conflict indicates strongly that translation does not take place primarily through word association but at the conceptual level" (Massey, 2016, p. 70).

TAPs were also employed by Kussmaul (2000) as dialogue protocols for students, and in particular with reference to creativity research. His dialogue protocols, too, support the idea that thought processes are happening at conceptual levels, although they are often triggered by metaphorical expressions and the initial attempt to provide a target-language equivalent for a linguistic metaphor. What remains unclear, however, is when exactly the conceptual level is accessed and what exactly triggers the need to access the conceptual level in real translation events.

Keystroke logging is a non-intrusive technology, which enables the researcher to collect realistic data similar to a genuine translation process. In keystroke logging, the software provides quantitative data about the process, i.e. cursor movements, changes and corrections, and position and length of pauses. It is thus a tool to acquire information about a translator's real-time decision-making processes. Hansen (2013), however, argues that the log files hardly "provide information about the translators' cognitive processes, i.e. what they are reflecting upon during the pauses and phases, or what resources or aids they refer to" (p. 91), and it therefore remains impossible to know the full mental processes. Nevertheless, the recorded pauses and revisions are indicators of interruptions to the flow of the translation process and "can be analysed in order to support hypotheses about comprehension, linguistic issues, problem solving, and formulation challenges" (Ehrensberger-Dow, 2018, p. 295). Using the keystroke logging software Translog, Jakobsen, Jensen, & Mees (2007), for example, measured processing time and noticed that idiomatic expressions (which are very often metaphorical) slow down the translation process.

Keystroke logging has also been used in experimental settings to test difficulties and cognitive load in metaphor translation, in particular by Sjørup (2011, 2013), Turkama (2017), and Förster Hegrenæs (2018). Turkama (2017) tested the cognitive translation hypothesis with reference to primary and complex metaphors in translating from English into Finnish. Her general aim was to find out which of these types of metaphor require more cognitive effort and to identify translation strategies. Her initial hypotheses were as follows:

> 1) primary metaphors are easier to translate than complex metaphors, 2) metaphors with shared mappings are easier to translate than metaphors with different mappings, 3) complex metaphors are easier to translate if primed by the translation of one of their assumed primary components.
>
> *Turkama, 2017, p. 50*

For investigating the cognitive effort of the translation process, fixation and time as recorded in the files collected by the Translog software were analysed. The notion of "fixation" describes the process of searching for a translation solution in which the original expression is copied literally. This is, however, a different use of "fixation" compared with eye-tracking studies (see later). The results of her experiments (conducted with students of translation) provided strong evidence that metaphorical expressions based on primary metaphors are easier to translate than those based on complex metaphors, thus providing support for the assumption that primary metaphors have a more universal experiential grounding, which facilitates their translation. It was also confirmed that different conceptual domains in the two languages make translation more difficult and increase the cognitive effort. Turkama also argues that "the conceptual domain or mapping used as the basis of the metaphorical expression was found to be a more significant factor in the difficulty and the amount of cognitive effort of translation than the primary vs. complex nature of metaphor" (p. 89). However, priming the translation of complex metaphors with the translation of primary metaphors did not facilitate the translation, and the third hypothesis was thus not verified.

Sjørup (2011) used keylogging to measure the production time of metaphorical expressions in TTs, relating them to specific translation strategies. Her basic assumption was that "the choice of translation strategy would have an effect on the cognitive effort involved in metaphor translation" (p. 201). However, she distinguished only three strategies: translating a metaphor to the same metaphorical image (M→M), use of another metaphorical phrase (M1→M2) or paraphrasing (M→P). In her study, professional translators translated two texts involving different conceptual domains. She noted a higher cognitive load for the paraphrase strategy and argues that this is due to two shifts involved: a shift from one domain to another, and a shift from metaphorical expressions to literal ones. This finding was slightly revised in her following study (see later).

Cognitive effort in dealing with metaphor is also the main interest of Förster Hegrenæs (2018), addressed from a developmental perspective. Her aim is to explore how cognitive effort is allocated when translating metaphors related to the linguistic and/or conceptual translation strategy chosen. Her investigation is based on experiments with students at different stages of their undergraduate programmes in both Norway and Germany. Production time of translating metaphorical expressions is seen as an indicator of cognitive effort and measured with keystroke logging (Translog II). Her investigation illustrated that translation strategies that involve a conceptual change (different conceptual mappings and different linguistic items; different conceptual mappings with partly similar linguistic items) increased production time. Overall, however, "no clearly definable development in terms of predicted production time" could be identified for the different strategies at different stages of competence development (p. 164). The analysis confirmed Mandelblit's hypothesis and also Sjørup's (2013) findings that strategies which require a shift of conceptual mapping are marked by longer production times. In respect of the selection of linguistic and conceptual translation strategies, all participant groups selected metaphorical strategy types more often than non-metaphorical ones. In other respects, however, her results differ from Sjørup's findings. For example, in respect of production time, Sjørup's suggestion that translators opt for strategies which require less time was not corroborated.

In some of her studies on cognitive effort in metaphor translation, Sjørup (2008, 2013) combines eye-tracking and keylogging methods. Eye tracking provides information about gaze activity, such as eye fixations, movements of the eye, gaze time, fixation duration and number of refixations. The use of eye tracking is based on the assumption that eye movements are indications of cognitive activity and that there is a correlation between the time readers fixate on a word and the amount of processing that takes place. Researchers try to infer what a translator is attending to at any particular moment of text processing (e.g. a specific word, the ST or the TT). One hypothesis in Sjørup's study of 2008 was that "translators dwell longer over the processing complexities involved in translating a metaphor than a non-metaphorical concept" (p. 53). Using naturally occurring text, she studied eye fixation time of professional translators and discovered that there was indeed a longer fixation time for metaphors compared with non-metaphorical language, which seemed to confirm her hypothesis.

The question of whether metaphors are processed differently compared with non-figurative language is also addressed in Sjørup (2013). This study combines an investigation of both the comprehension and the production phase. The overriding assumption is that comprehension mainly takes place during reading and is manifested in the eye movements, whereas production (mental and manifested) takes place mainly in the text production process and can be investigated through keylogging data. She analysed the cognitive effort involved in comprehending metaphors compared with non-metaphors when the text is processed for translation, and the cognitive effort involved in translating metaphorical expressions compared with translating non-metaphors. For investigating comprehension, total fixation time, total fixation number, and first pass fixation time were measured as dependent variables. A main finding in respect of comprehension was that even though "metaphors do not require a greater cognitive effort to read for translation than literal expressions", the "significant interaction between the variables Type and Task suggests a tendency that metaphors are in fact more cognitively effortful to read for translation than literal expressions, but less cognitively effortful to read for comprehension only" (p. 137). This result also shows that translation is a particular type of text processing and that measures which have been used in monolingual metaphor studies are not equally relevant when it comes to metaphor in translation. The measure chosen for analysing cognitive effort of the production stage is total production time. Sjørup (2013) also asks whether certain translation strategies required more effort than others, as indicated by gaze times and production time. An interesting result is that M→P (Metaphor into Paraphrase) is not more cognitively effortful than the direct translation strategy M→M (used most frequently by the participants of her study). M1→M2 is the least preferred strategy. For her analysis, effects of independent variables, such as the length of the area of interest, its position, the domain, the task, etc., are taken into account. However, the role of individual factors such as textual context and participant background are not considered, since her aim is to "identify commonalities of metaphor translation rather than investigate individual factors such as experience within different translation genres" (p. 9). Sjørup also acknowledges that it is impossible to distinguish exactly the comprehension aspect from the text production task. Although longer fixation times can be interpreted as a greater cognitive processing load, it is impossible to determine how this load is distributed between metaphor interpretation and the choice of a translation strategy and a target-text formulation. Moreover, longer gaze time could also be related to a translator reflecting on the consequences of local decisions for the overall function of the text—a point not problematized in depth by Sjørup.

What process research so far has revealed is that processing and translating metaphors do indeed seem to be linked to greater cognitive load, as reflected in longer pauses, total length of completing the translation task and more uncertainty (verbalized in TAPs and/or noticeable in Translog reports). Although the research illustrated earlier makes use of CMT, the analyses focus on metaphorical expressions. Scholars who used process methods have argued that their data have not been conclusive in arriving at a firm evaluation of whether the translators did indeed access the conceptual level or whether they were guided by the surface structure.

18.3 Recent developments

Translation process research has been a growing area in the discipline of TS, with progress in research methods and tools. That is, for data collection, researchers have made increasing use of several methods in combination (triangulation), such as keystroke logging, screen recording, eye tracking and retrospective interviews. In particular, keystroke logging and eye tracking are good indicators of processing effort in translation in real time. The studies devoted to metaphor in translation summarized earlier have signalled where processing effort is predominantly located, despite partly conflicting results. Such issues continue to be the concern of recent developments, although we also see an increase in multi-method approaches and an investigation of processes in different modes of translation as well as translation-related tasks. Some examples of such research will be summarized in the following.

18.3.1 Multi-methods in metaphor analysis

Triangulation is used by Massey (2016) and Massey and Ehrensberger-Dow (2017) in investigating how conceptual metaphor is re-conceptualized or re-mapped in the translation process. In their studies, they combine product-oriented approaches with techniques used to access translators' cognitive processes. Massey's exploratory study

> triangulates data from screen recordings with eye tracking, retrospective verbal commentaries and target-text products in an attempt to shed light on how translators at various levels of experience, and working in different language pairs, appear to map or remap conceptual metaphorical meanings as they work.
>
> *Massey, 2016, p. 71*

Screen recordings register all the changes that take place on the computer screen, including the use of resources (the Internet, electronic dictionaries). The data were collected from Bachelor (BA) and Master (MA) students and professional translators, with the translations (from German into English) being done in laboratory experimental settings. Two complex conceptual metaphors, an ontological one (a topographical personification: the German word *Hang*, i.e. "inclination", used in its psychological sense) and an orientational one (*zugrunde liegen*, literally, "to lie at the bottom of" as an example of causative orientation), were analysed in detail. At first, a product-oriented analysis categorized the translation procedures employed, i.e. metaphor into same metaphor, metaphor into different metaphor, metaphor into non-metaphor (or represented sense) and metaphor into zero. For example, for *zugrunde liegen*, all vertically oriented spatial metaphors in the TTs were treated as metaphor into same metaphor. Then, the retrospective verbal protocols (RVPs), gathered on the basis of the screen recordings overlaid with eye-tracking visualizations, were analysed for problem indicators, that is, for determining the participants' awareness of the conceptual metaphors. This is based on the assumption that mentioning the selected metaphors in their RVPs indicated the participants' awareness that the conceptual metaphors represented a translation issue which needed to be addressed. Finally, the "screen-recorded processes were examined for patterns of internal and external resource consultation" (p. 75). The analyses revealed correlative tendencies across experience levels. That is, in respect of both the translation products (i.e. translation procedures) and the process (i.e. problem awareness and resource-use behaviour), the results revealed differences between the three groups, with more similarities between the advanced MA students and the professionals. Massey (2016) argues that one "palpable conclusion to draw is that experience and/or training appears to be a central factor in handling conceptual metaphor" (p. 78).

In a follow-up study, Massey and Ehrensberger-Dow (2017) triangulated data from keystroke logs, RVPs and TTs produced by professionals, advanced MA students and beginner BA students, all translating into their L1 (English or German). The methodology for this study was very similar to the one used by Massey (2016). In a first stage, the translation procedures were identified in the products, followed by the analysis of process data. A pause of five seconds or more in the keystroke logs was taken to be a problem indicator (pauses are considered to indicate internal cognitive resource use). In the third stage, the RVPs were analysed in more detail to ascertain the participant groups' comprehension of the metaphors in the text. Aligning the product and process analyses, Massey and Ehrensberger-Dow (2017) noted differences in the behaviour of professional translators working into English (ProE) or into German (ProG). In the products, it was noted that most of the ProE translators

> remapped only partially, but maintained the generic metaphor structure of personification, [...] This seems to be accompanied by a very low degree of pausing and few conceptual clarity issues, as well as moderate transfer or formulation difficulties expressed in the RVPs.
>
> *Massey & Ehrensberger-Dow, 2017, p. 184–185*

However,

> two-thirds of the ProG translators re-mapped directly to the source-language target domain, but [...] this was accompanied by high levels of pausing and RVP mentions—which are commonly assumed by process researchers to indicate non-routine problem identification and cognitive processing.
>
> *Massey & Ehrensberger-Dow, 2017, p. 185*

This result makes them ask: "So could it be that re-mapping to source-language target domains, rather than cross-domain re-mapping, demands more attention, causes greater transfer and production uncertainty and, therefore, requires increased cognitive effort?" (p. 185). Investigating this question would need to deploy more direct elicitation methods, such as structured retrospective interviews.

18.3.2 Metaphor in translation-related tasks

Investigations of metaphor and cognition have also been conducted for different modes of translation and for translation-related activities. Zheng and Zhou (2018) used eye tracking for investigating processing time for metaphorical expressions (ME) during English–Chinese sight translation, with a particular interest in the eye-voice span. The processes in the sight translation task, conducted by students with no experience in professional translation or interpreting, were registered by eye tracker and audio recorder, supplemented by post-task retrospection data. The research focused on the eye movements and fixations during the pauses that preceded metaphorical expressions and during their oral translation time. Their study illustrated that metaphorical expressions increased cognitive effort, as indicated in a slowed-down pace of reading ahead and the existence of reading backwards or re-reading when sight translating. They acknowledge that their method of calculating fixation time "might have ignored the cognitive function of context (or co-text) in processing MEs" and that reading ahead "could also reasonably indicate some essential effort for working out the translation of ongoing ME, by seeking clues from the following co-text" (p. 756). They therefore conclude that there are "some grey zones that eye-tracking technology might not be possible to identify; or to be more precise, there are some intricate human cognitive processes which cannot be accurately investigated by eye-tracking data alone" (p. 756).

Recent developments in the translation industry mean that translators are presented with text already translated to a certain degree (as a result of translation memory systems or machine translation). These tools change the cognitive demands on and processes of translation (see also Jakobsen & Mesa-Lao, 2017). An aspect of technology is also addressed by Koglin (2015), who investigated cognitive effort required to post-edit machine-translated metaphors compared with translating metaphors manually. Her hypothesis was that manual translation is more cognitively demanding than post-editing. To investigate it, an experiment was conducted using eye tracking, keylogging and retrospective TAPs. Professional translators performed the translation task, whereas students post-edited two versions of the same newspaper ST produced by two different machine translation systems. Koglin, however, does not problematize her setup by using different groups of subjects to perform the two tasks. After each task, two TAPs were recorded. Participants were asked, first, "to think aloud while their full post-editing process was replayed on Translog-II screen", and for a following guided protocol, "they were asked two questions related to metaphor interpretation and its post-editing decision-making process" (pp. 130–131). In the analysis of cognitive effort, she focused on keylogged data in respect of pauses and on eye-tracking data related to total fixation duration for two specific metaphors selected as areas of interest in both the ST and the TT. She found that on average, more time was spent on manually translating the text compared with post-editing it. There were fewer deletions and insertions for post-editing in comparison to manual translation. For fixation duration, Koglin noticed variation among the participants of both tasks. For example, "participants had longer average fixation durations on the TT area in post-editing and longer average fixation durations on the ST area in manual translation" (p. 135), and the total of pauses was lower in post-editing than in manual translation. Koglin argues that her hypothesis was confirmed, although she found that there "has not been a significant difference between cognitive effort required to post-edit machine translated metaphors in comparison to manually translating them" (p. 137). She acknowledges, however, that due to her small sample size for the translation task and the high variation among the individual participants, her "findings might not be generalizable" (p. 136).

In respect of metaphors, there is much more research into translation than interpreting, another mode which involves decision making. Schäffner (2017b) investigated how interpreters deal with metaphors at international press conferences, looking at interpreting strategies (i.e. how the simultaneous interpreters dealt with metaphorical expressions) and at coping strategies (i.e. strategies that might signal a particular cognitive load, such as lengthening, doubling and hedging). However, since her analysis was based predominantly on the product (i.e. the transcripts of the press conferences) and not on the actual process, her assumptions about cognitive load cannot be empirically verified. Similarly, Saltalamacchia's (2014) investigation of how time metaphors in plenary speeches in the European Parliament were dealt with by simultaneous interpreters in the English, German, French, Italian and Portuguese booths was also based on the analysis of the official transcripts. Her aim, too, was to understand the cognitive processes involved when conveying highly complex metaphors. Although her analysis was based on transcripts, she also had audio recordings of the interpreters' performance on which we can hear interpreters breathe heavily, stammer or laugh, which also indicates cognitive effort. She, too, suggests that triangulating the analysis of transcripts with video recordings of the interpreters and eye tracking could be more useful in gaining insights into the cognitive processes. Research into the cognitive load in simultaneous interpreting has recently made increasing use of eye tracking (e.g. Seeber, 2013). In particular, the psycho-physiological method of pupillometry, i.e. a study of the pupil size and pupil dilation when carrying out a task, has emerged as a promising method to measure cognitive load during simultaneous interpreting in real time. Pupillometry is based on the assumption that eye movements and pupil size are indications of cognitive activity. However, it has not yet been employed for studying metaphor in translation and/or in interpreting.

Cognitive processes can also be measured by other objective psycho-physiological techniques, such as measures of brain or heart activity, e.g. usage of electroencephalography (EEG), positron emission tomography (PET) or functional magnetic resonance imaging (fMRI). Due to certain constraints, especially the (at least partly) intrusive nature of these methods, they have only rarely been used for research into cognitive aspects of translation and interpreting (for an evaluation of some such research, see Seeber, 2013). An interesting experiment was conducted by Rojo et al. (2014), who investigated the emotional impact of metaphors by measuring the participants' heart rate with a heart monitor. Heart rate is related to emotions, and heart rate acceleration or deceleration can be an indicator of the degree of physiological arousal. Rojo et al. (2014)

were interested in finding out whether a metaphorical or a non-metaphorical translation of the same figurative expression would result in differences in the emotional impact. An experiment was carried out which involved two Spanish translations of English short stories, with the final sentence containing either a metaphorical or a non-metaphorical expression of the four basic emotions happiness, sadness, rage and fear. Their results showed differences in the participants' heart rate: metaphorical expressions caused an increase in the participants' mean heart rate, whereas non-metaphorical ones caused a decrease for rage, fear and happiness expressions, with the opposite pattern for sadness expressions. They argue that "the loss of metaphorical image will most probably result in diminished emotional impact" and suggest that "translators should certainly be more sensitive to the implications of reproducing the meaning at the cost of the image" (p. 38). They acknowledge, however, that "heart rate can also be altered by other factors, such as participants' stress or tiredness" and plead for further studies, which should combine "heart rate with measurements of other indicators, such as galvanic skin response, subjective feelings or even retrospective interviews" (p. 38). In this study, the emotional impact metaphors had on recipients was measured. Translators, and interpreters too, are in a way recipients of the ST, and it would be interesting to investigate their own emotional reaction and heart rate when engaging with the metaphors in the ST and producing their TTs. By triangulating methods, e.g. heart rate measurement and eye tracking, it could be possible to see whether there are correlations between physiological, emotional and cognitive factors. Such research could also lend support to investigating translation as an embodied activity.

18.4 Concluding remarks

As illustrated in this chapter, a large amount of research into the relationship between metaphor in translation and cognition has been conducted in experimental settings. Professional translators, however, normally operate in real-life contexts, performing authentic assignments, using a variety of tools and cooperating with other agents. Translation as a socially situated, embodied and enacted activity means that multiple actors and factors influence a translator's decision-making process in a concrete translation event. As Martín de León (2013) argues, "[r]esearching distributed cognition in translation amounts to studying complex real-life translation projects" (p. 115). There has been a growing amount of research investigating how physical, organizational, environmental and other relevant ergonomic factors impact on translation practice (e.g. Ehrensberger-Dow & Massey, 2017; Risku 2010), and on how cognitive and situational levels interact in translation and interpreting (e.g. Ehrensberger-Dow & Englund Dimitrova, 2018). In respect of metaphor in translation, Massey and Ehrensberger-Dow (2017) point out that "[t]hose addressing translated conceptual metaphor need to bear [the essential situatedness of translation] in mind when collecting and analysing their data and when interpreting results" (p. 175). Although workplace studies have significant benefits, so far, real-life translation projects have not yet been used for investigating cognitive aspects in metaphor translation but should be considered for future research.

In addition, more process studies in experimental settings will also be relevant, since previous research has left a number of open questions as well as findings for which a firm explanation has not yet been found (see also Shuttleworth, 2019, on topics for future research). For example, previous research has provided evidence that thought processes are often triggered by metaphorical expressions and the initial attempt to provide a direct metaphorical equivalent in the target language, which seems to suggest that translating metaphor into the same metaphor is indeed the default procedure. This could be tested further by a closer qualitative analysis of the keylogging data. The log files can show whether translators initially typed the closest target-language equivalent and changed it at a later stage. Similarly, processes of revision and post-editing, including post-editing of machine translation output, can show which changes were made in respect of metaphors, e.g. whether or not an initially used metaphorical expression was changed into a paraphrase or deleted. Such research could also get us closer to answering the question: What exactly triggers the translator's need to access the conceptual level in real translation events? Another question that needs further exploration is: (How) can we determine how the cognitive load is distributed between metaphor interpretation, the choice of a translation procedure, and target-text production? For example, pauses in typing, before or after typing and/or corrections, as well as longer gaze times may be due to a translator reflecting on consequences of local decisions for the overall function of the text and may thus not be immediately relatable to a comprehension or production stage. Most scholars have been cautious in making generalizations about processes in metaphor translation, having discovered that subjects tend to differ enormously in their performance. This only confirms that there are several variables in the translation process.

Process-oriented studies of metaphor in translation can also provide insights with which theories of conceptual metaphor and cognitive linguistics can be tested. However, translation scholars who have referred to conceptual metaphor theories have not (yet) problematized them. There is also hardly any engagement with work in CMT beyond that of Lakoff and Johnson, and blending theory has not yet been used to any significant extent. There is thus much scope for TS research, including the study of metaphor in different types of mediated communication. That is, in addition to interpreting, briefly mentioned earlier, one could also investigate how verbal metaphors interact with metaphors in other modes, i.e. focusing on the multimodal nature of the translation and/or interpreting process. This could be studied for the audiovisual mappings as they are relevant in audiovisual translation. Metaphors in signed languages are also worthy of investigation. Roush (2011), for example, asked to what extent a specific metaphor exists in the conceptions of American sign language (ASL) signers, and what procedures a translator might use to handle potential differences between English and ASL. For the translation procedures, he used those suggested by Al-Harrasi (2001). With the increased interest in performance by non-professional translators, one could also investigate how they process metaphors in comparison to professional translators.

Triangulation of research methods will continue to be useful to investigate cognitive aspects of metaphor in translation. Moreover, multi-method approaches can be combined with an interdisciplinary orientation, drawing on different theoretical and methodological frameworks, such as Translation Studies, cognitive linguistics, sociology, network theory or neuropsychology (for an overview of neurocognitive research on translation and interpreting, see Muñoz et al., 2018). Translation Studies can benefit from interdisciplinary projects, since the various disciplines can provide complementary perspectives on understanding the complexity of translating and interpreting.

In respect of metaphor in translation, the process studies illustrated in this chapter tended to focus on cognitive load and the allocation of processing effort during task execution. The position and function of a metaphorical expression in a text, as well as specific linguistic features, deserve more attention too, since language processing is an essential element of cognitive processing in translation. As already indicated, the choice of a translation strategy can also be due to certain constraints on dealing with metaphor (e.g. culture-specific aspects, word form or genre). One such constraint can also be the use of a particular language by a speaker for whom this language is not the first language. Such texts could include metaphorical expressions reflecting cultural variation, which can have an influence on language processing of translators and interpreters. The potential processing problems that texts and speeches with English as a lingua franca (ELF) cause for translators and interpreters compared with those in standard English, the cognitive load associated with processing ELF, and the coping strategies used by translators and interpreters are being investigated in an interdisciplinary project, which draws on perspectives from translation and interpreting studies, ELF studies and neuroscience (CLINT, n.d.). It will be interesting to see whether linguistic and conceptual metaphors are among the ELF features that have an impact on the performance of translators and interpreters.

In acknowledging the value of multi- and interdisciplinary research, Shlesinger's argument, although related to interpreting research, is still very pertinent today: "To fully appreciate the merits and drawbacks of the available approaches, we apparently require more research into research; i.e., efforts to validate the relevance of our methodologies and to avoid counterproductive or misleading ones" (Shlesinger, 2000, p. 13).

Further reading

Kövecses, Z. (2005). *Metaphor in culture: Universality and variation*. Cambridge: Cambridge University Press.
An overview of a cognitive linguistic view of metaphor with a special focus on dimensions, components and causes of intra- and intercultural metaphor variation.

Miller, D. R., & Monti, E. (Eds.) (2014). *Tradurre figure / Translating figurative language*. Bologna: Bononia University Press.
The contributions in this volume address interlingual translation of figurative language, which often foregrounds the complexities of the translation process as well as cultural variation.

Rojo, A., & Ibarretxe-Antuñano, I. (Eds.) (2013). *Cognitive linguistics and translation. Advances in some theoretical models and applications*. Berlin/Boston: de Gruyter.
The chapters in this volume explore what cognitive linguistics can bring to Translation Studies, and vice versa, thus contributing to the development of a cognitive translation theory.

Semino, E., & Demjén, Z. (Eds.) (2017). *The Routledge handbook of metaphor and language*. Abingdon: Routledge.
A comprehensive overview of interdisciplinary research on metaphor and language, addressing theory, methodology and application of research.

Trim, R., & Śliwa, D. (2019). *Metaphor and translation*. Newcastle-upon-Tyne: Cambridge Scholars Press.
Contributions (in English or French) dealing with theoretical issues and methodology in translating metaphor as well as empirical studies of metaphor translation, illustrated with a variety of text types.

References

Al-Harrasi, A. (2001). Metaphor in (Arabic-into-English) translation with specific reference to metaphorical concepts and expressions in political discourse [unpublished doctoral dissertation]. Aston University, Birmingham.

Brdar, M., & Brdar-Szabó, R. (2017). Moving-time and moving-ego metaphors from a translational and a contrastive linguistic perspective. *Research in Language, 15*(2), 191–212. doi: 10.1515/rela-2017–0012

Burmakova, E. A., & Marugina, N. I. (2014). Cognitive approach to metaphor translation in literary discourse. *Procedia—Social and Behavioral Sciences, 154*, 527–533. doi: 10.1016/j.sbspro.2014.10.180

Charteris-Black, J. (2004). *Corpus approaches to critical metaphor analysis*. New York: Palgrave Macmillan.

Chilton, P. (1996). *Security metaphors. Cold War discourse from containment to common house*. New York: Lang.

Chilton, P. (2019). Cognitive linguistics. In W. Brekhus, & G. Ignatow (Eds.), *Oxford handbook of cognitive sociology* (pp. 243–270). New York: Oxford University Press.

CLINT (n.d.) Cognitive Load in Interpreting and Translation (CLINT). www.zhaw.ch/en/linguistics/institutes-centres/iued/research/clint/

Croft, W., & Cruse, D. A. (2004). *Cognitive linguistics*. Cambridge: Cambridge University Press.

Dagut, M. (1976). Can "metaphor" be translated? *Babel, 22*(1), 21–33.

Ehrensberger-Dow, M. (2018). Process research. In L. D'hulst, & Y. Gambier (Eds.), *A history of modern translation knowledge* (pp. 293–300). Amsterdam: John Benjamins.

Ehrensberger-Dow, M., & Englund Dimitrova, B. (Eds.) (2018). *Exploring the situational interface of translation and cognition*. Amsterdam/Philadelphia: John Benjamins. [Benjamins Current Topics 101].

Ehrensberger-Dow, M., & Massey, G. (2017). Socio-technical issues in professional translation practice. *Translation Spaces, 6*(1), 104–121. https://doi.org/10.1075/ts.6.1.06ehr

Förster Hegrenæs, C. (2018). Translation competence development and the distribution of cognitive effort: An explorative study of student translation behavior [Unpublished doctoral dissertation]. Norwegian School of Economics, Bergen.

Guldin, R. (2016). *Translation as metaphor*. Abingdon & New York: Routledge.

Hansen, G. (2013). The translation process as object of research. In C. Milán, & F. Bartrina (Eds.), *The Routledge handbook of translation studies* (pp. 88–101). London & New York: Routledge.

Jääskeläinen, R. (2009.) Think-aloud protocols. In M. Baker, & C. Saldanha (Eds.), *The Routledge encyclopedia of translation studies* (2nd ed., pp. 290–293). London & New York: Routledge.

Jakobsen, A. L., Jensen, K. T. H., & Mees, I. M. (2007). Comparing modalities: Idioms as a case in point. In F. Pöchhacker, A. L. Jakobsen, & I. M. Mees (Eds.), *Interpreting studies and beyond. A tribute to Miriam Shlesinger* (pp. 217–249). Copenhagen: Samfundslitteratur.

Jakobsen, A. L., & Mesa-Lao, B. (Eds.). (2017). *Translation in transition: Between cognition, computing and technology*. Amsterdam: John Benjamins.

Koglin, A. (2015). An empirical investigation of cognitive effort required to post-edit machine translated metaphors compared to the translation of metaphors. *Translation & Interpreting,* 7(1), 126–141. doi: ti.106201.2015.a06

Kövecses, Z. (2005). *Metaphor in culture: Universality and variation*. Cambridge: Cambridge University Press.

Kövecses, Z. (2014). Conceptual metaphor theory and the nature of difficulties in metaphor translation. In D. R. Miller, & E. Monti (Eds.), *Tradurre Figure / Translating figurative language* (pp. 25–39). Bologna: Bononia University Press.

Kussmaul, P. (2000). *Kreatives Übersetzen*. Tübingen: Stauffenburg.

Lacruz, I., & Jääskeläinen, R. (Eds.). (2018). *Innovation and expansion in translation process research*. Amsterdam: John Benjamins.

Lakoff, G. (1993). The contemporary theory of metaphor. In A. Ortony (Ed.), *Metaphor and thought* (pp. 202–251). Cambridge: Cambridge University Press.

Lakoff, G., & Johnson, M. (1980/2003). *Metaphors we live by*. Chicago & London: University of Chicago Press.

Lee, D. A. (2001). *Cognitive linguistics: An introduction*. Melbourne: Oxford University Press.

Levý, J. (1967/2000). Translation as a decision process. In L. Venuti (Ed.), *The translation studies reader* (pp. 148–159). London & New York: Routledge.

Mandelblit, N. (1995). The cognitive view of metaphor and its implications for translation theory. In M. Thelen, & B. Lewandowska-Tomaszczyk (Eds.), *Translation and meaning Part* 3 (pp. 482–495). Maastricht: Universitaire Press.

Martín de León, C. (2013). Who cares if the cat is on the mat? Contributions of cognitive models of meaning to translation. In A. Rojo, & I. Ibarretxe-Antuñano (Eds.), *Cognitive linguistics and translation. Advances in some theoretical models and applications* (pp. 99–122). Berlin/Boston: de Gruyter.

Massey, G. (2016). Remapping meaning: Exploring the products and processes of translating conceptual metaphor. In Ł. Bogucki, B. Lewandowska-Tomaszczyk, & M. Thelen (Eds.), *Translation and meaning* (Vol. 2(1), pp. 67–83). Frankfurt am Main: Peter Lang.

Massey, G., & Ehrensberger-Dow, M. (2017). Translating conceptual metaphor: The processes of managing interlingual asymmetry. *Research in language, 15*(2),173–189. doi: 10.1515/rela-2017-0011

Miller, D. R., & Monti, E. (Eds.) (2014). *Tradurre figure / Translating figurative language*. Bologna: Bononia University Press.

Muñoz, E., Calvo, N., & García, A. M. (2018). Grounding translation and interpreting in the brain: What has been, can be, and must be done. *Perspectives, 27*(4), 483–509. doi: 10.1080/0907676X.2018.1549575

Muñoz Martín, R. (2013a). More than a way with words: The interface between cognitive linguistics and cognitive translatology. In A. Rojo, & I. Ibarretxe-Antuñano (Eds.), *Cognitive linguistics and translation. Advances in some theoretical models and applications* (pp. 75–97). Berlin/Boston: de Gruyter.

Muñoz Martín, R. (2013b). Cognitive and psycholinguistic approaches. In C. Milán, & F. Bartrina (Eds.), *The Routledge handbook of translation studies* (pp. 241–256). London and New York: Routledge.

Musolff, A. (2016). *Political metaphor analysis. Discourse and scenarios*. London: Bloomsbury.

Newmark, P. (1981). *Approaches to translation*. Oxford: Pergamon.

Pragglejaz Group (2007). MIP: A method for identifying metaphorically used words in discourse. *Metaphor and Symbol, 22*(1), 1–39. doi: 10.1080/10926480709336752

Risku, H. (2010). A cognitive scientific view on technical communication and translation: Do embodiment and situatedness really make a difference? *Target, 22*(1), 94–111.

Rojo, A., & Ibarretxe-Antuñano, I. (Eds.). (2013). *Cognitive linguistics and translation. Advances in some theoretical models and applications*. Berlin/Boston: de Gruyter.

Rojo, A., Ramos, M., & Valenzuela, J. (2014). The emotional impact of translation: A heart rate study. *Journal of Pragmatics, 71*, 31–44. https://doi.org/10.1016/j.pragma.2014.07.006

Roush, D. (2011). Revisiting the conduit metaphor in American Sign Language. In C. B. Roy (Ed.), *Discourse in signed languages* (pp. 155–175). Washington DC: Gallaudet University Press.

Safarnejad, F., Ho-Abdullah, I., & Awal, N. M. (2014). Cultural basis of conceptual metaphors translation: Case of emotions in Persian and English. *Asian Social Science, 10*(7), 107–118. doi: 10.5539/ass.v10n7p107

Saltalamacchia, S. (2014). Neue Zeitmetaphern in der Energiedebatte—Herausforderung Simultandolmetschen. Ein Vergleich zwischen fünf Sprachen [unpublished Master's dissertation]. Zürcher Hochschule für Angewandte Wissenschaften, Zurich.

Schäffner, C. (2004). Metaphor and translation: Some implications of a cognitive approach. *Journal of Pragmatics, 36*(7), 1253–1269. https://doi.org/10.1016/j.pragma.2003.10.012

Schäffner, C. (2017a). Metaphor in translation. In E. Semino, & Z. Demjén (Eds.), *The Routledge handbook of metaphor and language* (pp. 247–262). Abingdon: Routledge.

Schäffner, C. (2017b). Self-awareness, norms and constraints: Dealing with metaphors in interpreter-mediated press conferences. In M. Biagini, M. S. Boyd, & C. Monacelli (Eds.), *The changing role of the interpreter. Contextualising norms, ethics and quality standards* (pp. 149–172). New York and London: Routledge.

Schäffner, C., & Shuttleworth, M. (2013). Metaphor in translation: Possibilities for process research. *Target, 25*(1), 93–106.

Schwieter, J. W., & Ferreira, A. (Eds.). (2017). *The handbook of translation and cognition*. Hoboken: Wiley Blackwell.

Seeber, K. (2013). Cognitive load in simultaneous interpreting. Measures and methods. *Target, 25*(1), 18–32.

Semino, E. (2008). *Metaphor in discourse*. Cambridge: Cambridge University Press.

Semino, E., & Demjén, Z. (2017a). Introduction: Metaphor and language. In E. Semino, & Z. Demjén (Eds.), *The Routledge handbook of metaphor and language* (pp. 1–10). Abingdon: Routledge.

Semino, E., & Demjén, Z. (Eds.). (2017b). *The Routledge handbook of metaphor and language*. Abingdon: Routledge.

Shlesinger, M. (2000). Interpreting as a cognitive process: How can we know what really happens? In S. Tirkkonen-Condit, & R. Jääskeläinen (Eds.), *Tapping and mapping the processes of translation and interpreting: Outlooks on empirical research* (pp. 3–15). Amsterdam: John Benjamins.

Shuttleworth, M. (2011). Translational behaviour at the frontiers of scientific knowledge: A multilingual investigation into popular science metaphor in translation. *The Translator, 17*(2), 301–323.

Shuttleworth, M. (2017). *Studying scientific metaphor in translation: An inquiry into cross-lingual translation practices*. Abingdon: Routledge.

Shuttleworth, M. (2019). Metaphor in translation: What do we know, and what do we still need to know? In R. Trim, & D. Śliwa (Eds.), *Metaphor and translation* (pp. 29–39). Newcastle-upon-Tyne: Cambridge Scholars Press.

Sjørup, A. (2008). Metaphor comprehension in translation: Methodological issues in a pilot study. In S. Göpferich, A. L. Jakobsen, & I. M. Mees (Eds.), *Looking at eyes. Eye-tracking studies of reading and translation processing* (pp. 53–77). Copenhagen: Samfundslitteratur.

Sjørup, A. (2011). Cognitive effort in metaphor translation: An eye-tracking study. In S. O'Brien (Ed.), *Cognitive explorations of translation* (pp. 197–214). London: Continuum.

Sjørup, A. (2013). Cognitive effort in metaphor translation: An eye-tracking and key-logging study [doctoral dissertation, Copenhagen Business School]. www.econstor.eu/bitstream/10419/208853/1/cbs-phd2013-18.pdf

St André, J. (Ed.). (2010). *Thinking through translation with metaphor*. Manchester: St Jerome.

Steen, G. (2017). Identifying metaphors in language. In E. Semino, & Z. Demjén (Eds.), *The Routledge handbook of metaphor and language* (pp. 73–87). Abingdon: Routledge.

Tcaciuc, L. S. (2014). The conceptual metaphor "money is a liquid" and "economy is a living organism" in Romanian translations of European Central Bank documents. In D. R. Miller, & E. Monti (Eds.), *Tradurre figure / Translating figurative language* (pp. 99–112). Bologna: Bononia University Press.

Teixeira, C., & O'Brien, S. (2017). Investigating the cognitive ergonomic aspects of translation tools in a workplace setting. *Translation Spaces, 6*(1), 79–103. https://doi.org/10.1075/ts.6.1.05tei

Tirkkonen-Condit, S. (2001). Metaphors in translation processes and products. *Quaderns. Revista de traducció, 6*, 11–15.

Toury, G. (1995). *Descriptive translation studies and beyond*. Amsterdam: John Benjamins.

Trim, R., & Śliwa, D. (Eds.). (2019). *Metaphor and translation*. Newcastle-upon-Tyne: Cambridge Scholars Press.

Turkama, K. (2017). Difficulty of the translation of primary and complex metaphors. An experimental study [doctoral dissertation, University of Eastern Finland]. Publications of the University of Eastern Finland. Dissertations in Education, Humanities, and Theology No. 95.

Turner, M., & Fauconnier, G. (2002). *The way we think. Conceptual blending and the mind's hidden complexities*. New York: Basic Books.

Van den Broeck, R. (1981). The limits of translatability exemplified by metaphor translation. *Poetics Today, 2*(4), 73–87. doi: 10.2307/1772487

Yan, D., Noël, D., & Wolf, H.-G. (2010). Patterns in metaphor translation: A corpus-based study of the translation of FEAR metaphors between English and Chinese. In Xiao, R. (Ed.), *Using corpora in contrastive and translation studies* (pp. 40–61). Newcastle: Cambridge Scholars Publishing.

Zheng, B., & Zhou, H. (2018). Revisiting processing time for metaphorical expressions: An eye-tracking study on eye-voice span during sight translation. *Foreign Language Teaching and Research, 50*(5), 744–759.

19

Translation, equivalence and cognition

Erich Steiner

19.1 Introduction: A historical perspective

An exploration of a core notion—and one at times controversial—such as "equivalence" usually gains from some historical grounding. In the following brief historical sketch focusing on recent developments, different notions of "equivalence" will be reviewed as well as some implications they have had for conceptualizing relationships between texts. One implication is a differentiation between *paraphrase*, *variation* and *translation* as related, yet different, notions. The questions will then be addressed of whether and in what sense "equivalence" between source texts (ST) and target texts (TT) has been regarded as an essential property of *translation*, and if so, equivalence with respect to *what*. Implications of multimodality for the concept of equivalence will be mentioned, before we focus on a brief overview of equivalence in models of cognition in translation up to now.[1]

19.1.1 Different notions of "equivalence" and different notions of "translation"

Different notions of equivalence, or even the outright dismissal of such a notion as being relevant, will lead to different notions of translation, especially in comparison with *paraphrase* and *variation*. "Equivalence" has often been left vague and/or has been given different meanings according to different approaches, but in its more precise formulations, it allows us to make some important distinctions between *translation* and other forms of multilingual text production. Such distinctions are a prerequisite for models of cognition in translation, if they are to be translation-specific and different from models of text understanding and production in language processing in general.

The concept of equivalence has a long and controversial history in Translation Studies (see, among many others, Halverson, 1997; House, 1997, p. 24, 2015, p. 5; Koller, 1995, 1997, p. 159; and for a more comprehensive historical overview, Munday, 2001, Ch. 3), but it is not easy to define translation entirely without it. Even in approaches shifting the domain of equivalence away from any level of textual structure to "equivalent effect" between the receptors and STs and TTs, respectively (Nida, 1964, p.159), the question arises of how we ensure identity or similarity between source and target contexts, because receptors and contexts cannot simply be presupposed in their common-sense meanings as tools for explicit models of translation. If, with relevance theory, translation is considered as "interlingual interpretive use" (Gutt, 1991, p. 100), we again move away from levels of linguistic encoding as such, but once more we must assume some sort of equivalence between contexts and between assumptions, implicatures or propositions in which these can be expressed. The concept of equivalence has thus sometimes been shifted away from structural encoding to processing and/or context, but without explicit models of the latter, this simply means shifting the burden of definition to other disciplines, or falling back on common-sense notions—not a viable strategy for a scientific discipline. Without a clear notion of the specifics of the object *translation*, we cannot expect models clear and specific enough for research hypotheses and empirical investigations.

Halverson (1997, p. 209) in a representative survey article reminds us that at least three points need to be clarified for any discussion of equivalence:

a) specification of entities, between which the relationship holds, and in what sense they are comparable
b) specification of the specific *nature* and the *degree* of "likeness/sameness/ similarity/equality" between entities compared
c) specification of the quality in terms of which the sameness is defined.

Bearing in mind Halverson's requirements, an attempt will be made to relate translation to other relevant forms of textual relationships through the role played by equivalence (see Steiner, 2001). *Paraphrase* is a truth-condition-preserving, and in that sense equivalent, relationship among sets of propositions intra- or interlingually. Paraphrases do not preserve the text-building patterns, such as information structure, and they have no very clear relationship towards the interpersonal semantics (mood, speech acts, illocutionary acts and appraisal) of the encoding. Paraphrases as sets of propositions are possible for texts within and across languages, preserving important aspects of their experiential and logical semantics (Halliday & Matthiessen, 2014, p.85; Matthiessen, 2001, p.100ff). Yet the notion of "equivalence" as paraphrase alone under- and over-determines a relevant notion of translation.

Variation as dealt with in variation theory and register theory (see Biber, 1993; Halliday & Hasan, 1989; House, 1997, 2015) is possible within sets of intertextually related texts, both intralingually and interlingually. But these texts are not translationally equivalent precisely because at least some aspect of their overall meaning (field, tenor, mode of discourse, or Biber's functional dimensions) must by definition vary.

Translation, for the present purposes, is taken as a relationship between STs and TTs (or smaller translation units) which approximates equivalence in a combination of the dimensions of field, tenor and mode of discourse, and ideational, interpersonal and textual meaning in terms of clause semantics and grammar. These dimensions will usually be ranked depending on the context of the translation brief (similar to Koller's 1997 model). The approximation thus is a multidimensional optimization task rather than one single right-or-wrong solution. Especially in a model oriented to cognitive processes, the TT may show additional traces of the process of understanding (e.g. *explicitation*) and other aspects of the translation process. Translation is different from paraphrase in respect to interpersonal and textual meanings in addition to experiential (propositional) and logical meaning. And translation is different from variation by keeping many of the parameters of variation maximally stable. In the end, we may say that translation is an approximation to a multi-functional paraphrase of the ST by some TT—rather than the mono-functional paraphrases of logic-oriented semantics—under the constraints of the process of understanding and of the typology of the language systems involved. Each individual translation, i.e. situated language (instantiation), is text production under the constraints of a ST. To what extent "instantiation" here is the same as "[translator's] reading of a text" is a question to which we shall return.

19.1.2 Equivalence between source texts and target texts as a defining property of translation?

Has equivalence so far been regarded as a defining property of translation, or has it been dismissed, given its inherent difficulties? The notion of equivalence has more recently been relativized (Halverson, 1997, p. 217f), and its domain has sometimes been shifted from levels of text representation, including context, to postulated translation units or even to the process itself. Catford (1965), in his classic account, already diversified the notion across an entire range of linguistic ranks and levels. Matthiessen (2001, 2014) took that approach substantially further within a highly developed model of systemic functional linguistics with a wider semiotic background. Frame-based approaches give a privileged place to the frame as a key locus of equivalence of meaning (Czulo, 2017). Nida (1964) and Gutt (1991) shifted it outside structural encoding, locating it in behavioural reactions and interpretations in the translation process. Transfer-based models of (machine) translation (MT) relied on notions of equivalence-based transfer on a hierarchy of levels (EUROTRA as in Durand et al., 1991), as did other MT systems based on stratificational models (see Carl & Schaeffer, 2017), or in the extreme case on an interlingua which would neutralize language-dependent differences. More recent statistics-based MT models incorporate equivalence under "adequacy" (vs. fluency) in MT evaluation (Banchs et al., 2015; Chungyu & Tak-ming, 2015), as do current models of neural-networks-based MT with some modifications (see Gupta et al., 2015). For a model of cognition in translation, it has been said that removing equivalence altogether simply leaves us with notions of text production, text comprehension, bilingual processing— or a combination of these at best (see Schwieter & Ferreira, 2017, p. 144; Shreve & Lacruz, 2017, pp.129, 134 on the notion of "transfer"). This may be true even if we extend the notion of translation across different modes of meaning (Bateman, 2008; Kress, 2010, introducing the notion of "transduction" across modes; Matthiessen, 2001, p. 50). Multimodal semiotic objects are objects of translation, and they pose a creative challenge to any form of text production (Hiippala, 2012), but the very different codes employed in the different modalities caution against a premature full-scale extension of the notion of equivalence to all of these. If we want to speak of "multimodal translation", though, a motivated notion of equivalence across modes needs to be developed.

19.1.3 Cognition and translation

To what extent have models of translation so far included a perspective on the wider context of human cognition? A number of fairly detailed studies exist by now on specific aspects of equivalence: for example, several of those brought together in Shreve and Angelone (2010). Yet, many publications have been programmatic rather than empirical and methodologically mature (Halverson, 2010; Muñoz Martín, 2010; Steiner, 2012). Even the more recent overview articles of Alves and Hurtado (2017) and Muñoz Martín (2017) show that the precise identification of the objects and above all, methods of research are not yet clear enough. In order to test anything empirically, in terms of corpus-based work (product) or of experiments (process), models are needed from which to derive sufficiently specific hypotheses. Too few, if any, such models exist so far, because the existing models of the translation process are—understandably in view of the recent history of empirical Translation Studies and in view of the overall complexity of the process of translation—not fine-grained enough to derive research hypotheses from. In order to arrive at such models, certain key issues need clarification, in particular[2]:

- Textuality: Levels of abstraction, syntagmatic vs. paradigmatic axes, intra-level ranks
- Potential vs. instance; Text vs. reading
- Explicitness vs. implicitness of meaning
- Models of translation and cognition

The place of equivalence along all these dimensions will be discussed later, as well as its potential implications for cognition.

19.2 Core issues

Some core issues arise out of recent debates and appear important for making progress towards models of translation, equivalence and cognition.

One of these is the issue of textuality, more specifically "what are STs and TTs, and what are translation units?" Languages, registers and individual texts can be represented on different levels of abstraction (lexicogrammar, semantics, context/register), on syntagmatic vs. paradigmatic axes, and within levels even on ranks (sentences, clauses, phrases, words and morphemes). The precise number and kind of levels of textuality depend on the model of language used, but equivalence and cognitive processes can be modelled on any of these levels. The ST in a translation relationship cannot be a pre-theoretically-given unanalysed object of everyday perception; it is a highly structured semiotic web of texture.

Another issue will be that of potential vs. actual (system and instance): it is a fundamental assumption that translation (studies) deals with texts (instances) rather than with language systems (potential). This needs some elaboration, and in particular, we need to ask whether there is a further distinction between the "text" (instance) and the "reading" of a text. Equivalence can be located on any of these dimensions, but the question is which cognitive aspects of the translation process are responsive to which dimension.

A third issue will be the distinction between explicitness and implicitness of meaning, another way of saying "encoded vs. inferred" meaning. Again, which is the domain of equivalence, and which of the two types of meaning lead to which cognitive processes?

A fourth issue directly addresses the question of where, in a model of translation (studies), cognition finds its place. The obvious place would seem to be the translation process and accordingly, process-related studies. Yet, there are more possible answers, such as, for example, a model of translation based on some version of cognitive linguistics (see Geeraerts & Cuyckens, 2007). Whichever of the approaches is taken, Translation Studies needs to make itself familiar with methodologies of empirical disciplines, both corpus-based and experimental, which means considerable methodological refinement. At the same time, a development of methodologies appropriate to its own object of research is necessary, rather than simply taking over methodologies from adjacent fields. One of the typical and unique properties of translations, once again, seems to be equivalence. Contrastive linguistics (see König and Gast, 2018) has developed the system side of equivalence (correspondence), but Translation Studies needs to develop the instance side, along with intercultural register studies.

19.2.1 Textuality: Levels of abstraction, syntagmatic vs. paradigmatic axes, intra-level ranks

An ST is not simply a sequence of letters, morphemes, words, phrases or clauses mapped onto some TT unit under some notion of sameness, or even equivalence. This was already recognized in versions of translation procedures and methods (see Fawcett, 1997 for an overview), it was programmatic in Catford's (1965) notion of translation diversified on different levels and ranks, and it is clearly recognized in approaches relating micro and macro translation units to grammatical units and shifts (Alves et al., 2010, p. 129ff). Yet, against the background of more elaborate semiotically based models (e.g. Halliday & Matthiessen, 2014), the semiotic text as an object of translation can be modelled on different levels of abstraction (phonology, lexicogrammar, semantics and context), each of these levels may be stratified by metafunctions (ideational, interpersonal and textual; or conceptual vs. procedural encoding in terms of relevance theory as applied in Alves et al., 2014), the focus may be on the syntagmatic axes (structure) or the paradigmatic axes (system of features), both system and structure can be dealt with in different degrees of specificity, and within the levels of abstraction, we may even assume different ranks. Questions of equivalence and cognition can be raised separately for each of these dimensions (see Matthiessen, 2001, 2014; Halverson, 1997, p. 212ff). And quite importantly, even with relatively stable ideational meaning between source and target, category change in the syntax-to-semantics mapping within and between languages plays a central role both in understanding and in translation (Serbina et al., 2017; Steiner, 2004, p.125ff). Another type of approach, relying on an elaborate but quite different architecture of textual structure, is cognitive-grammar-based approaches (Halverson, 2010, 2017; Sickinger, 2017). Elaborate notions of textuality such as those mentioned here allow us to ask the question of equivalence on different levels of representation. And depending on the particular type of translation context (including the translation brief, but also the level of expertise of the translator and constraints of time, resource, media, etc.), the search for equivalence by the translator and by the reviser can be expected to focus on different levels and ranks. And whatever shows up as macro- and micro-level translation units in processing can be expected to be directly responsive to the location of translation strategy in terms of levels and ranks (e.g. grammar-, semantics-and register-based translation strategies and methods).

19.2.2 Potential vs. instance; Text vs. reading

Another refinement of notions of "text" has to do with the opposition between "potential" and "instance": along each of the dimensions identified earlier, we can focus either on the system of possible options (e.g. the grammar, the lexicon or the inventory of registers) of a language, or, more typically for Translation Studies, on the instantiation of these systems (e.g. one text in its particular context of situation). Along this cline, Translation Studies clearly investigates instances that are in an assumed relationship of *translation*, ST and TT bound by some notion of equivalence. The potential (grammars, lexicons, repertoires of cohesive devices, genres and types of register) is important as a set of resources and constraints for the translator, and between units of these systems, there are relationships of correspondence, rather than the instantiated relationship of equivalence (see Munday, 2001, p. 47, following Koller), but the process itself deals with ST and TT as instances. It is these processes that are the object of research in studies using, for example, eye tracking, keylogging, (functional) magnetic resonance imaging ((f)MRI) patterns, etc. The patterns that we see there, hopefully reflecting relevant processes in the mind of the translator, are patterns of behaviour relating an ST instance to a TT instance—an extended version of the "eye-mind hypothesis".

As a further refinement, though, we need to distinguish "text" from "reading": even the individual text, e.g. as an object of pre-translational text analysis or of translation evaluation, is still full of ambiguities and vagueness; it allows different readings by the source-text audience, by the target-text audience and, particularly, by the translator. This is true for the mapping of frames/scripts to linguistic texts, for disambiguation of cohesive relationships in discourse, and even for merely sentential instantiations. Let me take here the problem of determining the semantic reading of a given lexical form, taken from a larger project on cohesion (Kunz et al., 2017; Steiner, 2015).

> THE HIDDEN GENETIC PROGRAM OF COMPLEX ORGANISMS
>
> Assumptions can be dangerous, especially in science. They usually start as the most plausible or comfortable interpretation of the available facts. But when their truth cannot be immediately tested and their flaws are not obvious, assumptions often graduate to articles of faith, and new observations are forced to fit them. Eventually, if the volume of troublesome information becomes unsustainable, the orthodoxy must collapse.
>
> *Mattick, 2004, 61–67*

When analysing this example, say in a pre-translational text analysis, the translator is faced with determining the reading of all of the lexical items. Importantly, in the example text, lexical items such as *assumptions*, *science*, *comfortable*, *interpretation*, *available*, *facts*, *truth*, *flaws*, etc. are vague and/or ambiguous on a simple dictionary look-up, and they need to be instantiated/disambiguated in the process of understanding and then translating, the latter also depending on the target language. This involves integrating the lexical items into relevant chains and fields of items. The result of this process is a "reading" of one textual instance such as the example text, and it is this reading that is input to the translation process, rather than the textual instance as a chain of words or phrases. Ideally, the translator might attempt to arrive at a TT version having exactly the same ambiguities as the ST—but this ideal can hardly ever be reached, other than in individual sample sentences. Each reading of such a text disambiguates, instantiates its meaning—and a model of translation ultimately needs to be clear about whether the object of a translation is the individual text or the reading of it by the translator. As Muñoz Martín (2010, p. 175) says, "It is interpretations, not texts or discourses, that are translated and interpreted" (see also Venuti, 2009, p. 162). And furthermore, readings can be governed by readers'/translators' dispositions and choices; they can be 'tactical, resistant, compliant' and thus subjectified readings (see Martin & Rose, 2003, p. 269ff) of instantiated texts, and so important for translators' choices in decision making (Munday, 2012). Obviously, this is a core area where the active role of the translator finds its place. And once more, in terms of cognitive processes, we need to look for traces of translators' choices of this type in our data, yet these will be choices under the constraints of equivalence between translation units, if they are translators' rather than other text producers' choices.

19.2.3 Explicitness vs. implicitness of meaning

Another debate under core issues has to do with explicitness vs. implicitness of meaning. If, against the background of relevance theory (Alves et al., 2014; Gutt, 1991), or the "iceberg metaphor" of textuality (see Linke & Nussbaumer, 2000, p. 435), we assume that the encoded meaning in a text is only a part of its overall meaning, then translators continuously need to make choices as to how much of the overall meaning they want to encode explicitly or else leave to inferencing (the "reading" referred to in Section 19.2.2). Translation Studies, like linguistics, needs to develop models to account for these types of meaning. To "account for" means to take semantic presuppositions, implications, implicatures, pragmatic presuppositions, conversational maxims, speech acts, etc. seriously by applying them to models of translation. The same applies for the interface between any text and its context, whether this context be modelled as script, frame or some form of contextual configuration underlying register choice. It is in the interaction between encoded meaning, implicit meaning and contexts that the overall meaning underlying translation lies. And it is a fascinating, yet at this stage not very well-modelled, issue how far the constraint of "equivalence" extends across all these types of meaning. The models to be developed for integrating explicit and implicit meaning in translation need to allow for both conscious choice, reflected in metacognition or awareness as revealed by recall protocols within a translation process approach, and non-conscious coding behaviour by the translator. This requirement arises not only if we want to empower translators, but also, and beyond this, as a descriptive requirement—translators make choices with regard to implicitness/explicitness, even if these choices are frequently unconscious. Work relying on relevance theory in process studies, as well as work investigating "explicitation"[3] in the product (Hansen-Schirra et al., 2012), provides important areas of development here.

19.2.4 Models of translation and cognition

There are two obvious further key questions here: (why) do we need the notion of "cognition" in models of translation, and how do we model it?

Traditionally, translation has often been studied with an emphasis either on translation theory or else on the product and function of translation. The "applied branch" of Translation Studies as conceived of in Holmes's "map" of Translation Studies (see Toury, 1995, p. 10) focused on training, translation aids or criticism. This was probably the dominant view on Translation Studies at least until the 1980s. Out of the types of Translation Studies, 13 altogether, mentioned in Holmes' taxonomy, only the "process-oriented descriptive" kind had a clear orientation towards cognition, but with an underdeveloped methodology compared with that in psycholinguistics. This process-oriented type of Translation Studies has since become much stronger and diversified, as shown in e.g. Shreve and Angelone (2010), in Schwieter and Ferreira (2017b) or in Hansen-Schirra et al. (2017). The notion of "equivalence" is often hidden in the procedures that are the prime object of study (see Shreve & Lacruz (2017, pp.129–134) on the notion of "transfer"), and cognition as such is their main object of study. We need to be aware, though, that process-oriented studies rely on observable behavioural data (eye movements, keylogging, reaction time, electroencephalography (EEG) patterns, fMRI patterns, etc.) to "infer" cognition. It is uncontroversial, for at least this growing family of approaches, that these data reflect cognition, and that cognition is a prime object of research for Translation Studies.

The second question is how we model translation and cognition against the broad background of existing models in cognitive science, psychology and linguistics (Czulo, 2017; Muñoz Martín, 2017, p. 558ff; Shreve & Angelone, 2010, p. 12). This raises questions not only of objects of research, but also of the methodologies involved. The history of research in cognitive processes in translation is relatively recent, although there has been a significant growth from the process studies of the 1980s, through models based on relevance theory (Alves & Gonçalves, 2003; Gutt, 1991), systemic functional linguistics (parts of Alves et al., 2010; Bell, 1991; Serbina et al., 2017) or frame theory (Czulo, 2017), to a much wider field by now. If, however, we look at relevant disciplines with an earlier start in researching language and cognition (linguistics, cognitive linguistics and psycholinguistics), we find a few central questions that need answers:

How realistic/naturalistic do we need our translational data and their elicitation to be? So far, data used in experimental work on language processing are generally created under very artificial (usually laboratory) conditions, and are very small scale, hardly ever more than one sentence, or certainly very few. Is that kind of data valid for the study of a process as complex and context dependent as translation? Data in current studies in translation process research still tend to be far too complex for classical accepted psycholinguistic methods but already too small scale and artificially controlled for classical accepted Translation Studies.

What is the temporal scale of experiments, a full translation session vs. a minute, a second or milliseconds of activity, and what is the level of activities (outwardly observable symbolic data vs. neurophysiological data vs. other behavioural data such as eye movements)? Classical psycholinguistic studies operate with milliseconds of activity, whereas most studies in translation process research still operate with longer time scales—which is a methodological problem. As far as the level of activities is concerned, process research in Translation Studies is not markedly different from cognitively oriented research in psycholinguistics, so there does not seem to be a major problem here.

How wide is the gap from corpus findings to possible cognitive explanations? Corpus findings have the advantage of working with naturalistic data, and if the sampling procedure is adequate—which has not always been the case in corpus-based translation research—then problems of naturalness or time scale do not arise. One problem with corpus data is, though, that there is initially a gap between observed phenomena in the data and cognitive explanations. Phenomena to be observed, such as, for example, explicitation (e.g. Hansen-Schirra et al., 2012), or register properties and translation direction (Evert & Neumann, 2017) can be diagnosed, yet it takes fairly elaborate statistical techniques to relate them to one or two specific variables. And finally, do current models of translation make specific predictions about any product or process data? The overview in Carl and Schaeffer (2017, especially p. 66) concludes that at least for process data, this is hardly the case as yet.

19.3 Recent developments pointing to the future

Finally, a few recent developments are identified, pointing to the future of Translation Studies. In the first place, we would like to mention here the *expanding methodologies* in cognitive translation research as hopefully enabling us to learn more about the highly complex process of translation. A second and important perspective is that of *integrating product and process* research, as they mutually complement each other rather than constituting rival candidate methodologies for translation research. The third important development has to be a considerable *strengthening of competence in empirical research* methods. This process has only just begun and can be expected to gain momentum. And finally, a perspective on *traces of "equivalence" in translation-related cognitive processes* will be discussed, because this is a distinctive area for Translation Studies and hence also an area where translation can teach us something about language processing that other phenomena cannot.

19.3.1 Triangulation—multiangulation

The first promising developments are combinations of different methods, and even methods from different disciplines, to study translation. Research into translation and cognition may have started with early think-aloud-protocols (TAPs), but since then, it has been developed further through "triangulation" of methods such as keylogging, eye tracking and verbal protocols in experiments on translation (e.g. Alves et al., 2010) combined with corpus data, further progressing through increasing use of a common time-stamp to enable multi-layer recording and analysis, and most recently, increasing interfaces with translation technology (Alves & Hurtado, 2017, p. 544; Carl et al., 2016). The general possibility is, of course, something like "multiangulation", because additional methods may be involved: linguistic analyses, MRI, EEG or electrocardiography (ECG) recordings, reaction time experiments and other classical psycholinguistic methods. They open a wide potential for interdisciplinary studies of translation in the sense of multiangulation, but they also require much more methodological competence than is usually included in the training of translation researchers. Hence, research-oriented teaching of students must adapt to these requirements. Evaluation of data obtained through these methods requires competence in standard techniques of statistical evaluation (Evert and Neumann 2017; Levshina, 2015). And there is a concern relevant to all the methods discussed here: cognition is a process of the mind, yet none of the methods discussed here observes such processes directly (see Schlesewsky, 2009; Steiner, 2012). Product data show the outcome of cognitive processes at best, and process data such as keylogging, eye tracking, EEG and (f)MRI data all show (hopefully) correlates of cognitive processes, but not these processes directly. And, at least for process data, the question of ecological validity remains: in order to conform to standards of experimental research, the realistic process of translation must be reduced to an observable configuration of explanatory (independent) and response (dependent) variables, and this means very artificial data. Corpus data are at an advantage here, because the data can be natural realistic data.

19.3.2 Process and product

The second of these recent developments has to do with recognizing translation not only as system and instance on different levels of abstraction, but also as product and process. Whereas investigations of the product have been very much at the heart of Translation Studies from its beginnings, and have gained in methodological refinement since the advent of corpus-based work (see Menzel et al., 2017), studies of translation processes have only been undertaken since the 1980s and have only recently reached any degree of methodological refinement. It is a widespread paradigm in empirical linguistics and psycholinguistics that hypotheses about language production are initially tested on product data (corpora), which usually yields correlations between situational variables and patterns in the product. Any further progress towards causal explanations involves experiments and predictions, and it is this combination of product and process data that brings us closer to causal in addition to correlational explanations (see Hansen-Schirra et al., 2017; Schlesewsky, 2009; Steiner, 2012)—if we have models of the translation process that make predictions specific enough to be tested through our methods. In product data, our particular interest as translation scholars is in data showing the typical constraint of equivalence, such as aligned ST-TT translation units, or histories of attempts at interim solutions before the translator arrives at a final solution (see also Serbina et al., 2017). This presupposes that we define relationships of (non-)equivalence on our aligned data, for example on typologically motivated shifts (grammatical metaphor) and on "crossing lines" and "empty links" in alignments (see Hansen-Schirra et al., 2012, p. 268ff). In process data, we can then test whether the observed phenomena from corpora (e.g. grammatical shifts, grammatical metaphor, word order changes, explicitation, simplification, etc.) are indeed arising out of specific hypothesized stages in the process, and which of these may be due to the search for equivalence by the translator, mirrored in parallel processing of STs and TTs, in specific histories of revisions and gaze patterns. It is this intersecting of data that begins to bring us closer to something like causal explanations—which are, after all, what we want in the end.

19.3.3 *Empirical methods*

The third of these recent developments has to do with the movement of at least some strands of Translation Studies in the direction of empirical disciplines: these involve techniques needed for corpus building, including annotation and querying, and for experimental methodologies, still underdeveloped in Translation Studies. Our sense of "empirical" here is more specifically oriented to corpus-based and experimental methodologies than the sense of "descriptive" in Holmes' map of Translation Studies (Toury, 1995, p. 9ff). Examples can be found in de Sutter et al. (2017) or in Hansen-Schirra et al. (2017). Techniques are being developed for corpora, involving sampling, representation, representativeness, annotation and consistency checking, querying, evaluation of results, multivariate analyses, isolating individual explanatory variables from several explanatory sources in the product, scales of measurement on data (nominal, ordinal, interval and ratio; see Levshina, 2015, p. 16) and interfaces with translation technology (see Lapshinova-Koltunski, 2017). We also need to move forward in transparent documentation of work that has been done, not least because studies need to be repeatable to be fully convincing.

For experiments, we need to move forward on questions such as naturalness of experimental situation, size of sampled text material, hypothesis formulation and interpretation of results. Progress in experimental design needs to include competence in handling tools such as eye tracking and multi-layer integrated corpus building, formulation of testable hypotheses, operationalization of relevant variables, interfacing with translation technology, and contacts with the relevant research communities in interdisciplinary contexts. For all this, the enormous complexity of the translation process needs to be controlled—which is difficult without making the object of study un-natural. There are frequently too many interdependent variables in translation research, and any successful attempts at reducing these for the purpose of a given experiment will help towards a better methodology. And as in the case of corpus-based work, we need to introduce transparent documentation and replicate experiments in order to increase the sustainability of research design in process studies (see Alves and Hurtado, 2017, p. 547). Once more, as a community of translation researchers rather than researchers in writing research or in traditional psycholinguistics research, we need to keep the equivalence constraints between source and target units in focus (explicit vs. implicit meanings, grammar vs. semantics vs. context, and translation units vs. linguistic units), and we need to specify precisely whether and where they show up in our data. Translation involves both understanding and production, but it is different from other forms of understanding and production through the double-bind of the translation relationship both to the ST and to the TT, and additionally through the involvement of at least two language systems, even if the latter is shared by research into bilingual communication.

A final point needs mentioning, which is the construction of models of the translation process that are specific enough to generate research hypotheses in sufficiently high granularity, and which can interface with models of text production, including writing, and understanding. These models will have to be a merging of models from linguistics and models from processing, plus the specific components of the translation process, and again "equivalence" will have to play a key role here. The neglect in branches of Translation Studies to work towards such models is a hindrance to making progress on that front.

19.3.4 *Tracing "equivalence" in translation-related cognitive processes*

Equivalence is not "sameness", so there cannot be a question of simply transferring any element of structure or features unchanged from STs to TTs. Equivalence may not even be located in any

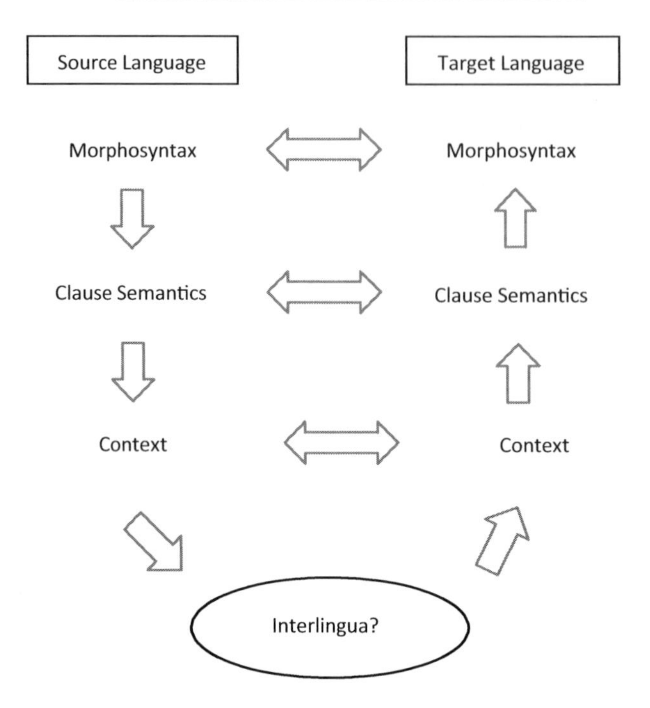

Figure 19.1 Models of translation, transfer and equivalence

level of encoding of the message, but instead may reside in patterns of interpretation or in contextually equivalent patterns of behaviour. If we subscribe to this notion of equivalence, then we must find ways of comparing patterns of behaviour in communicative forms other than language (see Section 19.1.2)—which is certainly not easy, at least in the case of "interpretations", which points us back towards linguistic encoding.

Most models of translation focusing on linguistic encoding are stratificational. They usually take the form of two parallel syntagmatic and paradigmatic axes and involve levels such as phonology, lexicogrammar, semantics, context (situation) and transfer between them. If an interlingua is involved, the model approaches a U-shape with the interlingua at the bottom (see Figure 19.1).

In such models, transfer happens intralingually between more formal and more contextual levels in analysis and production, and it happens interlingually between languages holding the level constant. "Equivalence" then means transfer between these levels, ideally without loss of meaning. Now, translation as interlingual transfer is the only form of text production that takes as an input a fully specified and instantiated linguistic text on all levels, with the ideal aim of producing a maximally complete "quotation" of it in a different language. This quotation is different from a paraphrase or a variant; it is like "quotation" but unlike "paraphrasing" and "retelling", in the sense of Martin (2006). Some models of translation operate with notions like "re-instantiation" (De Souza, 2013, following Martin, 2006; Matthiessen, 2014, p. 322) or "transfer", "reconstitution" and "translation competence" for the competence side (Shreve & Lacruz, 2017, p.127). Whichever metaphor we prefer for translation, even the wording indicates the specific property of the process: *equi*, *trans*, *re-*, *para*, all of which refer to something which is (relatively) unchanged.

Research in language understanding shows us traces of moving from a phonetic input to some form of deeper semantic representation, and the overall final interpretation of the message is the contextualized interpretation of all the levels involved (for one of the classical models, see van Dijk and Kintsch (1983) and later elaborations by Kintsch), its "reading" in the sense of what was presented in Section 19.2.2. Research in language production shows the reverse, and the end product of the production pipeline is the encoded text (see one of the classical models in Levelt, 1989; for models of writing see the summary in Shreve & Lacruz, 2017, p. 131), where the term *translation* is even used intralingually between adjacent levels of representation). Research in multilingual language processing shows all the above under conditions of two or more language systems. None of these processes, though, show patterns of "equivalence" in the stricter sense of moving from a linguistically fully specified input to an output in another linguistic system. In the sense of Figure 19.1, language understanding is the left-hand side of the schema, and language production is its right-hand side. Yet, transfer of readings, which is not simply the same as correspondence of structure, is the province of translation. And the processes arising out of transfer are unique to translation. My suggestion, therefore, is that this is a core area of promising research in translation-related cognition for the immediate future.

Where, then, do we locate traces of equivalence in natural language processing ((see Shreve & Lacruz, 2017) for a related, though different, perspective on this)? Reading for pre-translational text analysis is one case in point. It is to be expected that texts will be read differently depending on whether they are read for translation or for understanding without a translation brief. This should show up in various patterns of eye tracking, in tests involving recall of specific properties of the ST, in patterns of typing onset and pauses, and in neurophysiological patterns identifying problems during reading. Through the entire translation process, we would expect frequent and iterative coupling of ST and TT. Another very interesting source of relevant data should be histories of the development of translation units, both in the sense of behavioural units (Dragsted, 2010) and in the sense of aligned units of product data (for an attempt at integration, see Alves et al., 2010). The first type of data would essentially be process data, the second would be product data, and we would definitely expect histories of translation units to be different from histories of the production of text units in writing research, precisely in the sense that they reflect the search for equivalence between two linguistic encodings (ST-TT) rather than between meaning representations in other modalities (visual, propositional, auditory, etc.) and their linguistic encoding in non-translational forms of text production (e.g. audio description, forms of interpreting, and straight multilingual text production in multilingual technical writing). The search for equivalence should also show up in data from drafting and revision, either in straight annotated linguistic data or in advanced forms of keylogging and eye tracking. Above all, once we have available data in multi-level corpora, involving linguistic data, eye-tracking and keylogging data, magnetic resonance imaging and user-activity data of various types, and if these different levels of data are mapped onto one common time-stamp, and if we then compare translational data with data from other forms of multilingual text production, the specifics, in particular the search for equivalence, should show up clearly ((see Carl & Schaeffer, 2017, p. 63), though without a discussion of equivalence). It will, of course, be challenging, but highly interesting, to look at data that indicate which of all of these processes are governed by conscious problem solving, because it can be expected that precisely the search for equivalence in problematic cases requires translation-specific recourse to specific types of knowledge and of strategies. In all the process data we have mentioned here, we must, of course, presuppose that the behavioural data we are observing are direct traces of cognitive processes—a major assumption, indeed, but this risk we share with most methods of psycholinguistics and cognitive studies.

Notes

1 Some arguments in this chapter rely on concepts from systemic functional linguistics (see House, 2015:25f, 63f, 126; Steiner & Yallop, 2001). This applies in particular to the concepts of "potential vs. instance, text vs. reading, and multifunctional paraphrase" as used in Sections 19.1.2, 19.2.1 and 19.2.2. In SFL usage, "potential" refers to the linguistic system, as opposed to situated texts ("instance"). The "reading" of a "text" is the hearer/reader/translator's interpretation of a linguistic encoding, and a "multifunctional paraphrase" is a paraphrase which preserves not only truth conditions, but in addition interpersonal ("pragmatic") meaning and textual meaning (information structure and thematic progression).

2 To what extent information-theory-based models (e.g. Martínez Martínez & Teich, 2017; Teich et al., this volume) need all these clarifications is currently unclear to me. The notion of equivalence in such a kind of approach seems to be hidden under the low surprisal of a given translation solution. If the language model there is trained on a corpus of "good" translations, low surprisal of a given translation would also be an indicator of equivalence, without having any explicit model of the term.

3 For the distinction between explicitness and explicitation, cf. Hansen-Schirra et al. (2012, p. 59):

> We assume explicitation if a translation (or language-internally one text in a pair of register-related texts) realizes meanings [...] more explicitly than its source text—more precisely, meanings not realized in the less explicit source variant but implicitly present in a theoretically-motivated sense. The resulting text is more explicit than its counterpart.

Further reading

De Sutter, G., De Lefer, M.-A, & Delaere, I. (Eds.) (2017) *Empirical translation studies. New methodological and theoretical traditions*. Trends in linguistics. Studies and Monographs [TiLSM] 300. Berlin: De Gruyter Mouton.

A rich source for recent developments in empirical methodologies, albeit with a focus on product studies.

Halverson, S. (1997). The concept of equivalence in translation studies: Much ado about something. *Target, 9*(2), 207–233.

This is more than 20 years old, but it remains a valid source for fundamental issues concerning "equivalence" in Translation Studies.

Hansen-Schirra, S., Czulo, O., & Hofmann, S. (Eds.). (2017). *Empirical modelling of translation and interpreting*. Berlin: Language Science Press.

This presents a range of recent studies of cognition in translation and interpreting.

Koller, W. (1995). The concept of equivalence and the object of translation studies. *Target,* 7(2), 191–222.

This is more than 20 years old, but it remains a valid source for fundamental issues concerning "equivalence" in Translation Studies.

Schwieter, J., & Ferreira. A. (Ed.) (2017). *Handbook of translation and cognition*. Oxford: Wiley Blackwell.

A comprehensive and reasonably recent source for our topic as a whole.

Shreve, G., & Angelone, E. (Eds.) (2010). *Translation and cognition*. Amsterdam: John Benjamins.

This gives a representative survey of the field, at least for the time up to 2010, with much of it still not outdated.

References

Alves, F., & Gonçalves, J. L. (2003). A relevance theory approach to the investigation of inferential processes in translation. In F. Alves (Ed.), *Triangulating translation: Perspectives in process oriented research* (pp. 11–34). Amsterdam: John Benjamins.

Alves, F., Gonçalves, J. L., & Szpak, K. S. (2014). Some thoughts about the conceptual/ procedural distinction in translation: A key-logging and eye tracking study of processing effort. *MonTI, 1, Special issue: Minding Translation*, 151–175.

Alves, F., & Hurtado Albir, A. (2017). Evolution, challenges, and perspectives for research on cognitive aspects of translation. In J. Schwieter, & A. Ferreira (Eds.), *The handbook of translation and cognition* (pp. 537–554). Hoboken, NJ: Wiley Blackwell.

Alves, F., Pagano, A., Neumann, S., Steiner, E., & Hansen-Schirra, S. (2010). Translation units and grammatical shifts: Towards an integration of product- and process-based translation research. In G. Shreve, & Angelone, E. (Eds.), *Translation and cognition* (pp. 109–142). Amsterdam: John Benjamins.

Banchs, R. E., D'Haro, L. F., & Li, H. (2015). Adequacy - fluency metrics: Evaluating MT in the continuous space model framework. *IEEE/ACM Transactions on Audio, Speech and Language Processing, 23*(3), 472–482.

Bateman, J. A. (2008). *Multimodality and genre. A foundation for the systematic analysis of multimodal documents.* Basingstoke: Palgrave Macmillan.

Bell, R. T. (1991). *Translation and translating.* London: Longman.

Biber, D. (1993). *Dimensions of register variation.* Cambridge: Cambridge University Press.

Carl, M., & Schaeffer, M. (2017). Models of the translation process. In J. W. Schwieter, & A. Ferreira (Eds.), *The handbook of translation and cognition* (pp. 50–70). Hoboken, NJ: Wiley Blackwell.

Carl, M., Schaeffer, M., & Bangalore, S. (2016). The CRITT Translation Process Research Database. *Springer International Publishing*, 13–54.

Catford, J. C. (1965). *A linguistic theory of translation.* Oxford: Oxford University Press.

Chungyu, K., & Tak-Ming, B. W. (2015). Evaluation in machine translation and computer-aided translation. In C. Sin-Wai (Ed.), *The Routledge encyclopedia of translation technology* (pp. 213–236). London: Routledge.

Czulo, O. (2017). Aspects of a primacy of frame model of translation. In S. Hansen-Schirra, O. Czulo, & S. Hofmann (Eds.), *Empirical modelling of translation and interpreting* (pp. 465–490). Berlin: Language Science Press.

De Souza, L. M. F. (2013). Interlingual re-instantiation—a new systemic functional perspective on translation. *Text&Talk, 33*(4–5), 575–594.

De Sutter, G., De Lefer, M.-A, & Delaere, I. (Eds.). (2017). *Empirical translation studies. New methodological and theoretical traditions.* Trends in linguistics. Studies and Monographs [TiLSM] 300. Berlin: De Gruyter Mouton.

Dragsted, B. (2010). Coordination of reading and writing processes in translation: An eye on uncharted territory. In G. Shreve, & Angelone, E. (Eds.), *Translation and cognition* (pp. 41–62). Amsterdam: John Benjamins.

Durand, J., Bennett, P., Allegranza, V., van Eynde, F., Humphreys, L., Schmidt, P., & Steiner, E. (1991). The Eurotra linguistic specifications: An overview. In V. Allegranza, S. Krauwer, & E. Steiner (Eds.), *Machine Translation,* 6(2). *Special Issue Eurotra.*

Evert, S., & Neumann, S. (2017). The impact of translation direction on characteristics of translated texts. A multivariate analysis for English and German. In G. De Sutter, M.-A. De Lefer, & I. Delaere (Eds.), *Empirical translation studies. New methodological and theoretical traditions.* Trends in linguistics. Studies and Monographs [TiLSM] 300 (pp. 47–80). Berlin: De Gruyter Mouton.

Fawcett, P. (1997). *Translation and language. Linguistic theories explained.* Manchester: St. Jerome.

Geeraerts, D., & Cuyckens, H. (Eds.). (2007). *The Oxford handbook of cognitive linguistics.* Oxford: Oxford University Press.

Gupta, R., Orasan, C., & van Genabith, J. (2015). ReVal: A simple and effective machine translation evaluation metric based on recurrent neural networks. *Proceedings of the 2015 Conference on Empirical Methods in Natural Language Processing* (pp. 1066–1072). Lisbon: Association for Computational Linguistics.

Gutt, E. A. (1991). *Translation and relevance. Cognition and context.* Oxford: Blackwell.

Halliday, M. A. K., & Hasan, R. (1989). *Language, context and text: Aspects of language in a social-semiotic perspective.* Oxford: Oxford University Press.

Halliday, M. A. K., & Matthiessen, C. M. I. M. (2014). *An introduction to functional grammar.* London: Arnold (earlier versions by Halliday in 1985/1994, Halliday and Matthiessen, 2004).

Halverson, S. (1997). The concept of equivalence in translation studies: Much ado about something. *Target, 9*(2), 207–233.

Halverson, S. (2010). Cognitive translation studies. Developments in theory and method. In G. Shreve, & Angelone, E. (Eds.), *Translation and cognition* (pp. 349–370). Amsterdam: John Benjamins.

Halverson, S. (2017). Gravitational pull in translation. Testing a revised model. In G. De Sutter, M.-A. De Lefer, & I. Delaere (Eds.), *Empirical translation studies. New methodological and theoretical traditions.* Trends in linguistics. Studies and Monographs [TiLSM] 300 (pp. 9–46). Berlin: De Gruyter Mouton.

Hansen-Schirra, S., Czulo, O., & Hofmann, S. (Eds.). (2017). *Empirical modelling of translation and interpreting.* Berlin: Language Science Press.

Hansen-Schirra, S., Neumann, S., & Steiner, E. (2012). *Cross-linguistic corpora for the study of translations. Insights from the language pair English-German*. Berlin: Mouton de Gruyter.
Hiippala, T. (2012). The localisation of advertising print media as a multimodal process. In W. Bowcher (Ed.), *Multimodal texts from around the world: Cultural and linguistic insights* (pp. 97–122). Basingstoke: Palgrave Macmillan.
House, J. (1997). *Translation quality assessment. A model revisited*. Tübingen: Gunter Narr Verlag.
House, J. (2015). *Translation quality assessment. Past and present*. London: Routledge.
Koller, W. (1995). The concept of equivalence and the object of translation studies. *Target, 7*(2), 191–222.
Koller, W. (1997). *Einführung in die Übersetzungswissenschaft* (5th revised ed.). Wiesbaden: Quelle und Meyer.
König, E., & Gast, V. (2018). *Understanding English-German contrasts*. Berlin: Erich Schmidt Verlag (1st ed. 2007, 2nd revised ed. 2009, 3rd revised ed. 2012).
Kress, G. (2010). *Multimodality: A social semiotic approach to contemporary communication*. London: Taylor & Francis.
Kunz, K., Degaetano-Ortlieb, S., Lapshinova-Koltunski, E., Menzel, K., & Steiner, E. (2017). GECCo—An empirically-based comparison of English-German cohesion. In G. De Sutter, M.-A. De Lefer, & I. Delaere (Eds.), *Empirical translation studies. New methodological and theoretical traditions*. Trends in linguistics. Studies and Monographs [TiLSM] 300 (pp. 265–312). Berlin: Mouton de Gruyter.
Lapshinova-Koltunski, E. (2017). Exploratory analysis of dimensions influencing variation in translation. The case of register and translation method. In G. De Sutter, M.-A. De Lefer, & I. Delaere (Eds.), *Empirical translation studies. New methodological and theoretical traditions*. Trends in linguistics. Studies and Monographs [TiLSM] 300 (pp. 207–234). Berlin: Mouton de Gruyter.
Levelt, W. (1989). *Speaking. From intention to articulation*. Cambridge, MA: MIT Press.
Levshina, N. (2015). *How to do linguistics with R. Data exploration and statistical analysis*. Amsterdam: John Benjamins.
Linke, A., & Nussbaumer, M. (2000). Konzepte des Impliziten: Präsuppositionen und Implikaturen. In K. Brinker, G. Antos, W. Heinemann, & S. F. Sager (Eds.), *Text- und Gesprächslinguistik. Ein internationales Handbuch zeitgenössischer Forschung*. Berlin: Mouton de Gruyter.
Martin, J. (2006). Genre, ideology and intertextuality: A systemic functional perspective. *Linguistics and the Human Sciences, 2*(2), 275–298.
Martin, J., & Rose, D. (2003). *Working with discourse: Meaning beyond the clause*. London: Continuum.
Martínez Martínez, J., & Teich, E. (2017). Modeling routine in translation with entropy and surprisal: A comparison of learner and professional translations. In L. Cercel, M. Agnetta, & M. T. Amido Lozano (Eds.), *Kreativität und Hermeneutik in der Translation* (pp. 403–426). Tübingen: Narr Francke.
Matthiessen, C. M. I. M. (2001). The environments of translation. In E. Steiner, & C. Yallop (Eds.), *Exploring translation and multilingual text production: Beyond content* (pp. 41–126). Berlin: Mouton de Gruyter.
Matthiessen, C. M. I. M. (2014). Choice in translation: Metafunctional considerations. In K. Kunz, E. Teich, S. Hansen-Schirra, S. Neumann, & P. Daut (Eds.), *Caught in the middle—Language use and translation. A Festschrift for Erich Steiner on the occasion of his 60th birthday* (pp. 271–334). Saarbrücken: Saarland University Press.
Mattick, J. S. (2004, October). The hidden genetic program of complex organisms. *Scientific American,* 61–67.
Menzel, K., Lapshinova-Koltunski, E., & Kunz, K. (Eds.). (2017). *New perspectives on cohesion and coherence. Implications for translation*. Berlin: Language Science Press.
Munday, J. (2001). *Introducing translation studies. Theories and applications*. London and New York: Routledge (4th ed. 2016).
Munday, J. (2012). *Evaluation in translation. Critical points of translator decision-making*. Abingdon: Routledge.
Muñoz Martín, R. (2010). On paradigms and cognitive translatology. In G. Shreve, & Angelone, E. (Eds.), *Translation and cognition* (pp. 169–188). Amsterdam: John Benjamins.
Muñoz Martín, R. (2017). Looking toward the future of cognitive translation studies. In J. Schwieter, & A. Ferreira (Eds.), *The handbook of translation and cognition* (pp. 555–572). Hoboken, NJ: Wiley Blackwell.
Nida, E. A. (1964). *Toward a science of translating: With special reference to principles and procedures involved in Bible translation*. Leiden: E.J. Brill.
Schlesewsky, M. (2009). Linguistische Daten aus experimentellen Umgebungen: Eine multiexperimentelle und multimodale Perspektive. *Zeitschrift für Sprachwissenschaft, 28*(1), 169–178. doi:10.1515/ZFSW.2009.020
Schwieter, J. W., & Ferreira, A. (2017a). Bilingualism in cognitive translation and interpreting studies. In J. Schwieter, & A. Ferreira (Eds.), *The handbook of translation and cognition* (pp. 144–164). Hoboken, NJ: Wiley Blackwell.

Schwieter, J. W., & Ferreira, A. (Eds.). (2017b). *The handbook of translation and cognition*. Hoboken, NJ: Wiley Blackwell.
Serbina, T., Hintzen, S., Niemietz, P., & Neumann, S. (2017). Changes of word class during translation – Insights from a combined analysis of corpus, keystroke logging and eye-tracking data. In S. Hansen-Schirra, O. Czulo, & S. Hofmann (Eds.), *Empirical modelling of translation and interpreting* (pp. 177–208). Berlin: Language Science Press.
Shreve, G., & Angelone, E. (Eds.) (2010). *Translation and cognition*. Amsterdam: John Benjamins.
Shreve, G., & Lacruz, I. (2017). Aspects of a cognitive model of translation. In J. Schwieter, & A. Ferreira (Eds.), *The handbook of translation and cognition* (pp. 127–143). Hoboken, NJ: Wiley Blackwell.
Sickinger, P. (2017). Aiming for cognitive equivalence mental models as a tertium comparationis for translation and empirical semantics. *Research in Language, 15*(2), 213–236.
Steiner, E. (2001). Intralingual and interlingual versions of a text—how specific is the notion of translation. In E. Steiner, & C. Yallop (Eds.), *Exploring translation and multilingual text production: Beyond content* (pp. 161–190). Berlin: Mouton de Gruyter.
Steiner, E. (2004). *Translated texts: Properties, variants, evaluations*. Frankfurt: Peter Lang.
Steiner, E. (2012). Methodological cross-fertilization: Empirical methodologies in (computational) linguistics and translation studies. *Translation: Computation, Corpora, Cognition, 2*(1), 3–21. Special issue at the crossroads between contrastive linguistics, translation studies and machine translation. www.t-c3.org/vol-2-no-1-2012/
Steiner, E. (2015). Contrastive studies of cohesion and their impact on our knowledge of translation (English-German). *Target, 27*(3), 351–369.
Steiner, E., & Yallop, C. (Eds.). (2001). *Exploring translation and multilingual text production: Beyond content*. Berlin: Mouton de Gruyter.
Toury, G. (1995). *Descriptive translation studies and beyond*. Amsterdam: John Benjamins.
Van Dijk, T., & Kintsch, W. (1983). *Strategies of discourse comprehension*. New York: Academic Press.
Venuti, L. (2009). Translation, intertextuality, interpretation. *Romance Studies, 27*(3), 157–173.

20

Translation, information theory and cognition

Elke Teich, José Martínez Martínez and Alina Karakanta

20.1 Introduction

It is widely acknowledged that human language processing relies to a considerable degree on expectancy. In online processing, we make predictions about what comes next, and we choose particular linguistic encodings based on preceding events in the situation, the preceding discourse or assumptions about the addressee (e.g., shared knowledge or state of attention) (see Levy, 2008; Levy & Jaeger, 2007). Similarly, in offline (written) language use, we rely on the expectancy of particular linguistic encodings according to register, text type and/or genre. The formal basis for modelling predictability in context (linguistic, situational or register) is provided by probabilistic models of language use—so-called language models, which are widely used in natural language processing. While it has been suggested that probabilistic explanations should provide valuable insights into translators' behaviour and translational choice (Toury, 2004), in Translation Studies only a few attempts have been made to develop suitable formal approaches. Some earlier works are Levý's account based on game theory (Levý, 1967; see Cronin, 1998) and Gutt's account (Gutt, 1989) based on relevance theory (Sperber & Wilson, 1986).

The purpose of the present chapter is to sketch a formal basis for the probabilistic modelling of human translation that may provide a common basis for product-oriented and process-oriented research on translation and makes communicative explanations of translational choice and its linguistic traces possible: shining through, normalization, simplification, etc. (see Baker, 1996; Teich, 2003; Toury, 1995; Volansky et al., 2015). The approach is rooted in information theory, with Shannon's notion of information at the core (Shannon, 1948). We provide a definition of Shannon information applied to linguistic communication—called *surprisal* in research on comprehension and *information density* in production research—and discuss its relevance for modelling translation (Section 20.2.1).

According to Shannon, a crucial property of the communication channel is that it is noisy, i.e. the signal is always distorted, which may result in loss of information through transmission. We explain the concept of the noisy channel and the types of linguistic problems that are typically addressed with noisy channel models—including machine translation, which provides the link to modelling human translational choice (Section 20.2.2). Thirdly, we suggest that a number of

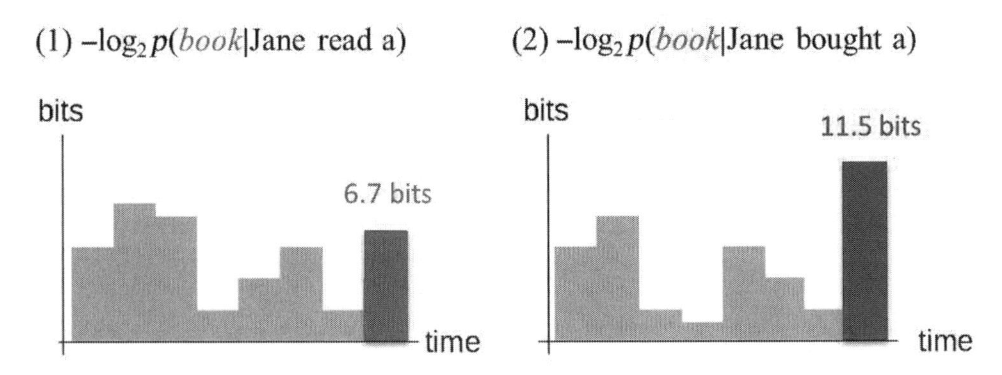

Figure 20.1 Surprisal

translation-relevant variables, notably (dis)similarity between languages, level of expertise and translation mode (i.e. interpreting vs. translation), may be appropriately indexed by entropy, which in turn has been shown to indicate production effort (Section 20.2.3). What is crucial about an information-theoretic approach is that it provides a link to cognition: Surprisal has been shown to correlate with a number of behavioural and neurophysiological indices (e.g. response latencies, reading time, pupil dilation and event-related potentials (ERP); see Crocker et al., 2016) and is therefore a suitable measure of cognitive effort in human language processing. Section 20.3 concludes with a brief summary, a discussion of methods to estimate probabilities from corpora (Section 20.3.1) and an outlook on rational explanations of translation (Section 20.3.2).

20.2 Core concepts

20.2.1 Shannon information and surprisal/information density

In human language processing, context exerts various constraints upon the kinds of linguistic unit that may come up. For example, at the level of syntax, given a preposition, the unit most likely to follow is a nominal phrase; or, at the level of words, given the word "read", a likely continuation is "book". Information theory allows us to define such probabilities in context on the basis of the amount of information that is conveyed by a given unit measured in bits. This is commonly formalized by the measure of *surprisal* (Equation 20.1), which estimates the probability of a *unit* (e.g. a word) given some *context* (e.g. the preceding *n* words) as the negative logarithm to the base 2 (alternatively, base 10).[1]

$$Surprisal = -\log_2 p(unit \mid context) \tag{20.1}$$

According to this model, linguistic events with high surprisal are low in probability and convey more information than events with low surprisal/high predictability in context. Crucially, predictability in context is inversely proportional to cognitive effort; i.e. higher surprisal incurs higher processing cost (Hale, 2001).

For illustration, consider the following two examples:

(1) Jane read a book.
(2) Jane bought a book.

In example (1), the item "read" in the preceding context of "book" makes "book" highly predictable (there aren't many other, similarly likely alternative completions); "book" has low information content, so surprisal is low. In example (2), in contrast, "buy" does not strongly license a particular continuation, so surprisal on "book" is relatively high: we get more information when we see/hear "book" in the context of "buy" than in the context of "read". Figure 20.1 illustrates why "book" needs more bits for encoding in (2) compared with (1), and so it incurs a higher processing cost.

For estimating probabilities in context, we need linguistic usage data. These can be obtained from corpora (corpus probabilities; see Section 20.2.2). While this involves some challenges (see Section 20.3.1 for discussion), in the following we focus on the potential of an information-theoretic approach for modelling human translational choice and contributing to the theory of translation at large.

20.2.1.1 Applications in linguistics

The perspective of predictability in context is extremely fruitful for the study of language use, variation and change. Apart from language comprehension, in language production it has been shown that shorter, more condensed linguistic variants are preferred in more predictive contexts and longer, more expanded variants when the context is less predictive. Examples are expanded/reduced vowel space size in speech (Schulz et al., 2016), fragments in syntax (Reich, 2017), condensed syntactic expression (e.g., coercion: *Jane began a book* (Delogu et al., 2017)) or optional marking of discourse relations (Rutherford et al., 2017). Thus, language users seem to strive for an optimal encoding of a given message depending on predictability in context by modulating the amount and rate of information in a message through specific linguistic choices. This seems to be a valid communicative explanation of certain types of linguistic variation to be observed in online communication, notably the choice of fully expanded vs. condensed linguistic forms such as relative pronoun or complementizer omission, syntactic fragments or shorter syllable durations. In the analysis of written, offline communication, surprisal can act as a measure of the (relative) complexity of a text as an alternative to type-token ratio, lexical density or the Fog index, which are based on simple frequency counts. Surprisal has the added value of being context aware and, more importantly, directly cognitively relevant. In a theoretical perspective, adopting an information-theoretic approach opens up the opportunity to explain language use on the basis of rational communication, according to which interlocutors want their interactions to be successful, as seen in Grice's (1975) maxims and in Sperber and Wilson's (1986) relevance theory, while at the same time keeping their cognitive effort at a reasonable level. There is now increasing empirical evidence that communicative concerns play an important role in language variation and change at large, as seen by Hume and Mailhot (2013), Degaetano-Ortlieb and Teich (2016), Rubino et al. (2016) or Baayen et al. (2017). The structure and evolution of the linguistic system may thus be explainable in terms of communicative concerns, striking a balance between expressiveness and efficiency (see e.g. Piantadosi et al. (2011) on the optimization of word lengths across languages).

20.2.1.2 Application to translation

The perspective of rational communication can be straightforwardly applied to translation. It is reasonable to assume that translators, too, want communication to be successful, and, as far as professional translation goes, they strive for a high-quality output. However, there are a number of specific constraints interfering with these goals. First, translators need to produce a translation that is true to the source-language text and conforms to the target-language expectations at the same time—the classic translation dilemma. Furthermore, compared with other linguistic processes (e.g. reading or shadowing), translation is a process with high resource limitations (time pressure and cognitive effort) (Hyönä et al., 1995), and it can therefore be assumed that translators, and even more so interpreters, have a vested interest in keeping their effort at a reasonable level. This includes efficient management of working memory, e.g. by maintaining only few hypotheses about a translation solution and trying to be as certain as possible about the best solution (Pym, 2008). Further interacting constraints are level of expertise (professional vs. learner), translation mode (translation vs. interpreting), language pair (i.e. (dis)similarity between source and target language) and translation direction (i.e. from/into native vs. non-native language).

Notwithstanding its specific constraints, translation can be modelled in probabilistic terms just like other processes of language use. Taking the production perspective, in the remainder of this chapter we will focus on how to model the following goals of translation and the linguistic traces that may be left by striving to realize these goals (see translationese: Baker, 1995; Gellerstam, 1986):

1. Be as true as possible to the source-language message.
2. Adhere to the norms of the target language.

In product-oriented Translation Studies, the first relates to the notion of equivalence, which defines a relation between two products, a source-language text and a target-language text; the second corresponds to the notion of adequacy or appropriateness (see e.g. Reiss, 1983). Thirdly, a condition rather than a goal, it is in the interest of translators to keep their efforts (cognitive and temporal as well as technical) at a reasonable level.

We assume that the translation output is optimal when goals (1) and (2) are reached, but typically they are compromised due to effort levelling and the interacting constraints discussed earlier. Also, one goal may be favoured over the other, possibly resulting in shining through/interference (overemphasizing goal 1) or normalization/standardization (overemphasizing goal 2).

Turning to probabilistic modelling, the successful outcome of goal 1 may be characterized as maximizing the probability of being able to retrieve the source-language expression given the chosen target-language translation; and goal 2 can be formalized in probabilistic terms as maximizing the probability of the chosen translation in the target-language context. To capture the gist of translation, we need to be able to model these two goals together. As will be explained in Section 20.2.2, this is exactly what the noisy channel model allows us to do.

Finally, regarding cognitive effort, we analyse the number and distribution of translation options considered at a given choice point as an important factor. Here, the hypothesis is that the more (equiprobable) options are being considered at a given choice point, the higher the cognitive effort involved. As will be explained in Section 20.2.3, the complexity of a choice (and associated cognitive effort) can be appropriately modelled by entropy.

20.2.2 Noisy channel

The noisy channel model of communication gives recognition to the fact that the transmission of a message is typically distorted. For example, an original input "first name" may be

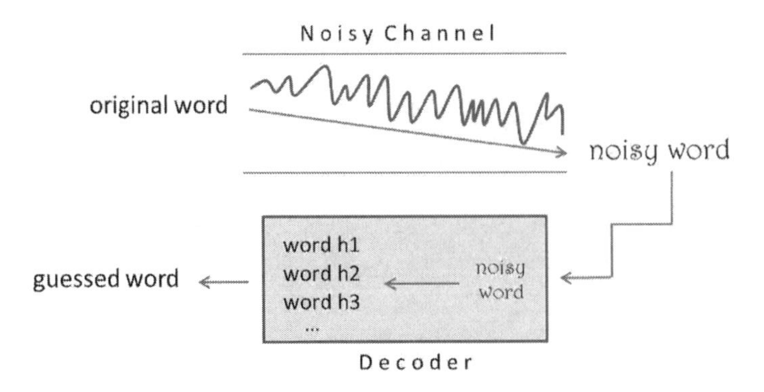

Figure 20.2 Noisy channel transmission

phonetically distorted in such a way that what is understood is "first time", or a given word may be orthographically distorted by misspelling, e.g. "embarrass" as "embarass".

The noisy channel model is applied successfully in a number of natural language processing tasks, including spelling and optical character recognition (OCR) correction, speech recognition and machine translation. In these tasks, the common goal is to restore the original input from the distorted output of a noisy channel. Formally, this is represented as

$$argmax_c\, p(c|x) = argmax_c\, p(x|c)\, p(c) \tag{20.2}$$

for all candidate matches c, where $p(x|c)$ is called the error model (how likely is it that x is a variant of c?) and $p(c)$ the language model (how likely is c in a given language?).

For the sake of illustration, consider an example from spelling error correction (see Figure 20.2; based on Jurafsky & Martin, 2018, p. 480). A sender sends a word through the noisy channel (e.g. "embarrass") which is distorted to a noisy word by misspelling (e.g. "embarass", "emberrass", "emberass"), and the receiver needs to "guess" what the original word was. In a noisy channel model, this guessing is done using knowledge about the original input *c* as well as knowledge about possible distortions in the channel. When receiving the word "emberrass", the receiver assesses how likely "a" is instead of "e" and how likely the word "embarrass" is. In many applications, this search process is performed with the help of a device called a decoder, which gathers candidate original words for the noisy word and decides on the basis of the underlying probabilistic model what the best match is.

To explain what the noisy channel model can do for us in modelling human translation, it is instructive to look at its application in machine translation.

20.2.2.1 Application in machine translation

The noisy channel is the underlying formal model in statistical machine translation. According to such a model, a translation itself is a distorted message; e.g. an intended message in a language *e* is output by an expression in another language *f* (see Figure 20.3). More technically, in statistical machine translation a text is translated according to a probability distribution $p(e|f)$, where e is an expression in a target language and *f* is an expression in a source language. For modelling, *e* and *f* are reversed to $p(f|e)$ (the latter is easier to estimate). Additionally, the probability of the expression *e* on its own (i.e. in original texts of target language *e*) is taken into account. $p(f|e)$ is the so-called *translation model* (= error model of the noisy channel) and $p(e)$ the *language model* (see Equation 20.3).

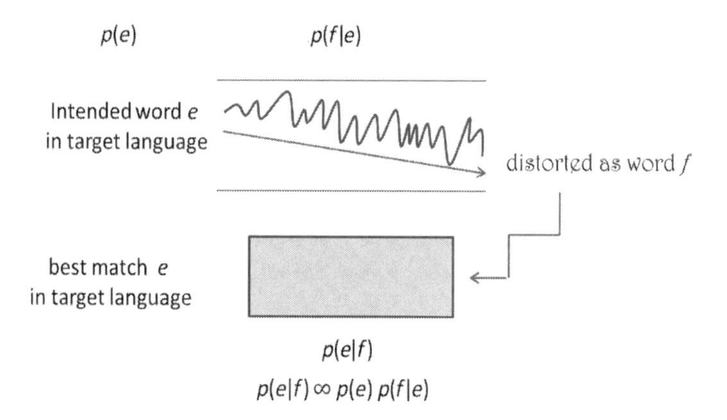

Figure 20.3 Translation as noisy channel

$$p(e|f) = p(f|e)\,p(e) \tag{20.3}$$

The ultimate goal in machine translation is, then, to find the optimal translation e, which maximizes the two probabilities:

$$argmax_e\,p(e|f) = argmax_e\,p(f|e)\,p(e) \tag{20.4}$$

where the *argmax* operation is a search process in the space of possible target translations—this is the decoding procedure as explained earlier for the task of spelling correction. Challenges involved in statistical MT are to come up with efficient approaches to decoding, to find the best translation unit granularity (word based vs. phrase based) and to determine the optimal parameter weights for a given translation task (e.g. giving more weight to $p(f|e)$ and less to $p(e)$ or vice versa). Also, the output quality of a statistical MT system rests very much on the size and quality of parallel corpora (see Section 20.3.1 for discussion).

20.2.2.2 Application to human translation

We can now conceptualize the components of the noisy channel model for machine translation in terms of human translation. The translation model and the language model can be used directly for representing the goals of human translation discussed earlier. Humans try to reconcile goals 1 and 2—in terms of a noisy channel model, they seek the optimal balance between $p(e|f)$ and $p(e)$. Goal 1, henceforth referred to as source-language (SL) fidelity, is represented by the translation model, i.e. maximizing the probability of retrieving the source-language expression s given the chosen target-language translation t. The type of data needed for modelling here is parallel corpora. In a statistical MT system, the probabilities based on the counts retrieved from the parallel corpus are represented in a so-called phrase table. For an example, see an extract of a phrase table in Figure 20.4, which lists the translations occurring in the underlying parallel corpus for the English source-language expression "prerequisite" into German with their translation probabilities.[2]

According to the example, the best match for "prerequisite" would be "Voraussetzung" ($p(s|t) = 0.516$).

Goal 2, henceforth referred to as target-language (TL) conformity, is represented by the target-language model, i.e. maximizing the probability of the target-language expression on its own. Here, we

SL	TL	$(s\|t)$
prerequisite	Bedingung	0.002433
prerequisite	Grundbedingung	0.002433
prerequisite	Grundvoraussetzung	0.096603
prerequisite	Voraussetzung dafür	0.002433
prerequisite	Voraussetzung	0.51612

Figure 20.4 Extract from a phrase table

need a monolingual corpus of the target language that is ideally as comparable as possible to the parallel corpus (register, domain). The language model is a measure of a well-formed expression in the target language. Moreover, it aids the translation model in difficult decisions by providing knowledge about the context of the expression (preceding words). For example, the translation model gives the highest probability to "Voraussetzung" as a translation of "prerequisite". However, depending on the context, the best match could be the option with the second highest probability, "Grundvoraussetzung". In this way, the language model is responsible for ensuring fluent output. See Section 20.3.1 for further discussion on giving more/less weight to language models or translation models.

Provided we are able to obtain empirically sound models, we may use them in a wide spectrum of relevant applications, ranging from translation quality assessment and translationese detection to compiling material for translation training and obtaining stimuli for experimental translation process research. For instance, in a given translation, we can assess whether it is more on the literal or on the adaptive end of the translation cline, i.e. emphasizing SL or TL. For example, in the EuroParl-UdS corpus we find two alternative translations for the English "development budget", "Entwicklungsbudget" and "Entwicklungshaushalt", and the preferred translation is "Entwicklungsbudget". However, in comparable original texts, "Entwicklungshaushalt" is clearly preferred. Choosing "Entwicklungsbudget" is thus an obvious case of literal translation, overemphasizing SL fidelity at the cost of TL conformity. Considering accumulated effects, noisy channel analysis provides a diagnostic tool for translationese: a tendency to favour high probabilities for $(s|t)$ is interpreted as SL shining through, while a tendency to favour high probabilities for (t) is indexical of TL normalization.

Furthermore, other methods from the same formal family of information-theoretic methods can be naturally connected. For instance, a translation's difference from a comparable target-language text (or text collection/corpus) can be measured with *relative entropy* (e.g. by Jensen–Shannon divergence or Kullback–Leibler divergence). Relative entropy is used as standard to compare probability distributions in terms of the number of additional bits needed for encoding when a non-optimal model is used (see Fankhauser et al. (2014) and Degaetano-Ortlieb & Teich (2016) for applications in contrastive and diachronic analysis). For an example, see Figure 20.5 showing word clouds of the most distinctive words for English translation vs. interpreting based on Kullback–Leibler divergence (KLD), (a) from German and (b) from Spanish. Size encodes the information gain (additional bits). In this example, we can observe, for instance, that the most distinctive features of interpreting (right) are markers of oral mode, such as interactant personal pronouns, contractions and grammatical coordination, whereas translation (left) exhibits clear

Figure 20.5 KLD-based word clouds for translation vs. interpreting. (a) Translation (left) and interpreting (right) from German into English; (b) translation (left) and interpreting (right) from Spanish into English

signs of written mode, such as topical words and the definiteness marker "the". Note that the difference between translation and interpreting outputs seems to be independent of the source language, as indicated by the example in Figure 20.5 (see Shlesinger & Ordan, 2012 for similar observations).

20.2.3 Entropy

Entropy quantifies the amount of uncertainty related to the outcome of an event. For illustration, let us consider a simple example. There are three bowls with four apples in each of them: in bowl A there are four green apples, bowl B contains one red and three green apples, and bowl C has two green and two red apples. With bowl A, we can be 100% certain that we get a green apple; with bowl B, there is a 75% chance that the apple we get is green and a 25% chance that it is red;

whereas with bowl C, green and red are equally probable (50%). Comparing the three cases, in A no uncertainty about the outcome is involved—therefore entropy is zero. C has the highest uncertainty, with both outcomes being equally likely, and entropy is 1. B is in between (0.8).

Technically, entropy measures the size of the search space of possible values of a random variable X and its associated probabilities (Manning & Schütze, 2003, p. 63). The mathematical formulation of entropy H is given in Equation 20.5, where $p(x)$ is the probability mass function of the random variable X (see preceding example: kind of apple) over all its possible values $x \in X$ (see example: green or red apple).

$$H(X) = \sum_{x \in X} p(x) \log_2 p(x) \tag{20.5}$$

Entropy, like surprisal, is measured in bits. When we are absolutely certain about the outcome, entropy is zero bits (as in A in the apple example). In contrast, the most uncertain situation, with the highest entropy possible, occurs when all outcomes are equiprobable (as in C). Then entropy equals $log_2 n$ (where n is the total number of possible outcomes). Like surprisal, the measure of entropy will always yield a positive number, with values ranging from 0 to $log_2 n$.

20.2.3.1 Application in linguistics

Entropy has been applied in linguistics to capture the complexity of linguistic choices. For instance, it has been shown that there is a relation between entropy and cognitive effort in language comprehension, which may be described as uncertainty on the part of readers or listeners about what comes next (e.g. a word). In sentence comprehension, entropy (uncertainty) typically decreases with each incoming word, and the amount of information that each word gives may be defined as the reduction in entropy due to that word (see Blache & Rauzy, 2011; Frank, 2010; Hale, 2001, 2003, 2006). To mention a more specific example, Linzen and Jaeger (2016) look at probabilities of complementation patterns of verbs with multiple subcategorization options (e.g., "He forgot my birthday", "He forgot about my birthday", "He forgot that it was my birthday"). When probability is evenly distributed across subcategorization frames, verbs with more frames have higher entropy; for the same number of frames, the less balanced the distribution, the lower the entropy. For processing, the prediction is that the first is more effortful than the second.

Also, language production has been shown to work with predictability (e.g. Jaeger, 2010). Here, entropy in the search space of linguistic options is an important indicator of processing complexity: as a tendency, the higher the entropy of the space of options, the higher the effort incurred in processing (measured e.g. by response latency or production time). In fact, this partly explains longer latencies in word search in older adults—due to lifelong experience, the space of options is simply much larger than in younger people. Hence, rather than general cognitive decline, a higher entropy in the lexicon may better explain word search problems in older people in some tasks (see e.g. Blanco et al. (2016) for relevant results from behavioural experiments).

20.2.3.2 Applications to human translation

If we define the task of translation as a search task in a space of alternative linguistic options in a target language (see $p(s|t)$ in the noisy channel model), we can apply entropy directly to formalize the set of possible translation outputs as:

$$H(T) = -\sum_{t \in T} p(t) \log_2 p(t) \tag{20.6}$$

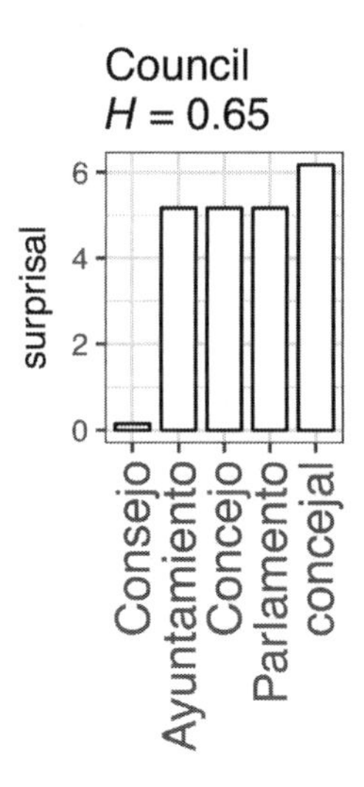

Figure 20.6 Low-entropy translation space for "Council" and its underlying probability distribution

where T stands for the *translation space*, i.e. the set of all possible translations t for a given source-text unit as found in a parallel corpus.

For illustration, consider two examples from an excerpt of the proceedings of the European Parliament in English translated into Spanish by translation trainees in Figures 20.6 and 20.7, showing two translation spaces with low and high entropy, respectively (see Martínez Martínez & Teich, 2017). The x-axis shows the individual options and the y-axis plots surprisal, i.e. how probable a translation is (low surprisal = high probability).[3] Figure 20.6 illustrates the translation space of "Council" showing a very low entropy due to few options (five), with one option being the clearly preferred translation ("Consejo"). We may infer that the translation of "council" into Spanish is quite straightforward (on the basis of the underlying parallel corpus) and not associated with high cognitive effort. In the other example (Figure 20.7), in contrast, there are many alternative options (35), and they show fairly similar surprisal scores; i.e. they are all similarly likely. Again, given that entropy is positively correlated with cognitive effort, it may be concluded that translational choice from English to Spanish regarding "gross breach" is associated with relatively high cognitive effort, as indicated by the underlying parallel corpus, while the translation of "Council" is associated with low effort.

Estimates of entropy based on parallel corpora may be directly used to assess translation difficulty from the production perspective, connecting up with existing work on translation in a natural way. The number of variants available to and considered by the translator is mentioned as an indicator of cognitive effort by both Krings (2001, pp. 536–537) and Englund Dimitrova (2005, p. 26). See also Angelone's work on uncertainty management in translation associated with problem solving (Angelone, 2010); or Campbell's choice network analysis, which confirms a correlation between the number of choices and cognitive effort (Campbell, 1999, 2001).

One of the most elaborate approaches using entropy in translation process research is word translation entropy, as initially proposed by Schaeffer and Carl (2014) and taken further in subsequent research as in Carl and Schaeffer (2017a, 2017b), Schaeffer et al. (2016) and Bangalore et al. (2016). This work emanates from exploiting data from the CRITT translation process database (Carl et al., 2016). Also, there are related tasks in lexical natural language processing similar to

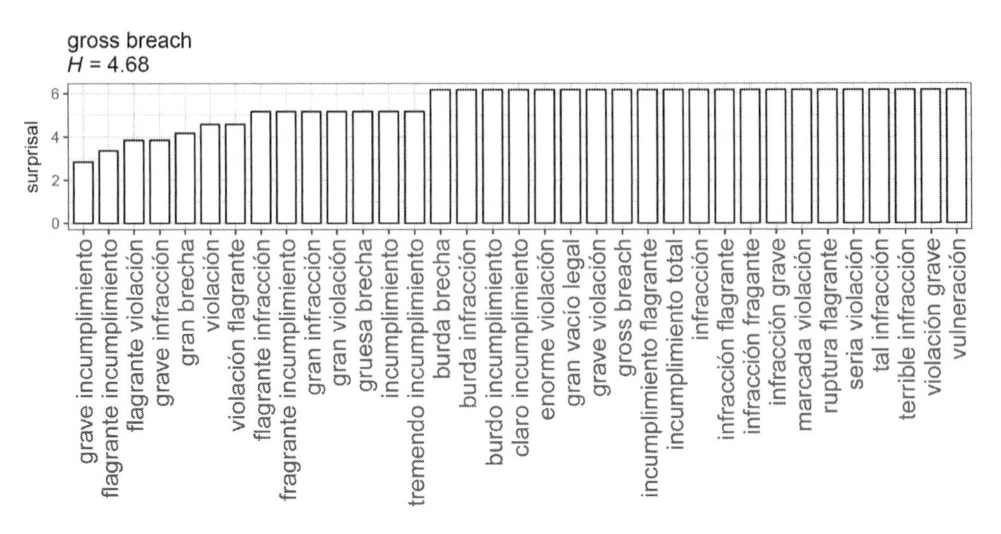

Figure 20.7 High-entropy translation space for "gross breach" and its underlying probability distribution

choice in translation, such as ambiguity detection based on semantic similarity between source and target text (see e.g. Cap, 2017; Villada Moirón & Tiedemann, 2006), that are relevant to consider for modelling human translation.

Other applications of entropy in Translation Studies include assessment of a given translation (or set of translations) in terms of degree of literalness or degree of expertise. For instance, Carl and Schaeffer (2017a) show that more literal translations are easier to produce than less literal translations, linking back to evidence from priming studies. Similar effects are shown by Bangalore et al. (2016) for syntactic choices (e.g. word order). Martínez Martínez and Teich (2017) show how entropy can be used to characterize learner vs. expert behaviour in translation, interpreting low entropy as an indicator of routine behaviour (high certainty about translational choice).

Entropy thus offers the possibility to capture effects of some important factors involved in translational choice, such as source- and target-language similarity or level of expertise, and it may be used for assessing translation output in terms of translation consistency (e.g. when multiple translators are involved). As a measure, entropy turns out to be reasonably stable across languages and modes of translation (e.g. from scratch vs. post-edited), thus providing a reliable instrument for comparative studies. Crucially, as we have shown with selected examples from online language processing, entropy provides a direct link to cognitive effort—the higher the entropy at a given choice point, the higher the effort incurred, as indicated by several behavioural measures (e.g. response latency and production time).

20.3 Discussion and future directions

In this chapter, we have sketched a formal basis for the probabilistic modelling of human translation based on information theory, adopting Shannon's noisy channel model and selected entropy-based measures of information. The new perspective that is opened up by applying information theory in Translation Studies is to consider translation as rational communication, according to which interlocutors aim to understand the sender's message and to be understood by the receiver while the cognitive effort spent remains reasonable. However rational language users may actually be in real life, it is clear that many constraints interact with interlocutors' assumed rational goals, including noise in the channel, uncertainty about the correct interpretation of a message or the adequacy of a produced message for a given audience, as well as cognitive resource limitations. Translation being a type of language use, albeit a special one, by implication operates under the same kinds of constraints. The specific goals of an interlocutor-as-translator impose additional, translation-specific constraints, notably to find a good balance between fidelity to the SL expression and conformity to TL expectations (see Chapter 2.4 on the notion of equivalence). We have shown how these objectives can be formally represented by the components of a noisy channel model and suggested a number of uses for such models (Section 20.2.2). Furthermore, we have shown that selected constraints (level of expertise and translation difficulty) may be captured by entropy as a suitable operationalization of the notion of (un)certainty about the best translational choice (Section 20.2.3).

The common assumption in applications of information theory to the study of language is that language use can be modelled as a probabilistic process, according to which interlocutors rely to a large extent on predictability in context (see Section 20.2.1). While probabilistic models of language use have been successfully applied in a number of natural language processing tasks, applications to linguistic or translatological questions are still rare.

Embarking on this line of research involves some methodological challenges (see Section 20.3.1) but bears a lot of promise for theoretical advancement (see Section 20.3.2).

20.3.1 Methods: Estimating probabilities

As in other data-driven, corpus-based accounts, there are a number of issues involved in estimating probabilities, such as representativeness of the data set, data sparseness, difference in size of data sets when comparing e.g. parallel and comparable monolingual corpora, as well as reliability of (pre-)processing, notably word alignment. Also, to enable rational explanations including cognitive interpretations, we want to integrate contextual constraints into our models (see surprisal, Section 20.2.1). Apart from the ambient linguistic context, which is known to have an immediate effect on online processing effort, there are many other relevant contextual variables that may exert an influence on the expectancy of a given linguistic unit, including discourse context, register/text type and world knowledge. With the approach sketched here, we can potentially incorporate any kind of contextual variable.

To start with, for approximating $p(unit|\, context)$, standard n-gram language models can be used, where $context$ is simply the preceding context of $n-1$ words. Other variables relevant for translation are brought into play by choosing specific corpora, e.g. interpreting vs. translation, different language pairs and translation directions, or levels of translation expertise. As in other data-driven approaches, very much rests upon the choice of adequate data sets. For a noisy channel model of translation, we need substantial amounts of translation data to estimate $p(s \mid t)$ as well as target-language data to estimate $p(t)$. A specific challenge is that the probability scores for the translation model and the language model will be in quite different ranges and not directly comparable. Comparing the rankings of the translation options based on probabilities of the translation model and of the language model can provide us with information on whether a given translation choice tends towards SL fidelity or TL conformity. If we want the two components to have equal influence on the output, we can experiment with different weights on the two components of the noisy channel model; i.e. we can induce greater TL conformity by giving more weight to the target-language model or greater SL fidelity by putting a larger weight on the translation model. In this way, we can explore the effects on translational choice of shifting between SL fidelity and TL conformity. See again the example of "Entwicklungsbudget" (higher probability according to translation model) vs. "Entwicklungshaushalt" (higher probability according to target-language model).

20.3.2 Theory: Translation as rational communication

Casting translation in terms of rational communication with information theory as a formal basis allows us to formalize some core notions of translation theory as well as provide the theoretical underpinnings for some less explained areas in Translation Studies. Theoretical notions such as equivalence (see Chapter 19, this volume), creativity (see Chapter 17, this volume), expertise (see Chapter 26, this volume) or competence (see Chapter 22, this volume) are all graded concepts and may adequately be represented in probabilistic terms. Furthermore, context is regarded as a crucial factor in translation, but it often remains too vaguely defined. In an information-based model, context is given in the very definition of information as the immediate linguistic context, register context and/or source- or target-language context (see discussion on methods in Section 20.3.1). Here again, probabilities are an adequate means for modelling the influence of context on translational choice. Finally, a number of translational phenomena that are hard to explain may be captured within a rational communicative framework. For instance, omission is a common choice in translation, and even more so in interpreting, but for a given occurrence it is often unclear when the choice is due to audience design (TL orientation) or to cognitive resource limitations on the part of the translator/interpreter. Here, the factors assumed to be involved in a given translational choice may be represented as multiple independent variables in a regression model.

In summary, while adopting an information-theoretic framework for modelling human translation presents some challenges, it bears the promise of integrating insights from descriptive corpus analysis and experimental results on selected cognitive processing aspects involved in translation for a more encompassing explanation of translation under the perspective of rational communication. The goals of rational interlocutors—successful message transmission with reasonable cognitive effort—are goals for translators, too. The formal apparatus we have described in this chapter provides the means to model rational communicative goals and the constraints acting upon them, including the specific constraints acting upon translation. With its link to evidence from cognitive processing, an information-based, rational communication approach provides an adequate level to formulate translation universals and a unifying framework for capturing translationese effects (see also Halverson, 2003 on the cognitive basis of translation universals). Finally, a rational communicative approach to translation opens up the opportunity of closer interaction with relevant linguistic research on language use, variation and change.

Notes

1 Formally, surprisal and information density are identical, but the term *surprisal* tends to be used in the context of language comprehension, while information density tends to be used in studies of language production. Commonly employed related measures are *Mutual Information*, a popular measure for characterizing collocations, and *Information Gain*, commonly used for comparing probability distributions.
2 The corpora used for illustration here and in other examples are EuroParl-UdS (Karakanta et al., 2018), available from CLARIN-D at http://fedora.clarin-d.uni-saarland.de/europarl-uds/, and the Translation and Interpreting Corpus (TIC) (Kajzer-Wietrzny, 2015).
3 We could have used frequency here instead, i.e. how often a translation is used.

Further reading

Grice, P. H. (1975). Logic and conversation. In P. Cole, & J. L. Morgan (Eds.), *Speech acts, syntax and semantics* (Vol. 3, pp. 41–58). New York: Academic Press.

In his pragmatic theory, Grice introduced four principles of linguistic interaction that describe the rational principles that people follow in effective communication. The four "maxims" of quality, quantity, relation (or relevance) and manner are the pillars of an overarching cooperative principle that Grice stipulated.

Harris, Z. (1991). *A theory of language and information. A mathematical approach.* Oxford: Clarendon Press.

The book introduces a formal theory of language on the basis of information theory. It includes an information-theoretically oriented account of register variation using the example of scientific language and offering a communicative explanation for register/sublanguage formation.

Jurafsky, D., & Martin, J. H. (2018). *Speech and language processing. An introduction to natural language processing, computational linguistics, and speech* (3rd ed.). India: Pearson.

This is a standard textbook in natural language processing (NLP). Different chapters introduce NLP techniques mentioned in the present article, including various kinds of computational language models.

Koehn, P. (2010). *Statistical machine translation* (1st ed.). New York: Cambridge University Press.

This is a standard textbook in statistical machine translation (SMT). Different chapters introduce the basic workings of language models in the context of translation, including word-based and phrase-based models, text-translation alignment and evaluation of MT output.

Kullback, S., & Leibler, R. A. (1951). On information and sufficiency. *The Annals of Mathematical Statistics, 22*(1), 79–86.

This article from the early 1950s provided an asymmetric measure of relative entropy, called Kullback–Leibler Divergence after the authors.

Shannon, C. E. (1948). A mathematical theory of communication. *The Bell System Technical Journal, 27*, 379–423, 623–656.

This is the original article by Claude E. Shannon on information theory, including the definition of general communication systems by a noisy channel model.

References

Angelone, E. (2010). Uncertainty, uncertainty management and metacognitive problem solving in the translation task. In G. Shreve, & E. Angelone (Eds.), *Translation and cognition* (pp. 17–40). Amsterdam: John Benjamins.

Baayen, H., Tomaschek, F., Gahl, S., & Ramscar, M. (2017). The Ecclesiastes principle in language change. In M. Hundt, S. Mollin, & S. E. Pfenninger (Eds.), *The changing English language* (pp. 21–48). Cambridge: Cambridge University Press.

Baker, M. (1995). Corpora in translation studies: An overview and some suggestions for future research. *Target, 2*, 223–243.

Baker, M. (1996). Corpus-based translation studies: The challenges that lie ahead. In H. Somers (Ed.), *Terminology, LSP and translation: Studies in language engineering in honour of Juan C. Sager* (pp. 175–186). Amsterdam: John Benjamins.

Bangalore, S., Behrens, B., Carl, M., Ghankot, M., Heilmann, A., Nitzke, J., Schaeffer, M., & Sturm, A. (2016). Syntactic variance and priming effects in translation. In M. Carl, S. Bangalore, & M. Schaeffer (Eds.), *New directions in empirical translation process research: Exploring the CRITT TPR-DB* (pp. 211–238). Cham: Springer.

Blache, P., & Rauzy, S. (2011). Predicting linguistic difficulty by means of a morpho-syntactic probabilistic model. In *Proceedings of the 25th Pacific Asia conference on language, information and computation* (pp. 160–167). Singapore: Institute of Digital Enhancement of Cognitive Processing, Waseda University.

Blanco, N. J., Love, B. C., Ramscar, M., Otto, A. R., Smayda, K., & Maddox, W. T. (2016). Exploratory decision-making as a function of lifelong experience, not cognitive decline. *Journal of Experimental Psychology: General, 145*(3), 284.

Campbell, S. (1999). A cognitive approach to source text difficulty in translation. *Target, 11*(1), 33–63.

Campbell, S. (2001). Choice network analysis in translation research. In M. Olohan (Ed.), *Intercultural faultlines: Research models in translation studies: Textual and cognitive aspects* (pp. 29–42). Manchester: St Jerome.
Cap, F. (2017). Show me your variance and I tell you who you are—Deriving compound compositionality from word alignments. In *Proceedings of the 13th workshop on multiword expressions (MWE 2017)* (pp. 102–107). Valencia, Spain: Association for Computational Linguistics.
Carl, M., & Schaeffer, M. (2017a). Why translation is difficult: A corpus-based study of non-literality in post-editing and from-scratch translation. *HERMES—Journal of Language and Communication in Business, 5*(6), 43.
Carl, M., & Schaeffer, M. (2017b). Sketch of a noisy channel model for the translation process. In S. Hansen-Schirra, O. Czulo, & S. Hofmann (Eds.), *Empirical modelling of translation and interpreting* (pp. 71–116). Berlin: Language Science Press.
Carl, M., Schaeffer, M., & Bangalore, S. (2016). The CRITT translation process research database. In M. Carl, S. Bangalore, & M. J. Schaeffer (Eds.), *New directions in empirical translation process research* (pp. 13–54). Cham: Springer.
Crocker, M., Demberg, V., & Teich, E. (2016). Information density and linguistic encoding (IDeaL). *KI—Künstliche Intelligenz, 30*(1), 77–81.
Cronin, M. (1998). Game theory and translation. In M. Baker (Ed.), *Routledge encyclopedia of translation studies* (pp. 91–93). London: Routledge.
Degaetano-Ortlieb, S., & Teich, E. (2016). Information-based modeling of diachronic linguistic change: From typicality to productivity. In *Proceedings of language technologies for the socio-economic sciences and humanities (LATECH'16), Association for Computational Linguistics (ACL)*. Berlin.
Delogu, F., Crocker, M. W., & Drenhaus, H. (2017). Teasing apart coercion and surprisal: Evidence from eye-movements and ERPs. *Cognition, 161*, 46–59.
Englund Dimitrova, B. (2005). *Expertise and explicitation in the translation process*. Amsterdam: John Benjamins.
Fankhauser, P., Knappen, J., & Teich, E. (2014). Exploring and visualizing variation in language resources. In *Proceedings of the ninth international conference on language resources and evaluation (LREC'14)*. Reykjavik.
Frank, S. L. (2010). Uncertainty reduction as a measure of cognitive processing effort. In *Proceedings of the 2010 workshop on cognitive modeling and computational linguistics* (pp. 81–89). Uppsala: Association for Computational Linguistics.
Gellerstam, M. (1986). Translationese in Swedish novels translated from English. In L. Wollin, & H. Lindquist (Eds.), *Translation studies in Scandinavia* (pp. 88–95).
Grice, P. H. (1975). Logic and conversation. In P. Cole, & J. L. Morgan (Eds.), *Speech acts, syntax and semantics* (Vol. 3, pp. 41–58). New York: Academic Press.
Gutt, E.-A. (1989). *Translation and relevance*. London: Routledge.
Hale, J. (2001). A probabilistic Earley parser as a psycholinguistic model. In *Proceedings of the second meeting of the North American chapter of the association for computational linguistics on language technologies* (pp. 1–8). Stroudsburg, PA: Association for Computational Linguistics.
Hale, J. (2003). The information conveyed by words in sentences. *Journal of Psycholinguistic Research, 32*(2), 101–123.
Hale, J. (2006). Uncertainty about the rest of the sentence. *Cognitive Science, 30*(4), 643–672.
Halverson, S. (2003). The cognitive basis of translation universals. *Target, 15*(2), 197–241.
Harris, Z. (1991). *A theory of language and information. A mathematical approach*. Oxford: Clarendon Press.
Hume, E., & Mailhot, F. (2013). The role of entropy and surprisal in phonologization and language change. In *Origins of sound patterns: Approaches to phonologization*, 29–47.
Hyönä, J., Tommola, J., & Alaja, A.-M. (1995). Pupil dilation as a measure of processing load in simultaneous interpretation and other language tasks. *The Quarterly Journal of Experimental Psychology Section A, 48*(3), 598–612.
Jaeger, T. F. (2010). Redundancy and reduction: Speakers manage syntactic information density. *Cognitive Psychology, 61*(1), 23–62.
Jurafsky, D., & Martin, J. H. (2018). *Speech and language processing. An introduction to natural language processing, computational linguistics, and speech* (3rd ed.). India: Pearson.
Kajzer-Wietrzny, M. (2015). Simplification in interpreting and translation. *Across Languages and Cultures, 16*(2), 233–255.
Karakanta, A., Vela, M., & Teich, E. (2018). Preserving metadata from parliamentary debates. In *Proceedings of the eleventh international conference on language resources and evaluation (LREC)*. Miyazaki, Japan.
Koehn, P. (2010). *Statistical machine translation* (1st ed.). New York: Cambridge University Press.

Krings, H. P. (2001). *Repairing texts: Empirical investigations of machine translation post-editing processes*. (G. S. Koby, Ed.). Kent: Kent State University Press.
Kullback, S., & Leibler, R. A. (1951). On information and sufficiency. *The Annals of Mathematical Statistics, 22*(1), 79–86.
Levý, J. (1967). Translation as a decision process. In J. Levý (Ed.), *To honor Roman Jakobson: Essays on the occasion of his seventieth birthday* (pp. 1171–1182). Hague: Mouton.
Levy, R. (2008). Expectation-based syntactic comprehension. *Cognition, 106*, 1126–1177.
Levy, R., & Jaeger, T. F. (2007). Speakers optimize information density through syntactic reduction. In B. Schölkopf, J. Platt, & T. Hofmann (Eds.), *Advances in neural information processing systems 19: Proceedings of the 2006 conference* (pp. 849–856). Cambridge, MA: The MIT Press.
Linzen, T., & Jaeger, T. F. (2016). Uncertainty and expectation in sentence processing: Evidence from subcategorization distributions. *Cognitive Science, 40*(6), 1382–1411.
Manning, C. D., & Schütze, H. (2003). *Foundations of statistical natural language processing* (6th ed.). Cambridge, MA: MIT Press.
Martínez Martínez, J. M., & Teich, E. (2017). Modeling routine in translation with entropy and surprisal: A comparison of learner and professional translations. In L. Cercel, M. Agnetta, & T. Amido Lozano (Eds.), *Kreativität und Hermeneutik in der Translation*. Tübingen: Narr Francke Attempto Verlag.
Piantadosi, S. T., Tily, H., & Gibson, E. (2011). Word lengths are optimized for efficient communication. *Proceedings of the National Academy of Sciences, 108*(9), 3526–3529.
Pym, A. (2008). Redefinindo competência tradutória em uma era eletrônica. Em defesa de uma abordagem minimalista. *Cadernos de Tradução, 1*(21), 9–40.
Reich, I. (2017). On the omission of articles and copulae in German newspaper headlines. *Linguistic Variation, 17*(2), 186–204.
Reiss, K. (1983). Adequacy and equivalence in translation. *The Bible Translator, 34*(3), 301–308.
Rubino, R., Lapshinova-Koltunski, E., & van Genabith, J. (2016). Information density and quality estimation features as translationese indicators for human translation classification. In *Proceedings of the 2016 conference of the North American chapter of the association for computational linguistics: Human language technologies* (pp. 960–970). San Diego, CA: Association for Computational Linguistics.
Rutherford, A., Demberg, V., & Xue, N. (2017). A systematic study of neural discourse models for implicit discourse relation. In *Proceedings of the 15th conference of the European chapter of the association for computational linguistics: Volume 1, long papers* (pp. 281–291). Valencia: Association for Computational Linguistics.
Schaeffer, M., & Carl, M. (2014). Measuring the cognitive effort of literal translation processes. In U. Germann, M. Carl, P. Koehn, G. Sanchis-Trilles, F. Casacuberta, R. Hill, & S. O'Brien (Eds.), *Proceedings of the EACL 2014 workshop on humans and computer-assisted translation* (pp. 29–37). Stroudsburg, PA: Association for Computational Linguistics.
Schaeffer, M., Dragsted, B., Hvelplund, K. T., Balling, L. W., & Carl, M. (2016). Word translation entropy: Evidence of early target language activation during reading for translation. In M. Carl, S. Bangalore, & M. Schaeffer (Eds.), *New directions in empirical translation process research: Exploring the CRITT TPR-DB* (pp. 183–210). Cham: Springer.
Schulz, E., Oh, Y. M., Malisz, Z., Andreeva, B., & Möbius, B. (2016). Impact of prosodic structure and information density on vowel space size. In *Proceedings of the international conference on speech prosody* (Vol. 2016–January, pp. 350–354). Boston: Boston University.
Shannon, C. E. (1948). A mathematical theory of communication. *The Bell System Technical Journal, 27*, 379–423, 623–656.
Shlesinger, M., & Ordan, N. (2012). More spoken or more translated?: Exploring a known unknown of simultaneous interpreting. *Target, 24*(1), 43–60.
Sperber, D., & Wilson, D. (1986). *Relevance: Communication and cognition*. Oxford: Blackwell.
Teich, E. (2003). *Cross-linguistic variation in system and text: A methodology for the investigation of translations and comparable texts*. Berlin: Walter de Gruyter.
Toury, G. (1995). *Descriptive translation studies and beyond*. Amsterdam: John Benjamins.
Toury, G. (2004). Probabilistic explanations in translation studies. In A. Mauranen, & P. Kujamäkipp (Eds.), *Translation universals: Do they exist?* (pp. 15–32). Amsterdam: John Benjamins.
Villada Moirón, B., & Tiedemann, J. (2006). Identifying idiomatic expressions using automatic word-alignment. *EACL-2006 workshop on multi-word-expressions in a multilingual context*, 33–40.
Volansky, V., Ordan, N., & Wintner, S. (2015). On the features of translationese. *Digital Scholarship in the Humanities, 30*(1), 98–118.

21

Translation, human–computer interaction and cognition[1]

Sharon O'Brien

21.1 Introduction

Translation is, without a doubt, a form of human–computer interaction (HCI). In a period of less than 30 years, technology has radically transformed the way in which professional translators work (Folaron, 2010, p. 429). Among other technologies, translation memory (TM) tools are now standard in many professional translation domains, and recent successes in machine translation (MT) have led to a significant increase in usage and commercial implementation, which is in turn touching on the lives of professional translators. The cognitive processes involved in translation must surely also be impacted by the significant use of technology.

Thinking about translation as a form of HCI requires a statement about the concept of "translation" underpinning the discussion. Tymoczko (2007) argues that the narrow English-language Western European concept of "translation" as a form of transfer between a written source-language text and a target-language one must be broadened into a concept of "*translation" as cross-cultural understanding that is not reliant on dominant Western European views or on restricted notions of what constitutes a text.[2] While Tymoczko's appeal for broadening the concept of translation within Translation Studies is acceptable, at the same time, it is legitimate also to consider translation as something more specific, especially when exploring the cognitive aspects. Otherwise, the scope would be too broad for a coherent discussion in this chapter. Therefore, the notion of translation considered here is that of bilingual, text-based translation destined for public consumption for which the translator is paid. While this may be a restricted concept of translation, it constitutes a significant global economic activity and is the type of translation from which many Translation Studies graduates earn their livelihoods.

The concept of translation under consideration here is one of which some repetition, high volume and time pressure are characteristic, making the task particularly suitable for computer-aided translation tools. However, other types of translation, or even *translation in Tymoczko's sense, are not explicitly excluded. For example, we might also include collaborative volunteer translation or subtitling and dubbing of audiovisual material, which are both also characterized by interaction with computers. Literary translation is not explicitly included in our discussion, but even the translation of literary text can also be a form of HCI, and the use of technology for literary translation is on the rise (see, for example, Toral & Way, 2015). While the primary focus here is on written translation as a form of HCI, it is important to acknowledge that interpreters use computer resources in their work, and that can therefore also be considered a form of HCI. Furthermore, speech is now becoming a more viable form of input, as will be discussed in Section 21.3.

This chapter is structured along the following lines. First, the core topics relating to HCI and translator–computer interaction (TCI) are explored, including the benefits and challenges presented. This is followed by a discussion of future directions.

21.2 Core topics

21.2.1 What is human–computer interaction?

HCI is defined as "the study of the interaction between people, computers and tasks" (Johnson, 1992, p. 1). It draws on the disciplines of science, engineering and art and has as a core concern the demands made by the computer on people's knowledge, tasks and learning. HCI is not just about the user interface of a software product. Two terms that are commonly used in the HCI domain are "human factors" and "ergonomics" (see Ehrensberger-Dow, this volume). Human factors focus on how people interact with tools and technology. While the term "ergonomics" traditionally referred to the ease with which hardware, such as keyboards, could be used, it has evolved to also include the "ease" with which software products can be used. A sub-domain within ergonomics is "cognitive ergonomics", which is concerned with the cognitive demands placed on users by the design and complexity of computer programs. In a description of the scope of a European Conference on Cognitive Ergonomics, it was stated that

> [r]ecent trends of cognitive ergonomics indicate that human interaction with IT-based systems is increasingly complex and thus needs more sophisticated social, cognitive, and affective support, and that diverse user groups should be considered from system requirements analysis and initial design stages, paying attention to personalization, care, and complexity.[3]

These days, translation requires ever-increasing, complex, physical and cognitive interaction with computers and computer programs. As described by Ehrensberger-Dow (this volume), the task of translation involves sitting at a workstation for extended periods of time interacting with various tools, and this has been both enabling and a source of malcontent in the translation profession.

21.2.2 Translation as human–computer interaction

The translation profession has changed over time, with the translator in some circumstances becoming almost symbiotic with the "machine" (used synonymously here with "computer"). Yet, TCI is not a new phenomenon. With the introduction of the electronic typewriter, with only two lines of memory, and the use of dictaphones, translation already became a computer-interaction task. This was followed by the introduction of word-processing software. Although the origins of word processing date to before the mid-1970s, word processing only started to become known globally in the mid-1970s and early 1980s (Haigh, 2006). This development required translators to interact with a computer for the first time. Not long after the mass embracing of word processing came the introduction of TM tools. In conjunction with this development came terminology management programs, which are ostensibly used to store terms and their corresponding translations in one or multiple languages, though it is well known that such programs are not restricted to the storage of terms but also store phrases and sometimes even sentences or larger chunks of text, therefore creating a fuzzy line between TM and terminology management tools. The information technology (IT) industry and, in particular, the software localization sector were the first to embrace TM tools. It is not surprising that TM tools grew out of IT companies (TM2 in the case of IBM) or out of technical translators who worked for IT companies (e.g. Trados Translator's Workbench), because this industry produces large volumes of repetitive text that is updated on a regular basis. Prior to the introduction of TM tools, content repetition was identified using compare features in word processors. Content that was identical was marked up by the word processor. The translator then had to locate that content in the previously translated document and copy and paste the relevant translated section into the new document. Needless to say, this was a cognitively tedious, time-consuming and error-prone task. The IT industry therefore had a problem, and TM tools were developed to solve it. Additionally, translation of the text in user interfaces (UI) was required. At first, this necessitated the extraction of UI strings of text into a contextless spreadsheet, but eventually, dedicated tools were developed for the translation of this specialized type of text.[4]

Access to the Internet and to personal computers grew in the early to mid-1990s, and this also had an impact on translators, who now had electronic dictionaries, encyclopaedias, and other sources of digital information at their fingertips. The IT industry deals in specialized terminology that is reproduced across different content types; it is important that the menu name in a program, for example, is reproduced consistently in the online help so as not to confuse and frustrate the user. Therefore, terminology management tools were introduced to solve another problem related specifically to specialized terminology management.

Not only has translation become a HCI task, but so has the task of running a translation business. E-mail and instant messaging have mostly replaced telephone conversations. Faxes have become redundant. Where once large-scale translation projects were delivered on disks or CD-ROMs in boxes, they are now downloaded from websites or accessed via specially designed workflow management tools.[5] Project team meetings are now done via online conferencing systems or team collaboration tools such as "Slack", and training is done via webinars. Purchase orders and invoicing are managed through ERP (Enterprise Resource Planning) systems.

The developments described can be categorized as technology that aids the human cognitive translation process. Initially, there was some resistance to the introduction of TM technology, because it meant a considerable change to the way translation was done, and translators also had well-founded fears that it would change how they were paid for their work. Although not all translators use it, it has become relatively standard in many professional domains.

Although introduced before TM, MT has taken a different path, and the professional community has been slower to adopt it (see Carl, this volume).

A report by Howard Taubman in the *New York Times* in 1967 stated:

> if you have begun to fear that there is no stopping the machine in its march to take over human duties, cheer up—at least for a while. A learned National Academy of Sciences has found that in one area, translation, man is not obsolescent.[6]

Taubman's words were prescient; MT did not make the translator obsolescent. However, recent developments in MT mean that the fear Taubman alluded to has returned. In the last decade, MT has been reinvigorated due to three key developments. First, large repositories of parallel translated data became available due to the use of TM technology for over 20 years. This gave birth to a new data-driven paradigm that produced better machine-translated output than the previous rules-based paradigm. Second, the World Wide Web provided a massive database of mostly free electronic text from which MT systems can learn. Third, recent developments in machine learning, coupled with greater computer processing power, have led to the successful application of neural networks to the problem of MT, which in turn has produced higher-quality (particularly in terms of target-language fluency) MT output. These key developments have led to a situation where the quality of machine-translated text is now at a level where it is a realistic aid to the cognitive process of human translators.

However, in the case of MT, human interaction is not just between translators and the machine, but also between end users who have an information need and the machine, or between volunteer translators, such as "fan-subbers" (O'Hagan, 2009), and the machine. The recent improvements in MT systems and their ease of access via the Internet have only increased the level of interaction between computers and the act or product of translation, leading to increased "machine translation literacy" (Bowker & Buitrago Ciro, 2019).

21.2.3 Translation–computer interaction—benefits

Development of translation as an HCI task has undoubtedly brought with it many challenges for humans. Before delving into these, however, it is important to enumerate the benefits. At least three groups benefit from HCI in translation: translation clients, end users (otherwise known as recipients of translation) and translators themselves.

For clients and end users, the use of translation technology theoretically speeds up the process, because a repeated sentence (normally) does not need to be retranslated. In turn, quality is improved through consistency, and costs are reduced because a client does not have to pay to retranslate text. Few experienced translators would deny the productivity increases brought about by the use of TM tools, assuming, of course, that the contents of the TM are of a high quality to begin with. Some have, however, questioned the contribution tools make to increased consistency and, ultimately, quality (e.g., Bowker, 2005). Nonetheless, it is mostly accepted that a quality-controlled deployment of terminology management and TM tools will contribute to translation consistency and quality. The third general advantage, reduced cost, is obviously a contentious one, with professional translators initially being very resistant to the reduction in word rates brought about by the introduction of TM tools. However, given that TM tools have now become mainstream, there is little doubt that cost advantages have been accrued, and these have not been limited only to translation clients or end users.

In addition to the three main advantages discussed, there are more subtle, process-based advantages for translators who interact with computer tools. For example, TM technology relieves a translator from having to translate the same sentence over and over again. Even when only a part of the sentence can be reused (as with a fuzzy match), the translator is saved from having to retype certain words or phrases. It has also replaced the error-prone and mind-numbing manual task of copying and pasting by a more intelligent and automatic search and replace and autocomplete tool. Thus, a TM tool could be seen as having relieved the human translator of a repetitive and boring task. Terminology management tools provide the translator with instant access to an approved term list, saving the translator from the effort of trying to remember how she or he translated a term previously, or having to look it up in several dictionaries, which effectively breaks the flow of the translation process, even if it is an essential part. If used correctly, both these tool types help contribute to translation expertise by supporting consistency and task flow.

Machine translation systems translate sentences at a speed that is significantly faster than a human translator can achieve. Even when post-editing (or fixing of MT errors) is necessary, research has shown that reasonable-quality raw MT output can enable the translator to work at speeds beyond what might otherwise be achievable and to translate a higher number of words per hour (Guerberof, 2009; O'Brien, 2007). The downward pressure on payment rates has somewhat been compensated for by higher throughput, supported by technology. Early research using data-driven MT engines suggested that novice translators, such as students and recent graduates, who are at the starting line with regard to their accumulation of professional expertise, might benefit from MT, while professionals with long-term experience might not benefit as much (or at all) (García, 2010). MT is also a useful tool for end users of translations—it can be used to decide whether the content of a document is interesting or relevant enough to have it either post-edited or translated by a human translator. Users can also use MT to get the gist of the message in a text written in a language they do not (fully) understand. Furthermore, MT is now being deployed in new areas, such as a writing aid for academic publication for non-native speakers of English (Bowker & Buitrago Ciro, 2019; Goulet et al., 2017; O'Brien et al., 2018a; Parra Escartín et al., 2017) or for educational purposes (Hu et al., 2019).

The uptake of MT also means that now even more information can be translated. It not only creates more translation-related work, but it can potentially have a positive impact on human rights. In the context of an explosion in user-generated content, it has been suggested that only 0.5% of the content being created today is translated (Vashee, 2010). Much of the translation done today is from English (content produced by multinationals who want to sell products) into the languages of the richest countries in the world. Very little content is formally translated into, from or between the many languages of Africa or India, for example. It has been suggested that machine translation can be the enabler of "translation as a human right" (Van Der Meer, 2010); that it will allow linguistic communities who do not have access to information to attain that access (O'Brien et al., 2018b).

In summary, the use of computers to aid translation creates a number of potential and real benefits, including faster throughput, increased consistency and lower costs for clients, possibly leading to higher volumes being translated as well as increased access to information in languages not normally seen as being commercially important.

21.2.4 Translator–computer interaction—challenges

Any person involved in translation, whether as a student, academic, professional translator, project manager, client or tools developer, will be only too aware, however, of the challenges that TCI introduces. While the introduction of technology to support translation has brought about many advantages, it has also introduced a number of significant challenges and raises some important questions about the future of the translation profession.

How the increasing use of technology impacts on the status of the translation profession has been, and continues to be, of considerable concern. Some translators feel dehumanized by the technology they are required to use. Having to fix the errors created by an MT system (or created by a human translator and propagated by a TM system) understandably irks some translators to such a degree that they refuse to interact with the technology. In the context of MT, not only can translators feel replaced by the machine, but the machine generates fundamental linguistic errors that a trained human translator would rarely generate. The professional translator is then demoted to the status of a fixer (Krings, 2001) of seemingly unintelligent errors. Cognitively, the task is boring and tedious, with translators often reflecting that they have been robbed of their creativity. That they are paid lower rates to fix such errors than to create their own translation adds to the feelings of negativity. There are, however, other dimensions to this complex debate. Cooper (2004) argues that

> [i]t doesn't require sophisticated tools to dehumanise your fellow human—a glance or a kick does it as well. It is not the technology that is dehumanising. It is the technologists, or rather the processes that technologists use, that create dehumanising products.
>
> *Cooper, 2004, p. 120*

Cooper's first point is peripheral; i.e. it is not just technology that can dehumanize; humans can too. The more relevant point here is that it is how the technology is created, or implemented, that has a dehumanizing effect. Technology created without consideration for the task or end users removes those end users from the equation. Karamanis et al. (2011) touch on this issue in their contextual-inquiry-based research into translators in the workplace. On the topic of machine translation, they note how translators see MT as a black box, something they do not quite understand and which removes them further from the task of translation, which, according to their observations, is a highly collaborative task, at least in the context they investigated. The lack of possibilities to collaborate with a machine (on the surface at least—but more about personalization in Section 21.3) leads to a level of mistrust and sometimes also to rejection of the technology. The more the professional translator is involved in the design, testing and implementation of translation technology, the more ownership she or he feels over the technology, and the more likely it is to be seen as an aid rather than a dehumanizing threat. In fact, the ability to have control over the use or non-use of MT has been shown to have an effect on feelings of "agency" among translators (Cadwell et al., 2016; Cadwell et al., 2017; Olohan, 2011). On the flipside of the dehumanizing debate, claims that TCI actually results in humanizing, or socializing, translation have been made (Pym, 2011). In the context of collaborative volunteer translation, candidate translations, whether created by a human or generated in some way by a computer program, are collaboratively assessed, negotiated, voted on and, finally, accepted. The many people involved in this process create a human translating network that is supported by technology. This is an interesting image that stands in quite stark contrast to that of machine as master and translator as a bored slave.

The tension between translators and computers is only one of many such frictions to have occurred over time. As Christian observes (2011, p. 84), the reshaping of job markets through automation and mechanization is centuries old. One side of the debate argues that machines take human jobs away, while the other side argues that increased mechanization has resulted in an economic efficiency that raises the standard of living for all, releasing humans from unpleasant tasks. Christian discusses a scenario that has some interesting parallels with translation. He describes how software programmers work directly on problems while at the same time trying to automate the solution to those problems. So, are software programmers programming their collective way out of a job? Christian concludes: "No, the consensus seems to be that they move on to progressively harder, subtler, and more complex problems, problems that demand more thought or judgement. They make their jobs, in other words, more human" (Christian, 2011, p. 88)—and potentially more cognitively fulfilling. Can translators make their jobs more human through HCI? Can we allow the machine to take over the boring, repetitive tasks and free ourselves up for the harder, subtler and more cognitively complex problems? And what are those problems that machines cannot solve, but human translators can? These are some of the important questions facing us today.

Psychological theories play a major role in HCI research (Johnson, 1992). Designers and developers of computer programs are often required to make assumptions about task structure, human behaviour during a task, user experience levels, a user's ability to learn, etc. The assumptions made by designers directly affect the experience of the user. Cooper (2004) and Kolko (2010) both appeal for software to be designed not by programmers but by interaction designers, suggesting the importance of understanding how the human interacts with the computer and specific task-supporting programs. Cooper talks about cognitive friction between users and devices, which he defines as "the resistance encountered by a human intellect when it engages with a complex system of rules that change as the problem changes" (Cooper, 2004, p. 19). He also points out that there is a tremendous difference between designing for function and designing for humans (Cooper, 2004, p. 90). Olohan (2011), exploring how sociologist of science Andrew Pickering's concept of the "mangle of practice" might be applied to translation and TM technology, also draws on the theme of resistance and echoes Cooper's sentiments when she points out:

> One argument to explain why systems sometimes fail is that system development is often regarded as technical change rather than socio-technical change; i.e., the human and organizational aspects are not addressed at all, or only implicitly, or in an ad-hoc fashion, when the system is being developed.
>
> *Olohan, 2011, p. 345*

Unfortunately, there is little evidence to suggest that tools that are proposed as aids to the translation process have been designed from the point of view of the humans who have to use them. That is not to say that all computer aids for translation are flawed. Without a doubt, features in many of the tools are useful and appreciated by translators. However, it is also clear that the tools are not all easy to learn or use, that they are not always stable, and that they have not been designed from the point of view of interaction with translators, as opposed to simply supporting functions within the translation task or supporting the managers of the translation business (Ehrensberger-Dow & O'Brien, 2015; O'Brien et al., 2017). While programmers know a lot about the functional design of software and have their own personal preferences regarding design, they rarely know about designing with the end user in mind (Cooper, 2004). This is probably true of computer aids for translation. What proportion of the programmers who have designed TM or terminology management tools have ever translated content? What proportion of MT system developers are translators? What do they know about the cognitive process involved in translation? This goes some way to explaining the friction that sometimes exists between translators and their computer aids.

As mentioned previously, TM tools were introduced to solve specific problems in the context of high-volume, high-repetition translation. It is only to be expected, then, that translators who work in this domain will engage in more revising and editing of other translators' work than in creating their own translations. The increasing use of MT further increases the editing component of the task, only in this case the editing is sometimes (but not always) of seemingly obtuse mistakes. A recent interesting development is the move to "neural machine translation" (NMT) (see Forcada, 2017, for a detailed explanation). NMT engines produce seemingly better output than previous types of engines (statistical or rules-based ones). Interestingly, they produce a new challenge for post-editors: the output is deemed to look and sound very fluent and sometimes almost like a human translation. However, the fluency can be misleading—the meaning might be quite incorrect, even though the translation sounds convincing. For instance, a city name might be replaced with a different city name from an entirely different country or continent. This is just a feature of how neural networks work. Nonetheless, it might be difficult to spot the error in "logic" if the grammar and meaning otherwise seem to be correct. At the time of writing this chapter, no substantial research had been conducted on the different cognitive demands of editing NMT output compared with output from older types of engines. However, it is reasonable to speculate that when the errors are not obvious and the output sounds fluent, the cognitive task will have changed from one of quickly spotting and fixing errors to a task that is closer to that of revision, by a human, of translation produced by a human. The reviser will have to pay close attention to the meaning of the source text in order to identify errors in the target text that are not so obvious at first glance. This leads us to wonder whether post-editing might become more of an accepted task among professionals, given that the task might now approximate that of traditional "revision".

At the same time, for many, editing is seen as a less creative task than translation (though this is certainly open to debate—can we really argue that improving or correcting what an author has written is "less creative" than translating another author's words?), and job satisfaction is further diminished by having to correct machine-generated mistakes or human mistakes propagated by the machine. A significant problem with the creativity argument is that the concept of creativity is very difficult to define and measure, and there are various definitions for the term. Recent research on creativity in the translation process has resulted in some operationalization of the construct in terms of cognitive shifts between source text and target text (Bayer-Hohenwarter, 2009, 2010, see also Bayer-Hohenwarter & Kussmaul, this volume). This is useful for the research domain, in particular for the study of the development of translation competence over time. However, in the field of professional translation, creativity is sometimes exactly what the client does not want, because it is associated (rightly or wrongly) with requiring more time and introducing inconsistency where consistency is valued more than creative (alternative) solutions. For the translator who prides him- or herself as working in a creative profession, this is difficult to accept. It is probably true that many professionals would like to think of their daily tasks as requiring some form of creativity, but the reality is that there are a great deal more humdrum than eureka moments, whatever the profession, and computerization has arguably added to this.

Until quite recently, translators translated texts, and some still do. However, in some domains the notion of a text, with a beginning, a middle and an end, has changed radically. Translators now frequently work with isolated "chunks", sentences or even "segments" and "sub-segments". This is a result not only of how translation tools broker text but also of the way in which information is now produced—we are moving more and more towards smaller chunks of information delivered in the form of SMS texts, tweets and blogs. Rather than having a simplifying effect on the task of translation, this radical change has resulted in making the task more complex. The linearity of the text, its cohesion, is disrupted (Pym, 2011, p. 3). Contextual clues are missing. Lay on top of this the fact that space limitations can also be imposed, that there is often no forgiveness for languages that happen to take more characters than English to communicate a message, or that the time available for reading can be restricted (e.g. for subtitles), and we have before us a rather complex puzzle that requires creative solutions! In his consideration of man vs. machine, Christian (2011) suggests that perhaps the greatest contribution of humans in an age where computers are automating many tasks will be the craft of coherence. Whereas artificial intelligence (AI) machines are successful at the word, phrase and segment level, they are less successful at text and discourse levels, at voice and register, levels which are, after all, also conduits for meaning. Pym (2011, p. 4) argues that the more technology is part of the equation, the less easy it is to make decisions about the linearity of the text. True, but translators are in an excellent position to compensate for the machine's failures in cohesion, coherence, register, voice and context generally.

21.3 Future directions

Translation as a human–computer interactive task has clearly brought many advantages, arguably to all stakeholders in the translation process and beyond, but this has not happened without significant changes to work practices and serious challenges for the translation profession and for translator trainers (for the latter topic, see also O'Brien & Rodríguez Vázquez, 2019). While it is always interesting to observe what has happened in the past, it is intriguing to contemplate what might happen in the future, and all the more because it would seem that we are living in a time of significant change in general, thanks to advances in AI.

Once upon a time, Kay (1980, p. 11) made the following prediction:

> I want to advocate a view of the problem in which machines are gradually, almost imperceptibly, allowed to take over certain functions in the overall translation process. First they will take over functions not essentially related to translation. Then, little by little, they will approach translation itself. The keynote will be modesty.
>
> At each stage, we will do only what we know we can do reliably. Little steps for little feet!

We have long passed the point Kay predicted. Little steps have turned into considerable leaps. What does the future hold?

Almost a decade ago, the general feeling among researchers was that translators would continue to play a central role in the production of high-quality translation by fine-tuning and repairing MT output. Expectations were that it was unlikely that there would be anything more than incremental advances in performance for the industry as a whole. However, the situation has changed substantially in the past five years or so, with the improvements in NMT, discussed earlier. Some view this as a "game changer" for the use of MT. Others caution that the advances should not be overstated and that there is still a significant need for professional translators for the future. What, exactly, the cognitive task of translation will involve in the future is open for discussion. Perhaps little will change, but the speed of technological advances in the past few years suggests that this is unlikely. Will more translators post-edit more content types for more language pairs? If MT produces reasonable quality for low-stakes content, does that mean that more text will be translated into more languages in the future, but professional translators will only handle high-stakes, highly creative or very sensitive content? These are open questions. What is certain for the next few years is that MT will become an even more dominant technology for professional translators, and we can expect an increased interaction between translators and computers, even for some genres that, heretofore, were deemed untouchable (see Toral & Way, 2015).

Ironically, the future of current TM technology, a long-standing translation aid, is now under question, fuelled by the recent success of MT. TM, as it was known ten years ago, is already changing, since it has now merged with MT. Within most TM tools these days, automatic translation via an MT engine is available as an option in addition to the traditional TM "matches". This means that the traditional differentiation between TM and MT technology is being eroded, as is the differentiation between "translating" with the help of a TM match and "post-editing" MT output. Cognitively, the tasks are getting closer. For instance, a translator can now be offered a TM match, part of which is then enhanced using MT—this is sub-segment machine translation. A translator may no longer know what part of the text comes from TM and what comes from MT, or even which MT engine generated the proposal. Other innovations also continue apace. For instance, "adaptive" MT now forms part of one of the major TM tools. When a sentence from MT is edited, the system learns from it and applies that learning to the sentences that follow. This tackles one of the major weaknesses of MT—generating the same or similar errors repeatedly. From a cognitive perspective, the innovation of "interactive" MT is perhaps even more interesting: traditionally, MT output is delivered as a *fait accompli* to the translator, who then assesses it and edits or retranslates, if necessary. With interactive MT, the translator accepts or rejects words produced on the fly by the MT system, like the functionality of an autocomplete text editor. Very little research has been conducted, from a cognitive perspective, on the implications of this novel way of working with MT. This mode of operation forces the translator to focus on the word more than ever and, operationally, demands that the translator either types a preferred word or just presses a key on the keyboard if they want to accept the proposal. Is this a step backwards from a HCI perspective? Does it reduce the translator to an operator that presses "go" or "no go" buttons? Or is it a very clever extension of the translator's brain? Early empirical research comparing traditional post-editing with interactive post-editing concludes that the interactive mode might be a viable alternative if productivity indicators and translators' qualitative feedback are considered (Sánchez Torrón, 2017), but obviously more research is required.

An area that is ripe for deployment is technology personalization. As was mentioned earlier, translation technology has not typically been developed in conjunction *with* translators, but rather in isolation *for* translators. Its impact on the cognitive task, on ergonomics, on job satisfaction and on the status of the profession has been largely ignored. "One-size-fits-all" has been the dominant approach. Now, in the era of machine learning, it is much more viable to expect AI-driven technology to learn from individual translators and to adapt on their behalf. This concept is known as personalization and is defined as "tailoring products and services to better fit the user" (Göker & Myrhaug, 2002, p. 1). As discussed in O'Brien & Conlan (2019), personalization of translation technology has, conceptually at least, much scope. Imagine a technological aid that learns about the specialized domain a translator is most interested in and finds only appropriate resources, discarding ones that are less relevant or less trustworthy. Imagine an aid that understands the context in which a translation is being produced and tailors its features accordingly. To elaborate, a context that demands very high quality might switch off the MT feature, whereas one that demands sufficient quality, but has an even greater demand on productivity, foregrounds the MT proposals. Or imagine one that learns about the uncertainty tolerances of each individual translator and only presents suggestions based on those tolerances. Imagine a technology that can detect, based on gaze information from eye trackers embedded in our personal devices, when a translator needs assistance and when she or he is in a cognitive "flow" and should not be interrupted with prompts. This is what personalized translation tools could potentially achieve. Personalization is not, however, trivial and does not succeed overnight. It requires the willingness of the user to train the personalization engine, and this takes time. Nonetheless, the potential for more sophisticated, useful HMI is considerable, should translation technology developers and translators decide to collaborate.

Since the invention of word processing, the mouse and keyboard have been the main mechanisms for TCI. Prior to that, the dictaphone was used. With current advances in speech recognition and increasing accuracy for speech to text conversion, it can be expected that new modes of interaction will emerge. This possibility has already gained some traction in our discipline, with researchers investigating user experience of multimodal input, including touch-enabled screens and speech recognition. Findings are reported as being promising, though inconclusive (Teixeira et al., 2019). Although we have some way to go with these new modes of interaction, it is quite likely that they will impact on TCI into the future. The positive aspect of this is that it may help to reduce some of the ergonomic challenges reported in the literature (e.g. Ehrensberger-Dow & Hunziker Heeb, 2016; Ehrensberger-Dow & O'Brien, 2015) by allowing translators to use voice rather than keyboard as input. The translator would thus not necessarily have to sit at his or her desk looking at a screen for many hours in the day. Also, using speech for input and as an output tool increases accessibility and opens up the task of translation to blind translators (Rodríguez Vázquez et al., 2018). It has to be noted, though, that speech input is still not entirely reliable, and it requires a quiet work setting.

All these changes—merging of TM and MT, adaptive MT, interactive MT and multi-modal input—are likely to impact on the process of translation from a cognitive perspective. Considerable research will be required in the coming years to help us understand what that impact is.

Notes

1 This chapter is a reworked and revised version of the following article: O'Brien, S. (2012). Translation as human-computer interaction. *Translation Spaces, 1*, 101–122.
2 Tymoczko deliberately uses the asterisk to differentiate the two concepts.
3 Retrieved 16 July 2019 from www.interaction-design.org/events/external-ux-events/ecce_2010_-_european_conference_on_cognitive_ergonomics
4 Examples of current visual localization tools are Alchemy Catalyst and SDL Passolo.
5 An example of such a tool is the GlobalSight product.
6 Reported by Andrew Joscelyne at the Translation Automation User Society (TAUS) conference, Portland, Oregon, 3–6 October 2010.

Further reading

Forcada, M. (2017). Making sense of neural machine translation. *Translation Spaces, 6*(2), 291–309.
This article explains how neural machine translation works in an accessible way for translation scholars.

O'Brien, S. (2016). Machine translation and cognition. In J. Schwieter, & A. Ferreira (Eds.), *The handbook of translation and cognition*. Oxford – UK: Wiley Blackwell.
This chapter discusses MT from a cognitive perspective, focusing in particular on two types of interaction: MT evaluation and post-editing.
See also in the present Handbook the chapter on *Translation, ergonomics and cognition* by Maureen Ehrensberger-Dow for a discussion of cognitive ergonomics and cognition in more detail as well as the chapter on *Translation, artificial intelligence and cognition* by Michael Carl for a coverage of machine translation in more depth.

References

Bayer-Hohenwarter, G. (2009). Translational creativity: Measuring the unmeasurable. In S. Göpferich, A. Lykke Jakobsen, & I. M. Mees (Eds.), *Behind the mind: Methods, models and results in translation process research* (pp. 39–59). Copenhagen: Samfundslitteratur.
Bayer-Hohenwarter, G. (2010). Comparing translational creativity scores of students and professionals: Flexible problem-solving and/or fluent routine behaviour? In S. Göpferich, F. Alves, & I. M. Mees (Eds.), *New approaches in translation process research* (pp. 83–111). Copenhagen: Samfundslitteratur.
Bowker, L. (2005). Productivity vs. quality: A pilot study on the impact of translation memory systems. *Localization Focus, 4*(1), 13–20.
Bowker, L., & Buitrago Ciro, J. (2019). *Machine translation and global research*. UK: Emerald Publishing.
Cadwell, P., Castilho, S., O'Brien, S., & Mitchell, L. (2016). Human factors in machine translation and post-editing among institutional translators. *Translation Spaces, 5*(2), 222–243.
Cadwell, P., O'Brien, S., & Teixeira, C. S. C. (2017). Resistance and accommodation: Factors for the (non-) adoption of machine translation among professional translators. *Perspectives: Studies in Translation Theory and Practice, 26*(3), 301–321.
Christian, B. (2011). *The most human human: A defence of humanity in the age of the computer*. London: Viking.
Cooper, A. (2004). *The inmates are running the asylum: Why hi-tech products drive us crazy and how to restore the sanity*. Indianapolis: SAMS.

Ehrensberger-Dow, M., & Hunziker Heeb, A. (2016). Investigating the ergonomics of a technologized translation workplace. In R. Muñoz Martín (Ed.), *Reembedding translation process research* (pp. 69–88). Amsterdam: John Benjamins.
Ehrensberger-Dow, M., & O'Brien, S. (2015). Ergonomics of the translation workplace: Potential for cognitive friction. *Translation Spaces, 4*(1), 98–118.
Folaron, D. (2010). Translation tools. In Y. Gambier, & L. Van Doorslaer (Eds.), *Handbook of translation studies*, 1 (pp. 429–436). Amsterdam: John Benjamins.
Forcada, M. (2017). Making sense of neural machine translation. *Translation Spaces, 6*(2), 291–309.
García, I. (2010). Is machine translation ready yet? *Target, 22*(1), 7–21.
Göker, A., & Myrhaug, H. I. (2002). User context and personalization. In M. H. Göker, & B. Smith (Eds.), *Workshop on case based reasoning and personalization*, 6th European conference on case based reasoning (ECCBR) (pp. 1–8). Aberdeen, Scotland.
Goulet, M.-J., Simard, M., Parra Escartín, C., & O'Brien, S. (2017). La traduction automatique comme outil d'aide à la rédaction scientifique en anglais langue seconde: résultats d'une étude exploratoire sur la qualité linguistique. *ASp—la revue du GERAS, 72*, 5–28.
Guerberof, A. (2009). Productivity and quality in the post-editing of outputs from translation memories and machine translation. *Localization Focus: The International Journal of Localization*, 7(1), 11–21.
Haigh, T. (2006). Remembering the office of the future: The origins of word processing and office automation. *Annals of the History of Computing, 28*(4), 6–31.
Hu, K., O'Brien, S., & Kenny, D. (2019). A reception study of machine translated subtitles for MOOCs. *Perspectives—Studies in Translation Theory and Practice*. doi: org/10.1080/0907676X.2019.1595069
Johnson, P. (1992). *Human computer interaction: Psychology, task analysis and software engineering*. Maidenhead: McGraw-Hill.
Karamanis, N., Luz, S., & Doherty, G. (2011). Translation practice in the workplace: A contextual analysis and implications for machine translation. *Machine Translation, 25*(1), 35–52.
Kay, M. (1980). *The proper place of men and machines in language translation*. Research Report CSL-80-11. Palo Alto, California: Xerox Palo Alto Research Center.
Kolko, J. (2010). *Thoughts on interaction design*. Burlington, MA: Morgan Kaufmann.
Krings, H.-P. (2001). *Repairing texts: Empirical investigations of machine translation post-editing processes* (G. S. Koby, Trans.). Kent, OH: Kent State University Press.
O'Brien, S. (2007). An empirical investigation of temporal and technical post-editing effort. *Translation and Interpreting Studies, 2*(1), 83–136.
O'Brien, S. (2016). Machine translation and cognition. In J. Schwieter, & A. Ferreira (Eds.), *The handbook of translation and cognition*. Oxford, UK: Wiley Blackwell.
O'Brien, S., & Conlan, O. (2019). Moving towards personalising translation technology. In H.V. Dam, M. N. Brøgger, & K. K. Zethsen (Eds.), *Moving boundaries in translation studies* (pp. 81–97). Oxford: Routledge.
O'Brien, S., Ehrensberger-Dow, M., Hasler, M., & Connolly, M. (2017). Irritating CAT tool features that matter to translators. *Hermes Journal of Language and Communication in Business, 56*, 145–162.
O'Brien, S., Federico, F., Cadwell, P., Marlowe, J., & Gerber, B. (2018b). Language translation during disaster: A comparative analysis of five national approaches. *International Journal of Disaster Risk Reduction, 31*, 627–636.
O'Brien, S., & Rodríguez Vázquez, S. (2019). Translation and technology. In S. Laviosa, & M. González-Davies (Eds.), *The Routledge handbook of translation and education*. London: Routledge.
O'Brien, S., Simard, M., & Goulet, M.-J. (2018a). Machine translation and self-post-editing for academic writing support: Quality explorations. In J. Moorkens, S. Castilho, S. Doherty, & F. Gaspari (Eds.), *Translation quality assessment: From principles to practice* (pp. 237–262). Germany: Springer.
O'Hagan, M. (2009). Evolution of user-generated translation: Fansubs, translation hacking and crowdsourcing. *Journal of Internationalization and Localization, 1*, 94–121.
Olohan, M. (2011). Translators and translation technology: The dance of agency. *Translation Studies, 4*(3), 342–357.
Parra Escartín, C., O'Brien, S., Goulet, M.-J., & Simard, M. (2017). Machine translation as an academic writing aid for medical practitioners. In *Proceedings of the machine translation summit XVI*, Nagoya, Japan, September, pp. 254–267.
Pym, A. (2011). What technology does to translating. *The International Journal for Translation and Interpreting Research, 3*(1), 1–9.
Rodríguez Vázquez, S., Fitzpatrick, D., & O'Brien, S. (2018). Is web-based computer-aided translation (CAT) software usable for blind translators? In K. Miesenberger, & G. Kouroupetroglou (Eds.), Lecture

notes in computer science (LNCS vol. 10896) (pp. 31–34). In *International conference on computers helping people with special needs (ICCHP 2018)*. Germany: Springer.
Sánchez Torrón, M. (2017). Productivity in post-editing and in neural interactive translation prediction: A study of English-to-Spanish professional translators [PhD thesis, University of Auckland, New Zealand]. https://researchspace.auckland.ac.nz/handle/2292/37205
Teixeira, C. S. C., Moorkens, J., Turner, D., & Vreeke, J. (2019). Creating a multimodal translation tool and testing machine translation integration using touch and voice. *Informatics, 6*(1), 13.
Toral, A., & Way, A. (2015). Machine-assisted translation of literary text: A case study. *Translation Spaces, 4*(4), 241–268.
Tymoczko, M. (2007). *Enlarging translation, empowering translators*. Manchester: St. Jerome.
Van der Meer, J. (2010). Translation in the 21st century. Retrieved 17 July 2019 from www.slideshare.net/TAUS/translation-in-the-21st-century-webinar
Vashee, K. (2010). What is holding the wider adoption of MT back? The empty pages blog, Retrieved 17 July 2019 from http://kv-emptypages.blogspot.com/2010/07/what-is-holding-wider-adoption-of-mt.html

22

Translation competence and its acquisition

Amparo Hurtado Albir

22.1 Introduction

In recent decades, Translation Studies has been concerned with describing the knowledge and abilities translators need to translate appropriately This has resulted in research evolving around the notion of translation competence and translation competence acquisition.

The notion of competence itself has a long history as the subject of analysis in other disciplines such as applied linguistics, work psychology and pedagogy. The concept of "communicative competence"[1] has been used in applied linguistics since the mid-1960s, with a long line of research by scholars such as Hymes (1971), Canale-Swain (1980), Canale (1983), Widdowson (1989), Spolsky (1989), Bachman (1990), etc. In the sphere of recruitment in the job market, research into professional competences dates as far back as the beginning of the 1970s in the field of work psychology with McClelland (1973) and the development of a behavioural approach to the study of competences. This was followed by various studies (Boyatzis, 1982, 1984; Spencer et al., 1994; etc.), which established competence models (known as "competency dictionaries") for specific job profiles based on studying professionals who perform well in the tasks required for these posts. Since the turn of the 21st century, a new pedagogical model known as competence-based training (CBT) has gained support. As Lasnier (2000, p. 22) points out, this is the logical continuation of the previous model (objectives-based learning). In CBT, competences are the core of curriculum design, and it advocates an integrated model of teaching, learning and evaluation. The basis for CBT can be found in cognitive-constructivist and socio-constructivist learning theories.

In Translation Studies there is no research tradition comparable to that in other disciplines. The study of the notion of translation competence (TC) started within the field of Translation Studies in the mid-1980s and began to figure more predominantly in the 1990s. Alongside this, research has also been developed on translation competence acquisition (TCA), although fewer models have been proposed than in the case of TC. In both cases, however, little empirical research has been developed.

22.2 Translation competence

Two major periods in the evolution of research into TC can be distinguished: the first until the end of the 1990s, which was the beginning of studies into TC; and the second from 2000 onwards, which was a period of consolidation and the beginning of empirical validation.[2]

22.2.1 Early studies on translation competence

With the exception of Wilss (1976) and Koller (1979), pioneers in this area, the study of the notion of TC began within the field of Translation Studies in the mid-1980s.

The 1980s and 1990s witnessed the first proposals for TC models; the majority of these were componential models, which focused on describing the components that make up TC. These were neither specific nor extensive studies on TC; most authors only dealt with this topic peripherally, and many of them derived from interest in curriculum design. Nevertheless, their major interest lies in the fact that they represent early reflections on the characteristic workings about TC and its components.

22.2.1.1 The components of translation competence

According to Wilss (1976, p. 120), translators must have three competences: source-language receptive competence (the ability to decode and understand the source text); target-language reproductive competence (the ability to use linguistic and textual resources in the target language); and super-competence (the ability to transfer messages between the source and target culture linguistic and text systems). Wilss defines TC as "the ability to integrate the two monolingual competences on a higher level, i.e. on the level of the text" (Wilss, 1982, p. 58).

Delisle (1980, p. 235) proposed that four competences were needed to know how to translate, although he did not make use of the term TC: linguistic competence, encyclopaedic competence, comprehension competence and reformulation competence.

Roberts (1984) differentiated five competences: linguistic competence—the ability to understand in source language and formulate in target language; transfer competence (*traductionnelle*)—the ability to grasp the meaning and to reformulate it, avoiding linguistic interference(s); methodological competence—the ability to find, document and assimilate suitable terminology; thematic competence; and technical competence—the ability to use different resources and tools to translate.

Hewson and Martin (1991, p. 52) distinguished three types of competences in a translator: acquired interlinguistic competence, i.e. linguistic competence in the two languages; a dissimilative competence, which consists of the aptitude to generate and dissimilate homologous statements and to define and recreate socio-cultural norms; and transferred competence, which includes not only what the translator knows but also that which is accumulated through dictionaries, data banks, etc.

In two studies, Nord distinguished three components in TC. In the first study, these are transfer competence, linguistic competence and cultural competence (Nord, 1988/1991, p. 161). In the second, Nord (1992, p. 47) sets out the following essential competences: reception competence and text analysis, research competence, transfer competence, text production competence, translation quality assessment competence, and linguistic and cultural competence—source and target.

Neubert (1994, p. 412) proposed three sub-competences in TC (which he called "translational competence"): linguistic competence, subject competence and transfer competence.

In particular, he stressed the importance of transfer competence as that which distinguishes translators from other professionals and governs the other competences.

Kiraly (1995, p. 108) offered an "integrated model" for translator competence, which incorporated: (1) knowledge concerning situational factors that may be involved in a given translation task; (2) translation-relevant knowledge that the translator possesses, i.e. linguistic knowledge in the source and target languages (syntactic, lexico-semantic, socio-linguistic and textual), cultural knowledge related to the source and target languages, and specialized knowledge; and (3) the translator's ability to begin the appropriate intuitive and controlled psycholinguistic processes to formulate the target text and monitor its adaptation to the original text.

Hurtado (1996a, p. 34; 1996b, p. 39) defined TC as the "ability to know how to translate"[3] and distinguished five sub-competences: (1) linguistic competence in the two languages, composed of source-language comprehension and target-language production—written for a translator, or oral for an interpreter; (2) extra-linguistic competence, i.e. encyclopaedic, cultural and thematic knowledge; (3) transfer competence, which consists of knowing how to work through the translation process correctly, in other words knowing how to understand the original text and reformulate it in the target language according to the purpose of the translation and characteristics of the target reader; (4) professional competence, which consists of knowing how to document and how to use new technologies and knowing the job market; and (5) strategic competence: conscious and individual procedures used by translators to solve problems encountered during the translation process according to their specific needs.

Presas (1996) highlighted the distinction between TC and bilingual competence, considering TC as specific to reception and text production. According to Presas, TC is founded in a "pre-translation competence", which consists of: knowledge of both languages, cultural awareness concerning these two languages, encyclopaedic knowledge, thematic knowledge and theoretical knowledge about translation. She pointed out two types of knowledge that make up and characterize TC: epistemic knowledge and operative knowledge. Epistemic knowledge includes knowledge of the two languages, such as cultural, encyclopaedic and thematic knowledge. As regards operative knowledge, she distinguished between nuclear, peripheral and tangential knowledge. Nuclear knowledge consists of reception of the source text for translating (identifying the "distances" and translation problems); setting up the translation project; and producing the translation. Peripheral knowledge refers to the specific instruments used by the translator and includes, among other things, the capacity to evaluate and use documentation sources; the capacity to acquire knowledge related to new or unfamiliar thematic areas; and the capacity to evaluate other translations. Finally, tangential knowledge refers to the ability to use standard work tools (text editing technology and desktop publishing). Presas underlines the importance of the relationships between these fields of knowledge.

Hansen (1997) made a distinction between implicit (automatized, unconscious) and explicit (conscious) knowledge and abilities. She distinguished three interacting sub-competences: translational competence; social, cultural and intercultural competence; and communicative competence. Translational competence consists of two competences: implicit and explicit. This author defines implicit translational competence as the ability to extract pertinent information from the source text considering the intention for which the translation has been commissioned, and to produce the target text so that it complies with the intended purpose. Explicit translational competence comprises the explicit knowledge of translation methods and the ability to choose the most appropriate, as well as strategies to recognize and solve translation problems. Social, cultural and intercultural competence include both implicit and explicit knowledge: implicit knowledge (socially and culturally conditioned) of one's own social and cultural context and of other contexts, as well as explicit knowledge of the social and cultural norms and differences. Finally, communicative competence includes pragmatic competence and linguistic competence.

Risku's (1998) is the most in-depth study on TC. This concerns a model along "pragmatic-cooperative" lines. Risku proposes a modular conception of TC, comprising four sub-competences which work together to construct meaning: setting up the macro-strategy, integrating information, planning and decision making, and self-organization. The purpose of the macro-strategy is to anticipate the communication context for the translation. Integrating information facilitates creating and contrasting representations of situations in the source text and the translation, as well as evaluating documentation. Planning and decision making ensure intratextual coherence and contrastivity. Finally, self-organization allows reflection on, and continual assessment of, decisions taken.

22.2.1.2 Translation "abilities" and "skills"

Some authors prefer to use the term "translation skill" or "translation ability". One such example is Lowe (1987), who used the term "translation skill" and distinguished eight categories of knowledge and skills which map out the ideal translator profile: source-language reading comprehension; ability to produce target-language texts; understanding the source-language style; mastery of the target-language style; comprehension of socio-linguistic and cultural aspects in the source language; mastery of socio-linguistic and cultural aspects in the target language; speed; and the "X factor", which according to Lowe refers to a quality that is difficult to define, but which renders a translation clearly superior to others given an equal rating (Lowe, 1987, p. 55).

Pym (1991, 1992), in turn, pointed out two "translation skills" that make up TC, which consist of the ability to generate different options for the source text, and the ability to choose one based on the specific end purpose and the target reader (Pym, 1992, p. 281).

The description by Hatim and Mason (1997, p. 205) of "translator abilities" is based on Bachman's model (1990) of "communicative ability": organizational competence, pragmatic competence and strategic competence. They envisage a three-phase translation process (processing the source text, transfer and processing the target text) and assign a series of skills to each of them, although they point out that during the translation process these skills interact. The source-text processing phase, depending on the estimated effect of the source text on the reader, requires recognizing intertextuality and situationality; inferring intentionality, organizing texture and the text structure; and evaluating the informativity depending on whether these are "static texts" (easy to process because they conform to pre-established text norms and with end reader expectations) or "dynamic texts" (difficult to process because they do not comply with text norms or reader expectations). In the transfer phase, a strategic re-negotiation is developed, in which adjustments are made regarding efficiency, effectiveness and the relevance of the communicative task of the translator concerning specifying this task in order to fulfil a given rhetorical purpose. Target-text processing requires establishing intertextuality and situationality, creating intentionality, organizing the texture and structure of the text, and balancing the informativity on the basis of its impact on the target-text reader.

Vienne (1998) criticized the tendency to reduce TC to linguistic competences (text analysis and production) and advocated a definition of the abilities a professional translator needs in a given translation situation from a functionalist approach. This particular focus on training translators deliberately omits linguistic competences and focuses on four aspects: ability to analyse different translation situations; ability to manage and process information; ability to discuss the decisions taken with whoever commissions the translation; and ability to cooperate with other experts.

22.2.1.3 Translation competence and translation proficiency

Working from Chomsky's distinction (1965) between competence and performance, Cao (1996) distinguished between TC and translation proficiency. She defined TC as many kinds of knowledge that are essential to the translation act (Cao, 1996, p. 326). Translation proficiency was defined as "the ability to mobilize translation competence to perform translation tasks in context for purposes of intercultural and interlingual communication" (Cao, 1996, p. 327). She considered translation proficiency as a global skill with various components for performing translation tasks.

She proposed three components for translation proficiency based on Bachman's model of communicative language ability (Bachman, 1990): translational language competence, translational knowledge structures and translational strategic competence.

Likewise, her description of translational language competence was based on Bachman and included SL and TL organizational competence, consisting of grammatical and textual competence, and SL and TL pragmatic competence, consisting of illocutionary and socio-linguistic competence.

Translational knowledge structures include general, special and literary knowledge, akin to Snell-Hornby's (1988) translation prototypology. General knowledge refers to knowledge about the world: ecology, material culture, social organization, etc. in the SL and TL language communities. Special knowledge includes specialist technical knowledge in different fields. Literary knowledge includes knowledge in areas including the Bible, stage, film, lyric, poetic and literary works, cultural history and literary studies.

Cao saw translational strategic competence as a mental ability that provides the means to relate the various components of translation proficiency during the translation process. This comprises two component parts: (1) assessment, planning and executive abilities inherent in all mental activities, including language use; (2) the skills demanded during the processing and non-verbal stage of reformulation and analogy by reasoning, specific to translation. This also includes psychological mechanisms: the cognitive aspect of human thought processes as well as the creative aspect.

Cao highlights the interaction between the various components of translation proficiency and the essential role of translational strategic competence. She concludes that translation is a special area of expertise that requires knowledge and specialization in many different areas, and points out that there are different levels of proficiency.

Cao's distinction between TC (knowledge that is essential to the translation act) and translation proficiency (ability to mobilize TC to perform translation tasks) is not shared by all those TC models that include abilities to perform translation tasks and highlight the fundamental role of strategic competence.

22.2.1.4 Translation competence and expertise

According to Bell (1991), there are three possible ways of characterizing TC. The first is in terms of ideal bilingual competence in the Chomskyan sense, which according to Bell is inadequate. The second is as an expert system, i.e. generalizations based on observations of how translators work, which entails two basic components: (1) a knowledge base in the source and target languages, text types, domain knowledge and contrastive knowledge; (2) an inference mechanism to decode and code texts. The third is related to the second and adopts a multi-component approach to communicative competence. He distinguishes between grammatical, socio-linguistic, discourse and strategic competence, along the lines of Canale and Swain (1980), concluding that a translator must have linguistic competence in the two languages and communicative competence in the two cultures. Bell (1991, p. 43) defines translator competence as the "knowledge and skills the translator must possess in order to carry it [the translation process] out".

Gile (1995, pp. 4–5) analyses the components of translation expertise required for interpreting and translation, differentiating between the following areas of knowledge and skills: good passive knowledge of their passive working languages; good command of their active working languages; enough knowledge of the subjects of the texts or speeches they process; and the know-how to translate, referring to the conceptual framework and the interpreting and translation technical skills (comprehension of principles of fidelity and of professional rules of conduct as well as techniques for knowledge acquisition, language maintenance, problem solving, decision making, etc.).

The PACTE research group was set up in 1997 to carry out experimental research on TC acquisition. PACTE considers this process in terms of the development from novice to expert knowledge (PACTE, 2000, p. 103). It puts forward a holistic TC model that was first presented in 1998.[4] In this first version of the model (PACTE, 2000), six interrelated, hierarchical sub-competences are proposed: (1) communicative competence in two languages; (2) extra-linguistic competence, including knowledge about translation as well as encyclopaedic and domain knowledge; (3) instrumental-professional competence, including both knowledge and skills related to the tools of the trade and the profession (knowledge and use of all kinds of documentation sources and new technologies; knowledge of the market and how a translator behaves professionally, especially in relation to professional ethics); (4) psycho-physiological competence, i.e. the ability to use all kinds of psychomotor, cognitive and attitudinal resources; (5) transfer competence, i.e. the core competence that integrates all the others, the ability to complete the transfer process from the ST to the TT; (6) strategic competence, including all procedures, conscious and unconscious, verbal and non-verbal, used to solve problems found during the translation process. Although all these competences are involved in TC, in PACTE's 1998 model, transfer and strategic competences played a predominant role in the interrelation between competences.

22.2.1.5 Translating into L2 (inverse translation)

In reference to translating into L2, Beeby (1996, pp. 91–92) talks of an "ideal translator communicative competence", which, broadly speaking, distinguishes four ideal sub-competences: (1) ideal translator grammatical competence—including linguistic knowledge and skills necessary to understand and express the literal meaning of utterances; (2) ideal translator socio-linguistic competence—the knowledge and ability necessary to produce and understand utterances adequately in the situational context of both cultures; (3) ideal translator discourse competence—the ability, in both languages, to produce formal cohesion and coherent meaning in different text genres; and (4) ideal translator transfer competence—mastering communication strategies that enable the transfer of meaning from the source to the target language and which may be used to improve communication or compensate for breakdowns. Beeby points out, however, the specificity of the translation towards the foreign language as regards the job market and capabilities.

Campbell (1998) also proposes a TC model for inverse translation, which comprises three relatively independent elements (1998, p. 152): target-language textual competence, "disposition" and "monitoring". For Campbell, target-language textual competence is a central consideration in inverse translation. He uses the term "disposition" to refer to the (non-linguistic) capacity, which consists of the way the translation task is approached. Monitoring competence refers to the capacity to monitor and supervise the product. Campbell points out that these three components are reflected in the following questions (1998, p. 155): Can a translator produce stylistically appropriate translations in the target language? Does the translator have the right personality to be able to translate? Is the translator capable of producing a text that requires only the minimum of proofing?

22.2.1.6 Key features of this period

The following summarizes the most characteristic features of this first period (Table 22.1).

1. Focus on the description of components: The transfer competence. The merit of these early studies is in highlighting that TC requires additional competences beyond linguistic competences. The majority focus on describing the components of TC, proposing different components: language skills; extra-linguistic knowledge, documenting skills and the use of tools; and transfer competence. Proposing transfer competence as a component of TC is characteristic of this period.
   These studies also illustrate that these components cover a wide range of areas: knowledge, abilities, skills and attitudes. In addition, some authors underline the procedural nature of TC and talk rather of abilities, distinguishing between declarative and operative components and stressing the importance of the strategic component (Beeby, 1996; Cao, 1996; Hatim & Mason, 1997; Hurtado, 1996a, 1996b; PACTE, 2000; Presas, 1996).
2. Translation competence as a form of expertise. A few authors (Bell, 1991; Cao, 1996; Gile, 1995; PACTE, 2000) relate TC to expertise, although during this period they seem to confuse both concepts and do not clearly distinguish between TC and translation expertise.
3. Specificity of translation into L2 (inverse translation). Some authors (Beeby, 1996; Campbell, 1998) take into consideration the specificity of TC in the case of translating into L2.
4. Lack of specific studies. With the exception of some studies, such as those by Cao (1996), Presas (1996) or Risku (1998), the majority of these initial TC proposals are isolated references, which deal with the issue tangentially.
5. Dearth of definitions and legions of labels. What first needs to be pointed out is the lack of definitions for TC; as we have seen, many authors cover the topic of TC, but only a few offer definitions: Wilss (1982); Bell (1991); Cao (1996); Hurtado (1996a, 1996b). A certain diversity can also be seen as regards the terms used to refer to TC: transfer competence (Nord, 1988/1991, p. 160), translational competence (Chesterman, 1997, p. 147; Hansen, 1997, p. 205; Pym, 1993, p. 26; Toury, 1995, p. 250), translator's competence (Kiraly, 1995, p. 108), translation ability (Lowe, 1987, p. 57; Stansfield et al., 1992), translation skills (Lowe, 1987, p. 57) and translation expertise (Gile, 1995, p. 4).
6. Lack of empirical studies. Empirical-experimental studies on written translation began towards the end of the 1980s and were developed in the 1990s. But rather than focusing on a holistic approach to TC, they deal with partial elements: linguistic knowledge of the translator, extra-linguistic knowledge, abilities and aptitudes such as creativity and emotivity, attention, the role of documentation, strategies used, etc.
   Orozco (2000, p. 113) points out that the study by Stansfield et al. (1992) is in fact the only effective attempt to operationalize TC, in their terms "translation ability". This research was commissioned by the FBI to create an instrument for determining the TC level of candidates applying for posts as translators. This was known as the "Spanish into English Verbatim Translation Exam" (SEVTE) and was subjected to validity and reliability tests. However, the authors themselves point out that the results cannot be used to make generalizations, given the small sample group used (seven FBI employees). In addition, Orozco (2000, p. 116) raises two objections to this instrument: there is no definition of TC, and so there is no way of knowing whether it manages to measure what they wanted to measure; and the instrument itself, since at no point were the subjects required to translate complete texts; rather, they were given words or segments from phrases, or phrases or paragraphs.

Table 22.1 Research into translation competence until 2000

Emergence of early translation competence models
Focus on the description of components: language skills; extra-linguistic knowledge; documentation skills and use of tools; transfer competence
Inclusion of the transfer component as a specific sub-competence
Initial proposals of translation competence as a form of expertise; confusion concerning both concepts
Lack of specific studies
Lack of definition and terminological diversity
Lack of empirical studies

22.2.2 Consolidation of research on translation competence

Moving forward to the new millennium, the number of studies on TC increases considerably and now plays a more important role in Translation Studies research, so TC is established as the object of specific studies. A significant publication is *Developing translation competence* (edited by Schäffner & Adab, 2000).

During this period, the approach is to consider TC as basically procedural knowledge (abilities, skills and strategies) and the importance of the strategic component. It is at this point that we see early attempts to establish differences between TC and translation expertise. A more interdisciplinary framework was established, since many of the proposals were based on research carried out in other disciplines. Furthermore, empirical validations were being designed.

22.2.2.1 Revisiting previous models

Neubert (2000) developed his 1994 proposal underlining the complexity and heterogeneity of TC, pointing out that TC comprises seven main characteristics:

(1) Complexity—translation is a complex activity and differs from the rest of the language-related professions.
(2) Heterogeneity—it implies developing abilities which are very disparate.
(3) Approximation—the impossibility of knowing all the thematic fields that can be translated and the need to have recourse to other disciplines.
(4) Open-endedness—the constant demand to be up to date.
(5) Creativity—to solve certain translation problems.
(6) Situationality—to adapt to new translation situations (purpose, commissions).
(7) Historicity, capacity to change, to be able to adapt oneself to other ways of focusing the translation, given space-time changes.

Neubert highlights that in order to accomplish this complex task, translators need expertise that distinguishes them from other language users (2000, p. 5). He establishes five parameters that make up TC: language competence, textual competence, subject competence, cultural competence and transfer competence. As he had already pointed out in his 1994 proposal, transfer competence is the competence which distinguishes translation from any other communicative activity.

Pym (2003) likewise reconsiders the question of TC. He states that proposals since the 1970s had approached TC from four perspectives: as some mode of bilingualism, subject to linguistic analysis; as a result of market demands, subject to social and historical changes; as a multi-component competence, comprising linguistic, cultural, technological and professional skills; and as a vague super-competence which transcends the other components. Pym is critical of the componential models of TC and advocates a minimalist concept based on producing and then eliminating alternatives. This draws on his earlier proposal in 1991: (1) the ability to generate a series of more than one viable target text (TTI, TT2 ... TTn) for a pertinent source text, and (2) the ability to choose only one viable TT from this series, quickly and with justified confidence (2003, p. 489).

Results from PACTE's exploratory studies (PACTE, 2002, 2003) with six professional translators resulted in modifications to the first version of the group's TC model. Changes in the model have a bearing on the sub-competences that make up TC; furthermore, the sub-competences are defined in terms of declarative and procedural knowledge to indicate their predominance in each sub-competence. As a result, PACTE (2003) adjusted the definition of the sub-competences as follows:

(1) Bilingual sub-competence. Predominantly procedural knowledge required to communicate in two languages (pragmatic, socio-linguistic, textual, grammatical and lexical knowledge).
(2) Extra-linguistic sub-competence. Predominantly declarative knowledge about the world in general and field specific (bicultural knowledge, encyclopaedic knowledge and subject knowledge).
(3) Knowledge of translation sub-competence. Predominantly declarative knowledge about what translation is and knowledge about the profession.
(4) Instrumental sub-competence. Predominantly procedural knowledge related to the use of documentation resources and information and communication technologies applied to translation.
(5) Strategic sub-competence. Procedural knowledge to guarantee the efficiency of the translation process and solve problems encountered. This is an essential sub-competence that affects all the others, since it creates links between the different sub-competences as it controls the translation process.
(6) Psycho-physiological components. Different types of cognitive and psycho-attitudinal components and mechanisms. They include cognitive components, such as memory, perception, attention and emotion; and attitudinal aspects, such as intellectual curiosity, perseverance, rigour, critical spirit, knowledge about and confidence in one's own abilities, the capacity to measure one's own abilities, motivation, etc.

22.2.2.2 Different approaches

The majority of the proposed models during this period are componential and deal with TC from different perspectives.

22.2.2.2.1 Didactic approach

From a didactic viewpoint, Kelly (2002, 2005, 2007) offers an integral approach to TC for curriculum design. Kelly (2005, p. 162) defines TC as the set of knowledge, skills, attitudes and aptitudes which a translator possesses in order to undertake professional activity in the field. She describes the components of TC (2005, pp. 32–33) as (1) communicative and textual competence in at least two languages and cultures; (2) cultural and intercultural competence; (3) subject area competence, i.e. basic knowledge of subject areas the future translator will/may work in; (4) professional and instrumental competence, i.e. the use of all manner of documentary resources, use of IT tools for professional practice, basic notions for managing professional activity, etc.; (5) attitudinal or psycho-physiological competence (self-concept, self-confidence, initiative, etc.); (6) interpersonal competence; and (7) strategic competence, i.e. organization and planning skills, problem identification and problem solving, monitoring, self-assessment and revision. The specificity of her proposal lies in introducing interpersonal competence as a separate competence: ability to work with other professionals and actors involved in the translation process, including teamwork, negotiation and leadership skills. Kelly (2002, p. 15) graphically illustrates her proposal by laying out the sub-competences that make up TC in the form of a pyramid model and emphasizing the role of the strategic sub-competence, located at the tip of the pyramid.

González Davies and Scott-Tennent (González Davies, 2004, pp. 74–75; González Davies & Scott-Tennent, 2005, p. 162) propose six aspects a translator should know: language work (source language/s and target language/s), subject matter (encyclopaedic knowledge related to different disciplines), translation skills (problem spotting and problem solving, creativity, self-confidence, etc.), resourcing skills (paper, electronic and human), computer skills and professional skills (translator's rights, contracts, etc.).

Katan (2008) puts forward a multi-component list of competences (2008, pp. 133–135) that can be employed in teaching specialized translation, based on previous proposals (PACTE, 2003; Pym, 2003; Schäffner, 2004). His proposal concerns two major blocks: lingua culture-specific competences and translation competences. The lingua and culture competences include competences related to textual competence (comprehension in the source language and production in the target language) and extra-linguistic competence (bicultural knowledge and knowledge of specialized topics). Translation competences include those related to general transfer/mediation competence (knowledge of the theories of translation, the ability to decide on a translation strategy, etc.); strategic transfer/mediation competence (special language related to specialist topics and linguistic/literary devices as compensation strategies, rhetorical strategies, etc.); instrumental/professional competence (knowledge and skills relating to professional translation practice); and attitudinal competence (flexibility, creativity, etc.).

In 2009, the European Master's in Translation (EMT)[5] established a translator competence profile framework and set out the competences translators need in order to work successfully in the professional translation market. It distinguishes six types of competences and sets out the corresponding components for each of them: translation service provision competence (interpersonal and production dimension); language competence (L1 and one's other working languages); intercultural competence (socio-linguistic and textual dimension in the comparison of and contrast between discursive practices in L1, L2 and L3); information mining competence; thematic competence; and technological competence (mastery of tools). In 2017, the model was revised and the following competences were proposed: language and culture (transcultural and socio-linguistic awareness and communicative skills); translation (strategic, methodological and thematic competence); technology (tools and applications); personal and interpersonal; and service provision.

The PACTE model (2003; see above) is also composed from a didactic perspective.

22.2.2.2.2 Relevance-theoretic approach

Gutt (2000), from a relevance-theoretic perspective of translation, defends a competence-oriented research of translation (CORT). The aim of CORT is to understand and explicate the mental faculties that enable human beings to translate in the sense of expressing in one language what has been expressed in another.

Also, from the relevance-theoretic perspective, and taking into consideration proposals from connectionist approaches, Gonçalves (2003, 2005) and Alves and Gonçalves (2007) differentiate between general translator competence and specific translator competence. General translator competence is defined as all knowledge, abilities and strategies a successful translator masters and which lead to adequate translation task performance. Specific translator competence, however, operates in coordination with other sub-competences and works mainly through conscious or metacognitive processes, being directly geared to the maximization of "interpretive resemblance".

22.2.2.2.3 Expertise studies approach

Shreve (2006) believes TC should be analysed within the scope of Expertise Studies. He focuses on TC as translation expertise and defines it as the multiple translation-relevant cognitive resources to perform a translation task (2006, p. 28). Shreve proposes that this competence could be seen as declarative and procedural knowledge from a variety of cognitive domains accumulated through training and experience and then stored and organized in a translator's long-term memory (2006, p. 28).

Shreve (2006, p. 40) maintains that an expertise-oriented model could assume that "knowing how to translate" implies having access to (1) L1 and L2 linguistic knowledge; (2) culture knowledge of the source and target culture, including knowledge of specialized subject domains; (3) textual knowledge of source and target textual conventions; and (4) translation knowledge—knowledge of how to translate using strategies and procedures including tools and information-seeking strategies. These four cognitive areas have to be integrated to satisfactorily complete the translation task. He adds that identifying these four sub-competences implies that translation expertise can be developed differently according to variations in how further experience in the domain of practice is acquired. He concludes that translation expertise is not "a homogeneous, easily describable set of uniform cognitive resources achieved by all translation experts" (Shreve, 2006, p. 40).

Göpferich (2008, p. 155) argues along the lines of expertise research and works from the PACTE model (2003), albeit with some modifications. Göpferich (2009) proposes a TC model as a point of reference for the TransComp project (a process-oriented longitudinal study that explores the development of TC). She distinguishes the following six sub-competences (Göpferich, 2009, pp. 21–23): communicative competence in at least two languages, which corresponds to PACTE's bilingual sub-competence; domain competence, which corresponds approximately to PACTE's extra-linguistic sub-competence; tools and research competence, which corresponds to PACTE's instrumental sub-competence; translation routine activation competence, which comprises the knowledge and the abilities to recall and apply certain (standard) transfer operations (or shifts), which frequently lead to acceptable target-language equivalents (corresponding to the ability to activate productive micro-strategies proposed by Hönig, 1991, 1995); psychomotor competence, which is the psychomotor abilities required for reading and writing (with electronic tools); and strategic competence, which corresponds to the PACTE's strategic competence and controls the application the other sub-competences.

She points out that employing and controlling sub-competences is determined by three factors: the translation brief and translation norms; the translator's self-concept/professional ethos; and the translator's psycho-physical disposition (intelligence, ambition, perseverance, self-confidence, etc.).

22.2.2.2.4 Knowledge management approach

Risku et al. (2010) work from a knowledge management perspective[6] and consider translation as expert knowledge. They pose the need for a new professional translator profile so that translators can take on their role in knowledge management endeavours and generate intellectual capital in the knowledge society. They classify the types of knowledge required of a translator into five categories (Risku et al., 2010, pp. 88–91): language, linguistic and text skills and translation competence; country and cultural knowledge; general and subject matter knowledge; client and business knowledge; and information technology and computer skills.

They assign a series of factors for each category to investigate the extent to which these types of knowledge are, or are not, codifiable and identify appropriate knowledge management tools and instruments: codifiable aspects (e.g. grammar and technology); knowledge management instruments for codifiable aspects (e.g. glossaries and translation memories); non-codifiable aspects (e.g. tacit understanding of context and variations in meaning); and Knowledge Management instruments for non-codifiable aspects (e.g. mailing lists and online communities).

22.2.2.2.5 Professional and behavioural approach

Gouadec, in various studies, has examined translation from the professional viewpoint (see, for example, 2002, 2005, 2007). He points out the following prerequisites and conditions for a good translator (Gouadec, 2007, p. 150): absolutely perfect mastery of the languages used, especially the target language; multicultural competence (in the broad sense, including technical culture, business culture, corporate culture, etc.); perfect familiarity with the domains they specialize in; absolute knowledge of what translation means, what it requires and what it implies; and doing the job as professionally as possible.

The behavioural approach has also been used in some TC studies, although it has had little impact to date.[7] Rothe-Neves (2005) put forward an empirical proposal based on the behavioural approach first introduced by McClelland (1973), with the aim of designing a model of competences for the translation profession (with its various areas of specialization) based on systematic observation of translators who perform well.

Surveys have also been carried out about which competences a good translator or interpreter requires according to their professional profile to satisfy the professional demands, and offers a repertoire of competences. For example, Mackenzie (2000) presents the results of the Practical Orientation of Studies in Translation and Interpreting (POSI) project carried out in Finland, aimed at users and providers of translation services; another example is Calvo Encinas (2004), who focuses on the profile of the community interpreter (carried out in the province of Toledo, Spain).

22.2.2.3 Key features of this period

As we have seen, with the crossing over to the 21st century, research into the functioning of TC is taking on greater importance in Translation Studies. There are various proposals concerning this, and various analysis perspectives. The following is a summary of the main characteristics of this period (Table 22.2):

1. Range of approaches. TC models have been designed with various aims in mind: to be used in curriculum design; with a view to performance in the job market; and with theoretical objectives to discover the function of the competences required of, and which identify, a translator. Most propose similar components for TC; however, they differ in their approach, the terminology used, and the distribution and importance given to these components. This disparity of criteria, however, serves to stress how complex it is to describe TC and the variety of sub-components it comprises.

   The majority of these models are cognitive in nature, but there are also some based on a behavioural approach. Both approaches to studying TC (what it is needed to "know how to do" to be a translator and what the translators "do") are complementary when it comes to describing the workings of TC.
2. The importance of the procedural component and of strategic competence. As opposed to the earlier period, the majority of models concern the procedural nature of TC and include strategic competence as an essential component to be able to resolve translation problems.

Table 22.2 Research into translation competence since 2000

Consolidation of translation competence as an object of study
Diversity of perspectives
- Minimalist approach
- Didactic approach
- Relevance-theoretic approach
- Expertise Studies approach
- Knowledge management approach
- Professional and behavioural approach
Consideration of the importance of the procedural component and the strategic competence
Linking the study of translation competence with Expertise Studies and establishing the differences between translation competence and translation expertise
Beginnings of empirical validation

3. Linking the study of TC with expertise studies and establishing the difference between TC and translation expertise. As was the case during the previous period, some authors relate TC to translation expertise (Göpferich, 2008, 2009; Shreve, 2006), establishing links between studies into TC and expertise studies. However, in this period, advances are made in establishing the features that characterize translation expertise and how they differ from TC. By way of example, PACTE carried out an additional study with the best translators from the sample in their experiment on TC. This further study (PACTE, 2017) clearly demonstrated that the results from this group, in the majority of indicators, were higher than among the rest of the translators. Features in this group were also found which, according to Expertise Studies, characterize experts and can serve as a basis to differentiate between TC and translation expertise (PACTE, 2017, pp. 293–294): superior performance; qualitative differences in the representation of knowledge; more highly developed structuring and interconnection of knowledge; more highly developed procedural knowledge; and more efficient use of documentation strategies.
4. The beginnings of empirical validation. It should be pointed out that the majority of the models proposed for TC have not been validated empirically, although there are a few cases of empirical research developed with this objective (Alves & Gonçalves, 2007; Gonçalves, 2003, 2005; PACTE, 2000, 2003).

   PACTE carried out an exploratory study on TC (PACTE, 2002, 2003) with six professional translators, a pilot study with three professional translators and three foreign language teachers (PACTE, 2005a, 2005b) and, finally, an experiment with 35 professional translators and 24 foreign language teachers comparing their performance (Hurtado, 2017b). The results of this study validated the proposed TC model.

   The TC model proposed by Gonçalves (2003, 2005) and Alves and Gonçalves (2007) is assessed in various empirical studies carried out by these authors with a range of subject types: four students of English, eight translation students and four professional translators (Gonçalves, 2003, 2005); 16 translation students (Alves & Gonçalves, 2003); 17 translation students (Alves & Magalhães, 2004); three professional translators (Alves, 2005a); and four professional translators (Alves, 2005b).

22.2.3 Essential characteristics of translation competence

The proposed models and emerging empirical research leave us with the following noteworthy essential TC characteristics.

- TC is a collection of knowledge, abilities and attitudes necessary in order to translate. It comprises declarative and operative knowledge, and is essentially operative knowledge, since it integrates the abilities to perform translation tasks.
- TC is comprised of a set of interrelated sub-competences: linguistic competence in at least two languages; extra-linguistic competence (encyclopaedic, cultural and thematic knowledge); knowledge of translation competence (principles which govern translation and professional aspects); instrumental competence (use of all kinds of documentary sources and information and communication technologies applied to translation); and strategic competence.
- Strategic competence plays a fundamental role in being able to guarantee the translation process efficacy and solve translation problems appropriately.
- TC affects the development of the translation process and its product (the quality of translations).
- There are differences in the functioning of TC depending on directionality (into L1 or L2) and the area of translation specialization in question (legal, technical, scientific, literary, audiovisual, localization, etc.), as well as differences of an individual nature.
- TC is an acquired competence, which is different from bilingual competence.

22.3 Translation competence acquisition

As some authors point out (Campbell, 1998, p. 18; Waddington, 2000, p. 135) no TC model would be complete without taking into consideration the process by which it is acquired; hence the need to also research the process of translation competence acquisition (TCA).

22.3.1 Models of translation competence acquisition

As opposed to the case for TC, there are few TCA model proposals.

22.3.1.1 Harris' natural translation

Harris (1973, 1977, 1980) and Harris and Sherwood (1978) point out that there is a "natural translation" ability, an innate ability of a universal nature, which all bilinguals have. Harris (1977) defines natural translation as translations by bilinguals (with no special training for this) in daily circumstances. Working from empirical studies, Harris and Sherwood (1978) demonstrate that natural translation is an innate ability, which appears at a very early age and which develops from a stage that they call "pre-translation" to a "semi-professional" stage.

Nevertheless, as pointed out by Toury (1986, 1995) and Presas (2000), this ability does not necessarily lead to TC. Toury (1986) adds that TC does not develop automatically and in parallel to natural bilingualism; a translator has to create a second competence in addition to linguistic competence, namely transfer competence, which requires transferring texts, implying knowledge structures that are not part of bilingualism. Presas (2000), in turn, contrasts the notion of a "natural translator" with that of a "trained translator", stressing that this natural translation ability is not enough to be a translator.

22.3.1.2 Toury's socialization of translation

Toury (1980, 1995) puts forward the concept of "native translator", a complementary notion to that of "native speaker" from the field of linguistics, which he defines as a person who has progressively accessed translation without any formal instruction.

Toury (1995, pp. 241–258) proposes a model of the process through which someone who is bilingual becomes a translator, which he calls "socialization as concerns translating". He proposes that the act of translating is always communicative production and hence, an interactive act in which environmental feedback plays a fundamental role. This feedback the translator receives is essentially normative and can be in the form of a "sanction" (if they have translated badly) or reward (the agent or receiver expresses their satisfaction with the translation). This happens particularly in the initial stages, since the novice translator is discovering how the social environment works and is not sure of what is expected from him/her or the criteria to judge the appropriateness of a range of translation solutions or use of alternative strategies. The translator later develops an internal control mechanism, which operates during the translation act. The more advanced the TC of the individual, the more s/he can begin to take potential responses to the normative pressure without the risk of being penalized; from this point on, a translator can not only contravene the established rules but can even bring about changes to them.

Toury points out that during this socialization process of translation, the translator is assimilating the feedback and, consequently, modifies her/his basic TC. Seen from this perspective, TC is, in each of the translator's phases of development, a mix of innate, assimilated and social mechanisms.

Another hypothesis proposed by Toury is that the greater the variety of translation situations encountered, the greater the range and flexibility of individual ability to perform in a socially adequate manner. So, what is acquired is "adaptability". He adds that "specialization" can conflict with the individual's adaptability, lowering their overall TC.

22.3.1.3 Shreve: From natural to constructed translation. The expertise trajectory

Shreve (1997) sees TC as a specialization of communicative competence, which not everyone has (unlike communicative competence); from this viewpoint, TC is not an innate ability.

For Shreve, TC develops along a continuum between "natural translation" and "constructed translation" (professional translation). However, this development is not automatic or linear, nor is there an established path. Shreve talks of a "three-dimensional polygon", which embraces the different translation forms and functions, translation experience and translation situations.

In the development of TC two types of variation can come about: (1) due to individual cognitive styles and (2) due to translation acquisition history (through teaching, mentoring from another translator or autonomously).

According to this author, the development of TC means changes in the nature of the translation process and the norms that govern translation. Thus, there are differences between the "natural translator" and the "professional translator". The natural translator produces culturally and stylistically inappropriate translations, translates in micro-units without taking into account questions of coherence and cohesion, and does not bear in mind the end purpose of the translation; lexicon takes precedence over other aspects, etc. Shreve stresses the influence exerted by the nature, level and frequency of translation tasks in the history of TCA; moreover, according to Shreve, this restructuring movement of TC does not happen if there are no changes in the translation tasks.

In a later study, Shreve (2006) relates TCA to the notion of the "expertise trajectory" (Lajoie, 2003). Shreve proposes that declarative knowledge (i.e. what is known about the task) is converted, with practice, into production rules leading to proceduralization and, therefore, less effortful processing and greater automaticity. Shreve argues that TCA can be developed differentially, depending on variations in how further practical experience is acquired.

22.3.1.4 Chesterman's five stages of translation expertise

Chesterman (1997), in turn, refers to the five stages that Dreyfus and Dreyfus (1986) identify in the acquisition of any skill.[8]

- Stage one: "novice". Recognition of predefined features and rules. The trainees learn to recognize objective facts and predefined relevant features, and acquire rules to determine actions related to these facts and features.
- Stage two: "advanced beginner". Recognition of non-defined but relevant features. The trainees begin to recognize features that are difficult to define (or are undefined), although relevant. Increased experience and level of recognition.
- Stage three: "competence". Hierarchical and goal-oriented decision making. Having more experience and greater recognition, it is necessary to develop a sense of priorities, that is, hierarchical decision-making procedures; at this stage they follow conscious rules, information is processed and decisions taken.
- Stage four: "proficiency". Intuitive understanding plus deliberative action. Decisions are taken more on the basis of personal experience and, to a lesser degree, following conscious rules; this is an intuitive understanding and also rational action.
- Stage five: "expertise". Fluid performance plus deliberative rationality. Involves acting fluently and deliberately; intuition takes priority and conscience becomes apparent in critical reflections on intuition.

Chesterman defines this development as a gradual process of automatization, which runs "from atomistic to holistic recognition, from conscious to unconscious responses, from analytical to intuitive decision-making, from calculative to deliberative rationality, from detached to involved commitment" (Chesterman, 1997, p. 150).

Chesterman states that the degree of deliberative rationality varies depending on the translation task (poetry, certificates, etc.), and that a trait of the expert translator is probably her/his ability to judge when this is needed and how to use it.

22.3.1.5 The PACTE dynamic and spiral model

PACTE (2000, 2014, 2015, 2019a; Hurtado, in press) conceives TCA as a spiral, non-linear process by which novice knowledge (Pre-TC) evolves into TC. According to the PACTE model, TCA is (1) a dynamic, spiral process that, like all learning processes, evolves from novice knowledge (pre-TC) to TC; (2) a process of restructuring and developing TC sub-competences; (3) a process in which both declarative and procedural types of knowledge are integrated, developed and restructured; (4) a process in which the development of procedural knowledge—and, consequently, the strategic sub-competence—is essential.

So, for PACTE, this is a process of developing from an initial stage (which only has bilingual and extra-linguistic competences and a rudimentary ability of natural translation) to the stage of TC.

According to PACTE, the sub-competences that play a part in the process (1) are interrelated and compensate for each other; (2) do not always develop in parallel (i.e. at the same time and rate); and (3) may vary depending on the translation direction (towards L1 or L2), language pairs, translation specialization (legal, technical, literary, etc.), learning context (formal training, self-learning, etc.) and teaching methodology used. There are also differences at the individual level (knowledge, abilities, cognitive styles, etc.).

22.3.1.6 Alves and Gonçalves' relevance-theoretic model

Alves and Gonçalves (2007) work from connectionist approaches and see the acquisition of TC as a gradual, systematic and recursive process of expanding neural networks between various units of the individual cognitive environment. Rather than a TCA model, they pose a scale of evolution. Working from empirical studies, they distinguish between two cognitive profiles, which would point to differentiating between translators with lower or higher levels of metacognitive activity:

- "Narrow-band translators"—those who mainly work on the basis of insufficiently contextualized cues (i.e., dictionary-based meaning of words instead of contextualized meaning) and fail to bridge the gap between procedurally, conceptually and contextually encoded information.
- "Broadband translators"—those who mainly tend to work on the basis of communicative cues provided by the ST and reinforced by the contextual assumptions derived from their cognitive environments. In this way, expert translators are able to integrate procedurally, conceptually and contextually encoded information into a coherent whole to encompass higher levels of metacognition.

22.3.1.7 Kiraly's four-dimensional model of the emergence of translator competence

Kiraly (2013, 2015) criticizes the two-dimensional TC models, as they are unable to capture the complexity involved. He proposes a three-dimensional TC model, to which he adds the time factor to turn it into a four-dimensional model of the emergence (rather than acquisition) of translator competence.

Kiraly's model reflects the complex interplay of competences and their non-parallel emergence over time, and represents uniqueness in that each individual's competence development is different. In his emergence model, the author describes TC as a complex network and sub-competences as sub-networks; however, he refuses to propose a list of specific sub-competences and justifies this by arguing that there is no consensus on which ones actually exist. He does underline the fact that a range of aspects influence the competence acquisition process, which is in constant evolution. Among these aspects are the translation tasks and projects in which the translator engages and which he or she learns from, their personal and interpersonal attitude towards translating, human and material resources available and employed, and also the influence of the learning environment (Kiraly, 2013, p. 212). The translator's disposition for learning, abstracting from experience, using language in a creative manner, adapting to norms, etc. are also included among the influential elements which can have a bearing on TCA and make every process unique.

In Kiraly's model, each sub-competence would appear near the lower (novice) level as a separate dynamic vortex but show complex links throughout the system towards the upper (expert) end of the model due to experience and learning, including learning from interpersonal interaction. Finally, the separate sub-competences merge into a highly integrated and mainly intuitive super-competence (Kiraly, 2015, p. 29).

22.3.2 Empirical studies on translation competence acquisition

Since the 1980s there has been a wide range of empirical studies, the majority with small samples, into questions related to TCA.[9] They focus on translation students at the same or different levels, or compare the performance of translation students with bilinguals or, for the main part, with professional translators. These studies deal with topics such as creativity, automatization processes, the process of comprehension, identifying problems, decision-making processes, using strategies, cultural competence, documenting sources, the influence of bilingualism, etc. The following studies are noteworthy as regards large sample groups:

- Séguinot (1991), which analyses a total of 195 students of specialized translation, who were tested at the beginning and end of each year. This study focuses on analysing translation strategies, comparing the difference between native and L2 strategies.

- Orozco (2000) and Orozco and Hurtado (2002) on developing instruments to measure the TCA process in written translation, which draws on a sample group of 235 first-year students from three Spanish universities.
- Lachat Leal (2003), which concerns the impact of experience and learning in the process of problem solving and using strategies. This was carried out on a sample group of 111 second-year undergraduate students, 98 fourth-year students—translation students in both cases—and 12 professional translators.
- Gregorio Cano (2014), which analyses the development of strategic competence for resolving problems of a cultural nature, with 1,046 undergraduate translation students from five Spanish faculties of translation and interpreting: 655 first-year students and 391 fourth-year students. It includes a longitudinal study with 37 students who were given the same test in their first and fourth years.
- Quinci (2014), who carried out a longitudinal study over three years with 53 undergraduate and master's students of translation (one group of BA students and two of MA students). This study also includes ten professional translators. The aim of this study is to observe whether different levels of competence reflect on different linguistic patterns.
- Massana-Roselló (2016), who performed an experimental study on TCA as regards resolving false friends (Portuguese–Spanish) with 30 undergraduate translation students from the second, third and fourth year at two Spanish universities and ten professional translators.
- Olalla-Soler (2017), who performed an experimental study on the acquisition of translator cultural competence with a sample group of 38 undergraduate translation students taking German as a second foreign language in the BA in Translation and Interpreting from the first, second, third and fourth years and ten professional translators.

However, these studies deal with partial aspects of TCA, and research into comprehensive monitoring of TCA as a whole is in short supply. Only three of the TCA proposed models have incorporated empirical validation. Harris and Sherwood (1978) base their analysis of how "natural translation" functions on data taken from studies of bilinguals (some of which are longitudinal) from birth to the age of 18, carried out in the US. Alves & Gonçalves (2007), in turn, work from empirical studies with translation students and professional translators (see Alves, 2005a, 2005b; Alves & Gonçalves, 2003; Alves & Magalhães, 2004) for their proposal of two TC profiles: narrow-band translators and broadband translators. PACTE (2014, 2015, 2019a; Hurtado, in press) carried out a study with 130 undergraduate students in translation and interpreting from the first, second, third and fourth year and those recently graduated (approximately 30 subjects from each of these groups). The results are compared with the group of translators from the TC experiment. The results have enabled four types of "evolution" to be identified, which vary according to the various competences (PACTE, 2019a, in press): (1) non-evolution—no difference in the values between successive groups between the first year and the end of training; (2) rising evolution—values rise between the first year and the end of training, with each value between successive groups being higher than or equal to the previous one; (3) falling evolution—values fall between the first year and the end of training, with each value between successive groups being lower than or equal to the previous one; and (4) mixed evolution—a combination of rising and falling evolution between the first year and the end of training.

There are two noteworthy research projects that have carried out longitudinal studies on TCA:

- The TransComp project (2008–2011, University of Graz) monitored the performance of 12 students of translation over a period of three years and compared it with the performance of ten professional translators (see Göpferich, 2009).

- The Capturing Translation Processes (CTP) project (2009–2011, ZHAW Institute of Translation and Interpreting), which compared translation students at various stages in their training (194 beginners and 112 advanced students) and 39 professional translators (see Ehrensberger-Dow, 2013; Ehrensberger-Dow & Massey, 2013; Massey & Ehrensberger-Dow, 2011).

Of another nature is the NACT project from the PACTE research group on Establishing Competence Levels in the Acquisition of Translation Competence in Written Translation (PACTE, 2018, 2019b), which aimed at drawing up a proposal for level descriptors that would be a first step in the direction of developing a common European framework in the academic and professional area of translation.

22.3.3 Essential characteristics of translation competence acquisition

The proposed models and empirical studies underscore the following essential characteristics of TCA.

- TCA is a cyclical process. As we have seen, all the models that have been put forward point out that TC is not an innate competence but rather, is acquired, and this acquisition is a cyclical process from an initial stage to a stage of consolidating competences. Given that it is a cyclical process, we can surmise that there are different TCA phases. We are, however, lacking in empirical information about what these different acquisition phases are and how they work.
- TCA is a process in which TC sub-competences are developed and restructured.
- TCA is a process in which it is essential to develop procedural knowledge and strategic competence in order to make progress in one's capacity to solve translation problems.
- TCA implies a gradual process of automatization. Some authors (Alves and Gonçalves, Chesterman, PACTE and Shreve) assimilate this acquisition process with the process of acquiring any knowledge and underline the gradual process of proceduralization and automatization.
- TCA involves a different process depending on the sub-competence. The results of PACTE's empirical research have demonstrated that not all sub-competences are developed in parallel, and there are different types of development depending on the case (non-evolution, rising evolution, falling evolution or mixed evolution).
- TCA is not a linear process. Some authors (Kiraly, PACTE and Shreve) give importance to the fact that, given the complexity of the TCA process, it does not develop in a linear fashion. The mixed evolution identified by PACTE in their TCA experiment serves as an illustrative example.
- TCA affects the translation process and its product, bringing about an evolution as regards the functioning of the translation process and as regards the quality of translations.
- There are variations in the TCA process. Some models (PACTE, Shreve and Kiraly) stress that TCA is a complex process in which we can find different types of variations: according to the way it is acquired (naturally and self-taught, guided by means of teaching-learning); according to the pedagogical context; according to directionality (towards L1 or L2); or according to professional profile (legal, technical, literary translator, etc.). And, of course, there are also variations at an individual level depending on the characteristics of each subject.

22.4 Research perspectives

22.4.1 Difficulties related to research on translation competence and translation competence acquisition

In 1998, Campbell had proposed some requirements for a TC model: (1) to show whether TC is divisible into components, and, if so, to describe those components and their relationships; (2) to describe the developmental pathway taken in learning how to translate; and (3) to include means for describing the differences between the performance of different translators (Campbell, 1998, p. 18). Waddington (2000) expresses himself along similar lines when discussing the problems involved in drawing up TC models: (1) it is difficult to know the number of components and to clearly identify them and the relationship between them; (2) a model that has been developed for one given level of competence is not necessarily valid for another; and (3) the competence model is therefore incomplete without a competence development model (Waddington, 2000, p. 135). Both, thus, agree on underlining the difficulties involved in describing TC and the need to describe the TCA process.

Advances have been made in recent decades as regards describing TC and TCA. Furthermore, major advances have been made in empirical research from the field of Translation Studies, and attempts have been made to deal with partial aspects of TC and TCA. However, most of the proposed TC and TCA models have not been validated empirically. This shortage of empirical research into TC and TCA as a whole could be due to various reasons:

(1) The complex nature of TC and TCA and the complexity of the relationship of their components. It should first be pointed out that research into TC is problematic because of its inherently complex nature, given the wide range of cognitive areas and activities involved and the complexity of the corresponding relationships.
(2) The procedural and automatized nature of TC and TCA. Proceduralization and automatization, characteristics that affect both the functioning of TC as well as its acquisition, make it difficult to analyse them, since procedural knowledge is more difficult to verbalize and observe.
(3) The heterogeneous nature of TC and TCA is another aspect that poses problems to research, since this implies a very diverse range of abilities, which in addition can vary depending on the subject.
(4) The diversification of TC and TCA. How TC functions and the relationships between its components are particularly difficult to observe, given the differences depending on the individual characteristics of the subjects (knowledge, experience, cognitive styles, etc.) and the way TC is acquired (guided, via teaching-learning; autonomously, through practice outside the teaching system). There are also differences depending on the translation direction (into L1 or L2). Furthermore, each specialized professional profile has its own specific characteristics.

22.4.2 Need for future advances in research on translation competence and translation competence acquisition

Although much has been achieved, there is still a long way to go in research on TC and TCA. Research should develop along four major lines.

(1) Empirical research on TC and TCA. As we have seen, most of the TC and TCA models have not been validated empirically, and there is a shortage of empirical research on TC and TCA as a whole. These studies should be based on large and representative

samples so that the results can be generalizable. Furthermore, studies already carried out should be replicated (e.g. those by PACTE on TC and TCA, TransComp on TCA) in different contexts, so that the results can be compared in each case, and we can deduce common and generalizable characteristics. As regards studying TCA, it is important to carry out longitudinal studies, which, by solving the inherent difficulties in performing this kind of study (controlling confounding variables, developing parallel instruments for each measure, subjects dropping out, etc.), are able to control all the possible external influences that could distort results.

(2) Research on the competences acquired in each professional profile and how they are acquired. The majority of studies on TC and TCA refer to generalist translator competences. So, what is lacking is advances in research into the competences required in each professional translation area and the process by which they are acquired: legal, business, financial, technical, scientific, literary, audiovisual, accessibility and localization.

(3) Research on possible variations which can crop up in TC and in TCA. It would be helpful to carry out studies that compare performance depending on translation direction (into L1 and into L2), according to the context (social, pedagogical), according to the language pair and according to the characteristics of the subjects, so that consistencies and differences can be identified.

(4) Establish levels of competences. As opposed to other disciplines, in translation there is no common base for describing levels of competences, as happens, say, in teaching languages. A description of this kind would provide a common framework, which would be of great use in the education and professional sectors of translation. This requires research into the TCA process, with large and representative samples, and also in different contexts in order to better understand the different phases in the TCA process. Furthermore, studies such as the PACTE group NACT project (see Section 22.3.2) would contribute to advances in describing and coming to a common agreement about TC levels.

Notes

1 The term was created by the anthropologist Hymes in 1966 in his paper entitled "On Communicative Competence" (published in 1971).

2 For previous descriptions of the TC model proposals, see Hurtado (2001/2011, pp. 383–392) and Hurtado (2017a, pp. 18–31).

3 Original quotation: "la habilidad de saber traducir".

4 This model was first presented on the poster "La competencia traductora y su aprendizaje: Objetivos, hipótesis y metodología de un proyecto de investigación" ("Translation competence and how it is learned: objectives, hypothesis and methodology behind a research project") at the *IV International Congress on Translation*, held at the Universitat Autònoma de Barcelona, 06/05/1998.

5 The EMT (http://ec.europa.eu/emt) is a European Commission partnership project in conjunction with higher-education institutions that offer translation programmes. The EMT sets out quality standards for translation programmes that meet professional standards and market needs. Approved centres are authorized to use their logo, which is a registered EU trademark.

6 Knowledge management is seen as a discipline that encourages an integrated approach to identifying, capturing, assessing, retrieving and sharing all the information assets of an enterprise, i.e. databases, documents, policies, procedures and previously un-captured expertise and experience in individual workers (Duhon, 1998).

7 Regarding the applications of the behavioural approach in Translation Studies, see Kuznik and Hurtado (2015).

8 Dreyfus and Dreyfus identify five stages but without awarding them a specific denomination; the denominations used here are those proposed by Chesterman for each stage.

9 See Massana-Roselló (2016, pp. 39–67) for a review of empirical studies into TCA up to 2015.

Further reading

Hurtado Albir, A. (2017). (Ed.) *Researching translation competence by PACTE Group*. Amsterdam: John Benjamins.

This book is a compendium of PACTE Group's experimental research in translation competence since 1997. The book is organized in four sections: Conceptual and methodological background, Research design and data analysis, Results of the PACTE translation competence experiment and Defining features of translation competence. It also includes eight appendices and a glossary.

Schäffner, Ch., & Adab, B. (Eds.). (2000). *Developing translation competence*. Amsterdam: John Benjamins.

This book presents a comprehensive study of what constitutes translation competence and its place in the translation training programmes. The book is organized in three sections: Defining translation competence, building translation competence and assessing translation competence.

References

Alves, F. (2005a). Ritmo cognitivo, metarreflexão e experiência: Parâmetros de análise processual no desempenho de tradutores novatos e experientes. In A. Pagano, C. Magalhães, & F. Alves (Eds.), *Competência em tradução: Cognição e discurso* (pp. 90–122). Belo Horizonte: Editora UFMG.

Alves, F. (2005b). Esforço cognitivo e efeito contextual em tradução: Relevância no desempenho de tradutores novatos e expertos. *Linguagem em (Dis)curso, 5*, 11–31.

Alves, F., & Gonçalves, J. L. (2003). A relevance theory approach to the investigation of inferential processes in translation. In F. Alves (Ed.), *Triangulating translation: Perspectives in process oriented research* (pp. 11–34). Amsterdam: John Benjamins.

Alves, F., & Gonçalves, J. L. (2007). Modelling translator's competence: Relevance and expertise under scrutiny. In Y. Gambier, M. Shlesinger, & R. Stolze (Eds.), *Translation studies: Doubts and directions. Selected papers from the IV Congress of the European Society for Translation Studies* (pp. 41–55). Amsterdam: John Benjamins.

Alves, F., & Magalhães, C. (2004). Using small corpora to tap and map the process-product interface in translation. *TradTerm, 10*, 179–211.

Bachman, L. F. (1990). *Fundamental considerations in language testing*. London: Oxford University Press.

Beeby, A. (1996). *Teaching translation from Spanish to English* [Didactics of Translation Series 2]. Ottawa: University of Ottawa Press.

Bell, R. T. (1991). *Translation and translating*. London: Longman.

Boyatzis, R. E. (1982). *The competent manager: A model for effective performance*. New York: Wiley-Interscience.

Boyatzis, R. E. (1984). *Identification of skill requirements for effective job performance*. Boston: McBer.

Calvo Encinas, E. (2004). La administración pública ante la interpretación social: Toma de contacto en la provincia de Toledo. *Puentes, 4*, 7–16.

Campbell, S. (1998). *Translation into the second language*. London: Longman.

Canale, M. (1983). From communicative competence to communicative language pedagogy. In J. C. Richards, & R. W. Schmidt (Eds.), *Language and communication* (pp. 2–27). London: Longman.

Canale, M., & Swain, M. (1980). Theoretical bases of communicative approaches to second language teaching and testing. *Applied Linguistics, 1*(1), 1–47.

Cao, D. (1996). Towards a model of translation proficiency. *Target, 8*(2), 325–340.

Chesterman, A. (1997). *Memes of translation*. Amsterdam: John Benjamins.

Chomsky, N. (1965). *Aspects of the theory of syntax*. Massachusetts: Institute of Technology Press.

Delisle, J. (1980). *L'Analyse du discours comme méthode de traduction*. Ottawa: University Press.

Dreyfus, H. L., & Dreyfus, S. E. (1986). *Mind over machine*. Oxford: Blackwell.

Duhon, B. (1998). It's all in our heads. *Inform, 12*(8), 8–13.

Ehrensberger-Dow, M. (2013) . *Capturing translation processes. Final report*. Zurich: ZHAW, Zurich University of Applied Sciences.

Ehrensberger-Dow, M., & Massey, G. (2013). Indicators of translation competence: Translators' self-concepts and the translation of titles. *Journal of Writing Research, 5*(1), 103–131.

EMT. (2009). Competences for professional translators, experts in multilingual and multimedia communication. Directorate-General Translation. Retrieved July 2017 from http://ec.europa.eu/dgs/translation/programmes/emt/key_documents/emt_competences_translators_en.pdf

EMT. (2017). European Master's in translation, Competence framework 2017. Retrieved 11 May 2019 from https://ec.europa.eu/.../emt_competence_fwk_2017_en_web.pdf

Gile, D. (1995/2009). *Basic concepts and models for interpreter and translator training*. Amsterdam: John Benjamins.
Gonçalves, J. L. (2003). O desenvolvimento da competência do tradutor: Investigando o processo através de um estudo exploratório-experimental [unpublished PhD dissertation]. Universidade Federal de Minas Gerais, Belo Horizonte.
Gonçalves, J. L. (2005). O desenvolvimento da competência do tradutor: Em busca de parâmetros cognitivos. In A. Pagano, C. Magalhães, & F. Alves (Eds.), *Competência em tradução: Cognição e discurso* (pp. 59–90). Belo Horizonte: Editora UFMG.
González Davies, M. (2004). Undergraduate and postgraduate translation degrees: Aims and expectations. In K. Malmkjaer (Ed.), *Translation as an undergraduate degree* (pp. 67–81). Amsterdam: John Benjamins.
González Davies, M., & Scott-Tennent, C. (2005). A problem-solving and student-centred approach to the translation of cultural references. *Meta, 50*(1), 160–179.
Göpferich, S. (2008). *Translationsprozessforschung: Stand- Methoden- Perspektiven*. Translationswissenschaft 4. Tübingen: Narr.
Göpferich, S. (2009). Towards a model of translation competence and its acquisition: The longitudinal study TransComp. In S. Göpferich, A. L. Jakobsen, & I. M. Mees (Eds.), *Behind the mind: Methods, models and results in translation process research* (pp. 12–38). Copenhagen: Samfundslitteratur.
Gouadec, D. (2002). *Profession: traducteur.* Paris: La Maison du Dictionnaire.
Gouadec, D. (2005). Modélisation du processus d'exécution des traductions. *Meta, 50*(2), 643–655.
Gouadec, D. (2007). *Translation as a profession*. Amsterdam: John Benjamins.
Gregorio Cano, A. (2014). Estudio empírico-descriptivo del desarrollo de la competencia estratégica en la formación de traductores [unpublished PhD dissertation]. Universidad de Granada, Granada.
Gutt, E. A. (2000). Issues of translation research in the inferential paradigm of communication. In M. Olohan (Ed.), *Intercultural faultlines—Research models in translation studies 1: Textual and cognitive aspects* (pp. 161–179). Manchester: St. Jerome Publishing.
Hansen, G. (1997). Success in translation. *Perspectives: Studies in Translatology, 5*(2), 201–210.
Harris, B. (1973). La traductologie, la traduction naturelle, la traduction automatique et la sémantique. *Cahiers de Linguistique, 10*, 11–34.
Harris, B. (1977). The importance of natural translation. *Working Papers on Bilingualism, 12*, 96–114.
Harris, B. (1980). How a three-year-old translates. In E. A. Afrendas (Ed.), *Patterns of bilingualism* (pp. 370–393). Singapore: National University of Singapore Press.
Harris, B., & Sherwood, B. (1978). Translating as an innate skill. In D. Gerver, & H. W. Sinaiko (Eds.), *Language, interpretation and communication* (pp. 155–170). Oxford: Plenum Press.
Hatim, B., & Mason, I. (1997). *The translator as communicator.* London: Routledge.
Hewson, L., & Martin, J. (1991). *Redefining translation. The variational approach.* London: Routledge.
Hönig, H. G. (1991). Holmes' "Mapping Theory" and the landscape of mental translation processes. In K. Van Leuven-Zwart, & T. Naajkens (Eds.), *Translation studies: The state of the art. Proceedings from the first James S. Holmes symposium on translation studies* (pp. 77–89). Amsterdam: Rodopi.
Hönig, H. G. (1995). *Konstruktives Übersetzen*. Tübingen: Stauffenburg.
Hurtado Albir, A. (1996a). La enseñanza de la traducción directa "general". Objetivos de aprendizaje y metodología. In A. Hurtado Albir (Ed.), *La enseñanza de la traducción* [Col. Estudis sobre la traducció 3] (pp. 31–55). Castellón: Universitat Jaume I.
Hurtado Albir, A. (1996b). La cuestión del método traductor. Método, estrategia y técnica de traducción. *Sendebar, 7*, 39–57.
Hurtado Albir, A. (2001/2011). *Traducción y traductología. Iniciación a la traductología.* Madrid: Cátedra.
Hurtado Albir, A. (2017a). Translation and translation competence. In A. Hurtado Albir (Ed.), *Researching translation competence by PACTE group* (pp. 3–33). Amsterdam: John Benjamins.
Hurtado Albir, A. (Ed.). (2017b). *Researching translation competence by PACTE group*. Amsterdam: John Benjamins.
Hymes, D. H. (1971). *On communicative competence.* Philadelphia: University of Pennsylvania Press.
Katan, D. (2008). University training, competencies and the death of the translator. Problems in professionalizing translation and in the translation profession. In M. Musacchio, & G. Henrot (Eds.), *Tradurre: Formazione e Professione* (pp. 113–140). Padova: CLEUP.
Kelly, D. (2002). Un modelo de competencia traductora: Bases para el diseño curricular. *Puentes,* 1, 9–20.
Kelly, D. (2005). *A handbook for translator trainers.* Manchester: St Jerome.
Kelly, D. (2007). Translator competence contextualized. Translator training in the framework of higher education reform: In search of alignment in curricular design. In D. Kenny, & K. Ryou (Eds.), *Across boundaries: International perspectives on translation studies* (pp. 128–142). Cambridge: Cambridge Scholars Publishing.

Kiraly, D. (1995). *Pathways to translation. Pedagogy and process*. Kent, OH: The Kent State University Press.

Kiraly, D. (2013). Towards a view of translator competence as an emergent phenomenon: Thinking outside the box(es) in translator education. In D. Kiraly, S. Hansen-Schirra, & K. Maksymski (Eds.), *New prospects and perspectives for educating language mediators* (pp. 197–224). Tubingen: Gunter Narr.

Kiraly, D. (2015). Occasioning translator competence: Moving beyond social constructivism toward a post-modern alternative to instructionism. *Translation and Interpreting Studies, 10*(1), 8–32.

Koller, W. (1979). *Einführung in die Übersetzungswissenschaft*. Heidelberg: Quelle und Meyer.

Kuznik, A., & Hurtado Albir, A. (2015). How to define good professional translators and interpreters: Applying the behavioural approach to studying competences in the field of translation studies. *Across Languages and Cultures, 16*(1), 1–27.

Lachat Leal, C. (2003). Estrategias y problemas de traducción [unpublished PhD dissertation]. Universidad de Granada, Granada.

Lajoie, S. P. (2003). Transitions and trajectories for studies of expertise. *Educational Researcher, 32*(8), 21–25.

Lasnier, F. (2000). *Réussir la formation par compétences*. Montreal: Guérin.

Lowe, P. (1987). Revising the ACTFL/ETS scales for a new purpose: Rating skill in translating. In M. G. Rose (Ed.), *Translation excellence: Assessment, achievement, maintenance* [American Translators Association Series 1] (pp. 53–61). New York: Suny Binghamton Press.

Massana-Roselló, G. (2016). La adquisición de la competencia traductora portugués-español: Un estudio en torno a los falsos amigos [unpublished PhD dissertation]. Universitat Autònoma de Barcelona, Barcelona.

Massey, G., & Ehrensberger-Dow, M. (2011). Investigating information literacy: A growing priority in translation studies. *Across Languages and Cultures, 12*(2), 193–211.

McClelland, D. (1973). Testing for competencies rather than for intelligence. *American Psychologist, 28*, 1–14.

Mackenzie, R. (2000). POSI-tive thinking about quality in translator training in Finland. In A. Beeby, D. Ensinger, & M. Presas (Eds.), *Investigating translation: Selected papers from the 4th International Congress on Translation, Barcelona, 1998* (pp. 213–222). Amsterdam: John Benjamins.

Neubert, A. (1994). Competence in translation: A complex skill, how to study and how to teach it. In M. Snell-Hornby, F. Pöchhacker, & K. Kaindl (Eds.), *Translation studies. An interdiscipline* (pp. 411–420). Amsterdam: John Benjamins.

Neubert, A. (2000). Competence in language, in languages, and in translation. In C. Schäffner, & B. Adab (Eds.), *Developing translation competence* (pp. 3–18). Amsterdam: John Benjamins.

Nord, C. (1988). *Textanalyse und Übersetzen*. Heidelberg: J. Groos Verlag (*Text analysis in Translation*. Amsterdam: Rodopi, 1991).

Nord, C. (1992). Text analysis in translator training. In C. Dollerup, & A. Lindegaard (Eds.), *Teaching translation and interpreting 1* (pp. 39–48). Amsterdam: John Benjamins.

Olalla-Soler, C. (2017). La competencia cultural del traductor y su adquisición. Un estudio experimental en la traducción alemán-español [unpublished PhD dissertation]. Universitat Autònoma de Barcelona, Barcelona.

Orozco, M. (2000). Instrumentos de medida de la adquisición de la competencia traductora: Construcción y validación [unpublished PhD dissertation]. Universitat Autònoma de Barcelona, Barcelona.

Orozco, M., & Hurtado Albir, A. (2002). Defining and measuring translation competence acquisition. *Meta,* 47(3), 375–402.

PACTE (2000). Acquiring translation competence: Hypotheses and methodological problems in a research project". In A. Beeby, D. Ensinger, & E. M. Presas (Eds.), *Investigating translation* (pp. 99–106). Amsterdam: John Benjamins.

PACTE (2002). Exploratory tests in a study of translation competence. *Conference interpretation and translation,* 4(2), 41–69.

PACTE (2003). Building a translation competence model. In F. Alves (Ed.), *Triangulating translation: Perspectives in process oriented research* (pp. 43–66). Amsterdam: John Benjamins.

PACTE (2005a). Primeros resultados de un experimento sobre la competencia traductora. In M. L. Romana García (Ed.), *II AIETI. Actas del II Congreso Internacional de la Asociación Ibérica de Estudios de Traducción e Interpretación. Madrid, 9–11 de febrero de 2005* (pp. 573–587). Madrid: AIETI.

PACTE (2005b). Investigating translation competence: Conceptual and methodological issues. *Meta, 50*(2), 609–619.

PACTE (2014). First results of PACTE group's experimental research on translation competence acquisition: The acquisition of declarative knowledge of translation. *MonTI. Monografías de Traducción e Interpretación. Special issue, 1*, 85–115.
PACTE (2015). Results of PACTE's experimental research on the acquisition of translation competence: The acquisition of declarative and procedural knowledge in translation. The dynamic translation index. *Translation Spaces, 4*(1), 29–35.
PACTE. (2017). The performance of the top-ranking translators. In A. Hurtado Albir (Ed.), *Researching translation competence by PACTE group* (pp. 269–280). Amsterdam: John Benjamins.
PACTE. (2018). Competence levels in translation: Working towards a European framework. *The Interpreter and Translator Trainer, 12*(2), 111–131.
PACTE. (2019a). Evolution of the efficacy of the translation process in translation competence acquisition. Results of the PACTE group's experimental research. *Meta, 64*(1), 242–265.
PACTE. (2019b). Establecimiento de niveles de competencias en traducción. Primeros resultados del proyecto NACT. *Onomázein. Revista de lingüística, filología y traducción, 43*, 1–25.
PACTE. (In press). Translation competence acquisition. Design and results of the PACTE group's experimental research. *The Interpreter and Translator Trainer, 14*(2). Special issue.
Presas, M. (1996). Problemes de traducció i competència traductora. Bases per a una pedagogia de la traducció [unpublished PhD dissertation]. Universitat Autònoma de Barcelona, Barcelona.
Presas, M. (2000). Bilingual competence and translation competence. In C. Schäffner, & B. Adab (Eds.), *Developing translation competence* (pp. 19–31). Amsterdam: John Benjamins.
Pym, A. (1991). A definition of translational competence, applied to the teaching of translation. In M. Jovanovic (Ed.), *Translation: A creative profession: 12th World Congress of FIT. Proceedings* (pp. 541–546). Belgrade: Prevodilac.
Pym, A. (1992). Translation error analysis and the interface with language teaching. In C. Dollerup, & A. Loddegaard (Eds.), *Teaching translation and interpreting. Training, talent and experience* (pp. 279–290). Amsterdam: John Benjamins.
Pym, A. (1993). *Epistemological problems in translation and its teaching*. Teruel: Caminade.
Pym, A. (2003). Redefining translation competence in an electronic age: In defence of a minimalist approach. *Meta, 48*(4), 481–497.
Quinci, C. (2014). Translators in the making: An empirical longitudinal study on translation competence and its development [PhD dissertation]. Trieste: Università degli studi di Trieste.
Risku, H. (1998). *Translatorische Kompetenz. Kognitive Grundlagen des Übersetzens als Expertentätigkeit*. Tübingen: Stauffenburg.
Risku, H., Dickinson, A., & Pircher, R. (2010). Knowledge in translation practice and translation studies: Intellectual capital in modern society. In D. Gile, G. Hansen, & N. K. Pokorn (Eds.), *Why translation studies matters* (pp. 83–96). Amsterdam: John Benjamins.
Roberts, R. P. (1984). Compétence du nouveau diplômé en traduction. In *Traduction et Qualité de Langue. Actes du Colloque Société des traducteurs du Québec/Conseil de la langue française* (pp. 172–184). Québec: Éditeur officiel du Québec.
Rothe-Neves, R. (2005). A abordagem comportamental das competências. Aplicabilidade aos estudos da tradução. In A. Pagano, C. Magalhães, & F. Alves (Eds.), *Competência em tradução. Cognição e discurso* (pp. 91–107). Belo Horizonte: Universidade Federal de Minas Gerais.
Schäffner, C. (2004). Developing professional translation competence without a notion of translation. In K. Malmkjaer (Ed.), *Translation as an undergraduate degree* (pp. 113–125). Amsterdam: John Benjamins.
Schäffner, C., & Adab, B. (Eds.). (2000). *Developing translation competence*. Amsterdam: John Benjamins.
Séguinot, C. (1991). A study of student translation strategies. In S. Tirkkonen-Condit (Ed.), *Empirical research in translation and intercultural studies* (pp. 79–88). Tübingen: Gunter Narr.
Shreve, G. M. (1997). Cognition and the evolution of translation competence. In J. H. Danks, G. M. Shreve, S. B. Fountain, & M. K. McBeath (Eds.), *Cognitive processes in translation and interpreting* (pp. 120–136). Thousand Oaks: Sage.
Shreve, G. M. (2006). The deliberate practice: Translation and expertise. *Journal of Translation Studies, 9*(1), 27–42.
Snell-Hornby, M. (1988). *Translation studies. An integrated approach*. Amsterdam: John Benjamins.
Spencer, L. M., McClelland, D. C., & Spencer, S. M. (1994). *Competency assessment methods: History and state of art*. New York: Hay/McBer Research Press.

Spolsky, B. (1989). Communicative competence, language proficiency, and beyond. *Applied Linguistics, 10*(2), 138–156.
Stansfield, C. W., Scott, M. L., & Kenyon, D. M. (1992). The measurement of translation ability. *The Modern Language Journal, 76*(4), 455–467.
Toury, G. (1980). The translator as a nonconformist-to-be, or: How to train translators so as to violate translational norms. In S. O. Poulsen, & W. Wilss (Eds.), *Angewandte Übersetzungswissenschaft: Internationales Übersetzungswissenschaftliches Kolloquium an der Wirtschaftsuniversität Århus/Dänemark, 19–21 Juni 1980* (pp. 180–194). Aarhus: Aarhus University.
Toury, G. (1986). Natural translation and the making of a native translator. *Textcontext, 1*, 11–29.
Toury, G. (1995/2012). *Descriptive translation studies and beyond*. Amsterdam: John Benjamins.
Vienne, J. (1998). Vous avez dit compétence traductionnelle? *Meta, 43*(2), 187–190.
Waddington, C. (2000). *Estudio comparativo de diferentes métodos de evaluación de traducción general (inglés-español)*. Madrid: Universidad Pontificia de Comillas.
Widdowson, H. G. (1989). Knowledge of language and ability for use. *Applied Linguistics, 10*(2), 128–137.
Wilss, W. (1976). Perspectives and limitations of a didactic framework for the teaching of translation. In R. W. Brislin (Ed.), *Translation applications and research* (pp. 117–137). New York: Gardner.
Wilss, W. (1982). *The science of translation. Theoretical and applicative aspects*. Tübingen: Gunter Narr.

23

Translation, the process–product interface and cognition

Silvia Hansen-Schirra and Jean Nitzke

23.1 Introduction: Why do we need a process–product interface?

According to Toury, discourse transfer is inherent in the mental processes involved in translation. From a product perspective, translation depends on a particular manner in which the source text is processed, so that "the more the make-up of a text is taken as a factor in the formulation of its translation, the more the target text can be expected to show traces of interference" (Toury, 1995, 276). Frawley defines translation as "a code in its own right" arising "out of the bilateral consideration of the matrix and target codes" (Frawley, 1984, p. 168). Baker (1996) pursues this argument and hypothesizes that translation might feature universal properties, which can be found among different translated texts but not in non-translated text. However, the question arises why these translation-specific properties exist. An answer might lie in the translation process itself, since these universal text properties might be triggered by universal processing taking place during translation.[1]

From a psycholinguistic perspective, several models of the translation process have been presented (for a summary of earlier models, see Koller, 2004; an overview of more recent models is given in Göpferich, 2008). Most of the models have in common that they split the translation process into a reception and a production phase (see Kautz, 2000; Steiner, 2001; or in machine translation, Vauquois, 1968) or a decoding and a coding phase (see also Schaeffer & Carl, 2013). In addition, some models suggest a further intermediary step, i.e. a transfer step or a conceptualization step (Levelt, 1989). This model has already been empirically tested in Translation Studies (see Carl & Dragsted, 2012; Tirkkonen-Condit, 2005). However, variations thereof have also been presented: Francis and Gallard (2005), for instance, suggest that translators always skip the conceptualization step; Tirkkonen-Condit (2005) assumes that translation is carried out by transcribing the source text into the target language with the help of self-monitoring. Steiner (2001) assumes grammatical deconstruction while simultaneously understanding the source text.

In the following, we argue that the same questions or very similar questions have been independently worked on from two different research branches in Translation Studies: from the product perspective, corpus linguistic techniques have been used to examine translation properties and universals, whereas the process perspective has evolved into the whole area of translation process research. The latter has adopted methods from psycholinguistics and cognitive sciences.

We postulate, however, that we have to bridge the gap between these research branches in order to get a holistic picture of the translation process, one that integrates statistics and explanatory power. Let's take the dichotomy of normalization and shining through: if we try to relate these translation universals to the process models, we assume that structures are primed without being conceptualized or, in contrast, that a monitoring or inhibition mechanism leads to the mental control of a translation equivalent before it is articulated or written (de Groot, 2011; Levelt, 1999). Accordingly, priming might lead to shining through, whereas monitoring might trigger normalization (Hansen-Schirra, 2017).

In order to investigate these possible correlations, we need to interface product- and process-based research. By doing so, we will get quantitative statistical evidence from the corpus data as well as explanations concerning translators' strategies, behaviour and cognition. In the following two sections, we discuss the product and process perspectives, outlining relevant methods and their advantages and drawbacks, before introducing the product–process interface in Section 23.4.

23.2 The product perspective: Corpus linguistics in Translation Studies

As mentioned earlier, different kinds of translation corpora have been built up in Translation Studies in order to empirically investigate translation properties or universals or to identify translation shifts triggered through specific translation strategies (see Olohan, 2004, or Hansen-Schirra & Teich, 2008 for an overview). The most important methodological considerations will be discussed in the following.

23.2.1 Corpus designs

Two types of corpus design are most commonly used in corpus-based Translation Studies: the parallel corpus and the monolingually comparable corpus. Parallel corpora consist of source-language texts and translations of those texts into a target language. They are employed in bilingual computational lexicography, machine translation and translation memories. In translation research, parallel corpora are used to provide information on language-pair-specific translation behaviour, to observe equivalence relations between lexical items or grammatical structures in the source and target languages or texts (see Hansen-Schirra, 2008), and to investigate translation problems and translation mistakes. Some corpus initiatives are moving towards more than one language pair, e.g. the Oslo Multilingual Corpus (OMC) of the SPRIK project (Johansson, 2002).

Monolingually comparable corpora (short: comparable corpora) are collections of translations (from one or more source languages) into one target language and original texts in the target language. Comparable corpora "should cover a similar domain, variety of language and time span, and be of comparable length" (Baker, 1995, p. 23); they have

> the potential to reveal most about features specific to translated text, i.e., those features that occur exclusively, or with unusually low or high frequency, in translated text as opposed to other types of text production, and that cannot be traced back to the influence of any one particular source text or language.
>
> *Kenny, 1997*

Among the features researchers have posited comparing between translations and texts originally produced in the target language are explicitation, simplification, normalization and conservatism (Baker, 1995, 1996; Hansen, 2003; Laviosa-Braithwaite, 1996; Kenny, 1998; Mauranen, 1997; Teich, 2003). The main application of comparable corpora is the investigation of the specific (and possibly universal) properties of translations. The fact that translations exhibit linguistic properties that distinguish them from texts that are not translations is sometimes also referred to as *translationese* (see Baroni & Bernardini, 2006).

The most recent type of corpus design is a combination of parallel and comparable corpora. Such a combination will automatically contain a third subcorpus: a multilingually comparable corpus (see Hansen-Schirra et al., 2012). This combination of corpora can be used for cross-linguistic comparison of original texts, cross-linguistic comparison of original and translated texts, cross-linguistic comparison of translated texts, and monolingual comparison of original and translated texts. Methodologically, the primary issue concerning comparable corpora, both monolingual and multilingual, is to define the notion of comparability (see Neumann & Hansen-Schirra, 2013). The favoured solution is to use the concept of register, i.e. functional variation or variation according to situational context. The Translational English Corpus, for example, has followed the design of the British National Corpus in terms of register distinctions.

More recently, empirical translation research has been based on multiple translation corpora. This architecture includes one source text in a given language and target texts translated by several translators into another language. The CRITT TPR DB is such a multiple corpus: several English source texts are translated by multiple translators (professional translators as well as students) into several languages (see Carl, Schaeffer, et al., 2016). Using this corpus, universal translation strategies can be investigated across the language pairs. Moreover, different phases of expertise can be analysed, since novice translation can be compared with expert translation.

23.2.2 Corpus annotation and alignment

In Translation Studies, most corpus-based research is carried out on the basis of raw text. Only the encoding of meta-information is a usual practice. More recently, however, the use of annotated corpora is becoming more common, because this is the only way of empirically investigating grammatical and semantic as well as discourse or register features (see Hansen-Schirra et al., 2012). Depending on how abstract the linguistic features to be analysed are, linguistic corpus annotation can be done automatically, semi-automatically or manually. There are a number of automatic annotation techniques. The most reliable ones are part-of-speech (PoS) taggers and morphological analysers, following either a rule-based or a statistical approach (see Hansen-Schirra & Teich, 2008). Also, there are a few treebanking efforts, i.e. annotation of parallel corpora in terms of syntactic structure (e.g. the Sofie Treebank; Samuelsson & Volk, 2005 or the CroCo Corpus; Hansen-Schirra et al., 2012). However, phrase structure and syntactic function as well as semantic annotations have to be corrected or carried out manually.

For the analysis of translation shifts in a parallel corpus, the units of translation (i.e. source-language text units and their translational equivalents) need to be aligned. There are various alignment programs freely available (Hofland, 1996); additionally, aligners are often incorporated in translation memories. The most commonly implemented technique is sentence-by-sentence alignment. Other alignment techniques apply to paragraphs (see Mihailov, 2001) or words (e.g. the word aligner GIZA++; Och & Ney, 2003), the latter being a prerequisite for statistical machine translation. The TreeAligner (Volk et al., 2006) can be used for parallel treebanking, i.e. to align bilingual sentence pairs already annotated in terms of syntactic structure.

Within the context of the CroCo Corpus, a combined query of annotation and alignment enables, for instance, the investigation of so-called "crossing lines" and "empty links" (Čulo et al., 2011): "Empty links" are units in the target text that do not have matches in the source text and vice versa on any alignment layers. The term "crossing lines" is used to denote units whose alignment crosses the alignment of a higher level. Instances of text contained in one sentence or phrase in the source text but spread over two sentences or phrases, respectively, in the target text can be detected. Such differences probably have implications for the information structure, the distribution of given and new information, or linguistic foci contained in the target text. These structures are very similar to the Cross values discussed in Section 23.4, which clearly shows the need for interfacing product and process research.

23.2.3 Statistics, data analytics and machine learning

Translation corpora can also be statistically exploited, which is the case for machine translation (MT) research. The basic idea of statistical MT is to generate a translation from a parallel training corpus by calculating the most likely equivalent in the target language. Statistical translation models are generated and trained on the corpora with the help of machine learning. Both mono- and bilingual corpora are used to capture the typical linguistic patterns of the languages—the monolingual corpora generate the language model; the bilingual parallel corpora generate the translation model. In addition, statistical MT uses word-aligned n-grams—sequences of words (usually $n \leq 7$)—that are assigned probabilities representing how probable the word or sequence is, based on its distribution in the training corpus. Attempts have been made to unite different approaches (usually rule based and statistical) in hybrid systems so that the advantages of the respective approaches can be combined (e.g. Eisele, 2007). The latest approach to MT is the use of neural networks. Neural MT systems try to build one large neural network for translation, while statistical MT systems are composed of many small subcomponents. If available, several language pairs are trained together in order to improve the quality and efficiency of the system.

Statistical methods can also be used in order to identify translationese or quantitative patterns in translation corpora (Diwersy et al., 2014; Neumann & Hansen-Schirra, 2013). Baroni and Bernardini (2006) apply text categorization methods and machine learning in order to distinguish translated from non-translated text. Based on support vector machines, they report an accuracy with nearly 90% precision and more than 80% recall for automatically detecting translations. Similar approaches have been proven successful for the automatic classification of translated text (Gaspari & Bernardini, 2008; Koppel & Ordan, 2011; Kurokawa et al., 2009; Nisioi, 2015; van Halteren, 2008; Volansky et al., 2015). Rabinovich et al. (2016) take this kind of research a step further: they use a clustering procedure on the basis of PoS tags in order to classify translated, non-translated and non-native text. Their results show that translations and non-native texts are more similar to each other than they are to original text production. This means that constrained texts, i.e. text varieties that are directly influenced by other languages, seem to share properties, which in turn seems to indicate that universal processes are at play, processes that have an impact on translation as well as non-native text production.

23.2.4 Challenges and problems

When working with multilingual corpora, text segmentation poses several challenges to the comparability of the corpus analyses: For example, French clitics have to be lemmatized in order to render the segmentation between French and German comparable and thus analysable. Above word level, similar problems arise, which concern, for instance, the diverging segmentation of non-finite clauses in German and English. In this case, a possible solution would be to base the recognition of segments on the comparison of the respective functions in each language. In order to investigate the relationship between the source-text corpus and the target-text corpus, the matching text units have to be parallelized. Here, the difficult question is precisely what has to be aligned: whole texts, sentences or smaller units (such as phrases, words or morphemes). For practical or technical reasons, the sentence is usually chosen as the alignment unit, since the automatic detection of sentence boundaries is relatively straightforward. For many research questions (especially the one on the translation unit), this choice is a compromise, the actual focus of the investigation being on the relationship between smaller units. Whereas word alignment is useful for machine translation and multilingual term extraction, this procedure is not sufficient for the analysis of translations, because the translation unit is not reflected in single words. The desirable units for alignment are located between words and sentences and should be flexible enough to cope with typological differences between source and target language. Other alignment problems arise from the fact that different languages have different patterns with respect to sentence length, and that during the process of translating source-language sentences into target-language sentences, the former can be split or merged (see Fabricius-Hansen, 1999).

Another major drawback for corpus-based Translation Studies is the fact that many researchers still rely on untagged, unlemmatized KWIC concordances for their analyses. On this basis, however, only raw text can be found. For instance, irregular language or inflected forms are neglected. A lemmatized or morphologically tagged corpus, where word stems can be explored for the generation of concordances, would offer a higher level of efficiency (i.e. higher precision); a semantically annotated corpus (including synonyms, hyponyms, etc.) would enable more comprehensive querying (i.e. higher recall). Furthermore, string-based approaches are rather lexically oriented. Raw text, concordances or word counts are inadequate for grammatical or semantic investigations, which call for deeper linguistic interpretation. A further problem emerges in connection with cross-linguistic comparisons, as string-based queries fail to take into account typological differences. The analysis of, for example, simplification on the basis of type-token ratio, lexical density and average sentence length, which entails word counts, is not particularly well suited to many language pairs, since typological differences in morphology bias the results (see Hansen & Teich, 1999 for the language pair English–German). Thus, alternative ways of operationalizing the testing of the rather abstract hypotheses both on a text basis and in multilingual environments have to be sought. Syntactic information enables the investigation of explicitation as well as simplification. Thus, deep linguistic annotation (e.g. syntactic, semantic or pragmatic annotation) helps to bridge the gap between the rather abstract hypotheses developed by translation scholars (e.g. on explicitation or simplification in translations) and their realizations in the source and target texts.

For the development of an annotation scheme that meets the requirements of a multilingual analysis, there seem to be two methods that take into consideration the typological characteristics of the involved languages. First, the multilingual corpus is split up into monolingual subcorpora, which are then annotated independently. The second method uses one language as a basis for compiling and analysing a multilingual corpus, whereas the others have to conform. Both methods, however, are rather problematic. The latter forces the adapted language to fit into the system of the language used as a fundament. In the former, questions of cross-linguistic comparison are merely postponed to a later stage and come into play in the interpretation of the data obtained from this research design. In order to avoid both problems, the annotation scheme has to abstract from the level of language-specific realizations to functional categories that are comparable across languages. In order to keep inter-annotator agreement and thus the annotation quality to an acceptable level, annotation schemes and alignment rules have to be very precise and transparent (see Brants & Hansen, 2002).

23.3 The process perspective: Translation process research

From the process perspective, analyses of the source text and the translation product are also very important. They are necessary to outline what might be happening in the translator's mind. To analyse cognitive translation processes, however, the following methods are well established in translation process research (see Göpferich, 2008; Krings, 2005).

23.3.1 Report data

Think-aloud protocols (TAPs) are used to record translators' thoughts during the translation process. Translators can be asked to verbalize their thoughts directly during the translation or retrospectively; the first variant, however, is more common. The transcriptions of these verbalizations are called think-aloud protocols. Using TAPs has various advantages and disadvantages. Among others, one major disadvantage of immediate verbalization is that studies have shown that verbalization changes the thought process (see Jakobsen, 2003) and therefore may also change the translation process. TAPs that are produced retrospectively do not change the translation process, because the translator is asked about specific translation units only after the whole translation has been produced. The problem is, however, that a lot of valuable thoughts might get lost between translation production and verbalization. Here, it is helpful to use a screen recording of the session—maybe with eye-tracking data—to help the participant recall the passages of interest. Further, only thoughts that are conscious can be uttered, and a high cognitive load during the task might prevent participants from verbalizing their thoughts. However, it is one of the few methods that actually purport to reproduce what is going on inside the heads of the participants, even if not completely (Jääskeläinen, 2010).

Questionnaires are usually distributed in a written form—either on paper or electronically—and contain a set of questions. These questions can be open—the participant decides which questions to answer and in how much detail—or closed—the participant can choose from a set of answers. Mixed questions elicit a number of possible answers. Participants can, however, add their own answers. Closed questions are much easier to assess, which saves time and money. Open questions, however, deliver more extensive and more differentiated data (Klöckner & Friedrichs, 2014). Compared with interviews, questionnaires are easier to distribute, and participants might be more willing to fill out a questionnaire whenever they have time than to schedule a date with the interviewer. Further, questionnaires are more discreet and more anonymous. On the other hand, participants must be able to read and write, which excludes some potential target groups (but this is not usually a problem in translation process studies). Written answers will probably be shorter and less detailed than answers in an interview. Finally, a questionnaire does not allow personal contact, so that participants are not able to ask questions and the examiner cannot get an impression of the person (see Döring & Bortz, 2014, pp. 398–399).

The requirements for a high-quality questionnaire are that the participant is able to answer the questions and that it is objective, valid and reliable. However, even a very sophisticated questionnaire can cause non-responses and different response qualities due to the individual characteristics of the participants. Furthermore, participants tend to follow certain response strategies, e.g. they prefer extreme categories or medium categories, they prefer the first or the last response choice (also called primacy or recency effect), or they answer according to social conventions (see Reinecke, 2014, pp. 612–613).

23.3.2 Behavioural data

Keylogging software allows the researcher to analyse the text production process and the associated mental processes. All key (and mouse) activities are recorded during text production, including typing processes, special key combinations and deleting activities. Further, pauses are recorded, which can elucidate reading processes and the segmentation of the text, which is done subconsciously by the participant and might indicate text passages that require high cognitive effort (see Alves & Vale, 2011). However, one has to keep in mind that the data only allow us to speculate on mental processes of participants engaged in text production. Other methods, especially eye tracking, can help to interpret participants' behaviour, e.g. during pauses (see Jakobsen, 2011, 37–38). Recording keylogging data with a keylogger is unobtrusive, because the program runs in the background and, hence, the recording process is not noticeable (see Carl, 2012, p. 4108).

The term *eye tracking* refers to the methodology with which human eye movement (called saccades) and fixations can be recorded and quantified. The basic assumption is that the human pays attention to the point (s)he fixates on (a concept also known as the eye-mind assumption, introduced in a reading study by Just & Carpenter, 1980), although this is not always the case. The eye movement data give us promising and important hints about what is going on in the translator's mind, although we can only interpret the data and cannot be entirely sure what is happening in the black box. This research method, however, brings not only advantages but also challenges to translation process research. For instance, the kind of eye-tracking system has to be considered. A remote eye tracker is considered ecologically more valid, because the participant can move comparably freely in front of the computer screen, while head-mounted eye trackers and eye trackers with chin or head rests produce more accurate data. Eye-tracking glasses, on the other hand, liberate the participants from the screen. In contrast to questionnaires, think-aloud protocols and keylogging technology, which involve only very low costs if the experimenter has a PC or laptop and an Internet connection, eye tracking requires expensive hardware and software. Further, eye-tracking experiments must be conducted in a consistent environment, e.g. with similar lighting conditions for all participants (see O'Brien, 2009, pp. 251–254; on detailed information on requirements for a suitable eye-tracking lab, see Rösener, 2016). The texts should be no longer than about 300 words in eye-tracking studies so that scrolling does not become necessary, because this makes the data harder to assess. Similarly, the texts should be in a font size of 16 or 18 and displayed with at least 1.5 line spacing so that the eye-tracking data can be mapped correctly (see O'Brien, 2009, pp. 261–262).

23.3.3 Neuroscientific data

Translation Studies have only in recent years slowly begun to use neuroscientific methods to ultimately tackle the problem of what is going on in the translator's "black box" while translating. These methods are typically used in very controlled experiments and cannot (yet) be used in authentic translation tasks. Electroencephalography (EEG) is used to record electrical activity in the brain. Depending on what needs to be measured and how precise the recordings need to be, 16 to 256 electrodes are distributed on the head. Artefacts in the EEG signals can be produced by external influences like head movement, blinking or other muscular activities. These artefacts cannot easily be avoided and have to be eliminated from recordings, if possible, before the signals are analysed and interpreted (see Freeman & Quian Quiroga, 2013, 5–6). The electrical activity in the brain produces oscillations, which have been associated with certain functions and pathological conditions of the brain. However, not only oscillations but also typical patterns can be studied with the help of EEG signals. An example of such an event-related potential (ERP) would be the N400. This amplitude peaks negatively about 400 ms after a participant has been presented with a stimulus. The N400 is widely acknowledged as a measure for semantic processing. If, for example, a sentence is presented that includes a semantically nonsensical stimulus, the negative amplitude will be easily visible 400 ms later (see Kutas & Federmeier, 2011). Translation Studies has used EEG, for example, to investigate priming, monitoring and inhibition (see Oster, 2018), cognitive effort in backward translation (García et al., 2015), conceptualization (e.g. Grabner et al., 2007) and expectancy violations (e.g. Elmer et al., 2010)—for an overview, see Hansen-Schirra (2017).

The applications of functional magnetic resonance imaging (fMRI) include all areas of brain imaging and have become a very important tool for neuroscience research, both clinical and cognitive (see Faro & Mohamed, 2006, pp. v–vi). Kim and Bandettini (2006, p. 3) observed "that functional brain mapping is possible by using the venous blood oxygen level-dependent (BOLD) magnetic resonance imaging (MRI) contrast". The BOLD method depends on the level of deoxyhaemoglobin, which can be seen in the signal intensity of magnetic resonance images when the level changes and can hence be used for human brain imaging. Compared with other brain imaging methods, fMRI is especially useful to show language areas in the brain, because it is non-invasive and produces images of good quality and with good localization (among other benefits). It is rather difficult to assign brain regions to single language processes such as phonetic, semantic or syntactic processes, because they often work together. Carefully selected research designs with contrasting conditions can, however, help in tackling these problems (see Binder, 2006, 245–248). Ahrens et al. (2010), Franceschini et al. (2003) and Kalderon (2017) were the first researchers to use fMRI to localize and visualize activated areas during translation and interpreting.

23.3.4 Challenges and problems

Despite the differences discussed with respect to the individual methods mentioned, they all have in common that the participants for a study must be chosen carefully. Although professional translators are often considered more valuable, as they have practical experience, they are harder to convince to participate in a study and often expect financial compensation, as they miss (part of) their workday. Students, on the other hand, are easier to recruit, as eye-tracking studies in translation process research are usually conducted at a university, and the study might even be credited in classes. However, students exhibit different cognitive processes and behavioural patterns than professionals, which might in turn affect the generalizability of the results. Further, some studies, like questionnaire studies, may be carried out at home, while other studies, especially those that require special equipment, as most eye-tracking, EEG or fMRI studies do, can only be conducted in special laboratories. In addition, one question is whether the participants will commit to an equal degree when they invest their free time into participating in studies as opposed to when they are paid or rewarded in a different manner for the task. Additionally, limited funding might restrict the number of participants that can be recruited for the study. However, it is doubtful whether small participant numbers, e.g. 12 participants or even fewer, can return generalizable and well-balanced results. Nonetheless, these studies are valuable to build hypotheses for larger studies. Finally, the professionalism of the participants has to be addressed, as not every translator with a degree in translation is equally capable of all tasks. Some issues may also occur which disqualify the participant for the study, e.g. typing skills, language competence or ability to follow instructions, or if the participant feels intimidated, judged or pressured during the session (see O'Brien, 2009, pp. 254–259, p. 262).

Another problematic issue concerning TPR studies concerns the authenticity of the translation situation. While some methods are quite authentic in their experimental setup—i.e. the participant can use a computer or laptop and online research for the translation of authentic texts—others are not natural at all, e.g. those involving an EEG cap or an fMRI scanner. The latter usually focus on single-word translation or translations, which are not written or verbalized at all, while the former also take place during a phase of observation. These experimental conditions are needed in order to figure out which effect is triggered by which stimulus. Furthermore, not only are many participants needed, but so are many repetitions of the stimuli, to improve statistical validity and significance. Thus, the more controlled an experiment is, the easier it will be to see a clear stimulus–effect relation. So, there is always a trade-off between authenticity of the translation situation and effect size concerning a given hypothesis.

23.4 The challenge of interfacing the two perspectives

Data triangulation—i.e. linking two or more sources of data, researchers, methodological approaches, theoretical ideas or analytical designs so that the different advantages of each methods can be exploited (see Thurmond, 2001, pp. 253–257)—has become more and more important in Translation Studies (e.g. Alves, 2003). Triangulation has the advantages that the research data become more reliable, inventive approaches are developed to comprehend and interpret research hypotheses, existing theories might be challenged or confirmed, and a phenomenon can be better understood. Every triangulation approach on its own, whether it is data, researcher, methodological, theory or analysis triangulation, has individual advantages, but also disadvantages. In general, these include

> [an] increased amount of time needed in comparison to single strategies, [...] difficult[ies] of dealing with the vast amount of data, [...] potential disharmony based on investigator biases, [...] conflicts because of theoretical frameworks, and [...] lack of understanding about why triangulation strategies were used.
>
> *Thurmond, 2001, p. 256*

In short, triangulation is valuable to gather findings from different perspectives that are complete and confirm each other, and to strengthen the findings. The researchers, however, must be able to explain why they used the triangulation method and why it was necessary. Many studies have adopted this approach in translation process research in recent years, especially concerning data triangulation. The danger, however, is that, among other possible problems, the amount of data becomes overwhelming. Some solutions on how to deal with large data sets might be to work in research teams, so that either the same research topic is analysed together or different researchers analyse various hypotheses on the same data (O'Brien, 2009, pp. 260–261, p. 264).

What we suggest here exceeds the triangulation of different perspectives, since we propose to interface different research branches, namely that of corpus linguistics with that of TPR. There are, however, two ways of achieving such interfacing:

- different data sets are used with different methods to answer one research question, or
- the same data set is used with different methods to answer one research question.

In the following, we will present both alternatives on the basis of selected examples.

23.4.1 Interfacing different data sets with different methods

The first studies in which different data sets were interfaced using different methods can be found in Hansen (2003) and Alves et al. (2010). The former examined the relation between normalization and translation expertise; the latter the relation between translation units and grammatical shifts occurring during translation. However, in both studies, the process-oriented task was rather example based and could not have been tested with statistical methods. The generalizability of the results is therefore limited. However, these studies can be regarded as pioneer work, given the methodological principle they adopt of interfacing product and process research.

For a more empirical approach, let's get back to our dichotomy of normalization vs. shining through mentioned in Section 23.1: Tirkkonen-Condit (2005) argues that literal translation, which may result in shining-through effects, is a default translation procedure, which is cognitively preferred to others. Chesterman (2011) and Halverson (2015) reintroduce the concept of literal translation, assuming that entrenchment effects strengthen the co-activation or priming effect of linguistic patterns and thus reduce the cognitive load during translation for literal renderings (see Schaeffer & Carl, 2014 for an empirical operationalization). Tirkkonen-Condit (2005) assumes that translators constantly monitor production, and as soon as a problem is encountered in their default translation routes, they stop the literal translation process and try to find a better solution. The monitor model has been empirically tested by Carl & Dragsted (2012).

The process-oriented continuum between monitoring and priming, i.e. literal translation, could be another way to perceive the product-based dichotomy between normalization and shining through. The monitor model, however, still exhibits some shortcomings. It is, for example, not precise enough to determine which factors influence priming and which linguistic levels are affected over others. In order to further clarify these open questions, Hansen-Schirra et al. (2017) investigated cognates (translation equivalents which share a similar form). Cognates are relatively easy to control in experimental settings and can be investigated in many language pairs. Using EEG methodology, Christoffels et al. (2006), de Groot (2011) and Oster (2018) show that cognates seem to be pre-activated during translation. This cognate facilitation effect indicates that priming takes place while participants are decoding the source text. On the other hand, Kußmaul (1989) argues that monitoring takes place while translators are translating cognates. In order to find out which factors trigger priming or monitoring processes, Hansen-Schirra et al. (2017) exploited large multilingual corpora to quantify cognate usage patterns, on the one hand, and carried out controlled experiments to explain translation-inherent factors for cognate translation, on the other. Using the Google n-gram viewer, they trace the impact of societal and technological development on the usage of selected cognates. They show cases where a cognate was first used in one language, but due to common language roots, it was accepted in other languages as well. In addition, typical preferences in terms of cognate usage could be identified for different text types in the German DWDS corpus. It seems to be the case that some text types, such as academic texts, are more receptive to cognates than others (e.g. newspaper texts). When comparing translation with post-editing corpora, they found that translations from scratch and post-edited target texts show similar cognate and non-cognate usage; however, the variety in non-cognate lexical choices is statistically higher in translations from scratch than in post-edited texts, for which the machine translation output was relied on in most cases. This shows that the usage of computer-aided translation may also have an influence.

Complementing the corpus results with experimental methods, Hansen-Schirra et al. (2017) contrasted single-word translation of cognates with translating them with context. The analysis showed that the use of cognates in translations is dependent on the context of the translation.

In general, the participants chose a cognate less frequently when they were translating a whole text than when they only had to find German equivalents in a word list. This might indicate that the cognate translation is the most obvious and the easiest without context, because the cognate is similar not only in meaning but also in form. When a cognate is embedded in context, however, the surrounding co-text of the word triggers different lexical choices. Another translation experiment included translation students with different levels of expertise. The comparison here suggests that monitoring or other control mechanisms develop with increasing translation competence, since the number of cognates decreases for more experienced translators. In addition, other studies have shown that the number of cognates in translations varies significantly depending on other factors, such as status of the respective languages (Vintăr & Hansen-Schirra, 2005), translation mode (Oster, 2017) or exposure to media in the mother tongue (Oster, 2018). In conclusion, it can be argued that some of the factors that might influence the production of cognates can only be isolated with corpus-based research, while others have to be approached from a more process-oriented perspective. A combination of both comprehensively shows how the translation of cognates can be predicted against the background of different constraints.

23.4.2 Interfacing the same data set with different methods

When interfacing the same data with information from several data collection methods, corpus enrichment becomes more complicated, since user-activity data is added to the text corpus. This means that each word, phrase or sentence is annotated not only with linguistic information, such as PoS tags, phrase structure, co-reference resolution, etc., but also with data concerning the translator's behaviour during the translation process. More concretely, data on eye movements, i.e. fixation times, saccades, regressions, etc., have to be mapped onto the source and target units. Information from the keylogs, i.e. on the production processes of words as well as pauses, also has to be added. The representation of data including all kinds of annotations and alignments can be quite complex. If the baseline is the word (it could also be the syllable or a time-stamp for spoken data), there are several alignment and annotation relations:

- product annotation with linguistic information (PoS, phrase structure, semantic relations, morphology, etc.)
- product alignment with source text (word alignment, sentence alignment, etc.)
- process annotation with production units from keylogs (production history, pauses, etc.)
- process alignment with eye-tracking data (fixation duration, fixation count, etc.)

While product annotation and alignment are statistical in nature, user-activity data from the process constitute dynamic information. Some of the linguistic annotations and alignments may be hierarchical in their structure (e.g. syntactic dependency trees), while others reflect temporal progress (e.g. eye data). Both phenomena complicate the corpus representation and the holistic integration of all data layers as well as querying procedures and corpus analyses. Moreover, the data may be further processed, resulting in more specified metrics, e.g. the Cross value (explained later), which in turn may again be added as annotation to each word of the corpus. This makes the corpus itself a dynamic resource.

The first dynamic corpora including user-activity data deal with individual language pairs and include either keylogging (e.g. Alves & Gonçalves, 2013; Jakobsen, 2002; Serbina et al., 2015) or eye-tracking data (e.g. Jakobsen & Jensen, 2008). The CRITT-TPR database is the first dynamic corpus for several language pairs and includes eye-tracking and keylogging data in addition to linguistic annotation. This multiple translation corpus includes different translation tasks (from scratch, post-editing MT, etc.) by professional and student translators (see Hvelplund, 2011). While some data categories represent "pure" eye-tracking and keylogging data, other parameters are provided which present processed data (e.g. the Cross values explained later). Finally, there is also additional information on the source and target-text units, such as PoS tags (see Carl & Schaeffer, 2013 or Carl, Schaeffer, et al., 2016; these papers also provide a detailed description of the parameters contained in the database). Schaeffer & Carl (2014) introduce a metric that operationalizes the above-mentioned literal translation hypothesis. Further, they evaluate the effort that is necessary for non-literal translation. To do so, they use the gaze behaviour of translators for different language pairs from the CRITT-TPR DB. Different lexical realizations of source words in the target language are counted, and for some words, higher variation is detected than for others. On a syntactic basis, crossing word alignments (similar to the crossing lines in Section 23.2) are introduced: the metric computes the Cross values for single words, based on their position in the source and the target language. Depending on the point of view, the Cross value can be realized from the source text as a reference and the target text as output ("CrossS") or the other way around ("CrossT"). The smaller the Cross value, the more similar the texts are in terms of structure; when the Cross value is high, syntax varies significantly. The authors show that higher Cross values strongly correlate with total reading time on source and target words and, therefore, prove that high syntactic variation takes more effort to produce. Furthermore, a strong correlation can be found between production time of the target-text word and number of alternative translations. "With few choices post-editors are quicker than translators, but this distance decreases as the number of translation choices increase" (ibid., p. 34). Additionally, a strong correlation was detected between total reading time on target-text word and number of alternative translations. This strongly supports the literal translation hypothesis by attesting higher cognitive load for non-literal translations—taking lexical as well as syntactic entropy into account. Bangalore et al. (2016) investigate the role of co-activation of languages in translation and its influence on the translator's behaviour, combining syntactic annotation with user-activity data.

23.5 Concluding remarks and outlook

In this chapter, we have:

- clarified the need for interfacing process and product data. However, we have also explained why it is not sufficient to triangulate different methods; rather, we need to combine different methods from different research branches in Translation Studies, i.e. corpus linguistics and translation process research.
- introduced the most common methods within the two research branches and discussed their strengths but also their shortcomings and problems.
- given examples of successful product–process interfaces—with a common data set or without.

We have shown that corpus linguistics leads us to the quantification of translation patterns. However, in order to explain the quantitative phenomena, it is necessary to complement them with cognitive experiments. For instance, product analyses may identify normalization or shining through, but only on the basis of process-based research can we explain how and why translators decided upon the strategies that trigger exactly these phenomena. The combination of corpus-based work on translated texts (i.e. product-oriented research) with experimental studies of translators' behaviour (i.e. process-oriented research) seems to be promising, since it enables not only the quantification and interpretation of translation properties and strategies but also their explanation concerning cognitive processes and resources. Recent advances in corpus architecture and corpus querying (including multi-layer annotation and alignment) and increasing incorporation of methods from psycholinguistics and cognitive science into process-oriented research (especially eye tracking and keystroke logging) point to a desired combination of corpus studies with a more direct, experimental insight into processing efforts.

Future work will include the integration of speech data such as translation dictation or sight translation, on the one hand (see Carl et al., 2016b), and deeper analysis as well as the meta-annotation of user-activity databases, on the other. With multimodal annotation tools, such as the web application developed by Alves and Vale (2009), it is possible to manually annotate the user-activity data and to query these corpora. Such tools facilitate translation-oriented annotations, such as the explicit labelling of translation procedures or the annotation of explicitation. On this basis, correlation between these annotations and the eye-tracking or keylogging data can be calculated. It can be clarified whether, for instance, explicitating ambiguous structures will influence source-text reading times.

Another development will be the integration of EEG and fMRI data into user-activity corpora. So far, these methods have been adopted from cognitive sciences in order to identify biological correlates of cognitive and motoric processes. However, they have not yet been integrated into multi-method approaches to translation. Focusing on single words as stimuli, or not being able to actually produce the translation orally or in written form, poses several problems for the investigation of cognitive translation processes: As translators usually translate whole texts in context and with respect to a given *skopos*, it is difficult to study complex translation processes or strategies using EEG or fMRI technology as described earlier. Decoding the source text and encoding the target text involve such a wide range of stimuli that it is difficult to isolate a single stimulus during an authentic translation task. This means that studies in which single-word translation is used or in which the translation has to be produced mentally are not ecologically valid, because they neglect the complex problem-solving mechanism employed during translation. Although the context problem has been relativized by van Hell and de Groot (2008), who found similar effects for context-free vs. sentence-context conditions, there is still a huge gap between existing translation-oriented theories and models and their operationalization with methods from cognitive science (see Göpferich, 2008). However, Nagels et al. (2013) integrate authentic text reception and production into fMRI methodology, which seems to be very promising for translation process research. Finally, it can be argued that it is beneficial to triangulate data collected in manipulated studies, on the basis of which clear effects can be derived with sentence-context studies that corroborate the results gained from the controlled setting. By interfacing different methods with different data sets, experimental control can be complemented with ecological validity.

Note

1 A critical discussion of the existence of translation universals can be found in Mauranen & Kujamäki (2004), Malmkjær (2005) or House (2008). Hansen-Schirra et al. (2012) reframed the concept and more cautiously call them translation-specific properties.

Further reading

Alves, F. (Ed.). (2003). *Triangulating translation: Perspectives in process oriented research*. Amsterdam: John Benjamins.

This collection of articles focuses on translation process research but also includes the integration of translation product analyses. The data triangulation is reflected from a theoretical and a methodological point of view.

Carl, M., Bangalore, S., & Schaeffer, M. (Eds.). (2016). *New directions in empirical translation process research*. London: Springer.

This volume focuses on the investigation of keylogging and eye-tracking data. Using multi-method approaches, "old" questions concerning, for instance, priming can be re-analysed.

Hansen-Schirra, S., Čulo, O., & Hofmann, S. (Eds.). (2017). *Empirical modelling of translation and interpreting* (Vol. 7). Berlin: Language Science Press. http://langsci-press.org/catalog/book/132

This book focuses on the data-driven modelling of translation and interpreting processes. By interfacing process and product data, more comprehensive models and theories can be formulated.

References

Ahrens, B., Kalderon, E., Krick, C., & Reith, W. (2010). fMRI for exploring simultaneous interpreting. In D. Gile, G. Hansen, & N. Pokorn (Eds.), *Why translation studies matters* (pp. 237–249). Amsterdam: John Benjamins.

Alves, F. (Ed.). (2003). *Triangulating translation: Perspectives in process oriented research*. Amsterdam: John Benjamins.

Alves, F., & Gonçalves, J. L. V. R. (2013). Investigating the conceptual-procedural distinction in the translation process: A relevance-theoretic analysis of micro and macro translation units. *Target, 25*(1), 107–124.

Alves, F., Pagano, A., Neumann, S., Steiner, E., & Hansen-Schirra, S. (2010). Translation units and grammatical shifts: Towards an integration of product- and process-based translation research. In G. M. Shreve, & E. Angelone (Eds.), *Translation and cognition*. Amsterdam: John Benjamins.

Alves, F., & Vale, D. (2009). Probing the unit of translation in time: Aspects of the design and development of a web application for storing, annotating and querying translation process data. *Across Languages and Cultures, 10*(2), 251–273.

Alves, F., & Vale, D. C. (2011). On drafting and revision in translation: A corpus linguistics oriented analysis of translation process data. *Translation: Corpora, Computation, Cognition, 1*(1), 105–122.

Baker, M. (1995). Corpora in translation studies: An overview and some suggestions for future research. *Target, 7*(2), 223–245.

Baker, M. (1996). Corpus-based translation studies: The challenges that lie ahead. In H. Somers (Ed.), *Terminology, LSP and translation: Studies in language engineering in honour of Juan C. Sager* (pp. 175–186). Amsterdam: John Benjamins.

Bangalore, S., Behrens, B, Carl, M., Ghankot, M., Heilmann, A., Nitzke, J., Schaeffer, M., & Sturm, A. (2016). Syntactic variance and priming effects in translation. In M. Carl, S. Bangalore, & M. Schaeffer (Eds.), *New directions in empirical translation process research* (pp. 211–238). Berlin: Springer.

Baroni, M., & Bernardini, S. (2006). A new approach to the study of translationese: Machine-learning the difference between original and translated text. *Literary and Linguistic Computing, 21*(3), 259–274.

Binder, J. (2006). fMRI of language systems: Methods and applications. In S. Faro, F. Mohamed, M. Law, & J. Ulmer (Eds.), *Functional neuroradiology* (pp. 245–277). Berlin: Springer.

Brants, S., & Hansen, S. (2002). Developments in the TIGER annotation scheme and their realization in the corpus. In *Proceedings of the Third International Conference on Language Resources and Evaluation (LREC-2002)*. Las Palmas, 1643–1649.

Carl, M. (2012). Translog-II: A program for recording user activity data for empirical reading and writing research. In *LREC, proceedings: Eighth International Conference on Language Resources and Evaluation* (pp. 4108–4112).

Carl, M., Aizawa, A., & Yamada, M. (2016). English-to-Japanese translation vs. dictation vs. post-editing. In *LREC, 2016 proceedings: Tenth International Conference on Language Resources and Evaluation* (pp. 4024–4031).

Carl, M., & Dragsted, B. (2012). Inside the monitor model: Processes of default and challenged translation production. *Translation: Corpora, Computation, Cognition, 2*(1), 127–145.

Carl, M., & Schaeffer, M. (2013). The CRITT translation process research database V1.4. https://odoko.cbs.dk/handle/10398/9058

Carl, M., Schaeffer, M., & Bangalore, S. (2016). The CRITT translation process research database. In M. Carl, S. Bangalore, & M. Schaeffer (Eds.), *New directions in empirical translation process research* (pp. 13–54). Berlin: Springer.

Chesterman, A. (2011). Reflections on the literal translation hypothesis. In C. Alvstad, A. Hild, & E. Tiselius (Eds.), *Methods and strategies of process research: Integrative approaches in translation studies* (pp. 23–35). Amsterdam: John Benjamins.

Christoffels, I. K., de Groot, A. M. B., & Kroll, J. F. (2006). Memory and language skills in simultaneous interpreters: The role of expertise and language proficiency. *Journal of Memory and Language, 54*, 324–345.

Čulo, O., Hansen-Schirra, S., Maksymski, K., & Neumann, S. (2011). Empty links and crossing lines: Querying multi-layer annotation and alignment in parallel corpora. *Translation: Computation, Corpora, Cognition, 1*(1), 75–104.

de Groot, A. M. B. (2011). *Language and cognition in bilinguals and multilinguals: An introduction*. New York: Psychology Press.

Diwersy, S., Evert, S., & Neumann, S. (2014). A weakly supervised multivariate approach to the study of language variation. In B. Szmrecsanyi, & B. Wälchli (Eds.), *Aggregating dialectology, typology, and register analysis. Linguistic variation in text and speech. Linguae & Litterae* (pp. 174–204). Berlin: de Gruyter.

Döring, N., & Bortz, J. (2014). *Forschungsmethoden und Evaluation*. 5. Berlin: Springer.

Eisele, A. (2007). Hybrid machine translation: Combining rule-based and statistical MT systems. First machine translation marathon, Edinburgh. http://mt-archive.info/MTMarathon-2007-Eisele.pdf

Elmer, S., Meyer, M., & Jancke, L. (2010) Simultaneous interpreters as a model for neuronal adaptation in the domain of language processing. *Brain Research, 1317*, 147–156.

Fabricius-Hansen, C. (1999). Information packaging and translation: Aspects of translational sentence splitting (German-English/Norwegian). In M. Doherty (Ed.), *Sprachspezifische Aspekte der Informationsverteilung* (pp. 175–214). Berlin: Akademie-Verlag.

Faro, S., & Mohamed, F. (2006). *Functional MRI: Basic principles and clinical applications*. New York: Springer Science & Business Media.

Franceschini, R., Zappatore, D., & Nitsch, C. (2003). Lexicon in the brain: What neurobiology has to say about languages. In J. Cenoz, B. Hufeisen, & U. Jessner (Eds.), *The multilingual lexicon* (pp. 153–166). New York: Kluwer Academic Publishers.

Francis, W. S., & Gallard, S. L. K. (2005). Concept mediation in trilingual translation: Evidence from response times and repetition priming patterns. *Psychonomic Bulletin & Review, 16*, 1082–1088.

Frawley, W. (1984). *Translation: Literary, linguistic & philosophical perspectives*. Newark: University of Delaware Press.

Freeman, W., & Quian Quiroga, R. (2013). *Imaging brain function with EEG: Advanced temporal and spatial analysis of electroencephalographic signals*. New York: Springer.

García, A. M., Mikulan, E., & Adolfi, F. (2015). *Reading and translation in bilinguals: A connectivity study with scalp and intracranial EEG recordings*. Buenos Aires: University of Buenos Aires.

Gaspari, F., & Bernardini, S. (2008). Comparing non-native and translated language: Monolingual comparable corpora with a twist. In *Proceedings of the international symposium on using corpora in contrastive and translation studies*. Zhejiang University, Hangzhou, 25–27 September. Lancaster University.

Göpferich, S. (2008). *Translationsprozessforschung: Stand—Methoden—Perspektiven*. Vol. 4. Translationswissenschaft. Tübingen: Narr.

Grabner, R., Brunner, C., Leeb, R., Neuper, C., & Pfurtscheller, G. (2007). Event-related EEG theta and alpha band oscillatory responses during language translation. *Brain Research Bulletin, 72*(1), 57–65.

Halverson, S. (2015). Cognitive translation studies and the merging of empirical paradigms. The case of "literal translation." *Translation Spaces, 4*(2), 313–343.

Hansen, S. (2003). *The nature of translated text: An interdisciplinary methodology for the investigation of the specific properties of translations*. Saarbrücken: DFKI/Universität des Saarlandes.

Hansen, S., & Teich, E. (1999). Kontrastive Analyse von Übersetzungskorpora: Ein funktionales Modell. In J. Gippert (Ed.), *Sammelband der Jahrestagung der GLDV 99* (pp. 311–322). Prague: Enigma Corporation.

Hansen-Schirra, S. (2008). Interactive reference grammars: Exploiting parallel and comparable treebanks for translation. In E. Yuste (Ed.), *Topics in language resources for translation and localisation*. Amsterdam: John Benjamins.

Hansen-Schirra, S. (2017). EEG and universal language processing in translation. In J. Schwieter, & A. Ferreira (Eds.), *The handbook of translation and cognition* (pp. 232–247). Oxford: Wiley-Blackwell.

Hansen-Schirra, S., Čulo, O., & Hofmann, S. (Eds.). (2017). *Empirical modelling of translation and interpreting* (Vol. 7). Berlin: Language Science Press. http://langsci-press.org/catalog/book/132

Hansen-Schirra, S., Neumann, S., & Steiner, E. (2012). *Cross-linguistic corpora for the study of translations—Insights from the language pair English-German. Text, translation, computational processing (11)*. Berlin: de Gruyter Mouton.

Hansen-Schirra, S., Nitzke, J., & Oster, K. (2017). Predicting cognate translation. In S. Hansen-Schirra, O. Čulo, & S. Hofmann (Eds.), *Empirical modelling of translation and interpreting* (pp. 3–22). Berlin: Language Science Press.

Hansen-Schirra, S., & Teich, E. (2008). Corpora in human translation. In A. Lüdeling, & M. Kytoe (Eds.), *International handbook on "corpus linguistics"*. Berlin: de Gruyter.

Hofland, K. (1996). A program for aligning English and Norwegian sentences. In S. Hockey, N. Ide, & G. Perissinotto (Eds.), *Creating and using English language corpora* (pp. 25–37). Amsterdam: Rodopi.

House, J. (2008). Beyond intervention: Universals in translation? *Trans-Kom, 1*(1), 6–19.

Hvelplund, K. (2011). Allocation of cognitive resources in translation: An eye-tracking and key-logging study [PhD dissertation]. Copenhagen: Copenhagen Business School. www.econstor.eu/bitstream/10419/208778/1/cbs-phd2011-10.pdf

Jääskeläinen, R. (2010). Think-aloud protocol. In Y. Gambier, & L. van Doorslaer (Eds.), *Handbook of translation studies 1* (pp. 371–374). Amsterdam: John Benjamins.

Jakobsen, A. L. (2002). Translation drafting by professional translators and by translation students. *Copenhagen Studies in Language, 27*, 191–204.

Jakobsen, A. L. (2003). Effects of think aloud on translation speed, revision and segmentation. In F. Alves, *Triangulating translation: Perspectives in process oriented research* (pp. 69–95). Amsterdam: John Benjamins.

Jakobsen, A. L. (2011). Tracking translators' keystrokes and eye movements with Translog. In C. Alvstad, A. Hild, & E. Tiselius (Eds.), *Methods and strategies of process research* (pp. 37–55). Amsterdam: John Benjamins.

Jakobsen, A. L., and Jensen, K. T. H. (2008). Eye movement behaviour across four different types of reading task. *Copenhagen Studies in Language, 36*, 103–124.

Johansson, S. (2002). Towards a multilingual corpus for contrastive analysis and translation studies. In E. Borin (Ed.), *Parallel corpora, parallel worlds* (pp. 47–59). Amsterdam: Rodopi.

Just, M., & Carpenter, P. (1980). A theory of reading: From eye fixations to comprehension. *Psychological Review, 87*(4), 329–354.

Kalderon, E. (2017). Neurophysiologie des Simultandolmetschens: eine fMRI-Studie mit Konferenzdolmetschern [PhD Dissertation]. Mainz: Gutenberg Qualify. https://publications.ub.uni-mainz.de/theses/volltexte/2017/100001169/pdf/100001169.pdf

Kautz, U. (2000). *Handbuch Didaktik des Übersetzens und Dolmetschens*. Munich: Iudicium.

Kenny, D. (1997). Creatures of habit? What collocations can tell us about translation. Poster presented at ACH/ALLC 97.

Kenny, D. (1998). Corpora in translation studies. In M. Baker (Ed.), *Routledge encyclopedia of translation studies* (pp. 50–53). London: Routledge.

Kim, S., & Bandettini, P. (2006). Principles of functional MRI. In S. Faro, & F. Mohamed (Eds.), *Functional MRI: Basic principles and clinical applications* (pp. 3–22). New York: Springer.

Klöckner, J., & Friedrichs, J. (2014). Gesamtgestaltung des Fragebogens. In N. Baur, & J. Blasius (Eds.), *Handbuch Methoden der Empirischen Sozialforschung* (pp. 675–685). Wiesbaden: Springer.

Koller, W. (2004). *Einführung in die Übersetzungswissenschaft*. Wiesbaden: Quelle & Meyer.

Koppel, M., & Ordan, N. (2011). Translationese and its dialects. *Proceedings of the 49th Annual Meeting of the Association for Computational Linguistics: Human Language Technologies*. Association for Computational Linguistics, 1318–1326.

Krings, H. (2005). Wege ins Labyrinth—Fragestellungen und Methoden der Übersetzungsprozessforschung im Überblick. *Meta: Translators' Journal, 50*(2), 342–358.

Kurokawa, D., Goutte, C., & Isabelle, P. (2009). Automatic detection of translated text and its impact on machine translation. *Proceedings of MT-Summit* XII, 81–88.

Kußmaul, P. (1989). Interferenzen im Übersetzungsprozeß—Diagnose und Therapie. In H. Schmidt (Ed.), *Interferenz in der Translation* (pp. 19–28). Leipzig: VEB Enzyklopädie.

Kutas, M., & Federmeier, K. (2011). Thirty years and counting: Finding meaning in the N400 component of the event related brain potential (ERP). *Annual Review of Psychology*, 62, 621–643.

Laviosa-Braithwaite, S. (1996). The English comparable corpus (ECC): A resource and a methodology for the empirical study of translation [PhD Dissertation]. Manchester: UMIST.

Levelt, W. J. M. (1989). *Speaking: From intention to articulation*. Cambridge, MA: MIT Press.

Levelt, W. J. M. (1999). Producing spoken language: A blueprint of the speaker. In C. M. Brown, & P. Hagoort (Eds.), *The neurocognition of language* (pp. 83–122). Oxford University Press. https://pure.mpg.de/rest/items/item_147935/component/file_196891/content

Malmkjær, K. (2005). Norms and nature in translation studies. *Synaps, 16*, 13–20.

Mauranen, A. (1997). Hedging in language revisers' hands. In R. Markkanen, & H. Schröder (Eds.), *Hedging and discourse: Approaches to the analysis of a pragmatic phenomenon in academic texts* (pp. 115–133). Berlin: de Gruyter.

Mauranen, A., & Kujamäki, P. (Eds.) (2004). *Translation universals. Do they exist?* Amsterdam: John Benjamins.

Mihailov, M. (2001). Two approaches to automated text aligning of parallel fiction texts. *Across Languages and Cultures, 2*(1), 87–96.

Nagels, A., Kauschke, C., Schrauf, J., Whitney, C., Straube, B., & Kircher, T. (2013). Neural substrates of figurative language during natural speech perception: An fMRI study. *Frontiers in Behavioral Neuroscience*, 7, 121.

Neumann, S., & Hansen-Schirra, S. (2013). Exploiting the incomparability of comparable corpora for contrastive linguistics and translation studies. In S. Sharoff, R. Rapp, P. Zweigenbaum, & P. Fung (Eds.), *Building and using comparable corpora* (pp. 321–335). Berlin: Springer.

Nisioi, S. (2015). Unsupervised classification of translated texts. In C. Biemann, S. Handschuh, A. Freitas, F. Meziane, & E. M'etais (Eds.), Natural language processing and information systems. *Proceedings of the 20th International Conference on Applications of Natural Language to Information Systems, NLDB* (pp. 323–334). Berlin: Springer.

O'Brien, S. (2009). Eye tracking in translation process research: Methodological challenges and solutions. In I. Mees, F. Alves, & S. Göpferich (Eds.), *Methodology, technology and innovation in translation process research* (pp. 251–266). Copenhagen: Samfundslitteratur.

Och, F., & Ney, H. (2003). A systematic comparison of various statistical alignment models. *Computational Linguistics, 29*(1), 19–51.

Olohan, M. (2004). *Introducing corpora in translation studies*. London: Routledge.

Oster, K. (2017). The influence of self-monitoring on the translation of cognates. In S. Hansen-Schirra, O. Čulo, & S. Hofmann (Eds.), *Empirical modelling of translation and interpreting* (pp. 3–22). Berlin: Language Science Press.

Oster, K. (2018). Lexical activation and inhibition of cognates among translation students [PhD dissertation]. Germersheim: Johannes-Gutenberg-Universität Mainz.

Rabinovich, E., Nisioi, S., Ordan, N., & Wintner, S. (2016). On the similarities between native, non-native and translated texts. *Proceedings of the 54th Annual Meeting of the Association for Computational Linguistics*, 1870–1881.

Reinecke, J. (2014). Grundlagen der Standardisierten Befragung. In N. Baur, & J. Blasius (Eds.), *Handbuch Methoden der Empirischen Sozialforschung* (pp. 601–617). Berlin: Springer.

Rösener, C. (2016). Eye tracking and beyond: The dos and don'ts of creating a contemporary usability lab. In S. Hansen-Schirra, & S. Gruzca (Eds.), *Eyetracking and Applied Linguistics* (pp. 143–162). Berlin: Language Science Press.

Samuelsson, Y., & Volk, M. (2005). Presentation and representation of parallel treebanks. *Proceedings of the Nodalida Workshop on Treebanks for Spoken Language and Discourse*. Joensuu, Finland. www.zora.uzh.ch/id/eprint/32953/2/Samuelsson_Volk_2005V.pdf

Schaeffer, M., & Carl, M. (2013). Shared representations and the translation process: A recursive model. *Translation and Interpreting Studies, 8*(2), 169–190.

Schaeffer, M., & Carl, M. (2014). Measuring the cognitive effort of literal translation processes. In U. Germann (Ed.), *Workshop on humans and computer-assisted translation. Association for computational linguistics* (pp. 29–37). Association for Computational Linguistics.

Serbina, T., Niemietz, P., & Neumann, S. (2015). Development of a keystroke logged translation corpus. In C. Fantinuoli, & F. Zanettin (Eds.), *New directions in corpus-based translation studies* (pp. 11–34). Translation and Multilingual Natural Language Processing. Berlin: Language Science Press.

Steiner, E. (2001). Translations English-German: Investigating the relative importance of systemic contrasts and of the text-type translation. In *SPRIKreport, 7, Reports from the project languages in contrast*. Oslo: University of Oslo.

Teich, E. (2003). *Cross-linguistic variation in system and text: A methodology for the investigation of translations and comparable texts*. Berlin: de Gruyter.

Thurmond, V. A. (2001). The point of triangulation. *Journal of Nursing Scholarship, 33*(3), 253–258.

Tirkkonen-Condit, S. (2005). The monitor model revisited: Evidence from process research. *Meta: Translators' Journal, 50*(2), 405–414.

Toury, G. (1995). *Descriptive translation studies and beyond*. Amsterdam: John Benjamins.

van Halteren, H. (2008). Source language markers in EUROPARL translations. In D. Scott, & H. Uszkoreit (Eds.), *Proceedings of the 22nd International Conference on Computational Linguistics*, 937–944.
van Hell, J. G., & de Groot, A. M. B. (2008). Sentence context modulates visual word recognition and translation in bilinguals. *Acta Psychologica*, 128, 431–451.
Vauquois, B. (1968). A survey of formal grammars and algorithms for recognition and transformation in mechanical translation. *Information Processing 68: Proceedings of IFIP* Congress 68, 1114–1123.
Vintar, S., & Hansen-Schirra, S. (2005). Cognates. Free rides, false friends or stylistic devices?: A corpus-based comparative study. In G. Barnbrook, P. Danielsson, & M. Mahlberg (Eds.), *Meaningful texts: The extraction of semantic information from monolingual and multilingual corpora, (Research in corpus and discourse)* (pp. 208–221). London: Continuum.
Volansky, V., Ordan, N., & Wintner, S. (2015). On the features of translationese. *Digital Scholarship in the Humanities, 30*(1), 98–118.
Volk, M., Gustafson-Capková, S., Lundborg, J., Marek, T., Samuelsson, Y., & Tidström, F. (2006). XML-based phrase alignment in parallel treebanks. *Proceedings of EACL Workshop on Multi-dimensional Markup in Natural Language Processing* (pp. 93–96). www.aclweb.org/anthology/W06-2717.pdf

24

Translation, multimodality and cognition

Jan-Louis Kruger

24.1 Introduction and background

There have been a number of publications on translation and multimodality over the past decade (cf. e.g. de Pedro Ricoy, 2012; González, 2014; Kaindl, 2013; O'Sullivan, 2013; Tuominen et al., 2018). On the whole, however, Translation Studies has not awarded much attention to multimodality, and even less to multimodality and cognition. When considering source and target texts, the emphasis traditionally has been on the linguistic level. Kaindl (2013, p. 257) points out that both diachronic and synchronic Translation Studies in the past focused mainly on the linguistic dimension, rendering the discipline monomodal.

This emphasis on language rather than multimodal meaning transfer has been a blind spot in translation research for much of the history of the discipline. But then, as Gambier (2006:6) points out, "[n]o *text* is, strictly speaking, monomodal. Traditional texts, hypertexts, screen texts combine different semiotic resources. Films and TV programs co-deploy gesture, gaze, movement, visual images, sound, colours, proxemics, oral and written language, and so on". And as Tuominen et al. (2018, p. 1) point out, most communication in contemporary society has become primarily multimodal, and this multimodality is therefore also a central concern in translation and gives rise to interesting questions about the optimal methodology to study multimodality in translation. In fact, according to Tuominen et al. (2018, p. 5), we could refer to multimodal Translation Studies as "a translation-specific derivation from the term 'multimodal studies', to foreground multimodally oriented translation research as a research orientation with a specific objective of addressing the presence and interplay of different modes in translational contexts".

The preceding reference to different text types also reveals the skewed emphasis in Translation Studies on the translation product. It is, however, what happens to this product when the target text is received and processed by the target-language audience, as well as how it comes into being (in other words, the production thereof), that makes it imperative to consider cognition.

Multimodality is a key concern in interlingual interpreting, where the interpreter has to provide not only a translation of the words of the source-language speaker but also an interpretation of how these words are presented (e.g. intonation) as well as of visual information that forms part of the context of what is spoken (e.g. gestures, notes, printed matter or slides). Sign language interpreting likewise mediates between an auditory spoken language (with a similar multimodal context as in interlingual interpreting) and a deaf audience who rely on a visual interpretation of the spoken language through a complex set of signs, including facial and bodily expressions. In audiovisual translation, the linguistic modes of spoken and written language (in signs and subtitles) always co-exist with various other modes such as moving and static images, film editing, camera angles, *mise-en-scène*, music and other sounds, presented in the medium of film, whether - in a theatre or digital on television or on the Internet. Even the translation of written text often has to consider extra-linguistic aspects such as typeface, layout and various forms of illustration or graphics that go beyond the single modality of written language.

Translation Studies has embraced cognitive approaches over the past decades, and indeed, according to Muñoz Martín, "cognitive approaches to translation and interpreting may be considered the oldest empirical research area of modern translation studies" (2016, p. 555). Cognitive Translation Studies deals with the processes in the production and reception of translation and interpreting products in an interdisciplinary manner, focusing on the cognitive processing of translators and interpreters as well as the users of translation and interpreting. In that sense, there is a strong link between multimodality and cognition in translation that becomes evident in empirical studies where ergonomics, the interaction between the translator and various modes and sources of information in different media, is investigated. Multimodality is also central in studies on the difference in processing between reading and writing, as well as between watching, listening and speaking. This emphasis on multimodality is evident in a number of recent books on Cognitive Translation Studies and translation process research (see Lacruz & Jääskeläinen, 2018; Muñoz Martín, 2017; Schwieter & Ferreira, 2017; Shreve & Angelone, 2010).

24.2 Multimodality

Before looking at cognition in the translation of multimodal texts, it might be useful to clarify what multimodality means. This is by no means a clear concept, but there seems to be some convergence on meaning in work on multimodality in the context of social semiotics (cf. Halliday, 1978). In this regard, it is important to distinguish between mode and medium. According to Kress and van Leeuwen (2001), mode is related to resources for making meaning that employ channels of communication or representation. It therefore includes linguistic elements, such as spoken dialogue and written text, as well as static and dynamic images, sound, gesture, gaze, facial and bodily expressions or posture. This conceptualization of mode and multimodality therefore takes the emphasis away from simply linguistic components. Medium, on the other hand, relates to the material form (see Littau, 2011) that conveys the message contained in the mode, such as paper, ink, film or hyperspace. Kress and van Leeuwen (2001) argue that medium has traditionally been neglected in both linguistics and semiotics, just as non-verbal modes have been neglected.

Fundamentally, however, looking at translation from a multimodal perspective poses a number of questions about the very nature and self-image of a discipline that has been conceptualized primarily on the basis of monomodality and linguistic texts and speech units (see Kaindl, 2013, p. 257). This is not to say that Translation Studies has been focused exclusively on the linguistic aspects of "monomodal" texts. Translation Studies has, in fact, been concerned with systems and agents that shape the translation product for many years. This is evident in Jakobson's (1959/2000) identification of intersemiotic translation as a category of translation; in Holmes's concept of function-oriented descriptive Translation Studies (or DTS), which is concerned with context rather than text (Holmes, 1988/2000); in polysystems theory, which considers literary translation within a broader literary system (Even-Zohar, 1978/2000); and in Reiss's (1981/2000) addition of audiomedial as a text type in addition to informative, expressive and operative texts, among others. This broader conceptualization of multimodality foregrounds the fact that, regardless of our definition of text, the linguistic content is always framed in, supplemented by and informed by context, co-texts and non-verbal modes that determine the translation as well as the reception of the translated product (see, for example, Kruger & Kruger, 2017).

Interestingly, when defining multimodal discourse in their theory of multimodal communication, Kress and van Leeuwen (2001, p. 21) saw modes as "semiotic resources which allow the simultaneous realization of discourses and types of (inter)action. Designs then use these resources, combining semiotic modes, and selecting from the options which they make available according to the interests of a particular communication situation". Kaindl (2013, p. 258) interprets this as meaning that modes are not in the first instance products, "but cultural processes which manifest themselves as discourses and the functions of which constitute texts in relation to other modes".

Klaus Kaindl's lucid and comprehensive contribution on "Multimodality in translation studies" in *The Routledge handbook of translation studies* (2013) provides an excellent critical overview of the concept of multimodality and how this has been approached in Translation Studies. Many of the problems in these approaches can be related to the confusion between mode and medium, and the use of semiotic systems in defining multimodality.

Toury (1994) defines intersemiotic translation as the translation between different codes, whereas interlingual translation is considered to be intrasemiotic translation (divided into intrasystemic or intralingual and intersystemic translations). Kaindl (2013, p. 261) therefore argues that the criteria of mode and medium should be clearly distinguished from each other and that the semiotic dimension is problematic for a translation-relevant text typologization such as those presented by Jakobson and Toury.

Having identified mode and medium as the categories that have to be distinguished in translation, Kaindl (2013, pp. 261–262) suggests a distinction between intramodal and intermodal translation in the dimension of mode or semiotic code, and between intramedial and intermedial translation (or transfer aspects) in the dimension of medium referring to the materialities of translation. In intramodal translation, one mode is translated with the same form of mode (i.e. linguistic to linguistic, such as in translation between dialects of the same language in a drama; image to image, such as in a translation of Disney cartoons into manga with an emphasis on the styles of animation; or one musical genre into another, etc.). Intermodal translation involves a change in mode, such as a translation from a linguistic mode to an image mode or vice versa (as in audio description). Intramedial translation maintains the same medium while transferring media elements in accordance with cultural conventions related to the medium (i.e. adhering to medium-related conventions in drama for different cultures in translation). Intermedial translation, by contrast, concerns translation across media barriers (such as the translation of a novel into a film or a play into a musical). Kaindl (2013) draws attention to a range of areas in Translation Studies that have engaged with multimodality (such as audiovisual translation, the translation of children's literature, the translation of pragmatic texts like advertisements, the localization of websites or video games). He also points out the lack of attention to multimodality in the areas of specialized texts (such as technical communication) and interpreting (where surprisingly little work has been done on multimodality up until recently—see Kaindl, 2013, p. 265 for some exceptions). Kaindl (2013, pp. 264–265) concludes by identifying a need for an extension not only of the definition of text as basis for translation to include multimodality, but also for an extension of the instruments used for analysis of these texts in Translation Studies. This centres primarily on text analysis, where he distinguishes between, firstly, an analysis of the composition and functioning of different modes, and, secondly, the correlation and interaction modalities between different modes beyond the verbal modes.

In the short period since Kaindl's seminal chapter in 2013, multimodality in translation has received significantly more attention (see, for example, O'Sullivan, 2013; Tuominen et al., 2018), particularly in audiovisual translation and in translation process research. In the field of audiovisual translation, for example, the increased focus on audio description has seen a number of studies and large projects aimed at investigating multimodality in translating audiovisual texts and contexts. In translation process research, multimodality has taken centre stage in the investigation of the way in which translators and interpreters process source texts and contexts and produce target texts, and significantly, on how audiences process these target texts and contexts. This has necessarily also seen a growing emphasis on the investigation of cognition. In the rest of this chapter, these areas will be discussed in more detail in order to show how multimodality and the cognitive processing of multimodality have shaped recent developments in Translation Studies.

24.3 Core issues in multimodality and cognition

Measuring the impact of multimodality on the production and processing of translation is an important step towards understanding how different modes impact on different aspects of cognition, such as knowledge, beliefs, attention, emotions and memory, in this complex environment. But what makes multimodality such a complicated issue to study? Part of the answer lies in the fact that the human mind has the capacity to process and integrate various different sources of information not only in sequence but also simultaneously. Multimodality means that we are not only forced to engage with the linguistic aspects of a multimodal text, but we also have to interrogate constantly how the different modes contribute to the creation of meaning. This is complicated by the fact that we process some elements of a multimodal text automatically: when text appears on a screen we automatically begin to read it (see d'Ydewalle and De Bruycker, 2007; d'Ydewalle and Gielen, 1992); when there is movement on a screen, our eyes are drawn to the movement; we tend to begin to process a video by looking at the centre of the screen or page. These automatic impulses are referred to as bottom-up processing and compete with top-down processing, where we consciously direct our attention to salient elements of a text.

The efficiency with which humans engage with multimodal contexts is obvious when we look at the astonishing speed with which infants pick up on cues around them and begin to interact with their surroundings. It is also evident in the way in which we can view a fiction film consisting of a series of distinct scenes and shots that are edited together, ignoring the discontinuities in these texts. We constantly pick up on overt and covert cues around us, including when we translate or process a translated text.

The relevance of all of this lies in the fact that multimodal texts consist of a number of modes that contribute to the creation of meaning. Much of this is redundant or interdependent. When we see an emotional scene in a film, we know a character is sad because of the words in the spoken dialogue, the intonation in the voices, the words in the subtitles, the expressions on faces, the musical score, the camera angles and close-up shots, and so forth (all of which, of course, also carry culturally defined meaning). In many cases, any one of these codes would be sufficient to give us the information we need. In that sense, the different sources could be considered redundant.

What makes redundancy such an interesting concept in the context of multimodality and cognition is that different users will process codes in a different order, or will process different codes to arrive at the same understanding. Redundancy has a number of functions. It often provides confirmation in one mode of what is presented in another. In that respect, it could be supplementary—adding more nuance to our understanding. In other cases, the redundancy could be partial, meaning that an interdependence will exist between two sources of information. But in some cases, redundancy could result in competition, introducing a complexity that forces the user to process a number of redundant sources of the same information in different modes simultaneously (see also Lautenbacher, 2018). Much of what we most need to understand, therefore, is how humans process multimodality cognitively.

In educational psychology and instructional design, the redundancy effect is associated with a negative impact of redundant information on comprehension and learning. This results in a potential overloading of the working memory that has a negative impact on performance (see Kalyuga, 2012; Moreno & Mayer, 2002). However, when there is an interdependence between information in verbal and visual modes, the redundancy effect does not occur—there is no dramatic increase in the cognitive load associated with processing both sources of information.

However, the impact of different sources of information varies from individual to individual. For this reason, it is essential to study the cognitive processing of multimodal texts—not only to provide us with a better understanding of how information presented in different modes is processed, but also to help us understand how translators code this multimodality. Audiovisual translation provides the ideal context for studying multimodality and the processing of various sources of information simultaneously.

24.4 Audiovisual translation (AVT)

In AVT, multimodality has long been acknowledged as a fundamental concern. Kaindl (2013) acknowledges that this field is at the cutting edge of the comprehensive and systematic description of "the role and function of non-verbal modes in transcultural communication" (2013, p. 263). In recent years, this focus has deepened (see e.g. Tuominen et al., 2018). For example, Taylor (2016) explores the multimodal approach in AVT by focusing on multimodal text analysis (including narrative, linguistic, semiotic and cultural considerations) before turning to multimodal transcription, which provides a useful methodology for engaging with the various modes that impact on AVT. He points out, however, that such transcription becomes too unwieldy when it comes to whole films, calling rather for phasal analyses that enable the translator "to identify homogeneous 'phases,' both continuous and discontinuous, within a multimodal text and to recognize register changes, character traits, and elements of cohesion and coherence that, if ignored, could lead to inconsistencies in translation" (2016, p. 230). This sense of economy in dealing with multimodality in AVT provides a scalable solution to an enterprise that could easily become bogged down in microtextual elements.

Audio description (AD) is the one form of AVT that engages with the multimodality of the text beyond the linguistic content of the dialogue to a much greater extent than any other form of AVT. Here, the dialogue is the one element of the text that is left undisturbed in the spoken form, while the various other modes that constitute film are mediated for the blind audience. Interestingly, much recent research on AD has centred on the difference between a focus on the unimodally visual component of what can be seen on screen in terms of characters, objects and action and a focus on the interaction between sound effects, visual presentation (in camera angles, editing and *mise-en-scène*) and verbal modes.

The foregrounding of multimodality in AD is also evident in a series of studies dealing with sound, film style and aesthetics. In a collection edited by Maszerowska, Matamala and Orero (2014), Elisa Perego (2014), for example, discusses the importance of visual composition and editing techniques for AD. This resonates with the work by Fryer and Freeman (2013), which also deals with multimodality in grappling with whether or not to include filmic language (descriptions of style) in AD. Mazur (2014) discusses interpretative description and shows that the interpretation of the visual modes of gestures and facial expressions is sometimes essential in order to convey the filmic narrative as well as its aesthetic dimension. Matamala (2014) considers the transfer of visual verbal text to spoken word in describing text on screen, before Szarkowska and Orero (2014) discuss the importance of describing sounds in AD in order to disambiguate sounds for the target audience. These studies, while by no means the only ones engaging with multimodality in AVT, signal an important shift in this branch of Translation Studies away from the linguistic mode in isolation.

In the context of multimodal analysis, Jiménez Hurtado and Soler Gallego (2013) explore the possibility of applying corpus-based methods to the analysis of film for AD, which enables the narratological structure of the film, the filmic language and the linguistic elements of the film to be analysed in relation to the AD. This multimodality is also addressed by Wilken and Kruger (2016) in looking at the impact of filmic elements such as *mise-en-shot* (the way in which visual elements are shown to an audience) on the processing of the film and particularly the immersion of the audience in the film. The multimodal transcription of scenes from a film forms the basis for their analyses.

In addition to text analysis, the processing of multimodal texts by audiences has received ample attention in AVT. I will present a few examples from the fields of AD, subtitling and dubbing, with a distinction between the offline and online measurement of the impact of multimodal features on immersion and cognitive load, with online measures mostly triangulated with offline measures (cf. Doherty & Kruger, 2018 and Kruger et al., 2015 for an overview, and Kruger & Doherty, 2016 for a multimodal methodology to measure cognitive load).

24.5 Offline measurements of the impact of AVT on immersion and cognitive load

Although the individual components of the multimodality of AVT products have received little overt attention, a number of studies have investigated the impact of adding subtitles to film on immersion and cognitive load, and the difference between the effects of subtitles and dubbing on film processing. Perego et al. (2015) provide a strong case showing that subtitles do not have cognitive costs when compared with dubbing and in fact, boost lexical acquisition. They use a range of offline cognitive measures, including a comprehension questionnaire, a dialogue recognition questionnaire, a face-name association test, and a visual scene recognition test. Kruger and Steyn (2014), and Kruger, Doherty, et al. (2017) include offline measures in their studies on the processing of subtitled video. Based on the principles of cognitive load theory, they use the items proposed by Leppink and colleagues (2014, 2015), which identify specific items that can be used in measuring different components of cognitive load. Extraneous cognitive load is of particular relevance for the measurement of the impact of subtitles and other multimodal elements on cognitive load. Extraneous cognitive load relates to the way in which information is presented and is therefore impacted when information is presented in different modes, as is the case with subtitles as well as AD.

Kruger et al. (2016) and Kruger, Soto-Sanfiel, et al. (2017) use a suite of offline measures of immersion (including items on presence, transportation and character identification). Immersion refers to the degree to which an audience becomes immersed, or lost, in a fictional reality. In their study, Kruger, Soto-Sanfiel, et al. (2017) find that immersion (and in particular transportation as the sense of being transported into a fictional reality) increases for a second-language audience watching a television drama with English same-language subtitles. In contrast, however, when comparing dubbing and subtitling, Wissmath et al. (2009) find that dubbing is more immersive than subtitles. However, they do not compare the dubbed and subtitled versions with a version without dubbing or subtitling, and their audience is accustomed to dubbing rather than subtitling.

The field of AD has produced a number of studies that investigate the interaction of different modes in creating an AVT product that will be accessible to an audience deprived of full access to the visual modes. As mentioned earlier, Wilken and Kruger (2016) investigate the impact of *mise-en-shot* and elements of film perspective on immersion. They find that AD does not always include a description of modes such as perspective provided by camera angles, shot length and editing, which has a negative impact on the ability of a sighted audience (and by extension, a blind or visually impaired audience) to immerse. Fryer and colleagues investigate the related concept of presence in a number of studies dealing with AD. They find that AD enables the blind audience to experience at least as high a sense of presence as a sighted audience, and in some cases an even higher sense of presence (Fryer & Freeman, 2012, 2014; Fryer et al., 2013).

Romero-Fresco and Fryer (2013) and Fryer and Romero Fresco (2014) transfer the theatrical convention (of providing blind audiences with an introduction to the theatre) to film. They investigate the use of audio introductions (10 minutes of description preceding the film that provides information on various elements of the visual modes, such as film style, and descriptions of the context, including characters and setting). The blind audience responds very positively to these introductions.

More recently, these offline measures have been triangulated with more objective online measures of the impact of certain aspects of AVT on the cognitive processing of audiences. This provides a promising avenue for research on the cognitive processing of multimodality in the context of translation.

24.6 Online measures of cognition and triangulation

The online measurement of the processing of multimodal texts has developed rapidly over the past decade (see, for example, Holsanova, 2014). The majority of this work in the context of Translation Studies has been done in AVT and in particular on the processing of subtitles by viewers, although a handful of studies also use eye tracking to investigate film processing for audio description using eye tracking. One example is the study by Vilaró et al. (2012) on the impact of sound on the visual processing of film. Another example is the study by Kruger (2012) on the way in which eye movements can provide information on viewer construction of narrative based on multimodal codes.

According to Doherty and Kruger (2018, p. 47), "empirical research on subtitling and captioning has understandably focused on examining their processing and reception by diverse audiences as part of a rich multimodal experience that spans various genres and formats". Originating in the 1980s in the work by d'Ydewalle and colleagues in Belgium, eye tracking has been used extensively and increasingly to study the visual processing of subtitles in order to get closer to an understanding of the cognitive processing of subtitles together with other visual modes in film. Kruger and Doherty (2018) provide an overview of more than 30 eye-tracking studies conducted in the context of subtitling (the majority of these studies in the past decade). Many of these studies deal primarily with attention distribution between the linguistic mode of subtitles and the visual modes on screen. Perego et al. (2010), however, investigate the cognitive effectiveness of the processing of subtitles by also testing the extent to which viewers manage to process other visual modes. Their study therefore grapples very specifically with cognition in the presence of multimodality.

Fox (2016) engages with another dimension of multimodality by using eye tracking to study the way in which viewers process "subtitles" that were created as an integrated element of the multimodal text. She calls this "integrated titles" and shows how placing the text according to aesthetic principles as well as according to principles of visual saliency (where the eyes would automatically be directed due to the other visual modes) facilitates the overall processing of the film. Caffrey (2012) likewise uses eye tracking to measure how experimental subtitles impact on viewer perception.

Szarkowska et al. (2011) and Szarkowska et al. (2016), as well as Krejtz et al. (2013), make use of eye tracking to investigate the impact of presentation speed, degree of editing and shot changes on subtitle processing, thereby making a contribution to our knowledge of the effects of these multimodal elements on the cognitive processing of subtitles. Although eye tracking provides a valuable measure for investigating cognition in the processing of multimodal translation products, particularly by providing evidence of the visual processing of written language and other visual codes such as images or movement, it has some limitations when it comes to investigating anything more than visual attention distribution. Measures of cognitive load, such as fixation duration and fixation count, are only meaningful when similar modes are compared, such as non-verbal visual modes. Due to the difference in visual processing required when reading text (i.e. many, short, linear fixations) in comparison to scanning a scene or exploring a face for cues about emotion (fewer, longer, distributed fixations), differences in fixation count or duration become meaningless. The nature of film also means that constant changes in the luminosity and location of fixations that alter the shape of the pupil make changes in pupil diameter less reliable than in studies on static stimuli. In this respect, measures such as revisits to different elements of the screen become more important. These limitations mean that triangulation with offline measures is essential and also make it important to explore other online measures.

Electroencephalography (EEG) is one such online measure. It measures electrical brain activity on the scalp to study cognitive processing. In the context of Translation Studies and AVT, EEG offers exciting possibilities for measuring cognitive processing during the reception and production of translation products. However, EEG data has to be triangulated with eye-tracking data and also with offline measures in order to validate what is as yet a largely uncharted methodology.

One early study attempting such triangulation is that by Kruger, Doherty, Fox, & de Lissa (2017). They triangulate alpha power with subjective measures of cognitive load, showing promising results in analysing EEG data as time course data to show how levels of cognitive load change over time. Kruger, Doherty, & Ibrahim (2017) also investigate beta coherence between the prefrontal cortex and the posterior parietal cortex as an online measure of psychological immersion as it fluctuates over the course of a film.

24.7 Concluding remarks

A focus on multimodality and cognition in translation research represents a shift away from the purely linguistic components of translation. This not only changes the discipline fundamentally but has also been necessitated by the proliferation of multimodality in the texts translators have to translate. This is evidenced by the central role online texts play in our lives today, as well as the proliferation of media. This chapter provides a brief (and selective) overview of the way in which multimodality impacts on the production as well as the reception of translation products, and of ways in which to study the impact of multimodality on cognition.

As translation scholars become more versed in the methodologies of disciplines such as psychology, psycholinguistics and cognitive science, the rigour and replicability of research increase. This opens up exciting possibilities for arriving at a better understanding of the way humans engage with the multimodal reality around them, particularly when this reality involves the complexities of two or more languages in interaction, while building on decades of research on linguistic and cultural aspects. The challenge we face is to disentangle the impact of various modes in varying relationships of redundancy, supplementarity and interdependence on cognition, not only for the translators producing target texts but also for the audiences who have to process the texts.

Further reading

Doherty, S., & Kruger, J. L. (2018). The development of eye tracking in empirical research on subtitling and captioning. In J. Sita, T. Dwyer, S. Redmond, & C. Perkins (Eds.), *Seeing into screens: Eye tracking and the moving image* (pp. 46–64). London: Bloomsbury.

In this chapter, the authors provide an overview of the development of eye tracking in the study of subtitling. They outline some of the main measures used in eye tracking in this context and identify the main uses and limitations of these measures.

Kaindl, K. (2013). Multimodality and translation. In C. Millán, & F. Bartrina (Eds.), *The Routledge handbook of translation studies* (pp. 257–269). London: Routledge.

This chapter traces the development of the concept of multimodality in Translation Studies. In particular, it engages with the implications for the discipline of viewing translation as a multimodal activity.

Kress, G., & Van Leeuwen, T. (2001). *Multimodal discourse: The modes and media of contemporary communication.* London: Hodder Arnold.

This is a foundational text on multimodality, outlining a theory of communication in the context of interactive multimedia. It provides a comprehensive delineation of the boundaries between modes and media of communication.

Kruger, H., & Kruger, J. L. (2017). Cognition and reception. In J. W. Schwieter, & A. Ferreira (Eds.), *The handbook of translation and cognition* (pp. 71–89). Malden, MA/Oxford, England: Wiley-Blackwell.

The authors of this chapter provide a perspective on cognitive Translation Studies that focuses on the reception of translations rather than the production of translation. As such, the emphasis is on the cognitive processing in the minds of readers, listeners and viewers who receive translation, interpreting and audiovisual translation products.

Muñoz Martín, R. (2017). Looking toward the future of cognitive translation studies. In J. W. Schwieter, & A. Ferreira (Eds.), *The handbook of translation and cognition* (pp. 555–572). Malden, MA/Oxford, England: Wiley-Blackwell.

This chapter argues for internal coherence among cognitive approaches to translation and interpreting, and identifies current and future trends in cognitive Translation Studies.

Tuominen, T., Jiménez Hurtado, C., & Ketola, A. (2018). Why methods matter: Approaching multimodality in translation research. *Linguistica Antverpiensia, New Series: Themes in Translation Studies, 17*, 1–21.

In their introduction to the special issue on multimodality in translation research, the authors engage with the interdisciplinary nature of translation research on multimodality that requires the crossing of various boundaries. In doing so, they highlight the methodological challenges this field encounters.

References

Caffrey, C. (2012). Using an eye-tracking tool to measure the effects of experimental subtitling procedures on viewer perception of subtitled AV content. In E. Perego (Ed.), *Eye tracking in audiovisual translation* (pp. 223–258). Rome: Aracne.

de Pedro Ricoy, R. (2012). Multimodality in translation: Steps towards socially useful research. *Journal Multimodal Communication, 1*(2), 181–203. doi:10.1515/mc-2012-0012

Doherty, S., & Kruger, J. L. (2018). The development of eye tracking in empirical research on subtitling and captioning. In J. Sita, T. Dwyer, S. Redmond, & C. Perkins (Eds.), *Seeing into screens: Eye tracking and the moving image* (pp. 46–64). London: Bloomsbury.

d'Ydewalle, G., & de Bruycker, W. (2007). Eye movements of children and adults while reading television subtitles. *European Psychologist, 12*(3), 196–205. doi: 10.1027/1016-9040.12.3.196

d'Ydewalle, G., & Gielen, I. (1992). Attention allocation with overlapping sound, image, and text. In K. Rayner (Ed.), *Eye movements and visual cognition* (pp. 415–427). New York: Springer.

Even-Zohar, I. (1978/2000). The position of translated literature within the literary polysystem. In. L. Venuti (Ed.), *The translation studies reader* (pp. 192–197). London: Routledge.

Fox, W. (2016). Integrated titles: An improved viewing experience? In S. Hansen-Schirra, & S. Grucza (Eds.), *Eye tracking and applied linguistics* (pp. 5–30). Berlin: Language Science Press.

Fryer, L., & Freeman, J. (2012). Cinematic language and the description of film: Keeping ad users in the frame. *Perspectives: Studies in Translatology, 21*(3), 412–426. doi:10.1080/0907676X.2012.693108

Fryer, L., & Freeman, J. (2013). Visual impairment and presence: Measuring the effect of audio description. *Proceedings of the 2013 Inputs-Outputs Conference: An interdisciplinary conference on engagement in HCI and performance*. Brighton, United Kingdom, 26 June.

Fryer, L., & Freeman, J. (2014). Can you feel what I'm saying? The impact of verbal information on emotion elicitation and presence in people with a visual impairment. In A. Felnhofer, & O. D. Kothgassner (Eds.), *Challenging presence: Proceedings of the 15th international conference on presence* (pp. 99–107). Wien: facultas.wuv.

Fryer, L., Pring, L., & Freeman, J. (2013). Audio drama and the imagination: The influence of sound effects on presence in people with and without sight. *Journal of Media Psychology: Theories, Methods, and Applications, 25*(2), 65–71. doi:10.1027/1864-1105/a000084

Fryer, L., & Romero-Fresco, P. (2014). Audiointroductions. In A. Maszerowska, A. Matamala, & P. Orero (Eds.), *Audio description: New perspectives illustrated* (pp. 11–28). Amsterdam: John Benjamins.

Gambier, Y. (2006) Multimodality and audiovisual translation. In M. Carroll, H. Gerzymisch-Arbogast, & S. Nauert (Eds.), *Audiovisual translation scenarios: Proceedings of the second MuTra conference in Copenhagen 1–5 May*. www.euroconferences.info/proceedings/2006_Proceedings/2006_Gambier_Yves.pdf

González, L. P. (2014). Multimodality in translation and interpreting studies: Theoretical and methodological perspectives. In S. Bermann, & C. Porter (Eds.), *A companion to translation studies* (Ch. 9, pp. 119–131). Malden, MA/Oxford, England: Wiley-Blackwell.

Halliday, M. A. K. (1978). *Language as social semiotic: The social interpretation of language and meaning*. London: Edward Arnold.

Holmes, J. S. (1988/2000). The name and nature of translation studies. In L. Venuti (Ed.), *The translation studies reader* (pp. 172–185). London: Routledge.

Holsanova, J. (2014). Reception of multimodality: Applying eye tracking methodology in multimodal research. In C. Jewitt (Ed.), *Routledge handbook of multimodal analysis* (2nd ed., pp. 285–296). London: Routledge.

Jakobson, R. (1959/2000). On linguistic aspects of translation. In L. Venuti (Ed.), *The translation studies reader* (pp. 113–118). New York: Routledge.

Jiménez Hurtado, C., & Soler Gallego, S. (2013). Multimodality, translation and accessibility: A corpus-based study of audio description. *Perspectives: Studies in Translatology, 21*(4), 577–594. doi:10.1080/0907676X.2013.831921

Kaindl, K. (2013). Multimodality and translation. In C. Millán, & F. Bartrina (Eds.), *The Routledge handbook of translation studies* (pp. 257–269). London: Routledge.

Kalyuga, S. (2012). Instructional benefits of spoken words: A review of cognitive load factors. *Educational Research Review,* 7(2), 145–159. doi: 10.1016/j.edurev.2011.12.002

Krejtz, I., Szarkowska, A., & Krejtz, K., (2013). The effects of shot changes on eye movements in subtitling. *Journal of Eye Movement Research, 6*(5), 1–12.

Kress, G., & Van Leeuwen, T. (2001). *Multimodal discourse: The modes and media of contemporary communication*. London: Hodder Arnold.

Kruger, H., & Kruger, J. L. (2017). Cognition and reception. In J. W. Schwieter, & A. Ferreira (Eds.), *The handbook of translation and cognition* (pp. 71–89). Malden, MA/Oxford, England: Wiley-Blackwell.

Kruger, J. L. (2012). Making meaning in AVT: Eye tracking and viewer construction of narrative. *Perspectives: Studies in Translatology, 20*(1), 67–86. doi: 10.1080/0907676X.2011.632688

Kruger, J. L., & Doherty, S. (2016). Measuring cognitive load in the presence of educational video: Towards a multimodal methodology. *Australasian Journal of Educational Technology,* 32(4), 19–31. doi:10.14742/ajet.3084

Kruger, J. L., & Doherty, S. (2018). Triangulation of online and offline measures of processing and reception in AVT. In E. Di Giovanni, & Y. Gambier (Eds.), *Reception studies and audiovisual translation* (pp. 91–109). Amsterdam: John Benjamins.

Kruger, J. L., Doherty, S., Fox, W., & de Lissa, P. (2017). Multimodal measurement of cognitive load during subtitle processing: Same-language subtitles for foreign-language viewers. In I. Lacruz, & R. Jääskeläinen (Eds.), *Innovation and expansion in translation process research* (Ch. 12, pp. 267–294). Amsterdam: John Benjamins. doi: 10.1075/ata.18.12kru

Kruger, J. L., Doherty, S., & Ibrahim, R. (2017). Electroencephalographic beta coherence as an objective measure of psychological immersion in film. *International Journal of Translation (RITT Rivista internazionale di tecnica della traduzione), 19*, 99–112.

Kruger J. L., Soto-Sanfiel, M. T., & Doherty S. (2017). Original language subtitles: Their effects on the native and foreign viewer. *Comunicar, 25*(50), 23–32.

Kruger, J. L., Soto-Sanfiel, M. T., Doherty, S., & Ibrahim, R. (2016). Towards a cognitive audiovisual translatology: Subtitles and embodied cognition. In R. Muñoz Martín (Ed.), *Reembedding translation process research* (pp.171–194). Amsterdam: John Benjamins.

Kruger, J. L., & Steyn, F., (2014). Subtitles and eye tracking: Reading and performance. *Reading Research Quarterly, 49*(1), 105–120. doi:10.1002/rrq.59

Kruger, J. L., Szarkowska, A., & Krejtz, I. (2015). Subtitles on the moving image: An overview of eye tracking studies. *Refractory: A Journal of Entertainment Media, 25*, n.p.

Lacruz, I., & Jääskeläinen, R. (2018) (Eds.). *Innovation and expansion in translation process research.* Amsterdam: John Benjamins.

Lautenbacher, O. P. (2018). The relevance of redundancy in multimodal documents. *Linguistica Antverpiensia, New Series–Themes in Translation Studies, 17*, 215–530.

Leppink, J., Paas, F., Van der Vleuten, C. P. M., Van Gog, T., & Van Merrienboer, J. J. G. (2014). Effects of pairs of problems and examples on task performance and different types of cognitive load. *Learning and Instruction, 30*, 32–42. doi: 10.1016/j.learninstruc.2013.12.001

Leppink, J., & van den Heuvel, A. (2015). The evolution of cognitive load theory and its application to medical education. *Perspectives on Medical Education, 4*(3), 119–127. doi:10.1007/s40037-015-0192-x

Littau, K. (2011). First steps towards a media history of translation. *Translation Studies, 4*(3), 261–281. doi:10.1080/14781700.2011.589651

Maszerowska, A., Matamala, A., & Orero, P. (Eds.) (2014). *Audio description: New perspectives illustrated.* Amsterdam: John Benjamins.

Matamala, A. (2014). Audio describing text on screen. In A. Maszerowska, A. Matamala, & P. Orero (Eds.), *Audio description: New perspectives illustrated* (pp. 103–120). Amsterdam: John Benjamins.

Mazur, I. (2014). Gestures and facial expressions in audio description. In A. Maszerowska, A. Matamala, & P. Orero (Eds.), *Audio description: New perspectives illustrated* (Ch. 10, pp. 179–198). Amsterdam: John Benjamins.

Moreno, R., & Mayer, R. E. (2002). Verbal redundancy in multimedia learning: When reading helps listening. *Journal of Educational Psychology, 94* (1), 156–163. doi:10.1037/0022-0663.94.1.156

Muñoz Martín, R. (2016). *Reembedding translation process research.* Amsterdam: John Benjamins.

Muñoz Martín, R. (2017). Looking toward the future of cognitive translation studies. In J. W. Schwieter, & A. Ferreira (Eds.), *The handbook of translation and cognition* (pp. 555–572). Malden, MA/Oxford, England: Wiley-Blackwell.

O'Sullivan, C. (2013). Introduction: Multimodality as challenge and resource for translation. *Jostrans, 20*, 2–14.

Perego, E. (2014). Film language and tools. In A. Maszerowska, A. Matamala, & P. Orero (Eds.), *Audio description: New perspectives illustrated* (Ch. 5, pp. 81–102). Amsterdam: John Benjamins.

Perego, E., Del Missier, F., & Bottiroli, S. (2015). Dubbing versus subtitling in young and older adults: Cognitive and evaluative aspects. *Perspectives: Studies in Translatology, 23*(1), 1–21. doi: 10.1080/0907676X.2014.912343

Perego, E., Del Missier, F., Porta, M., & Mosconi, M. (2010). The cognitive effectiveness of subtitle processing. *Media Psychology, 13*(3), 243–272. doi: 10.1080/15213269.2010.502873

Reiss, K. (1981/2000). Type, kind and individuality of text: Decision making in translation (S. Kitron, Trans.). In L. Venuti (Ed.), *The translation studies reader* (pp. 160–171). New York: Routledge.

Romero-Fresco. P., & Fryer, L. (2013). Could audio-described films benefit from audio introductions? An audience response study. *Journal of Visual Impairment and Blindness, 107*(4), 287–285. doi:10.1177/0145482X1310700405

Schwieter, J. W., & Ferreira, A. (Eds.). (2017). *The handbook of translation and cognition.* Hoboken: John Wiley & Sons.

Shreve, G. M., & Angelone, E. (Eds.). (2010). *Translation and cognition.* Amsterdam: John Benjamins.

Szarkowska, A., Krejtz, I., Klyszejko, Z, & Wieczorek, A. (2011). Verbatim, standard, or edited? Reading patterns of different captioning styles among deaf, hard of hearing, and hearing viewers. *American Annals of the Deaf, 156*(4), 363–378. doi:10.1353/aad.2011.0039

Szarkowska, A., Krejtz, I., Pilipczuk, O., Dutka, Ł., & Kruger, J. L. (2016). The effects of text editing and subtitle presentation rate on the comprehension and reading patterns of interlingual and intralingual subtitles among deaf, hard of hearing and hearing viewers. *Across Languages and Cultures, 17*(2), 183–204. doi:10.1556/084.2016.17.2.3

Szarkowska, A, & Orero, P. (2014). The importance of sound for audio description. In A. Maszerowska, A. Matamala, & P. Orero (Eds.), *Audio description: New perspectives illustrated* (pp. 121–140). Amsterdam: John Benjamins.

Taylor, C. (2016). The multimodal approach in audiovisual translation. *Target, 28*(2), 222–236. doi: 10.1075/target.28.2.04tay

Toury, G. (1994). A cultural-semiotic perspective. In T. A. Sebeok (Ed.), *Encyclopedic dictionary of semiotics* (Vol. 2N–Z, pp. 1111–1124). Berlin and New York: Mouton de Gruyter.

Tuominen, T., Jiménez Hurtado, C., & Ketola, A. (2018). Why methods matter: Approaching multimodality in translation research. *Linguistica Antverpiensia, New Series: Themes in Translation Studies, 17*, 1–21.

Vilaró, A., Duchowski, A. T., Orero, P., Grindinger, T., Tetreault, S., & Di Giovanni, E. (2012). How sound is the Pear Tree Story? Testing the effect of varying audio stimuli on visual attention distribution. *Perspectives: Studies in Translatology, 20*(1), 55–65. doi: 10.1080/0907676X.2011.632682

Wilken, N., & Kruger, J. L. (2016) Putting the audience in the picture: *Mise-en-shot* and psychological immersion in audio described film. *Across Languages and Cultures, 17*(2), 251–270. doi: 10.1556/084.2016.17.2.6

Wissmath, B., Weibel, D., & Groner, R. (2009). Dubbing or subtitling? Effects on spatial presence, transportation, flow, and enjoyment. *Journal of Media Psychology, 21*(3), 114–125. doi: 10.1027/1864-1105.21.3.114

25

Translation, risk management and cognition

Anthony Pym

25.1 Introduction

Risk management is a relatively new set of concepts in the study of translation. It can be applied to all kinds of decision making, at many different levels, from the running of translation companies through to the translator's cognitive management of uncertainty while translating. One of the potential advantages of the approach is thus that it can extend from the cognitive to the social (and back), bringing many disciplinary perspectives to bear on translation phenomena.

One of the possible drawbacks, though, is that since the term "risk" is not bound to cognitive science in any strong way, many disciplines use it in quite different ways, inviting as much confusion as insight. The term probably comes from the Ancient Greek ῥιζικόν ("root", used as a metaphor for an obstacle to be avoided at sea) or ῥίζα ("cliff"), similarly to be avoided at sea. This negative sense of a danger to be avoided was carried through in the Latin forms *resicum*, *risicum*, *riscus* and the vernaculars (Luhmann, 1991, p. 8; Skjong, 2005). The first occurrence of the term with a positive meaning ("to dare, to undertake, enterprise, hope for economic success") may have been in Middle High German in 1507 (Skjong, 2005)—the story that the older Chinese characters危机 ("crisis") mean both "danger" and "opportunity" is apocryphal. From the mid-seventeenth century, theories of probability gave risk a mathematical sense, with the mortality risk of diseases being calculated by John Graunt in 1662. This mathematical sense later provided the basis for applications to business models, most clearly in the distinction made by Knight (1921) between risk as a "known chance" and uncertainty as an "immeasurable probability". Risk management, as a specific term, then developed from the 1940s as an extension of "market management" (Dionne, 2013). It has since become a rationalism of business decision making, where the ideal is to quantify and prioritize risks in terms of magnitude and frequency while taking steps to minimize the negative consequences. The presupposition of rationality has nevertheless long been tempered by studies in the psychology of risk, for example in experiments where subjects who are made to feel confident about their risk management tend to take more risks (Krueger & Dickson, 1994).

Our task here is to relate those different senses to existing studies on various aspects of translation (cognitive and otherwise), and to propose areas in which risk management can help solve problems in cognitive studies.

25.2 Preludes to cognitive risk management in Translation Studies

Not surprisingly, some of the first applications of risk management to translation concerned setting up and running a translation business, drawing on the kind of theory that can be found in any course on business practices. Stoeller (2003), for example, offers sound advice to translation project managers, basically providing checklists that help project managers to plan what to do when things go wrong. This means categorizing the kinds of risks: "tigers" have high probability and high impact, while "kittens" have low probability and low impact—so you plan for the tigers and learn to live with the kittens. A series of commentators (e.g. Akbari, 2009; Canfora & Ottmann, 2015, 2018; Lammers, 2011) similarly apply business models to translation, without prolonged attention to what might be *specific* to translation as a business and without attempting to enter any cognitive or psychological dimension involved in translating language. Like the MBA graduates who tend to run translation companies in the US, these commentators basically see translation as just another business.

The potential for a more cognitive application of risk management nevertheless came from early forays into the psychology of translators. Henderson (1987), in attempting to compare the personalities of translators and interpreters, touches on the particular importance of interpreters' confidence and "tolerance of ambiguity", both traits that could be seen in terms of a greater propensity for risk taking. These and similar psychological factors were then investigated using think-aloud protocols. Fraser (1996) proposed that translators with more experience become good at finding ways to "live with" uncertainty in the start text—in the sense that one translation problem can be solved in several different ways (this "indeterminacy of translation" was formulated by Quine, 1960). Fraser (2000, p. 123) further investigated translators' particular "tolerance of ambiguity and uncertainty", finding that professionals generally have more tolerance than do novices. Tirkkonen-Condit (2000) similarly uses think-aloud protocols to observe the way translators use "uncertainty management", basically by becoming proficient at advancing tentative solutions. House (2000) looks at student translators' use of reference books and not surprisingly suggests that high-frequency use is related to low-risk personalities, whereas high-risk takers are confident with and without reference books. More intriguingly, House further finds that low-risk learners may enjoy being denied the support of dictionaries if other means of support are available, such as group work. That is, there is an emotional aspect to risk taking—there can be enjoyment, as all gamblers know.

These think-aloud studies, working on data from very small groups, do not mention risk management as such. Their concern is more explicitly with *uncertainty* management, as has been the case in later process studies that employ a wider variety of methods (cf. Angelone, 2010; Angelone & Shreve, 2011). Their frequent concern is with comparing the performances of novices and experts, in search of an empirical basis for a model of translation competence or expertise. They thus tend to bundle variables together, for example by insisting that trainers should boost the learners' confidence, since enhanced confidence is associated with greater emotional investment in the task, encourages tolerance of uncertainty, and tends to result in higher-quality translations. All good things come together. In the meantime, the most influential models of translation competence (PACTE, EMT) have been based on committees agreeing around a table rather than any bottom-up process research (despite appearances of firm empirical bases in Hurtado, 2017 and Toudic, 2012), and the calls for enhanced confidence thus remained difficult to separate from simple guesswork, where the novice's reduction of uncertainty can equally tend to be associated with a heightened risk of failure.

These initial forays into the general field of risk and uncertainty managed to touch on most of the topics that have been picked up in more recent research. They did not, however, look closely at the way risk is discussed in economics or business studies (where the distinction between risk and uncertainty had been operative for a long time), and they tended to forget the initial research interest in relating risk to the translator's personality. Further, the almost exclusive focus of attention was the translator's relation with the start text (source text) and occasionally with "reference books". The question of risk changes considerably when the scene includes machine translation, group translation, relations with clients and relations with end users. Risk is involved in all those relationships, since people other than the lone translator are taking risks whenever they choose to trust a particular translator.

When one loses sight of those other relations, one also loses the potential to have the risk management connect cognitive processes with social and economic constraints, with a reduced set of concepts addressing related problems at all those levels.

25.3 Core issues

The first attempts to apply risk management to translation in a more general way might be dated from Wilss (2005), who saw that the basic concepts could become a whole approach to translation. In the same years, Pym (2004/2005) argued that attention to risk management could explain translation processes in ways that mainstream linguistics and uncertainty analysis (i.e. the numbers of possible translation solutions) were unable to: three noun phrases may have exactly the same status as linguistic problems (let's say, each with three possible renditions), yet only one of them might be high risk for achieving the translator's communicative purpose. Sentence-level linguistics and uncertainty counts cannot see that risk.

This commonsensical insistence on communicative pragmatics also allows one to address issues of effort distribution: all else being equal, translators should invest more effort in solving the high-risk problem and should not spend too long on the low-risk problems. Some support for this proposition is found in empirical observations that more experienced translators have a more uneven distribution of effort than do novices: they tend to automatize some complex tasks but also shift between automatized routine tasks and conscious problem solving (Jääskeläinen & Tirkkonen-Condit, 1991; Krings, 1988; see Englund Dimitrova, 2005, p. 15; Pym, 2008a). Novices, on the other hand, appear to be more prone to not investing sufficient effort in high-risk items (i.e. guessing) and investing too much effort in low-risk items (i.e. over-translating).

There is, however, more at stake than getting students to rationalize their distribution of effort.

25.3.1 What is specific to translation?

One of the reasons for turning to risk management does not concern cognitive studies in the close sense of observing what goes on in translators' brains. It has more to do with solving a problem of definition in what one might call pure translation theory, on the level of what goes on when commentators and theorists write.

Broadly since the Early Modern period, the Western translation form has operated through claims of fidelity to the start text, variously conceptualized in the twentieth century as "adequacy" and then "equivalence". Not surprisingly, cognitive studies on uncertainty and risk have mostly focused on the same relation with the start text. That focus, however, has been challenged in two ways. For a range of approaches spanning from hermeneutics to deconstruction, the meaning of a text is indeterminate, allowing different understandings by different readers. This means that there is, in theory, little stability in the start text to which the translator could be faithful, adequate or equivalent. That is, there is uncertainty at the source, even prior to the translator's attempts to render the text into other languages. A further challenge has then come from German-language *skopos* theory, which proposes that the translator's priority should be to fulfil the target-side purpose, breaking with equivalence claims as required (Reiss & Vermeer, 1984, p. 100). This effectively throws several other spanners into the works. First, it increases the potential uncertainty in the translator's decision-making process. (What to do, for example, when the client's instructions contradict the start text and the receiver's expectations?) And then, if the translator now has the right to adapt everything, can the product still be called a translation?

These challenges have effectively undone the conceptual basis on which the translation form had been institutionalized for several centuries. They constitute a theoretical problem in dire need of a solution. Risk management can provide something like a way out.

The first step is to abandon the presumption of substantially reproduced meaning. A translation has the status of a translation because it is *received* as a translation, that is, because it creates the illusory supposition that the value of the translated text is on some level the same as that of an anterior text in another language. The terms "illusory" and "value" are motivated here. Think of the way the monetary system depends on the illusion that banknotes or electronic numbers actually represent something like gold, when they do not do so in any full way. It is the shared illusion that creates the value. Break that illusion, and the monetary system falls down. Any social system requires this element of *illusio* if it is to function (Bourdieu, 1980, pp. 103–104). Translation may be no different.

What this means is that the translator's work is going to be socially valued as such for as long as it maintains *credibility* in the eyes of those receiving the translation, including whoever is paying for the translation. From this perspective, the greatest risk is not necessarily misrepresenting the start text or failing to resolve uncertainties but losing the *trust* of the other parties involved in the communication act. What risk must translators manage? First and foremost, it is their credibility.

This is the level, quite removed from the intricacies of text-bound cognition, where risk management can offer a way of rethinking whole translation forms without abandoning the specificity of translation altogether. For each historical or cultural translation form, one can describe the things that credibility entails (a language border, the alien-I, and quantitative co-variance in the case of the Western translation form, but those features can vary). The important point is that the translator's cognitive processes are, on this view, guided by a framing need to maintain trust, and that this general disposition is to be analysed in terms of each specific translation form.

25.3.2 What types of risk are pertinent to translation?

Adopting this broad approach, Pym (2015) attempts to distinguish between three different kinds of risk that are involved in the translating process:

- *Credibility risk*, as we have seen, concerns the need to gain and maintain trust relationships with the other parties involved in the communicative act. Its nature varies in accordance with historical translation forms. In negative terms, this is the risk of losing trust.
- *Uncertainty risk*, then, concerns the translator's doubts about how to render a start-text item. This is the kind of risk that has been dealt with in think-aloud protocol studies under the aegis of "uncertainty management". In negative terms, it is the risk of rendering a start-text item in a non-optimal way.
- *Communicative risk* is then the risk of the translation not fulfilling the desired communicative function, no matter how that specific aim might be established.

These three kinds of risk are obviously related. In an ideal situation, assessment of the communicative risk should define the kind of optimal rendition required, and thus regulate uncertainty risk, while credibility risk would define the kind of communicative risk pertinent to the translation as a translation. The risks are nevertheless quite different when seen in terms of frames for decision making. Two start-text items can have exactly the same linguistic status and entail exactly the same uncertainty risk (a throw-away example would be the names of the mother and the midwife in a birth certificate) yet be radically different in terms of communicative risk (a mistake in the name of the mother would be high risk; in the name of the midwife, not as high). Or again, a Chinese government translator might adopt some radical rewriting in order to make official jargon understandable to the Western English-language reader, thus reducing communicative risk, yet in so doing they incur the risk of losing the trust of their government employers ("this is not what we expect of our translators") and possibly of informed China-watchers ("this does not read like an official Chinese translation"). Such discrepancies between the three kinds of risk thus open up pathways for several further kinds of analysis.

Most usefully, the discrepancy between uncertainty and communicative risk enables an analysis of how much *effort* should be invested in reducing uncertainty. Here we adopt a principle of effability, which fictitiously posits that all uncertainties produced in the translatorial reading of the start text could be resolved if the translating subject simply worked hard enough. This is indeed the fiction that many beginner translators fall into, and that some of their instructors are strangely pleased to encourage. Attention to communicative risk, however, tells us that some start-text problems involve higher risks than others. A simple rationalism then proposes that the translator should work most on the problems that involve the greatest communicative risks, and less on those that do not. This is impeccable common-sense advice that obviously needs no fancy theory to sustain it. Yet it is overlooked by the widespread fiction that all uncertainty can and should be resolved. Further, as noted earlier, it allows learners to be warned of two particularly ubiquitous sins: when the translator invests high effort in a low-risk problem, the result is over-translation—recognized as a professional shortcoming since Mounin (1963)—and when the translator invests low effort in a high-risk problem, the result is guesswork, which can have rather more serious consequences. The more general principle is obviously that the degree of effort should be adjusted to correspond to the degree of communicative risk, not the degree of uncertainty.

This kind of modelling finds some justification in empirical studies, mentioned earlier, that show professional translators having a more irregular distribution of effort in their problem solving, as opposed to novices, who tend to have a more even distribution, working hard to solve every problem that comes along.

25.3.3 What types of risk management are pertinent to translation?

A more traditional categorization of risk is in terms of what can be done with it: the translator can reduce it (risk reduction), transfer it to parties like clients or readers (risk transfer), accept it (risk taking), reduce it by introducing a risk of lesser proportions (risk mitigation), or use combinations of all these possibilities. In theory, such strategies can be applied to all three kinds of risk (credibility, uncertainty and communication). The following examples concern uncertainty and communicative risk:

- *Risk reduction* accounts for all the preliminary processing, documentation, checking, revision and reviewing that can be done by the translator, and possibly pre-editing, then post-editing, when machine translation is used. So much effort is invested in these processes that one might falsely assume that translators can only reduce risk.

- *Risk transfer* is when a party other than the translator is in some way made responsible for a decision. For instance, the translator may not know whether to use the intimate or formal second person when promoting a computer product, and consultation of parallel texts fails to resolve the issue (both cases can be found), so the client is consulted. Once the client decides, the risk of communicative failure has been transferred to them. (Another strategy for risk reduction here would be to avoid the second person altogether.) In a wonderful think-aloud study on how students and professionals rendered a semi-colon, Künzli (2004) found that some professionals considered adding a note for the client (risk transfer), whereas the student translators were more prone to guess (risk taking).
- *Risk taking* is where the translator is aware that a decision may involve serious risks but makes the decision anyway. In some cases, such decisions are uneducated enough to count as guesses: student think-aloud studies abound in self-justifications such as "It sounds better" (Pavlović, 2010), where intuition becomes a reason for choosing between alternative solutions. Robinson (1991) argues that many translation decisions are based on "gut feelings"—we translate as much with our body as we do with our mind. Extreme examples of risk taking might include magnificent works like Ezra Pound's *Cathay* (1915), using poetic intuition to render Chinese poetry in the absence of direct knowledge of the Chinese language.
- *Risk mitigation* in a translation context is a hybrid strategy employed when a minor risk is incurred in order to reduce a major risk (Pym & Matsushita, 2018). For example, the use of untranslated start-text terms in the translation can be seen as transferring the risk of interpretation to the reader, although at the hopefully lower risk of confusing some readers. The giving of multiple solutions might be viewed in the same way: some of those solutions might muddy the reception, but one will hopefully succeed (Matsushita, 2016).

By combining these strategies with the various kinds of risk, a fairly rich and varied metalanguage can be developed for describing the ways translators solve problems, with reference to translation as a social relation (credibility), as a particular kind of textual product (uncertainty) and as an effective social action (communication). This is also a metalanguage that can prove useful in the training of translators, where it can be taught through practical activities and examples, making students reflect on their own assessments of relative risks.

More importantly, perhaps, this is a conceptual framework that does *not* make some of the major assumptions that have proved precarious over time: it does not espouse equivalence or fidelity; it does not assume just one set of translation norms; and although it certainly does assume a rational subject in its theoretical calculations of comparative risks and corresponding effort distributions, actual human calculations are inescapably subjective, made by the translator in the act, and conceptual space can and should be made for non-rationalist models of intuition, inspiration and confidence (see later).

25.3.3.1 Is there a translator personality?

When one goes over the think-aloud protocol studies in search of examples of risk management, the overwhelming majority of instances concern risk aversion. This conclusion can also be reached from textual studies of "translation tendencies" (after Levý, 1963/2011), where it is generally found that, when compared with non-translations, translations tend to be simpler, less varied in lexicon, more explicit, less given to extreme writtenness or spokenness (Shlesinger, 1989, in Pym, 2007), and so on (Pym, 2008b). All these tendencies would indicate risk aversion. One might therefore legitimately wonder whether risk aversion forms part of a particular translator personality. Do particularly risk-averse people become translators? Or does translating perhaps turn you into a particularly risk-averse person?

As mentioned, some of the think-aloud studies have asked similar questions, generally concluding that experienced translators develop something called "tolerance of ambiguity and uncertainty" (Fraser, 2000) and positing that development of such psychological traits should be an important part of translator training (Tirkkonen-Condit, 2000). Research on this tolerance has continued, in part because it can be measured with established psychometric instruments. Bolaños Medina (2015) finds that students who do not think they are cut out to become professional translators tend to obtain low scores for ambiguity tolerance (although this might simply mean they express underlying anxiety on both counts, and perhaps in many other instances as well). Rosiers and Eyckman (2017, p. 52) find that interpreters have greater ambiguity tolerance than do translators; they propose that "the nature of the interpreter's job aids the development of tolerance of ambiguity" (although the personality trait might have been there first). Other research could be mustered along similar lines: some forms of emotional intelligence (Hubscher-Davidson, 2013) could be key in avoiding panic (and thus unwarranted risk taking), and higher self-esteem and emotional stability could similarly help manage risks in sign language interpreting (Bontempo et al., 2014), although the exact connections with risk seem not to have been investigated. Ongoing research by Pirouznik (see Pirouznik, 2014) finds no significant correlation between translators' personalities and risk taking, although the research does not compare translators with non-translators or novices.

Some of the research on ambiguity tolerance seems interestingly coloured by the political origins of the term. Frenkel-Brunswik (1949) first developed the construct to study ethnocentrism in children, with low ambiguity intolerance being associated with "the authoritarian personality" (Adorno et al., 1950). Since translation is supposed to work across cultures, the last thing one ideologically wants to find is ethnocentric authoritarianism. Yet that understandable desire seems to be a long way from the situation where the translator is uniquely aware of several legitimate ways to solve a problem, and usually has to opt for one of them. Decisions are made; some kind of authority is assumed. The ability to take underdetermined decisions might well be a correlative of ambiguity tolerance (since that is what the studies find), but its strictly cognitive dimension does not necessarily connect with any disposition to risk aversion.

Indeed, the connection with risk aversion, if there is any, may not be a consequence of cognitive processes or personality traits at all. It seems more likely to ensue from the way translators are employed and viewed in certain social situations. In a translation culture where the intermediary is supposed to be subservient and invisible, a translator has little motivation to take risks: there are no rewards to be gained. As Leonardo Bruni (1405/1969, pp. 102–103) complained, if something in the translation is good, the author gets the credit, and if something is bad, the translator gets the blame. The problem may be that most cognitive studies and personality tests have been carried out in situations where translators are culturally conditioned to be risk averse.

25.3.3.2 How is risk related to the reception of translations?

In an early paper, Fraser (1996) compared how "commercial translators" and "community translators" expressed uncertainty about how much explanatory information they should add to a government brochure. These doubts concerned not just the limits of the translation task, but also the translators' images of who their receivers were—the community translators were marginally more concerned with envisaging the receiver.

This observation is of some import for general translation theory, and indeed for the pragmatics of reception. Theories of "audience design", dating from Bell (1984/1997, 2001) and working from the "participation framework" modelled by Goffman (1981), broadly propose that text producers adjust their style to suit particular kinds of audience (basically addressees, auditors, overhearers and eavesdroppers). The idea has been applied to translation (particularly in Mason, 2000) and has been allied with the general theory that the way one translates depends primarily on the purpose (*skopos*) of the translation (Reiss & Vermeer, 1984, p. 100): "Essentially the audience design in translation is what the Skopos Theory is about" (Schubert, 2005, p. 132). The operative assumption underlying most of these approaches is that the translator can know, should know and sometimes does in fact know where the translation is headed and who is going to use it. In terms of applied risk management, however, that assumption is often rather precarious.

Bell (1984/1997, p. 162) actually has two major categories in his tree of audience types: "responsive audience design", when the text producer knows who the receiver is and adjusts appropriately, and "initiative audience design", when that knowledge is not in place and the producer uses a certain style in the hope that the corresponding people will be reached—in a sense, the translation goes looking for its readership. It is not entirely clear how these two types map onto the translator's risk management. If the translator knows where the text is going, "responsive" design can justify some intelligent risk management, potentially using all available strategies. On the other hand, if the translator has significant *doubts* about the possible audiences, Bell's "initiative design" would suggest that the translator adopts a risk-taking attitude, opting for choices that may not suit all but will eventually be agreeable to someone. And yet all the evidence we have, from admittedly limited cultural situations, is that risk aversion dominates.

Tellingly, some of the studies that apply the notion of audience design actually reveal unclear or multiple reception. Suojanen et al. (2015), advocating an approach called "user-centered translation", admit that the translator's audience can be a "murky, faceless entity", but they then claim that "audience design has potential as a tool that clarifies the target audience" (2015, p. 69). This clarification is supposed to come from having the translator reflect on a checklist of possible audiences: addressees, auditors, overhearers, etc. Confrontation with actual checklists nevertheless shows that the translator often has only a faint idea of who could be out there and has limited resources for adjustments to the various possibilities (see Suojanen et al., 2015, pp. 26, 136–137). When the purpose and the readership are in doubt, and there are no rewards for taking risks, the translator has several reasons to produce a bland lowest-common-denominator text. Who could blame them for being risk averse?

One might lament (as does Nord, 2014) that translators are not more courageous in their decisions. But if we want them to take intelligent risks, we have to make sure they are aware of who their communication partners are and what rewards the translator will reap from communicative success.

25.4 Recent developments

Risk management remains a minor avenue of research in Translation Studies, with only occasional mention in empirical cognitive studies. It nevertheless offers a powerful instrument for interpreting and explaining the results of cognitive studies. Further, it can do so in ways that connect to some of the ongoing general discussions in the discipline. Here I look at three such discussions: attention to group translation work, the claims of neural machine translation, and increasing awareness that emotions play a role in translation. All three cases concern more than the lone translator.

25.4.1 Risk in distributed translation tasks

Although much of the research on cognitive processes looks at situations where the lone translator is confronting a text, translation scholars are increasingly aware that other people are also involved: co-translators, revisers, reviewers, editors, clients, censors, distributors and receivers can all play roles in the creation of meaning. Machine translation and translation memories effectively enact similar extensions, adding the work of previous translators to the mix. This extended scene obliges us to consider not just the risk relations between the translator and the text, but also the relations between the translator and the other actors in the scene.

We know remarkably little about how readers construe translations. If losing credibility is the greatest risk to the translator, it is in the process of reception that the game is played. At the same time, receivers clearly take risks when they decide to trust one translator rather than another. So what kinds of signals indicate that a translator is trustworthy? A certain economic analysis of the signalling is possible (Pym et al., 2012), but what behavioural or textual features trigger the loss of trust? There is a whole set of cognitive processes waiting to be discovered on the receiver's side.

The categories of risk management invite studies that assess the strategies of all participants in a translation event and use that matrix to try to explain translator decisions. An example would be Hui's 2012 study of a simulated training activity, where risk analysis is applied at all levels, from translator–client relations and end-user analysis right through to think-aloud protocols of the translation process and clients' evaluations of translator performance, including a risk disposition profile for each translator (based on the risk strategies evinced in each solution contemplated and the solutions finally selected). A more extreme case of a whole-situation analysis would be the multi-party risk taking surrounding an instance of effective non-translation by a military interpreter in Afghanistan (Pym, 2016), where the interpreter's actions would break all the codes of professional ethics but can nevertheless be explained as a rational instance of risk management in a particularly high-risk environment.

Such analysis can involve factors that are not commonly seen in a narrow focus on the intermediary's choices; they operate more on the level of language policy (Pym, 2018). In studies on the use of interpreters in asylum-seeker centres (Fiedler & Wohlfarth, 2018; Pokorn & Čibej, 2018), researchers were surprised to find that the recent migrants often did not appreciate the presence of the interpreters, who were ostensibly there to help them. The interpreters were implicitly not trusted, perhaps because the migrants placed more value on independence and learning the host language, but also possibly because the migrants were from cultures where mediation is likely to be by an informant of some kind. Trust was not established, and so the migrants instead calculated the risks, accepting to work with interpreters only in high-risk situations. Similarly, in many healthcare encounters doctors choose *not* to use interpreters in low-risk situations where the expense (in time, more than money) is not merited by the potential risk (Diamond et al., 2009). Both these cases suggest that professional mediation is most called for in particularly high-risk situations. This may be a further reason for adopting risk management as a frame of analysis.

25.4.2 Risk in the use of machine translation

As the quality of machine translation improves, the translator's work increasingly involves various degrees of post-editing, which is a distributed task that potentially involves a special kind of risk management. Krings (1995/2001) does talk about uncertainty in the post-editing process, but we know relatively little about how those processes differ from the revision of a human translation. Once again, the more important game could be played out on the side of reception.

The greatest perceived risk is no doubt the idea that machine translation will replace human translators. Why should a human mediator be called for when developers claim that neural machine translation has now reached "parity" with fully human translation? Microsoft researchers make this claim on the basis of 18 million bilingual sentence pairs being evaluated by "bilingual crowd workers", who were asked whether each candidate translation conveyed "the semantics of the source text" (Hassan et al., 2018, p. 3). On this measure, the MT outputs were found to be statistically as good as the reference human translations. From the perspective of risk management, though, the important point is that the human testing was of context-free sentence pairs being assessed in terms of content, not form. To appreciate the problem, imagine you are a client being told that there are only two mistakes per page of output, and that this probability is the same in both machine and human translation. So is there the same communicative risk? If one of the things we know about professional translators is that they can distinguish intuitively between low-risk and high-risk textual items, and they can invest more effort in solving the high-risk problems, then the chances of those two errors being calamitous should be much lower for the human translation, at least until the neural algorithms can detect communicative risks and redistribute effort.

Given the continued presence of communicative risk, the role of human translators may be expected to become that of the guarantor or certifier of (post-edited) machine translations. Rather than sell numbers of words, translators will sell trustworthiness.

25.4.3 Risk as a play of emotions

Many areas of the humanities have been turning to the role of emotions in what once appeared to be rational decision-making processes. Risk management is no exception. As rationalist risk management theory was developing within economics, Keynes (1936/1962, pp. 161–162) remained aware that economies also run on 'animal spirits'. It has been found, for example, that investors overprice stocks on cloudy days (Goetzmann et al., 2015), and it seems that a country's economy is boosted by similar emotions if they win the World Cup in football (although no reliable economic study on this seems to be available). Such connections with emotions have rarely been dealt with in process studies on translation, but they have never been far away. When Frenkel-Brunswik (1949, p. 108) formulated "ambiguity tolerance" as an instrument, she was partly working from "emotional ambivalence". More recently, as noted earlier, the numerical variables of cognitive studies are paralleled by serious calls for attention to "emotional intelligence" (Hubscher-Davidson, 2013), while Robinson has gone from the somatics of the translator (Robinson, 1991) to the wider "socioaffective ecologies" by which civilizations work and interrelate (for example, Robinson, 2017).

Most of these openings to emotion are nevertheless hard to relate to risk management as such. Their broad claims and checklists of psychological traits seem somehow either too sweeping or too static to relate to the very specific kinds of decisions that concern risk. One might consider building on more particular aspects like the nervous excitement of consciously taking some kinds of risks (as noted by House, 2000), the emotional balancing act of presenting swear words to an undefined audience (Hjort, 2017), the quite different anxious optimism of trusting a translator or a client for the first time, the sense of commitment and responsibility of being trusted for the first time, the gut feeling of deception when trust is betrayed, and even the feeling of intense relief that comes at the end of an interpreting assignment yet is not so available to the written translator, for whom there is always the lingering risk that a mistake will be discovered. Such emotions might also hopefully be related to the kinds of happiness that translators experience at work (Liu, 2013) and to enjoyment of linguistic difference as a kind of aesthetic pleasure rather than a trigger for guilt (Malmkjær, 2007).

To reach some of these particularly positive emotions, translators would have to break from regimes of subservience to other social actors, of inadequacy in the face of impossible goals, and of guilt ensuing from a job that could always have been done better. A major step in that direction could be the recognition that risk taking can be justified, pleasurable and socially rewarded.

25.5 Concluding remarks: Ways to mediate in risk society

Modernity has been described in term of a developing "risk society" (Beck, 1986; Giddens, 1999), in part because technological complexity and environmental threats mean that cause and consequence cannot be grasped and controlled directly. We once trusted a multitude of intermediaries (doctors, scientists, banks, electricity companies, and so on) to control cause and effect, but as that trust breaks down, each of those relations becomes a calculation of risk, with attendant anxiety. The work of translators is no different in this regard: the illusory certainties of "*the* translation" delivered by "*the* professional" have largely given way to relative probabilities of communicative success and calculated distributions of effort. In the meantime, the free availability of machine translation means that everyone can do some kind of translation, so all translators can be second-guessed. This opens at least two possible avenues for translators' management of risks.

On the one hand, as mentioned, the role of the professional translator might become that of the consecrated post-editor, the authority who can legitimize one among many variants, the professional who sells degrees of trustworthiness rather than numbers of words. (People still go to doctors for calming advice, since self-diagnosis on the Internet just makes people feel nervous and sick; they might turn to professional translators for similar reasons.) Translators thus become figures of authority, taking on the risks of those who turn to them; they learn to look and sound trustworthy, in denial of the indeterminism at the core of their task.

On the other hand, the technologies of machine translation, of cheap publication and of crowdsourcing open new possibilities for the social dimension of translation. The kind of secret "insider" translation knowledge that print culture long reserved for the lone translator can now be made more public: all can potentially participate in the excitement and pleasures of variant solutions, engaging in and producing multiple different translations, sharing rather than transferring risks, extending and enhancing what has technically been called tolerance of ambiguity but has always been the opposite of the authoritarian personality and cultural closure.

Exactly how such technological possibilities affect translators' cognitive processes remains to be discovered. Before we can even begin to tackle such questions, however, we need some kind of empirical answers to a few of the basic questions raised by the concept of risk management. For example, do human translators consistently distinguish between low-risk and high-risk problems? Do they really expend less effort on low-risk problems (in a way that machine translation does not)? Do readers and/or clients actually choose to trust a translation on the level of the whole text and/or appropriateness to the communicative situation? Do readers trust a translation or an image of the translator? Does the social status of the translator actually affect the way a translation is received? To what extent are readers with field expertise able to correct machine translation errors as they read and thus mitigate high risks? Is it more efficient for them to read in this way or to employ a human translator? To what extent are certain personality types given to enjoy the indeterminacy of translation problems, either as translators or as readers? And do the answers to these questions depend in the first place on a particular historical translation form (which can change)?

When we have data rather than brave assumptions on these points, we might be able to indicate a way forward.

Further reading

Matsushita, K. (2016). Risk management in the decision-making process of English–Japanese news translation [unpublished PhD thesis]. Rikkyo University, Tokyo. https://tinyurl.com/y67h6r5y
This ground-breaking empirical research demonstrates how the concepts of risk management can account for translation decisions that might otherwise appear to be ethically aberrant. The cases include not just print news translation but also the work of media interpreters at press conferences, where they are called upon to deal with highly sensitive topics on the fly.

Pym, A. (2015). Translating as risk management. *Journal of Pragmatics, 85*, 67–80.
This article attempts to bring together and name the various levels on which risk management can be applied to translation, mustering evidence from cognitive and textual studies. It nevertheless fails to include risk mitigation as a category, which was added to the conceptual framework later (Pym & Matsushita, 2018).

References

Adorno, T., Frenkel-Brunswik, E., Levinson, D., & Sanford, N. (1950). *The authoritarian personality. Studies in prejudice*. New York: Harper & Row.
Akbari, M. (2009). Risk management in translation. In H. C. Omar, H. Haroon, & A. Ghani (Eds.), *The sustainability of the translation field: The 12th International Conference on Translation* (pp. 509–518). Kuala Lumpur, Malaysia: Persatuan Penterjemah Malaysia.
Angelone, E. (2010). Uncertainty, uncertainty management and metacognitive problem solving in the translation task. In G. Shreve, & E. Angelone (Eds.), *Translation and Cognition* (pp. 17–40). Amsterdam: John Benjamins.
Angelone, E., & Shreve, G. M. (2011). Uncertainty management, metacognitive bundling in problem solving, and translation quality. In S. O'Brien (Ed.), *Cognitive explorations of translation* (pp. 17–40). London: Bloomsbury.
Beck, U. (1986). *Risikogesellschaft. Auf dem Weg in eine andere Moderne*. Frankfurt a.M.: Suhrkamp.
Bell, A. (1984/1997). Language style as audience design. In N. Coupland, & A. Jaworski (Eds.), *Sociolinguistics: A reader and coursebook* (pp. 240–250). New York: St Martin's Press.
Bell, A. (2001). Back in style: Reworking audience design. In P. Eckert, & J. R. Rickford (Eds.), *Style and sociolinguistic variation* (pp. 139–169). Cambridge: Cambridge University Press.
Bolaños Medina, A. (2015). La tolerancia a la ambigüedad y los procesos cognitivos del traductor. *Babel, 61*(2), 147–169.
Bontempo, K., Napier, J., Hayes, L., & Brashear, V. (2014). Does personality matter? An international study of sign language interpreter disposition. *Translation & Interpreting, 6*(1), 23–46.
Bourdieu, P. (1980). *Le sens pratique*. Paris: Minuit.
Bruni, L. (1405/1969). Praefatio in Vita M. Antonii ex Plutarcho traducta. In H. Baron (Ed.), *Leonardo Bruni Aretino. Humanistisch-philosophische Schriften mit einer Chronologie seiner Werke und Briefe* (2nd ed., pp. 102–104). Stuttgart: Teubner.
Canfora, C., & Ottmann, A. (2015). Risikomanagement für Übersetzungen. *Trans-Kom, 8*(2), 314–346.
Canfora, C., & Ottmann, A. (2018). Of ostriches, pyramids, and Swiss cheese—Risks in safety-critical translations. *Translation Spaces, 7*(2), 167–201.
Diamond, L. C., Schenker, Y., Curry, L., Bradley, E. H., & Fernandez, A. (2009). Getting by: Underuse of interpreters by resident physicians. *Journal of General Internal Medicine, 24*(2), 256–262.
Dionne, G. (2013). Risk management: History, definition and critique. *Risk Management and Insurance Review, 16*(2), 147–166.
Englund Dimitrova, B. (2005). *Expertise and explicitation in the translation process*. Amsterdam: John Benjamins.
Fiedler, S., & Wohlfarth, A. (2018). Language choices and practices of migrants in Germany: An interview study. *Language Problems and Language Planning, 42*(3), 267–287.
Fraser, J. (1996). Mapping the process of translation. *Meta, 41*(1), 84–96.

Fraser, J. (2000). What do real translators do? Developing the use of TAPs from professional translators. S. Tirkkonen-Condit, & R. Jääskeläinen (Eds.), *Tapping and mapping the processes of translation and interpreting: Outlooks on empirical research* (pp. 111–122). Amsterdam: John Benjamins.
Frenkel-Brunswik, E. (1949). Intolerance of ambiguity as an emotional and perceptual personality variable. *Journal of Personality, 18*(1), 108–143.
Giddens, A. (1999). *Runaway world: How globalization is reshaping our lives*. London: Profile.
Goetzmann, W., Kim, D., Jumar, A., & Wang, Q. (2015). Weather-induced mood, institutional investors, and stock returns. *The Review of Financial Studies, 8*(1), 73–111.
Goffman, E. (1981). *Forms of talk*. Philadelphia, PA: University of Pennsylvania Press.
Hassan, H., Aue, A., Chen, C., Chowdhary, V., Clark, J., Federmann, C., Huang, X., Junczys-Dowmunt, M., Lewis, W., Li, M., Liu, S., Liu, T.-Y., Luo, R., Menezes, A., Qin, T., Seide, F., Tan, X., Tian, F., Wu, L., ...Zhou, M. (2018). Achieving human parity on automatic Chinese to English news translation. www.microsoft.com/en-us/research/uploads/prod/2018/03/final-achieving-human.pdf.
Henderson, J. A. (1987). *Personality and the linguist: A comparison of the personality profiles of professional translators and conference interpreters*. Bradford: University of Bradford Press.
Hjort, M. (2017). Affect, risk management and the translation of swearing. *Rask. Internationalt tidsskrift for sprog og kommunikation, 46*, 159–180.
House, J. (2000). Consciousness and the strategic use of aids in translation. In S. Tirkkonen-Condit, & R. Jääskeläinen (Eds.), *Tapping and mapping the processes of translation and interpreting: Outlooks on empirical research* (pp. 149–162). Amsterdam: John Benjamins.
Hubscher-Davidson, S. (2013). Emotional intelligence and translation studies: A new bridge. *Meta, 58*(2), 324–346.
Hui, M. T. T. (2012). Risk management by trainee translators: A study of translation procedures and justifications in peer-group interaction [unpublished PhD thesis.] Universitat Rovira i Virgili. www.tdx.cat/handle/10803/83497
Hurtado Albir, A. (Ed.). (2017). *Researching translation competence by PACTE group*. Amsterdam: John Benjamins.
Jääskeläinen, R., & Tirkkonen-Condit, S. (1991). Automatised processes in professional vs. non-professional translation: A think-aloud protocol study. In S. Tirkkonen-Condit (Ed.), *Empirical research in translation and intercultural studies* (pp. 89–110). Tübingen: Narr.
Keynes, J. M. (1936/1962). *The general theory of employment, interest and money*. New York: Harcourt Brace.
Knight, F. (1921). *Risk, uncertainty and profit*. Boston & New York: Houghton Mifflin.
Krings, H. P. (1988). Blick in die "Black Box"—eine Fallstudie zum Übersetzungsprozeß bei Berufsübersetzern. In R. Arntz (Ed.), *Textlinguistik und Fachsprache* (pp. 393–411). Hildesheim: Olms.
Krings, H. P. (1995/2001). Texte reparieren: empirische Untersuchungen zum Prozeß der Nachredaktion von Maschinenübersetzungen [Habilitation thesis]. Hildesheim: University of Hildesheim. Translated as *Repairing texts: Empirical investigations of machine translation post-editing processes*, G. S. Koby (Ed.), Kent OH & London: The Kent State University Press.
Krueger, N., & Dickson, P. R. (1994). How believing in ourselves increases risk taking: Perceived self-efficacy and opportunity recognition. *Decision Sciences, 25*(3), 385–400.
Künzli, A. (2004). Risk taking: Trainee translators vs professional translators. A case study. *Journal of Specialised Translation, 2*, 34–49. www.jostrans.org/issue02/art_kunzli.pdf
Lammers, M. (2011). Risk management in localization. In K. J. Dunne, & E. S. Dunne (Eds.), *Translation and localization project management: The art of the possible* (pp. 211–232). Amsterdam: John Benjamins.
Levý, J. (1963/2011). *Umění překladu*. P. Corness (Trans.), Z. Jettmarová (Ed.), *The art of translation*. Amsterdam: John Benjamins.
Liu, F. M. C. (2013). Revisiting the translator's visibility: Does visibility bring rewards? *Meta, 58*(1), 25–57.
Luhmann, N. (1991). *Soziologie des Risikos*. Berlin & New York: de Gruyter.
Malmkjær, K. (2007). Translation competence and the aesthetic attitude. In A. Pym, M. Shlesinger, & D. Simeoni (Eds.), *Beyond descriptive translation studies: Investigations in homage to Gideon Toury* (pp. 293–310). Amsterdam: John Benjamins.
Mason, I. (2000). Audience design in translating. *The Translator, 6*(1), 1–22.
Matsushita, K. (2016). Risk management in the decision-making process of English-Japanese news translation [unpublished PhD thesis]. Rikkyo University, Tokyo. https://researchmap.jp/7000009916/published_papers/21633146
Mounin, G. (1963). *Les problèmes théoriques de la traduction*. Paris: Gallimard.
Nord, C. (2014). *Hürden-Sprünge. Ein Plädoyer für mehr Mut beim Übersetzen*. Berlin: BDÜ-Fachverlag.

Pavlović, N. (2010). What were they thinking?! Students' decision making in L1 and L2 translation processes. *Hermes, 44*, 63–87.
Pirouznik, M. (2014). Personality traits and personification in translators' performances. Report on a pilot study. In E. Torres-Simón, & D. Orrego-Carmona (Eds.), *Translation research projects 5* (pp. 93–112). Tarragona: Intercultural Studies Group.
Pokorn, N., & Čibej, J. (2018). "It's so vital to learn Slovene": Mediation choices by asylum seekers in Slovenia. *Language Problems and Language Planning, 42*(3), 286–304.
Pound, E. (1915). *Cathay*. London: Elkin Mathews.
Pym, A. (2004/2005). Text and risk in translation. In M. Sidiropoulou, & A. Papaconstantinou (Eds.), *Choice and difference in translation. The specifics of transfer* (pp. 27–42). Athens: University of Athens. Extended version in Aijmer, K., & Alvstad, C. (Eds.). (2005). *New tendencies in translation studies* (pp. 69–82). Göteborg: Göteborg University.
Pym, A. (2007). On Shlesinger's proposed equalizing universal for interpreting. In F. Pöchhacker, A. L. Jakobsen, & I. Mees (Eds.), *Interpreting studies and beyond: A tribute to Miriam Shlesinger* (pp. 175–190). Co penhagen: Samfundslitteratur Press.
Pym, A. (2008a). On omission in simultaneous interpreting: Risk analysis of a hidden effort. In G. Hansen, A. Chesterman, & H. Gerzymisch-Arbogast (Eds.), *Efforts and models in interpreting and translation research* (pp. 83–105). Amsterdam: John Benjamins.
Pym, A. (2008b). On Toury's laws of how translators translate. In A. Pym, M. Shlesinger, & D. Simeoni (Eds.), *Beyond descriptive translation studies: Investigations in homage to Gideon Toury* (pp. 311–328). Amsterdam: John Benjamins.
Pym, A. (2015). Translating as risk management. *Journal of Pragmatics, 85*, 67–80.
Pym, A. (2016). Risk analysis as a heuristic tool in the historiography of interpreters. For an understanding of worst practices. In J. Baigorri-Jalón, & K. Takeda (Eds.), *New insights in the history of interpreting* (pp. 247–268). Amsterdam: John Benjamins.
Pym, A. (2018). Introduction: Why mediation strategies are important. *Language Problems and Language Planning, 42*(3), 255–265.
Pym, A., Grin, F., Sfreddo, C., & Chan, A. L. J. (2012). *The status of the translation profession in the European Union*. Luxembourg: European Commission.
Pym, A., & Matsushita, K. (2018). Risk mitigation in translator decisions. *Across Languages and Cultures, 19*(1), 1–18.
Quine, W.V. O. (1960). *Word and object*. Cambridge, MA: MIT Press.
Reiss, K., & Vermeer, H. J. (1984). *Grundlegung einer allgemeinen Translationstheorie*. Tübingen: Niemeyer.
Robinson, D. (1991). *The translator's turn*. Baltimore & London: The Johns Hopkins University Press.
Robinson, D. (2017). *Exorcising translation. Towards an intercivilizational turn*. New York & London: Bloomsbury Academic.
Rosiers, A., & Eyckman, J. (2017). Investigating tolerance of ambiguity in novice and expert translators and interpreters: An exploratory study. *Translation & Interpreting, 9*(2), 52–66.
Schubert, K. (2005). Translation studies: Broaden or deepen the perspective? In H.V. Dam, J. Engberg, & H. Gerzymisch-Arbogast (Eds.), *Knowledge systems and translation* (pp. 125–148). Berlin: De Gruyter.
Shlesinger, M. (1989). Simultaneous interpretation as a factor in effecting shifts in the position of texts on the oral-literate continuum [unpublished MA thesis]. Tel Aviv University. www.researchgate.net/publication/329371864_Simultaneous_interpretation_as_a_factor_in_effecting_shifts_in_the_position_of_texts_in_the_oral-literate_continuum
Skjong, R. (2005). Etymology of risk. http://research.dnv.com/skj/papers/etymology-of-risk.pdf
Stoeller, W. (2003). Risky business. Risk management for localization project managers. www.translationdirectory.com/article462.htm
Suojanen, T., Koskinen, K., & Tuominen, T. (2015). *User-centered translation*. London & New York: Routledge.
Tirkkonen-Condit, S. (2000). Uncertainty in translation processes. In S. Tirkkonen-Condit, & R. Jääskeläinen (Eds.), *Tapping and mapping the processes of translation and interpreting: Outlooks on empirical research* (pp. 123–142). Amsterdam: John Benjamins.
Toudic, D. (2012). Employer consultation synthesis report. OPTIMALE Academic network project on translator education and training. Université Rennes 2, Rennes.
Wilss, W. (2005). Übersetzen als Sonderform des Risikomanagements. *Meta, 50*(2), 656–664.

Part IV

Taking Cognitive Translation Studies into the future

26

Translation, expert performance and cognition

Igor A. L. da Silva

26.1 Introduction

The relationship between expertise and both translation and interpreting has not only intrigued scholars since ancient times but has also become a subject of investigation in disciplines of their own in the last century. This has led to a myriad of definitions for each term, ranging from commonsensical notions to more sophisticated accounts of what expertise entails or what translation is all about. This chapter takes up the challenge of exploring a combination of such terms: expertise in translation/interpreting, focusing on the expertise framework and its implications for the study of translation and interpreting.

Expertise may be used as a buzzword in our daily lives, especially when it comes to pointing out those who (seemingly) excel in a domain or those who are (seemingly) skilful or well informed in a domain. This chapter particularly draws on the notion of expertise as developed from an expert-performance approach, as it seems to have paved the way for a new trend of investigations in Cognitive Translation Studies since Ericsson's (2000) breakthrough article on expertise in interpreting and Shreve's (2002, 2006) discussion of the impact of an expert-performance approach on the investigation of translation as a cognitive activity. Both Ericsson and Shreve have been amply cited in papers and articles related to translation at the interface with expertise.

The first principled investigations of the topic started nearly 50 years ago, with Chase and Simon (1973) and Simon and Chase (1973) focusing on expertise in chess. Investigations of expertise have had a great impact on cognitive psychology (see Ericsson et al., 2006), but expertise is also relevant because of its societal importance (see Bereiter & Scardamalia, 1993). This holds true for expertise in translation and expertise in interpreting, which are of great relevance in a number of cultural encounters and commercial businesses (Cronin, 2006).

This chapter is divided into four sections including this Introduction. Section 26.2 describes expertise in any given domain, making the point that Cognitive Translation Studies should resort to the notion of expertise as a more "elucidating" substitute for the notion of competence. Section 26.2 also discusses whether expertise in translation and expertise in interpreting are one single construct, focusing primarily on expertise in translation and providing a brief account of expertise in interpreting. Section 26.3 points to future directions and to approaches to expertise that could somehow find their way into Cognitive Translation Studies. Finally, Section 26.4 concludes with some remarks on the impact of the expertise framework on future endeavours in Cognitive Translation Studies.

26.2 Core topics

26.2.1 Expertise in a domain

Because the word "expert" shares the same origin as the words "experience" and "experiment", "expertise" may at first refer to the characteristics, skills and/or knowledge of someone who has learned from experience (see Ericsson, 2006, 2018). This definition is misleading in that one is led to assume that 1) expertise, measured mostly in years of practice, requires experience and also that 2) experience will necessarily result in expertise. While the first part of the assumption is true, the second is not (see Ericsson et al., 2006).

Ericsson (2006, pp. 10–14) describes five major theoretical conceptions of expertise, namely: 1) the result of individual differences in mental capacities (e.g. Galton, 1869/1979), 2) the extrapolation of everyday skills to extended experience (e.g. de Groot, 1946/1978), 3) qualitatively different representation and organization of knowledge (e.g. Chi, 2006b; Chi et al., 1981), 4) elite achievement resulting from superior learning environments (e.g. Roe, 1952; Zuckerman, 1977), and 5) reliably superior (expert) performance on representative tasks (e.g. Ericsson & Smith, 1991; van der Maas & Wagenmakers, 2005). Apart from conception 1), which has failed tests over the twentieth century, such major frameworks consider expertise as an acquired skill rather than the result of basic endowment (e.g. innate talent, mental capacities, intelligence or memory) and constrain potential innate genetic predictors of expertise to body mass and height in some domains (e.g. basketball and ballet) (Ericsson, 2000).

Following Ericsson's (2000) seminal approach to expertise in interpreting, it seems that scholars have been mostly committed to conception 5 (Ericsson et al., 2006), also known as the expert-performance approach, especially when it comes to theoretical accounts of expertise in translation, with some rare interfaces with conception 3 (e.g. da Silva, 2007, 2012; Pagano & da Silva, 2008). In both cases, the aim is to understand skills, knowledge and/or characteristics that distinguish experts from less skilled people.

A working definition of expertise within conception 5 could relate to consistently superior, or outstanding, or exceptional, performance in a given domain, with a domain being some kind of skilled activity (Shreve, 2006), which can be either informal (e.g. sewing or cooking) or formal (e.g. biology or chess) (Chi, 2006a). The demonstration of such consistently superior performance can be attained on a set of reproducible representative tasks in any given domain and is assumed to be the result of deliberate practice, i.e. the engagement in training activities especially designed for developing and consistently maintaining high performance levels in that domain (Ericsson & Charness, 1997).

The crucial aspect of deliberate practice is that unlike the mere accumulation of experience in performing a regular activity, it is focused and well planned, necessarily requiring both regular execution of well-defined tasks designed to be performed sequentially, with clear goals and adequate, gradually increased levels of difficulty for the individual, and informative feedback followed by opportunities for peer observation, repetition and error correction (Ericsson, 2006) over a significant period of time—ten years or so (Ericsson & Crutcher, 1990). In other words, regular engagement in an activity with a view to improving performance is a prerequisite of deliberate practice, which entails conscious performance monitoring (Horn & Masunaga, 2006), avoidance of arrested development associated with automaticity, and acquisition of metacognitive skills to support continued learning and improvement (Ericsson, 2000, 2006). Performers' practice is assessed either by themselves or by their peers or coaches against the predetermined goals or expected levels of achievement (Horn & Masunaga, 2006).

A combination of characteristics is required for one to attain expertise (Ericsson & Smith, 1991). Drawing on Tiselius (2013b), they may be summarized as follows: 1) regular outstanding performance in the domain, which rules out single top performances, 2) access to expert knowledge when needed; i.e. both novices and experts may have similar performances in routine tasks but not in difficult situations, 3) extensive experience in the domain, which can be roughly estimated to be ten years or 10,000 hours of practice and training, 4) engagement in deliberate practice, which consists of specific tasks, often coached, particularly designed to be completed at a time set aside only for practice, and solely aimed at improving a given skill, 5) definition of clear goals, with final goals usually divided into reachable part-time goals on both macro- and micro-levels, which is partly connected with deliberate practice, and 6) openness to feedback.

Further characteristics have also been associated with experts, especially when it comes to their expanded working memory (e.g. Atkinson & Shiffrin, 1968; Baddeley & Hitch, 1994; Ericsson, 2018)—the capacity of temporarily holding information available for processing, usually in three to seven chunks of information (e.g. Baddeley, 1986; Miller, 1967). Besides, compared with non-experts, experts tend to 1) know more, 2) produce the best solutions more quickly and more accurately, 3) represent a task and see problems at a deeper level and analyse them qualitatively, 4) have more efficient monitoring abilities to detect errors, 5) more successfully choose appropriate strategies to solve problems, 6) make more efficient use of information sources available while solving a problem and 7) have better recall as well as retrieve relevant knowledge and strategies with minimum cognitive effort (Chi, 2006a, 2006b; Glaser & Chi, 1988; Swanson & Holton III, 2001). Such positive qualities can also be overcome by some pitfalls and shortcomings that are common among experts, including (see Chi, 2006b) 1) poor calibration of skills required to perform a task because of overconfidence, 2) negligence of details at the surface level, 3) need for contextual clues to solve problems, 4) inflexibility and difficulty in adapting to changes, and 5) bias towards providing results based on what is already known, thereby neglecting other possibilities.

There have been two major traditions in the study of expertise, which Chi (2006a) names the "absolute approach" and the "relative approach". The first approach builds on studying "truly exceptional" individuals to understand "how they perform in their domain of expertise" (Chi, 2006a, p. 21). Two major methods to identify such individuals are the retrospective method, i.e. by "looking at how well an outcome or product is received" (Chi, 2006a, p. 21), and the concurrent measure method, i.e. by using a rating system or measures of how well the experts perform a task. The second approach draws on studying experts in comparison to novices. The definition of expertise used in this approach is to some extent loose or generic, with experts being those individuals with more knowledge or better performance than novices. Expertise in this sense can be either grossly assessed through such measures as academic qualifications, consensus among peers, seniority or years performing a task, or more thoroughly assessed through domain-specific knowledge or performance tests (Chi, 2006a, pp. 22–23). One should notice, however, that in some domains, such as sports, more declarative knowledge does not necessarily lead to superior performance (Ericsson & Lehmann, 1996; Williams & Davids, 1995).

Drawing on Chase and Simon's (1973) and Simon and Chase's (1973) pioneering work on chess, which has set the ground for modern laboratory research on expertise, Ericsson and Smith (1991) tend to favour the first approach but admit the possibility of adopting the second approach in step 2 of their three-step model for investigating expertise in domains where no external ranking is available to define outstanding performance, especially in complex domains.

The first step consists of finding or collecting a set of standardized, representative tasks to capture superior performance in the domain. The second step is to examine the expert performance in a laboratory by using methods of analysis available in cognitive psychology, which include 1) methods of inferring mediating processes, as one cannot directly observe such processes, 2) expertise–novice comparisons, 3) extensive studies of single subjects, and 4) studies of particular aspects of expert performance, rather than its entirety. The underlying assumption for the second step is that the mediating mechanism for expert performance is stable and not much influenced by the additional experience in the laboratory, which is only a fraction of the expert's total experience. The third step involves accounting for superior performance by experts, i.e. providing theoretical and empirical accounts of how the mechanisms identified in step 2 can be acquired through training and practice.

Following this general framework of the studies of expertise and expert performance, the next section delves into how expertise has been approached within Cognitive Translation Studies.

26.2.2 Expertise in Cognitive Translation Studies

Building on Halverson (2010), what is now termed Cognitive Translation Studies seems to fit well into what Holmes (1972/2000) envisioned as "process-oriented descriptive translation studies", a kind of research within Translation Studies that would profit from psychology to gain access to the "complex mental processes" underlying the "process or act of translation itself" (Holmes, 1972/2000, p. 177). The field has evolved far beyond this interaction with psychology and has been influenced by a number of other disciplines, including linguistics, neuroscience and cognitive science, to name a few (see Alves, 2015; Alves & Hurtado, 2010; Alvstad et al., 2011). However, as O'Brien (2013) points out, such influence does not necessarily entail "interdisciplinarity", as the field has been more of a borrower than a lender and has had very little impact on the other domains. O'Brien's observation adds to Alves and Hurtado's (2010) contention that cognitive approaches to translation still borrow extensively from other disciplines while striving to establish their own tradition.

This observation and the call for reciprocity between Cognitive Translation Studies and other disciplines are perceptive and well timed. However, O'Brien does not account for why it has been so. While this chapter is not intended to uncover all such reasons, a starting point seems to be how Cognitive Translation Studies appropriates concepts and methods from other disciplines without a clear understanding of the epistemological consequences (Marín, 2017, 2019). This holds particularly true when it comes to the notion of expertise and expert performance, originally introduced in cognitive psychology (see Ericsson & Crutcher, 1990; Ericsson & Smith, 1991; Scardamalia & Bereiter, 1991), and how several scholars have been at odds concerning how to relate it to the notion of translation competence or vice versa (see Alves, 2015; Muñoz Martín, 2014; PACTE, 2003, 2005, 2017; Pym, 2003; Schäffner & Adab, 2000; Shreve, 2002, 2006; Sirén & Hakkarainen, 2002; Tiselius, 2013a, 2013b).

Competence may be basically understood in two different ways: either as underlying knowledge, i.e. the cause or prerequisite of performance, or as aptitude, i.e. the result of performance (see Rothe-Neves, 2007). In their struggle to relate competence and expertise, translation scholars have 1) roughly equated expertise with translator's competence (e.g. Muñoz Martín, 2014; PACTE, 2003, 2005, 2017; Rodrigues, 2018a), 2) implicitly considered expertise either as 2.1) the final stage or result of translation competence (e.g. Englund Dimitrova, 2005) or as 2.2) the underlying factor or cause of competence (e.g. Shreve, 2002, 2006), or 3) even explicitly seen expertise as a broader term encapsulating competence (e.g. Tiselius, 2013a). Alongside the fact that the very notion of competence remains unclear within Translation Studies, as it is interchangeably associated with underlying knowledge, qualities of a good translator, aptitude, performance or the result of performance (see Malmkjaer, 2009; Pym, 2003; Rothe-Neves, 2007; Schäffner & Adab, 2000), the notion of competence seems to have brought little profit to cognitive approaches to translation, especially if one considers its influence or gradual detachment from Chomsky's (1965) generative grammar. Even though Alves (2015) argues that cognitive approaches to translation have explicitly claimed affiliation with the studies of expertise and expert performance, such affiliation, which has been dependent on how expertise relates to competence, has not entailed a rethinking of the epistemological position of the notion of expertise within Cognitive Translation Studies (Marín, 2017, 2019) and how it can feed back into the studies of expertise and expert performance.

Following Shreve et al.'s (2018) argument, competence models (e.g. Alves & Gonçalves, 2007; Bell, 1991; Hatim & Mason, 1997; Kelly, 2002; Neubert, 2000; PACTE, 2003, 2005, 2017), especially the multicomponential models, seem to have undeniable pedagogical value but tend to be less relevant to understanding cognition and mental processes, for which "the concept of expertise could be a robust and more enlightening substitute" (Shreve et al., 2018, p. 37). Besides, if Cognitive Translation Studies truly means to be reciprocal to other disciplines, it is by using the notion of expertise instead of that of competence that it may find a theoretical apparatus not only to feed into Cognitive Translation Studies but also to feed back into the studies of expertise and expert performance—for instance, by employing experimental methods, accommodating findings, and expanding theoretical frameworks with evidence from translation (Shreve et al., 2018). In Ericsson's (2000, p. 214) words, both the domains of interpreting and translation offer a unique window on real-time comprehension and seem to provide a sufficiently constrained task for the study of the most elusive phenomena in skilled activities, i.e. comprehension and production of verbal messages.

Shreve et al. (2018) draw their conclusions from seeds planted by such authors as Sirén & Hakkarainen (2002), Shreve (2002, 2006), Tiselius (2013a), and Muñoz Martín (2014), to name a few, who anticipated issues in relating expertise to competence. Their original, rather strong claim that the concept of expertise may be more "robust" and "enlightening" in Cognitive Translation Studies points to some maturity in how to conceive of and tap into expertise in translation and expertise in interpreting.

Having provided some information on how Cognitive Translation Studies has incorporated the concept of expertise, the next section discusses whether expertise in translation and expertise in interpreting should be regarded as one single construct.

26.2.3 Expertise in translation and expertise in interpreting: One single construct?

Muñoz Martín (2014, p. 4) mentions that "expertise in translation and expertise in interpreting are often implicitly assumed to be different constructs" and argues for an understanding of "expertise in translation" as a construct referring to both translation and interpreting tasks. He contends that 1) both translating and interpreting skills are "concrete, adaptive developments of various levels of bilingual proficiency to specific task configuration" (Muñoz Martín, 2014, p. 4), 2) they share some features, including an optimized bilingual lexicon and inhibitory rules for competing linguistic forms in different languages, 3) people often engage in both translation and interpreting activities, thereby gaining experience from both, and 4) "research constructs must include all cases in the definition while excluding all others" (Muñoz Martín, 2014, p. 4).

Reasons 1) and 2) seem to be valid, but it is quite debatable whether people often do engage in both translation and interpreting tasks (reason 3) and whether a research construct including all cases in the definition while excluding all others (reason 4) is feasible, considering that expertise is domain specific (Ericsson, 2000; Ericsson & Smith, 1991). If the underlying assumptions in the theoretical framework of the studies of expertise and expert performance remain true, especially when it comes to identifying the set of reproducible representative tasks in a given domain (Ericsson & Smith, 1991), it follows that an all-encompassing research construct named "expertise in translation" is inconsistent with the epistemological and theoretical frameworks of the expert-performance approach. There are some points of overlap, particularly considering Muñoz Martín's reason 2, but the representative tasks for translation are extensively different from those for interpreting languages orally, which in turn are also quite different from those for interpreting signed languages. Such differences are also reflected in the fact that studies of translation have traditionally focused on several variables and parameters that are quite different from those used in studies of interpreting. Indeed, one cannot expect an expert translator to necessarily be an expert interpreter, since expertise does not transfer across domains, however much it may help to perform tasks in other domains (Chi, 2006b; Scardamalia & Bereiter, 1991). The notion of domain can be even stricter, including text types and subject areas, in such a way that one cannot even expect, for instance, an expert literary translator to be an expert scientific-technical translator (Shreve, 2002, p. 153).

The fundamental, all-encompassing research construct should remain that of expertise, as specified, especially because translation scholars tend to explicitly state their affiliation with the studies of expertise and expert performance and usually relate their research to the notion of expertise as consistently superior performance—see, for instance, Shreve et al.'s (2018) account of and approach to the matter. In other words, even though researchers can find characteristics that are specific to translators and interpreters alike, it seems there is no need for such a specific research construct as expertise in translation for Cognitive Translation Studies. As a matter of fact, this would eventually prevent the field from feeding back into the studies of expertise and expert performance (see Section 26.2.2).

As such, the following sections address expertise in translation and expertise in interpreting separately. According to Pym (2003), the adoption of the expertise framework in the Translation Studies dates back to Holz-Mänttäri (1984), but the following section focuses on expertise studies that build on cognitive psychology, more specifically on the expert-performance approach. The use of such specific framework started in interpreting studies (e.g. Ivanova, 1999; Moser-Mercer, 2000), and seemingly gained momentum with Ericsson (2000), a specialist from cognitive psychology.

26.2.4 Expertise in translation

This section is subdivided into two: the first addresses the theoretical framework that has already been introduced into Cognitive Translation Studies, while the second addresses how this framework has been translated into empirical studies.

26.2.4.1 A theoretical account

Expertise in translation has long intrigued translation scholars, institutions and professionals (see Pym, 1996). Several questions have been raised, such as: How have translators become experts? Are *all* translators experts?

However, it was only in 2000 that Ericsson's article paved the way for the study of interpreting (and translation) from an expert-performance approach. It did not take long for Shreve (2002) to provide what is perhaps the first comprehensive account of a potential interface between such an approach and Cognitive Translation Studies.[1] Shreve (2002) discusses the theoretical issues around the notion of expertise in translation, how it relates to the notion of translation competence, and the cognitive changes that take place in translation experts. Particularly, he observes that "making the concept of expertise in translation operational would require a performance model for translation" (Shreve, 2002, p. 152). Following the studies of expertise and expert performance, such a model requires consistency, with translation experts consistently displaying expert performance.

By then, probably because of the novelty of an expert-performance approach to translation, Shreve (2002, p. 154) declared that "[o]ur interest in expertise in translation is an extension of our research in translation competence". Apparently, this perspective remained until 2018 and was notably present in Shreve's (2006) much-cited paper, in which he particularly focused on the implications of the concept of deliberate practice in translation. He wisely pointed out that a performance model "must fully describe the empirical characteristics that define superiority" (Shreve, 2006, pp. 28–29). As performance in translation cannot be easily measured, unlike in such domains as sports, Shreve (2002, 2006) seemed to be aware that performance in translation has yet to be properly defined, and Cognitive Translation Studies does need to come up with measurable, well-defined, translation-specific tasks if it is to qualitatively differentiate superior performances from simply acceptable or good performances.

Focusing on four conditions that are necessary for deliberate practice to sustain expertise—1) a well-defined task, 2) appropriate difficulty of the task for the individual, 3) informative feedback, and 4) opportunities for repetition and error corrections—Shreve (2006) accounts for the challenges that Cognitive Translation Studies may face if it is to truly adopt the notion of deliberate practice. As for condition 1), he argues that translation per se is an extremely complex task, which requires performance in a number of other related tasks, including reading and writing, which vary greatly according to experiential parameters, including pragmatic circumstances, text types, subject domain and translation *skopos*. In fact, one could contend that translation is an ill-defined, problem-solving task (see Nye et al., 2016), whose final outcome is open to some variation. However, following Pym's (2003) minimalist approach, Shreve (2006, p. 31) seems to suggest that the notion of a well-defined task in translation could be a combination of the following two activities, performed under specific experiential circumstances: "the ability to generate a series of more than one viable target text (TT1, TT2, ..., TTn) for a pertinent source text (ST)" and "the ability to select only one viable TT from this series, quickly and with justified confidence" (Pym, 2003, p. 489).

As for the appropriate difficulty condition, Shreve (2006) places it at the intersection between the specific textual properties and the existing, possibly deficient, cognitive resources of the translators. According to him, textual and contextual causes may precipitate specific difficulties in a number of stages during the translation task, including input processing, comprehension and target-text production. For instance, complex, highly dense texts may require re-reading and refreshing of working memory.

When it comes to feedback, Shreve (2006) stresses that feedback mechanisms usually work well in translation learning settings but are rare in the workplace. Still, feedback is crucial because of the interdependency with the capacity for self-regulation. One potential feedback mechanism in the workplace can be error correction in the form of editing and revising processes, when they truly provide room for the translator to perform changes and understand their inefficiency. Further compensatory error correction schemes could include mentoring or coaching, which are not rare in other domains, as well as Internet-enabled collaboration. Besides, repetition in such contexts would not be necessarily related to translating the same text over and over but to specializing in a language pair, in particular subject areas, and in particular text types. This could be a useful way to recognize, analyse and schematize linguistic, textual and conceptual patterns.

Shreve (2006) argues that expertise in translation includes performance in four distinct areas, namely: 1) L1 and L2 linguistic knowledge, 2) culture knowledge of the source and target culture, including domain knowledge of specialized domains, 3) textual knowledge of text conventions in L1 and L2, and 4) specific knowledge of translation, including strategies and procedures, such as use of translation tools and application of information-seeking strategies. According to the author, translation experts excel in their respective domains not only because of their number of cognitive resources but, most importantly, because of the quality and constitution of such resources.

Besides, consistently with the studies of expertise and expert performance, Hurtado and Alves (2009) summarize expert performance in translation as an acquired skill, which 1) requires a high level of metacognitive activity, 2) involves proceduralization of knowledge related to domain specificities, 3) requires self-regulatory behaviour, i.e. monitoring, resource allocation and planning, and 4) is not necessarily related to general cognitive capacities such as memory or intelligence. However, despite these theoretical attempts to describe expertise in translation, empirical cognitive translation investigations have not been able to address such skills under the expert-performance framework because of the unstable relationship between expertise in translation and translation competence. The following section addresses this troublesome relationship.

26.2.4.2 An account of empirical studies

As mentioned in Section 26.2.1, two major traditions stand out in the study of expertise, namely, the "absolute approach" and the "relative approach". The theoretical account described earlier calls for an absolute approach, but empirical cognitive translation investigations have primarily adopted the relative approach.

This imbalance between the theoretical account and the empirical practices may explain why scholars have mistakenly brought the notion of competence to bear on the expert-performance framework. In assuming that professional, experienced, usually (but not necessarily) trained translators are "expert enough" in the translation domain, since they have the necessary competence to perform a translation task, competence-based studies most often take it for granted that it is valid to carry out experiments with participants screened and selected on the basis of the numbers of years they have worked as professional translators. Such a procedure is consistent with a "relative approach", provided the studies limit their findings to an understanding of how an individual with more experience than another seems to perform differently, sometimes better, in a task.

Much as the relative approach tradition is valid within its own theoretical framework and arguably allows translation competence to be equated with expertise in translation, it comes at a price. Experts, loosely defined as such, do not necessarily perform better than non-experts or novices, nor do they necessarily perform better consistently. For example, the PACTE group started out with 35 professional translators but ended up with only nine top-ranked translators. PACTE (2003, p. 44) stated that they "started from the concept of translation as a communicative activity [... that] requires expert knowledge. In Translation Studies, this expert knowledge is called Translation Competence". However, PACTE (2017) had to take a relative approach to account for the performance of the nine top-ranked translators, even though all participants had originally been selected because of their assumed competence/expertise resulting from years of experience (see especially PACTE, 2017, pp. 269–302). Besides, several professional translators performed similarly to non-translators, and some were even outperformed by them in at least one of the tasks. On top of that, the participants performed only two tasks, which does not necessarily guarantee that the nine top-ranked translators would consistently perform at superior levels. PACTE (2017, p. 294) insists that translation competence *is* expert knowledge, but it turns out that translation competence may be at best a prerequisite for expertise: if one assumes that the nine top-ranked translators were indeed experts from the perspective of a relative approach to expertise, they were actually found because of their alleged competence.

The notion of competence seems to be dear to Translation Studies, and competence models do have pedagogical value (e.g. Esqueda, 2020). Besides, studies following this notion may well contribute to the relative approach to expertise if they recognize themselves as such. They may also contribute to the absolute approach to expertise if researchers learn from their task applications and how they can serve the purpose of deliberate practice and identification of consistently superior performance, especially considering steps 1 and 2 in Ericsson and Smith's (1991) three-step model for investigating expertise.

To the best of this author's knowledge, no scholar within Cognitive Translation Studies has followed the three steps fully in empirical studies, especially when it comes to the third step: accounting for consistently superior performance by translation experts. In fact, the field has yet to come up with a collection of representative tasks, which, to date, may only be speculatively derived from a combination of the individual tasks used in several empirical studies. Most empirical investigations have focused on comparisons between experts, loosely defined as such, and non-experts (e.g. Jakobsen, 2002, 2003), or between performance in an expertise-specific domain and performance in a slightly different domain (e.g. different text type, subject matter, language pair, directionality or number of text features) (e.g. da Silva, 2007, 2012; Schmaltz, 2015), or even both (e.g. Alves et al., 2011; Dragsted, 2005). Few studies have focused on single expert individuals (e.g. Jakobsen, 2005). Methods have usually included verbal protocols (e.g. da Silva, 2015; Englund Dimitrova, 2005), keylogging (e.g. Jakobsen, 2002, 2003), eye tracking (e.g. Jakobsen & Jensen, 2008; O'Brien, 2006) or triangulation of different sources of data elicitation (e.g. Carl & Dragsted, 2012; Hvelplund, 2011), usually concerned with ecological validity (see Alves, 2003; Shlesinger, 2000). More recent studies have also used resources from the neurosciences, including functional magnetic resonance imaging (e.g. García, 2013; Szpak, 2017).

A systematic expert-performance approach to translation has yet to emerge in empirical cognitive translation investigations, especially when it comes to defining or singling out expert individuals and having them participate in reproducible tasks. Nonetheless, clear attempts to qualify expert performance have been made. The analyses have focused on a range of dependent, process-related variables of temporal, cognitive and technical nature (Krings, 2001), some of which have evolved independently within Translation Studies, Cognitive Translation Studies or even other fields, including working memory, pauses, segmentation, micro-units, recursiveness, renditions, fixations and metarepresentation. The results of the individual empirical studies are not generalizable, but they seem to indicate that within their domain of expertise, experts, loosely defined as such, tend 1) to operate in an integrated processing mode when dealing with familiar, easy texts, i.e. a mode featuring long average segment sizes, high production speed and short pauses, with most processing at the clause/sentence level, and 2) to make decisions more quickly than novices and stick to them once they write them, with recursiveness and changes in renditions having an impact on the macro-level of the text rather than being mere changes of words on the micro-level. Also, it is worth noticing Jakobsen's (2005) attempt to define an objective, systematic metric to identify expert performance, aka peak performance, i.e. producing extremely long segments by pressing 60 keys or more sequentially without any interruption longer than 2.4 seconds.

In addition to the traditional expert–novice comparison, a few studies have also compared "expert" translators with field specialists in the translation of technical-scientific texts (e.g. Alves et al., 2011; Pagano & da Silva, 2008). Field specialists, such as physicists and physicians in non-English-speaking countries, often develop translation skills as an integral part of their expertise in their respective fields. This allows them to participate in various academic activities (e.g. to attend conferences and publish in peer-reviewed journals). Studies of such field specialists have attempted to uncover the impact of domain knowledge on task execution. Findings seem to show that field specialists cannot operate in an integrated processing mode, regardless of their domain knowledge of the subject matter addressed in the text. In other words, their domain knowledge, including source-text comprehension and vocabulary, does not allow them to perform to standards that would be expected from expert translators when they are dealing with source texts in domains with which they are familiar, as reflected in the production of long segments with few and short pauses in between (e.g. Dragsted, 2005; Jakobsen, 2005). Even so, some field specialists can perform tasks at similar speeds as translators and can represent meaning at higher linguistic levels (see Chi, 2006b), including those of the clause and sentence (da Silva, 2015), with some of their target texts ranking better than the translators' when assessed by a panel including field specialists, linguists and translators (Braga, 2012). These findings, which seem to challenge some current notions of expertise in translation (e.g. are exceptionally long segments and few, short pauses truly indicative of expertise in all cases? Do they always translate into fast performance and production of quality target texts?), may be used in the future to improve the collection of reproducible tasks that point to expertise in translation.

26.2.5 Expertise in interpreting

Moser-Mercer appears to have introduced the expert-performance approach to the interpreting community when she invited Ericsson to the Ascona workshops in 1997 and 2000 (see Tiselius, 2013b). However, it was Ericsson's (2000) article that formally proposed an expert-performance approach to interpreting to substitute those he named "traditional approaches", including the pioneering work by Gerver (1971). Furthermore, even though Shreve (2006) does not explicitly mention "interpreting", it seems that his account of translation is all-encompassing and does include interpreting, as he mentions "listening" once (Shreve, 2006, p. 32).

Early studies in interpreting assumed that expertise in interpreting was dependent on some innate special skill (see Dillinger, 1989). Subsequent attempts were built on introspective descriptions of mediating mechanisms or on basic processes such as word recognition and categorization, but they lacked independent verification and experimental validation, and therefore did not meet the standards of laboratory research (see Massaro & Shlesinger, 1997; Shlesinger, 2000). According to Shlesinger (2000), there is evidence against the decomposability of interpreting into components and sub-processes, since the complex semantic context at the level of discourse is relevant for generating the meaning of a word or a sentence.

In his proposal for an expert-performance approach to interpreting, Ericsson (2000) draws on the three-step model described in Section 26.2.1. The author also makes a strong case for the potential of such an approach to interpreting. However, as he could not find studies that met the criteria for the expert-performance approach at that time, he did not demonstrate how it could be applied.

According to Tiselius (2013b, p. 19), the expert-performance approach has since found its way into studies of expertise in interpreting. For instance, Moser-Mercer has investigated expertise from the learners' perspective (e.g. Moser-Mercer, 2000; Moser-Mercer et al., 2000). Besides, some dissertations have explored expertise in interpreting: Ivanova (1999) addressed problem-solving strategies, Liu (2001) focused on working memory, and Vik-Tuovinen (2006) included preparation in her study of expertise from a wider perspective.

It seems that, as in the studies involving translation, a systematic expert-performance approach has yet to emerge in empirical studies of interpreting. Still, Liu (2008) reports several findings in interpreting that can be correlated with expertise (e.g. Dillinger, 1989; Goldman-Eisler, 1972; Isham, 1994; Liu, 2001). She points to three major processes in interpreting—comprehension, translation and production—and further divides them into subskills and cognitive abilities (concurrent articulation, articulatory suppression, working memory and attention shift). Building on such processes and subskills, she reports some common features among experienced interpreters compared with non-experienced interpreters that might be indicative of expertise in interpreting, such as 1) better semantic processing, 2) better selection of the most important meaning units,

3) less disturbance by delayed auditory feedback, 4) larger digit span when it comes to working memory, and 5) rapid attention switching between listening and speaking. These skills usually entail fewer errors, faster responses and less effort.

These findings were obtained in studies involving oral languages in either simultaneous or consecutive interpreting. To the best of the author's knowledge, Cognitive Translation Studies is still in its infancy when it comes to interpreting sign languages (Rodrigues, 2018a, 2018b). This is even more the case regarding studies of expertise in sign language interpreting. In one of the few studies, Rodrigues (2018a), who arguably equates competence with expertise, advocates the inclusion of an intermodal component if a translation competence model is to be "universal" or at least capable of accounting for interpreting sign language. In a related work, Rodrigues (2018b) states that there are at least two language modalities—an auditory-vocal modality and a visual-gesture modality—and contends that modality impacts on interpreting from oral language to sign language because of 1) the use of code blending, 2) the necessary body-visual performance, and 3) the prevalence of translation into the L2. According to Rodrigues (2018a, 2018b), expertise in interpreting sign languages requires both a kinaesthetic-bodily ability to produce signs and a visual-cognitive ability to understand signs, both of which are linked to the linguistic competence and the communicative competence.

26.3 Future directions

Besides the expert-performance approach, other approaches to expertise, not necessarily mutually exclusive, can likewise contribute to Cognitive Translation Studies. Several translation scholars have claimed that translation requires "competence" in reading and writing (see Muñoz Martín, 2014; Shreve, 2006, among others). Therefore, it seems promising to understand the developments in studies of expertise in such related domains. For instance, Scardamalia and Bereiter (1991) suggest that expertise in writing is a dialectical process whereby individuals both deduce from their domain knowledge and their discourse knowledge to solve a particular case and infer from a particular case to reformulate their domain knowledge and their discourse knowledge. The authors point out that there are two different writing processes. In the *knowledge transfer* process, the text tends to reflect the order in which the individual thinks of something, rather than an order imposed by the content as a result of planning. In the *knowledge transformation process*, two spaces are interconnected: the content space, where there are problems related to domain (or content, or subject-matter) knowledge, and the rhetorical space, where there are problems related to writing (discourse knowledge).

Scardamalia and Bereiter (1991) also state that, in several domains, experts need to develop skills in their core tasks while also developing skills in writing and reading. They suggest that expertise in domains requiring reading and writing skills may be related to how such skills are incorporated into the professional performance and to some habits of the individuals. Some scholars in Cognitive Translation Studies have mentioned Scardamalia and Bereiter's work, especially when dealing with comparisons between translators and field specialists (e.g. da Silva, 2007, 2012; Pagano & da Silva, 2008), but they have not explicitly focused on the translators' habits and how their reading and writing skills are incorporated in their performance. Besides, studies of translation have not investigated the implications of fast performance as an unreliable predictor of writing and reading expertise (see Scardamalia & Bereiter, 1991); i.e. unlike in most domains, expert writers and expert readers are not necessarily significantly faster than novices, which may also hold true for expert translators to some extent.

From a sociological perspective, Collins and Evans (2007) posit several different types of expertise, two of which seem to be of relevance to studies of translation and interpreting. The authors suggest that specialized tacit knowledge can be associated with two different types of specialized expertise; namely, contributory expertise and interactional expertise. In general terms, contributory expertise is what allows individuals to perform their activities in their domains (e.g. a surgeon operating on a patient), while interactional expertise is what allows reviewers, journalists, sociologists and translators to perform a great deal of their tasks upon conversations and interactions with those who have contributory expertise; i.e. it is co-created between the individuals. In other words, the former allows individuals to perform directly in their own domain, whereas the latter implies encyclopaedic knowledge and mastery of the language of another domain without knowledge of actual practice in that domain. Drawing on the interactional expertise paradigm, da Silva and Silveira (2017) carried out a two-year longitudinal study with novice translators who periodically interacted with physicians and performed tasks aimed at deliberate practice. The authors showed that interactional expertise enabled the novice translators by themselves to pinpoint and solve problems in the writing of the source text before producing an adequate translation with improved confidence in their own work. The focus in such an approach is, therefore, not exclusively on the individual performer but, rather, on their interactions with others, as elaborated on by da Silva (2019).

26.4 Concluding remarks

The present chapter is necessarily a selective approach, which neglects several aspects, theories and historical accounts of expertise in general and expertise in translation/interpreting in particular. It provides a general overview of some aspects of several different definitions of expertise, both concerning studies of expertise and expert performance and regarding Cognitive Translation Studies (see Muñoz Martín, 2014; Tiselius, 2013b, Marín, 2017, 2019, for further discussions). There seem to be competing definitions of what expert performance is in interpreting and in translation; besides, they may be dependent on text type, *skopos*, among other intervening factors.

By explicitly affiliating with some paradigms, traditions or approaches from other disciplines, while also understanding their impact on the tradition of empirical-experimental research in Translation Studies in general and Cognitive Translation Studies in particular, researchers can overcome several shortcomings and most likely improve their understanding of translation and interpreting while also feeding back into other disciplines. For instance, Tiselius (2013b) affiliated with the expert-performance approach and found results that challenge the notion of deliberate practice in interpreting. While she did not deny the validity of the concept, Tiselius (2013b) pleaded for a more precise, unambiguous understanding of it, both from within the discipline as well as from the perspective of the expert-performance approach. Evidence of expertise in translation or in interpreting is mostly similar to that in other domains (Jääskeläinen, 2011), and a closer dialogue between Cognitive Translation Studies and studies of expertise and expert performance could be fundamental, not only to improve our understanding of translation as a highly demanding cognitive act and of expert professional behaviour, but also to better place translation and Cognitive Translation Studies on the map of other disciplines.

Cognitive Translation Studies is paving its way to reaching maturity in its investigation of expertise in translation and expertise in interpreting. Perhaps, the next step is to truly overcome the unstable relationship, at times dichotomous, at others close, between expertise in translation and translation competence, which has long trapped Cognitive Translation Studies in its theoretical and methodological accounts of expertise in translation and expertise in interpreting. Hopefully, this chapter has provided a contribution to this debate by shedding light on the differences between an absolute and a relative approach to expert performance with respect to translational activity. In doing so, researchers will most likely create a bulk of empirical studies that are comparable because they share methodological procedures, core definitions and theoretical frameworks, not only in

Cognitive Translation Studies but also in neighbouring disciplines. In this way, expert performance and expertise are likely to become strong constructs rather than buzzwords indistinguishably associated with competent, professional and/or experienced individuals. The field may also come up with objective metrics of what absolute expertise in translation or in interpreting really is, which could also feed back into the studies of expertise and expert performance.

Acknowledgements

The author is thankful to Prof. K. Anders Ericsson for his valuable contributions to early drafts and for his insightful questions, which have triggered a significant amount of reflection about the status of the expert-performance approach in Cognitive Translation Studies. He is also thankful to CAPES, Coordination for Improvement of Higher Education Personnel (grant 88887.375027/2019-00), for its support.

Note

1 Sirén and Hakkarainen (2002) also provided a comprehensive account of the potential of investigating expertise in translation, but not from an expert-performance approach.

Further reading

Alvstad, C., Hild, A., & Tiselius, E. (Eds.). (2011). *Methods and strategies of process research: Integrative approaches in translation studies*. Amsterdam: John Benjamins. doi: 10.1075/btl.94

A volume with 19 chapters on innovative methodological approaches to investigating translation and interpreting.

Ericsson, K. A., Hoffman, R. R., Kozbelt, A., & Williams, A. M. (Eds.). (2018). *The Cambridge handbook of expertise and expert performance* (2nd ed.). Cambridge: Cambridge University Press. doi: 10.1017/9781316480748

As an updated edition of the 2006 *Cambridge handbook of expertise and expert performance*, this handbook provides vast documentation of expertise in 16 major domains, including theoretical and methodological discussions.

Lacruz, I., & Jääskeläinen, R. (Eds.). (2018). *Innovation and expansion in translation process research* (pp. 37–54). Amsterdam: John Benjamins. doi: 10.1075/ata.xviii

A collection of 12 chapters on translation process that report on methods, research topics, and interactions of the Translation Studies with other disciplines.

Shreve, G. M, & Angelone, E. (Eds.). (2010). *Translation and cognition: Recent developments*. Amsterdam: John Benjamins. doi: 10.1075/ata.xv

A collection of 15 chapters that assesses the state of the art in Cognitive Translation Studies by focusing on methodological innovation, research design and research issues.

References

Alves, F. (Ed.). (2003). *Triangulating translation: Perspectives in process oriented research*. Amsterdam: John Benjamins. doi: 10.1075/btl.45

Alves, F. (2015). Translation process research at the interface. Paradigmatic, theoretical, and methodological issues in dialogue with cognitive science, expertise studies, and psycholinguistics. In A. Ferreira, & J. W. Schwieter (Eds.), *Psycholinguistic and cognitive inquiries into translation and interpreting* (pp. 17–40). Amsterdam: John Benjamins. doi: 10.1075/btl.115.02alv

Alves, F., &. Gonçalves, J. L. V. R. (2007). Modelling translator's competence: Relevance and expertise under scrutiny. In Y. Gambier, M. Shlesinger, &. R. Stolze (Eds.), *Translation studies: Doubts and directions*. (pp. 41–55). Amsterdam: John Benjamins. doi: 10.1075/btl.72.07alv

Alves, F., & Hurtado Albir, A. (2010). Cognitive approaches to translation. In Y. Gambier, & L. van Doorslaer (Eds.), *The John Benjamins handbook of translation studies* (pp. 28–35). Amsterdam: John Benjamins. doi: 10.1075/hts.1.cog1

Alves, F., Pagano, A., & da Silva, I. A. L. (2011). Modeling (un)packing of meaning in translation: Insights from effortful text production. In B. Sharb, M. Zock, M. Carl, & A. L. Jakobsen (Eds.), *Proceedings of the 8th international NLPCS workshop* (pp. 153–162). Copenhagen: Samfundslitteratur.

Alvstad, C., Hild, A., & Tiselius, E. (Eds.). (2011). *Methods and strategies of process research: Integrative approaches in translation studies.* Amsterdam: John Benjamins. doi: 10.1075/btl.94

Atkinson, R. C., & Shiffrin, R. M. (1968). Human memory: A proposed system and its control processes. In K. W. Spence, & J. T. Spence (Eds.), *The psychology of learning and motivation*, vol. 8 (pp. 115–118). London: Academic Press. doi: 10.1016/S0079-7421(08)60422–3

Baddeley, A. D. (1986). *Working memory*. Oxford: Clarendon Press.

Baddeley, A. D., & Hitch, G. J. (1994). Developments in the concept of working memory. *Neuropsychology, 8*(4), 485–493. doi: 10.1037/0894-4105.8.4.485

Bell, R. T. (1991). *Translation and translating*. London/New York: Longman.

Bereiter, C., & Scardamalia, M. (1993). *Surpassing ourselves: An inquiry into the nature and implications of expertise*. Michigan: Open Court Publishing Company.

Braga, C. N. O. (2012). O texto traduzido sob a perspectiva do avaliador: Um estudo exploratório [Unpublished doctoral dissertation]. Universidade Federal de Minas Gerais, Belo Horizonte, Brazil.

Carl, M., & Dragsted, B. (2012). Inside the monitor model: Processes of default and challenged translation production. *Translation: Computation, Corpora, Cognition, 2*(1), 127–143.

Chase, W. G., & Simon, H. A. (1973). The mind's eye in chess. In W. G. Chase (Ed.), *Visual information processing* (pp. 215–281). New York: Academic Press. doi: 10.1016/B978-0-12-170150-5.50011-1

Chi, M. T. H. (2006a). Two approaches to the study of experts' characteristics. In K. A. Ericsson, N. Charness, P. J. Feltovich, & R. R. Hoffman (Eds.), *The Cambridge handbook of expertise and expert performance* (pp. 21–30). Cambridge: Cambridge University Press. doi: 10.1017/CBO9780511816796.002

Chi, M. T. H. (2006b). Laboratory methods for assessing experts' and novices' knowledge. In K. A. Ericsson, N. Charness, P. J. Feltovich, & R. R. Hoffman (Eds.), *The Cambridge handbook of expertise and expert performance* (pp. 167–184). Cambridge: Cambridge University Press. doi: 10.1017/CBO9780511816796.010

Chi, M T. H., Feltovich, P. J., & Glaser, R. (1981). Categorization and representation of physics problems by experts and novices. *Cognitive Science, 5*, 121–152. doi: 10.1207/s15516709cog0502_2

Chomsky, N. (1965). *Aspects of the theory of syntax*. Cambridge: MIT Press. doi: 10.21236/AD0616323

Collins, H., & Evans, R. (2007). *Rethinking expertise*. Chicago/London: University of Chicago Press. doi: 10.7208/chicago/9780226113623.001.0001

Cronin, M. (2006). *Translation and identity*. Oxford/New York: Routledge. doi: 10.4324/9780203015698

da Silva, I. A. L. (2007). Conhecimento experto em tradução: Aferição da durabilidade de tarefas tradutórias realizadas por sujeitos não-tradutores em condições empírico experimentais [Unpublished master's thesis]. Universidade Federal de Minas Gerais, Belo Horizonte, Brazil.

da Silva, I. A. L. (2012). (Des)compactação de significados e esforço cognitivo no processo tradutório: Um estudo da metáfora gramatical na construção do texto traduzido [Unpublished doctoral thesis]. Universidade Federal de Minas Gerais, Belo Horizonte, Brazil.

da Silva, I. A. L. (2015). On a more robust approach to triangulating retrospective protocols. In A. Ferreira, & J. W. Schwieter (Eds.), *Psycholinguistic and cognitive inquiries into translation and interpreting* (pp. 175–201). Amsterdam: John Benjamins. doi: 10.1075/btl.115.08sil

da Silva, I. A. L. (2019). An interactional expertise-based approach to specialized inverse translation. *Tradução em Revista, 26*(1), 86–98.

da Silva, I. A. L., & Silveira, F. A. (2017). A expertise por interação como condicionante da competência do tradutor de textos técnicos e científicos. *Domínios de Lingu@gem, 11*(5), 1746–1763. doi: 10.14393/DL32-v11n5a2017-19

De Groot, A. D. (1946/1978). *Thought and choice in chess.* The Hague: Mouton Publishers.

Dillinger, M. L. (1989). Component processes of simultaneous interpreting [Unpublished doctoral dissertation]. McGill University, Montreal, Canada.

Dragsted, B. (2005). Segmentation in translation differences across levels of expertise and difficulty. *Target, 17*(1), 49–70. doi: 10.1075/target.17.1.04dra

Englund Dimitrova, B. (2005). *Expertise and explication in the translation process.* Amsterdam: John Benjamins. doi: 10.1075/btl.64

Ericsson, K. A. (2000). Expertise in interpreting: An expert-performance perspective. *Interpreting, 5*(2), 187–220. doi: 10.1075/intp.5.2.08eri

Ericsson, K. (2006). An introduction to *The Cambridge handbook of expertise and expert performance*: Its development, organization and content. In K. A. Ericsson, N. Charness, P. J. Feltovich, & R. R. Hoffman (Eds.), *The Cambridge handbook of expertise and expert performance* (pp. 3–19). Cambridge: Cambridge University Press. doi: 10.1017/CBO9780511816796

Ericsson, K. A. (2018). Superior working memory in experts. In K. A. Ericsson, R. R. Hoffman, A. Kozbelt, & A. M. Williams (Eds.), *The Cambridge handbook of expertise and expert performance* (2nd ed., pp. 696–713). Cambridge: Cambridge University Press. doi: 10.1017/9781316480748.036

Ericsson, K. A., & Charness, N. (1997). Cognitive and developmental factors in expert performance. In P. J. Feltovich, K M. Ford, & R. R. Hoffman (Eds.), *Expertise in context: Human and machine* (pp. 3–41). Cambridge: MIT Press.

Ericsson, K. A., Charness, N., Feltovich, P. J., & Hoffman, R. R. (Eds.) (2006). *The Cambridge handbook of expertise and expert performance*. Cambridge: Cambridge University Press. doi: 10.1017/CBO9780511816796

Ericsson, K. A., & Crutcher, R. J. (1990). The nature of exceptional performance. In P. B. Baltes, D. L. Featherman, & Richard M. Lerner (Eds.), *Lifespan development and behavior* (pp. 188–218). Hillsdale: Lawrence Erlbaum Associates.

Ericsson, K. A., Hoffman, R. R., Kozbelt, A., & Williams, A. M. (Eds.). (2018). *The Cambridge handbook of expertise and expert performance* (2nd ed.). Cambridge: Cambridge University Press. doi: 10.1017/9781316480748

Ericsson, K. A., & Lehmann, A. (1996). Expert and exceptional performance: Evidence on maximal adaptations on task constraints. *Annual Review of Psychology, 47*, 273–305. doi: 10.1146/annurev.psych.47.1.273

Ericsson, K. A., & Smith, J. (1991). *Toward a general theory of expertise: Prospects and limits*. Cambridge: Cambridge University Press.

Esqueda, M. D. (2020). *Ensino de tradução: Proposições didáticas à luz das competências em tradução*. Uberlândia: EDUFU.

Galton, F. (1869/1979). *Hereditary genius: An inquiry into its laws and consequences*. London: Julian Friedman Publishers. doi: 10.1037/13474-000

García, A. M. (2013). Brain activity during translation: A review of the neuroimaging evidence as a testing ground for clinically-based hypotheses. *Journal of Neurolinguistics, 26*, 370–383. doi: 10.1016/j.jneuroling.2012.12.002

Gerver, D. (1971). Aspects of simultaneous interpretation and human information processing [Unpublished doctoral dissertation]. Oxford University, Oxford, United Kingdom.

Glaser, R., & Chi, M. T. H. (1988). Overview. In M. T. H. Chi, R. Glaser, & M. J. Farr (Eds.), *The nature of expertise* (pp. xv–xxviii). Hillsdale: Lawrence Erlbaum Associates.

Goldman-Eisler, F. (1972). Segmentation of input in simultaneous translation. *Journal of Psycholinguistic Research, 1*(2), 127–140. doi: 10.1007/BF01068102

Halverson, S. L. (2010). Cognitive translation studies: Developments in theory and methods. In G. M. Shreve, & E. Angelone (Eds.), *Translation and cognition: Recent developments* (pp. 349–369). Amsterdam: John Benjamins. doi: 10.1075/ata.xv.18hal

Hatim, B, & Mason, I. (1997). *The translator as communicator*. London/New York: Routledge.

Holmes, J. S. (1972/2000). The name and nature of translation studies. In L. Venuti (Ed.), *The translation studies reader* (pp. 172–185). London: Routledge.

Holz-Mänttäri, J. (1984). *Translatorisches Handeln: Theorie und Methode*. Helsinki: Suomalainen Tiedeakatemia.

Horn, J., & Masunaga, H. (2006). A merging theory of expertise and intelligence. In K. A. Ericsson, N. Charness, P. J. Feltovich, & R. R. Hoffman (Eds.), *The Cambridge handbook of expertise and expert performance* (pp. 587–611). Cambridge: Cambridge University Press. doi: 10.1017/CBO9780511816796.034

Hurtado Albir, A., & Alves, F. (2009). Translation as a cognitive activity. In J. Munday (Ed.), *The Routledge companion to translation studies* (pp. 210–234). London: Routledge.

Hvelplund, K. T. (2011). Allocation of cognitive resources in translation: An eye-tracking and key-logging study [Unpublished doctoral dissertation]. Copenhagen Business School, Copenhagen, Denmark.

Isham, W. P. (1994). Memory for sentence form after simultaneous interpretation: Evidence both for and against deverbalization. In S. Lambert, &. B. Moser-Mercer (Eds.), *Bridging the gap. Empirical research in simultaneous interpretation* (pp. 191–211). Amsterdam: John Benjamins. doi: 10.1075/btl.3.15ish

Ivanova, A. (1999). Discourse processing during simultaneous interpreting: An expertise approach [Unpublished doctoral dissertation]. University of Cambridge, United Kingdom.

Jääskeläinen, R. (2011). Studying the translation process. In K. Malmkjaer, & K. Windle (Eds.), *The Oxford handbook of translation studies* (pp. 123–135). Oxford: Oxford University Press.

Jakobsen, A. L. (2002). Orientation, segmentation, and revision in translation. In G. Hansen (Ed.), *Empirical translation studies: Process and product* (pp. 191–204). Copenhagen: Samfundslitteratur.

Jakobsen, A. L. (2003). Effects of think aloud on translation speed, revision and segmentation. In F. Alves (Ed.), *Triangulating translation: Perspectives in process-oriented research* (pp. 69–95). Amsterdam: John Benjamins. doi: 10.1075/btl.45.08jak

Jakobsen, A. L. (2005). Instances of peak performance in translation. *Lebende Sprachen, 50*(3), 111–116. doi: 10.1515/LES.2005.111

Jakobsen, A. L., & Jensen, K. T. H. (2008). Eye movement behaviour across four different types of reading task. In S. Göpferich, A. L. Jakobsen, & I. Mees (Eds.), *Looking at eyes: Eye-tracking studies of reading and translation processing* (pp. 103–124). Copenhagen: Samfundslitteratur.

Kelly, D. (2002). Un modelo de competencia traductora: Bases para el diseño curricular. *Puentes, 1*, 9–20.

Krings, H. P. (2001). *Repairing texts: Empirical investigations of machine translation post- editing process.* Ohio: The Kent State University Press.

Lacruz, I., & Jääskeläinen, R. (Eds.). (2018). *Innovation and expansion in translation process research* (pp. 37–54). Amsterdam: John Benjamins. doi: 10.1075/ata.xviii

Liu, M. (2001). Expertise in simultaneous interpreting [Unpublished doctoral dissertation]. University of Texas at Austin, Texas, United States.

Liu, M. (2008). How do experts interpret? Implications from research in interpreting studies and cognitive science. In G. Hansen, A. Chesterman, &. H. Gerzymisch-Arbogast (Eds.), *Efforts and models in interpreting and translation research. A tribute to Daniel Gile* (pp. 159–178). Amsterdam: John Benjamins. doi: 10.1075/btl.80.14liu

Malmkjaer, K. (2009). What is translation competence? *Revue Française de Linguistique Appliquée, 14*, 121–134.

Marín García, A. (2017). Theoretical hedging: The scope of knowledge in translation process research (Unpublished doctoral dissertation). Kent State University, Kent, USA.

Marín García, A. (2019). The opportunities of epistemic pluralism for Cognitive Translation Studies. *Translation, Cognition & Behavior, 2*(2), 147–168. doi: doi.org/10.1075/tcb.00021.mar

Massaro, M. W., & Shlesinger, M. (1997). Information processing and a computational approach to the study of simultaneous interpretation. *Interpreting, 2*(1/2), 13–53. doi: 10.1075/intp.2.1-2.02mas

Miller, G. A. (1967). *The psychology of communication: Seven essays.* New York/London: Basic Books Inc.

Moser-Mercer, B. (2000). The rocky road to expertise in interpreting: Eliciting knowledge from learners. In M. Kadric, K. Kaindl, & F. Pöchhacker (Eds.), *Translationswissenschaft* (339–352). Tübingen: Stauffenburg Verlag.

Moser-Mercer, B., Fraunfelder, U., Casado, B., & Künzli, A. (2000). Searching to define expertise in interpreting. In B. Englund Dimitrova, & K. Hyltenstam (Eds.), *Language processing and simultaneous interpreting: Interdisciplinary perspectives* (pp. 1–21). Amsterdam: John Benjamins. doi: 10.1075/btl.40.09mos

Muñoz Martín, R. (2014). Situating expertise in translation. In J. W. Schwieter, & A. Ferreira (Eds.), *The development of translation competence: Theories and methodologies from psycholinguistics and cognitive science* (pp. 2–56). Newcastle upon Tyne: Cambridge Scholars Publishing.

Neubert, A. (2000). Competence in language, in languages, and in translation. In C. Schäffner, & B. Adab (Eds.), *Developing translation competence* (pp. 3–18). Amsterdam: John Benjamins.

Nye, B. D., Boyce, M. W., & Sottilare, R. A. (2016). Defining the ill-defined: From abstract principles to applied pedagogy. In R. A. Sottilare, A. Graesser, X. Hu, A. Olney, B. B. Nye, & A. M. Sinatra (Eds.), *Design recommendations for intelligent tutoring systems: Domain modelling*, vol. 4 (pp. 19–38). Orlando: US Army Research Laboratory.

O'Brien, S. (2006). Eye-tracking and translation memory matches. *Perspectives: Studies in Translatology, 14*(3), 185–205. doi: 10.1080/09076760708669037

O'Brien, S. (2013). The borrowers: Researching the cognitive aspects of translation. *Target, 25*(1), 5–17. doi: 10.1075/target.25.1.02obr

PACTE. (2003). Building a translation competence model. In F. Alves (Ed.), *Triangulating translation: Perspectives in process oriented research* (pp. 43–66). Amsterdam: John Benjamins. doi: 10.1075/btl.45.06pac

PACTE. (2005). Investigating translation competence: Conceptual and methodological issues. *Meta, 50*(2), 609–619. doi: 10.7202/011004ar

PACTE. (2017). *Researching translation competence by PACTE group.* Amsterdam: John Benjamins.

Pagano, A., & da Silva, I. A. L. (2008). Domain knowledge in translation task execution: An analysis of academic researchers performing as translators. In International Federation of Translators (Ed.), *Proceedings of the XVIII FIT world congress* (pp. 1–12). Beijing: Foreign Language Press.
Pym, A. (1996). Ideologies of the expert in discourses on translator training. In M. Snell-Hornby, & Y. Gambier (Eds.), *Problems and trends in the teaching of interpreting and translation* (pp. 139–149). Misano: Instituto San Pellegrino.
Pym, A. (2003). Redefining translation competence in an electronic age. In defense of a minimalist approach. *Meta, 48*(4), pp. 481–497. doi: 10.7202/008533ar
Rodrigues, C. H. (2018a). Competência em tradução e línguas de sinais: A modalidade gestual-visual e suas implicações para uma possível competência intermodal. *Trabalhos em Linguística Aplicada, 57*(1), 287–318. https://doi.org/10.1590/010318138651578353081
Rodrigues, C. H. (2018b). Interpretação simultânea multimodal: Sobreposição, performance corporal-visual e direcionalidade inversa. *Revista da ANPOLL, 44*(1), 111–119. doi: 10.18309/anp.v1i44.1146
Roe, A. (1952). *The making of a scientist*. New York: Dodd, Mead & Company.
Rothe-Neves, R. (2007). Notes on the concept of "translator's competence". *Quaderns, 14*, 125–138.
Scardamalia, M., & Bereiter, C. (1991). Literate expertise. In K. A. Ericsson, & J. Smith (Eds.), *Toward a general theory of expertise* (pp. 171–194). Cambridge: Cambridge University Press.
Schäffner, C., & Adab, B. (2000). Developing translation competence: Introduction. In C. Schäffner, & B. Adab (Eds.), *Developing translation competence* (pp. vii–xv). Amsterdam: John Benjamins. doi: 10.1075/btl.38.01sch
Schmaltz, M. (2015). Resolução de problemas na tradução de metáforas linguísticas do chinês para o português: Um estudo empírico-experimenta [Unpublished doctoral dissertation]. University of Macau, Macau, China.
Shlesinger, M. (2000). Interpreting as a cognitive process: How can we know what really happens? In S. Tirkkonen-Condit, & R Jääskeläinen (Eds.), *Tapping and mapping the processes of translation and interpreting* (pp. 3–15). Amsterdam: John Benjamins. doi: 10.1075/btl.37.03shl
Shreve, G. M. (2002). Knowing translation: Cognitive and experiential aspects of expertise in translation from the perspective of expertise studies. In A. Riccardi (Ed.), *Translation studies: Perspectives on an emerging discipline* (pp. 150–171). Cambridge: Cambridge University Press.
Shreve, G. M. (2006). The deliberate practice: Translation and expertise. *Journal of Translation Studies, 9*(1), 27–42.
Shreve, G. M, & Angelone, E. (Eds.). (2010). *Translation and cognition: Recent developments*. Amsterdam: John Benjamins. doi: 10.1075/ata.xv
Shreve, G. M., Angelone, E., & Lacruz, I. (2018). Are expertise and translation competence the same? Psychological reality and the theoretical status of competence. In Lacruz, I., & Jääskeläinen, R. (Eds.), *Innovation and expansion in translation process research* (pp. 37–54). Amsterdam: John Benjamins. doi: 10.1075/ata.18.03shr
Simon, H. A., & Chase, W. G. (1973). Skill in chess. *American Scientist, 61*(4), 394–403.
Sirén, S., & Hakkarainen, K. (2002). Expertise in translation. *Across Languages and Cultures, 3*(1), 71–82. doi: 10.1556/Acr.3.2002.1.5
Swanson, R. A., & Holton III, E. F. (2001). *Foundations of human resource development*. San Francisco: Berrett-Koehler.
Szpak, K. (2017). A atribuição de estados mentais em atividades de tradução: Um estudo conduzido com rastreamento ocular e ressonância magnética funcional [Unpublished doctoral dissertation]. Universidade Federal de Minas Gerais, Belo Horizonte, Brazil.
Tiselius, E. (2013a). Same, same but different? Competence and expertise in translation and interpreting studies [Trial Lecture for the Doctoral Degree]. University of Bergen, Bergen, Norway.
Tiselius, E. (2013b). Experience and expertise in conference interpreting [Unpublished doctoral dissertation]. University of Bergen, Bergen, Norway.
Van der Maas, H. L. J., & Wagenmakers, E.-J. (2005). A psychometric analysis of chess expertise. *American Journal of Psychology, 118*(1), 29–60.
Vik-Tuovinen, G.-V. (2006). Tolkning på olika nivåer av professionalitet [Unpublished doctoral thesis]. University of Vaasa, Vaasa, Finland.
Williams, M., & Davids, K. (1995). Declarative knowledge in sport: A by-product of experience or a characteristic of expertise? *Journal of Sport & Exercise Psychology, 17*(3), 259–275. doi: 10.1123/jsep.17.3.259
Zuckerman, H. (1977). *Scientific elite: Nobel laureates in the United States*. New York: Free Press.

27

Translation and situated, embodied, distributed, embedded and extended cognition

Hanna Risku and Regina Rogl

27.1 Introduction

In the last few decades, cognitive science has experienced some major conceptual and methodological changes in the way intelligence, especially the function of the human mind (i.e. for instance perception, attention, memory, learning, problem solving, affect and creativity), is studied and understood. This chapter offers a short overview of recent, situated approaches to cognitive science—namely situated, embodied, distributed, embedded and extended cognition—and seeks both to sketch their historical roots and motivations and to demonstrate their consequences for the study of translation and interpreting, in particular the translation and interpreting process. As a single chapter in a thematic handbook, it can only provide a general orientation for further reading, but it nonetheless aims to show where these new concepts and approaches differ or overlap. Overall, these approaches have led to the emergence of new ideas in Cognitive Translation Studies, which have in turn raised new questions and introduced new terms and methods. It should, however, be noted that the examples of their application within Translation Studies provided in this chapter often cannot be assigned exclusively and unambiguously to one particular situated approach. Indeed, they frequently combine multiple situated frameworks (e.g. situated, embodied cognition) or draw inspiration from the basic assumptions of a specific approach, yet do not always follow it strictly to its conclusion. Accordingly, much of what is presented in this chapter is essentially still a vision for the future. Indeed, given the many intersections between the recent cognitive approaches, different situated frameworks might be useful to tackle a specific research question, depending on the actual focus of the study. We will, however, attempt to demonstrate the underlying theory behind them, their conceptual foci and the consequences of their application to specific research questions, thus embedding them in the Cognitive Translation Studies context.

To illustrate the initial motivations behind the development of the current approaches in cognitive science, the chapter will begin with a brief overview of their predecessors and their respective influences on early translation process research. We will then present the basic tenets and concepts of modern-day cognitive science, relate them to the status quo in Cognitive Translation and Interpreting Studies, and provide examples of their application. Finally, we will point to their potential for the development of cognitive translation and interpreting studies.

27.2 The roots: Early cognitive science and translation process research

Even though the question of the relationship between the brain and the mind has long been discussed in philosophy, and especially in the philosophy of the mind, cognitive science's history in the theoretical and empirical study of the workings of the mind is still relatively short. It started in the middle of the 20th century, when different disciplines, such as informatics, psychology, biology and linguistics, began to model human intelligence from an interdisciplinary perspective. These initial models were mainly oriented along the lines of the newest technological achievement at that time: the metaphor of computers as automatic information processing systems using algorithms to carry out complex calculations and thereby produce an output to a given input according to programmed rules. These symbol- and rule-based systems became the basis of the first generation of cognitive scientific theories, the computational-representational understanding of the mind (CRUM; Thagard, 2005, or "good old-fashioned artificial intelligence"—GOFAI—to its critics; cf. Haugeland, 1985). Cognition was seen here as a process in which sensory input is transformed into internal symbols that act as representations of the environment, which are then manipulated in the brain.

The symbolic view of the mind is also the one that gained ground in early cognitive translation and interpreting research (see e.g. Danks et al., 1997; Massaro & Shlesinger, 1997). Symptomatic of this approach is the view of translation as rational problem solving and a process of trying to learn and apply rules (strategies, procedures) with which to reach a desired goal. Given the theoretical conceptualization of translation as information processing, early translation process research experiments typically involved solving tasks in laboratory or didactic settings. The predefined tasks and limited segments of translation work included in such experiments—e.g. short, mostly informative general language texts with no specific constraints and no consideration of contacts with clients, proofreaders or editors—can be described as micro-worlds (cf. Walter, 2014, p. 21), far removed from the simulation of authentic translation as a professional practice in the workplace. The information processing approach proved influential, especially on the methodological level, as many early translation process researchers relied mainly on think-aloud protocols (TAPs) to make translation processes visible and analysable, often combining these with the assumption that translation is essentially a verbalizable problem-solving process (Gerloff, 1988; Jääskeläinen, 1993; Königs, 1987; Krings, 1986; Tirkkonen-Condit, 1989).

In the 1980s, cognitive scientists increasingly began to question whether human cognition could really be captured as a calculation based solely on symbolic representations and started to look for alternative explanatory models on a sub-symbolic level. Connectionist models (also often referred to as "parallel distributed processing"; Rumelhart, McClelland, et al., 1986), the second generation of cognitive science theories, were born. They were inspired by the structure of the brain with its complex network of neurons, synapses and axons, which they drew upon as a metaphor to model cognition. Parallel activations spread as impulses are conducted from neurons to other neurons, modifying the strength of the specific neural pathways and resulting in experience-based, action-oriented patterns of activation. In the connectionist cognitive models, this process of modifying and developing neural activation patterns was modelled as learning.

The notion of non-symbolic, experience-based and, thus, subjective learning formed the basis for the development of several cognitive scientific approaches within the connectionist framework (e.g. schema theory, cf. Rumelhart, Smolensky, et al., 1986; prototype theory, cf. Rosch, 1973, 1975; scenes-and-frames semantics, cf. Fillmore, 1977; and mental models, cf. Johnson-Laird, 1983), all of which assumed that relatively stable, internal, tacit mental patterns enable humans to make sense of their environment and to act accordingly. In these second-generation models, cognition was still seen as a process in which sensory input is transformed into internal activation patterns in units that act as representations of the environment. However, in the connectionist view, these units are not linguistically transparent symbols for rule-based manipulation. Instead, they are sub-symbolic activation patterns whose structures depend on the experience of the individual network/brain itself. According to connectionists, it is through such experience that internal representations gain sense and significance, whereas the abstract symbolic representations of first-generation cognitive science are well structured but meaningless.[1] In addition, connectionist network models are able to accommodate the continuity and time structure of cognitive processes, i.e. their dependency on previous activations and present states of a cognitive system ("recurrent neural networks"; see Elman, 1990). Human decision making was thus modelled as contextualized and path dependent instead of isolated and historically decontextualized.

In the 1980s and 1990s, the philosophical underpinnings of connectionism were adopted as central elements in some of the translation process models of the time. Notions such as schemas, prototypes, scenes and frames, scripts, and bottom-up and top-down processes were increasingly applied to the translation process. Vannerem and Snell-Hornby (1986) were among the first to introduce Fillmore's scenes-and-frames semantics to Translation Studies in the hope that it would provide a better understanding of texts as products of culture (see also Snell-Hornby, 1988; Vermeer & Witte, 1990). Another early contribution was provided by Hönig (1995), who, inspired by Tannen's (1979) model of "structures of expectation" (i.e. schemata, scripts and frames), developed a model of translators' mental processes that depicted them as occurring in both a controlled workspace (acquired knowledge about rules, principles and facts) and an uncontrolled workspace (intuition and creativity). With his model, Hönig tried to provide translators with an understanding of their own internal processes that would allow them to develop a macro-strategic approach to translating (see also Hönig, 1997). These frameworks were used extensively in Translation Studies until at least the 1990s to reframe and shed light on numerous aspects of the translation process which it had not previously been possible to explain adequately (see e.g. Kußmaul, 1991, 2000, on creativity in the translation process; Risku, 1998, for a situated account of connectionist schemas as the basis of translation competence; Padilla, Bajo, and Padilla's (1999) schema theoretical proposal for a theory of translation and interpreting; and Chesterman's 2000 reference to the Dreyfus and Dreyfus (1986) model of the stages of expertise). Connectionist approaches to cognition remain influential in current-day translation process research, albeit with due consideration of recent scientific findings. A good example is the competence model formulated by Alves and Gonçalves (2007), which combines connectionist notions with relevance theory. Connectionist-inspired approaches to the translation process bring contextuality, creativity and individual experience to the fore. From the connectionist perspective, translation does not consist of reproduction but of the creation of a possibility to construct meaning (Risku, 2016, p. 84).

Before we move on to our overview of recent cognitive approaches to translation, we would first like to make a short comment on theoretical compatibility. To date, there is no agreement in Translation Studies on whether classical cognitive science theories and more recent frameworks could be used in a complementary way. Following Bermúdez (2014), O'Brien (2017) suggests that the difference between computational, connectionist and situated approaches is more one of level of abstraction than of irreconcilable theoretical positions, and that this would allow a combination of views from different frameworks (see also Shreve, 2018, on levels of explanation). Others, however, contend that computational, connectionist and situated frameworks could be considered as paradigmatic (in line with Kuhn, 1970) and thus require a radical reframing of our view of cognition in translation theory and its empirical implementation in translation research.

In a classical understanding of the term "paradigm", this would mean that findings obtained in the earlier computational (and connectionist) research traditions would need to be rejected entirely (see Marín García, 2017, for a critical analysis of these conflicting voices in translation process research). However, as Muñoz Martín (2010a, pp. 154ff.) shows, many classical views of cognition (implicitly) continue to persist in current translation process research, either because researchers lag behind new developments in cognitive science or because newer concepts only gradually seep into translation research and become increasingly accepted by a larger group. This disagreement as to the status of the different cognitive research traditions is not only to be found where translation scholars borrow from cognitive science. Similar debates also exist in cognitive science itself, where this issue remains to be resolved (see e.g. Dominey et al., 2016, pp. 336f.).

27.3 Recent reorientations in cognitive science: Situated, embodied, distributed, embedded and extended cognition

At least since the 1990s, the above-mentioned computational-representational and connectionist approaches to cognition have been challenged by theories that attribute a central role in human cognition to the body and to physical and social interaction rather than to the notion of mental representation. Within cognitive science, there has been a growing interest in the body as more than just a device to transport signals to the brain and to carry out the decisions that are made there. A similar level of interest has been shown in the brain as more than just a device to transform sensory input into internal, mental representations that can be manipulated to produce motor output. One of the core ideas of the current approaches is thus to extend cognition beyond the individual brain to situationally contingent interaction with and cognitive exploitation of the environment.

In the following sections, we will try to explain how the so-called situated approaches to cognition (also often referred to as 4EA cognition, i.e. "embodied, embedded, extended, enacted, affective cognition") have increasingly gained momentum in cognitive science. Whereas cognitive processes were depicted in the traditional view of cognition as mental operations on representations inside the brain, and perception and action were not seen as being cognitive in nature (but as transport and application processes), situated approaches maintain that intelligent action cannot be explained without according a role to the body and the relevant physical and social environment as integral constituents of the cognitive process.

The situated approaches do not constitute a unified theory of cognition but, rather, a cluster of different "intellectual threads" (Beer, 2014, p. 136) that originally stem from different disciplinary areas such as developmental psychology, human–computer interaction and anthropology. They all address—with different conceptual positions and emphases—the embodied, embedded nature of cognition, extend the notion of the cognitive system to extra-cranial entities like the body and the environment, and see cognition as enactive instead of representational.

In this article, we will outline five orientations that are committed to the situatedness of cognition: 1) situated cognition, 2) embodied cognition, 3) distributed cognition, 4) embedded cognition and 5) extended cognition. These are increasingly gaining a place in the Translation Studies discourse, albeit for the most part more as research desiderata or in the form of individual, frequently isolated presuppositions and less as comprehensively and consistently applied approaches. Accordingly, and in addition to providing current examples of their application in Translation Studies, we will also endeavour in each case to include research objects and questions to which we feel they can be applied with particularly beneficial effect and which would serve to extend previous research work.

27.3.1 Situated cognition

In the 1980s, Suchman (1987) was one of the scholars who prepared the ground for a situated reorientation of cognitive science. With her observations and analyses of human–machine interaction, she was able to profoundly reconceptualize cognitive theories of the time. According to Suchman, mental representations ("plans") do not determine and control intelligent action; they are only one of the resources involved in it. Internal representations develop as memorized patterns of activation in the course of human activities and serve as individual and (due to social interaction in similar environments and activities) cultural resources in the process of further activities. Other resources used in cognition are the physical and social elements that are perceived *in situ* and reacted to depending on the current point in the dynamic path of neural, sensory and motor activity, thus resulting in actions that deviate from internal representations as plans or goals that precede the activity. Trying to observe or report on our own mental representations, for example in TAPs, is thus a learned, cultural activity, based on the assumption—or rather, illusion—that these internal processes would be accessible to conscious attention and that they actually steer and control our actions. Suchman suggests, instead, that describing and explaining action as a result of rational problem solving and decision making in the brain enables us to indicate to others and ourselves what we did, what we are going to do and why. Instead of steering intelligent action, mental representations provide us with metacognitive capabilities like self-reflection and self-evaluation ("second-order cognitive dynamics"; see Clark, 1998, p. 208). These descriptions and activities are, however, no less situated and contextually contingent than any other human activities (Suchman, 2007, p. 13).

As an anthropologist, Suchman (1987, 2007) also considers the role of language and culture in cognition. She notes that the significance of an expression in any actual occasion depends on the circumstances of its use. Language is thus an indexical resource, just like the plans "in the head" mentioned earlier, whose significance is dependent on the world in which they are embedded (Suchman, 2007, pp. 77ff.). From an ethnomethodological perspective, her interest also lies in how the feeling of objective grounding and shared understanding of social situations is accomplished despite their context dependency. According to Suchman (2007, p. 77), language has an important, typifying role in the practices and methods of establishing common ground because we assume common conventional meanings.

The notion of situatedness is the element in this framework that is so far most clearly established in Translation Studies. First introduced to Translation Studies by the functional approaches to translation, and explicitly by Vienne (1994), it has become a central principle not only in different cognitive approaches but also in other translation research traditions, where researchers emphasize the situatedness and context dependency of translation work. Examples here include sociological translation research (e.g. on practice theory; see Olohan, 2017) and workplace research (e.g. on ergonomics; see Ehrensberger-Dow & Hunziker Heeb, 2016; and on the sociology of work; see Kuznik, 2016). As Marín García (2017, pp. 35f.) acknowledges, even researchers who base their work on earlier cognitive approaches would not in fact reject the influence of situative factors either. The difference lies in whether they are seen for methodological reasons as a negligible and abstractable context or as a central constituent of cognition and thus an essential focus of empirical enquiries.

We believe that situated action in Suchman's (2007) sense provides a solid general basis and inspiration for research agendas (including the definition of research objects and questions) and research designs (including the choice of data acquisition and analysis methods) that are committed to a non-computational understanding of the cognitive processes in translation. It is moderate in its position towards representations, since it does not deny their existence or relevance altogether but merely relativizes their role in the process and emphasizes their experience-based origins.[2] As Clark (1998, p. 149) summarizes, representations are seen here as "local and action-oriented rather than objective and action-independent". Suchman's (and Clark's) view on language and culture seems productive for any approach in contemporary Translation Studies that adopts a contextual or constructivist view on meaning and translation or deals with the creative and situated nature of translation as localization and transcreation.

This view differs from computationalist and connectionist approaches in that it places a stronger focus on the relevance of concrete action in a situative context, assumes a development perspective on competences and situations, and includes external factors like space, time and social interaction in the analysis of cognitive processes. The approach initially attracted the interest of Translation Studies scholars primarily on a theoretical level (see e.g. Krüger, 2015a, 2015b; Martín de León, 2008; Muñoz Martín, 2010a, 2010b; Risku, 2002, 2010; Risku et al., 2013), and the new concepts it contained did not initially find their way into empirical studies. However, its potential in the study of work processes, social interaction and the use of artefacts in translation and translation project management has been demonstrated in recent field studies (Risku, 2014, 2016). The concept of situated cognition is currently being discussed and/or applied in numerous Translation Studies publications—albeit frequently not as a fully-fledged framework, but nonetheless as a bundle of presuppositions, as an analogy or in further deliberations. The range of translation practice fields studied thereby (e.g. specialized translation, Krüger, 2015a, 2015b; literary translation, Kolb, 2017; translation project management, Risku, 2016; interpreting project management, Dong & Turner, 2016) and research questions discussed in this context (e.g. work processes, Kolb, 2017; work content, Kuznik & Verd, 2010; situated knowledge, Risku et al., 2010; Olohan, 2017; expertise, Angelone & Marín García, 2017; social interaction and cooperation, Risku & Pircher, 2006; use of tools, Risku, 2016; Teixeira & O'Brien, 2017; creativity, Risku, Milošević, et al., 2017; affect, Hokkanen, 2017; Hokkanen & Koskinen, 2016) underline the framework's application potential. Situated cognition can also provide a model for further studies of translation as human–machine/computer interaction. Following Suchman's orientation towards applied research on the usability and optimization of technological artefacts, it can also help in the design of projects relating to the development and assessment of translation technologies. The recent publication of a series of dedicated special issues on situated views of translation likewise shows that the application of situated views on translation is not necessarily restricted to translation process research, but that they can be used advantageously at the interface between cognitive, sociological, didactic and ergonomic issues in translation and interpreting. A special issue of the journal *Translation Spaces* edited by Ehrensberger-Dow and Englund Dimitrova (2016), for instance, looked at the cognitive space of translation and interpreting as situational interface; Risku, Rogl, et al. (2017) brought together a variety of field studies with socio-cognitive research foci in another volume of *Translation Spaces*; and González-Davies and Enríquez Raído (2016) dedicated a special issue of the journal *The Interpreter and Translator Trainer* to situated learning in translation and interpreting, in which they combine situated cognition and social constructivist (e.g. Kiraly, 2000) perspectives on learning and translation pedagogics.

27.3.2 Embodied cognition

Both parallel and subsequent to the reorientation of many cognitive scientists towards the situatedness of cognitive processes, the realization that the body might play an irreducible role in cognition began to gain ground. Vygotsky's (1978) theory of developmental psychology, which originated in the 1930s, modelled how children accomplish new tasks with adult help, e.g. being able to walk when physically supported by an adult. Conceptualized as "scaffolding" by Wood et al. (1976), this mechanism of physically acting in and being supported by the given environment was shown to exist not only for bodily actions but also for cognitive abilities and formed one of the cornerstones of the embodied cognition approach.

The notion behind embodied cognition maintains that we will not be able to understand cognition as long as we only look at the processes in the brain, since these derive their significance as parts of bodily activities. Indeed, according to Heidegger (1953, originally published in 1927), we grasp the world by acting in it: The bodily practice of using, manipulating and navigating in the material world is primary to and the basis for human cognition.[3] Our sense of relevance is procedural "know-how" rather than propositional "know-what" and is founded on our interaction with the world (see also Dreyfus, 1972). No matter how detailed our data on or knowledge of them might be, abstract representations or neural patterns will only become meaningful and explain cognitive processes if they are analysed as parts of bodily interactions in and with the surrounding circumstances and context of action.

Lakoff and Johnson (1980) showed how our early sensory and motor experience provides us with a basic first understanding of the world that is equally valid no matter how abstract the concept we are trying to understand might be. The bodily grounding of language enables us to understand both concrete and abstract language as well as its situated, contingent and metaphoric use: We can, for instance, understand and accept the use of the word "table" to refer to a tree stump if that is the function given to it (or the "affordance" that it has; Gibson, 1977) on that specific occasion. In accordance with phenomenology of perception (Merleau-Ponty, 1965, originally published in 1945), it is not the consciousness but the body through which we experience and gain knowledge about the world and that gives meaning to signs and perceptions.

Wilson (2002) points out that embodied cognition is tightly knit with the situated, embedded, distributed and extended approaches to cognition and that it houses a number of theoretical positions. These range from the rather non-revolutionary acknowledgement that cognition is closely tied to action, that abstract (offline) cognition is body based or that the body influences cognitive processes to the more substantial and consequential position that bodily interactions with the environment actually constitute cognition. At the very least, Menary (2010a, p. 460) allows us to conclude that "cognitive capacities have turned out to be highly embodied and embedded [...] and that, therefore, we need explanatory resources that go beyond the dynamics of neural networks".

Beer (2014, pp. 132f.) mentions three to some extent different perspectives in the embodiment context, namely the *conceptual*, *physical* and *biological* levels. Of these, *conceptual* embodiment is the one that has been most widely discussed and applied in Translation Studies and centres on the notion that "even when engaged in pure ratiocination, our most abstract concepts are still ultimately grounded in our bodily experiences and body-oriented metaphors" (Beer, 2014, p. 132). On a theoretical level, Martín de León (2017) provides an overview of different conceptualizations of embodied representations and mental simulations in cognitive science, makes specific mention of the concept of radical embodied cognition (Chemero, 2009) and describes how these approaches have been applied to Translation Studies. The notion of conceptual embodiment also implies a new notion of language and initially comes into play in terminology research. Faber (2011), for instance, addresses the problem that dynamicity cannot be adequately represented in terminology resources, since these usually rest on traditional cognitive science theories; i.e. they are "based on the abstract, amodal representation of entities, events, and processes" (Faber, 2011, p. 25). She outlines the implications of drawing on newer cognitive frameworks in terminology research and suggests ways of developing new terminological resources accordingly. Tercedor Sánchez (2011) looks at the extent to which terminological variation can be viewed and empirically studied as a situated, dynamic phenomenon. In her view, situated and embodied conceptualizations of cognition also serve as the basis for a dynamic representation, in her case of terminological variation. Other good examples of the application of the concept of embodiment to terminology include Tercedor Sánchez et al.'s (2011) study of blind children's drawings, which—so the authors found—were derived from embodiment or perceptual symbolic simulations that had their parallel in language, and Prieto Velasco and Tercedor Sánchez's (2014) research on the embodied nature of medical concepts such as pain.

Physical embodiment, the second perspective on embodiment described by Beer, focuses on

> the uniquely physical aspects of an agent's body [that] are crucial to its behavior, including its material properties, the capabilities for action provided by the layout and characteristics of its degrees of freedom and effectors, the unique perspective provided by the particular layout and characteristics of its sensors, and the modes of sensorimotor interaction that the sensors and effectors collectively support.
>
> *Beer, 2014, p. 132*

According to Beer, this perspective could be seen as a specific form of situatedness. While the idea of situatedness comprises any interaction with the environment, physical embodiment, as a special case of situatedness, focuses on the physical and mediated interactions of a body. In contrast, *biological* embodiment, the third strand of the embodiment concept described by Beer, emphasizes a body's "biological facts", "including the relevant neuroscience, physiology, development, and evolution" (Beer, 2014, p. 132).

These frameworks could provide fertile theoretical ground for a new conceptualization of bodily factors in translation research or provide complementary insights into research that has already been undertaken, especially in the field of physical ergonomics. While translation workplace research has hitherto approached the translation process primarily from an interaction, efficiency or production perspective, the ergonomics approach is now increasingly looking at the physiological aspects of translation workplaces and thus in turn at the question of the health and wellbeing of translators at their workplaces (Ehrensberger-Dow et al., 2016; Meidert et al., 2016; Pineau, 2011).

Recognition and awareness of the physical aspects of work are already more firmly anchored in interpreting research. This could be attributed to the interaction- and dialogue-oriented approaches that gained relevance in such research (see e.g. Baraldi & Gavioli, 2012; Mason, 2001; Wadensjö, 1998). From these starting points, it was not a big leap to questions pertaining to physical position, bodily presence or bodily movements such as gaze, gestures, head movements, changes in posture, etc. (see e.g. Mason, 2012, on the role of gaze and positioning in interpreted interaction, or Kinnunen, 2017, on the role of spatial and embodied aspects of interaction in court interpreting). Interesting new questions are currently also being raised in the field of sign language interpreting. Richardson (2017), for instance, combines a discussion of both Merleau-Ponty's (1965) phenomenological approach to perception (see earlier) and mirror neurons to illustrate his critique of existing research into sign language theatre interpreting, which he claims shows a lack of consideration for factors such as "bodily perception and the pre-conscious empathic response that comes from experiencing onstage physicality" (Richardson, 2017, p. 56). In their study of interpreted encounters with deafblind persons, Raanes and Berge (2017) provide a micro-level analysis of interpreted interactions that sheds light on embodied actions and the interpreters' use of haptic signs.

The above-mentioned bodily aspects might at first seem interesting largely from an interactionist perspective or could be seen, in line with Beer's definition, as a special case of situatedness.

However, the addition of an embodied cognition perspective could enrich and nuance this discussion on a cognitive level, because it can account for the bi-directional relation between the external world and the physiology and morphology of the human body that mediates our experience of the environment.

In any case, the methodological repertoire for exploring such research in Translation Studies is still in its infancy. Ethnographic translation and interpreting research would seem to be a good starting point. However, studying embodied cognition also requires methods that allow the collection of data with a far higher level of granularity than can be obtained through classical participant observation. It requires micro-analytical approaches and new instruments that enable more detailed data collection in situated translation and interpreting contexts. At the same time, it is important to look at the whole body of a translator/interpreter in its embeddedness in and interaction with both a physical environment and the associated collaborators, a view which exceeds the possibilities of many traditional process research methods (such as screen recording, keystroke logging, eye tracking or other physiological measures). Micro-ethnography using video data of translation and interpreting activities could prove to be a promising future methodological avenue (Hirvonen & Tiittula, 2018; Risku et al., forthcoming). Scholars could also draw on the expertise in the empirical micro-analysis of bodily presence in social situations and multimodal aspects of communication that has been gained in other areas of Translation Studies research, e.g. in interpreting research (see e.g. Davitti, 2012; Pasquandrea, 2011; or Biagini et al.'s 2017 thematic section of the *Journal of Pragmatics* on "Participation in interpreter-mediated interaction: Shifting along a multidimensional continuum") or in research on audiovisual translation (Hirvonen & Tiittula, 2010; Lautenbacher et al., 2015; Taylor, 2003, 2004, 2016).

27.3.3 Distributed cognition

While the embodied cognition approach expanded the cognitive scientific research object from the brain to the individual as a whole (acting in a specific environment), some of the parallel developments in cognitive science went a step further. Hutchins (1995a, 1995b, 2001), for instance, states that we cannot understand cognitive systems by only studying individuals: No matter how detailed our data and our knowledge of the intra- and extra-cranial activities of an individual agent, we would then only be looking at a part of a cognitive system that, as such, does not suffice as an explanatory resource. According to Hutchins (1995b), cognitive systems are comprised of individuals and artefacts (instruments, technologies and tools of any kind), e.g. an airplane cockpit with the pilots, the flight instruments and the controls on the instrument panel. Essential in his approach is the notion that these elements contain internal and external representations and that there are constant transformations from external to internal representation (internalization, e.g. in perception and learning) and from internal to external representation (externalization, e.g. in communication and in writing a reminder to oneself). We operate on, manipulate, design and interact with external structures and social agents to remember, navigate and accomplish other cognitive tasks.

In addition to his prototypical example of the cockpit, Hutchins also studied the navigation of large naval vessels. He noted that when we study cognition in real-world contexts ("in the wild"), i.e. outside the laboratory, it becomes apparent that cognition does not take place in the head or in an agent, but in the world. It can be distributed across members of a social group, involve internal and external structures, and stretch its effects over time so that changes and outputs of one event transform later events. Similar to Norman's (1993) "things that make us smart", humans manipulate and develop their thinking by manipulating and changing their environment.

The distributed cognition approach thus expanded the boundaries of the unit of analysis of cognition to include events beyond the skin and skull of an individual. Through this shift, Hutchins not only departs from the notion that cognition happens in the logical engine of the brain; he also brings social and cultural processes—such as the development of artefacts, language and social groups as elements of human intelligence—into the focus of cognitive science, with a special emphasis on how cognitive processes can be distributed between these instances (see Hutchins, 2001).

In contrast to situated cognition, where context and social situation are studied first and foremost as factors that influence cognitive processes, the focus in distributed cognition lies more on the distributedness of cognition and action. Potential fields of application in Translation Studies therefore include all forms of complex systems, structures or frameworks in which translators act. This could in essence include any form or constellation of translation activity or translation work reality. Indeed, studies indicate that a whole range of people are involved even in apparently small translation projects (Kolb, 2017; Risku, 2016; Risku et al., 2016). Socially distributed cognition therefore lends itself well to the analysis of the distributed nature of translation activities and the related sub-processes as performed in different kinds of translator networks and communities, be they production or knowledge networks, geographically restricted or distributed groups, or online or offline constellations. With its close ties to the debate on "the collective mind" or "collective/swarm intelligence", distributed cognition, in this context, could thus serve as a valuable complement to existing research on translators' professional networks and online platforms (McDonough, 2007; Mihalache, 2008; Risku & Dickinson, 2009); on different kinds of collaborative (online) networks (McDonough Dolmaya, 2017; O'Hagan, 2011); and on fan or crowdsourced translation (Jiménez-Crespo, 2017; McDonough Dolmaya, 2012; O'Hagan, 2011; Orrego-Carmona & Lee, 2017). Applied to such research objects, a distributed cognition perspective allows a micro-level analysis of social interaction and cooperation, and helps to reframe problem-solving—a classic research question in cognitive translation research—not from an individual and inherently mental perspective but as a collaborative endeavour (Jiménez-Crespo, 2017, pp. 106f.).

The distributed cognition approach is also useful in the identification and analysis of socio-technical constellations. It can be applied to all forms of workplace, network or interaction research that focus on the social, material and physical interplay in a translation or interpreting constellation. This perspective has already been used to some extent in studies of distributed processes in translation settings in order to study the cognitive aspects of the use of tools at heavily computerized workplaces (cf. research on human–computer interaction and cognitive ergonomics, e.g. O'Brien, 2012; Teixeira & O'Brien, 2017). Micro-analyses of the use of artefacts (e.g. hardware equipment, desktop setups, specific software configurations, translation memories and post-editing tools) allow us to paint a clearer picture of the distributedness of translation activity across a range of social, physical and virtual entities. Christensen, in particular, views situated, embodied and distributed cognition as promising frameworks to provide more descriptive research on cognitive aspects of translator–computer interaction:

> In particular research on the mental consequences of the latest developments within TM technology seems desirable, for instance the integration of MT into the TM-assisted translation process and the adoption of side-by-side environments. Also, studies which combine the investigation of what happens in the individual translator's mind (internal processes) with what happens in the translator's environment (external processes) are needed in order to learn more about situated and distributed translation and the increasingly relevant role played by tools and teams.
>
> *Christensen, 2011, p. 155*

In this respect, and also as a hypothesis on the hybridity of cognitive systems, the notion of distributed cognition is compatible with other socio-cognitive approaches. It shows parallels, for example, with Latour's actor-network theory (as stated by Latour himself, 2005, p. 11[4]), which has already assumed a fixed place in the theoretical and methodological repertoire of sociological approaches to translation (cf. Abdallah, 2014; Buzelin, 2005, 2007).

The distributed cognition approach is closely linked with the methodological commitment to ethnographic observation—going "into the wild" and following people and activities in the field. Hutchins (1995a, p. 371) calls this methodological approach "cognitive ethnography" in reference to the method of observing cognitive processes in real-life settings, focusing on the social and material distribution of cognition and acknowledging the interrelations of cognition, culture and history. The implications of Hutchins' methodological considerations for application in Translation Studies were discussed, for instance, by Martín de León (2008) and followed up in an empirical setting by Risku and Windhager (2013). According to Jiménez-Crespo, the same methodological argumentation is also valid for the study of crowdsourced translation:

> [The] agenda to study translation "in the wild" resonates in crowdsourcing approaches. This is not only due to the dichotomy of controlled experimental vs. real professional practices. It also resonates in the relative [sic] unpredictable, free and uncontrolled way in which translation processes are carried out, or planned to be carried out, in volunteer initiatives.
>
> *Jiménez-Crespo, 2017, p. 107*

Studying the distributedness of activities in larger and geographically distributed networks confronts researchers with new methodological challenges that also cannot be adequately overcome using traditional ethnographic assumptions and instruments. They can benefit here from drawing increasingly on methodological approaches like multi-sited ethnography (Marcus, 1995), micro-ethnography (Garcez, 1997; LeBaron, 2008) or virtual ethnography (Hine, 2000, 2015) with its focus on the embodied, embedded and everyday aspects of the Internet, or borrowing from social sciences network research (e.g. Hollstein, 2006, on qualitative network analysis) and developing these approaches for application in Translation Studies.

27.3.4 Embedded cognition

As already mentioned, the different approaches to studying the situatedness, embodiment and embeddedness of cognition have fuzzy boundaries and slightly different foci. They also overlap greatly. A very broad notion of embeddedness assumes that cognitive processes can only be understood when an individual's environment is taken into account. However, in this chapter, we draw on a narrower understanding of embedded cognition, namely the theoretical position on cognition that relinquishes the idea and function of internal, mental representations altogether. Such non-representational approaches argue that intelligent action does not need an internal "image" of the outer world, because the outer world is there to be operated on. Instead, it is traces of our own actions and experiences that have been generated in the brain—traces that do not have a representational function (Chemero, 2009).

The non-representational, embedded view of cognition knits perception and action tightly together: They form the ingredients of cognition, without an intermediate central steering unit that would only need them as its input and output devices. The continuous flow of activity enables the cognitive system to react adequately to contingent circumstances—an ability that has been equated with intelligence from the very beginnings of cognitive science but has proved hard to simulate for any systems that depend on centralized, amodal information processing. The crucial difference here from simply stating (trivially) that the environment influences cognitive processes is that the inclusion of the environment into the cognitive process makes internal representation dispensable. Paradigmatic examples of such cognitive processes include "epistemic actions" (Kirsh & Maglio, 1994), which use the world to solve problems, e.g. in the manipulation of external elements by hand when playing games like Tetris (Kirsh & Maglio, 1994) or Scrabble (Walter, 2014) instead of processing the information in the head and then executing the solution.

The idea of having the world as "its own best model" (Brooks, 1991, p. 148) emphasizes the fluent, *in situ* interaction with the environment instead of the attempt to memorize new and retrieve prior information. External structures are "scaffolds" (Clark, 1998) onto which we "off-load" processing load and with which we (inter)act. This continuous, instantaneous link to the environment explains why we do not always adhere to plans but act instead according to the possibilities provided, why we are so easily distracted in our activities and how heavily we rely on our customary material and social environment and tools like computer programs, calendars and notebooks to function or perform as usual. Consequently, changing the social or material environment can have positive (and negative) effects, for instance, on creativity and innovation ("enabling spaces"; Peschl, 2007).[5]

Non-representational stances on cognition can be mathematically modelled in dynamic systems theory (DST) terms, i.e. the notion of continuous activations of interdependent units that result in dynamics that cannot be explained by the properties of the individual units. The dynamics of the system thus result in properties that are referred to as "emergent" and rely on decentralized and self-organizing processes. In this framework, human cognition is modelled as a complex adaptive system that changes in line with positive or negative feedback from the environment (see e.g. Smith & Thelen, 2003).

As Walter (2014) notes, DST provides a mathematically exact and explicit model for describing and analysing cognitive processes. It suggests that cognitive processes are not comprised of distinct steps but can be described instead as continuous changes in status, a trajectory in a present activation space. Phenomena such as behavioural phase transitions and convergence towards attractors are patterns of activity that emerge from the self-organizational dynamics of the system itself and can thus be explained without reference to a central steering unit. As Beer (2014, p. 135) points out, it is important to keep in mind that DST is "a body of mathematics, and not itself a scientific theory of the natural world", which is why it is not explicitly aligned with one particular cognitive research tradition. Indeed, in Translation Studies, O'Brien (2017, p. 324; see also Elman, 1998) claims close theoretical compatibility between connectionist positions and DST, while other scholars (e.g. Chemero, 2009) stress the proximity between (radical) embodied cognition and DST.

In any case, taking a dynamic view of a real-world phenomenon entails a particular perspective that shapes our research questions and theoretical assumptions as well as our data analysis and interpretation (Beer, 2014, p. 135). Given that cognitive systems are complex systems coupled with many other changing and interdependent systems, systems theory is better able to describe cognitive behaviour than the sequential information processing models encountered in first-generation cognitive science. However, due to the complexity involved in describing the mutual influences of continually changing systems and their effects on the superordinate system, the dynamic systems perspective has been criticized for its inability to explain higher cognitive functions. In Translation Studies, it will probably be used more as a powerful theoretical metaphor than as a genuine calculation or simulation instrument. Göpferich (2013), for example, has already attempted to bring it to bear on Translation Studies by discussing empirical data from a longitudinal research project on translators' competence in a DST context.

27.3.5 Extended cognition

The hypothesis of extended cognition can refer to two perspectives. It points, on the one hand, to the cognitive science position, not only that the environment *influences* and *interacts* with cognitive processes, but also that relevant parts of the social and material environment can actually be *constitutive* of cognitive processes. On the other hand, it can refer to the "extended mind" approach proposed by Clark and Chalmers (1998), which views the brain, the body and the environment as a "coupled system". This is the view we will discuss in this section. Here, neural and non-neural elements constitute potentially proper parts of the mind. Clark and Chalmers refer to the necessary criterion for actually constituting a part of cognition as the "parity principle":

> If, as we confront some task, a part of the world functions as a process which, *were it done in the head*, we would have no hesitation in recognizing as part of the cognitive process, then that part of the world *is* [...] part of the cognitive process.
>
> *Clark & Chalmers, 1998, p. 8*

For instance, cognitive phenomena such as remembering and calculating have been (phylogenetically and ontogenetically) developed in interaction with the environment and can be carried out in the head or with the help of a notebook, a computer or a friend. In the extended cognition view, there is no reason to rule out these extra-cranial and social parts of the process. In this respect, there is an obvious overlap with the embedded cognition approach. However, extended cognition is also a separate conceptual framework that clearly puts its emphasis on the idea that it is typical of humans as "natural-born cyborgs" (Clark, 2003) to utilize external artefacts as active constituents of cognitive systems.

The extended mind hypothesis deals mainly with the location, the "where", of cognition: It looks at the boundaries of cognitive systems. These are, of course, also extended in other approaches to cognition—from the brain to the body in embodied cognition, and from the individual to the social units in their environments by the distributed cognition view. However, the extended mind approach advocates an "active role of the environment in driving cognitive processes" (Clark & Chalmers, 1998, p. 7), which it refers to as "active externalism". Clark and Chalmers emphasize that this sort of externalism seeks to go a step further than simply stating that the meanings of words are not "in the head" (i.e. that they are indexical and dependent on the situation) by proposing that meaning is constituted by the whole "loop" of the brain, the body and the relevant parts of the environment. It is not only cognitive processing that is extended, but the mind itself, including seemingly inner mental functions such as beliefs, which, according to Clark and Chalmers (1998), can be constituted partly by features of the material and social environment when these drive the cognitive process *in situ*. Clark and Chalmers (1998, pp. 12ff.) maintain that believing, for example, that the Museum of Modern Art is located on a specific street because you have written it down in a notebook that you always carry with you is as reliable a belief as if you had memorized it in the brain. If a person constantly uses a notebook as a matter of course, dismissing it as a non-cognitive part of the process would mean missing the way in which the notebook entries play the kind of role that internal beliefs play in human cognition.

Clark and Chalmers (1998) accept that we might intuitively think that only inner mental cognitive processes are reliable portable resources that can be put into use in different situations to master different tasks. However, they argue that learned bodily actions and tool use also enable reliable coupling with the environment. They see language as one of the main resources that extend the cognitive into the world, e.g. when we memorize things by writing notes, think through writing or solve problems by discussing them with others. The extended mind hypothesis thus requires a reconceptualization of action and communication as thought and reasoning processes. Applied to translation, an extended cognition approach could provide a radically new view of the unit that is actually doing the translating. For example, translation competence and expertise would thus not be features of individual "naked" brains, but of the hybrid unit of humans and their social and artefact infrastructure.

Very little translation research has so far tried to accommodate extended cognition as a theoretical backing for empirical research. The first such studies have been mostly explorative in nature and can only point to possible implications that should be further investigated in larger research projects and from a longitudinal perspective (see e.g. Risku, 2014; Risku & Windhager, 2013).

Viewing cognition as extended makes a difference for the methodological choices when studying cognitive processes, since the whole loop has to be included in scientific investigations with corresponding explanatory power—as the following statement indicates:

> [A]s a methodological consequence, extended cognition studies inevitably follow "leaking minds" into their social and technical environments, thereby including process, interaction and artefact analysis into a combined and linked view on dynamic complexity.
>
> *Risku & Windhager, 2013, p. 36*

Adopting an extended cognition perspective also changes our moral criteria, with particular implications for research ethics: Since environmental and social features can be parts of minds, changing the environmental or social circumstances can be seen as interventions to minds.

27.4 Future developments

It seems likely that the insights emerging from research on situated frameworks will help to avoid a disembodied, ahistorical, reductionist view of translators or interpreters and of the cognitive processes in translation and interpreting. We might end up declining a view of them as defective computational machines after all. Putting translators and interpreters back in their bodies and worlds will open up new research questions and avenues for translation and interpreting process research and for Cognitive Translation and Interpreting Studies. However, as Jakobsen claims for translation process research, this area of research will firmly retain its focus on its research object of the generation of translations (and the same will probably apply to interpreting):

> The focus will continue to be on how a translator cognitively interacts with an external ST, an external TT, and writing tool, but will cast its net wider to be able to take more account of individual, emotional, interactional, technological, and institutional factors.
>
> *Jakobsen, 2017, p. 40*

Including these aspects will give us a more balanced view of the translation and interpreting process and do justice to the complexity of the cognitive processes involved, thus legitimating both micro- and macro-perspectives on the research object. After all, phenomena ranging from the neural and hormonal to organizational and social factors all affect translation and interpreting performance. In a similar vein, Muñoz Martín (2016) questions and deconstructs—or rather reconnects—the concepts of translation acts, events and practice: From a cognitive point of view, we cannot separate cognitive, sociological and cultural, historical and anthropological processes. The social/cognitive divide seems to be less drastic than it is often portrayed to be. The usual demarcation lines between cognition, culture and practice, for example, might have to be rethought:

> Where a narrow and a broader approach inevitably overlap is in their attention to how a translation emerges from another text and the degree of ease, effortlessness, pleasurableness, speed, efficiency, and quality with which this happens.
>
> *Jakobsen, 2017, p. 40*

Viewing cognition as situated is not only a terminological position; it also raises some serious demands on the methodological level. We will increasingly need research designs that can cope with cognition as this kind of self-organizing navigation of a hybrid network in a real-world setting. However, situated frameworks do not determine methodological choices—there are still questions to be answered, both with qualitative and quantitative as well as with field and laboratory research. While situated frameworks will require more than just experimental laboratory research, many explanations of how translators exploit their social and material environment will also have to be identified in a laboratory setting. As Jakobsen (2017, p. 40) maintains, "in research there is a place for the microscope as well as for the telescope, and also for unaided observation".

The new orientations in cognitive science outlined in this chapter open up exciting new perspectives for Translation Studies. New research fields will include, for instance, the increasing use of cognitive approaches in ergonomics research, a greater interest in the physicality or distributedness of cognition and action, or the increased study of affective factors and emotions as part of a situated, embodied translation activity. Current developments in cognitive sciences that could prove interesting for such research in the future include the concepts of the "predictive mind" or "predictive coding" (Clark, 2013, 2016) and in particular the "enacted cognition" approach (Hutto & Myin, 2013; Thompson, 2007; Varela et al., 1995), which also opens up exciting new perspectives, not least for interpreting research.

To this end, Cognitive Translation and Interpreting Studies will need to collaborate in future not only with the cognitive sciences and related disciplines (neuroscience, neuropsychology and social psychology) but also more strongly with the social sciences. At the same time, it will—as always—be important to keep abreast of developments in cognitive science and process the new insights obtained in this field—with all their theoretical implications—for application in cognitive approaches to Translation Studies and empirical translation and interpreting research.

Notes

1 See Harnad (1990) for a detailed account of the "symbol grounding problem".
2 See Martín de León (2017) for a detailed account of the role given or denied to mental representations in different cognitive research traditions and some deliberations on what this means for an application in a Translation Studies context.
3 See also Piaget's (1952) position that human intelligence is based on the individual's manipulation of and interaction with its environment.
4 See also Giere and Moffatt (2003) and Toon (2014) for more on the debate regarding the existence of a divide between or an increasing intertwining of sociological and cognitive approaches.
5 Similar links between translation workplaces, organizational frameworks, team constellations and translator creativity were observed, for instance, by Risku, Milošević, et al. (2017) in the translation department at a public institution.

Further reading

Clark, A. (2008). *Supersizing the mind: Embodiment, action, and cognitive extension.* Oxford: Oxford University Press.

With this work, Andy Clark continues his critical endeavour of putting brain, body and world back together again. He combines current research from robotics to psychology, linguistics, neuroscience and beyond, sketching a landscape of cognitive research that shifts from brain-centred, mental approaches to an embodied, extended view of cognition. As an added bonus to the foundational scientific insights, Clark's writing is as compelling as ever.

Hutchins, E. (1995). *Cognition in the wild.* Cambridge: MIT Press.

Edwin Hutchins' now classic empirical study of ship navigation provides the theoretical and methodological foundations for a distributed approach to cognition. Using anthropological methods, the navigation team with its specific interactive and communicative environment is described as a cultural and cognitive system, thus producing a new understanding of cognition and of cognitive science.

Menary, R. (Ed.). (2010). *The extended mind.* Cambridge: MIT Press.

In this edited volume, Richard Menary collects essays which build on a landmark 1988 paper by the philosophers Andy Clark and David Chalmers and which discuss how intelligent action can be scientifically explained as an emergent property of cognitive processing that extends the boundaries of the skin and the skull.

Shapiro, L. A. (2010). *Embodied cognition.* London: Routledge.

This monograph is an excellent introduction to the foundations of embodied cognition. It starts with a summary of the commitments of traditional, propositional-representational cognitive science and then goes on to describe how different disciplines and authors have contributed to the development of the embodied approach to cognition.

Suchman, L. (2007). *Human–machine reconfigurations.* Cambridge: Cambridge University Press.

This classic work includes Lucy Suchman's ground-breaking *Plans and situated actions* (1987) and weaves it into a coherent, current view on situated cognition. Through an empirical analysis of human–machine interaction, the renowned anthropologist shows how social and material practices constitute technical systems and offers a seminal, non-computational approach to human agency and cognition.

References

Abdallah, K. (2014). The interface between Bourdieu's Habitus and Latour's Agency: The work trajectories of two Finnish translators. In G. M. Vorderobermeier (Ed.), *Remapping habitus in translation studies* (pp. 111–132). Amsterdam: Rodopi.

Alves, F., & Gonçalves, J. L. (2007). Modelling translator's competence: Relevance and expertise under scrutiny. In Y. Gambier, M. Shlesinger, & R. Stolze (Eds.), *Doubts and directions in translation studies: Selected contributions from the EST Congress, Lisbon 2004* (pp. 41–55). Amsterdam: John Benjamins.

Angelone, E., & Marín García, Á. (2017). Expertise acquisition through deliberate practice: Gauging perceptions and behaviors of translators and project managers. *Translation Spaces, 6*(1), 122–158. https://doi.org/10.1075/ts.6.1.07ang

Baraldi, C., & Gavioli, L. (Eds.). (2012). *Coordinating participation in dialogue interpreting.* Amsterdam: John Benjamins.

Beer, R. D. (2014). Dynamical systems and embedded cognition. In K. Frankish, & W. M. Ramsey (Eds.), *The Cambridge handbook of artificial intelligence* (pp. 128–150). Cambridge: Cambridge University Press.

Bermúdez, J. L. (2014). *Cognitive science: An introduction to the science of the mind* (2nd ed.). Cambridge: Cambridge University Press.

Biagini, M., Davitti, E., & Sandrelli, A. (Eds.). (2017). Special section. Participation in interpreter-mediated interaction: Shifting along a multidimensional continuum [Special issue]. *Journal of Pragmatics, 107.*

Brooks, R. A. (1991). Intelligence without representation. *Artificial Intelligence, 47*(1–3), 139–159. https://doi.org/10.1016/0004-3702(91)90053-M

Buzelin, H. (2005). Unexpected allies: How Latour's network theory could complement Bourdieusian analyses in translation studies. *The Translator, 11*(2), 193–218. https://doi.org/10.1080/13556509.2005.10799198

Buzelin, H. (2007). Translation studies, ethnography and the production of knowledge. In P. St-Pierre, & P. C. Kar (Eds.), *In translation: Reflections, refractions, transformations* (pp. 39–56). Amsterdam: John Benjamins.
Chemero, A. (2009). *Radical embodied cognitive science*. Cambridge, MA: MIT Press.
Chesterman, A. (2000). Teaching strategies for emancipatory translation. In C. Schäffner, & B. Adab (Eds.), *Developing translation competence* (pp. 77–89). Amsterdam: John Benjamins. https://doi.org/10.1075/btl.38.09che
Christensen, T. P. (2011). Studies on the mental processes in translation memory-assisted translation: The state of the art. *Trans-Kom, 4*(2), 137–160.
Clark, A. (1998). *Being there: Putting brain, body, and world together again*. Cambridge: MIT Press.
Clark, A. (2003). *Natural-born cyborgs: Minds, technologies, and the future of human intelligence*. Oxford: Oxford University Press.
Clark, A. (2013). Whatever next? Predictive brains, situated agents, and the future of cognitive science. *Behavioral and Brain Sciences, 36*(3), 181–253. https://doi.org/10.1017/S0140525X12000477
Clark, A. (2016). *Surfing uncertainty: Prediction, action, and the embodied mind*. Oxford: Oxford University Press.
Clark, A., & Chalmers, D. J. (1998). The extended mind. *Analysis, 58*(1), 7–19.
Danks, J. S., Shreve, G. M., Fountain, S. B., & McBeath, M. K. (Eds.). (1997). *Cognitive processes in translation and interpreting*. London: SAGE.
Davitti, E. (2012). Dialogue interpreting as intercultural mediation: Integrating talk and gaze in the analysis of mediated parent-teacher meetings [unpublished PhD thesis]. University of Manchester, Manchester.
Dominey, P. F., Prescott, T. J., Bohg, J., Engel, A. K., Gallagher, S., Heed, T., Hoffmann, M., Knoblich, G., Prinz, W., & Schwartz, A. (2016). Implications of action-oriented paradigm shifts in cognitive science. In A. K. Engel, K. J. Friston, & D. Kragic (Eds.), *The pragmatic turn: Toward action-oriented views in cognitive science* (pp. 333–361). Cambridge, MA: MIT Press.
Dong, J., & Turner, G. H. (2016). The ergonomic impact of agencies in the dynamic system of interpreting provision: An ethnographic study of backstage influences on interpreter performance. *Translation Spaces, 5*(1), 97–123. https://doi.org/10.1075/ts.5.1.06don
Dreyfus, H. L., & Dreyfus, S. E. (1986). *Mind over machine: The power of human intuition and expertise in the era of the computer*. Oxford: Blackwell.
Dreyfus, H. L. (1972). *What computers can't do: The limits of artificial intelligence*. New York: MIT Press.
Ehrensberger-Dow, M., & Englund Dimitrova, B. (Eds.). (2016). Cognitive space: Exploring the situational interface [Special issue]. *Translation Spaces,* 5(1).
Ehrensberger-Dow, M., & Hunziker Heeb, A. (2016). Investigating the ergonomics of a technologized translation workplace. In R. Muñoz Martín (Ed.), *Reembedding translation process research* (pp. 69–88). Amsterdam: John Benjamins.
Ehrensberger-Dow, M., Hunziker Heeb, A., Massey, G., Meidert, U., Neumann, S., & Becker, H. (2016). An international survey of the ergonomics of professional translation. *ILCEA (Revue de l'Institut des langues et cultures d'Europe, Amérique, Afrique, Asie et Australie), 27*. http://ilcea.revues.org/4004
Elman, J. L. (1990). Finding structure in time. *Cognitive Science, 14*(2), 179–211. https://doi.org/10.1207/s15516709cog1402_1
Elman, J. L. (1998). Connectionism, artificial life, and dynamical systems. In W. Bechtel, & G. Graham (Eds.), *A companion to cognitive science* (pp. 488–505). Malden: Blackwell.
Faber, P. (2011). The dynamics of specialized knowledge representation: Simulational reconstruction or the perception-action interface. *Terminology, 17*(1), 9–29.
Fillmore, C. J. (1977). Scenes-and-frames semantics. In A. Zampolli (Ed.), *Linguistic structures processing* (pp. 55–81). Amsterdam: North-Holland Publishing.
Garcez, P. M. (1997). Microethnography. In N. H. Hornberger, & D. Corson (Eds.), *Encyclopedia of language and education* (2nd ed., pp. 187–196). Dordrecht, Boston, London: Kluwer.
Gerloff, P. (1988). From French to English: A look at the translation process in students, bilinguals, and professionals [unpublished dissertation]. Harvard University, Cambridge.
Gibson, J. J. (1977). The theory of affordances. In R. Shaw, & J. Bransford (Eds.), *Perceiving, acting, and knowing: Toward an ecological psychology* (pp. 67–82). Hillsdale: Erlbaum.
Giere, R., & Moffatt, B. (2003). Distributed cognition: Where the cognitive and the social merge. *Social Studies of Science, 33*(2), 301–310.
González-Davies, M., & Enríquez Raído, V. (Eds.). (2016). Situated learning in translator and interpreter training: Bridging research and good practice [Special issue]. *The Interpreter and Translator Trainer, 10*(1).

Göpferich, S. (2013). Translation competence: Explaining development and stagnation from a dynamic systems perspective. *Target, 25*(1), 61–76.
Harnad, S. (1990). The symbol grounding problem. *Physica D, 42*(1–3), 335–346. https://doi.org/10.1016/0167-2789(90)90087-6
Haugeland, J. (1985). *Artificial intelligence: The very idea.* Cambridge, MA: MIT Press.
Heidegger, M. (1953). *Sein und Zeit.* Tübingen: Niemeyer.
Hine, C. (2000). *Virtual ethnography.* London, Thousand Oaks: SAGE.
Hine, C. (2015). *Ethnography for the internet: Embedded, embodied and everyday.* London: Bloomsbury.
Hirvonen, M., & Tiittula, L. (2010). A method for analysing multimodal research material: Audio description in focus. *MikaEl,* 4. www.sktl.fi/@Bin/40698/Hirvonen%26Tiittula_MikaEL2010.pdf
Hirvonen, M., & Tiittula, L. (2018). How are translations created? Using multimodal conversation analysis to study a team translation process. *Linguistica Antverpiensia, 17,* 157–173.
Hokkanen, S. (2017). Emotions of involvement and detachment in simultaneous church interpreting. *Translation Spaces, 6*(1), 62–78. https://doi.org/10.1075/ts.6.1.04hok
Hokkanen, S., & Koskinen, K. (2016). Affect as a hinge: The translator's experiencing self as a sociocognitive interface. *Translation Spaces, 5*(1), 78–96. https://doi.org/10.1075/ts.5.1.05hok
Hollstein, B. (2006). Qualitative Methoden und Netzwerkanalyse. In B. Hollstein, & F. Straus (Eds.), *Qualitative Netzwerkanalyse: Konzepte. Methoden. Anwendungen* (pp. 11–35). Wiesbaden: VS Verlag für Sozialwissenschaften.
Hönig, H. (1995). *Konstruktives Übersetzen.* Tübingen: Stauffenburg.
Hönig, H. (1997). Translating: The constructive way. *Ilha Do Desterro, 33,* 11–23.
Hutchins, E. (1995a). *Cognition in the wild.* Cambridge, MA: MIT Press.
Hutchins, E. (1995b). How a cockpit remembers its speeds. *Cognitive Science, 19*(3), 265–288.
Hutchins, E. (2001). Cognition, distributed. In N. Smelser, & P. Baltes (Eds.), *The international encyclopedia of the social and behavioral sciences* (4th ed., pp. 2068–2072). New York: Elsevier.
Hutto, D. D., & Myin, E. (2013). *Radicalizing enactivism: Basic minds without content.* Cambridge: MIT Press.
Jääskeläinen, R. (1993). Investigating translation strategies. In S. Tirkkonen-Condit, & J. Laffling (Eds.), *Recent trends in empirical translation research* (pp. 99–120). Joensuu: University of Joensuu, Faculty of Arts.
Jakobsen, A. L. (2017). Translation process research. In J. W. Schwieter, & A. Ferreira (Eds.), *The handbook of translation and cognition* (pp. 21–49). Oxford: Wiley Blackwell.
Jiménez-Crespo, M. A. (2017). *Crowdsourcing and online collaborative translations: Expanding the limits of translation studies.* Amsterdam: John Benjamins.
Johnson-Laird, P. N. (1983). *Mental models: Towards a cognitive science of language, inference and consciousness.* Cognitive science series: Vol. 6. Cambridge: Harvard University Press.
Kinnunen, T. (2017). Körperlich-räumliche Aspekte gedolmetschter Interaktion im Gericht. *Trans-Kom, 10*(1), 45–74.
Kiraly, D. (2000). *A social constructivist approach to translator education: Empowerment from theory to practice.* Manchester, UK: St. Jerome.
Kirsh, D., & Maglio, P. (1994). On distinguishing epistemic from pragmatic action. *Cognitive Science, 18*(4), 513–549. https://doi.org/10.1207/s15516709cog1804_1
Kolb, W. (2017). "It was on my mind all day": Literary translators working from home—some implications of workplace dynamics. *Translation Spaces, 6*(1), 27–43. https://doi.org/10.1075/ts.6.1.02kol
Königs, F. G. (1987). Was beim Übersetzen passiert: Theoretische Aspekte, empirische Befunde und praktische Konsequenzen. *Die Neueren Sprachen, 86*(2), 182–185.
Krings, H. (1986). *Was in den Köpfen von Übersetzern vorgeht: Eine empirische Untersuchung zur Struktur des Übersetzungsprozesses an fortgeschrittenen Französischlernern.* Tübingen: Narr.
Krüger, R. (2015a). Fachübersetzen aus kognitionstranslatologischer Perspektive: Das Kölner Modell des situierten Fachübersetzers. *Trans-Kom, 8*(2), 273–313.
Krüger, R. (2015b). *The interface between scientific and technical translation studies and cognitive linguistics: With particular emphasis on explicitation and implicitation as indicators of translational text-context interaction.* Berlin: Frank & Timme.
Kuhn, T. S. (1970). *The structure of scientific revolutions* (2nd ed.). Chicago: University of Chicago Press.
Kußmaul, P. (1991). Creativity in the translation process: Empirical approaches. In K. M. van Leuven-Zwart, & T. Naaijkens (Eds.), *Translation studies: The state of the art. Proceedings of the 1st James S. Holmes Symposium in Translation Studies* (pp. 91–101). Amsterdam: Rodopi.

Kußmaul, P. (2000). *Kreatives Übersetzen.* Tübingen: Stauffenburg.
Kuznik, A. (2016). La traduction comme travail: Perspectives croisées en ergonomie, sociologie et traductologie. *ILCEA (Revue de l'Institut des langues et cultures d'Europe, Amérique, Afrique, Asie et Australie), 27.* http://ilcea.revues.org/4036
Kuznik, A., & Verd, J. M. (2010). Investigating real work situations in translation agencies: Work content and its components. *Hermes: Journal of Language and Communication Studies, 44*, 25–43.
Lakoff, G., & Johnson, M. (1980). *Metaphors we live by.* Chicago, London: University of Chicago Press.
Latour, B. (2005). *Reassembling the social: An introduction to Actor-Network-Theory.* Oxford: Oxford University Press.
Lautenbacher, O. P., Tiittula, L., Hirvonen, M., Laaksonen, J., & Kurimo, M. (2015). Towards reliable automatic multimodal content analysis. In A. Belz, L. Coheur, V. Ferrari, M.-F. Moens, K. Pastra, & I. Vulić (Eds.), *Proceedings of the Fourth Workshop on Vision and Language* (pp. 6–7). Stroudsburg, PA: Association for Computational Linguistics. https://doi.org/10.18653/v1/W15-2803
LeBaron, C. (2008). Microethnography. In W. Donsbach (Ed.), *The international encyclopedia of communication* (pp. 3120–3124). Malden, Oxford, Victoria: Blackwell Publishing.
McDonough, J. (2007). How do language professionals organize themselves? An overview of translation networks. *Meta, 52*(4), 793–815. https://doi.org/10.7202/017697ar
McDonough Dolmaya, J. (2012). Analyzing the crowdsourcing model and its impact on public perceptions of translation. *The Translator, 18*(2), 167–191.
McDonough Dolmaya, J. (2017). Expanding the sum of all human knowledge: Wikipedia, translation and linguistic justice. *The Translator, 23*(2), 143–157. https://doi.org/10.1080/13556509.2017.1321519
Marcus, G. E. (1995). Ethnography in/of the world system: The emergence of multi-sited ethnography. *Annual Review of Anthropology, 24*, 95–117.
Marín García, Á. (2017). Theoretical hedging: The scope of knowledge in translation process research [unpublished dissertation]. Kent State University, Kent.
Martín de León, C. (2008). Translation in the wild: Traductología y cognición situada. In L. Pegenaute, J. DeCesaris, M. Tricás, & E. Bernal (Eds.), *La traducción del futuro: Mediación lingüística y cultural en el siglo XXI. Vol. II: La traducción y su entorno* (pp. 55–64). Barcelona: PPU.
Martín de León, C. (2017). Mental representations. In J. W. Schwieter, & A. Ferreira (Eds.), *The handbook of translation and cognition* (pp. 106–126). Oxford: Wiley Blackwell.
Mason, I. (Ed.). (2001). *Triadic exchanges: Studies in dialogue interpreting.* Manchester: St. Jerome.
Mason, I. (2012). Gaze, positioning and identity in interpreter-mediated dialogues. In C. Baraldi, & L. Gavioli (Eds.), *Coordinating participation in dialogue interpreting* (pp. 177–200). Amsterdam: John Benjamins.
Massaro, D. W., & Shlesinger, M. (1997). Information processing and a computational approach to the study of simultaneous interpretation. *Interpreting, 2*(1–2), 13–53. https://doi.org/10.1075/intp.2.1-2.02mas
Meidert, U., Neumann, S., Ehrensberger-Dow, M., & Becker, H. (2016). Physical ergonomics at translators' workplaces: Findings from ergonomic workplace assessments and interviews. *ILCEA (Revue de l'Institut des langues et cultures d'Europe, Amérique, Afrique, Asie et Australie), 27.* http://journals.openedition.org/ilcea/3996
Menary, R. (2010a). Introduction to the special issue on 4E cognition. *Phenomenology and the Cognitive Sciences, 9*(4), 459–463.
Merleau-Ponty, M. (1965). *Phenomenology of perception* (C. Smith, Trans.) London: Routledge. Originally published in 1945.
Mihalache, I. (2008). Community experience and expertise: Translators, technologies and electronic networks of practice. *Translation Studies, 1*(1), 55–72.
Muñoz Martín, R. (2010a). Leave no stone unturned: On the development of cognitive translatology. *Translation and Interpreting Studies, 5*(2), 145–162. https://doi.org/10.1075/tis.5.2.01mun
Muñoz Martín, R. (2010b). On paradigms and cognitive translatology. In G. Shreve, & E. Angelone (Eds.), *Translation and cognition* (pp. 169–189). Amsterdam: John Benjamins. https://doi.org/10.1075/ata.xv.10mun
Muñoz Martín, R. (2016). Processes of what models? On the cognitive indivisibility of translation acts and events. *Translation Spaces, 5*(1), 145–161.
Norman, D. A. (1993). *Things that make us smart: Defending human attributes in the age of the machine.* Boston: Addison-Wesley Longman.
O'Brien, S. (2012). Translation as human–computer interaction. *Translation Spaces, 1*, 101–122.

O'Brien, S. (2017). Machine translation and cognition. In J.W. Schwieter, & A. Ferreira (Eds.), *The handbook of translation and cognition* (pp. 313–331). Oxford: Wiley Blackwell.
O'Hagan, M. (2011). Community translation: Translation as a social activity and its possible consequences in the advent of Web 2.0 and beyond. *Linguistica Antverpiensia New Series—Themes in Translation Studies, 10*, 11–23.
Olohan, M. (2017). Knowing in translation practice: A practice-theoretical perspective. *Translation Spaces, 6*(1), 159–180. https://doi.org/10.1075/ts.6.1.08olo
Orrego-Carmona, D., & Lee, Y. (Eds.). (2017). *Non-professional subtitling*. Newcastle upon Tyne: Cambridge Scholars Publishing.
Padilla, P., Bajo, M.T., & Padilla, F. (1999). Proposal for a cognitive theory of translation and interpreting: A methodology for future empirical research. *The Interpreter's Newsletter, 9*, 61–78. www.openstarts.units.it/bitstream/10077/2213/1/04Padilla.pdf
Pasquandrea, S. (2011). Managing multiple actions through multimodality: Doctors' involvement in interpreter-mediated interactions. *Language in Society, 40*(4), 455–481. https://doi.org/10.1017/S0047404511000479
Peschl, M. F. (2007). Enabling Spaces. Epistemologische Grundlagen der Ermöglichung von Innovation und knowledge creation. In N. Gronau (Ed.), *Professionelles Wissensmanagement. Erfahrungen und Visionen* (pp. 362–372). Berlin: GITO.
Piaget, J. (1952). *The origins of intelligence in children* (M. Cook, Trans.). New York: International Universities Press.
Pineau, M. (2011). La main et le clavier: Histoire d'un malentendu. *ILCEA (Revue de l'Institut des langues et cultures d'Europe, Amérique, Afrique, Asie et Australie), 14*. http://ilcea.revues.org/1067
Prieto Velasco, J. A., & Tercedor Sánchez, M. (2014). The embodied nature of medical concepts: Image schemas and language for pain. *Cognitive Processing, 15*(3), 283–296. https://doi.org/10.1007/s10339-013-0594-9
Raanes, E., & Berge, S. S. (2017). Sign language interpreters' use of haptic signs in interpreted meetings with deafblind persons. *Journal of Pragmatics, 107*, 91–104. https://doi.org/10.1016/j.pragma.2016.09.013
Richardson, M. (2017). Sign language interpreting in theatre: Using the human body to create pictures of the human soul. *TranscUlturAl, 9*(1), 45–64.
Risku, H. (1998). *Translatorische Kompetenz: Kognitive Grundlagen des Übersetzens als Expertentätigkeit*. Studien zur Translation: Vol. 5. Tübingen: Narr.
Risku, H. (2002). Situatedness in translation studies. *Cognitive Systems Research, 3*(3), 523–533. https://doi.org/10.1016/S1389-0417(02)00055-4
Risku, H. (2010). A cognitive scientific view on technical communication and translation: Do embodiment and situatedness really make a difference? *Target, 22*(1), 94–111. https://doi.org/10.1075/target.22.1.06ris
Risku, H. (2014). Translation process research as interaction research: From mental to socio-cognitive processes. *MonTi*, 331–353. https://doi.org/10.6035/MonTI.2014.ne1.11
Risku, H. (2016). *Translationsmanagement: Interkulturelle Fachkommunikation im Informationszeitalter* (3rd ed.). Tübingen: Narr Francke Attempto.
Risku, H., & Dickinson, A. (2009). Translators as networkers: The role of virtual communities. *Hermes: Journal of Language and Translation in Business, 22*(42), 49–70.
Risku, H., Dickinson, A., & Pircher, R. (2010). Knowledge in Translation Studies and translation practice: Intellectual capital in modern society. In D. Gile, G. Hansen, & N. Pokorn (Eds.), *Why translation studies matters* (pp. 83–94). Amsterdam: John Benjamins.
Risku, H., Hirvonen, M., Rogl, R., & Milošević, J. (forthcoming). Ethnographic research. In F. Zanettin, & C. Rundle (Eds.), *Routledge handbook of translation and methodology*. Oxford: Routledge.
Risku, H., Milošević, J., & Rogl, R. (2017). Creativity in the translation workplace. In L. Cercel, M. Agnetta, & M. T. Amido Lozano (Eds.), *Kreativität und Hermeneutik in der Translation* (pp. 455–469). Tübingen: Narr Francke Attempto.
Risku, H., & Pircher, R. (2006). Translatory cooperation: Roles, skills and coordination in intercultural text design. In M. Wolf (Ed.), *Übersetzen—translating—traduire: Towards a "social turn"?* (pp. 253–264). Wien, Berlin: LIT.
Risku, H., Rogl, R., & Milošević, J. (Eds.). (2017). Translation practice in the field: Current research on socio-cognitive processes [Special issue]. *Translation Spaces, 6*(1).
Risku, H., Rogl, R., & Pein-Weber, C. (2016). Mutual dependencies: Centrality in translation networks. *The Journal of Specialised Translation, 25*, 232–253.

Risku, H., & Windhager, F. (2013). Extended translation: A socio-cognitive research agenda. *Target, 25*(1), 33–45.
Risku, H., Windhager, F., & Apfelthaler, M. (2013). A dynamic network model of translatorial cognition and action. *Translation Spaces, 2*, 151–182. https://doi.org/10.1075/ts.2.08ris
Rosch, E. (1973). Natural categories. *Cognitive Psychology, 4*, 328–350.
Rosch, E. (1975). Cognitive representation of semantic categories. *Journal of Experimental Psychology, 104*(3), 192–233.
Rumelhart, D. E., McClelland, J. L., & the PDP Research Group (Eds.). (1986). *Parallel distributed processing: Explorations in the microstructure of cognition*. Cambridge: MIT Press.
Rumelhart, D. E., Smolensky, P., McClelland, J. L., & Hinton, G. E. (1986). Schemata and sequential thought processes in PDP models. In J. L. McClelland, D. E. Rumelhart, & the PDP Research Group (Eds.), *Parallel distributed processing: Explorations in the microstructure of cognition* (pp. 7–57). Cambridge, London: MIT Press.
Shreve, G. M. (2018). Levels of explanation and translation expertise. *Hermes: Journal of Language and Communication in Business, 57*, 97–108.
Smith, L. B., & Thelen, E. (2003). Development as a dynamic system. *Trends in Cognitive Sciences, 7*(8), 343–348. https://doi.org/10.1016/S1364-6613(03)00156-6
Snell-Hornby, M. (1988). *Translation studies: An integrated approach*. Amsterdam: John Benjamins.
Suchman, L. (1987). *Plans and situated action: The problem of human–machine communication*. Palo Alto Research Centre.
Suchman, L. (2007). *Human–machine reconfigurations*. Cambridge: Cambridge University Press.
Tannen, D. (1979). What's in a frame? Surface evidence for underlying expectations. In R. O. Freedle (Ed.), *New directions in discourse processing* (pp. 157–181). Norwood: Ablex.
Taylor, C. (2003). Multimodal transcription in the analysis, translation and subtitling of Italian films. *The Translator, 9*(2), 191–205. https://doi.org/10.1080/13556509.2003.10799153
Taylor, C. (2004). Multimodal text analysis and subtitling. In E. Ventola, C. Charles, & M. Kaltenbacher (Eds.), *Perspectives on multimodality* (pp. 153–172). Amsterdam: John Benjamins.
Taylor, C. (2016). The multimodal approach in audiovisual translation. *Target, 28*(2), 222–236. https://doi.org/10.1075/target.28.2.04tay
Teixeira, C., & O'Brien, S. (2017). Investigating the cognitive ergonomic aspects of translation tools in a workplace setting. *Translation Spaces, 6*(1), 79–103. https://doi.org/10.1075/ts.6.1.05tei
Tercedor Sánchez, M. (2011). The cognitive dynamics of terminological variation. *Terminology, 17*(2), 181–197. https://doi.org/10.1075/term.17.2.01ter
Tercedor Sánchez, M., Faber, P., & D'Angiulli, A. (2011). The depiction of wheels by blind children: Preliminary studies on pictorial metaphors, language, and embodied imagery. *Imagination, Cognition and Personality, 31*(1), 113–128. https://doi.org/10.2190/IC.31.1-2.j
Thagard, P. (2005). *Mind: Introduction to cognitive science* (2nd ed.). Cambridge, MA: MIT Press.
Thompson, E. (2007). *Mind in life: Biology, phenomenology, and the sciences of mind*. Cambridge, London: Belknap Press.
Tirkkonen-Condit, S. (1989). Professional vs. non-professional translation: A think-aloud protocol study. In C. Séguinot (Ed.), *The translation process* (pp. 73–85). Toronto: H.G. Publications.
Toon, A. (2014). Friends at last? Distributed cognition and the cognitive/social divide. *Philosophical Psychology, 27*(1), 112–125. https://doi.org/10.1080/09515089.2013.828371
Vannerem, M., & Snell-Hornby, M. (1986). Die Szene hinter dem Text. "Scenes-and-frames semantics" in der Übersetzung. In M. Snell-Hornby (Ed.), *Übersetzungswissenschaft—eine Neuorientierung: Zur Integrierung von Theorie und Praxis* (pp. 184–205). Tübingen: Francke.
Varela, F. J., Thompson, E., & Rosch, E. (1995). *Embodied mind: Cognitive science and human experience* (4th ed.). Cambridge, MA: MIT Press.
Vermeer, H. J., & Witte, H. (1990). *Mögen Sie Zistrosen? Scenes & frames & channels im translatorischen Handeln*. Heidelberg: Groos.
Vienne, J. (1994). Toward a pedagogy of translation in situation. *Perspectives, 2*(1), 51–59. https://doi.org/10.1080/0907676X.1994.9961222
Vygotsky, L. (1978). *Mind in society: The development of higher psychological processes*. Cambridge: Harvard University Press.

Wadensjö, C. (1998). *Interpreting as interaction*. London: Longman.
Walter, S. (2014). *Kognition*. Stuttgart: Reclam.
Wilson, M. (2002). Six views of embodied cognition. *Psychonomic Bulletin & Review, 9*(4), 625–636. https://doi.org/10.3758/BF03196322
Wood, D., Bruner, J. S., & Ross, G. (1976). The role of tutoring in problem-solving. *Journal of Child Psychology and Psychiatry, 17*, 89–100.

28

Translation, artificial intelligence and cognition

Michael Carl

28.1 Introduction

Artificial intelligence (AI) and machine translation (MT) have a closely entangled history. Both fields of research originated in the United States in the 1950s and were ignited during the Cold War by the attempt to use computers to automatically translate Russian scientific journals into English (Camburn, 2013). While the early success of MT triggered an enthusiasm for fully automatic high-quality translation, it was quickly recognized that—in order to scale up MT—one had to understand the grammar of the source and the target language, the morphology (the grammar of word forms), and the syntax (the grammar of sentence structure). But to understand syntax, one also had to understand semantics (the meaning of words and sentences) and the pragmatics of language use. The attempt to translate texts automatically from one language into another thus quickly developed into a much bigger endeavour, which involved all areas of Computational Linguistics. According to the Stanford Encyclopedia of Philosophy (2018, n.p.), Computational Linguistics is "the scientific and engineering discipline concerned with understanding written and spoken language from a computational perspective [...] [and] a computational understanding of language [that] provides insight into thinking and intelligence".

While the idea of using computers to simulate intelligent human behaviour (i.e. to provide insight into thinking and intelligence) goes back to Turing (see Section 28.2), the term "Artificial Intelligence" was coined at a workshop in 1956, when a small group of ten scientists met for six weeks to discuss how "to make machines use language, form abstractions and concepts, solve kinds of problems now reserved for humans, and improve themselves" (McCarthy et al., 1955, p. 1). The underlying conviction was that "every aspect of learning or any other feature of intelligence can in principle be so precisely described that a machine can be made to simulate it" (McCarthy et al., 1955, p.1). Based on this assumption, Simon and Newell subsequently developed a theory of the general problem solver in 1963 and the physical symbol systems hypothesis (PSSH) in 1972, which is the basis of several computational cognitive models and brought about a huge push in AI (Newell & Simon, 1963, 1972). The human mind was conceived as a machine that manipulates symbols, and the manipulation operations were identified as computational processes of a program which could equally well run in the human brain or in a computer. It was believed that "[a] physical symbol system has the necessary and sufficient means for general intelligent action" (Newell & Simon, 1976, p. 116).

Accordingly, in the 1970s and the early 1980s, natural language processing (NLP) approached language explicitly as a cognitive process. These NLP systems were based on intuitions about how people "understand", and linguistic theories were thought to constitute the building blocks for mental representations and models of human thinking. For instance, Nagao (1984, p.173 proposed to model MT in line with human behaviour, stating that "we have to think about the mechanism of human translation, and have to build a model based on the fundamental function of language processing in the human brain". Nagao suggested a cognitive approach to translation, which became known as example-based machine translation (EBMT), deemed to mimic the human translation process. Based on this model, a large number of EBMT systems have been developed (see e.g. Carl & Way, 2003, for an overview); however, none of them were based on actual psycholinguistic investigation taking into account how humans really translate.

Until the late 1980s, most work in MT was concerned with symbolic systems, involving large collections of handwritten rules to model linguistic phenomena. These kinds of "knowledge-based" systems had a number of bottlenecks, mainly due to the unexpected complexity and ambiguity of natural languages. Since the late 1980s, there has been a significant shift towards statistical methods, whereby rules and probabilistic generalizations are learned from data rather than being produced manually. This "data-driven" development became possible due to the availability of texts in electronic form and later from the World Wide Web and increased computer processing power, which is required to derive the information with necessary precision from large data sets. In addition, in his Chinese room experiment, Searle (1980) showed that intelligent behaviour (e.g. translation) can be simulated by machines without proper understanding of the meaning. Thus, intelligent behaviour is possible by manipulating symbols without insights into the underlying causal relations and implications. Together, this has led to a path-breaking rise of statistical MT, and more recently to machine learning and Big Data technologies with unprecedented translation quality, which have been developed without reference to human–machine analogy or psycholinguistic insights, hence decoupling intelligent behaviour from deeper understanding. However, very recently—within the discussion of human-machine parity in translation (Läubli et al., 2020)—we see a resurgence of interest in psycholinguistics in the context of interactive translation, translation reception and evaluation.

Crocker (2013) argues that the divergence between computational linguistics and computational psycholinguistics (and thus computational cognitive modelling) already started in the 1970s. He sees the reasons for this separation in divergent goals. While psycholinguistics is concerned with questions of linguistic competence (such as the processes and representations that allow humans to arrive at incremental sentence understanding), computational linguistics, and in particular its engineering branch, NLP, became more concerned with linguistic performance. This includes how linguistic knowledge can be used—among other things—for spell checking, keyword and information extraction, text summarization, document classification, question answering, MT, etc. Questions that would address how *humans* process and understand language in real time became increasingly uninteresting for NLP applications, so that MT and computational cognitive modelling are now only marginally overlapping fields of research. Current NLP applications are limited tasks that need to be performed accurately and robustly, often without real "understanding" (e.g. spam filters, information retrieval, document clustering, summarization, etc.). In its present form, modern NLP has shifted emphasis onto coverage and efficiency with little concern about cognitive plausibility.

However, with the increasing quality and usability of MT output, Cognitive Translation Studies has recently become concerned with human–computer interaction as a form of extended cognition (Muñoz Martín, 2010). As O'Brien (2017, p. 313) points out, "[w]ith the current state of the art, machine translation (MT) cannot show us how translators work at a cognitive level. We can, however, investigate how translators and end users [...] interact with MT through cognitive process studies". Despite increased translation quality, MT output is in many cases not good enough for at least some end users. Recent years have witnessed increased research into how cognitive effort in translation can be assessed and measured, and how humans and machines can work in tandem as a hybrid intelligence system, driven by the aim of minimizing human translation effort while maintaining the required translation quality. These studies reveal that the success of human–computer interaction—for instance interactive MT (e.g. Daems & Macken, 2019; Knowles et al., 2019) and MT post-editing—depends on various parameters, such as the quality of the MT output, the type and domain of the texts, the expertise of the translators, the availability of suitable interfaces and visualizations of translation options, etc. These studies are indicative of a renewed interest in the conditions of the human translating mind, which often have the goal of designing "machines to augment human expertise (*intelligence augmentation*)" so as to "leverage the best of both machine intelligence and human expertise rather than exclude one or the other out of hubris or habit"[1] (Dillinger, 2018, emphasis in original).

Section 28.2 traces some historical aspects in the relation between artificial intelligence and cognition. It outlines the PSSH, which claims that humans and computers are symbol-manipulating machines. The PSSH was challenged in the 1980s by a connectionist alternative, which today is experiencing a powerful, ubiquitous revival. Section 28.3 traces the development of interactive machine translation (IMT) in relation to embedded, extended and situated forms of cognition (Clarke, 2008; Muñoz Martín, 2010, 2017), where the translator engages in a dialogue with the MT system during translation production. We point to the particularly innovative and disruptive potential of IMT—as it models a tight integration of humans and machines—within different incarnations, presupposing different conceptions of AI, the functioning of the human mind, and technological frameworks. Section 28.4 summarizes the discussion with a view on dual-process/dual-system theory as a possible framework for integrating the various approaches, and the potential of Cognitive Translation Studies to advance AI approaches and MT.

28.2 Core topics at the interface between artificial intelligence and cognition

In 1936, Alan Turing developed a (theoretical) machine to formalize the notion of computation. Turing showed how this machine was capable of manipulating symbols and could thus solve problems that would up till then have required human intelligence. Turing's formalization of the notion of computation implied that machines could mimic the human mind—thus providing the technology and conceptual backbone of what would later be called "artificial intelligence".

Starting from the late 19th century, behaviourism was the dominant method for understanding human behaviour. Behaviourism postulates that only measures of observable behaviours and events, and in later incarnations also feelings, states of mind and introspection, can be objects of scientific investigation. From the 1950s, behaviourism was heavily criticized on the basis that it would not examine the role of mental processes, and by the 1970s there was a consensus that the processes and mechanisms of the mind could not be understood purely on the basis of behavioural experiments. It became obvious that behaviourism could not explain the intricate details of the human mind and its manifestation in behavioural flexibility. As there are very many parameters that influence the results in any experiment involving the human mind, observations could not be explained with sufficient precision.

28.2.1 The physical symbol systems hypothesis

Computational cognitive sciences emerged under the assumption that computational models are necessary for understanding a system as complex and as diverse as the human mind. According to Sun (2008), "[a] cognitive architecture helps to narrow down possibilities, provides scaffolding structures, and embodies fundamental theoretical postulates." (Sun, 2008, p. 9). Newell and Simon (1976), as well as Marr (1982), suggest cognitive architectures in which cognitive processes can be formulated on three more or less independent levels of description: (1) at a computational or knowledge level, to determine which are the goals and purposes of the cognitive processes; (2) at the symbol representation level, to identify the tasks, symbol structures and manipulation procedures; and (3) at the physical, implementation level, by any available physical means. The idea was that these levels of description are to a large degree independent and compositional. Cognition (thinking, perception, decision making, etc.) consists of manipulating the syntax of symbols, and their analysis would enable the prediction of behaviour. In this view, the mind is a syntactically driven machine in which meaning derives from the manipulation of symbols. A number of cognitive architectures, such as Soar (Laird, 2012), ACT-R (Anderson & Bower, 1973) and CLARION (Sun, 2017), have been developed, which provide a framework for implementing and testing the PSSH, aiming to provide conceptual clarity and precision of the predictions.

28.2.2 Strong vs. weak AI

The PSSH has generated controversial discussions. It stipulates the *strong AI hypothesis*, by which machines can be built that represent the human mind and the activities humans perform. The *weak AI hypothesis*, in contrast, states that machines can be built that simulate (but do not represent) human behaviour. Searle (1980, p. 417) states that "[i]n strong AI, because the programmed computer has cognitive states, the programs are not mere tools that enable us to test psychological explanations; rather, the programs themselves are the explanations". In this view, the brain is a computer, and the mind is the result of the program that the brain runs. Searle further explains that "instantiating a formal program with the right input and output is a sufficient condition of, indeed is constitutive of, intentionality [...] the essence of the mental is the operation of a physical symbol system" (Searle, 1980, p. 421) Strong AI protagonists thus maintain that Marr's (1982) level (1) and level (2) descriptions are sufficient to represent the human mind.

In his *Gedankenexperiment*, the Chinese Room, Searle (1980) attempts to reject the strong AI hypothesis through a paradox. He argues that perfect answers (and thus e.g. perfect translations) are possible without an understanding of the questions (i.e. a source text), just by following a sufficiently precise set of formal rules. Searle uses the Chinese Room experiment to show that the human mind is more than a syntactically driven machine which runs in the brain, and that no Turing machine (i.e. no usual computer) can ever achieve human-like understanding. For Searle (1980, p. 446), "the whole idea of strong AI is that we don't need to know how the brain works to know how the mind works"; and for weak AI it is not important to know how the brain works. While there was—and still is—a debate about under what conditions understanding can emerge in a computer, there is little doubt that perfect automatic translation is possible without proper understanding. For instance, an MT system may produce perfect translations for some sentences or texts, but we do not assume that the system "understands" what is being translated. Accordingly, the development of MT systems—and of computer linguistics or AI applications in general—has not paid much attention to psycholinguistic research in recent years. As Crocker (2013, p. 484) points out, computational linguistics has "a greater interest in optimizing the computational properties of parsing algorithms, such as their time and space complexity. Computational psycholinguistics, in contrast, places particular emphasis on the incremental processing behaviour of the parser", which is crucial in the understanding of how humans process language.

28.2.3 Connectionist modelling

Artificial neural networks (NNs) are the most common type of connectionist networks, which provide an alternative computational framework for computational cognitive modelling. There is a wide variety of NN architectures, which are derived from an abstraction of how the brain works. The earliest implementation of a simple artificial neuron goes back to McCulloch & Pitts (1943) and the perceptron (Rosenblatt, 1958). However, it was not until the mid-1980s, due to increased computer processing power, that the usage of multi-layer perceptrons (which include one or more hidden layers) and the discovery of back-propagation algorithms (Rumelhart et al., 1986) resulted in its breakthrough. NNs are interconnected simple processing units that operate in parallel and that spread activation through layers of connected nodes. A learning algorithm changes the behaviour of NNs over time (i.e. back-propagation adjusts activation weights between the individual nodes), but the numeric properties of the connections and node activation make an interpretation of the inner working of NNs notoriously difficult.

There was strong opposition to connectionism from proponents of the classical computational theory of mind, due to the fact that it is usually impossible to know what activity patterns inside NNs actually mean. Both connectionists and defenders of the PSSH maintain that the brain implements a computer. However, the latter state in addition that the mind operates with symbols, which have representational and syntactic properties, and that mental processes (e.g. thinking) manipulate sequences of those representations defined by the combinatorial structure of the representations, while the former argue that symbols cannot (easily) be isolated in distributed NNs—and usually not without loss of information—which undermines the basic assumptions of the PSSH.

In defence of the language of thought (LOT), Fodor and Pylyshyn (1988) launched a debate stating that connectionists cannot provide a scientific explanation of cognition, as connectionism lacks a representational system and a combinatorial syntax and semantics. Connectionist models tend to focus on a single cognitive phenomenon or process, while cognitive architectures (such as ACT-R; e.g. Anderson & Bower, 1973) cover a large range of cognitive activities. In connectionist systems, nodes are labelled, but "the operation of the machine is unaffected by the syntactic and semantic relations that hold among the expressions that are used as labels" (Fodor & Pylyshyn, 1988, p. 296). In addition, some connectionists claim that NNs can learn representations that have a combinatorial syntax and semantics and produce structure-sensitive transformations on those representations (e.g. Chorowski & Zurada, 2011). These structures, they claim, can be extracted in the form of rules, in the following manner: "Given a trained neural network and the examples used to train it, produce a concise and accurate symbolic description of the network" (Roy, 2002, p. 269). However, it is unlikely that any human could understand the "symbolic description" of a moderately intelligent system (in a reasonable amount of time), which undermines the notion of symbol altogether.

Most connectionists follow Chalmers (1993), for whom the "deepest philosophical commitment of the connectionist movement is [...] the rejection of the atomic symbol as the bearer of meaning", as symbolic descriptions "do not carry enough information with them to be useful in modeling human cognition". Other attempts develop two-factor theories, which "explain certain 'low-level' aspects of cognition without resort to representations, but [...] representational hypotheses will still be needed to account for the intentionality-based features of cognition and [...] higher level processes" (Von Eckardt, 2012, p. 45; see also Kahneman, 2011, later).

28.2.4 Neo-connectionism

While connectionist architectures in the 1980s were developed and discussed in the light of how they simulate human cognition, this aspect is largely lacking in the discussion of newer NN architectures. The recent success of Neural Machine Translation (NMT) systems[2] started around 2014 and, to a large extent, it is due to (1) increased computer processing power (memory and speed) and (2) a technique that encodes words in vector spaces (word embeddings). Vector models are continuous space representations, which have been used since the 1990s. These vectors capture the semantic similarities between words based on the distribution of collocational properties in large corpora words that are used in the same contexts tend to carry similar meanings. Earlier vector models include Latent Dirichlet Allocation (LDA) and Latent Semantic Analysis (LSA). More recently, NNs have been used to embed words as vectors into a lower dimensional space. For instance, an NMT system with a vocabulary of 50,000 words would initially encode each word as a "one-hot" vector of 50,000 bits, in which only one bit is set. This one-hot vector is then mapped into a continuous vector space of 100 to 500 or more nodes, which represent the contextualized meaning of the encoded word. The main benefit of vector space models and word embeddings is that they can be learned as an unsupervised task, which does not require pricey annotation of the data. Word embeddings make it possible to measure the semantic similarity between words, e.g. using the cosine similarity. As Koehn (2017, p. 36) points out, they also allow for semantic inference such as:

queen = king + (woman – man)
queens = queen + (kings – king)

The "meaning" of the nodes that make up word embeddings is, however, very different from the micro-features in a sub-symbolic system, as conceived by Fodor and Pylyshyn (1988). Fodor and Pylyshyn assume that concepts can be automatically decomposed into micro-features from the samples in the training set, so that, e.g., BACHELOR corresponds to a vector in a space of features that includes {+ADULT, +HUMAN, +MALE, −MARRIED} or CUP consists of a set of nodes which contains {+has-a-handle}, etc. Accordingly, they argue, sub-symbolic connectionist states do have semantics, though it is not the semantics of representations at the "conceptual level". Hence, symbol-based cognition emerges from the sub-symbols, so that connectionist systems are merely an implementation of the classical symbol systems hypothesis.

However, despite observations that in current vector space encodings "countries are clustered close together and syntactically similar words occupy similar locations in the vector space",[3] it is more than questionable whether it will be possible to assign a compositional meaning to each of the 500 or so nodes (i.e. sub-symbolic micro-features) of a word embedding. Current NMT systems are first of all technological solutions, and the architecture of the networks is inspired by pragmatic considerations, not through computational cognitive insights.

NMT systems consist of an input layer (the word embedding), an output layer and several hidden layers.[4] The nodes in the output layer can be decoded into symbols (e.g. words), for instance with support of the softmax transformation, by which representations of words emerge on the surface of the output. It is thus possible to compose and decompose connectionist input and output representations into words (i.e. symbols at the conceptual level), but the mapping mechanisms that are responsible for reordering and re-phrasing the input vectors into an output representation—which are learned during the training phase of the NMT system, and which are encoded in the hidden layers—are implicit (non-concatenative) realizations, which may be difficult or impossible to tokenize, extract and represent in some meaningful compositional way.

28.2.5 Connectionism and cognitive theories

Clarke (2008) has suggested a similar view to understand human language processing: words (and languages) are "inputs [to the mind] (whether externally or internally generated) that drive, sculpt, and discipline the internal representational scheme" (Clarke, 2008, p. 54) They are "basic tools to discipline and stabilize dynamic processes of reason and recall", which give us and others the "ability to reliably follow trajectories in representational space" and create the "stable attendable structure to which subsequent thinking can attach" (Clarke, 2008, pp. 53–59) However, the encoding and processing of words and language in the human mind does not amount to an LOT, as suggested by Fodor and Pylyshyn (1988).

In order to create an attendable and attachable structure in NMT systems, and to make them flexible and better adaptable to new domains or text types, methods need to be developed that in some way interfere in the translation process to adjust systematic shortcomings of the NMT systems. For instance, the infusion of specific translations (e.g. terminology) or fixed, idiomatic translations into an NMT system during the translation phase or the controlled production of preferred target structures would probably require some sort of online access into the internal structure of an NMT system. Research efforts are being made to infuse higher-order monitoring and intervention processes as extensions in the NN architecture, which allow targeted manipulations of the NMT processes that may lead to reliable output in a more predictive manner.

Dual-process and dual-system theories (e.g. Kahneman, 2011) provide a conceptual framework, which accounts for similar processes in the human cognitive system. According to Kahneman (2011), while one system (System 1) is automatic, fast and realizes non-conscious processes, the other system (System 2) is controlled, slow and processes conscious forms of thinking. These two processes are implemented in two different but connected systems. The integration of higher-order monitoring processes and lower-order automatized routines has also been the architecture in human-computer interaction in translation. In the next section, we will discuss how MT systems have been embedded in translation workflows and used as an instrument for embedded and extended cognition.

28.3 Recent developments concerning machine translation and cognition

Numerous models have been elaborated to describe cognitive processes in translation, and several of these models provide evidence of a dual process/dual system of reasoning in translation (e.g. Hönig, 1991; Ivir, 1981; Schaeffer & Carl, 2013; Tirkkonen-Condit, 2005). The dual-process/dual-system theory assumes that there are two types of systems (system 1 and system 2), which result in two different, overlapping processes which can be disentangled to some extent. "The implicit system [system 1] is non-conscious or pre-conscious, rapid, parallel, low effort, high capacity and shaped by biologically constrained, domain-specific learning. The explicit system [system 2], by contrast, is conscious, slow, serial, high effort, limited capacity and responsive to verbal instruction" (Frankish, 2010, p. 920). For Frankish (2010, p. 919), the implicit system 1 is "associated with parallel, connectionist processing; the [explicit system 2] [...] with serial, rule-governed processing".

The dual-system approach has also been extended and criticized. Instead of a dual system, Muñoz Martín (2017, p. 564), for instance, presupposes a mesh of interconnected processes, which are "conscious and unconscious, logical and analogical, rational and emotional to different degrees, often at once". Some incarnations of the dual-system theory and its one-system alternative could be understood as existing on a continuum that could plausibly be characterized as either one system or a dual system (Mugg, 2015). However, none of these theories is in principle incompatible with the assumption that the human mind is *embedded* within a social-cultural milieu and *extended* into the body and the environment. The mind is *embedded*, as it is "designed to function in tandem with the environment" (Rowlands, 2010), and it is also *extended*, as it offloads tasks onto the external environment to free up limited cognitive resources (e.g. using a notepad as external memory storage).

MT and IMT document workflows and their technological solutions are part of the socio-cultural milieu in which translators operate. Since the 1950s, attempts have been made to integrate the human translator and automated assistance (e.g. an MT system) into one hybrid cognitive system, in which a translator is either *embedded*—to give advice whenever a decision cannot be taken automatically by the MT system with high enough precision—or *extended* with automated help to free up her/his limited cognitive resources. The aim of any hybrid integration is to eliminate the weakness of each component by the (relative) strength of the other component so as to increase the performance of the integrated human–machine interactive system. This is possible if the weaknesses and strengths are unevenly distributed in the two (or more) sub-components (e.g. in system 1 and system 2). According to the conceptualization of the human mind as a device that manipulates symbols and the connectionist approach, these strengths have been differently defined. This becomes particularly apparent in the changing design of IMT over the past decades, which will be traced in this section.

Bruderer (1978) suggests four principles by which MT systems can cooperate with humans:

1. fully automatic MT, which is possible for easy texts, restricted domains and/or limited quality expectations
2. pre-edition, mainly to reduce ambiguities and to simplify the source text for better MT results
3. post-edition, the process to amend grammar and style of the MT output
4. IMT, where the translator intervenes in dialogue with the MT system

In contrast to the other forms of human–MT cooperation, IMT presumes MT operations preceding *and* following human intervention, thus *embedding* or *extending* human translation activities. While the concept of pre- and post-editing has remained—to a large extent—similar over the past decades, irrespectively of whether the underlying MT technology has been rule-based, statistical or neural concepts of IMT have dramatically changed with evolving translation technology in the past 30 years and accordingly have taken different views on embedded/extended cognition.

28.3.1 Interactive translation aids

For Hutchins and Somers (1992), the antecedents of MT are, in fact, mechanical dictionaries, which would be used interactively to support translators in their work and reduce their mechanical and cognitive workload. They report that in 1933 two independent teams, led by George Artsrouni and Petr Smirnov-Troyanskii, worked on two different interactive translation devices that can be seen as predecessors of MT. Artsrouni proposed to store a dictionary on a paper-tape-based general-purpose storage device that would be queried automatically to show word translations that could be used by a translator. Smirnov-Troyanskii, in turn, suggested a multi-step MT system in which the machine and a human would collaborate interactively. A human operator would first transform each source word into its basic form using a specific editor. The machine would then translate each word into a target-language basic word form, and finally a reviser would generate the final sentence form in the target language.

Such simple hybrid systems can be represented as simple box-and-arrow models, which make explicit the cognitive structure of the integrated system. The different processing steps can be represented by boxes and the relationships between them by arrows. Such diagrams—similar to a computer flowchart—facilitate a detailed understanding of an extended/embedded human translator and her/his machine interaction and make it possible to design the workflow in a way that compensates their mutual weaknesses and accumulates their strengths.

28.3.2 Interactive embedded translation

A next step in the development of IMT systems was related to the introduction of rule-based MT (RBMT) systems in the 1970s. RBMT systems typically first analyse the syntactic structure of the source text (ST). The ST structure and the lexical items are then transferred and mapped onto the target language, and a target-language surface form is generated. Due to the large number of possible analyses for an ST string, it was considered the most difficult task in early RBMT systems to generate (or select) a correct ST structure. In the case a decision was too difficult for the machine to take (such as prepositional phrase (PP) attachment), interactive RBMT systems would allow the MT system to "consult a human when it does not know what to do" (Bisbey & Kay, 1972, p. 9). Bisbey and Kay developed the MIND system, which embeds a "monolingual consultant to resolve ambiguities in the [machine] translation process" (Bisbey & Kay, 1972, p. 9). The human component was considered an additional *embedded* resource for the MT system with easy access to additional (world-)knowledge resources that are required to resolve linguistic ambiguities that the computer could not (yet) solve. In line with the predominant view at the time that cognitive processes in humans and machines were symbolic by nature, the human–machine communication was assumed to take place on a symbolic metalanguage level. It was "hoped that interaction could be restricted to analysis [...] [but] it was found that some interaction was also required in transfer" (Melby et al., 1980 p. 424).

According to Weaver (1988), the aim of interaction in RBMT systems was then considered to be helping the computer produce better draft translations, which would reduce the amount of successive post-editing. Weaver mentions several forms of interaction, the most common being lexical interaction by which an ST parser is provided with additional (semantic) constraints that help the system find better translations. A disadvantage of early RBMT interaction was that the "on-line interaction requires specially trained operators" (Melby et al., 1980). To overcome this shortcoming, Boitet et al. (1995) suggested a dialogue-based RBMT system whereby untrained editors would choose a disambiguating item from a pre-processed list in order to disambiguate the source linguistic analysis of the text to be translated.

28.3.3 Interactive extended translation

An alternative perspective on human-computer interaction evolved as a consequence of these difficulties. Instead of embedding the translator into a computational system to resolve remaining ambiguities that RBMT systems cannot solve alone, Martin Kay (1980) called for a re-examination of the role of humans and machines (including his own MIND system) with his Translator Amanuensis and envisioned a gradual extension of the translator, i.e. a human–machine partnership to assist translators. With his Translator Amanuensis, Kay suggests an incremental, pragmatic approach to human–machine interaction to produce not only a translation of an ST but also a device that constantly accumulates experiences that have been agreed upon between the human translator and the machine. By doing so, it *extends* the translator's memory, as it memorizes the translator's previous solutions, and it reduces motor activity (typing effort), thus enhancing its translation performance. Kay's vision of interactive extended translation can be considered a precursor of computer-assisted translation (CAT tools) and translation memory systems (TMS) as we know them today, being incrementally built through translation activities into which translators "offload" their translation solutions, which become the basis of future translations.

28.3.4 Interactive statistical translation

A further step in extending translators with AI systems arrived with data-driven technology. By the 1990s, statistical MT (SMT) systems were developed, which would focus on target-text generation, turning the focus away from SL analysis. In their seminal paper, Brown et al. (1990, p. 79) proposed SMT in which "every sentence in one language is a possible translation of any sentence in the other language", and they subsequently elaborated a mathematical framework to assign the (hypothetical) probabilities with which a translator would produce the various different target sentences for a given SL sentence (Brown et al., 1993). The emphasis in SMT shifted from selecting the best (or appropriate) ST analysis to the question of how to filter the best out of a large number of possible translations.

SMT systems have been further developed into interactive SMT systems, e.g. TransType (Foster et al., 2002) and TransType2 (Macklovitch, 2006), which offer an interaction with the translator. TransType proposes translation completion at the translator's cursor position, thus providing situated assistance at the time when a translator needs it. Not only the translator's memory, but also the human translation process, is extended into the translation environment, which provides the translator with potentially new translation solutions, but "the translator retains full control over their output" (Macklovitch & Valderrábanos, 2001) rather than being an appendix to an MT system. Foster et al. (1997) show how TransType can predict up to 70% of the characters in the translator's intended target text. Experiments with TransType (Foster et al., 2002) show a 10% gain in translation time. Later experiments with TransType2 (Macklovitch, 2006) showed a productivity increase from 15% to 55%, depending on the text to be translated, with professional translators in a translation agency.

28.3.5 Interactive cloud-based translation

A continuation of TransType2 was carried out in the context of the CASMACAT[5] project (Alabau et al., 2013). The objective of CASMACAT was to devise computer interfaces based on the outcome of cognitive studies and translators' behaviour. CASMACAT implemented a browser-based system, featuring a number of innovative techniques, such as online learning and various visualization options.

CASMACAT makes use of cloud-based translation servers, which is by now a common practice in the translation industry. Cloud-based solutions process computationally heavy translation tasks with remote, powerful hardware and communicate the results to a client, usually in real time. Sophisticated software solutions allow real-time updating of the translation models, and enhanced translation solutions are immediately available for successive interactive translations, e.g. of the same document. While CASMACAT is a non-commercial prototype, Lilt[6] is the first commercial solution that uses browser-based interface and online learning (Green et al., 2014).

28.3.6 Interactive crowd-based translation

Cloud-based techniques enable novel business models whereby source texts are fragmented into many small translation jobs, which can be distributed over a large crowd of translators, while the shared resources on the MT server help ensure consistency of the translated text segments within a distributed translation project. These techniques may be complemented by *active learning* strategies, in which the MT system re-orders the source-language segments to be translated and interactively edited in order to maximize its expected learning effect. Thus, the text is presented in an order that suits the machine, so that it can learn most quickly or efficiently from the human corrections. Interactive crowd-based MT supports a further move from "content being rolled out in a static, sequential manner" to translated content being "integrated into a dynamic system of ubiquitous delivery" (Cronin, 2013, p. 498).

Companies such as Unbabel[7] are experimenting with interactive translation editing tools for hand-held devices in which segments of limited length are post-edited out of context by the crowd and re-assembled into coherent text in the server back-end. Similarly, MotaWord[8] produces translations on a collaborative cloud platform "coordinated efficiently through a smart back end" in which, according to their website, over 5,600 translators participate. Such distributed translation platforms allow a translator to request new segments on the fly so that they can assign and change their workload at will, while simultaneous co-sharing of translation resources and sophisticated quality assessment routines make it possible to maintain the minimum requested translation quality.

28.3.7 Interactive machine translation teaching

An extension of this trend toward decontextualization of the translation job is put forward by Dillinger (2018). For Dillinger,[9] there are (at least) 50 shades of collaboration in which IMT systems are designed as *augmented intelligence*, where translators become "machine teachers". Dillinger promotes a "New Paradigm for Building Machine Learning Systems" (Simard et al., 2017), in which the role of a domain expert, becomes that of a "teacher [...] who transfers concept knowledge to a learning machine". According to this paradigm, a "concept is a mapping from an example to a label value", whereby the teacher selects examples that exemplify useful aspects of the concept under consideration. Concepts can be decomposed and re-factored, subconcepts can be created, and each concept might provide its own mechanisms for example selection, labels, schemas (i.e. hierarchies of concept relations) and features, etc. Such concepts with their examples—Simard claims—are important in AI systems to train a learning system for analysing pictures, speech, text, or the relationship between people, things and texts; they are necessary for named entity recognition, document classification, sentiment or intent detection, etc., etc.

The *augmented intelligence* approach assumes that the "teacher has access to feature concepts not available to the training set"; i.e. experts have access to knowledge that the learning agent (the machine) has not, and teachers can "use their domain knowledge to pick the right examples and counterexamples for a concept and explain why they differ".

For instance, according to Dion Wiggins,[10] the word "run" is the most ambiguous English word, with 197 meanings, including "I ran for office", "I went for a run", "I ate bad food and got the runs", "I scored a home run", "The dry run went well", etc., etc. Such sentences are likely to produce translation errors, which could be addressed in a systematic manner. In an MT teacher scenario, the MT system could first produce all translations that it is uncertain of (for instance, translations that include the word "run") and ask a teacher—through an interactive information exchange with a learning system—to provide a disambiguating label, probably best in the form of a correct translation. The learning agent would then store and generalize these examples and re-use them in similar contexts.

Note the similarity of this scenario to the role of the translator in the MIND and Dialogue systems discussed in Section 28.3.3. However, in contrast to embedded IMT in the 1970s, the actual interactive interface might just look like a well-known translation workbench. In line with such a vision, Alonso and Vieira (2017) project a vision of the Translator's Amanuensis into the year 2020 (TA2020), where they anticipate a multimodal (text, speech and picture) translation workstation, to be used by experts and by non-experts, that would allow ubiquitous translation production and translation retrieval, embedded in any kind of hand-held or other device.

28.4 Summary and outlook

From the late 1950s until about the 1980s, the predominant framework of AI was based on the Physical Symbol Systems Hypothesis (PSSH) which postulates that intelligent computer programs represent human cognition, information processing, memory, learning and decision processes. According to this hypothesis, the human mind was thought to process sequences of symbols, and computational models were built not (only) to explain but instead to represent intelligent behaviour, such as translation. A number of techniques were suggested (see Section 28.3.2) to interactively embed a translator into an RBMT system to provide additional knowledge resources at run time.

By the mid-1980s, the PSSH was challenged by connectionist architectures of massively parallel cooperative and competitive neuron-like units. Inspired by the functioning of biological systems, sets of neurons would accumulate potential and emit spikes of activation, and connections between them could be learned and the activation adjusted based on a training corpus and a learning algorithm. The PSSH was increasingly undermined as artificial NN (and other machine learning algorithms) were able to simulate intelligent behaviour without manipulating symbols but rather reacting directly to the input signal.

In addition, it was shown that human understanding is more than just the ability to manipulate symbols. Instead of aiming at fully automatic, all-purpose MT solutions, the ALPAC report (1966) suggested that tools should be developed that would support translators in their work. Such computer-assisted translation (CAT) tools would support and extend the translators' mechanical and cognitive capacities; humans and machines would collaborate in the production of translations in novel workflows.

Clarke (2008, pp. 39–42) makes the distinction between "a user in command of a new tool" and a "reconfigured user", in which a user and the tool form a brand-new systemic whole. In the former case, the agent "would *always* use tools the way we typically *begin* to use them", in a way that requires conscious allocation of cognitive resources and effort. In the latter case, the tool character disappears and becomes a new (cognitive or bodily) extension of the human agent. The agent "learns a complex problem-solving routine that makes a variety of deep implicit commitments to the robust bioexternal availability of certain operations". He argues that our "plastic neural resources become recalibrated [...] so as to *automatically* take account of new [...] opportunities". This recalibration is possible because our minds are "forever testing and exploring the possibilities for incorporating new resources and structures deep into [our] embodied acting and problem-solving regimes". A translator will thus recalibrate, automatize and offload translation tasks and translation decisions in line with the opportunities offered by the Computer Assisted Translation environment.

Kahneman (2011) makes a distinction between more automatized and more conscious processes within the dual-process/dual-system theory. An implicit system [system 1] is preconscious, rapid, parallel, with high capacity and low effort, whereas the explicit system [system 2] is slow, serial, conscious and involves high effort. Tirkkonen-Condit (2005) adapts this theory in a monitor model of translation production, which consists of a "literal translation automaton" (i.e. system 1) that operates on a lexical as well as a syntactic level during the production of translations. This literal translation automaton is quick with a high capacity, it does not require working memory and it "generates literal or formally corresponding linguistic material as long as the material thus produced is semantically and syntactically acceptable" (Tirkkonen-Condit, 2005, p. 412). System 2 processes consist of a self-aware monitoring mechanism, which controls and intervenes in unwanted literal rendering.

As discussed in Section 28.3, different IMT frameworks address the two systems in a different way: The interactive RBMT paradigm (see Section 28.3.2), addresses the conscious system 2, in which an *embedded* translation assistant is asked to disambiguate the ST analysis (e.g. whether an item is animated, countable or liquid, how a PP should be attached in a syntactic analysis, etc.). While the human translation assistant adjusts to the environment (gets used to the graphical user interface and the kinds of question to be solved), there is no intention (and maybe no chance) for the translator to automatize the decision process and thus to reduce the cognitive load.

The interactive SMT paradigm, by contrast, together with its cloud- and crowd-based translation scenarios (Sections 28.3.4–28.3.6), starts out with the attempt to *extend* the translators' cognitive process. It provides the translator with bilingual terminologies, translation memories (Section 28.3.3), and automatically generated translation proposals and translation completions. This IMT framework aims at mapping otherwise cognitively effortful operations (i.e. system 2 operations, such as internal or external search for translation alternatives or terminology, etc.) onto automatized system 1 processes in which the translator is invited to accept or select translation suggestions. The interactive SMT (or more recently also NMT) paradigm is thus likely to reduce the translators' workload, allowing "genuine deep implicit commitments" (see Clarke, 2008), in which the translator and the MT system become a brand-new systemic whole.

Risku (2014, p. 339) defines cognition in translation as consisting of interconnected and self-organizing processes, which "includes all operations that work on internal and external representations with the aim of creating translations". Working with CAT and IMT is thus a form of extended cognition, because the mind offloads tasks into the environment. As Muñoz Martín (2017, p. 564) puts it, "the brain uses (parts of) the environment as a tool for thought—such as rereading the original on the screen instead of memorizing it [...] thereby rendering the distinction internal/external irrelevant". In an extended and situated translation environment—as discussed e.g. in Section 28.3—it is crucial to understand what pieces of information can (or should) be offloaded, how and which translation processes can be best supported by the IMT system, and how the offloaded information should be retrieved and/or textually integrated at the right point in time.

An underlying prerequisite for this to work is the complementarity of the human and machine component in the integrated IMT process. However, this complementarity is undermined as MT errors become increasingly "human-like" and therefore difficult to detect and eventually impossible or unnecessary for humans to improve. Läubli et al. (2020) discuss the possibility of human–machine parity in translation and show that only with advanced evaluation methods and/or professional raters a significant difference can be measured between human and machine translation. Yamada (2019, p. 87), finds that NMT errors are similar to those of human translators, making them more difficult to spot and correct. He concludes that "translation training is necessary for students to be able to shift their attention to the right problems (such as mistranslation) and be effective post-editors". Novel methods and visualization tools might be required that highlight potential IMT errors or highly translation-ambiguous words so as to notify the human agent about the doubts of the (N)MT. By post-editing and even more so through IMT, the translator has already become a "machine teacher" who instructs the machine as to which translation(s) should be preferred under what conditions and in which context. Cognitive research in translation studies may need to address this new role of translation and the translator and investigate the impact of AI on subliminal translation and translation reception processes. Muñoz Martín (2017, p. 563) notes that cognition "works in tandem with the environment and [the various sub-systems] cannot be analyzed in isolation". This suggests that cognitive translation studies ought to develop methods to determine and assess the factors of interaction and co-variation between these subsystems aiming at better understanding the nature and functions of the various internal and environmental components and to cope with the biases that new AI technologies (re)produce.

Notes

1 www.linkedin.com/pulse/50-shades-ai-mike-dillinger-phd/
2 Currently almost all major MT providers (Google, Bing, Systran, etc.) make use of some sort of NMT system.
3 http://ruder.io/word-embeddings-1/
4 Current Google translate architecture contains eight hidden layers (www.quora.com/How-many-nodes-and-layers- does-Googles-machine-translation-neural-network-have).
5 Cognitive Analysis and Statistical Methods for Advanced Computer Aided Translation (www.casmacat.eu/).
6 http://labs.lilt.com/
7 https://unbabel.com/
8 www.motaword.com
9 www.linkedin.com/pulse/50-shades-ai-mike-dillinger-phd/
10 https://omniscien.com/wp-content/uploads/2016/02-new/M1-Dion-Wiggins-Found-In-Translation-Language-Meets-Technology.pdf

Further reading

Lilt. (2019). Your next million customers don't speak English. https://lilt.com/
Up-to-date interactive machine translation system.
On techniques of IMT as well as multimodal interfaces and adaptive learning in IMT.
Peris, A., Domingo, M., & Casacuberta, F. (2017). Interactive neural machine translation. *Computer Speech & Language, 45*, 201–220. www.sciencedirect.com/science/article/pii/S0885230816301000

Schwieter, W., and Ferreira, A. (2017). *The handbook of translation and cognition*. Hoboken, NJ: Wiley Blackwell.
An earlier handbook of cognitive aspects of translation.

Overview of research topics and applications of interactive machine translation:
Toselli, A. H., Vidal, E., & Casacuberta, F. (2011) Interactive machine translation. In A. H. Toselli, E. Vidal, & F. Casacuberta (Eds.), *Multimodal interactive pattern recognition and applications* (pp. 47–59). London: Springer.

References

Alabau, V., Bonk, R., Buck, C., Carl, C., Casacuberta, F., García-Martínez, M., González-Rubio, J., Koehn, P., Leiva, L., Mesa-Lao, B., Ortiz-Martínez, D., Saint-Amand, H., Sanchis-Trilles, G., & Tsoukala, C. (2013). CASMACAT: An open source workbench for advanced computer aided translation *Prague Bulletin of Mathematical Linguistics, 100*, 101–112.

Alonso, E., & Nunes Vieira, L. (2017). The Translator's Amanuensis 2020. *The Journal of Specialised Translation, 28*, 345–361.

ALPAC. (1966). *Languages and machines: Computers in translation and linguistics. A report by the Automatic Language Processing Advisory Committee, Division of Behavioral Sciences, National Academy of Sciences, National Research Council.* Washington, DC: National Academy of Sciences, National Research Council. (Publication 1416.) 124pp.

Anderson, J. R., & Bower, G. H. (1973). *Human associative memory.* Washington, DC: Winston and Sons.

Bisbey, R. L., & Kay, M. (1972). *The MIND translation system: A study in man-machine collaboration.* Santa Monica, CA: RAND Corporation. www.rand.org/pubs/papers/P4786.html

Boitet, C., Blanchon, H., Planas, E., Blanc, E., Guilbaud, J. P., Guillaume, P., Lafourcade, M., & Sérasset, G. (1995). LIDIA-1.2, une maquette de TAO personnelle multicible, utilisant la messagerie usuelle, la désambiguisation interactive, et la rétrotraduction. Rapport Final, GETA-UJF, 30 pp.

Brown, P. F., Cocke, J., Della Pietra, S. A., Della Pietra, V. J., Jelinek, F., Lafferty, J. D., & Watson, T. J. (1990). A statistical approach to machine translation. *Computational Linguistics, 16*(2), 79–85.

Brown, P. F., Della Pietra, S. A., Della Pietra, V. J., & Mercer, R. L. (1993). The mathematics of statistical machine translation: Parameter estimation. *Computational Linguistics, 19*(2), 263–312.

Bruderer, H. (1978). Sprache—Technik—Kybernetik. In H. Bruderer, *Aufsätze zur Sprachwissenschaft, maschinelle Sprachverarbeitung, künstlichen Intelligenz und Computerkunst* (pp. 113–126). Müsslingen bei Bern: Verlag Linguistik.

Carl, M., & Planas, E. (2019). Advances in interactive translation technology. In E. Angelone, M. Ehrensberger-Dow, & G. Massey (Eds.), *The Bloomsbury companion to language industry studies.* London: Bloomsbury.

Carl, M., & Way, A. (2003). *Recent advances in example-based machine translation.* Dordrecht/Boston/London: Kluwer Academic Publishers.

Camburn, R. (2013). *The development of natural language processing.* California State University, Fresno, CA.

Chalmers, D. J. (1993). Connectionism and compositionality: Why Fodor and Pylyshyn were wrong. *Philosophical Psychology, 6*, 305–319.

Chorowski, J., & Zurada, J. M. (2011). Extracting rules from neural networks as decision diagrams. *IEEE Transactions on Neural Networks, 22*(12), 2435–2446. doi: 10.1109/TNN.2011.2106163

Clarke, A. (2008). *Supersizing the mind: Embodiment, action, and cognitive extension.* New York: Oxford University Press.

Crocker, M. W. (2013). Computational psycholinguistics. In A. Clark, C. Fox, & S. Lappin (Eds.), *The handbook of computational linguistics and natural language processing* (pp. 482–512). Hoboken, NJ: Wiley-Blackwell.

Cronin, M. (2013). Translation and globalization. In C. Millán, & F. Bartrina (Eds.), *The Routledge handbook of translation studies* (pp. 491–502). London, UK: Routledge.

Daems, J., & Macken, L. (2019). Interactive adaptive SMT versus interactive adaptive NMT: A user experience evaluation. *Machine Translation, 33*, 117. https://doi.org/10.1007/s10590-019-09230-z

Dillinger, M. (2018). www.linkedin.com/pulse/50-shades-ai-mike-dillinger-phd/

Fodor, J. A., & Pylyshyn, Z. W. (1988). Connectionism and cognitive architecture: A critical analysis. *Cognition, 28*(1–2), 3–71.

Foster, G., Isabelle, P., & Plamondon, P. (1997). Target text mediated interactive machine translation. *Machine Translation, 12*(1), 175–194.

Foster, G., Langlais, P., & Lapalme, G. (2002). User-friendly text prediction for translators. *EMNLP '02 Conference on Empirical Methods in Natural Language Processing*, 148–155.

Frankish, K. (2010). Dual-process and dual-system theories of reasoning. *Compass, 5*(10), 914–926.

Green, S., Chuang, J., Heer, J., & Manning, C. (2014). Predictive translation memory: a mixed-initiative system for human language translation. In *Proceedings of the 27th annual ACM symposium on user interface software and technology October 2014* (pp. 177–187). https://doi.org/10.1145/2642918.2647408

Hönig, H. G. (1991). Holmes' "Mapping Theory" and the landscape of mental translation processes. In K. van Leuven-Zwart, & T. Naajkens (Eds.), *Translation studies: The state of the art. Proceedings from the first James S. Holmes symposium on translation studies* (pp. 77–89). Amsterdam: Rodopi.

Hutchins, J., & Somers, H. L. (1992). *An introduction to machine translation*. London: Academic Press.

Ivir, V. (1981). Formal correspondence vs. translation equivalence revisited. *Poetics Today, 2*(4), 51–59. doi: 10.2307/1772485

Kahneman, D. (2011). *Thinking, fast and slow*. London: Allen Lane.

Kay, M. (1997) The proper place of men and machines in language translation. *Machine Translation, 12*(1–2), 3–23. www.mt-archive.info/Kay-1980.pdf

Knowles, R., Sanchez-Torron, M., & Koehn, P. (2019). A user study of neural interactive translation prediction. *Machine Translation, 33*, 135–154. https://doi.org/10.1007/s10590-019-09235-8

Koehn, P. (2017). Neural machine translation. arXiv.org. https://arxiv.org/abs/1709.07809

Laird, J. E. (2012). *The Soar cognitive architecture*. Boston: MIT Press.

Läubli, S., Castilho, S., Neubig, G.. Sennrich, R., Shen, Q., & Toral, A. (2020). A Set of Recommendations for Assessing Human–Machine Parity in Language Translation. *IJCAI, 67*, pp. 653–672.

Lilt. (2019). Your next million customers don't speak English. https://lilt.com/

McCarthy, J., Minsky, M., Rochester, N., & Shannon, C. E. (1955). A proposal for the Dartmouth summer research project on artificial intelligence. http://raysolomonoff.com/dartmouth/boxa/dart564props.pdf

McCullough, W. S., & Pitts, W. (1943). A logical calculus of the ideas immanent in nervous activity. *The Bulletin of Mathematical Biophysics, 5*(4), 115–133.

Macklovitch, E. (2006) TransType2: The last word. *Proceedings of LREC 2006, Genoa, Italy* (pp. 167–172). http://rali.iro.umontreal.ca/rali/?q=en/node/955

Macklovitch, E., & Valderrábanos, A. S. (2001). Rethinking interaction: The solution for high-quality MT? *MT Summit, Santiago*. http://mt-archive.info/MTS-2001-TOC.htm

Marr, D. (1982). *Vision*. San Francisco: W. H. Freeman and Co.

Melby, A. K., Smith, M. R., & Peterson, J. (1980). ITS: Interactive translation system. In *COLING '80 Proceedings of the 8th conference on Computational linguistics*, 424–429. http://dl.acm.org/citation.cfm?id=990251

Mugg, J. (2015). Two minded creatures and dual-process theory. *Journal of Cognition and Neuroethics, 3*(3), 87–112.

Muñoz Martín, R. (2010). On paradigms and cognitive translatology. In G. M. Shreve, & E. Angelone (Eds.), *The handbook of translation and cognition* (pp. 169–187). Hoboken, NJ: John Wiley & Sons. 10.1075/ata.xv.10mun

Muñoz Martín, R. (2017). Looking toward the future of cognitive translation studies. In J. W. Schwieter, & A. Ferreira (Eds.), *The handbook of translation and cognition* (pp. 555–572). Hoboken, NJ: John Wiley & Sons.

Nagao, M. (1984). A framework of mechanical translation between Japanese and English by Analogy Principle. In A. Elithorn, & R. Banerji (Eds.), *Artificial and human intelligence*. (Ch. 11, n.p.). Elsevier Science Publishers. B.V. www.mt-archive.info/Nagao-1984.pdf

Newell, A., & Simon, H. A. (1963). GPS: A program that simulates human thought. In E. A. Feigenbaum, & J. Feldman (Eds.), *Computers and thought* (pp. 279–293). New York: McGraw-Hill.

Newell, A., & Simon, H. A. (1972). *Human problem solving*. Englewood Cliffs, NJ: Prentice-Hall.

Newell, A., & Simon, H. A. (1976). Computer science as empirical inquiry: Symbols and search. *Communications of the ACM, 19*(3), 113–126, doi:10.1145/360018.360022

O'Brien, S. (2017). Machine translation and cognition. In J. W. Schwieter, & A. Ferreira (Eds.), *The handbook of translation and cognition*. (pp. 311–331). Hoboken, NJ: Wiley Blackwell.

Peris, A., Domingo, M., & Casacuberta, F. (2017). Interactive neural machine translation. *Computer Speech & Language, 45*, 201–220. www.sciencedirect.com/science/article/pii/S0885230816301000

Risku, H. (2014). Translation process research as interaction research: From mental to socio-cognitive processes. *MonTI 1, Special Issue—Minding Translation*, 331–353.

Rosenblatt, F. (1958). The Perceptron: A probabilistic model for information storage and organization in the brain. *Psychological Review, 65*(6), 386–408. doi:10.1037/h0042519

Rowlands, M. (2010). *The new science of the mind: From extended mind to embodied phenomenology*. Cambridge, MA: MIT Press. http://dx.doi.org/10.7551/mitpress/9780262014557.001.0001

Roy, A. (2002) On connectionism and rule extraction. In R. Tagliaferri, & M. Marinaro (Eds.), *Neural Nets WIRN Vietri-01*. (Perspectives in Neural Computing series) (pp. 269–287). London: Springer.

Rumelhart, D. E., Hinton, G. E., & Williams, R. J. (1986). Learning representations by back-propagating errors. *Nature, 323*, 533–536.

Schaeffer, M., & Carl, M. (2013). Shared representations and the translation process: A recursive model. *Translation and Interpreting Studies, 8*(2), 169–190.

Schwieter, W., and Ferreira, A. (2017). *The handbook of translation and cognition*. Hoboken, NJ: Wiley Blackwell.
Searle, J. R. (1980), Minds, brains, and programs. *Behavioral and Brain Sciences, 3*, 417–424.
Simard, P.Y., Amershi, S., Chickering, D. M., Pelton, A. E., Ghorashi, S., Meek, C., Ramos, G., Suh, J., Verwey, J., Wang, M., & Wernsing, J. (2017). *A new paradigm for building machine learning systems*. https://arxiv.org/pdf/1707.06742.pdf
Stanford Encyclopedia of Philosophy. (2018). Retrieved 18 March 2020 from https://plato.stanford.edu/entries/computational-linguistics/
Sun, R. (2008). Introduction to computational cognitive modeling. In R. Sun (Ed.), *The Cambridge handbook of computational psychology* (pp. 3–19). Cambridge: Cambridge University Press.
Sun, R. (2017). The CLARION Cognitive architecture: Toward a comprehensive theory of the mind. In S. E. F. Chipman (Ed.), *The Oxford handbook of cognitive science*. (117–145). New York: Oxford University Press. doi: 10.1093/oxfordhb/9780199842193.013.11
Thouin, B. (1982). The Meteo system. In V. Lawson (Ed.), *Practical experience of machine translation* (pp. 39–44). © ASLIB. Amsterdam: North-Holland Publishing Company.
Tirkkonen-Condit, S. (2005). The monitor model revisited: Evidence from process research. *Meta: Translators' Journal, 50*(2), 405–414.
Toselli, A. H., Vidal, E., & Casacuberta, F. (2011) Interactive machine translation. In A. H. Toselli, E. Vidal, & F. Casacuberta (Eds.), *Multimodal interactive pattern recognition and applications* (pp. 47–59). London: Springer.
Von Eckardt, B. (2012). The representational theory of mind. In K. Frankish, & W. Ramsey (Eds.), *The Cambridge handbook of cognitive science*. Cambridge: Cambridge University Press.
Weaver, A. (1988). Two aspects of interactive machine translation. In M. Vasconcellos (Ed.), *Technology as translation strategy* (pp. 116–124). Amsterdam: John Benjamins.
Wiggins, D. https://omniscien.com/wp-content/uploads/2016/02-new/M1-Dion-Wiggins-Found-In-Translation-Language-Meets-Technology.pdf
Yamada, M. (2019). The impact of Google neural machine translation on post-editing by student translators. *Journal of Specialised Translation, 31*(1), 87–106.

29

Translation, multilingual text production and cognition viewed in terms of systemic functional linguistics[1]

Christian M. I. M. Matthiessen

29.1 Introduction

This handbook is concerned with different aspects of translation in relation to cognition, a fairly recent technology-enabled addition to scholarly perspectives on translation (see e.g. Rojo, 2015; Schwieter & Ferreira, 2017; Shreve & Angelone, 2010; and also García, 2019). Complementing this approach to translation as a *cognitive phenomenon*, we can approach translation as a *semiotic phenomenon*, foregrounding *meaning* rather than *knowledge*. As we argue in Halliday and Matthiessen (1999), drawing on *systemic functional linguistics* (SFL), the cognitive perspective and the semiotic one are not mutually exclusive alternatives; rather, they *complement* one another, and should ideally together provide a more holistic account of the phenomena under investigation.

This chapter focuses on SFL as a framework for engaging with translation, identifying contributions SFL can make, as a holistic theory of language, as a resource for making meaning in context. Thus, translation is construed as the recreation of meaning in context, investigated as choices in the different modes of meaning "embodied" in language (ideational (logical and experiential), interpersonal and textual), and the primary domain of translation is seen as texts theorized as units of meaning in context.

Translation is a linguistic phenomenon in the first instance, so that is how it is theorized in SFL; but at the same time, SFL locates it within an ordered typology of systems (see below). It can be viewed, observed and investigated alongside other linguistic phenomena that are in some sense "meta" to an original text, including revising and editing, abridging and summarizing, and even performing (e.g. a play)—as an enhanced version of reading aloud; but it differs from these in presupposing that language users engaging in translation have mastered a *multilingual meaning* potential (see Bateman et al., 1999; Matthiessen, 2018), and as we argue in Matthiessen et al. (2008), the study of translation can be included in *multilingual studies*, where it enters into dialogue with language typology and comparison, studies of second/foreign language development and also machine translation (MT), since the communities of researchers involved in MT and in Translation Studies have tended to remain quite separate, even though there is considerable potential for collaboration and cross-fertilization.[2]

I will address systemic functional contributions to the study of translation in this chapter, but first I will take a step back to locate the focus of translation—i.e. language—in an ordered typology of systems of increasing complexity in order to engage with "cognition" and also to discuss the different systemic orders within which translation operates:

In Section 29.2, I will explore the systemic location of translation in reference to the conception of language as a higher-order system in an ordered typology of systems;
In Section 29.3, I will focus on translation as a fourth-order, semiotic phenomenon, and suggest what SFL can contribute to the study of this phenomenon by offering an account of the "architecture" of language in context;
In Section 29.4, I will deal with the local dimensions of this architecture—dimensions that are manifested as fractals in the organization of all local sub-systems of language in context;
In Section 29.5, I will turn to the global dimensions of this architecture;
In Section 29.6, against the background of the two previous architectural chapters, I will highlight the notion of choice in the proposed conception of translation as the recreation of meanings in context;
In Section 29.7, I will provide a summary of the significance of the systemic functional architecture of language in context in relation to translation.

29.2 Language as a higher-order semiotic system in an ordered typology of systems

Translation is a linguistic phenomenon in the first instance—one involving different linguistic systems and processes in their contexts; but like other linguistic phenomena, it is also social, biological and material. Two or more of these different orders of manifestation of translation may be reflected in Translation Studies, but typically not all of them, so I will draw on the ordered typology of systems operating in different phenomenal realms proposed within SFL (see e.g. Halliday, 1996, 2005; Halliday & Matthiessen, 1999, 2006; Matthiessen, 2007, forthcoming).

29.2.1 Translation viewed in terms of four orders of system

We observe and theorize translators and translation as system and process in terms of an ordered typology of systems operating in different phenomenal realms. There are four types of system, ordered in increasing complexity from physical to semiotic systems. They are shown schematically in Figure 29.1 and can be characterized as follows:

- First-order systems—*physical systems*. These are omnipresent, extending through the whole universe, having emerged with the Big Bang on the order of 13.5 billion years ago and being subject to the laws of physics.
- Second-order systems—*biological systems* (physical systems + "life"). As far as we know, biological systems have only emerged under very specific conditions on our own planet (conditions that James Lovelock (1991) has called "the narrow window of life") on the order of 3.5 billion years ago (see Wilson, 2019). In addition to being physical, they have properties transcending physical systems: they are self-replicating and are subject to evolution and natural selection, with cross-generational memory and individuation.
- Third-order systems—*social systems* (biological systems + "value" (social order: roles in social networks)). Social systems have the properties of biological systems, through which they are manifested; but, in addition, they impose *value* or social order on biological organisms and

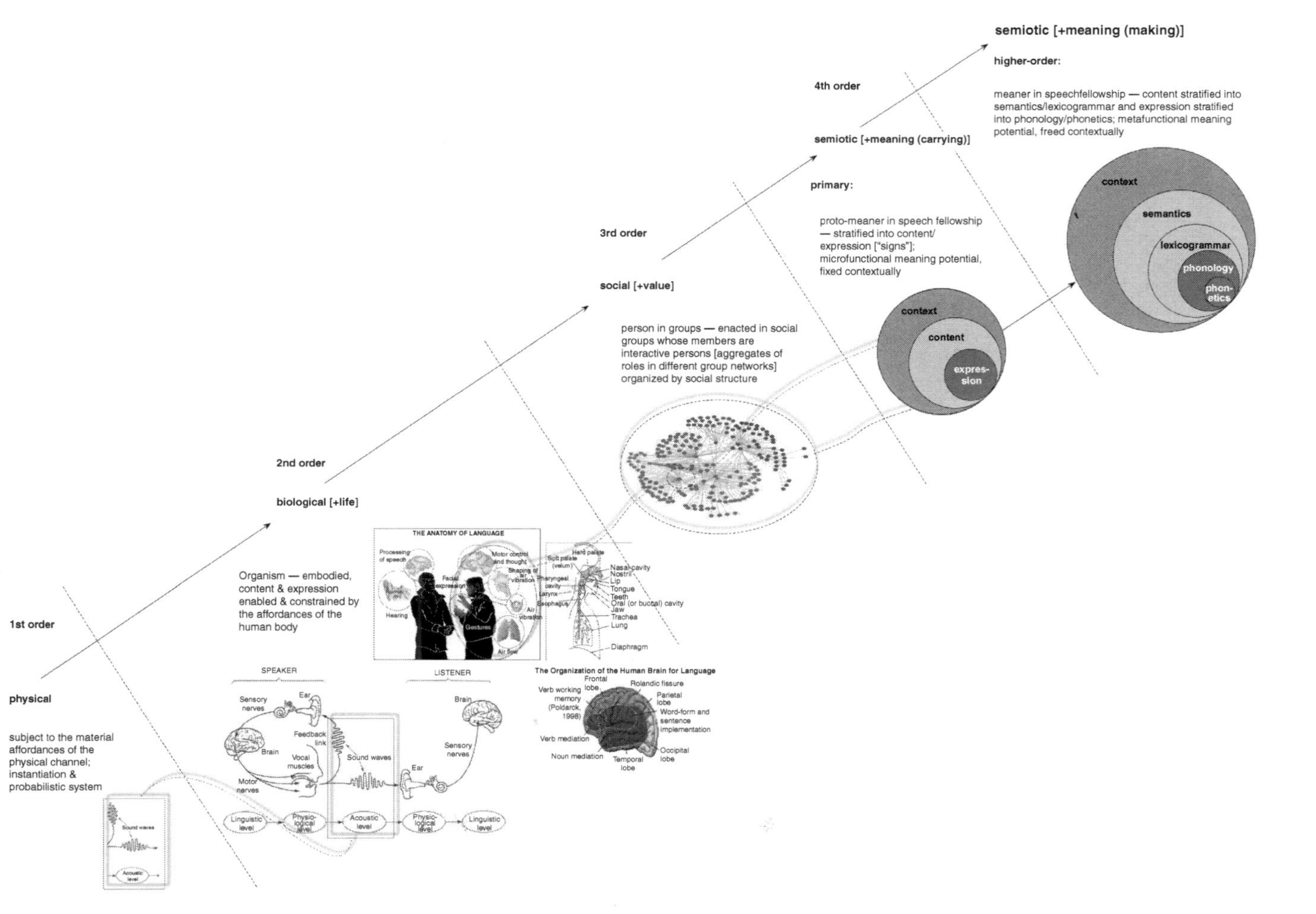

Figure 29.1 The ordered typology of systems operating in different phenomenal realms

populations through the organization into social role networks. They must have emerged on numerous occasions in different biological populations since the beginning of life (see Wilson, 2019). In social systems, biological individuals (organisms) take on multiple social roles, operating in the distinct role networks that characterize different social groups. In taking on different social roles (or personae), individuals become persons (see Butt, 1991; Firth, 1950; Halliday, 1978).

- Fourth-order systems—*semiotic systems* (social systems + "meaning"). Semiotic systems have the properties of social systems (so also of biological systems, and of physical ones); but, in addition, they carry or even create *meaning*. To have the property of meaning, they must be stratified into content and expression planes; and in the case of language, both these planes are further stratified: content into semantics and lexicogrammar, and expression into phonology and phonetics (or graphology and graphetics, or their analogues in the sign languages of deaf communities).

Physical and biological systems constitute *material systems*; they are "made of" matter, so they are subject to the laws of nature. In contrast, social and semiotic systems constitute *immaterial systems*; they are not "made of" matter, but, rather, impose different kinds of order on material systems—social order ("value") and semiotic order ("meaning").

The four orders are hypothesized to have emerged in the sequence of their order of increasing complexity; this is consistent with accounts by various scholars concerned with systems in general, such as Layzer's (1990) model of *cosmogenesis*). Once higher-order systems emerge, they then develop together with lower-order systems in a kind of dialectic, as put forward e.g. by Deacon (1992, 1997) in his hypothesis of the *co-evolution* of language and the brain and by Dunbar (1993, 1996) in his hypothesis of the correlation in evolution between language, brain and group size.

To advance our understanding of how systems of different orders relate to one another, we need to spell out explicitly how higher-order patterns are manifested as lower-order ones—of critical importance in the study of translation, e.g. in order to relate observations focused on translation as a neural activity and observations dealing with translation as a semiotic activity. In the case of SFL, this task has only recently been put on the research agenda; but Lamb (2013) suggests that the relational networks developed by him and his colleagues within relational network theory (see Lamb, 1999; García et al., 2017) can serve as an analytical interface between system networks (semiotic order) and neural networks (biological order), since relational networks have been developed as a model of neural networks.

To summarize, translation is in the first instance a semiotic phenomenon (as both system and process)—the recreation of meaning in context, and translators are in the first instance semiotic beings—multilingual meaners recreating meaning in context. But like other semiotic phenomena, they are enacted socially and embodied biologically, as shown in Table 29.1, which indicates how translation and translators can be studied at the different orders; but, in addition, correlations across the orders are an important focus of investigation. In fact, we need the semiotic, social and biological orders of systemic interpretation of translators to articulate what has come to be called "translation competence" (see PACTE, 2003).

29.2.2 The fourth order: The complementarity of semiotic and cognitive interpretations

Most approaches to linguistic phenomena in general, and to translation in particular, would probably agree on the nature and significance of the material orders of systems—i.e. on

Table 29.1 Properties and organizational features of the four orders of system (modes of study in italics)

Systemic order		*Properties*		*Translation*	*Translator*
immaterial	fourth order: semiotic	+meaning	stratified systems: content—expression	translation as recreation of meaning in context—*text (discourse) analysis, descriptions of multilingual meaning potentials*	translator as multilingual meaner—*think-aloud protocols (a kind of meta-discourse about translation choices)*
	third order: social	+value	role networked systems	translation as social behaviour—(professional) service—*questionnaires, ethnographic interviews, focus groups*	translator as professional person—*questionnaires, ethnographic interviews, focus groups*
material	second order: biological	+life	individuation, self-replication, evolution, natural selection	translation as neural activity (in the first instance) related to sensory and motor systems	translator as biological organism—*eye tracking, keystroke logging, brain scanning*
	first order: physical		subject to "laws of nature", extended in space-time	translation in (physical) workspace	translator in habitat—affordances of workspace

physical and biological systems; and while they would pay different degrees of attention to social systems, they would all acknowledge them.[3] However, the highest order of immaterial systems, fourth-order systems, have proved to be more challenging, even at the basic stage of interpreting what kinds of phenomena we are observing and thus need to theorize. Scholars have developed two interpretations of fourth-order systems, either as *cognitive systems* or as *semiotic systems* (Figure 29.2).

In Halliday and Matthiessen (1999), we argued for the *complementarity* of the semiotic and cognitive interpretations of phenomena of the fourth order. For example, when researchers offer models of "cognitive processing" and explore "cognitive load" in investigations of translation, we can also develop models of semiotic processing and investigate semiotic load (related to neurological activity).

The development of cognitive science as what I would call a macro-discipline, i.e. as a discipline made out of a number of separate disciplines, is well known and well documented (see e.g. Gardner, 1985, and for a personal perspective by one of the founders, Miller, 2003), as are criticisms (see Edelman's, 1992, early critique of the computational view of the brain). I would just add a reminder of Whorf's (1956) much earlier contributions from the 1930s and 1940s, which are of fundamental importance in explorations of translation (see also Grace, 1981). So since cognitive science is well documented, I will focus on a complementary angle of approach to fourth-order systems—social semiotics.

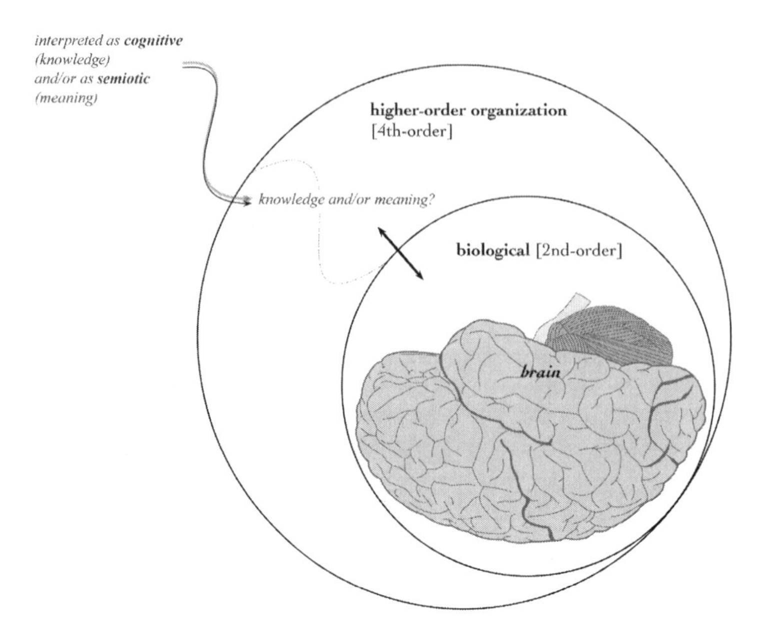

Figure 29.2 Interpretation of fourth-order systems—as systems of knowledge, cognitive systems, or as systems of meaning, semiotic systems

While early cognitive science was being developed in the US, M. A. K. Halliday was working along very different lines, originally in Europe and then in Australia. He was busy creating SFL, starting in the 1950s and taking off in the 1960s, drawing on the British linguistic tradition known variously as Firthian linguistics, system-structure theory and prosodic analysis, and also referring to other functional traditions (including the Prague School, noted for its contributions to Translation Studies) and to his experience studying with Wang Li in China in the late 1940s. Like J. R. Firth, his British teacher, Halliday included social systems in his account of language, drawing on social sciences, initially in particular on Bronislaw Malinowski's functional-contextual theory. This is important, since Malinowski had addressed the problem of translating Kiriwinan texts in their contexts into English, starting with his fieldwork in the Trobriand Islands in the 1910s (see Steiner, 2005; 2015, and Macdonald, this volume).

Halliday himself made some early contributions to the study and computational modelling of translation based on the theory that was later to emerge as systemic functional theory—scale and category theory (see Halliday, 1956, 1962; Halliday et al., 1964); and, using the same early framework, Catford (1965) published a short monograph on translation.

As SFL was developing in the 1960s, cognitive science began to take shape, one aspect of which was the emergence of psycholinguistics as a clearly recognizable branch within linguistics. In Halliday's view, there were two problems here in relation to his development of SFL: (1) social systems were almost entirely left out in the development of "mainstream" linguistic theories, and (2) cognitive claims about language were not grounded in empirical studies of the brain (cf. Halliday, 1995).[4] He needed a much more holistic conception of language in context, and referring to continental European semiotics, he introduced the notion of *social semiotics* as a way of framing the study of language and also of other systems of meaning—semiotic systems, taking the term "social semiotics" from Greimas. A number of publications by him in this area were collected and published as Halliday (1978), *Language as social semiotic* (LASS), which turned out to have a huge impact.

LASS foregrounded "the social interpretation of language and meaning", and with it Halliday had not only provided the foundation for *Social Semiotics*, which became an increasingly active area of investigation, paved the way for multimodal studies, and took on a life of its own (see Kress & Hodge, 1988; Maagerø & Andersen, 2015; van Leeuwen, 2005). LASS also constituted the beginning of an alternative to cognitive science—an approach where fourth-order phenomena are interpreted semiotically as meaning rather than cognitively as knowledge.

Building on our joint work in the 1980s, Halliday and I began to articulate this line of theorizing more clearly in the 1990s, partly in the context of our work on the modelling of meaning in computational linguistics, more specifically in work on text generation by computer (which later also became relevant in work on multilingual text generation and on machine translation: Halliday & Matthiessen, 1999), but also more generally as part of positing and exploration of the ordered typology of systems presented in Figure 29.1: Halliday (1996; 2005), Matthiessen (2007, forthcoming).

In Halliday and Matthiessen (1999), we emphasized that we conceived of knowledge and meaning as interpretations of fourth-order phenomena as complementary interpretations rather than as competing alternatives. But the semiotic interpretation of phenomena of this highest order of complexity enables us to bring out features of such phenomena that are arguably not foregrounded in cognitive interpretations (or only with an effort, as when "mainstream" cognitive scientists encountered Vygotsky's work in the 1930s; see Byrnes, 2006).

For the study of translation, the ordered typology of systems (Figure 29.1) is fundamental, since it makes it possible to identify precursors to semiotic systems that are social systems with the added feature of "meaning"; these precursors are *bio-semiotic systems* (Halliday & Matthiessen, 1999)—more specifically, sensory systems for construing our experience of the world as neural models and motor systems for enacting our intentions as muscular activities. Thus, we can compare the "translation" of visual images into meaning in bio-semiotic systems to translation between texts in two or more languages.

But there are also more specific advantages that follow from adopting the interpretation of fourth-order systems as systems of meaning rather than as systems of knowledge (Halliday & Matthiessen, 1999):

- explanation of the co-evolution of language and the brain (see e.g. Deacon, 1992, 1997; Halliday, 1995);
- understanding of the empowering role of language in individual development (e.g. Halliday, 2003a; Painter, 1999);
- addition of the interpersonal—the exchange of meaning, and so of the collective meaning potential;
- integration of language and culture both as systems of meaning, as semiotic systems, related to one another in explicit terms[5];
- the interpretation of individual and community multilingualism by reference to a multilingual meaning potential (Bateman et al., 1999; Matthiessen, 2018);
- and, at a more practical level, the extensive coverage of the description of fourth-order systems as semiotic systems.

While all are relevant to translation, the last four points have the most immediate impact, and the last point has significant implications for training translators by empowering them to use linguistic tools and in terms of undertaking Translation Studies equipped by comprehensive descriptions of the languages in their contexts of cultures that are in focus. (Naturally, the focus on interpersonal considerations is crucial in particular for engagement with interpreting and interpreting studies.)

29.3 Translation as a semiotic phenomenon, construed systemic-functionally

29.3.1 Translation as semiotic order of systems: The recreation of meaning in context

Adopting the semiotic interpretation of fourth-order systems as a complement to cognitive ones, we can say that translation is the *recreation of meaning in context* (see Matthiessen, 2001); we can call this *remeaning* (alongside rewording and rewriting).[6] This process presupposes a *multilingual meaning potential* (see Bateman et al., 1999; Matthiessen, 2018)—a potential that represents the meaning-making resources of the two or more languages that any translator must have mastered in order to be able to translate effectively in such a way that each language has its own integrity but is at the same time related to the other languages so as to support not only multilingual semiosis, such as translating and interpreting, but also code-switching and code-mixing.

The translator interprets the patterns of meaning in the source-language text in its context, making choices in the source-language meaning potential, and recreates these meanings by producing a text in the target language, making choices in the target-language meaning potential. Thus, central to the process of translation is *choice*—choice in meaning, as explored by e.g. Halliday (2013) and Matthiessen (2014); the recreation of meaning in contexts involves choices being made among options in the translator's multilingual meaning potential. These choices are, naturally, informed by contextual considerations (as well as considerations from below, i.e. lexicogrammar in the first instance); and they are *probabilistic* in nature (see Halliday, 1993; Matthiessen, 2015a; Toury, 2004) and thus need to be investigated in terms of corpora, as in Hansen-Schirra et al. (2012).

29.3.2 Examples of systemic functional studies of translation

The understanding of translation as the recreation of meaning in context is central to the systemic functional approach to translation as a phenomenon and also to the study of this phenomenon, i.e. to Translation Studies; but all aspects of the holistic systemic functional theory of language in contexts are relevant to the engagement with translation—that is, all aspects of the architecture of language in context. To bring this point out, I have given some examples of aspects of SFL that have been applied in Translation Studies in Table 29.2 (although not including the growing number of contributions written in a number of languages other than English, especially in Chinese).

I will focus on some key aspects of the SFL architecture of language, indicating their significance in relation to translation. I will be concerned with the semiotic order of systems, noting that I have already indicated that in SFL all four orders of systems operating in different phenomenal realms are relevant to the study of translation, as summarized in Table 29.1, and to the theorization of translators as "multi-layered" individuals (Figure 29.1).

29.3.3 Language in context organized multidimensionally

Language in context is organized in terms of a number of *semiotic dimensions*, each of which is the domain of a unique type of *relation*. It is modelled theoretically as a vast *multidimensional*

Table 29.2 Aspects of SFL that have been applied in Translation Studies

Aspect of SFL		*Applied to Translation Studies*
all dimensions:	SFL as resource in Translation Studies	Catford (1965) (pre-SFL); Wang & Ma (in press)
	nature of translated vs. original texts	Teich (2003)
stratification:	contextualism	Steiner (2005, 2015); Matthiessen (2001)
	context in translation evaluation	House (1997, 2001, 2015)
instantiation:	register variation (functional variation in context)	Teich (1999); Lavid (2000); Murcia-Bielsa (2000); Steiner (2004); Hansen-Schirra et al. (2012)
	particular registers, e.g. poetry and drama	Wang and Ma (forthcoming); Ma and Wang (forthcoming); Macdonald (2019, this volume)
	process and product	Alves et al. (2010)
metafunction:	all metafunctions (in relation to equivalence and shift)	Matthiessen (2014a)
	textual	McCabe (1999); Munday (2000); Kim and Matthiessen (2015); Kunz et al. (2017); Wang and Ma (2018)
	experiential	Mason (2012); Matthiessen et al. (forthcoming)
	interpersonal	Teich (1999); Lavid (2000); Murcia-Bielsa (2000); Wang (2004); Munday (2012, 2018); Yu and Wu (2016)
axis:	systemic (paradigmatic) organization: choice	Halliday (2010); Matthiessen (2014a)
	probabilistic choice	Toury (2004)
rank:	"level" of translation	Halliday (1962, 2009); Halliday et al. (1964)

network of relations, and all semiotic phenomena are defined in terms of the relations that they enter into—they are not things in themselves but, rather, nodes in networks of relations. This relational conception of language in context goes back to the 1960s, and it is entirely resonant with the general network science that has emerged in the last couple of decades since then (see Figure 29.1); see e.g. Barabási (2016). Translators operate in terms of multilingual multidimensional networks; in fact, being a multilingual speaker means having developed additional relational links, and becoming a translator means gradually growing further links.

The overall organization of language is determined by *global semiotic dimensions*—that is, the hierarchy of stratification, the cline of instantiation and the spectrum of metafunction; and the hierarchy of stratification organizes language in context into a number of sub-systems, each of which is organized in terms of *local semiotic dimensions*—that is, the hierarchies of axis and/or rank:

global semiotic dimensions

the hierarchy of stratification: the organization of language in context into an ordered series of sub-systems, that is, context—language, and within language: content (semantics—lexicogrammar)—expression (phonology—phonetics, or graphology—graphetics, or the two analogous expression strata in sign languages of deaf communities).

the cline of instantiation: the extension of language in context along a continuum from the potential pole (the meaning potential of a language operating in the context of culture (or cultural potential)) to the instance pole (texts unfolding in their contexts of situation), with intermediate patterns that have been identified in the study of language, including Translation Studies, as registers, sub-languages and text types.

the spectrum of metafunction: the diversification of meaning into a range of different modes within context and the content plane of language, that is, field—ideational meaning (the construal of experience as meaning) / tenor—interpersonal meaning (the enactment of roles, relations and values as meaning) / mode—textual meaning (the creation of ideational and interpersonal meaning as a flow of text in context).

local semiotic dimensions

the hierarchy of axis: the organization of each stratal sub-system into systems of choice among options (paradigmatic axis) and structures realizing options (syntagmatic axis).

the hierarchy of rank: the organization of each stratal sub-system into an ordered series of domains or units in terms of composition, e.g. in the lexicogrammars of many languages, clause—group—word—morpheme.

Together, these dimensions define what I have called the *environments of translation* (Matthiessen, 2001). We can formulate this as an ecological principle:

Ecological Principle:

the higher the unit that you translate in terms of (i) stratification and (ii) within a given stratum, the more information you will have access to in order to make an informed choice.

Thus, if we translate texts as semantic units in their contexts of situation, we have access to maximal contextual and semantic information as we make choices in translation; but if we translate clauses as grammatical units within the lexicogrammatical stratum, we haven't got access to semantic information, nor have we got access to contextual information—one reason why Malinowski emphasized the importance of translating texts in their contexts (see Steiner, 2005, 2015).

I will start with the local semiotic dimensions, discussing their implications for the study of translation, and then take a step back to bring the global dimensions into view.

29.4 The local semiotic dimensions

The local semiotic dimensions organize a stratal sub-system of language in context; they are the *hierarchy of rank* and the *hierarchy of axis*. The hierarchy of rank is like a local hierarchy of stratification except that it is based on the extension of units within each stratum, from most extensive to least extensive, e.g. from clause to group to word to morpheme in many languages (though not all).

29.4.1 The hierarchy of rank (rank scale)

The ecological principle was first established by Halliday in relation to the hierarchy of rank within lexicogrammar: the higher up the rank scale we move when we translate, the more information we have access to, so the more likely it is that the translation will be effective and acceptable (see e.g. Halliday, 1959–60, 1962; Halliday et al., 1964, pp. 126–128, on this "progressive selection", based on the grammatical scale of rank), illustrated by an example adapted from Halliday (1959–60) showing the progressive approximation in the

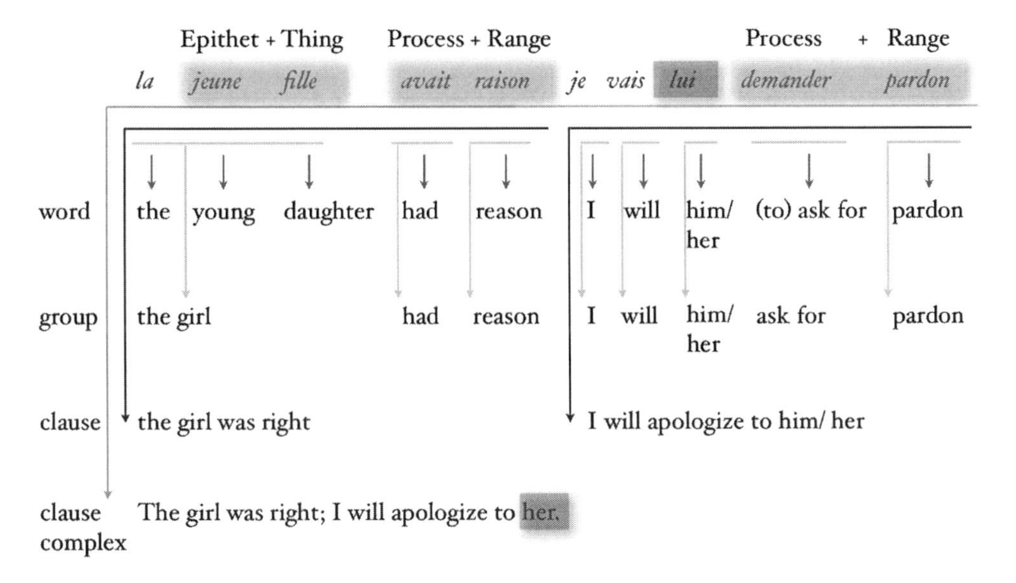

Figure 29.3 Progressive translation of a French clause complex into an English one as a move up the rank scale, with an increase in information available to translators at each rank (and finally at the combination of clauses as a clause complex)

translation of a French clause complex into an English one as the translation process moves up the rank scale from word rank to clause rank and the combination of clauses into clause complexes. (The same principle applies as we move from lexicogrammar to semantics, and from semantics to semantics in relation to context; the domain of translation can be expanded from a lexicogrammatical domain, i.e. the clause complex, to a semantic one, i.e. the text, and by another step by text in context.)

In general, translators will translate within the most extensive grammatical domain—that of the clause complex, and also, as already pointed out, ascending from lexicogrammar into semantics—i.e. translators typically focus on meaning rather than wording when they translate, except in special cases such as *interlinear glossing*.[7]

Staying within the lexicogrammatical stratum, we can thus say that maximally informed choices will be made within maximally extensive domains—in other words, within clause complexes. Using the rank scale, we can also set up a unit- or domain-based framework for studying empirically when translation choices lead to *shifts* in domain and when they do not, i.e. when the domains remain constant in the course of translation choices: see Table 29.3.

In this matrix, the domains of the source text are set out as column headings, and the domains of the translated text as row headings. If the two languages being investigated are typologically fairly similar, we can expect the defaults to be that the domains remain the same, as indicated by bold in the cells of the matrix.[8] But when we investigate original and translated texts, we certainly find that choices lead to incongruence in domains; domains may be "up-ranked", such as when a clause is translated by a clause complex, or "down-ranked", such as when a clause is translated by a phrase.

Within the ranks in Table 29.3, we can further differentiate grammatical classes, e.g. specifying groups as nominal, verbal or adverbial; and this will allow us to compile more detailed information about equivalences and shifts between pairs of languages such as French and English. While such detailed text-based investigations have perhaps not been a major focus in studies of texts translated by human translators, such detail has been essential in the development of machine translation systems, a classic contribution being Dorr (1994).

Text-based empirical investigations will continue to tell us more about similarities and differences between pairs, or sets, of languages. Some grammatical areas seem to be more likely than others to vary across languages. For example, while all languages almost certainly have a system of POLARITY ("positive"/"negative"), and "negative" is probably always the marked option, in that it is the one needing a special marker and in that the probability of it being chosen over "positive" is 0.1 vs. 0.9 (see Halliday & James, 1993; Toury, 2004), it turns out that the realizations of "negative" are quite varied around the languages of the world in terms of rank and primary class within a given rank: see Matthiessen (2004).

29.4.2 The hierarchy of axis

The reference to POLARITY immediately above serves as an introduction to the other semiotic dimension that is local to the organization of a stratal sub-system of language in context: the hierarchy of axis. The differentiation in linguistics, drawing on the insights from major linguists in the European tradition from the first half of the 20th century (including Saussure, Mathesius, Firth and Hjelmslev), between the two axes of organization in language—the paradigmatic axis and the syntagmatic one—has been widely accepted in the study of various linguistic phenomena, including translation. But thanks to Halliday's work in the 1960s (see Halliday, 1966; Matthiessen, 2015b, for a review), SFL has become unique in positing the paradigmatic axis as primary and the syntagmatic axis as "derived" from it by means of realization statements.

Treating the paradigmatic axis as primary foregrounds *choice among options in meaning* and follows from Halliday's conception of language as a *resource* for making meaning. This is obviously absolutely central to the conception and study of translation (see Halliday, 2013; Matthiessen, 2014): translated texts are the "product" of innumerable choices, and the process of translation is a process of making innumerable choices. Translating text is primarily a matter of choosing among the options of the source-language meaning potential and those of the target language.

One of the consequences of making the paradigmatic axis primary is that it becomes possible to differentiate paradigmatic similarities and differences among languages from syntagmatic ones. As far as translators are concerned, it is the choice that is important (see Halliday, 2010, on "pinpointing the choice" in translation); the syntagmatic marker is simply an automatic realization.

The translated text may thus be syntagmatically different—there is, as it were, a *shift in syntagmatic realization*; but paradigmatically it will be "equivalent". However, differences that emerge in the course of translation are, obviously, very often systemic; translators diverge in the choices they make within the meaning potential of the language that they translate into, which may, of course, happen even when the systems of the two languages appear to be similar at that point, as in the systems of MOOD in different languages explored by e.g. Teich (1999), Murcia-Bielsa (2000) and Lavid (2000), who identified registerial differences among languages in the deployment of the system of MOOD. For another illustration, involving the systems of MODALITY in English and German, see Matthiessen (2014).

One interesting finding coming out of systemic analyses of translations is that the more translations of the same source texts are produced, the more of the meaning potential of the target language is revealed. Different translators often make systemically related choices as they recreate the source-text meanings in the target language, but as we examine an increasing number of translations, we are likely to find a greater range of related choices. (Of course, this principle also applies to the translators' choices in the analysis and interpretation of the source text.)

Unless scholars undertake research projects where multiple translators are asked to translate one source text, multiple translations of the same text tend to be limited to texts that are highly valued in a given culture—certainly in the "target culture"; and this tends to happen with texts regarded as socio-politically important, sacred or belonging to high literature. As an illustration of this phenomenon (which is certainly well known but not usually construed in systemic functional terms), see my discussion of multiple translations of Charles Baudelaire's poem *L'Albatros* in Matthiessen (2014).

29.5 The global semiotic dimensions

The global semiotic dimensions organize the overall system of language in context; and, importantly, they also define the interfaces between language and other (denotative) semiotic systems and, by another step, between language and systems of lower phenomenal orders (see Figure 29.1).

29.5.1 The hierarchy of stratification

The *hierarchy of stratification* is central to translation because it can be used to specify the location of translation: as a process of recreating meaning in context, translation is located within the *content plane* of language operating within *context*. Stratification is, in a sense, the rank scale writ large, except that while the rank scale is based on the relation of composition (e.g. clauses are composed of groups, groups are composed of words, and words are composed of morphemes), the hierarchy of stratification is based on the more abstract relation of *realization* (e.g. meanings are realized by wordings, within the content plane of language).

The hierarchy of stratification orders *strata* (stratal sub-systems) of language in context from the stratum of context to that stratum that provides the interface to the material manifestation of meaning, which in the case of language is phonetics, graphetics or sign. Language is, of course, both the prototypical human semiotic and the most powerful one, and it is typically the focus of translation (allowing inter-semiotic types of translation; cf. Jakobson, 1959 ("intersemiotic translation or transmutation"); Matthiessen, 2001); but language operates in context together with other semiotic systems, such as gesture and other somatic semiotic systems in the case of spoken language and pictorial systems in the case of written language. Context provides the semiotic environment for all such denotative semiotic systems, and serves to determine the division of semiotic labour among them and coordinate them so that they complement one another in the making of meaning. This is highly relevant to types of translation involving technologically enabled channels of meaning making such as subtitling, and is related to multimodal studies informed by SFL (see e.g. Taylor, 2003, 2013).

Prototypically, translation is a semantic process—one that always references the stratum immediately above, i.e. context. There are certain exceptions (see Figure 29.3), including interlinear glossing, mentioned earlier. This is still within the content plane of language (semantics—lexicogrammar), but if translation is broadened to include transliteration, then this would be "translation" within the expression plane (in this case, graphology—graphetics).

Once we introduce the consideration of the global semiotic dimension of organization of stratification, we can construe the traditional distinction between "literal" and "free" translation as a cline whose outer poles are low-ranking translation within lexicogrammar to translation within semantics in reference to context—or even, by another step, translation focusing not on

Table 29.3 Translation shifts characterized in terms of rank (shaded cells = downgrading of unit from source to target)

		From source text			
		Clause complex	**Clause**	**Phrase**	**Group**
to target text	**clause complex**	**clause complex > clause complex**	clause > clause complex		
	clause	*clause complex > clause*	**clause > clause**	phrase > clause	
	phrase		*clause > phrase*	**phrase > phrase**	
	group			*phrase > group*	**group > group**

text per se but, rather, on comparable contexts of situation that may require quite different types of text (see Steiner, 2004) or even different (mixtures of) semiotic systems.

But here I need to take one step further in order the clarify the implications of characterizing translation as the "recreation of meaning in context". As I have noted, this implicates the semantic stratum within the content plane of language. However, once we recognize that context is in fact also a system of meaning, as explained by Halliday (1978), we can begin to explore the differentiation between translation focused on meaning within semantics operating in context and translation focused on meaning within context, i.e. focused on cultural meanings.

Translation is usually of the first kind, i.e. focused on meaning within semantics operating in context; but the focus of translation may be located within context in the first instance rather than within semantics. This means that as long as translators recreate the contextual meanings, there may be significant divergence in the semantics of the "translated" text. One obvious example is advertising (see e.g. Steiner, 2004). Here, translators may serve as mediators between two different contexts of culture, as also in Bible translation (see Nida, 2001). The focus on context will inform the semantic choices they make, even if there is a considerable shift from the meaning of the original text (see Anwyl et al., 1991); and here investigations of comparable texts are very helpful (see Abelen et al., 1993).

29.5.2 The cline of instantiation

While the global dimension of stratification is a hierarchy ordering sub-systems of language from the most "abstract" (context) to the ones that interface with the human body most directly (phonetics or graphetics), the *cline of instantiation* is a continuum—one extended from the overall meaning potential of a language to acts of meaning instantiating this potential as texts unfold in their contexts of situation. In 20th-century linguistics, this continuum has usually been misconstrued as a dichotomy, like the Saussurean dichotomy of *langue* and *parole* and the original Chomskyan dichotomy of competence and performance.[9]

But what has been conceptualized as a dichotomy is, in fact, a cline or continuum, as shown by Halliday—a cline that is one of the global dimensions of the organization of language in context; this is the *cline of instantiation* (see Halliday, 2002, 2003b; Matthiessen, 2007, 2019). The outer poles of this cline are the potential pole and the instance pole. In the case of language, these are the meaning potential of a given language and the texts instantiating this meaning potential. Intermediate between these two poles, there are patterns of variation in instantiation. Looked at from the potential pole, these are sub-potentials or *registers* (characterized as "sub-languages" in the work on machine translation, e.g. Kittredge & Lehrberger, 1982); looked at from the instance pole, they are *text types*.

In the engagement with, and study of, translation, the cline of instantiation is of central importance for two related reasons.

First, translators translate texts located at the instance pole of the cline of instantiation; they recreate patterns of meaning in the source text as patterns of meaning in the target text.[10] This *process of recreation* is thus (among other things) a process of reinstantiation, and since it involves movement up and down the cline of instantiation, it is, of course, a phased one: translators begin to translate, keep translating and stop translating once done. As researchers, we can study the different phases. Traditionally, the focus has been on the initial and final phases—on the source text as a "product" and on the translated text as a "product";[11] they can be explored using the tools and techniques of text analysis (including both "discourse analysis" and "corpus analysis"). But thanks to theoretical and, importantly, technological developments, it has become increasingly possible to investigate and shed light on the process of translating, the move in instantiation from source to target: see Jakobsen (2014, 2017), Saldanha and O'Brien (2014, Ch. 4) and see also Alves (2003).

The technological developments have enabled researchers to study translation as process in particular "from below" in terms of the ordered typology of systems presented earlier (see Figure 29.1 and Table 29.1). That is, the translation process can now be studied as a material phenomenon by tracking eye movements, logging keystrokes and even scanning the brain in action. One fascinating challenge is to relate the findings shedding light on translation as a material process to interpretations of translation as process within the semiotic order of systems. Theoretically, it has long been known in SFL that texts unfolding in their contexts of situation can be modelled as ongoing *choices* in the system networks that represent the potential or one or other sub-potential of a language (see Halliday, 1977; Matthiessen, 2002).

There have also been advances in the development of computational models of the process of instantiation, in particular in text generation (see Matthiessen & Bateman, 1991; Patten, 1988; Teich, 2009). But a great deal of work remains to be done in order to enable us to model and study semiotic processes of instantiation in relation to the significant findings "from below", from observations of translation as a material process. In particular, we need more work on the computational modelling—or at least pre-modelling—of translation, and the computational models must be interpreted theoretically as a step towards incorporating semiotic processes in the overall theory of language in context. Still, we can be guided by important findings from the work on text generation, including the in-principle demonstration that the unfolding of text can be interpreted as massively parallel (see Matthiessen et al., 1988; Matthiessen & Bateman, 1991). This is important partly because it enables us to explore competing motivations that translators must entertain as they make choices in the course of translation, e.g. trying to optimize the combination of distinct metafunctional considerations.

Secondly, when translators translate texts, they obviously have to refer to the source-language meaning potential instantiated by the source text and to the target-language meaning potential instantiated as they make choices from it to generate the target text. This means that they move from the instantial pole of the cline of instantiation towards the potential pole, and from the potential pole to the instantial pole. But how far up the cline of instantiation towards the potential pole do they move? This will depend on the extent to which they can stay within certain source and target *registers* (sub-potentials).

When they are translating texts that instantiate conventional registers, they can access the registerial "routines" instead of moving all the way up to the overall meaning potentials,[12] even undertaking some degree of automated preparatory translation. In fact, this was, of course, the discovery made within the machine translation community in the second half of the 1970s—that automatic translation would be more effective if the MT systems could be "tuned" to particular registers, or "sub-languages" as they came to be called in computational linguistics, such as weather forecasts or stock market reports (see e.g. Kittredge & Lehrberger, 1982).

Viewed in terms of work on problem solving within AI, registers can be interpreted as "precompiled" solutions that have evolved to address recurrent problems (see Matthiessen, 1993; Patten, 1988). In fact, translators need such registerial adaptations embodied, because registers will differ from one language to another—as shown in SFL by e.g. Teich (1999), Lavid (2000), Murcia-Bielsa (2000), and Steiner (2004).

29.5.3 Stratification × instantiation

By intersecting stratification and instantiation, we can examine particular registers located mid-region along the cline of instantiation and view them stratally from the point of view of context ("from above"), from the point of view of semantics—the semantic strategies that constitute a given register ("from roundabout"), and from the point of view of lexicogrammar—the patterns for wording realizing the semantic strategies ("from below"). In this way, we can study comparable registers in two or more languages, drawing on parallel or comparable texts instantiating these registers.

When we examine the systems of the languages involved at the potential pole of the cline of instantiation, they may appear quite similar up to a certain point (in delicacy); but when we move down the cline of instantiation to study registerial sub-systems, we may find that there are significant differences in the way that languages have adapted to comparable contextual goals in their registers.

29.5.4 Metafunction

Metafunction is organized as a spectrum of different modes of meaning—the metafunctions that organize the content plane of language. The metafunctions organize the systems of the content plane into simultaneous strands,[13] and they correlate with different parameters within context:

The *ideational* metafunction provides the resources for construing our experience of the world, organized either logically into chains of related units or experientially into configurations within units. It resonates with the *field* parameter within context—what's going on in the context (in terms of the field of activity, the field of experience being created by or accompanying this activity).

The *interpersonal* metafunction provides the resources for enacting our roles and relations in the engagement with each other. It resonates with the *tenor* parameter within context—who are taking part in what's going on (in terms of their roles and relations and value systems).

Table 29.4 Translation shifts characterized in terms of metafunction

		From source text			
		Textual	**Ideational: logical**	**Ideational: experiential**	**Interpersonal**
To target text	**textual**	textual > textual: e.g. thematic shift	logical > textual: e.g. complex to cohesive sequence		
	ideational: logical	textual > logical: e.g. cohesive sequence to complex	logical > logical: e.g. tactic shift		interpersonal > logical: e.g. mood or modality represented by verbal or mental clause in clause complex of projection
	ideational: experiential		logical > experiential: e.g. clause (in complex) > phrase (as circumstance)	experiential > experiential: e.g. process-type shift	
	interpersonal				interpersonal > interpersonal: e.g. mood type shift

The *textual* metafunction provides the resources for encapsulating ideational and interpersonal meanings as a flow of information unfolding as text in context, providing speakers with the strategies for enabling their listeners in the processing of text, and listeners with the strategies for interpreting text according to the guidance they are provided with. It resonates with the *mode* parameter within context—the role that language is playing within the context in which text unfolds (along with other semiotic systems, and also social systems).

As translators recreate meaning in the course of translation, all three modes of meaning are in principle equally important for them to pay attention to.[14] However, there seems to be a tendency for translators to focus more on ideational meaning than on interpersonal and textual meaning, and arguably more on interpersonal meaning than on textual meaning (see Kim & Matthiessen, 2015). In other words, translators would seem to tend to prioritize ideational equivalence over interpersonal and textual equivalence; and if this generalization is on the right track, we can expect to find more interpersonal and textual translation shifts than ideational ones.[15] But this may depend on register (see Steiner, 2004).

We can frame translation shifts (and equivalences) in terms of the metafunctions. To do this, we can use a matrix analogous to the one designed for the study of rank-based translation shifts in Table 29.3: see Table 29.4. Here the metafunctions of the source language are set out as column headings and the metafunctions of the target language as row headings. I have separated the two modes of construing experience within the ideational metafunction—the logical and the experiential modes, since I have found over the years that interesting shifts often take place between them, one key reason being that languages vary in how they deploy these two complementary modes of construing experience (see Matthiessen, 2004, 2015c).

In the matrix set out in Table 29.4, the diagonal from top left to bottom right represents choices where translators have stayed within the same metafunction. This does not mean that the original and translated versions are equivalent; they may be (to the extent that equivalence is possible for the two or more languages being compared), but they need not be—the shifts may be internal to the relevant metafunction. The cells located outside this diagonal all represent shifts where translators make choices that involve changes in metafunction.

The matrix is used in the interpretation of translation shifts (see Matthiessen, 2014); but many more corpus-based studies ranging over a significantly larger number of language pairs are needed to show what shifts are common. Tentatively, shifts between the logical and the textual ways of indicative rhetorical relations are fairly common, and similarly, shifts between the logical and the experiential modes of construing experience are fairly common (predictably; see Grace, 1981; Halliday & Matthiessen, 1999, Ch. 7; Matthiessen, 2004). But what about shifts between the interpersonal metafunction and the other metafunctions? They may be less likely, but they do occur—at least in cases involving *grammatical metaphor* of the interpersonal kind, where the ideational metafunction may be "co-opted" in one language but not in another (see Halliday & Matthiessen, 2014, Ch. 10).

Even where there are translation shifts, they usually appear within one and the same metafunction; and, as noted earlier, it would seem that translators are least aware of textual systems, so it is quite possible than they make "uninformed" choices within textual systems (see Kim, 2007; Kim & Matthiessen, 2015).

29.6 Translation choices

29.6.1 Translation choices to recreate meaning

Choices in the recreation of meanings turn out to be the central phenomenon (within the semiotic order of systems). Translators have to make choices as they interpret the source text as an instance of the meaning potential of the source language "embedded" in its cultural potential, and they have to make choices as they generate the "target" text as an instance of the meaning potential of the "target" language embedded in its cultural potential.

So we can investigate these choices, identifying the shifts that translators make—keeping in mind that the limiting case of shift is simply "equivalence" (and that equivalence is always relative to the multilingual meaning potential that translators have mastered—which, as translation researchers, we must describe).

29.6.2 Translation choices leading to shifts

By adopting a multidimensional theory of the architecture of language along the lines of SFL, we can systematically intersect the dimensions of the languages involved, to locate translation choices and to characterize the patterns of *shifts* and equivalences that we find. I have used two dimensions to illustrate this method of investigation: the hierarchy of rank (rank scale), set out in Table 29.3; and the spectrum of metafunction, set out in Table 29.4. But we can extend this approach to other semiotic dimensions. For example, we can investigate translation shifts in reference to axis (paradigmatic—syntagmatic), and within the paradigmatic axis in reference to the cline of delicacy (cf. Matthiessen et al., forthcoming). In this way, our exploration of translation shifts is empowered by our linguistic theory descriptions.

By tracking translation shifts, we can identify those shifts that appear to involve *explicitation* (e.g. Teich, 2003). Such shifts are likely to include those involving upgrading in terms of rank, i.e. those where the translated unit is of a higher rank than that of the unit in the source text (e.g. group > phrase, phrase > clause in Table 29.3); but they are also likely to include certain metafunctional shifts, such as when "implicit" orientation within the interpersonal metafunction is translated as "explicit" orientation by means of clause complexes of projection (see Table 29.4). And when translators "unpack" grammatical metaphors in the source text as part of the recreation of meaning in the target language, they are likely to make choices that involve shifts in terms of both rank and metafunction.

29.6.3 Semantic choices: Problem solving

Translation choices are semantic choices in context in the first instance. Lexicogrammatical ones will, in principle, follow automatically, since choices in meaning are realized by choices in wording; and choices in wording are realized automatically by choices in sounding (phonology) or in writing (graphology). So we can say that as far as content is concerned, lexicogrammatical choices are *automatized* (see Halliday, 1982); and, by the same token, lexicogrammatical shifts follow automatically from semantic shifts.

However, while translation of texts within fairly restricted registers, or even within less restricted registers that translators have become very familiar with, can be quite straightforward, translation always involves *problem solving* (see Halliday, 2010: "the basic problem for the translator is the problem of choice"); it is just that when the relevant registers are familiar, translators are likely to have ready-made solutions. When such solutions are not readily available, translators may have to *de-automatize* lexicogrammatical choices, and more consciously consider the realizational dialectic between semantics and lexicogrammar (and possibly between these two strata and phonology, as in the translation of poetry).

29.6.4 Systemic orders of choices

To investigate the way that translators make choices under these different conditions, we can profitably study the translation process in terms of its biological manifestation, logging keystrokes and tracking eye movements. For example, if we have undertaken the huge tasks of developing descriptions of the multilingual meaning potentials and wording potentials of the two or more languages a given translator is working with, we can certainly identify areas where there is such a degree of incongruence between or among the languages that it is very likely that translators will have to go into serious problem-solving mode—for example, the task of translating English existential clauses into German (see Matthiessen, 2001) or the more challenging task of translating choices in the English system of TENSE as choices in the Chinese system of ASPECT (see Matthiessen, 2018).

The latter is an area where the fundamentally important work by Christiane von Stutterheim and her colleagues has shed light on the way that the different systems for construing our experience of the flow of events through time influence not only conceptualization but even perception (see Carroll & von Stutterheim, 2011); so it is surely an area that would be fascinating to study "from below", from the vantage biological point of translators being tracked through keystroke logging, eye tracking or brain scanning.

29.7 Concluding remarks

Here I have explored translation as a *semiotic process*, taking place in the first instance within fourth-order systems in an ordered typology of systems operating in different phenomenal realms (see Figure 29.1). Translation is, fundamentally, the *recreation of meaning*; and our interpretation of it can be informed by systemic functional *theory*—the theory of language as a general human system common to all of us—and informed by systemic functional *descriptions* of particular languages. Outlining the architecture of language according to systemic functional theory, I have deconstructed the conception of translation as the "recreation of meaning in context" in terms of the semiotic dimensions that this architecture is based on. Thus, the process of recreating meaning can be illuminated in terms of all these *semiotic dimensions*:

- locally:
  in terms of the *hierarchy of axis*, the recreation of meaning involves systemic choice (the paradigmatic axis), with structural realizations (the syntagmatic axis) as automatic realizations of the choices being made;
  in terms of the *rank scale*, the recreation of meaning involves a downward move, from more extensive domains to less extensive ones (context: downwards from context of situation; semantics: downwards from text; lexicogrammar: downwards from clause);
- globally:
  in terms of the *hierarchy of stratification*, the recreation of meaning involves a focus on the stratum of semantics in relation to context in the first instance, with patterns at the lower strata as automated realizations of higher-stratal patterns;
  in terms of the *cline of instantiation*, the recreation of meaning involves reinstantiation of meaning, including the phases of analysis of the source text and the generation of the translated text;
  in terms of the *spectrum of metafunction*, the recreation of meaning involves all the metafunctional modes of meaning, although translators will face the challenge not only of taking all of them into account but also of optimizing competing motivations.

By interpreting translation as a semiotic process in the first instance, we are locating this phenomenon in the *ordered typology of systems* operating in different phenomenal realms set out in Figure 29.1. Like other linguistic phenomena—including other *metalinguistic phenomena* concerned with operations on existing texts in context, such as revising (see Bowen, 2019), editing, summarizing and abbreviating, and adapting (say to different groups of readers)—translation is enacted socially (in groups), embodied biologically (in organisms) and ultimately manifested physically. This is, in fact, one of the reasons why there have been a number of "turns" in the approach to and study of translation (see Snell-Hornby, 2010). So we need to investigate it in its various phenomenal "guises", adopting different vantage points and viewing it in different perspectives (through different "lenses"). This has been a reason for institutionalizing "Translation Studies" as a distinct discipline, one that must engage with a number of distinct disciplines; but this is equally true of all other linguistic phenomena (see Halliday, 1978). As the science of language, linguistics must be in constant interaction with other disciplines that engage with language in some way, as in the hyphenated branches of linguistics such as socio-linguistics, psycholinguistics and neurolinguistics—each of which comes with methods and methodologies also used in other disciplines. Thus, the investigation of translation is no different from the study of any other linguistic—or, indeed, semiotic—phenomena. And if we construe and enact it as a distinct discipline, we may run the danger of losing connection, on the one hand, with other metalinguistic phenomena and, on the other, with other multilingual phenomena (see Matthiessen et al., 2008).

As areas of research become more mature and more specialized, they are likely to drift apart, their researcher communities becoming insulated from one another and gradually forming quite distinct communities with different programmes, venues of publication, associations and conferences. Observing this tendency, a natural one in the evolution of scientific knowledge, in physical sciences, Bohm (1979) warned against the "fragmentation of knowledge". Such fragmentation is, in a very real sense, one consequence of Cartesian Analysis—a scientific approach that has been dominant as modern science gradually emerged during the last few centuries; but it will obscure the more holistic understanding that we need to develop (see Capra, 1996). My concern here has been with SFL as a resource for developing a well-rounded understanding of and engagement with translation. It is only by deploying a holistic theory of language such as SFL that we can show the complementarity of, say, eye tracking and brain scanning, social network analysis, text (discourse) analysis, and investigations of contexts of culture.

Notes

1 Acknowledgement: I am very grateful to the two editors, Fabio Alves and Arnt Lykke Jakobsen, not only for inviting me to contribute to this volume in the first place, but also for their patience and generous comments, helping me shorten an overlong manuscript—a task that my colleague and collaborator Wang Bo has also kindly helped me with.
2 The disciplinary trend has arguably been in the opposite direction—that is, the direction of institutionalizing "Translation Studies" as a distinct discipline, going back to James Holmes' frequently referenced map from the 1970s (e.g. Holmes, 1988; Toury, 1995), and it is important to recognize what this has meant in terms of university programmes, degrees and even department, and also in terms of conferences and publication channels. At the same time, the notion of "Translation Studies" has tended to move away from other relevant linguistic areas of research (such as those mentioned) and even from the highly relevant work on machine translation and arguably also from studies of other metalinguistic phenomena such as editing and summarization. The contribution of linguistics is—or rather should be—to provide a holistic account of all the related phenomena.
3 One among innumerable examples is the stance Jackendoff (1997, pp. 2–3) makes explicit:

> What about the abstract and social aspects of language? One can maintain a mentalist stance without simply dismissing them, as Chomsky sometimes seems to. [...] The mentalist stance would say, though, that we eventually need to investigate how such properties are spelled out in the brains of language users, so that people can use language. It then becomes a matter of where you want to place your bets methodologically: life is short, you have to decide what to spend your time studying. The bet made by generative linguistics is that there are some important properties of human language that can be effectively studied without taking account of social factors.

4 This was to be addressed much later, one important contribution being Varela et al. (1991).
5 Thus, while the importance of culture has been illuminated in descriptive translation theory, e.g. Toury (1995) (and in polysystem theory), in SFL, language has been theorized as "embedded in" the context of culture from the start (an insight which, of course, goes back to the work by Malinowski, and which was developed within linguistics first by Firth and then by Halliday and other systemic functional linguists).
6 This characterization of translation needs to be interpreted in terms of the architecture of language in context developed within SFL and sketched here. Importantly, it does not involve the *transfer metaphor* commonly used in definitions and characterizations of translation; the fundamental problems with this metaphor were identified a long time ago by Reddy (1979) in his study of the common conduit metaphor in explorations of language. Compare House's (2016) characterization of translation: "translation is a text-processing and text-reproducing activity which leads from a source text to a resulting text".

Obviously, the notion of the recreation of meaning in context also includes cultural meanings, since culture is interpreted as a (connotative) semiotic system, as outlined by Halliday (1978). The recreation of cultural meanings has, of course, been foregrounded as part of the "cultural turn" in Translation Studies; cf. e.g. Bassnett and Lefevere (1990), Snell-Hornby (2010), Marinetti (2011) and Katan (2009, 2014). Translation has been called "rewriting"; this term is intended to bring out the cultural aspects of translation (e.g. Lefevere, 1992). But the "recreation of meaning in context", or "remeaning", serves to locate the process stratally within semantics in context in the first instance, highlighting the semiotic nature of translation as one of meaning.

7 The most widely used being the "Leipzig Glossing Rules": www.eva.mpg.de/lingua/pdf/Glossing-Rules.pdf

8 The situation may be significantly different for languages that are, in terms of certain systems, typologically quite different; for a discussion of English and Kalam, see Halliday & Matthiessen (1999, Ch. 7) and references therein to relevant work in language typology and Translation Studies (including Grace, 1981).

9 Chomsky's focus on competence as opposed to performance is echoed in further elaborations of the notion of competence—"communicative competence" and also "translator competence". But the distinction between competence and performance was never accepted in SFL, and it was explicitly rejected by Halliday (1973). See immediately below.

10 When researchers investigate translation as process, there is perhaps a tendency to view it biologically in the first instance—that is, in terms of bodily activities that can be tracked through keystroke logging, eye tracking and even brain scanning. But it is equally a semiotic process—translation as (making) meaning, and, of course, a social one as well—translation as behaviour. As far as translation as a semiotic process is concerned, one constraint has simply been that process models have tended to be developed within computational linguistics rather than linguistics, so they have often not been interpreted within linguistic theory (cf. Matthiessen & Bateman, 1991), although there have been interesting explorations, such as the work on "dynamic syntax", e.g. Kempson et al. (2001).

11 In their discussion of product-based research in Translation Studies, Saldanha and O'Brien (2014, Ch. 3) focus on translated texts; but it is equally important to investigate the original products: translation can be illuminated by comparing carefully analysed original and translated texts, as e.g. Teich's (2003) corpus-based study shows.

12 The SFL theory of register and register variation was developed originally in the 1960s, drawing on the notion of context from Malinowski and Firth and generalizing Firth's notion of restricted languages (see e.g. Halliday, 1978; Halliday et al., 1964). Catford (1965) noted the significance of register in translation, writing e.g. (p. 90):

> In translation, the selection of an appropriate register is often important. Here, if the TL has no equivalent register, untranslatability may result. One of the problems of translating scientific texts into certain languages which have recently become National Languages, such as Hindi, is that of finding, or creating, an equivalent scientific register.

Since then, quite a substantial body of work on the significance of register in translation has emerged in SFL; and scholars in other traditions have also emphasized the need to characterize registers in the context of work on translation, often under the heading of "text type", e.g. Snell-Hornby (1995) and Nord (2005).

13 In SFL, stratification and metafunction are thus treated as separate semiotic dimensions that intersect; but in a number of other approaches they are, as it were, conflated.

14 I've written "in principle" because depending on the nature of the context in which registers operate, they may be oriented towards communicative goals that are associated with tenor in the first instance or with field. For example, in contexts where the field of activity is one of promoting goods-&-services, the goals can be characterized primarily in terms of tenor: in promotional registers such as advertisements, writers try to shift the positions of their readers towards greater readiness to purchase a commodity—one type of persuasion. But in contexts where the field of activity is one of expounding knowledge, the goals can be characterized in terms of field of experience in the first instance: for example, in explanations, writers try to produce texts that will enable readers to build new knowledge. Naturally, the translator's role also varies, depending on the mixture of mediation and reproduction as goals for a given translation.

15 Here it is important to note that in her classic translation textbook, Baker (1992) highlights the textual metafunction (chs. 5 and 6).

Further reading

Halliday, M. A. K. (2009). The gloosy ganoderm: Systemic Functional Linguistics and translation. *Chinese Translators Journal,* 1 (An introduction to theories of interpretation between English and Chinese), 17–26. Reprinted as Ch. 6 in Halliday, M. A. K. (2013). *Halliday in the 21st century*. Vol. 11 in the Collected Works of M. A. K. Halliday (J. J. Webster, Ed.). London: Bloomsbury Academic.

This chapter provides a brief overview of studies of translation informed by systemic functional linguistics and of the environments of translation, and offers examples from two texts of the study of translations between English and Chinese.

Halliday, M. A. K., & Matthiessen, C. M. I. M. (2006). *Construing experience through meaning: A language-based approach to cognition*. London & New York: Continuum.

This book presents a language-based approach to "cognition" interpreted as meaning; it provides part of the semiotic foundation upon which the current chapter is based.

Matthiessen, C. M. I. M. (2014). Choice in translation: metafunctional consideration. In Kunz, K., Teich, E., Hansen-Schirra, S., Neumann, S., & Daut, P. (Eds.), *Caught in the middle—language use and translation: A festschrift for Erich Steiner on the occasion of his 60th birthday* (pp. 271–333). Saarbrücken: Universaar, Saarland University Press.

This chapter explores choice as a central phenomenon in translation in terms of the different metafunctional modes of meaning.

Steiner, E., & Yallop, C. (Eds.). (2001). *Beyond content: Exploring translation and multilingual text production*. Berlin & New York: de Gruyter.

While published almost two decades ago, this book still contains contributions that are relevant to and have been influential in systemic functional work on translation.

Wang, B., & Ma, Y. (in press). *Systemic functional translation studies: Theoretical insights and new directions*. Sheffield: Equinox.

This is a new introductory overview of systemic functional linguistics as a resource in Translation Studies.

References

Abelen, E., Redeker, G., & Thompson, S. A. (1993). The rhetorical structure of US-American and Dutch fund-raising letters. *Text, 13*(1), 323–350.

Alves, F. (Ed.). (2003). *Triangulating translation: Perspectives in process oriented research*. Amsterdam: John Benjamins.

Alves, F., Pagano, A., Steiner, E., Neumann, S., & Hansen-Schirra, S. (2010). Translation units and grammatical shifts: Towards an integration of product- and process-based translation research. In G. M. Shreve, & E. Angelone (Eds.), *Translation and cognition* (pp. 109–142). Amsterdam: John Benjamins.

Anwyl, P., Matsuda, T., Fujita, K., & Kameda, M. (1991). Target-language driven transfer and generation. In *Proceedings of The 2nd Japan-Australia Joint Symposium on Natural Language Processing (JAJSNLP '91), October 2–5 1991* (pp. 234–243). Kyushu Institute of Technology, Iizuka City, Japan.

Baker, M. (1992). *In other words: A coursebook in translation* (3rd ed. 2018). London: Routledge.

Barabási, A. L. (2016). *Network science*. Cambridge: Cambridge University Press.

Bassnett, S., & Lefevere, A. (1990). *Translation, history and culture*. London: Pinter.

Bateman, J. A., Matthiessen, C. M. I. M., & Licheng, Z. (1999). Multilingual language generation for multilingual software: A functional linguistic approach. *Applied Artificial Intelligence: An International Journal, 13*(6), 607–639.

Bohm, D. (1979). *Wholeness and the implicate order*. London: Routledge.

Bowen, N. (2019). Unfolding choices in digital writing: A functional perspective on the language of academic revisions. *Journal of Writing Research, 10*(3), 465–498.

Butt, D. G. (1991). Some basic tools in a linguistic approach to personality: A Firthian concept of social process. In F. Christie (Ed.), *Literacy in social processes: Papers from the Inaugural Australian Systemic Functional Linguistics Conference, Deakin University, January 1990* (pp. 23–44). Darwin: Centre for Studies of Language in Education, Northern Territory University.

Byrnes, H. (Ed.). (2006). *Advanced instructed language learning: The complementary contribution of Halliday and Vygotsky*. London: Continuum.

Capra, F. (1996). *The web of life*. New York: Doubleday.
Carroll, M., & von Stutterheim, C. (2011). Event representation, time event relations, and clause structure: A crosslinguistic study of English and German. In J. Bohnemeyer, & E. Pedersen (Eds.), *Event representation in language and cognition* (pp. 68–83). Cambridge: Cambridge University Press.
Catford, J. C. (1965). *A linguistic theory of translation*. London: Oxford University Press.
Deacon, T. (1992). Brain-language coevolution. In J. A. Hawkins, & M. Gell-Mann (Eds.), *The evolution of human languages* (pp. 49–85). Redwood City, CA: Addison-Wesley (Proceedings Volume XI, Santa Fe Institute Studies in the Sciences of Complexity).
Deacon, T. (1997). *The symbolic species: The co-evolution of language and the human brain*. Harmondsworth: Penguin Books.
Dorr, B. J. (1994). Machine translation divergences: A formal description and proposed solution. *Computational Linguistics, 20*(4), 597–633.
Dunbar, R. I. M. (1993). Coevolution of neocortical size, group size and language in humans. *Behavioral and Brain Sciences, 16*(4), 681–735.
Dunbar, R. I. M. (1996). *Grooming, gossip, and the evolution of language*. London: Faber & Faber.
Edelman, G. (1992). *Bright air, brilliant fire: On the matter of the mind*. New York: Basic Books.
Firth, J. R. (1950). Personality and language in society. *Sociological Review, 42*(1), 37–52. Reprinted in Firth, J. R. (1957). *Papers in linguistics 1934–1951* (pp. 177–189). London: Oxford University Press.
García, A. M. (2019). *The neurocognition of translation and interpreting*. Amsterdam: John Benjamins.
García, A. M., Sullivan, W., & Tsiang, S. (2017). *An introduction to relational network theory: History, principles, and descriptive applications*. London: Equinox.
Gardner, H. (1985). *The mind's new science: A history of the cognitive revolution*. New York: Basic Books.
Grace, G. W. (1981). *An essay on language*. Columbia, SC: Hornbeam Press.
Halliday, M. A. K. (1956). The linguistic basis of a mechanical thesaurus, and its application to English preposition classification. *Mechanical Translation, 3*(3), 81–88. Reprinted in Halliday, M. A. K. (2004). *Computational and quantitative studies*. Volume 6: The Collected Works of M. A. K. Halliday (J. Webster, Ed.) (pp. 6–19). London: Continuum.
Halliday, M. A. K. (1959–60). Typology and the exotic. Combination of two lectures, one delivered at the Linguistics Association Conference, Hull, in May 1959, the other to the St. Andrews Linguistic Society, in May 1960. In M. A. K. Halliday, & A. McIntosh (1966). *Patterns of language: Papers in general, descriptive and applied linguistics* (Ch. 10, pp. 165–182) London: Longman.
Halliday, M. A. K. (1962). Linguistics and machine translation. *Zeitschrift für Phonetik, Sprachwissenschaft und Kommunikationsforschung, 15*(1–2), 145–158.
Halliday, M. A. K. (1966). Some notes on "deep" grammar. *Journal of Linguistics, 2*(1), 57–67. Reprinted in Halliday, M. A. K. (2002). *On grammar*. Volume 1 in the Collected Works of M. A. K. Halliday, J. Webster (Ed.) (pp. 106–117). London: Continuum.
Halliday, M. A. K. (1973). *Explorations in the functions of language*. London: Edward Arnold.
Halliday, M. A. K. (1977). Text as semantic choice in social contexts. In T. van Dijk, & J. Petöfi (Eds.), *Grammars and descriptions* (pp. 176–225). Berlin: Walter de Gruyter. Reprinted in Halliday, M. A. K. (2002). *Linguistic studies of text and discourse*. Volume 2 in the Collected Works of M. A. K. Halliday, J. Webster (Ed.) (Ch. 2, pp. 23–81). London: Continuum.
Halliday, M. A. K. (1978). *Language as social semiotic: The social interpretation of language and meaning*. London: Edward Arnold.
Halliday, M. A. K. (1982). The de-automatization of grammar: From Priestley's "An Inspector Calls". In J. M. Anderson (Ed.), *Language form and linguistic variation: Papers dedicated to Angus McIntosh* (pp. 129–159). Amsterdam: John Benjamins. Reprinted in Halliday, M. A. K. (2002). *Linguistic studies of text and discourse*. Volume 2 in the Collected Works of M. A. K. Halliday, J. Webster (Ed.) (pp. 126–148). London: Continuum.
Halliday, M. A. K. (1993). Quantitative studies and probabilities in grammar. In M. Hoey (Ed.), *Data, description, discourse: Papers on the English language in honour of John McH. Sinclair* (pp. 1–25). London: Harper Collins. Reprinted in Halliday, M. A. K. (2005). *Computational and quantitative studies*. Volume 6 in the Collected Works of M. A. K. Halliday, J. Webster (Ed.) (Ch. 7, pp. 130–156). London: Continuum.
Halliday, M. A. K. (1995). On language in relation to the evolution of human consciousness. In S. Allén (Ed.), *Of thoughts and words: Proceedings of Nobel Symposium 92 "The relation between language and mind"*, Stockholm, 8–12 August 1994 (pp. 45–84). Singapore, River Edge NJ & London: Imperial College

Press. Reprinted in Halliday, M. A. K. (2003) *On language and linguistics*. Vol. 3 in the collected works of M. A. K. Halliday, J. Webster (Ed.) (pp. 390–432). London: Continuum.

Halliday, M. A. K. (1996). On grammar and grammatics. In R. Hasan, C. Cloran, & D. Butt (Eds.), *Functional descriptions: Theory into practice* (pp. 1–38). Amsterdam: John Benjamins. Reprinted in Halliday, M. A. K. (2002). *On grammar*. Volume 1 of Collected Works of M. A. K. Halliday, J. Webster (Ed.) (Ch. 15, pp. 384–417). London: Continuum.

Halliday, M. A. K. (2002). Computing meanings: Some reflections on past experience and present prospects. In G. Huang, & Z. Wang (Eds.), *Discourse and language functions* (pp. 3–25). Shanghai: Foreign Language Teaching and Research Press. Reprinted in Halliday, M. A. K. (2005). *Computational and quantitative studies*. Volume 6 in the Collected Works of M.A.K. Halliday, J. Webster (Ed.) (pp. 239–267). London: Continuum.

Halliday, M. A. K. (2003a). *The language of early childhood*. Volume 4 of Collected Works of M. A. K. Halliday, J. Webster (Ed.). London: Continuum.

Halliday, M. A. K. (2003b). On the "architecture" of human language. In J. J. Webster (Ed.), *On language and linguistics* (pp. 1–29). Volume 3 in the Collected Works of M. A. K. Halliday. London and New York: Continuum.

Halliday, M. A. K. (2005). On matter and meaning: The two realms of human experience. *Linguistics and the Human Sciences, 1*(1), 59–82.

Halliday, M. A. K. (2009). The gloosy ganoderm: Systemic functional linguistics and translation. *Chinese Translators Journal* [中国翻译], *1*, 17–26. Reprinted in Halliday, M. A. K. (2013). *Halliday in the 21st century*. Volume 11 in the Collected Works of M. A. K. Halliday, J. Webster (Ed.). (pp. 105–126). London: Bloomsbury.

Halliday, M. A. K. (2010). Pinpointing the choice: Meaning and the search for equivalents in a translated text. In J. Webster (Ed.) *Halliday in the 21st century. Collected Works of M. A. K. Halliday*. (Vol. 11, Ch. 8, pp. 143–154). London: Bloomsbury Academic.

Halliday, M. A. K. (2013). Meaning as choice. In L. Fontaine, T. Bartlett, & G. O'Grady (Eds.), *Systemic functional linguistics: Exploring choice* (pp. 15–36). Cambridge: Cambridge University Press.

Halliday, M. A. K. & James, Z. L. (1993). A quantitative study of polarity and primary tense in the English finite clause. In J. M. Sinclair, M. Hoey, & G. Fox (Eds.), *Techniques of description: Spoken and written discourse (A Festschrift for Malcolm Coulthard)* (pp. 32–66). London: Routledge. Reprinted in Halliday, M. A. K. (2005). *Computational and quantitative studies*. Volume 6 in the Collected Works of M. A. K. Halliday, J. Webster (Ed.) (Ch. 6, pp. 93–129). London: Continuum.

Halliday, M. A. K., McIntosh, A., & Strevens, P. (1964). *The linguistic sciences and language teaching*. London: Longman.

Halliday, M. A. K., & Matthiessen, C. M. I. M. (1999). *Construing experience through meaning: A language-based approach to cognition*. London: Cassell.

Halliday, M. A. K., & Matthiessen, C. M. I. M. (2006). *Construing experience through meaning: A language-based approach to cognition*. London & New York: Continuum.

Halliday, M. A. K., & Matthiessen, C. M. I. M. (2014). *Halliday's introduction to functional grammar* (4th ed.). London: Routledge.

Hansen-Schirra, S., Neumann, S., & Steiner, E. (2012). *Cross-linguistic corpora for the study of translations: Insights from the language pair English–German*. Berlin: de Gruyter.

Holmes, J. S. (1988). *Translated! Papers on literary translation and translation studies*. Amsterdam: Rodopi.

House, J. (1997). *Translation quality assessment: A model revisited*. Tübingen: Gunter Narr.

House, J. (2001). Translation quality assessment: Linguistic description versus social evaluation. *Meta: Translators' Journal, 46*(2), 243–257.

House, J. (2015). *Translation quality assessment: Past, present and future*. Oxon & New York: Routledge.

House, J. (2016). *Translation as communication across languages and cultures*. London: Routledge.

Jackendoff, R. (1997). *The architecture of the language faculty*. Cambridge, MA: The MIT Press.

Jakobsen, A. L. (2014). The development and current state of translation process research. In E. Brems, R. Meylaerts, & L. van Doorslaer (Eds.), *The known unknowns of translation studies* (pp. 65–88). Amsterdam: John Benjamins.

Jakobsen, A. L. (2017). Translation process research. In J. Schwieter, & A. Ferreira (Eds.), *The handbook of translation and cognition* (pp. 19–49). Hoboken, NJ: Wiley-Blackwell.

Jakobson, R. (1959). On linguistic aspects of translation. In R. A. Bower (Ed.), *On translation* (pp. 232–239). New York: Oxford University Press.

Katan, D. (2009). Translation as intercultural communication. In J. Munday (Ed.), *The Routledge companion to translation studies* (pp. 74–92). London: Routledge.
Katan, D. (2014). *Translating cultures: An introduction for translators, interpreters and mediators* (2nd ed.). London: Routledge.
Kempson, R., Meyer-Viol, W., & Gabbay, D. (2001). *Dynamic syntax: The flow of language understanding.* Oxford: Blackwell.
Kim, M. (2007). Using systemic functional text analysis for translator education: An illustration with a focus on textual meaning. *The Interpreter and Translator Trainer, 1*(2), 223–246.
Kim, M., & Matthiessen, C. M. I. M. (2015). Ways to move forward in translation studies: A textual perspective. *Target, 27*(3), 335–350.
Kittredge, R., & Lehrberger, L. (Eds.) (1982). *Sublanguage: Studies of language in restricted semantic domains.* Berlin: de Gruyter.
Kress, G., & Hodge, B. (1988). *Social semiotics.* Ithaca: Cornell University Press.
Kunz, K., Degaetano-Ortlieb, S., Lapshinova-Koltunski, E., Menzel, K., & Steiner, E. (2017). English–German contrasts in cohesion and implications for translation. In G. de Sutter, M.-A. Lefer, & I. Delaere (Eds.), *Empirical translation studies: New methodological and theoretical traditions* (pp. 265–311). Berlin: Mouton de Gruyter.
Lamb, S. M. (1999). *Pathways of the brain: The neurocognitive basis of language.* Amsterdam: John Benjamins.
Lamb, S. M. (2013). Systemic networks, relational networks, and choice. In L. M. Fontaine, T. A. M. Bartlett, & G. N. O'Grady (Eds.), *Systemic functional linguistics: Exploring choice* (pp. 137–160). Cambridge: Cambridge University Press.
Lavid, J. (2000). Cross-cultural variation in multilingual instructions: A study of speech act realisation patterns. In E. Ventola (Ed.), *Discourse and community: Doing functional linguistics* (pp. 71–85). Tübingen: Günter Narr Verlag.
Layzer, D. (1990). *Cosmogenesis: The growth of order in the universe.* Oxford: Oxford University Press.
Lefevere, A. (1992). *Translation, rewriting, and the manipulation of literary fame.* London: Routledge.
Lovelock, J. (1991). *Gaia: The practical science of planetary medicine.* Sydney: Allen & Unwin.
Ma, Y., & Wang, B. (forthcoming). *Translating Tagore's Stray Birds into Chinese: Applying systemic functional linguistics to Chinese poetry translation.* Under contract. To be submitted to Routledge in March 2021.
Maagerø, E., & Andersen, T. H. (Eds.). (2015). *Social semiotics: Key figures, new directions.* London: Routledge.
McCabe, A. (1999). Theme and thematic patterns in Spanish and English history texts [unpublished PhD dissertation]. Aston University.
Macdonald, K. (2019). Construing musical discourses: axial reasoning for a contrastive description of habitual ideational resources in English and Korean, with reflection on translation [unpublished PhD thesis]. Hong Kong Polytechnic University.
Marinetti, C. (2011). Cultural approaches. In Y. Gambier, & L. van Doorslaer (Eds.), *Handbook of translation studies* (Vol. 2, pp. 26–30). Amsterdam: John Benjamins.
Mason, I. (2012). Text parameters in translation: Transitivity and institutional cultures. In L. Venuti (Ed.), *The translation studies reader* (3rd ed., pp. 399–410). London: Routledge.
Matthiessen, C. M. I. M. (1993). Register in the round: Diversity in a unified theory of register analysis. In M. Ghadessy (Ed.), *Register analysis: Theory and practice* (pp. 221–292). London: Pinter.
Matthiessen, C. M. I. M. (1998). Construing processes of consciousness: From the commonsense model to the uncommonsense model of cognitive science. In J. R. Martin, & R. Veel (Eds.), *Reading science: Critical and functional perspectives on discourses of science* (pp. 327–357). London: Routledge.
Matthiessen, C. M. I. M. (2001). The environments of translation. In E. Steiner, & C. Yallop (Eds.), *Beyond content: Exploring translation and multilingual text* (pp. 41–124). Berlin: Mouton de Gruyter.
Matthiessen, C. M. I. M. (2002). Lexicogrammar in discourse development: Logogenetic patterns of wording. In G. Huang, & Z. Wang (Eds.), *Discourse and language functions* (pp. 91–127). Shanghai: Foreign Language Teaching and Research Press.
Matthiessen, C. M. I. M. (2004). Descriptive motifs and generalizations. In A. Caffarel, J. R. Martin, & C. M. I. M. Matthiessen (Eds.), *Language typology: A functional perspective* (pp. 537–673). Amsterdam: John Benjamins.
Matthiessen, C. M. I. M. (2007). The "architecture" of language according to systemic functional theory: Developments since the 1970s. In R. Hasan, C. M. I. M. Matthiessen, & J. Webster (Eds.), *Continuing discourse on language* (Volume 2, pp. 505–561). London: Equinox.
Matthiessen, C. M. I. M. (2014). Choice in translation: Metafunctional consideration. In K. Kunz, E. Teich, S. Hansen-Schirra, S. Neumann, & P. Daut (Eds.), *Caught in the middle—language use and translation:*

A festschrift for Erich Steiner on the occasion of his 60th birthday. Saarbrücken: Universaar, Saarland University Press. http://universaar.uni-saarland.de/monographien/volltexte/2014/122/pdf/Kunz_etal_Festschrift_Steiner.pdf

Matthiessen, C. M. I. M. (2015a). Halliday's probabilistic theory of language. In J. Webster (Ed.), *The Bloomsbury companion to M.A.K. Halliday* (pp. 203–241). London: Bloomsbury Academic.

Matthiessen, C. M. I. M. (2015b). Halliday on language. In J. Webster (Ed.), *The Bloomsbury companion to M.A.K. Halliday* (pp. 137–202). London: Bloomsbury Academic.

Matthiessen, C. M. I. M. (2015c). The language of space: Semiotic resources for construing our experience of space. *Japanese Journal of Systemic Functional Linguistics, 8*, 1–64.

Matthiessen, C. M. I. M. (2018). The notion of a multilingual meaning potential: A systemic exploration. In A. Baklouti, & L. Fontaine (Eds.), *Perspectives from systemic functional linguistics*. (Ch. 6, pp. 90–120). London: Routledge. Version with additional figures to be available at www.syflat.tn

Matthiessen, C. M. I. M. (2019). Register in Systemic Functional Linguistics. *Register Studies, 1*(1), 10–41.

Matthiessen, C. M. I. M. (forthcoming). *The architecture of language according to systemic functional theory*. Book MS.

Matthiessen, C. M. I. M., Arús-Hita, J., & Teruya, K. (forthcoming). Translations of representations of moving and saying from English into Spanish. To be submitted to *Word*.

Matthiessen, C. M. I. M., & Bateman, J. A. (1991). *Systemic linguistics and text generation: Experiences from Japanese and English*. London: Frances Pinter.

Matthiessen, C. M. I. M., Sondheimer, N., & Tung, Y.-W. (1988). *On parallelism and the Penman Natural Language Generation System*. University of Southern California/ Information Sciences Institute.

Matthiessen, C. M. I. M., Teruya, K., & Canzhong, W. (2008). Multilingual studies as a multi-dimensional space of interconnected language studies. In J. Webster (Ed.), *Meaning in context* (pp. 146–221). London: Continuum.

Miller, G. A. (2003). The cognitive revolution: A historical perspective. *Trends in Cognitive Sciences, 7*(3), 141–144.

Munday, J. (2000). Using systemic functional linguistics as an aid to translation between Spanish and English: Maintaining the thematic development of the ST. *Revista Canaria de Estudios Ingleses, 40*, 37–58.

Munday, J. (2012). *Evaluation in translation: Critical points of translator decision-making*. London: Routledge.

Munday, J. (2018). A model of appraisal: Spanish interpretations of President Trump's inaugural address 2017. *Perspectives: Studies in translation theory and practice, 26*(2), 180–195.

Murcia-Bielsa, S. (2000). The choice of directives expressions in English and Spanish instructions: A semantic network. In E. Ventola (Ed.), *Discourse and community: Doing functional linguistics. Language in Performance*, 21 (pp. 117–146). Tübingen: Gunter Narr Verlag.

Nida, E. A. (2001). *Contexts in translating*. Amsterdam: John Benjamins.

Nord, C. (2005). *Text analysis in translation: Theory, methodology, and didactic application of a model for translation-oriented text analysis*. Amsterdam: Rodopi.

PACTE (2003). Building a translation competence model. In F. Alves (Ed.), *Triangulating translation: Perspectives in process oriented research* (pp. 43–66). Amsterdam: John Benjamins.

Painter, C. (1999). *Learning through language in early childhood*. London: Cassell.

Patten, T. (1988). *Systemic text generation as problem solving*. Cambridge: Cambridge University Press.

Reddy, M. J. (1979). The conduit metaphor: A case of frame conflict in our language about language. In A. Ortony (Ed.), *Metaphor and thought* (pp. 284–310). Cambridge: Cambridge University Press.

Rojo, A. (2015). Translation meets cognitive science: The imprint of translation on cognitive processing. *Multilingua: Journal of Cross-Cultural and Interlanguage Communication, 34*(6), 721–746.

Saldanha, G., & O'Brien, S. (2014). *Research methodologies in translation studies*. London: Routledge.

Schwieter, J. W., & Ferreira, A. (Eds.). (2017). *The handbook of translation and cognition*. Hoboken, NJ: Wiley-Blackwell.

Shreve, G. M., & Angelone, E. (Eds.). (2010). *Translation and cognition*. Amsterdam: John Benjamins.

Snell-Hornby, M. (1995). *Translation studies: An integrated approach* (2nd ed.). Amsterdam: John Benjamins.

Snell-Hornby, M. (2010). The turns of translation studies. In Y. Gambier, & L. van Doorslaer (Eds.), *Handbook of translation studies* (Vol. 1, pp. 366–370). Amsterdam: John Benjamins.

Steiner, E. (2004). *Translated texts: Properties, variants, evaluations*. Frankfurt: Peter Lang.

Steiner, E. (2005). Hallidayan thinking and translation theory—enhancing the options, broadening the range, and keeping the ground. In R. Hasan, C. M. I. M. Matthiessen, & J. Webster (Eds.), *Continuing discourse on language: A functional perspective* (Vol. 1, pp. 481–500). London: Equinox.

Steiner, E. (2015). Translation. In J. Webster (Ed.), *The Bloomsbury companion to M.A.K. Halliday* (pp. 412–426). London: Bloomsbury Academic.

Steiner, E., & Yallop, C. (Eds.). (2001). *Beyond content: Exploring translation and multilingual text production*. Berlin & New York: de Gruyter.
Taylor, C. (2003). Multimodal transcription in the analysis, translation and subtitling of Italian films. *The Translator, 9*(2), 191–205.
Taylor, C. (2013). Multimodality and audiovisual translation. In Y. Gambier, & L. van Doorslaer (Eds.), *Handbook of translation studies* (Vol. 4, pp. 98–104). Amsterdam: John Benjamins.
Teich, E. (1999). System-oriented and text-oriented comparative linguistic research: Cross-linguistic variation in translation. *Languages in Contrast, 2*(2), 187–210.
Teich, E. (2003). *Cross-linguistic variation in system and text: A methodology for the investigation of translations and comparable texts*. Berlin: Mouton de Gruyter.
Teich, E. (2009). Computational linguistics. In M. A. K. Halliday, & J. Webster (Eds.), *A companion to systemic functional linguistics* (pp. 113–127). London: Continuum.
Toury, G. (1995). *Descriptive translation studies and beyond*. Amsterdam: John Benjamins.
Toury, G. (2004). Probabilistic explanations in translation studies: Welcome as they are, would they qualify as universals? In A. Mauranen, & P. Kujamäki (Eds.), *Translation universals: Do they exist?* (pp. 15–32). Amsterdam: John Benjamins.
van Leeuwen, T. (2005). *Introducing social semiotics*. New York: Routledge.
Varela, F. J., Thompson, E., & Rosch, E. (1991). *The embodied mind: Cognitive science and human experience* (revised ed. 2016). Cambridge, MA: The MIT Press.
Wang, B., & Ma, Y. (2018). Textual and logical choices in the translations of dramatic monologue in *Teahouse*. In A. Sellami-Baklouti, & L. Fontaine (Eds.), *Perspectives from systemic functional linguistics* (pp. 140–162). Oxon & New York: Routledge.
Wang, B., & Ma, Y. (in press). *Systemic functional translation studies: Theoretical insights and new directions*. Sheffield: Equinox.
Wang, B., & Ma, Y. (forthcoming). *Lao She's Teahouse and its two English translations: Exploring Chinese drama translation with systemic functional linguistics*. Under contract. To be submitted to Routledge in January 2020.
Wang, P. (2004). "Harry Potter" and its Chinese translation: An examination of modality system in Systemic Functional approach [unpublished PhD thesis].
Whorf, B. L. (1956). *Language thought and reality: Selected writing of Benjamin Lee Whorf*, J. B. Carroll (Ed.). Cambridge, MA: The MIT Press.
Wilson, E. O. (2019). *Genesis: The deep origin of societies*. New York: Liveright Publishing Corporation.
Yu, H., & Wu, C. (2016). Recreating the image of Chan master Huineng: The roles of mood and modality. *Functional Linguistics, 3*(4), 1–21.

30

Grounding Cognitive Translation Studies: Goals, commitments and challenges

Fabio Alves and Arnt Lykke Jakobsen

30.1 Introduction

With this chapter, we come to the end of *The Routledge handbook of translation and cognition*. In the Introduction and the preceding 29 chapters, we have attempted to develop the three main goals we had for the volume: to help lay the epistemological, paradigmatic and interdisciplinary foundations for a sub-discipline still in the making. As we wrote in the Introduction, we aimed at openly acknowledging the lack of unity we presently see in the study of translation and cognition and have therefore striven to create a space in which a variety of conceptualizations, approaches and topics could be reflected that would collectively represent the current state of the art in Cognitive Translation Studies (CTS). Although the Handbook does not offer a perfectly integrated theory of how CTS could be finally grounded, we hope that it nevertheless provides a state-of-the-art basis for taking the last steps towards a more unified theory of this exciting new sub-discipline within Translation Studies.

As we come to a close, we see three main issues that remain to be discussed: namely the question of how we refer to the sub-discipline, its main commitments and the paradigmatic ground(s) on which it stands. In the following section, we will draw on thoughts raised by some of the authors who have contributed to this volume and use some of their voices to consubstantiate some of our own thinking. We end with a reflection on some challenges ahead for CTS.

30.2 Core issues

30.2.1 The name

As we have seen throughout the previous chapters, the historical development of the study of translation and cognition has always drawn on disparate disciplinary affiliations. If one looks back at the seminal work of Seleskovitch (1968), one can identify the prominence given to meaning construction and representation, which was of paramount importance in her vision of how translation unfolds as a cognitive activity. However, there has also been a parallel tendency to look at translation and cognition from the perspective of problem solving and verbalization. In the mid-1980s, researchers began an attempt to tap into the black box of translation (see Königs, 1987; Krings, 1986) by drawing on the information processing paradigm to provide an account of how the translation process unfolded in real time.

Within that tradition, the work of the TRAP Group at the Copenhagen Business School (CBS) in the 1990s, and the development of the Translog software (Jakobsen & Schou, 1999) in particular, led to a line of research that has been widely recognized as translation process research (TPR) (see Jakobsen, 2011). With the growth of TPR, first drawing on keylogged data and later on a combination of keylogging and eye-tracking data, the discussion about the translation unit, aspects of segmentation in translation, phases of the translation process and instances of peak performance, among other topics, gained prominence in research. With the creation of the Centre for Research in Translation and Translation Technology (CRITT) at CBS, TPR established itself as a widely used methodological paradigm for empirical research in translation and cognition at several university centres around the world.

Around 2010, this unifying, technology-based paradigm for studying translation from a cognitive perspective began to be questioned. Instead, a different approach suggested that translation as a cognitive activity should be regarded as situated action rather than cognition in the head. The term cognitive translatology (Muñoz, 2010) appeared as an alternative term to translation process research, claiming a clear differentiation between an approach which drew on the information processing paradigm and an approach which promoted a vision of translation in terms of embodied, enacted, embedded, extended and affective, so-called 4EA, cognition.

As we wrote in the introduction, cognitive translatology aimed at recontextualizing, or "reembedding", translation process research (Muñoz, 2016), thus broadening the scope of research to include considerations which were not taken onboard by the information processing paradigm implied in the TPR approach. By drifting away from an analogy with computational modelling, cognitive translatology rejected the dualism separating mind and body and proposed grounding mind in matter (see Maturana & Varela, 1987; Varela et al., 1991). Several chapters in the *Routledge handbook of translation and cognition* have contributed to taking this discussion to a higher level (see Section 30.2.3), addressing methodological consequences that follow from this disciplinary recontextualization.

Around the same time as cognitive translatology was presented as an alternative to TPR, Halverson (2010) proposed the term Cognitive Translation Studies (CTS) as a terminological alternative in view of the broad engagement of cognitive translation scholars with cognitive theories. For Halverson (2010, p. 353), "a cognitive theory of translation must build on cognitive theories of language". In our opinion, CTS is a broad enough term to accommodate both the computational information processing and the situated-action paradigms, whether supported methodologically by experiments, introspection, concept analysis, think-aloud, corpus analysis or field studies. We believe that these paradigmatic and methodological differences are best addressed from a complementary, integrative approach. This is in line with recent views voiced e.g. by García (2019, p. 26) and Kotze (2019, p. 333). Therefore, while acknowledging the relevance of translation process research and cognitive translatology for the historical development of the study of translation as a cognitive activity, CTS is our preferred terminological label for the sub-discipline described in the present Handbook.

The discussion about the name of the sub-discipline would not be complete without some considerations concerning interpreting as a cognitive activity. Throughout the previous 29 chapters, although several references have been made to interpreting, the main focus of the volume has fallen on translation. However, there is no doubt that translation and interpreting have enough in common, cognitively, to be brought under a single disciplinary framework. Defrancq et al. (2020) speak of "reuniting the sister disciplines of translation and interpreting studies" (p. 1), and the forthcoming publication of a *Routledge handbook of interpreting and cognition* offers another opportunity to unite Cognitive Studies of Translation and Interpreting. This would lead to extending the current denomination to cognitive translation and interpreting studies (CTIS), a term that has already appeared in several recent publications and in the present Handbook, as a signal to future theoretical development.

30.2.2 Commitment(s)

The primary commitment of CTS, in our view, is specifically to how *meaning* is communicated across languages. Meaning is the sense we make of reality with all our senses, emotions and intellect. Meaning is therefore inherently multimodal, and increasingly we see multimodal representation of meaning being used in everyday communication as well as in translation. The verisimilitude and situatedness achievable with multimodal, technology-supported representation of meaning may explain its astounding popularity, but spoken and written forms of language remain very important communicative components of multimodal communication for their ability to allow people to align and share thoughts. Language is so integrated with our perception of reality that it co-shapes our sense of reality and co-socializes us to a specific culture. Video images and audio sounds (even if technically digital) are analogical media and therefore travel the world more easily than language communications, although video and audio communications also need to be interpreted. Language communications are symbolic representations of meaning and therefore coded. This means that languages build boundaries around communities of language users, excluding speakers of other languages. That is why translation and interpreting are necessary and why the primary commitment of CTS is to explain how meaning travels across language boundaries. The ancient metaphor of translation as a bridge across a divide between two languages and cultures reflects this fundamental commitment.

A second commitment, therefore, is to a theory of how *language* mediates or helps mediate multimodal meaning: How language users use language to represent their meaning, and how they create meaning from other persons' language representations. Halverson (Chapter 2) argues that a theory of translation must build on a clearly articulated commitment to a view of language and language processing in translation to provide a solid and coherent account of CTS. Her commitment is to cognitive linguistics. Carl (Chapter 28) attempts to integrate connectionist and 4EA ideas of embeddedness and extended cognition in his computational modelling of translation, but such modelling still relies heavily on the physical symbol systems hypothesis, i.e. on linguistic representations of meaning.

Kotze (Chapter 6) shows how language contact may have similar effects to those of translation on language use. Researchers working from large language corpora, including Hansen-Schirra (2011) and Neumann (2013, 2014), have shown how sets of linguistic features demonstrate complex patterns of over- and under-representation in translations compared with original texts, reflecting both cross-linguistic influence on cognition and (over)adjustment to target-language norms. In parallel, researchers working on translation process data have examined links between cross-linguistic units in source and target texts and have found patterns that show how different types of transfer can point to a default strategy for reducing cognitive load (see Tirkkonen-Condit, 2005) or to more complex strategies when meaning has to be unpacked and repacked in distinct cross-linguistic units in source and target texts (see Alves et al, 2010; Carl & Dragsted, 2012; Schaeffer & Carl, 2013). Shreve & Lacruz (2017) present an overview of research from psycholinguistics and cognitive science showing how factors like task schemas interact with bilingual language production in the context of translation. The approaches represented in Chapters 9, 10, 11, 19, 20 and 23 also clearly prioritize linguistically represented meaning. Finally, Matthiessen's commitment (Chapter 29), which is to a semiotic-based meaning approach to translation rooted in systemic functional grammar, is also rooted in language.

Muñoz Martín and Martín de León's commitment (Chapter 3) is to a 4EA view of meaning as articulated in cognitive science. This commitment is not necessarily language bound, or bound to processes in the brain only, but extends into the body and the environment. Their prioritized commitment is most clearly to *cognition*. The same can be said of Chapters 7, 14, 15, 16, 17 and 18, all of which explore cognitive effects of translation either from linguistic representations or from behaviour associated with translation (and interpreting). Matthiessen (Chapter 29) considers that his commitment is complementary to what he considers to be a fundamentally knowledge-based cognitive approach. In our view, the various commitments listed here are indeed complementary, for whether or not one takes meaning or cognition or language to be primary, CTS is necessarily committed to all three. Regardless of priority, there is no escape from the inseparableness of meaning, language and cognition or, in the words of Macdonald (Chapter 5), from the "inextricable tripartite of thought, language and culture".

This means that the questions we asked in the introduction about translation and cognition remain open: Is language, translated or non-translated, a way to understanding the mind? Do we construe experience through linguistic meaning? Or do we construe experience *as* meaning, pre-linguistically, from interaction with our environment? Can we know about the mind from studying behaviour? From studying the brain? From studying culture? We believe the oscillation between the primary focus being on meaning, language or cognition as a means of discovery about translation is of benefit for CTS. They constitute a creative and productive trinity.

Chesterman (Chapter 1) raises concern about the scope of CTS with the situated-action approach. "If everything is connected to everything else, and the agent is seen as embedded in a complex environment, it will not be easy to isolate elements for close study; and if one has to study the whole network, where does the network end?", he asks. This suggests a final commitment of an ethical kind for CTS researchers to always explicate, to the extent possible, the *scope* of their inquiries.

30.2.3 The ground

The epistemological and ontological ground on which CTS is based is thoroughly explained in the Introduction and the four chapters in Part I. Corresponding to the triple commitment, CTS will be grounded in theories of semiosis (meaning making) and linguistics (language use) and on cognitive science (neurocognition and situated-action cognition).

Methodologically, we believe there is virtue in both inductive and deductive approaches and that empirical studies should go hand in hand with conceptual development, theory building and computational modelling. This appears to be generally acknowledged by all contributors to the present Handbook. Therefore, there are multiple ways of methodologically grounding CTS. CTS is already a large, some might say sprawling, construct with great methodological diversity. Empirical, experimental, corpus, ethnographic, computational and neurophysiological methods are all being used, and it is not surprising that many chapters advocate a multi-method approach. CTS even welcomes explorative approaches based on intuition or curiosity. They can sometimes lead the way to new discoveries.

In the Introduction, we made it clear that the Handbook would focus on exchanges with disciplines CTS has interacted with most vigorously, such as anthropology, psychology and cognitive science. This interdisciplinary dialogue has most often focused on conceptual aspects. As O'Brien (2013) rightly put it, within CTS, borrowing has been much more common than lending. Alves (2015) also explored interfaces of CTS with neighbouring disciplines. Alves and

Hurtado (forthcoming) differentiate between disciplines with direct and indirect impact on the study of translation as a cognitive activity. Muñoz Martín and Martín de León (Chapter 3) raise considerations about the implications of drawing on cognitive science for the growth of CTS. All these authors agree that interaction with cognitive science is currently highly profitable and productive and offers CTS a firm cognitive grounding. This means that although CTS is its own sub-discipline, its key tenets should be consistent with key tenets in cognitive science. Inconsistencies arising from new findings in one of the disciplines will have repercussions in the other, leading eventually to adjustment and growth. This applies more broadly to tenets in the family of disciplines CTS relates to. Being interdisciplinary, CTS is bound to operate in a multi-relational conceptual and methodological research environment, in which emerging inconsistency may well be a sign of disciplinary growth resulting from interdisciplinary dialogue.

Shreve (Chapter 4) defines CTS as "a cover term referring to a research tradition within Translation Studies that focuses on explaining the cognitive foundations of translating and other language mediation tasks like interpreting". He places CTS within the boundaries of cognitive science and points out that cognitive science is eminently interdisciplinary. Like CTS, it has also struggled to integrate disparate constructs stemming from multiple disciplinary frameworks. "In our understanding, as Shreve further suggests, many of the most pressing issues of Cognitive Translation Studies today are reflections of issues facing cognitive science broadly". For him, both CTS and cognitive science face the challenge of "how to integrate more traditional computationalist (information processing) models of cognition with both connectionist and extended cognition frameworks such as distributed and situated cognition".

If CTS was driven mainly by the information processing paradigm in the early phases, the most recent development has been more clearly driven by theories of mind, cognition and meaning developed by cognitive scientists who have insisted that cognition is situated and distributed as well as embodied, embedded, enacted, extended and affective. In Chapter 27, Risku and Rogl offer an overview of recent, situated approaches to cognitive science—namely situated, embodied, distributed, embedded and extended cognition. They argue that "there is no agreement in Translation Studies on whether classical cognitive science theories and more recent frameworks could be used in a complementary way". They also suggest that the differences between computational, connectionist and situated approaches to cognition occur more on the level of abstraction rather than on the level of irreconcilable theoretical positions. For Risku and Rogl, this potentially makes a combination of views from different frameworks possible, with an impact on levels of explanation. We agree that recent developments in cognitive science require reformulation of previously used concepts and paradigms in CTS. This, however, does not necessarily invalidate the findings of previously undertaken studies drawing on the information-processing paradigm, but earlier findings may need to be put to the test and reformulated in the light of more recent cognitive approaches to language and translation.

In their account of situated, embodied, distributed, embedded and extended approaches to cognition, Risku and Rogl (Chapter 27) argue that "embodied cognition is tightly knit with the situated, embedded, distributed and extended approaches to cognition". For them, the distributed cognition approach has expanded the boundaries of the cognitive unit of analysis. While in the situated approach to cognition, context and social situation are fundamental factors that influence cognitive processes, the distributed approach to cognition focuses on the distributedness of cognition and action, including all forms of complex systems, structures or frameworks. To this we would add that the distributed approach to cognition can also accommodate aspects of computational modelling and artificial intelligence that have already changed the way humans and machines now interact in translation-related matters. Neural machine translation, for instance, works on the basis of a distributed complex systems network based on tenets postulated by connectionism. In their account of approaches to cognition, Risku and Rogl point out that "situatedness, embodiment and embeddedness of cognition have fuzzy boundaries and slightly different foci". They add that these three approaches also overlap greatly. There may be fine-grained aspects that differentiate them, such as the rejection of internal, mental representations in the embedded approach to cognition. Nevertheless, for the purposes of CTS, we would like to suggest that situatedness can also accommodate both embodiment and embeddedness.

Shreve (Chapter 4, this volume) proposes to include CTS under the umbrella of complex adaptive systems (CAS) theory. For Shreve, CTS is "multi-scale", and, as a complex adaptive system, it can be studied at all levels, from the level of cell activity or below to the social and cultural level or beyond. For this reason, Shreve adds, it is important that the scope of research within CTS is always specified. Shreve notes that Marais (2014) already detailed the implications of complexity thinking for Translation Studies in general and focused on translation as an emergent phenomenon. Shreve's proposal, however, is primarily targeted at providing CTS with a theoretical ground. As Shreve makes a case for CAS, he includes situated, distributed and extended aspects of cognition into the boundaries of the complex adaptive system pertaining to translational action. There are concrete implications of such an approach for applied aspects of CTS, including the acquisition of translation competence (Chapter 22), the path to translation expertise (Chapter 26) and the traits of risk management (Chapter 25), as well as the allocation of cognitive resources to translation such as effort, attention, emotion and creativity (Chapters 14, 15, 16, 17 and 18).

Finally, the extended approach to cognition, according to Risku and Rogl, refers to the "extended mind" approach proposed by Clark and Chalmers (1998), considering the brain, the body and the environment as a "coupled system". In the extended approach to cognition, neural and non-neural elements constitute potentially proper parts of the mind. Cognition is extended, "as it offloads tasks onto the external environment to free up limited cognitive resources (e.g. using a notepad as external memory storage)" (Carl, Chapter 28).

In closing the handbook, we would like to take the discussion a step further by proposing what we are tentatively calling a SDE (situated, distributed and extended) approach to cognition within CTS. We suggest that several of the labels used to describe 4EA cognition (embodied, embedded, extended, enacted and affective) as well as situated and distributed cognition are somewhat redundant for CTS. To avoid the many terminological overlaps among these various construals and approaches, we propose an account in terms of only situated, distributed and extended (SDE) approaches to translation and cognition. The situated approach brings specific context and social factors into focus, the distributed approach can add a systemic view of interconnectedness at different levels, and the extended approach, embracing information theory (Chapter 20), artificial intelligence (Chapter 28), cognitive ergonomics (Chapter 8) and human–computer interaction (Chapter 21), among other relevant factors, accounts for intricate and complex affordances in the relationship of human beings with physical, virtual or digital artefacts. In line with Shreve (see Chapter 4), situated, distributed and extended aspects of cognition pertain to translational action and constitute fundamentally important pillars for CTS. We would like to argue that a SDE approach to cognition can provide a cognitive science ground for CTS and offer the field an opportunity to improve the balance between—and the internal coherence within—the cognitive translatological approach and its computational information processing counterpart.

Computers have inspired our thinking about human cognition for a long time, as is clear from the metaphors we have used and are still using. Some have humanized computers, e.g. into electronic brains with artificial intelligence, and some have computerized humans, e.g. into robots or cyborgs. Information processing and problem-solving activity was frequently modelled in the 1980s and 1990s in dehumanized flowcharts with binary options and loops. A more recent metaphor frequently used both in reference to computational and more human-oriented views is the network, often a hugely intricate, global network enabling everything to connect to everything. It is used both in reference to humans and machines to illustrate the connectivity of the brain or a computer, especially a processor, and also to illustrate people's social connectivity, including their virtual, technology-enabled connectivity. With its broad and suggestive application, the network metaphor often tends to underprivilege humans, often viewing them as "network agents" on a par with artefacts and ideas. In Latour's actor-network theory (ANT) (Latour, 1987), anything in existence can be an agent (or "actant"), with no particular importance given to human agents. ANT, incidentally, has borrowed the concept of translation and appropriated it for its own purposes to describe a network process, which reminds us that CTS is concerned with human meaning making.

Our hope and prediction are that metaphors more suggestive of a living, creative and interactive human activity will soon emerge. The jungle metaphor cited by Chesterman (Chapter 1) nicely suggests life and complexity and also danger and risk like other complex bio- or ecosystems one may think of. We will leave it as one of the challenges for the future to find an appropriate metaphor for capturing the living, adaptable, human complexity from which meaning, language, cognition and translation all emerge.

30.3 Challenges ahead

As we have seen throughout the volume, CTS is now in a position to claim the status of a sub-discipline within Translation Studies with challenging ideas for an epistemological, paradigmatic and methodological framework, from which CTS can grow and establish itself as a reference for scientific work on translation as a cognitive activity. We envisage a general need for research to continue in many of the areas covered in this Handbook but also several challenges:

- There is still a need to strengthen the epistemological grounding of CTS, which will enable the sub-discipline to establish itself in its own right (see Chesterman, Chapter 1).
- There is a need to reinforce the commitment to mind, meaning and language within the sub-discipline, bringing together language and knowledge-oriented cognitive approaches and a meaning-oriented semiotic approach to CTS (Halverson, Chapter 2, and Matthiessen, Chapter 29).
- There is also a need to further strengthen an already productive dialogue with cognitive science so that CTS becomes a discipline that can both benefit from and contribute to the development of multilingual aspects of cognitive science (Muñoz Martín & Martín de León, Chapter 3; Shreve, Chapter 4; García & Muñoz, Chapter 13, Risku & Rogl, Chapter 27).
- There is a need to maintain and strengthen an interdisciplinary dialogue around cognition, meaning and language with disciplines and research fields such as anthropology (Macdonald, Chapter 5); contact linguistics (Kotze, Chapter 6); pragmatics (Alves, Chapter 7); ontology (Pagano, Chapter 9); corpus linguistics (Neumann & Serbina, Chapter 10); linguistics (Malmkjær, Chapter 11); psycholinguistics (Chmiel, Chapter 12); and neuroscience (García & Muñoz, Chapter 13).
- A remaining particular challenge is theorizing the cognitive comparability of cross-linguistic, cross-cultural meaning, perhaps in the kind of ethnolinguistics envisaged by Franz Boas (Macdonald, Chapter 5), perhaps by updating our construal of the troubled concept of equivalence (Steiner, Chapter 19). Kotze (Chapter 6) reminds us that, as far as intercultural communication is concerned, cognitive processes and personal language usage are both affected by exposure to and processing of linguistic features, subject to language contact and sufficient

cognitive entrenchment. In the end, she writes, "the cognitive representations of the two languages syncretize and converge, becoming less representationally distinct."
- Technological developments, e.g. in biometrics, scanning and wearables, create a need to further investigate translation-specific phenomena in relation to CTS, addressing topics such as effort (Gile & Lei, Chapter 14); attention (Hvelplund, Chapter 15); emotions (Lehr, Chapter 16); creativity (Bayer-Hohenwarter & Kußmaul, Chapter 17); metaphor (Schäffner & Chilton, Chapter 18); and the product–process interface (Hansen-Schirra & Nitzke, Chapter 23).
- Multimodality (Kruger, Chapter 24) is an increasingly important field of study in CTS. As stated in the Introduction, it has already expanded the concept and scope of translation to include description of action represented in moving pictures, as in audio description. In a wider perspective, this expansion highlights the element of translation and subjective interpretation both, for instance, in news reporting and, more fundamentally and epistemologically, in our entire perception of reality. The huge power of the concepts of translation and interpreting constitutes a challenge to all CTS research, making it important to define how translation is construed in a given study.
- The accelerating use of translation technology continues to change the nature of translational activity. This affects the ways in which translators interact with technology and with their working environment. It also affects translators and their products. All of this needs to be continuously reassessed, as in ergonomics (Ehrensberger-Dow, Chapter 8) and human–computer interaction (O'Brien, Chapter 21). Technology's contribution to more and more effectively modelling human translation computationally, as in information theory (Teich et al, Chapter 20) and artificial intelligence (Carl, Chapter 28), is another continuing challenge.
- A final challenge is to continue to demonstrate the practical relevance of CTS for learning/teaching translation and for doing it competently and expertly with professional risk assessment; see *Translation competence and its acquisition* (Hurtado, Chapter 22), *Risk management* (Pym, Chapter 25), and *Expert performance* (da Silva, Chapter 26). The exact construal of competence and expertise is still under scrutiny (Marín García, 2017) and remains a challenge for the future.

Jääskeläinen and Lacruz (2018, p. 1) state: "Cognitive research in translation and interpreting has reached a critical threshold of maturity that is triggering rapid expansion along several innovative paths." This threshold of maturity is witnessed by the rapidly growing number of publications of various kinds, including handbooks of the present type, as well as dedicated journals, conferences and competitively funded projects. An interesting parallel to this growth is the slow emergence of clearer lines of demarcation between major approaches to the study of cognition in general and to the study of translational cognition in particular.

If the general framework suggested in this volume is accepted and adopted by the research community in CTS at large, the challenges ahead will be responded to under better-aligned paradigmatic frameworks or even a unified umbrella. If the SDE approach we have outlined here is widely adopted, this might allow CTS to consolidate the view of translation as an ontologically situated, distributed and extended cognitive activity. And if discussions with the cognitive interpreting studies community prove fruitful, we may envisage the development of a unified framework for the whole field of Cognitive Translation and Interpreting Studies (CTIS).

Further reading

The best general recommendation we can give is to read the chapters in the present handbook.

Halliday, M. A. K., & Matthiessen, C. M. I. M. (1999) *Construing experience through meaning: A language-based approach to cognition*. London: Continuum.

This is the fullest presentation of systemic functional grammar. A rich and demanding book, which presents language as a fourth-order social-semiotic system by means of which we construe our experience of the world as meaning. Matthiessen (Chapter 29 this volume) is a perfect introduction.

Marais, K. (2014). *Translation theory and development studies: A complexity theory approach*. New York and Abingdon: Routledge; Marais, K., & Meylaerts, R. (Eds.). (2018). *Complexity thinking in translation studies: Methodological considerations*. Abingdon & New York: Routledge.

The first attempts to apply complexity theory to Translation Studies. The two volumes introduce an epistemological complexity paradigm opposed to the binary thinking in the paradigm of reduction which is claimed to have characterized Western thinking for the last three centuries. A preprint of the introductory chapter to the 2018 publication (by Marais & Meylaerts) is available at www.academia.edu/38528885/20190311-Introduction-Complexity_thinking_in_translation.pdf

Varela, F. J., Thompson, E., & Rosch, E. (1991, revised ed. 2016). *The embodied mind: Cognitive science and human experience*. Cambridge, MA: MIT Press.

This is the classic book that launched the idea of embodied cognition, which sees both the viewer and the viewer's environment as jointly "enacting" meaning. In a general phenomenological framework, reality is understood as emerging, in the form of lived experience, as a result of this embodied interaction. The revised edition includes important introductions by Thompson and Rosch.

References

Alves, F. (2015). Translation process research at the interface: Paradigmatic, theoretical, and methodological issues in dialogue with cognitive science, expertise studies, and psycholinguistics. In A. Ferreira, & J. Schwieter (Eds.), *Psycholinguistic and cognitive inquiries into translation and interpreting* (pp. 17–40). Amsterdam: John Benjamins.

Alves, F., & Hurtado Albir, A. (forthcoming). *Translation as a cognitive activity. Introducing theories, models and methods for empirical research*. London: Routledge.

Alves, F., Pagano, A., Neumann, S., Steiner, E., & Hansen-Schirra, S. (2010). Translation units and grammatical shifts: Towards an integration of product- and process-based translation research. In G. M. Shreve, & E. Angelone (Eds.), *Translation and cognition* (pp. 109–142). Amsterdam: John Benjamins.

Carl, M., & Dragsted, B. (2012). Inside the monitor model: Process of default and challenged translation production. *TC3 Translation: Computation, Corpora, Cognition, 2*(1), 127–145.

Clark, A., & Chalmers, D. J. (1998). The extended mind. *Analysis, 58*(1), 7–19.

Defrancq, B., Daems, J., & Vandevoorde, L (2019). Reuniting the sister disciplines of translation and interpreting studies. In L. Vandevoorde, J. Daems, & B. Defrancq (Eds.), *New empirical perspectives on translation and interpreting* (pp. 1–9). Milton Park, Abingdon: Routledge.

García, A. (2019). *The neurocognition of translation and interpreting*. Amsterdam: John Benjamins.

Halliday, M. A. K., & Matthiessen, C. M. I. M. (1999) *Construing experience through meaning: A language-based approach to cognition*. London: Continuum.

Halverson, S. (2010). Cognitive translation studies: Developments in theory and method. In G. Shreve, & E. Angelone (Eds.) *Translation and cognition* (pp. 349–369). Amsterdam: John Benjamins.

Hansen-Schirra, S. (2011). Between normalization and shining-through: Specific properties of English–German translations and their influence on the target language. In S. Kranich, V. Becher, S. Höder, & J. House (Eds.), *Multilingual discourse production: Diachronic and synchronic perspectives* (pp. 135–162). Amsterdam: John Benjamins.

Jääskeläinen, R., & Lacruz, I. (Eds.) (2018). *Innovation and expansion in translation process research*. Amsterdam: John Benjamins.

Jakobsen, A. L. (2011). Tracking translators' keystrokes and eye movements with Translog. In C. Alvstad, A. Hild, & E. Tiselius (Eds.), *Methods and strategies of process research. Integrative approaches in translation studies* (pp. 37–55). Amsterdam: John Benjamins.

Jakobsen, A. L., & Schou, L. (1999). Translog documentation. In G. Hansen (Ed.), *Probing the process in translation: Methods and results* (pp. 151–186). Copenhagen: Samfundslitteratur.

Königs, F. G. (1987). Was beim Übersetzen passiert: Theoretische Aspekte, Empirische Befunde und praktische Konsequenzen. *Die neueren Sprachen, 86*, 162–185.

Kotze, H. (2019). Converging what and how to find out why: An outlook on translation studies. In L. Vandevoorde, J. Daems, & B. Defrancq (Eds.), *New empirical perspectives on translation and interpreting* (pp. 333–371) Milton Park, Abingdon: Routledge.

Krings, H. (1986). *Was in den Köpfen von Übersetzern vorgeht.* Tübingen: Narr.
Latour, B. (1987). *Science in action: How to follow scientists and engineers through society.* Cambridge, MA: Harvard University Press.
Marais, K. (2014). *Translation theory and development studies: A complexity theory approach.* New York and Abingdon: Routledge.
Marais, K., & Meylaerts, R. (Eds.). (2018). *Complexity thinking in translation studies: Methodological considerations.* Abingdon & New York: Routledge.
Marín García, Á. (2017). Theoretical hedging: The scope of knowledge in translation process research [unpublished PhD dissertation]. Kent State University.
Maturana, H. R., & Varela, F. J. (1987). *The tree of knowledge: The biological roots of human understanding.* Boston: New Science Library.
Muñoz Martín, R. (2010). On paradigms and cognitive translatology. In G. M. Shreve, & E. Angelone (Eds.), *Translation and cognition* (pp. 169–187). Amsterdam: John Benjamins.
Muñoz Martín, R. (Ed.). (2016). *Reembedding translation process research.* Amsterdam: John Benjamins.
Neumann, S. (2013). *Contrastive register variation: A quantitative approach to the comparison of English and German.* Berlin: de Gruyter.
Neumann, S. (2014). Cross-linguistic register studies: Theoretical and methodological considerations. *Languages in Contrast, 14*, 35–57.
O'Brien, S. (2013). The borrowers. Researching the cognitive aspects of translation. *Target, 25*(1), 5–17.
Schaeffer, M., & Carl, M. (2013). Shared representations and the translation process: A recursive model. *Translation and Interpreting Studies, 8*(2), 169–190.
Seleskovitch, D. (1968). *L'interprète dans les conférences internationales: Problèmes de langage et communication.* Paris: Minard (*Interpreting for international conferences: Problems of language and communication.* Washington: Pen and Booth).
Shreve, G. M., & Lacruz, I. (2017). Aspects of a cognitive model of translation. In J. W. Schwieter, & A. Ferreira (Eds.), *The handbook of translation and cognition* (pp. 127–143). Hoboken, NJ: Wiley Blackwell.
Tirkkonen-Condit, S. (2005). The monitor model revisited: Evidence from process research. *Meta, 50*(2), 405–414.
Varela, F. J., Thompson, E., & Rosch, E. (1991). *The embodied mind: Cognitive science and human experience.* Cambridge, MA: MIT Press.

Index

Figures and tables are denoted by *italic* and **bold** text respectively, and notes by "n" and the number of the note after the page number e.g., 306n1 refers to note number 1 on page 306. The acronyms "CTS" and "SFL" refer to Cognitive Translation Studies and systemic functional linguistics respectively.

Printed in Great Britain
by Amazon